# Graphs from Rings

David F. Anderson • T. Asir • Ayman Badawi
T. Tamizh Chelvam

# Graphs from Rings

Springer

David F. Anderson
Department of Mathematics
University of Tennessee
Knoxville, TN, USA

Ayman Badawi
Department of Mathematics, Nab 262
American University of Sharjah
Sharjah, United Arab Emirates

T. Asir
Department of Mathematics
Madurai Kamaraj University
Madurai, Tamil Nadu, India

T. Tamizh Chelvam
Department of Mathematics
Manonmaniam Sundaranar University
Tirunelveli, Tamil Nadu, India

ISBN 978-3-030-88409-3 ISBN 978-3-030-88410-9 (eBook)
https://doi.org/10.1007/978-3-030-88410-9

Mathematics Subject Classification (2020): 13xx, 20xx, 05xx, 68Rxx



This Springer imprint is published by the registered company Springer Nature Switzerland AG
The registered company address is: Gewerbestrasse 11, 6330 Cham, Switzerland

**Dedicated**

*To my wife Konnie, my favorite collaborator for the past 48 years,*
*our children Sarah and Jonathan, and our grandchildren*
*Brooke, Sophia, Nehemiah, Judson, Caedmon, Luke, Boaz, and Adeline*
David F. Anderson

*In memory of my father Thangaraj, to my mother Victoria, wife Selva Jeba,*
*daughter Ainsley Priscilla and son Anston Peter*
T. Asir

*In memory of my parents Rateb and Fawziya, to my wife Rawya, son Nadeem*
*and daughter Raneem*
Ayman Badawi

*To my father P. Thirugnanam, mother T. Sivagami, wife R. Ezily*
*and son T.Kuralamudhan*
T. Tamizh Chelvam

# Preface

In this book, we introduce several graphical representations of a ring. In the past three decades, graphs constructed from algebraic structures have been studied extensively by many authors and have become a major field of research. Assigning graphs to rings has played an important role in the study of the structure of rings. The benefit of studying these graphs is that one may find some results about the algebraic structures and vice versa.

Algebra and Discrete Mathematics are two vital disciplines of mathematics. These two disciplines interact in many ways by extending tools from one to the benefit of the other. The first example of a group construction from a graph comes from the study of automorphism groups of a graph. This raised the fundamental question of which groups can be realized as the automorphism group of some graph. In the other direction, this study aims for a graph construction from a group. The first graph construction from a given group was the construction of graphs from finite groups by Arthur Cayley in 1878; it was called the pictorial representation of groups. These Cayley graphs now play a vital role as a discrete structure in parallel computing and are used to solve routing problems in parallel computing.

Another important graph construction is the construction of graphs from rings. The study of graphs from rings contributes to the interplay between the ring structure and the derived graph structure. This study of graphs from rings starts from the most well-studied zero-divisor graphs from commutative rings. Here, one defines a graph whose vertices are a set of elements or ideals of the ring, and the edges are defined with respect to an algebraic condition on the elements of the vertex set. These well-studied classes of graphs include zero-divisor graphs, total graphs, annihilating graphs, annihilator graphs, comaximal graphs, unit graphs, Cayley graphs, Jacobson graphs, generalized total graphs, Cayley sum graphs, and trace graphs of matrices from commutative rings. These graphs constructed from commutative rings help researchers to study the algebraic properties of commutative rings using graph-theoretical tools. One can sometimes translate algebraic properties of commutative rings to graph-theoretic language, and then the geometric properties of the graphs can help explore some interesting results related to commutative rings.

In the literature, there are many papers on assigning a graph to a ring, group, semigroup, or module. One may address three major problems in this area: characterization of the resulting graphs, characterization of rings with isomorphic graphs, and realization of the connection between the algebraic structures and the corresponding graphs. The hope is that the graph-theoretic properties of the graph from the ring will help us better understanding the ring-theoretic properties of the ring. Not only have graphs from rings been studied by professional mathematicians, but they have also been the focus of undergraduate research, masters theses, and doctoral dissertations.

We have taken an essentially "historical" approach in this book and have tried to cover as many topics and be as complete as possible. The literature on associating graphs to rings is vast; it, and even the number of new graphs associated to rings, is growing at a rapid rate. In order to keep this book to a reasonable length, we have chosen to omit many of the proofs, but instead to give detailed comments and references.

We thank the reviewers for their constructive assessment of the manuscript and suggestions on many aspects. Finally, we would like to express our sincere thanks to the professional staff at Springer, and in particular to Elena Griniari and Leonie Kunz, for their help in producing this book.

September 2021

David F. Anderson
T. Asir
Ayman Badawi
T. Tamizh Chelvam

# Notation

Throughout, unless otherwise stated, our notations will be as given below:

| | |
|---|---|
| $\mathbb{N}$ | Set of natural numbers (positive integers) |
| $\mathbb{R}$ | Set of real numbers |
| $\mathbb{Q}$ | Set of rational numbers |
| $\mathbb{C}$ | Set of complex numbers |
| $\mathbb{Z}$ ($\mathbb{Z}^+$) | Set of integers (nonnegative integers) |
| $R$ | Ring |
| $\mathfrak{m}$ | Maximal ideal of $R$ |
| $R^*$ | $R \setminus \{0\}$ |
| $\mathbb{Z}_n$ | Congruence classes of the integers modulo $n$ |
| $F$ | Field |
| $\mathbb{F}_{p^n}$ or $GF(p^n)$ | Field with $p^n$ elements |
| $det(A)$ | Determinant of the matrix $A$ |
| $\|A\|$ | Cardinality of the set $A$ |
| $A \setminus B$ or $A - B$ | Set-theoretic difference |
| $(a)$ or $\langle a \rangle$ | Ideal generated by $a \in R$ |
| $Ann(a)$ or $ann(a)$ | Annihilator ideal of $a \in R$ |
| $U(R)$ | Group of units of $R$ |
| $Z(R)$ | Set of zero-divisors of $R$ |
| $Reg(R) = R \setminus Z(R)$ | Set of regular elements of $R$ |
| $Nil(R)$ | Nilradical of $R$ |
| $J(R)$ | Jacobson radical of $R$ |
| $Max(R)$ | Set of maximal ideals of $R$ |
| $Min(R)$ | Set of minimal prime ideals of $R$ |
| $Spec(R)$ | Set of prime ideals of $R$ |
| $Soc(R)$ | Sum of minimal ideals of $R$ |
| $Ass(R)$ | Set of associated prime ideals of $R$ |
| $\mathbb{I}(R)$ | Set of nonzero ideals of $R$ |
| $\mathbb{A}(R)$ | Set of annihilating-ideals of $R$ |

| | |
|---|---|
| $R/I$ or $\frac{R}{I}$ | Quotient or factor ring |
| $R_S$ | Ring of fractions of $R$ with respect to $S$ |
| $T(R) = R_{R\setminus Z(R)}$ | Total quotient ring of $R$ |
| $dim(R)$ | (Krull) Dimension of $R$ |
| $R_1 \times R_2$ | Direct product of two rings $R_1$ and $R_2$ |
| $M$ | $R$-module |
| $G = (V, E)$ | Graph with vertex set $V$ and edge set $E$ |
| $deg_G(v)$ or $deg(v)$ | Degree of the vertex $v \in V(G)$ |
| $\Delta(G)$ | Maximum degree of $G$ |
| $\delta(G)$ | Minimum degree of $G$ |
| $K_n$ | Complete graph on $n$ vertices |
| $K_{m,n}$ | Complete bipartite graph with partitions having $m$ and $n$ vertices |
| $\overline{K}_{m,3}$ | Join of $K_{m,3}$ and $K_{1,m}$ |
| $P_n$ | Path on $n$ vertices |
| $C_n$ | Cycle on $n$ vertices |
| $kG$ | $k$ copies of the graph $G$ |
| $\langle S \rangle$ | Subgraph induced by $S \subseteq V$ |
| $CG$ | Complement graph of $G$ |
| $N_G(v)$ or $N(v)$ | Open neighborhood of $v \in V(G)$ |
| $N_G(S)$ or $N(S)$ | Open neighborhood of $S \subseteq V(G)$ |
| $N[v] = N(v) \cup \{v\}$ | Closed neighborhood of $v \in V(G)$ |
| $e(v)$ | Eccentricity of $v \in V(G)$ |
| $rad(G)$ | Radius of $G$ |
| $diam(G)$ | Diameter of $G$ |
| $Aut(G)$ | Automorphism group of $G$ |
| $\chi(G)$ | Vertex chromatic number of $G$ |
| $\chi'(G)$ | Edge chromatic number of $G$ |
| $\omega(G)$ | Clique number of $G$ |
| $\alpha(G)$ | Independence number of $G$ |
| $\beta'(G)$ | Edge independence number of $G$ |
| $\kappa(G)$ | Vertex connectivity number of $G$ |
| $\lambda(G)$ | Edge connectivity number of $G$ |
| $gr(G)$ | Girth of $G$ |
| $g(G)$ | Genus of $G$ |
| $\tilde{g}(G)$ | Crosscap of $G$ |
| $\gamma(G)$ | Domination number of $G$ |
| $\gamma_i(G)$ | Independent domination number of $G$ |
| $\gamma_p(G)$ | Perfect domination number of $G$ |
| $\gamma_t(G)$ | Total domination number of $G$ |

| | |
|---|---|
| $\gamma_c(G)$ | Connected domination number of $G$ |
| $\gamma^{'}(G)$ | Inverse domination number of $G$ |
| $\gamma_s(G)(\gamma_w(G))$ | Strong (weak) domination number of $G$ |
| $d(G)$ | Domatic number of $G$ |
| $d_i(G)$ | Independent domatic number of $G$ |
| $d_p(G)$ | Perfect domatic number of $G$ |
| $b(G)$ | Bondage number of $G$ |
| $G\square H$ | Cartesian product of the graphs $G$ and $H$ |
| $G \times H$ | Direct product of the graphs $G$ and $H$ |
| $G \wr H$ | Wreath product of the groups $G$ and $H$ |
| $\Gamma(R)$ | Zero-divisor graph of a commutative ring $R$ |
| $\Gamma_I(R)$ | Ideal-based zero-divisor graph of $R$ |
| $\Gamma_E(R)$ | Compressed zero-divisor graph of $R$ |
| $\overline{\Gamma}(R)$ | Extended zero-divisor graph of $R$ |
| $\overrightarrow{\Gamma}(R)$ | Directed zero-divisor graph of a noncommutative ring $R$ |
| $\Gamma(S)$ | Zero-divisor graph of a semigroup $S$ |
| $\Gamma(R)$ | Zero-divisor graph of a semiring $R$ |
| $\Gamma_I(R)$ | Ideal-based zero-divisor graph of a semiring $R$ |
| $\Gamma(M)$ | Zero-divisor graph of a module $M$ |
| $T(\Gamma((R))$ | Total graph of a commutative ring $R$ |
| $IT(R)$ | Intersection graph of the total graph of $R$ |
| $IG(R)$ | Intersection graph of ideals of $R$ |
| $T_0(\Gamma((R))$ | Total graph without zero of a commutative ring $R$ |
| $Reg(\Gamma(R))$ | Regular graph of $R$ |
| $CT(\Gamma(R))$ | Complement of the total graph of $R$ |
| $\Gamma(R, G, S)$ | Generalized total graph of $R$ |
| $\overline{\Gamma}(R, G, S)$ | Generalized total graph with loops of $R$ |
| $L(T(\Gamma(R)))$ | Line graph of the total graph of $R$ |
| $T(\Gamma_I(R))$ | Ideal-based total graph of $R$ |
| $GT_S(R)$ | Total graph of a multiplicatively closed subset $S$ of $R$ |
| $GT_H(R)$ | Total graph of a multiplicatively prime subset $H$ of $R$ |
| $T(\Gamma(M))$ | Total graph of a module $M$ |
| $T(\Gamma(R))$ | Total graph of a commutative semiring $S$ |
| $IS(R)$ | Identity summand graph of $R$ |
| $\Gamma_c(R)$ | Co-maximal graph of $R$ |
| $\Gamma'(R)$ | Cozero-divisor graph of $R$ |
| $\mathfrak{J}_R$ | Jacobson graph of $R$ |
| $AG(R)$ | Annihilator graph of $R$ |
| $\mathrm{AG_E}(R)$ | Compressed annihilator graph of $R$ |
| $\Gamma_\sim(R)$ | Congruence-based zero-divisor graph of $R$ |

| | |
|---|---|
| $ZD(R)$ | Zero-divisor dot product graph of $R$ |
| $TD(R)$ | Total dot product graph of $R$ |
| $UD(R)$ | Unit dot product graph of $R$ |
| $EUD(R)$ | Equivalence unit dot product graph of $R$ |
| $\Gamma_t(M_n(R))$ | Trace graph of matrices over $R$ |
| $C\Gamma_t(M_n(R))$ | Complement of the trace graph of matrices over $R$ |
| $\Gamma_{I^t}(M_n(R))$ | Ideal-based trace graph of matrices over $R$ |
| $\Gamma_C(R)$ | Commuting graph of $R$ |
| $\Gamma_N(R)$ | Nilpotent graph of $R$ |
| $I(R)$ | Idempotent graph of $R$ |
| $\overrightarrow{IT}(R)$ | Cayley ideal graph of $R$ |
| $G(x)$ | Irreducible divisor graph of $R$ |
| $\overline{G}(x)$ | Reduced irreducible divisor graph of $R$ |
| $\mathrm{Cay}(G, \Omega)$ | Cayley graph of a group $G$ |
| $\overrightarrow{\Gamma}_H(G)$ | Subgroup graph of a group $G$ |

# Contents

# Chapter 1

# Introduction

The purpose of this chapter is to familiarize the reader with the terminology and notation that we shall use in the book. For the sake of completeness, we state some definitions and notation used throughout to keep this book as self-contained as possible. This chapter includes some basic definitions and results which are needed for the subsequent chapters. A few other definitions and notation will be given as needed throughout the book. Unless otherwise noted, all definitions relating to algebra are from Kaplansky [228], and all definitions relating to graph theory are from Chartrand and Zhang [160]. The interested reader can find information on commutative algebra in Atiyah and Macdonald [105], and for a graph-theoretic perspective, in Balakrishnan and Ranganathan [132], while additional information about algebraic graph theory can be found in Biggs [146]. We let $\mathbb{N}, \mathbb{Z}^+, \mathbb{Z}, \mathbb{Q}, \mathbb{R}, \mathbb{C}, \mathbb{Z}_n$, and $\mathbb{F}_q$ denote the positive integers, nonnegative integers, integers, rational numbers, real numbers, complex numbers, integers modulo $n$, and the finite field with $q$ elements, respectively. For $X$ a subset of a ring or module, let $X^* = X \setminus \{0\}$.

## 1.1 Basics in Algebra

Much of abstract algebra involves properties of sets. The term group was coined by Évariste Galois (1811-1832) about 200 years ago to describe sets of one-to-one functions on finite sets that could be combined together to form a closed set. As is the case with most fundamental concepts in mathematics, the modern definition of a group that follows is the result of a long evolutionary process.

**Definition 1.1.1.** A nonempty set $G$ together with an operation "$\cdot$" is called a *group* if the following four properties are satisfied:

(i) *Closure*: $a \cdot b \in G$ for all $a, b \in G$;
(ii) *Associativity*: $(a \cdot b) \cdot c = a \cdot (b \cdot c)$ for all $a, b, c \in G$;
(iii) *Identity*: There is an $e \in G$ such that $a \cdot e = e \cdot a = a$ for all $a \in G$;
(iv) *Inverses*: For $a \in G$, there is an $a^{-1} \in G$ such that $a \cdot a^{-1} = a^{-1} \cdot a = e$.



D. F. Anderson et al., *Graphs from Rings*, https://doi.org/10.1007/978-3-030-88410-9_1

If a group has the property that $a \cdot b = b \cdot a$ for every pair of elements $a$ and $b$, then the group is called an *abelian group*. In this case, we often use "+" (resp., $0$, $-a$) rather than "$\cdot$" (resp., $e$, $a^{-1}$). If a subset $H$ of a group $G$ is itself a group under the operation of $G$, then $H$ is called a *subgroup* of $G$. Equivalently, $\emptyset \neq H \subseteq G$ is a subgroup of $G$ if $a \cdot b^{-1} \in H$ whenever $a$ and $b$ are in $H$. For more about groups, see [199]. We will usually just write $ab$ for $a \cdot b$. A nonempty set with an associative binary operation is called a *semigroup*, and a semigroup with an identity element is called a *monoid*.

Many sets are naturally endowed with two binary operations called addition and multiplication. Examples that quickly come to mind are the integers, real numbers, matrices, and polynomials. When considering these sets as groups, we simply used addition and ignore multiplication. In many instances, one needs to take into account both addition and multiplication. One abstract concept that does this is a ring. The term ring was coined by the German mathematician David Hilbert (1862-1943). This notion originated in the mid nineteenth century with Richard Dedekind (1831-1916).

**Definition 1.1.2.** A *ring* $R$ is a nonempty set $R$ with two binary operations, addition "+" and multiplication "$\cdot$", such that

(i) $(R, +)$ is an abelian group;

(ii) *Associativity*: $(ab)c = a(bc)$ for all $a, b, c \in R$;

(iii) *Distributivity*: $a(b + c) = ab + ac$ and $(a + b)c = ac + bc$ for all $a, b, c \in R$.

Note that multiplication need not be commutative. When it is, we say the ring is *commutative*. Also, a ring need not have an identity element under multiplication. When a ring other than $\{0\}$ has an identity element under multiplication, we say the ring is a ring with *identity*. The multiplicative identity is denoted by 1. A nonzero element of a ring with identity need not have a multiplicative inverse. When it does, we say it is a unit element. Thus an element $x \in R$ is called a *unit* if there exists a $y \in R$ with $xy = yx = 1$; $y$ is called the *inverse* of $x$ and is often denoted by $x^{-1}$. The collection of all units in $R$ is denoted by $U(R)$. It is easy to check that $U(R)$ is a group under multiplication and is called the multiplicative *group of units* of $R$. For $a$ and $b$ in a commutative ring $R$, we say $a$ *divides* $b$ and write $a \mid b$, if there exists an element $c$ in $R$ such that $b = ac$. If $a$ does not divide $b$, then we write $a \nmid b$. Also, for $A$ a subset of an algebraic system with a 0 element, we let $A^* = A \setminus \{0\}$. Unless mentioned otherwise, we assume that all rings are commutative with identity. However, in several places, the proofs make no use of the identity.

Recall that in the case of groups, the identity and inverses are unique. What about rings? The same is true for rings, provided they exist. An element $a$ in a commutative ring $R$ is called a *zero-divisor* if there is a $0 \neq b \in R$ such that $ab = 0$. The set of all *zero-divisors* of $R$ is denoted by $Z(R)$, i.e., $Z(R) = \{a \in R \mid ab = 0 \text{ for some } b \neq 0\}$, and $Z(R)^* = Z(R) \setminus \{0\}$. A commutative ring with identity is said to be an *integral domain* if it has no nonzero zero-divisors. Thus, in an integral domain, a product is zero only when one of the factors is zero, that is, $ab = 0$ if and only if $a = 0$ or $b = 0$. A commutative ring with identity is called a *field* if every nonzero element is a unit. We have numerous examples of fields: the complex numbers, the real numbers, the rational numbers. Group theory, ring theory, field theory, and module theory are the four main branches of abstract algebra. The abstract theory of fields was initiated by Heinrich Weber (1842-1913) in 1893.

The *characteristic* of a ring $R$, denoted by $char(R)$, is the smallest positive integer $m$ such that $\underbrace{1+1+\cdots+1}_{m-times} = 0$ in $R$. If no such integer exists, the characteristic of $R$ is 0. Note that the characteristic of an integral domain is 0 or prime.

A subset $S$ of a ring $R$ with identity is a *subring* of $R$ if $S$ is itself a ring with the operations of $R$. A subset $I$ of a commutative ring $R$ is called an *ideal* of $R$ if $(I, +)$ is a subgroup of $R$ and $ar \in I$ for all $a \in I$ and $r \in R$. Let $R$ be a commutative ring with identity and let $a \in R$. The set $(a) = Ra = \{ra \mid r \in R\}$ is an ideal of $R$ called *the principal ideal* generated by $a$. We will often denote $\{0\}$ by 0; so $0 = \{0\} = (0)$. A proper ideal $I$ of a commutative ring $R$ is said to be a *prime ideal* of $R$ if whenever $ab \in I$ for $a, b \in R$, then $a \in I$ or $b \in I$. A proper ideal $I$ of $R$ is said to be a *maximal ideal* of $R$ if whenever $J$ is an ideal of $R$ and $I \subseteq J \subseteq R$, then $J = I$ or $J = R$. So, the only ideal that properly contains a maximal ideal is the entire ring. The set of all maximal ideals of $R$ is denoted by $Max(R)$. The set of all prime ideals of $R$ is called the *spectrum* of $R$ and is denoted by $Spec(R)$.

A commutative ring $R$ with identity is called a *local ring* (or *quasilocal ring*) if it has a unique maximal ideal. Note that if $R$ is a local ring with maximal ideal $\mathfrak{m}$, then every element of $R \setminus \mathfrak{m}$ is a unit. Conversely, if $R$ is a commutative ring with identity in which the set of nonunits forms an ideal $\mathfrak{m}$, then $R$ is a local ring with unique maximal ideal $\mathfrak{m}$. An ideal $I$ of $R$ is said to be *irreducible* if whenever $I = J \cap K$ for ideals $J$ and $K$ of $R$, then $I = J$ or $I = K$. An ideal $I$ of $R$ is called *dense* if $r \in R$ with $rI = \{0\}$ implies $r = 0$.

It is easy to verify that $Z(R)$ is an ideal of $R$ if and only if $Z(R)$ is closed under addition. According to [228, p. 3], the set of all zero-divisors of a commutative ring $R$, $Z(R) = \cup P_i$, where each $P_i$ is a prime ideal of $R$. Thus, we have $xy \in Z(R)$ for $x, y \in R$ implies $x \in Z(R)$ or $y \in Z(R)$, and hence if $Z(R)$ is an ideal of $R$, then $Z(R)$ is actually a prime ideal of $R$.

A $0 \neq x \in R$ is said to be *regular* [220, p. 447] if $x$ is not a zero-divisor in $R$. The set of all regular elements of $R$ is denoted by $Reg(R)$, i.e., $Reg(R) = R \setminus Z(R)$. In the case of a finite commutative ring $R$, one can easily verify that $Reg(R) = U(R)$, the set of units in $R$. If $a \in R$, then $\{x \in R \mid ax = 0\}$ is called the *annihilator* of $a$ and is denoted by $Ann_R(a)$, or just $Ann(a)$. It is well known that $Z(R) = \bigcup_{0 \neq a \in R} Ann(a)$. An element $r \in R$ is called *quasiregular* if there is an $s \in R$ such that $r + s - sr = 0$.

A prime ideal $P$ of $R$ is said to be an *associated prime ideal* if $P$ is the annihilator of some element $a$ of $R$. The collection of associated prime ideals of $R$ is denoted by $Ass(R)$. An element $x \in R$ is said to be *nilpotent* if $x^m = 0$ for some positive integer $m$. The set of all nilpotent elements of $R$, denoted by $Nil(R)$, forms an ideal of $R$ and is called the *nilradical* of $R$. Moreover, $Nil(R)$ is the intersection of all the prime ideals of $R$. The *Jacobson radical* of $R$, denoted by $J(R)$, is the intersection of all the maximal ideals of $R$. Moreover, $J(R) = \{x \in R \mid 1 - rx \in U(R) \text{ for all } r \in R\}$. The *socle* of $R$ is the sum of all the (nonzero) minimal ideals of $R$ and is denoted by $Soc(R)$.

Let $R$ be a ring and $I$ an ideal of $R$. Since $R$ is a group under addition and $I$ is a normal subgroup of $R$, we can form the factor group $R/I = \{r + I \mid r \in R\}$. The natural question at this point is how can we make a ring from this group of cosets? The addition is already taken care of by the group of cosets (i.e., $(a+I)+(b+I) = (a+b)+I$); we define the product of two cosets $a+I$ and $b+I$ as $ab + I$. The definition works as long as $I$ is an ideal and not just a subring of $R$. Actually, $R/I$ is a ring if and only if $I$ is an ideal of $R$. The ring $R/I$ is called the *quotient ring* or *factor ring* of $R$ by $I$. Note that the factor ring $R/I$ is an integral domain if and only if $I$ is prime and $R/I$ is a field if and only if $I$ is maximal. Thus every maximal ideal is a prime ideal.

A *(ring-)homomorphism* $f$ from a ring $R$ to a ring $S$ is a mapping from $R$ to $S$ that preserves the two ring operations; that is, $f(a + b) = f(a) + f(b)$ and $f(ab) = f(a)f(b)$ for all $a, b \in R$. We also assume that $f(1) = 1$. A one-to-one (ring-)homomorphism that is also onto is called an *(ring-)isomorphism*. Let $f$ be a homomorphism from a ring $R$ to a ring $S$. Then, the *kernel* of $f$, $Ker(f) = \{r \in R \mid f(r) = 0\}$, is an ideal of $R$.

Let $R$ be a commutative ring. The set of formal symbols $R[x] = \{a_n x^n + a_{n-1}x^{n-1} + \cdots + a_1 x + a_0 \mid a_i \in R, n \in \mathbb{Z}^+\}$ is called the *ring of polynomials* over $R$ in the indeterminate $x$. Let $D$ be an integral domain. Then $D[x]$ is an integral domain with $U(D[x]) = U(D)$. An $f(x) \in D[x]$ that is neither the zero polynomial nor a unit in $D[x]$ is said to be *irreducible* over $D$ if whenever $f(x)$ is expressed as a product $f(x) = g(x)h(x)$ with $g(x), h(x) \in D[x]$, then either $g(x)$ or $h(x)$ is a unit in $D[x]$. A nonzero, nonunit element of $D[x]$ that is not irreducible over $D$ is called *reducible* over $D$. The *content* of a polynomial $a_n x^n + a_{n-1}x^{n-1} + \cdots + a_0 \in \mathbb{Z}[x]$ is the greatest common divisor of the integers $a_n, a_{n-1}, \ldots, a_0$. A *primitive polynomial* is an element of $\mathbb{Z}[x]$ with content 1.

A *principal ideal domain* (PID) is an integral domain $R$ in which every ideal has the form $(a) = \{ra \mid r \in R\}$ for some $a$ in $R$. An integral domain $D$ is a *unique factorization domain* (UFD) if (*i*) every nonzero element of $D$ that is not a unit can be written as a product of irreducible elements of $D$; and (*ii*) the factorization into irreducible elements is unique up to associates (i.e., unit multiples) and the order in which the factors appear. An integral domain $D$ is called a *Euclidean domain* (ED) if there is a function $d$ from the nonzero elements of $D$ to the nonnegative integers such that (i) $d(a) \leq d(ab)$ for all $0 \neq a, b \in D$; and (ii) if $a, b \in D$, $b \neq 0$, then there are $q, r \in D$ such that $a = bq + r$, where $r = 0$ or $d(r) < d(b)$. Note that every Euclidean domain is a principal ideal domain and every principal ideal domain is a unique factorization domain. Also, if $F$ is a field, then $F[x]$ is a principal ideal domain, in fact, a Euclidean domain.

A commutative ring $R$ is said to be *Noetherian* if every ideal of $R$ is finitely generated, or equivalently, if every strictly ascending chain of ideals in $R$ is finite. This designation honors Emmy Noether (1882-1935), whose revolutionary ideas inaugurated the use of chain conditions in algebra. If every prime ideal in a ring $R$ is finitely generated, then $R$ is Noetherian [228, Theorem 8]. A commutative ring $R$ is said to be *Artinian* if every strictly descending chain of ideals in $R$ is finite. This designation honors Emil Artin (1898-1962). The structure theorem for Artinian rings says that an Artinian ring $R$ is uniquely (up to isomorphism) a finite direct product of Artinian local rings, i.e., $R = R_1 \times \cdots \times R_n$, where each $R_i$ is a local ring. Three well-known facts are that every ideal of the product ring $R = R_1 \times \cdots \times R_n$ has the form $I_1 \times \cdots \times I_n$ for ideals $I_i$ of $R_i$, that $I_1 \times \cdots \times I_n$ is a prime (resp., maximal) ideal of $R$ if and only if $I_i$ is a prime (resp., maximal) ideal of $R_i$ for some $1 \leq i \leq n$ and $I_j = R_j$ for all $j \neq i$, and that an Artinian commutative ring $R$ (e.g., $R$ is finite) is isomorphic to a finite product of Artinian local rings.

Note that in an Artinian ring, every prime ideal is maximal and the nilradical is equal to the Jacobson radical. By a chain of prime ideals of $R$, we mean a finite strictly increasing chain $P_0 \subsetneq P_1 \subsetneq \cdots \subsetneq P_n$ of prime ideals; the length of this chain is $n$. The *(Krull) dimension* of $R$, denoted by $dim(R)$, is the supremum of the lengths of all chains of prime ideals of $R$. A field has dimension 0 and $\mathbb{Z}$ has dimension 1. Note that a commutative ring $R$ is Artinian if and only if $R$ is Noetherian and $dim(R) = 0$. Also, we use the simple fact that $dim(R) = 0$ if $R$ is a finite ring. A ring $R$ is a *chained ring* if the (principal) ideals of $R$ are linearly ordered (by inclusion), equivalently, if either $x|y$ or $y|x$ for all $x, y \in R$.

A commutative ring $R$ is said to be *reduced* if $Nil(R) = (0)$, where $Nil(R)$ is nilradical of $R$. For an ideal $I$ of $R$, we define $\sqrt{I} = \{r \in R \mid r^k \in I \text{ for some } k \in \mathbb{N}\}$. Thus $Nil(R) = \sqrt{(0)}$. We say that an ideal $I$ is a *radical ideal* if $I = \sqrt{I}$, i.e., $x \in I$ whenever $x^n \in I$ for some $x \in R$ and integer $n \geq 1$. It is easy to see that $R/I$ is a reduced ring if and only if $I$ is a radical ideal of $R$. A commutative ring $R$ is called *von Neumann regular* if for each $x \in R$, there is a $y \in R$ with $x^2y = x$ (equivalently, $R$ is reduced and $dim(R) = 0$). The simplest examples of von Neumann regular rings are direct products of fields. A ring $R$ is a *Boolean ring* if $x^2 = x$ for all $x \in R$, i.e., every element of $R$ is idempotent. The simplest example of a Boolean ring is the power set of a set with symmetric difference as addition and intersection as multiplication (i.e., a direct product of $\mathbb{Z}_2$'s). Clearly a Boolean ring is von Neumann regular, but not conversely. For example, let $\{F_i\}_{i\in I}$ be a family of fields. Then $\prod_{i\in I} F_i$ is always von Neumann regular, but it is Boolean if and only if $F_i \cong \mathbb{Z}_2$ for all $i \in I$. Note that $R$ is Boolean if and only if its quotient ring $T(R)$ is Boolean, and in this case, $T(R) = R$. It is well known that Boolean rings are commutative and reduced with characteristic 2. For a commutative ring $R$, the set of idempotents of $R$ forms a *Boolean algebra*, denoted by $B(R)$, under the operations $a \wedge b = ab$ and $a \vee b = a + b - ab$, with largest element 1, smallest element 0, and complement given by $a' = 1 - a$. A ring $R$ is called *Bézout* if every finitely generated ideal of $R$ is principal. Note that every von Neumann regular ring is Bézout.

A *simple ring* is a nonzero ring that has no nonzero proper ideals. A commutative ring $R$ is called *semisimple* if it is a direct sum of minimal ideals. If $I$ is a proper ideal of a semisimple ring, then $R/I$ is also semisimple. A *semilocal* ring is a ring for which $R/J(R)$ is a semisimple ring, where $J(R)$ is the Jacobson radical of $R$. The definition of semilocal for commutative rings is usually taken to be "having finitely many maximal ideals". Some literature refers to a commutative (semi)local ring in general as a *quasi(semi)local ring.* A semilocal ring is thus more general than a local ring. The rings $\mathbb{Z}_n$ and a finite direct product of fields are semilocal rings.

Note that the intersection of all the subfields of a field is itself a subfield, and so every field has a smallest subfield (that is, a subfield that is contained in every subfield). This subfield is called the *prime subfield* of the field. Also, the prime subfield of a field of characteristic $p$, a prime, is isomorphic to $\mathbb{Z}_p$, while the prime subfield of a field of characteristic 0 is isomorphic to $\mathbb{Q}$. Although the integral domain $\mathbb{Z}$ is not a field, it is at least contained in a field, namely, the field of rational numbers. Note that the field of rational numbers is nothing more than quotients of integers. Actually, if $D$ is an integral domain, then there exists a field $q(D) = F$ (called the *field of quotients* of $D$) that contains a subring isomorphic to $D$.

A *semiring* is similar to ring, but without the requirement that every element must have an additive inverse. That is, a semiring $R$ is a semigroup $(R, \cdot)$ and a commutative monoid $(R, +)$ in which 0 is the additive identity and $r\cdot 0 = 0\cdot r = 0$ for all $r \in R$, and both operations are connected by ring-like distributivity. A *commutative semiring* is a semiring with $a \cdot b = b \cdot a$ for all $a, b \in R$. A nonempty subset $I$ of a semiring $R$ is called an *ideal* if $a, b \in I$ and $r \in R$ implies $a + b \in I$ and $ra \in I$. A *prime ideal* of $R$ is a proper ideal $P$ of $R$ in which $x \in P$ or $y \in P$ whenever $xy \in P$. Note that by [210, Corollary 6.11], any maximal ideal of a semiring $R$ is a prime ideal. A *subtractive ideal* (or *k-ideal*) is an ideal $K$ such that if $x, x + y \in K$, then $y \in K$ (so $\{0\}$ is a $k$-ideal of $R$). Note that in [210], Golan uses the term "subtractive ideal", but in the literature of semirings, authors use the term "$k$-ideal" as well. For some examples of nonsubtractive ideals in a semiring, see [210, Chapter 5]. An ideal $I$ is said to be a *strongly k-ideal* if and only if $a + b \in I$ for $a, b \in R$ implies that $a \in I$ and $b \in I$. Clearly, every strongly $k$-ideal of a semiring $R$ is a $k$-ideal of $R$. But the converse is not true in general.

For example, the set $2\mathbb{Z}^+$ of all nonnegative even integers is a subtractive ideal of the semiring $\mathbb{Z}^+$ of all nonnegative integers. But it is not a strongly $k$-ideal since $3+5 \in 2\mathbb{Z}^+$, while neither 3 nor 5 belongs to $2\mathbb{Z}^+$.

Let $R$ be a semiring and $K \subseteq R$. The *k-closure* of $K$, defined by $cl(K) = \{a \in R \mid a + c = d \text{ for some } c, d \in K\}$, is an ideal of $R$ satisfying $K \subseteq cl(K)$ and $cl(cl(K)) = cl(K)$. Thus an ideal $K$ of $R$ is a $k$-ideal if and only if $K = cl(K)$. If $I$ is an ideal of $R$, then the *radical* of $I$, denoted by $\sqrt{I}$, is the set of all $x \in R$ for which $x^n \in I$ for some positive integer $n$. This is an ideal of $R$, contains $I$, and is the intersection of all the prime ideals of $R$ that contain $I$. A *primary ideal* $I$ of $R$ is a proper ideal of $R$ such that, if $xy \in I$ and $x \notin I$, then $y^n \in I$ for some positive integer $n$. If $I$ is primary, then $\sqrt{I} = P$ is a prime ideal of $R$. In this case, we also say that $I$ is a $P$-primary ideal of $R$. An ideal $I$ of $R$ is called a *radical ideal* if $I = \sqrt{I}$. Let $I$ be an ideal of a semiring $R$. A prime ideal $P$ of $R$ is called an *associated prime ideal* of $I$ if $P$ is the annihilator $(0 : x)$ of some $x \in I$ (so $P$ is a $k$-ideal).

A nonzero element $x$ of a semiring $R$ is said to be a *semi-unit* in $R$ if there are $r, s \in R$ such that $1 + rx = sx$. A semiring $R$ is said to be a *local semiring* if it has a unique maximal $k$-ideal. If $R$ is a semiring, then $R$ is *Noetherian* (resp., *Artinian*) if every nonempty set of $k$-ideals of $R$ has a maximal element (resp., minimal element) with respect to set inclusion. This definition is equivalent to the ascending chain condition (resp., descending chain condition) on $k$-ideals of $R$. A semiring $R$ is called *reduced* if it contains no nonzero nilpotent elements. For more details regarding semirings, see [210].

**Definition 1.1.3.** Let $R$ be a commutative ring with identity 1. An *R-module* over $R$ is a set $M$ together with

(1) a binary operation "+" on $M$ under which $M$ is an abelian group, and
(2) an action of $R$ on $M$ (that is, a map $R \times M \to M$), denoted by $rm$ for all $r \in R$ and $m \in M$, which satisfies
    (i) $(r+s)m = rm + sm$ for all $r, s \in R,\ m \in M$;
    (ii) $(rs)m = r(sm)$ for all $r, s \in R,\ m \in M$;
    (iii) $r(m+n) = rm + rn$ for all $r \in R,\ m, n \in M$;
    (iv) $1m = m$ for all $m \in M$.

Let $S$ be a multiplicatively closed subset of a ring $R$ (i.e., $st \in S$ if $s, t \in S$). (It is a good idea to assume that $0 \notin S$; otherwise everything in sight will collapse to 0.) Let $M$ be an $R$-module. We define the *localization* of $M$ at $S$, denoted by $S^{-1}M = M_S$, to be the set of equivalence classes $m/s$ of pairs $(m, s)$ with $m \in M,\ s \in S$, the equivalence relation being: $(m, s) \sim (m_1, s_1)$ if there exists $s_2 \in S$ such that $s_2(s_1 m - s m_1) = 0$. Then $M_S$ is an $R$-module via $m/s + n/t = tm + sn/st$ and $r \cdot m/s = rm/s$. Let $R$ be a commutative ring. Then $R_S$ becomes a commutative ring with multiplication given by $a/s \cdot b/t = ab/st$, and $M_S$ is also an $R_S$-module with scalar multiplication given by $r/s \cdot m/t = rm/st$. For $P$ a prime ideal of $R$, we write $M_P$ (resp., $R_P$) for $M_S$ (resp., $R_S$), where

$S = R \setminus P$. The ring $R_P$ has exactly one maximal ideal $P_P$, and thus $R_P$ is quasilocal. Note that the ring $S^{-1}R = R_S$ is the smallest ring in which the elements of $S$ become units. Also, the map $\varphi : R \to S^{-1}R$ is an injection if and only if $S$ contains no zero-divisors of $R$, and $S^{-1}R = 0$ if and only if $0 \in S$, hence if and only if $S$ contains nilpotent elements.The ring $T(R) = R_S$, where $S = R \setminus Z(R)$, is called the *total quotient ring* of $R$. An *overring* of $R$ is a subring of $T(R)$ containing $R$. For more information about rings, modules, and localizations, one can refer to [228].

Let $M$ be an $R$-module. The set $Z(M) = \{r \in R \mid rm = 0 \text{ for some } 0 \neq m \in M\}$ will denote the set of zero-divisors of $M$. For $x \in M$, the *residual* of $Rx$ by $M$, denoted by $(x : M)$, is a set of elements $r \in R$ such that $rM \subseteq Rx$. The *annihilator of an $R$-module $M$*, denoted by $Ann_R(M)$, is $(0 : M)$. An $R$-module $M$ is called a *reduced module* if $rm = 0$ implies that $rM \cap Rm = 0$, where $r \in R$ and $m \in M$. (We will often write 0 for $\{0\}$ in a ring or module.) It is clear that $M$ is a reduced module if and only if $r^2m = 0$ for $r \in R$, $m \in M$ implies that $rm = 0$. A proper submodule $N$ of $M$ is called a *prime submodule* of $M$ if $rm \in N$ implies that $m \in N$ or $r \in (N : M)$, where $r \in R$ and $m \in M$. A prime submodule $N$ of $M$ is called a *minimal prime submodule* of a submodule $H$ of $M$ if it contains $H$ and there is no smaller prime submodule with this property. A minimal prime submodule of the zero submodule is also known as a minimal prime submodule of the module $M$.

An $R$-module $M$ is said to be a *multiplication module* if for every submodule $K$ of $M$, there is an ideal $I$ of $R$ such that $K = IM$. The notation $Spec(M)$ will denote the set of the prime submodules of $M$, and $Nil(M) = \cap_{N \in Spec(M)} N$ will denote the nilradical of $M$. Note that a multiplication $R$-module $M$ is reduced if and only if $Nil(M) = 0$. Also, every nonzero multiplication $R$-module has a maximal submodule, and thus has a minimal prime submodule. An $R$-module $M$ is called a *von Neumann regular module* if every cyclic submodule of $M$ is a direct summand of $M$. If $N$ is a direct summand of $M$, then $N$ is pure, but not conversely, and so every von Neumann regular module is reduced. An $R$-module $M$ is called a *cancellation* (resp., *weakly cancellation*) module if $IM = JM$ implies that $I = J$ (resp., $I + Ann(M) = J + Ann(M)$) for any two ideals $I, J$ of $R$. An $R$-module $M$ is called a *simple $R$-module* if $M \neq 0$ and $M$ has no $R$-submodules other than 0 and $M$. An $R$-module $M$ is said to be *semisimple* if every $R$-submodule of $M$ is an $R$-module direct summand of $M$.

## 1.2 Basics in Graph Theory

This section gives a gentle, yet concise, introduction to graph theory for most of the terminology used later in the book. The basic concepts of graph theory are extraordinarily simple and can be used to express problems from many different subjects. The word graph is Latin for "picture".

The Pregolya River passes through a city once known as Königsberg. In the 1700's, seven bridges were situated across this river. The city's residents enjoyed strolling on these bridges, but as hard as they tried, no resident of the city was ever able to walk a route that crossed each of these bridges exactly once. The Swiss mathematician Leonhard Euler (1707-1783) learned of this frustrating phenomenon, and in 1736, he wrote an article about it. His work on the "Königsberg Bridge Problem" is considered by many to be the beginning of the field of graph theory.

Many real-world situations can conveniently be described by means of a diagram consisting of a set of points together with lines joining certain pairs of these points. For example, the points could represent people, with lines joining pairs of friends; or the points might be communication centers, with lines representing communication links. Notice that in such diagrams one is mainly interested in whether two given points are joined by a line; the manner in which they are joined is immaterial. A mathematical abstraction of situations of this type gives rise to the concept of a graph.

A *graph* $G = (V, E)$ is a nonempty set $V$ of objects called vertices together with a set $E$ of unordered pairs of distinct vertices of $G$ called edges. The vertex set and the edge set of $G$ are denoted by $V(G)$ and $E(G)$, respectively. If $e = \{u, v\}$ is an edge, then we write $e = uv$ (or $u - v$); we say that $e$ joins the vertices $u$ and $v$, $u$ and $v$ are *adjacent* vertices, or $u$ and $v$ are *incident* with $e$. If two vertices are not joined, then we say that they are *nonadjacent*. If two distinct edges are incident with a common vertex, then they are said to be adjacent to each other. The number $p = |V(G)|$ is called the *order* of $G$ and $q = |E(G)|$ is called the *size* of $G$. A graph of order $p$ and size $q$ is called a $(p, q)$ graph. By abuse of notation, we will sometimes write $|G|$ for $|V(G)|$.

A graph $H$ is called a *subgraph* of $G$ if $V(H) \subseteq V(G)$ and $E(H) \subseteq E(G)$. A *spanning subgraph* $H$ of $G$ is a subgraph of $G$ with $V(H) = V(G)$. For any subset $S$ of vertices of $G$, the *induced subgraph* generated by $S$, denoted by $\langle S \rangle$, is the maximal subgraph of $G$ with vertex set $S$. Thus two vertices of $S$ are adjacent in $\langle S \rangle$ if and only if they are adjacent in $G$.

The *complement* $CG$ of a graph $G$ is the graph whose vertex set is $V(G)$ and such that for a pair $u, v$ of vertices of $G$, $uv$ is an edge of $CG$ if and only if $uv$ is not an edge of $G$. The *line graph* $L(G)$ of a nonempty graph $G$ is the graph with vertex set $E(G)$, and two vertices $e_1$ and $e_2$ of $L(G)$ are adjacent if and only if $e_1$ and $e_2$ are adjacent edges in $G$.

A *multigraph* $M$ is a nonempty set of vertices, every two of which are joined by a finite number of edges. Two or more edges that join the same pair of distinct vertices are called *parallel edges*. An edge joining a vertex to itself is called a *loop*. Structures that permit both parallel edges and loops (including parallel loops) are called *pseudographs*. A *digraph* (or *directed graph*) $D$ is a nonempty set $V$ of vertices and a set $E$ of ordered pairs of distinct vertices. The elements of $E$ are called *directed edges* or *arcs*. An undirected graph with no multiple edges or loops is often called a *simple graph.*

A graph $G$ is said to be *complete* if every pair of distinct vertices are adjacent, and a complete graph on $n$ vertices is denoted by $K_n$. A graph $G = (V, E)$ is said to be *bipartite* if the vertex set $V$ can be partitioned into two nonempty disjoint subsets $A$ and $B$ such that every edge in $G$ has one end in $A$ and the other end in $B$, denoted by $V(G) = (A, B)$. In this case, the sets $A$ and $B$ are called a *bipartition* of $G$. If every vertex of $A$ is adjacent with all the vertices of $B$, then $G$ is said to be a *complete bipartite* graph, and it is denoted by $K_{|A|,|B|}$. A graph is called a *star graph* if it is a complete bipartite graph with one partition a singleton set.

The *open neighborhood of a vertex* $v$ in a graph $G$ is the set of vertices of $G$ which are adjacent with $v$, and it is denoted by $N_G(v)$ or $N(v)$. The *closed neighborhood of* $v$ is defined by $N[v] = N_G[v] = N(v) \cup \{v\}$. For a subset $S \subseteq V$, the open neighborhood of $S$ is defined by $N(S) = N_G(S) = \bigcup_{v \in S} N(v)$, and the closed neighborhood by $N[S] = N_G[S] = N(S) \cup S$.

The *degree* of a vertex $v$ in a graph $G$, denoted by $deg_G(v)$ or $deg(v)$, is the number of vertices adjacent with $v$ in $G$. Note that $deg(v)$ may be infinite, and in fact, in most situations we consider the infinite case in this book. If $deg(v) = 0$, then $v$ is called an *isolated vertex*, and if $deg(v) = 1$, then $v$ is called an *end vertex* (or *pendent vertex*) of $G$. The *maximum degree* of $G$ is defined by $\Delta(G) = \sup\{|N(v)| \mid v \in V\}$, and the *minimum degree* of $G$ is defined by $\delta(G) = \inf\{|N(v)| \mid v \in V\}$. A graph $G$ is said to be *regular of degree* $k$ if every vertex of $G$ has degree $k$, and such graphs are called $k$*-regular* graphs.

Let $u$ and $v$ be (not necessarily distinct) vertices of a graph $G$. A $u-v$ *walk* of $G$ is a finite, alternating sequence $u = u_0, e_1, u_1, e_2, \ldots, e_n, u_n = v$ of vertices and edges beginning with vertex $u$ and ending with vertex $v$ such that $e_i = u_{i-1}u_i$, $i = 1, 2, \ldots, n$. The number $n$ is called the length of the walk. A $u - v$ walk is said to be *open* if $u$ and $v$ are distinct vertices; otherwise, it is *closed*. A walk $u_0, e_1, u_1, e_2, \ldots, e_n, u_n$ is denoted by $(u_0, u_1, \ldots, u_n)$. A $u - v$ *trail* of $G$ is a $u - v$ walk in which no edge is repeated. A walk in which $u_0, u_1, \ldots, u_n$ are distinct is called a *path*. A path on $n$ vertices is denoted by $P_n$.

A graph $G$ is called *connected* if there is a path between any two distinct vertices in $G$; otherwise, $G$ is said to be *disconnected*. At the other extreme, we say that $G$ is *totally disconnected* if no two vertices of $G$ are adjacent. There are some graphs that are so slightly connected that they can be disconnected by the removal of a single vertex or a single edge. Let $v$ be a vertex and $e$ an edge of a graph $G$. If $G - v$ has more components than $G$, then $v$ is a *cut-vertex* of $G$; while if $G - e$ has more components than $G$, then $e$ is a *bridge* of $G$. In particular, if $v$ is a cut-vertex of a connected graph $G$, then $G - v$ is disconnected; and if $e$ is a bridge of a connected graph $G$, then $G - e$ is disconnected and is necessarily a graph with exactly two components. A subgraph $H$ of a graph $G$ is said to be a *block* if $H$ contains no cut-vertices. A *vertex-cut* of a graph $G$ is a subset $S$ of $V(G)$ such that $V(G) \setminus S$ is disconnected. A vertex-cut of minimum cardinality in $G$ is called a *minimum vertex-cut* of $G$, and this cardinality is called the *vertex-connectivity* (or simply, the *connectivity*) of $G$ and is denoted by $\kappa(G)$. So the connectivity of a graph $G$ is the smallest number of vertices whose removal from $G$ results in either a disconnected graph or a trivial graph. A graph $G$ is *$k$-connected*, for $k \geq 1$, if $\kappa(G) \geq k$. An *edge-cut* of a graph $G$ is a subset $X$ of $E(G)$ such that $E(G) \setminus X$ is disconnected. An edge-cut of minimum cardinality in $G$ is a *minimum edge-cut*, and this cardinality is the *edge-connectivity* of $G$, which is denoted by $\lambda(G)$.

A closed path is called a *cycle* of $G$. A cycle on $n$ vertices is denoted by $C_n$. A graph $G$ is called *acyclic* if it has no cycles. A connected acyclic graph is called a *tree*. A graph is called *Eulerian* if it contains a closed trail containing every edge. It is well known that a connected graph $G$ is Eulerian if and only if every vertex of $G$ has even degree.

The problems proposed by Sir William Rowan Hamilton (1805-1865) in his Icosian game gave rise to concepts in graph theory which eventually became a popular subject of study by mathematicians. A *Hamiltonian cycle* of $G$ is a cycle that contains every vertex of $G$. A graph is *Hamiltonian* if it contains a Hamiltonian cycle. A graph with just one cycle is called a *unicyclic* graph. A graph $G$ in which each cycle of length at least four has a chord is called a *triangulated* graph. A graph $G$ of order $m \geq 3$ is *pancyclic* if $G$ contains cycles of all lengths from 3 to $m$. Also, $G$ is called *vertex-pancyclic* if each vertex $v$ of $G$ belongs to a cycle of every length $\ell$ for $3 \leq \ell \leq m$.

The *clique number* $\omega(G)$ of a graph $G$ is the number of vertices in a largest complete subgraph of $G$. If $K_n \subset G$ for every integer $n \geq 1$, we set $\omega(G) = \infty$. A set of vertices in $G$ is said to be *independent* if no two vertices in the set are adjacent. The *independence number* $\alpha(G)$ is the maximum cardinality of an independent set in $G$. The *edge independence number* $\beta'(G)$ of a nonempty graph $G$ is the maximum number of edges in an independent set of edges of $G$. Furthermore, if the order of $G$ is $n$, then $\beta'(G) \leq n/2$. For additional notations and definitions of graph theory, the reader is referred to [160] or any standard text on graph theory.

Next, we discuss the coloring of a graph. Beginning with the origin of the Four Color Problem in 1852, the field of graph colorings has developed into one of the most popular areas of graph theory. If the countries in a map of South America are to be colored in such a way that every two countries with a common boundary are colored differently, then this map could be colored using only four colors. Is this true for every map? While it is not difficult to color a map of South America with four colors, it is not possible to color this map with fewer than four colors. In fact, every two of Brazil, Argentina, Bolivia, and Paraguay are neighboring countries, and so four colors are required to color only these four countries. The problems in graph colorings that have received the most attention involve coloring the vertices of a graph. Furthermore, the problems in vertex colorings that have been studied most often are those referred to as proper vertex colorings.

A *proper vertex coloring* of a graph $G$ is an assignment of colors to the vertices of $G$, one color to each vertex, so that adjacent vertices are colored differently. A *k-vertex coloring* of a graph $G$ is an assignment of $k$ colors $\{1, \ldots, k\}$ to the vertices of $G$ such that no two adjacent vertices have the same color. Thus a $k$-vertex coloring of a graph $G$ can be described as a function $f : V(G) \to \{1, 2, \ldots, k\}$ such that $f(v_1) \neq f(v_2)$ for every two adjacent vertices $v_1$ and $v_2$ in $G$. The *vertex chromatic number* of a graph $G$, denoted by $\chi(G)$, is the minimum $k$ for which $G$ has a $k$-vertex coloring. The chromatic number of a graph $G$ is the minimum number of independent sets into which $V(G)$ can be partitioned. For every graph $G$, we have $\chi(G) \geq \omega(G)$.

An *edge coloring* of a graph $G$ is an assignment of colors to the edges of $G$, one color to each edge. If adjacent edges are assigned distinct colors, then the edge coloring is a *proper edge coloring.* A proper edge coloring that uses colors from a set of $k$ colors is a *k-edge coloring.* Thus a $k$-edge coloring of a graph $G$ can be described as a function $f : E(G) \to \{1, 2, \ldots, k\}$ such that $f(e_1) \neq f(e_2)$ for every two adjacent edges $e_1$ and $e_2$ in $G$. A graph G is *k-edge colorable* if there exists a $k$-edge coloring of $G$. The *edge chromatic* $\chi'(G)$ of a graph $G$ is the minimum positive integer $k$ for which $G$ is $k$-edge colorable. Furthermore, $\chi'(G) = \chi(L(G))$ for every nonempty graph $G$. Vizing's Theorem says that $\chi'(G) \leq 1 + \Delta(G)$ for every nonempty graph $G$, where $\Delta(G)$ denotes the maximum degree of $G$. Thus $\chi'(G)$ is either $\Delta(G)$ or $1 + \Delta(G)$. For all other coloring concepts, we direct the reader to [161].

Next, we discuss the distance concepts in graphs. For vertices $u$ and $v$ of $G$, the length of a shortest path from $u$ to $v$ is denoted by $d(u, v)$ if a path from $u$ to $v$ exists. Also, we define $d(u, u) = 0$ and $d(u, v) = \infty$ if there is no path between $u$ and $v$. The *diameter of* $G$ is $diam(G) = sup\{d(u, v) \mid u, v \in V(G)\}$. The *eccentricity of a vertex* $u$ in $G$ is defined by $e(u) = max\{d(u, v) \mid v \in V(G)\}$, and the *radius of* $G$ is given by $rad(G) = min\{e(v) \mid v \in V(G)\}$. A graph $G$ is said to be *self-centered* if $e(v) = rad(G)$ for all $v \in V(G)$. The *girth of* $G$, denoted by $gr(G)$, is the length of a smallest cycle in $G$, and $gr(G) = \infty$ if $G$ contains no cycles.

Graphs $G$ and $H$ are said to be *isomorphic*, written $G \cong H$, if there exists a one-to-one correspondence $f : V(G) \to V(H)$ such that for every pair $x, y$ of vertices of $G$, $xy \in E(G)$ if and only if $f(x)f(y) \in E(H)$. An *automorphism* of a graph is an isomorphism of the graph to itself. In the case of a simple graph, an automorphism is just a permutation $\alpha$ of its vertex set which preserves adjacency, i.e., if $uv$ is an edge if and only if $\alpha(u)\alpha(v)$ is an edge. The automorphisms of a graph reflect its symmetries. We denote the set of all automorphisms of a graph $G$ by $Aut(G)$. It can be verified that $Aut(G)$ is a group under the operation of composition, called the *automorphism group* of $G$. The automorphism group of $K_n$ is the symmetric group $S_n$, consisting of all permutations of its vertex set. In general, for any simple graph $G$ on $n$ vertices, $Aut(G)$ is a subgroup of $S_n$. For instance, the automorphism group of the cycle $C_n$ is $D_n$, the dihedral group on $n$ elements.

The *direct product* of two graphs $G$ and $H$, denoted by $G \times H$, is the graph in which $V(G\times H) = V(G)\times V(H)$. That is, the vertex set is the set of ordered pairs of vertices from $G$ and $H$. The edge set of $G \times H$ consists of all $\{(u_1, v_1), (u_2, v_2)\}$ such that $u_1, u_2 \in V(G)$ and $v_1, v_2 \in V(H)$, $\{u_1, u_2\} \in E(G)$ and $\{v_1, v_2\} \in E(H)$. The *Cartesian product* of two graphs $G$ and $H$, denoted by $G\Box H$, is the graph with vertex set $V(G\Box H) = V(G) \times V(H)$ and edge set $E(G\Box H) = \{((u_1, v_1), (u_2, v_2)) \mid (u_1, u_2) \in E(G)$ with $v_1 = v_2$, or $(v_1, v_2) \in E(H)$ with $u_1 = u_2\}$.

Recall that two distinct vertices $a$ and $b$ of a graph $G$ are said to be *orthogonal*, written $a\perp b$, if $a$ and $b$ are adjacent and there is no vertex $c$ of $G$ which is adjacent to both $a$ and $b$, i.e., the edge $a - b$ is not part of any triangle in $G$. We say that $G$ is *complemented* if for each vertex $a$ of $G$, there is a vertex $b$ of $G$ (called a *complement* of $a$) such that $a\perp b$, and that $G$ is *uniquely complemented* if $G$ is complemented and whenever $a\perp b$ and $a\perp c$, then $N(b) = N(c)$, where $N(x)$ denotes the open neighborhood of $x \in V(G)$ (i.e., any two complements of a vertex are adjacent to the same vertices).

One of the fastest growing areas in graph theory is the study of domination in graphs. The following problem can be said to be the origin of the study of dominating sets in graphs. According to the rules of chess, a queen can, in one move, advance any number of squares horizontally, vertically, or diagonally (assuming that no other chess piece lies in its way). In the 1850's, chess enthusiasts in Europe considered the problem of determining the minimum number of queens that can be placed on a standard $8 \times 8$ chessboard so that all squares are either attacked by a queen or are occupied by a queen. It was correctly thought in the 1850's that five is the minimum number of queens that can dominate all of the squares of an $8\times 8$ chessboard. The problem of dominating the squares of a chessboard can be stated more generally as the problem of dominating the vertices of a graph.

A nonempty subset $S$ of $V$ is called a *dominating set* if every vertex in $V \setminus S$ is adjacent to at least one vertex in $S$. A subset $S$ of $V$ is called a *total dominating set* if every vertex in $V$ is adjacent to some vertex in $S$. A dominating set $S$ is called a *connected* (resp., *clique*) *dominating set* if the subgraph induced by $S$ is connected (resp., complete). A dominating set $S$ is called an *independent dominating set* if no two vertices of $S$ are adjacent. A dominating set $S$ is called a *perfect dominating set* if every vertex in $V\setminus S$ is adjacent to exactly one vertex in $S$. A dominating set $S$ is called an *efficient dominating set* if $S$ is both independent and perfect. A dominating set $S$ is called a *strong* (resp., *weak*) *dominating set* if for every vertex $u \in V \setminus S$, there is a vertex $v \in S$ with $deg(v) \geq deg(u)$ (resp., $deg(v) \leq deg(u)$) and $u$ is adjacent to $v$.

The *domination number* of $G$, denoted by $\gamma(G)$, is defined to be the minimum cardinality of a dominating set in $G$, and the corresponding dominating set is called a $\gamma$-set of $G$. In a similar way, we define the *total dominating number* $\gamma_t$, *connected dominating number* $\gamma_c$, *clique dominating number* $\gamma_{cl}$, *independent dominating number* $\gamma_i$, *perfect dominating number* $\gamma_p$, *efficient dominating number* $\gamma_{eff}$, *strong dominating number* $\gamma_s$ and the *weak dominating number* $\gamma_w$. A graph $G$ is called *excellent* if for every vertex $v \in V(G)$, there is a $\gamma$-set $S$ containing $v$. A *domatic partition* of $G$ is a partition of $V(G)$ into dominating sets of $G$. The maximum number of sets in a domatic partition of $G$ is called the *domatic number* of $G$ and is denoted by $d(G)$. In a similar way, we define the *perfect domatic number* $d_p(G)$, *independent domatic number* $d_i(G)$, and the *total domatic number* $d_t(G)$. A graph $G$ is called *domatically full* if $d(G) = \delta(G) + 1$, which is the maximum possible order of a domatic partition of $V$. For results concerning domination parameters, see Haynes et al. [214].

The prime objective of topological graph theory is to draw a graph on a surface so that no two edges cross, an intuitive geometric problem that can be enriched by specifying symmetries or combinatorial side-conditions. To solve the problem and its variants, techniques are adapted from a broad range of mathematics, especially from algebraic topology and group theory, and more recently, from enumerative combinatorics and the analysis of algorithms. Graphs on surfaces form a natural link between discrete and continuous mathematics. They enable us to understand both graphs and surfaces better.

Let $S_k$ denote the sphere with $k$ handles, where $k$ is a nonnegative integer, that is, $S_k$ is an oriented surface with $k$ handles. The (orientable) *genus of a graph* $G$, denoted by $g(G)$, is the smallest integer $n$ such that the graph can be embedded in $S_n$. Intuitively, $G$ is embedded in a surface if it can be drawn in the surface so that its edges intersect only at their common vertices. We say a graph $G$ is *planar* if $g(G) = 0$, i.e., $G$ can be drawn in the plane in such a way that no edges intersect, except of course at a common vertex; and *toroidal* if $g(G) = 1$. A *minor* of $G$ is the graph obtained from $G$ by contracting edges in $G$ or deleting edges and isolated vertices in $G$. A classical theorem due to Wagner [419] states that a graph $G$ is planar if and only if $G$ does not have $K_5$ or $K_{3,3}$ as a minor. It is well known that if $G'$ is a minor of $G$, then $g(G') \leq g(G)$. A nonorientable compact surface is a connected sum of finite copies of projective planes; the number of copies of projective planes is called the *crosscap number* of this surface. For example, *projective space* is of crosscap one, and the *Klein bottle* is of crosscap two. A nonplanar graph is said to be *projective* if it can be embedded in the projective plane. For details on the graph genus, see [429].

# Chapter 2

# Distances in Zero-divisor Graphs

Let $G$ be a graph. If $u$ and $v$ are distinct vertices in a connected graph $G$, then there is a $u-v$ path in $G$. In fact, there may very well be several $u-v$ paths in $G$, possibly of varying lengths. This information can be used to provide a measure of how close $u$ and $v$ are to each other or how far from each other they are. The most common definition of distance between two vertices in a connected graph is the following. The *distance* $d(u,v)$ from a vertex $u$ to a vertex $v$ in a connected graph $G$ is the minimum of the lengths of the $u-v$ paths in $G$. The *eccentricity* $e(v)$ of a vertex $v$ in a connected graph $G$ is the distance between $v$ and a vertex farthest from $v$ in $G$. The *diameter* $diam(G)$ of $G$ is the greatest eccentricity among the vertices of $G$, while the *radius* $rad(G)$ is the smallest eccentricity among the vertices of $G$. The diameter of $G$ is also the greatest distance between any two vertices of $G$. A vertex $v$ with $e(v) = rad(G)$ is called a *central vertex* of $G$. The *girth* of a graph $G$ with cycles is the length of a smallest cycle in $G$. These concepts are the considerations of this chapter. Actually, we present results concerning the diameter, girth, and center of the zero-divisor graph of a commutative ring. We begin the chapter by introducing and discussing some basic concepts of the zero-divisor graph of a commutative ring.

## 2.1 Zero-divisor Graphs

The idea of a graph, denoted by $\Gamma_0(R)$, associated to the zero-divisors of a commutative ring $R$ was introduced by I. Beck [137] in 1988, where he studied colorings of such graphs. After introducing this graph, Beck conjectured that the clique number and chromatic number of $\Gamma_0(R)$ are equal. In 1993, D. D. Anderson and M. Naseer [60] settled Beck's conjecture in the negative by providing a counterexample. The definition, along with the name for the zero-divisor graph, was introduced by D. F. Anderson and P. S. Livingston [83] in 1999, after modifying Beck's definition (also, see [264]). They also investigated the interplay between the ring-theoretic properties of a commutative ring and graph-theoretic



D. F. Anderson et al., *Graphs from Rings*, https://doi.org/10.1007/978-3-030-88410-9_2

properties of its zero-divisor graph. In the original definition, Beck [137] took all elements of the commutative ring $R$ as vertices of the graph $\Gamma_0(R)$. The modified definition of the zero-divisor graph is as follows.

**Definition 2.1.1.** ([83]) Let $R$ be a commutative ring with identity and $Z(R)$ its set of zero-divisors. The *zero-divisor graph* of $R$, denoted by $\Gamma(R)$, is the simple graph with vertex set $Z(R)^* = Z(R) \setminus \{0\}$, and two distinct vertices $x$ and $y$ are adjacent if $xy = 0$.

As mentioned earlier, when Beck [137] presented the idea of associating a "zero-divisor" graph with a commutative ring, he used a slightly different definition for $\Gamma(R)$. We denote Beck's zero-divisor graph of $R$ by $\Gamma_0(R)$ (Beck just used $R$). In $\Gamma_0(R)$, the vertex 0 is adjacent to every other vertex, but non-zero-divisors are adjacent only to 0. Note that $\Gamma(R)$ is a (induced) subgraph of $\Gamma_0(R)$ and $\Gamma(R)$ is the empty graph if and only if $R$ is an integral domain. Results for $\Gamma(R)$ have natural analogs for $\Gamma_0(R)$; however, the definition of $\Gamma(R)$ is a better illustration of the zero-divisor structure of $R$. The following example illustrates the difference between the graphs due to Beck (the corresponding graph is $\Gamma_0(R)$) and Anderson and Livingston (the corresponding graph is $\Gamma(R)$).

**Example 2.1.2.** If $R = \mathbb{Z}_6$, then $Z(R) = \{0, 2, 3, 4\}$. The two graphs $\Gamma_0(\mathbb{Z}_6)$ and $\Gamma(\mathbb{Z}_6)$ are given in Figure 2.1.

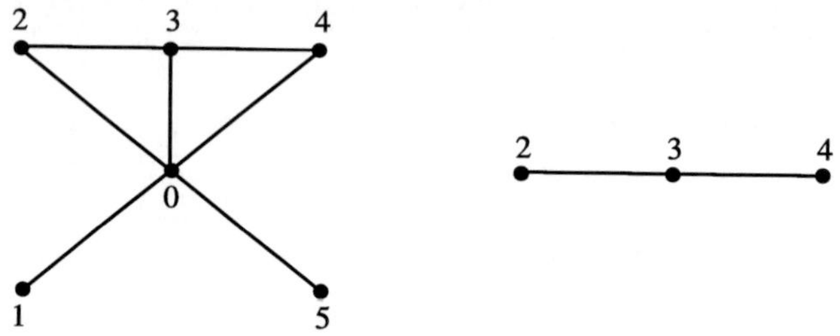


Figure 2.1: $\Gamma_0(\mathbb{Z}_6)$ and $\Gamma(\mathbb{Z}_6)$

Also, it is easy to see that $\Gamma_0(R)$ is complete if and only if $R \cong \mathbb{Z}_2$, and for $R$ not an integral domain, $\Gamma_0(R)$ is complete bipartite if and only if $R \cong \mathbb{Z}_4$ or $\mathbb{Z}_2[x]/(x^2)$. In view of these, $\Gamma(R)$ has a much richer and more appealing structure than $\Gamma_0(R)$ and better reflects properties of $Z(R)$.

Now we start with some examples of the zero-divisor graph $\Gamma(R)$ of a commutative ring $R$ which motivate later results. Also, we discuss the completeness of the zero-divisor graph. In the following example, easy details (including labeling the vertices) are left to the reader.

$\Gamma(\mathbb{Z}_4)$ or $\Gamma(\mathbb{Z}_2[x]/(x^2))$

$\Gamma(\mathbb{Z}_9)$, $\Gamma(\mathbb{Z}_2 \times \mathbb{Z}_2)$, or $\Gamma(\mathbb{Z}_3[x]/(x^2))$

$\Gamma(\mathbb{Z}_6)$, $\Gamma(\mathbb{Z}_8)$, or $\Gamma(\mathbb{Z}_2[x]/(x^3))$

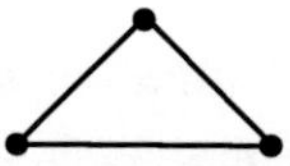

$\Gamma(\mathbb{Z}_2[x,y]/(x^2,xy,y^2))$ or $\Gamma(\mathbb{F}_4[x]/(x^2))$

Figure 2.2: Zero-divisor graphs of order less than four

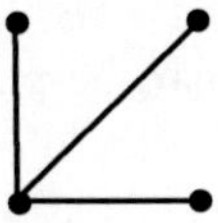

$\Gamma(\mathbb{Z}_2 \times \mathbb{F}_4)$

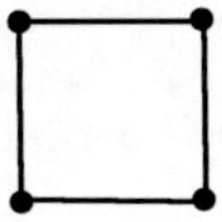

$\Gamma(\mathbb{Z}_3 \times \mathbb{Z}_3)$

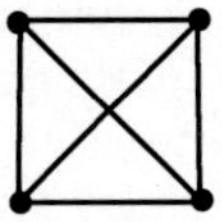

$\Gamma(\mathbb{Z}_{25})$ or $\Gamma(\mathbb{Z}_5[x]/(x^2))$

Figure 2.3: Zero-divisor graphs of order four

**Example 2.1.3.** ([83, Example 2.1], [75, Example 2.1]) (a) In Figure 2.2, zero-divisor graphs for several rings are given. Note that these examples show that non-isomorphic rings may have the same zero-divisor graph and that the zero-divisor graph does not detect nilpotent elements.

(b) By Figure 2.2, all connected graphs with fewer than four vertices may be realized as $\Gamma(R)$. Of the eleven graphs with four vertices, only six are connected. Of these six, only the three graphs given in Figure 2.3 can be realized as $\Gamma(R)$.

(c) Now, let us see the possible zero-divisor graphs $\Gamma(R)$ with $|V(\Gamma(R))| \leq 4$. Up to isomorphism, each graph may be realized as $\Gamma(R)$ by precisely the following rings:

(i) $\mathbb{Z}_4$, $\mathbb{Z}_2[x]/(x^2)$;
(ii) $\mathbb{Z}_9$, $\mathbb{Z}_2 \times \mathbb{Z}_2$, $\mathbb{Z}_3[x]/(x^2)$;
(iii) $\mathbb{Z}_6$, $\mathbb{Z}_8$, $\mathbb{Z}_2[x]/(x^3)$, $\mathbb{Z}_4[x]/(x^2-2,2x)$;
(iv) $\mathbb{F}_4[x]/(x^2)$, $\mathbb{Z}_4[x]/(x^2+x+1)$, $\mathbb{Z}_4[x]/(2,x)^2$, $\mathbb{Z}_2[x,y]/(x,y)^2$;
(v) $\mathbb{Z}_2 \times \mathbb{F}_4$;
(vi) $\mathbb{Z}_3 \times \mathbb{Z}_3$; and
(vii) $\mathbb{Z}_{25}$, $\mathbb{Z}_5[x]/(x^2)$.

**Remark 2.1.4.** ([83, Example 2.1]) (a) Now we sketch a proof that the graph $P_4$, the path of length 3, cannot be realized as $\Gamma(R)$. Let $P_4$ be the graph with vertices $\{a,b,c,d\}$ and edges $a-b, b-c, c-d$ . Suppose that there is a ring $R$ with $Z(R) = \{0,a,b,c,d\}$, $ab = bc = cd = 0$, and no other product is zero. Then $a+c \in Z(R)$ since $(a+c)b = 0$. Hence $a+c$ must be either $0, a, b, c$, or $d$. A simple check yields $a+c = b$ as the only possibility. Similarly, one can see that $b+d = c$. Hence $b = a+c = a+b+d$; so $a+d = 0$. Thus $bd = b(-a) = 0$, a contradiction. The proofs for the other two connected graphs on four vertices as nonrealizable are similar.

(*b*) We have seen in Example 2.1.3 that $\Gamma(R)$ can be a triangle or a square. But, $\Gamma(R)$ cannot be an $n$-gon for any $n \geq 5$ (the proof is similar to that in part($a$) above). However, for every $n \geq 3$, there is a zero-divisor graph with an $n$-cycle as a subgraph.

Given $n \geq 3$, let $R_n = \mathbb{Z}_2[X_1, \ldots, X_n]/I = \mathbb{Z}_2[x_1, \ldots, x_n]$, where $I = (X_1^2, \ldots, X_n^2, X_1X_2, X_2X_3, \ldots, X_nX_1)$. Then $\Gamma(R_n)$ is finite and has a cycle of length $n$, namely, $x_1 - x_2 - \cdots - x_n - x_1$.

(*c*) Let $A$ and $B$ be integral domains and $R = A \times B$. Then $\Gamma(R)$ is the complete bipartite graph $K_{|A|-1,|B|-1}$ since $V(\Gamma(R))$ can be partitioned into two disjoint vertex sets $V_1 = \{(a, 0) \mid a \in A^*\}$ and $V_2 = \{(0, b) \mid b \in B^*\}$ and two vertices $x$ and $y$ are adjacent if and only if they are in distinct vertex sets. If $A = \mathbb{Z}_2$, then $\Gamma(R)$ is a star graph. For example, $\Gamma(\mathbb{F}_p \times \mathbb{F}_q) = K_{p-1,q-1}$ and $\Gamma(\mathbb{Z}_2 \times \mathbb{F}_q) = K_{1,q-1}$. Two such specific examples are given in Figure 2.4.

Figure 2.4: Complete bipartite zero-divisor graphs

Next, we provide all zero-divisor graphs and their associated commutative rings with up to 14 vertices. These tables are due to Redmond [325, 326].

| Vertices | Ring $R$ | $\lvert R\rvert$ | Graph | Type |
|---|---|---|---|---|
| 1 | $\mathbb{Z}_4$ | 4 | $K_1$ | local |
| | $\mathbb{Z}_2[x]/(x^2)$ | 4 | $K_1$ | local |
| 2 | $\mathbb{Z}_2 \times \mathbb{Z}_2$ | 4 | $K_2$ | reduced |
| | $\mathbb{Z}_9$ | 9 | $K_2$ | local |
| | $\mathbb{Z}_3[x]/(x^2)$ | 9 | $K_2$ | local |
| 3 | $\mathbb{Z}_6$ | 6 | $K_{1,2}$ | reduced |
| | $\mathbb{Z}_8$ | 8 | $K_{1,2}$ | local |
| | $\mathbb{Z}_2[x]/(x^3)$ | 8 | $K_{1,2}$ | local |
| | $\mathbb{Z}_4[x]/(2x, x^2 - 2)$ | 8 | $K_{1,2}$ | local |
| | $\mathbb{Z}_2[x, y]/(x, y)^2$ | 8 | $K_3$ | local |
| | $\mathbb{Z}_4[x]/(2, x)^2$ | 8 | $K_3$ | local |
| | $\mathbb{F}_4[x]/(x^2)$ | 16 | $K_3$ | local |
| | $\mathbb{Z}_4[x]/(x^2 + x + 1)$ | 16 | $K_3$ | local |

| Vertices | Ring $R$ | $\lvert R\rvert$ | Graph | Type |
|---|---|---|---|---|
| 4 | $\mathbb{Z}_2 \times \mathbb{F}_4$ | 8 | $K_{1,3}$ | reduced |
| | $\mathbb{Z}_3 \times \mathbb{Z}_3$ | 9 | $K_{2,2}$ | reduced |
| | $\mathbb{Z}_{25}$ | 25 | $K_4$ | local |
| | $\mathbb{Z}_5[x]/(x^2)$ | 25 | $K_4$ | local |
| 5 | $\mathbb{Z}_2 \times \mathbb{Z}_5$ | 10 | $K_{1,4}$ | reduced |
| | $\mathbb{Z}_3 \times \mathbb{F}_4$ | 12 | $K_{2,3}$ | reduced |
| | $\mathbb{Z}_2 \times \mathbb{Z}_4$ | 8 | Fig.2.5(a) | mixed |
| | $\mathbb{Z}_2 \times \mathbb{Z}_2[x]/(x^2)$ | 8 | Fig.2.5(a) | mixed |
| 6 | $\mathbb{Z}_3 \times \mathbb{Z}_5$ | 15 | $K_{2,4}$ | reduced |
| | $\mathbb{F}_4 \times \mathbb{F}_4$ | 16 | $K_{3,3}$ | reduced |
| | $\mathbb{Z}_2 \times \mathbb{Z}_2 \times \mathbb{Z}_2$ | 8 | Fig.2.5(b) | reduced |
| | $\mathbb{Z}_{49}$ | 49 | $K_6$ | local |
| | $\mathbb{Z}_7[x]/(x^2)$ | 49 | $K_6$ | local |
| 7 | $\mathbb{Z}_2 \times \mathbb{Z}_7$ | 14 | $K_{1,6}$ | reduced |
| | $\mathbb{F}_4 \times \mathbb{Z}_5$ | 20 | $K_{3,4}$ | reduced |
| | $\mathbb{Z}_3 \times \mathbb{Z}_4$ | 12 | Fig.2.6(a) | mixed |
| | $\mathbb{Z}_3 \times \mathbb{Z}_2[x]/(x^2)$ | 12 | Fig.2.6(a) | mixed |
| | $\mathbb{Z}_{16}$ | 16 | Fig.2.6(b) | local |
| | $\mathbb{Z}_2[x]/(x^4)$ | 16 | Fig.2.6(b) | local |
| | $\mathbb{Z}_4[x]/(x^2+2)$ | 16 | Fig.2.6(b) | local |
| | $\mathbb{Z}_4[x]/(x^2+2x+2)$ | 16 | Fig.2.6(b) | local |
| | $\mathbb{Z}_4[x]/(x^3-2, 2x^2, 2x)$ | 16 | Fig.2.6(b) | local |
| | $\mathbb{Z}_2[x,y]/(x^3, xy, y^2)$ | 16 | Fig.2.7(a) | local |
| | $\mathbb{Z}_8[x]/(2x, x^2)$ | 16 | Fig.2.7(a) | local |
| | $\mathbb{Z}_4[x]/(x^3, 2x^2, 2x)$ | 16 | Fig.2.7(a) | local |
| | $\mathbb{Z}_4[x,y]/(x^2-2, xy, y^2, 2x, 2y)$ | 16 | Fig.2.7(a) | local |
| | $\mathbb{Z}_4[x]/(x^2+2x)$ | 16 | Fig.2.7(b) | local |
| | $\mathbb{Z}_8[x]/(2x, x^2+4)$ | 16 | Fig.2.7(b) | local |
| | $\mathbb{Z}_2[x,y]/(x^2, y^2-xy)$ | 16 | Fig.2.7(b) | local |
| | $\mathbb{Z}_4[x,y]/(x^2, y^2-xy, xy-2, 2x, 2y)$ | 16 | Fig.2.7(b) | local |
| | $\mathbb{Z}_4[x,y]/(x^2, y^2, xy-2, 2x, 2y)$ | 16 | Fig.2.8 | local |
| | $\mathbb{Z}_2[x,y]/(x^2, y^2)$ | 16 | Fig.2.8 | local |
| | $\mathbb{Z}_4[x]/(x^2)$ | 16 | Fig.2.8 | local |
| | $\mathbb{Z}_4[x]/(x^2)$ | 16 | Fig.2.8 | local |
| | $\mathbb{Z}_2[x,y,z]/(x,y,z)^2$ | 16 | $K_7$ | local |
| | $\mathbb{Z}_4[x,y]/(x^2, y^2, xy, 2x, 2y)$ | 16 | $K_7$ | local |
| | $\mathbb{F}_8[x]/(x^2)$ | 64 | $K_7$ | local |
| | $\mathbb{Z}_4[x]/(x^3+x+1)$ | 64 | $K_7$ | local |

| Vertices | Ring $R$ | $\|R\|$ | Graph | Type |
|---|---|---|---|---|
| 8 | $\mathbb{Z}_2 \times \mathbb{F}_8$ | 16 | $K_{1,7}$ | reduced |
| | $\mathbb{Z}_3 \times \mathbb{Z}_7$ | 21 | $K_{2,6}$ | reduced |
| | $\mathbb{Z}_5 \times \mathbb{Z}_5$ | 25 | $K_{4,4}$ | reduced |
| | $\mathbb{Z}_{27}$ | 27 | Fig.2.9(a) | local |
| | $\mathbb{Z}_9[x]/(3x, x^2-3)$ | 27 | Fig.2.9(a) | local |
| | $\mathbb{Z}_9[x]/(3x, x^2-6)$ | 27 | Fig.2.9(a) | local |
| | $\mathbb{Z}_3[x]/(x^3)$ | 27 | Fig.2.9(a) | local |
| | $\mathbb{Z}_3[x,y]/(x,y)^2$ | 27 | $K_8$ | local |
| | $\mathbb{Z}_9[x]/(3,x)^2$ | 27 | $K_8$ | local |
| | $\mathbb{F}_9[x]/(x^2)$ | 81 | $K_8$ | local |
| | $\mathbb{Z}_9[x]/(x^2+1)$ | 81 | $K_8$ | local |
| 9 | $\mathbb{Z}_2 \times \mathbb{F}_9$ | 18 | $K_{1,8}$ | reduced |
| | $\mathbb{Z}_3 \times \mathbb{F}_8$ | 24 | $K_{2,7}$ | reduced |
| | $\mathbb{F}_4 \times \mathbb{Z}_7$ | 28 | $K_{3,6}$ | reduced |
| | $\mathbb{Z}_2 \times \mathbb{Z}_2 \times \mathbb{Z}_3$ | 12 | Fig.2.9(b) | reduced |
| | $\mathbb{Z}_4 \times \mathbb{F}_4$ | 16 | Fig.2.10(a) | mixed |
| | $\mathbb{Z}_2[x]/(x^2) \times \mathbb{F}_4$ | 16 | Fig.2.10(a) | mixed |
| 10 | $\mathbb{Z}_3 \times \mathbb{F}_9$ | 27 | $K_{2,8}$ | reduced |
| | $\mathbb{F}_4 \times \mathbb{F}_8$ | 32 | $K_{3,7}$ | reduced |
| | $\mathbb{Z}_5 \times \mathbb{Z}_7$ | 35 | $K_{4,6}$ | reduced |
| | $\mathbb{Z}_{121}$ | 121 | $K_{10}$ | local |
| | $\mathbb{Z}_{11}[x]/(x^2)$ | 121 | $K_{10}$ | local |
| 11 | $\mathbb{Z}_2 \times \mathbb{Z}_{11}$ | 22 | $K_{1,10}$ | reduced |
| | $\mathbb{F}_4 \times \mathbb{F}_9$ | 36 | $K_{3,8}$ | reduced |
| | $\mathbb{Z}_5 \times \mathbb{F}_8$ | 40 | $K_{4,7}$ | reduced |
| | $\mathbb{Z}_2 \times \mathbb{Z}_9$ | 18 | Fig.2.10(b) | mixed |
| | $\mathbb{Z}_2 \times \mathbb{Z}_3[x]/(x^2)$ | 18 | Fig.2.10(b) | mixed |
| | $\mathbb{Z}_5 \times \mathbb{Z}_4$ | 20 | Fig.2.11(a) | mixed |
| | $\mathbb{Z}_5 \times \mathbb{Z}_2[x]/(x^2)$ | 20 | Fig.2.11(a) | mixed |
| | $\mathbb{Z}_2 \times \mathbb{Z}_8$ | 16 | Fig.2.11(b) | mixed |
| | $\mathbb{Z}_2 \times \mathbb{Z}_2[x]/(x^3)$ | 16 | Fig.2.11(b) | mixed |
| | $\mathbb{Z}_2 \times \mathbb{Z}_4[x]/(2x, x^2-2)$ | 16 | Fig.2.11(b) | mixed |
| | $\mathbb{Z}_2 \times \mathbb{Z}_2[x,y]/(x,y)^2$ | 16 | Fig.2.12(a) | mixed |
| | $\mathbb{Z}_2 \times \mathbb{Z}_4[x]/(2,x)^2$ | 16 | Fig.2.12(a) | mixed |
| | $\mathbb{Z}_4 \times \mathbb{Z}_4$ | 16 | Fig.2.12(b) | mixed |
| | $\mathbb{Z}_4 \times \mathbb{Z}_2[x]/(x^2)$ | 16 | Fig.2.12(b) | mixed |
| | $\mathbb{Z}_2[x]/(x^2) \times \mathbb{Z}_2[x]/(x^2)$ | 16 | Fig.2.12(b) | mixed |
| 12 | $\mathbb{Z}_3 \times \mathbb{Z}_{11}$ | 33 | $K_{2,10}$ | reduced |
| | $\mathbb{Z}_5 \times \mathbb{F}_9$ | 45 | $K_{4,8}$ | reduced |
| | $\mathbb{Z}_7 \times \mathbb{Z}_7$ | 49 | $K_{6,6}$ | reduced |
| | $\mathbb{Z}_2 \times \mathbb{Z}_2 \times \mathbb{F}_4$ | 16 | Fig.2.13(a) | reduced |
| | $\mathbb{Z}_{169}$ | 169 | $K_{12}$ | local |
| | $\mathbb{Z}_{13}[x]/(x^2)$ | 169 | $K_{12}$ | local |

| Vertices | Ring $R$ | $\lvert R\rvert$ | Graph | Type |
|---|---|---|---|---|
| 13 | $\mathbb{Z}_2 \times \mathbb{Z}_{13}$ | 26 | $K_{1,12}$ | reduced |
| | $\mathbb{F}_4 \times \mathbb{Z}_{11}$ | 44 | $K_{3,10}$ | reduced |
| | $\mathbb{Z}_7 \times \mathbb{F}_8$ | 56 | $K_{6,7}$ | reduced |
| | $\mathbb{Z}_2 \times \mathbb{Z}_3 \times \mathbb{Z}_3$ | 18 | Fig.2.13(b) | reduced |
| | $\mathbb{Z}_2 \times \mathbb{Z}_2 \times \mathbb{Z}_4$ | 16 | Fig.2.14(a) | mixed |
| | $\mathbb{Z}_2 \times \mathbb{Z}_2 \times \mathbb{Z}_2[x]/(x^2)$ | 16 | Fig.2.14(a) | mixed |
| 14 | $\mathbb{Z}_3 \times \mathbb{Z}_{13}$ | 39 | $K_{2,12}$ | reduced |
| | $\mathbb{Z}_5 \times \mathbb{Z}_{11}$ | 55 | $K_{4,10}$ | reduced |
| | $\mathbb{Z}_7 \times \mathbb{F}_9$ | 63 | $K_{6,8}$ | reduced |
| | $\mathbb{F}_8 \times \mathbb{F}_8$ | 64 | $K_{7,7}$ | reduced |
| | $\mathbb{Z}_2 \times \mathbb{Z}_2 \times \mathbb{Z}_2 \times \mathbb{Z}_2$ | 16 | Fig.2.14(b) | reduced |
| | $\mathbb{Z}_3 \times \mathbb{Z}_9$ | 27 | Fig.2.15 | mixed |
| | $\mathbb{Z}_3 \times \mathbb{Z}_3/(x^2)$ | 27 | Fig.2.15 | mixed |

(a) $\Gamma(\mathbb{Z}_2 \times \mathbb{Z}_4)$ and $\Gamma(\mathbb{Z}_2 \times \mathbb{Z}_2[x]/(x^2))$ (b) $\Gamma(\mathbb{Z}_2 \times \mathbb{Z}_2 \times \mathbb{Z}_2)$

Figure 2.5: Zero-divisor graphs on 5 and 6 vertices

(a) $\Gamma(\mathbb{Z}_3 \times \mathbb{Z}_4)$ and $\Gamma(\mathbb{Z}_3 \times \mathbb{Z}_2[x]/(x^2))$ (b) $\Gamma(\mathbb{Z}_{16})$, $\Gamma(\mathbb{Z}_2[x]/(x^4))$, $\Gamma(\mathbb{Z}_4[x]/(x^2+2))$, $\Gamma(\mathbb{Z}_4[x]/(x^2+2x+2))$, and $\Gamma(\mathbb{Z}_4[x]/(x^3-2, 2x^2, 2x))$

Figure 2.6: Zero-divisor graphs on 7 vertices

It is natural to ask for which positive integers $n$ is there a commutative ring $R$ whose zero-divisor graph has $n$ vertices, equivalently, when is there a commutative ring $R$ with $|Z(R)| = n+1$? Using a formula for the number of zero-divisors in a direct product of rings by R. Gilmer [206], Redmond [327] used computer calculations to show that there are no reduced commutative rings (with identity) with $1206, 1210, 1806, 3342, 5466, 6462, 6534, 6546,$ or $7340$ zero-divisors. Additional work showed that there are no commutative rings $R$ (with identity) with

(a) $\Gamma(\mathbb{Z}_2[x,y]/(x^3,xy,y^2))$, $\Gamma(\mathbb{Z}_8[x]/(2x,x^2))$, $\Gamma(\mathbb{Z}_4[x]/(x^3,2x^2,2x))$, and $\Gamma(\mathbb{Z}_4[x,y]/(x^2-2,xy,y^2,2x,2y))$.

(b) $\Gamma(\mathbb{Z}_4[x]/(x^2+2x))$, $\Gamma(\mathbb{Z}_8[x]/(2x,x^2+4))$, $\Gamma(\mathbb{Z}_2[x,y]/(x^2,y^2-xy))$, and $\Gamma(\mathbb{Z}_4[x,y]/(x^2,y^2-xy,xy-2,2x,2y))$

Figure 2.7: Zero-divisor graphs on 7 vertices

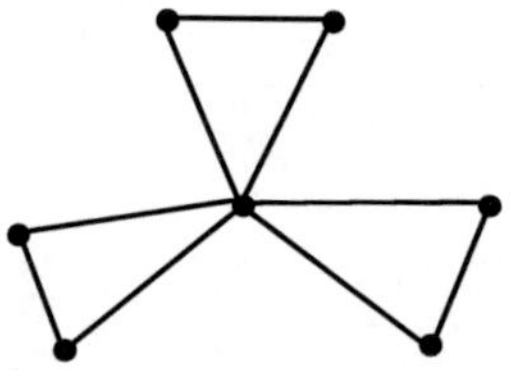

$\Gamma(\mathbb{Z}_4[x,y]/(x^2,y^2,xy-2,2x,2y))$, $\Gamma(\mathbb{Z}_2[x,y]/(x^2,y^2))$, and $\Gamma(\mathbb{Z}_4[x]/(x^2))$

Figure 2.8: Zero-divisor graphs on 7 vertices

1210, 3342, or 5466 zero-divisors. Thus, there is no commutative ring $R$ (with identity) such that $\Gamma(R)$ has 1209, 3341, or 5465 vertices. However, see Section 2.6 for commutative rings without identity.

Beck's focus was on rings that could be finitely colored, i.e., $\chi(\Gamma_0(R)) < \infty$, where $\chi(G)$ denotes the chromatic number of $G$, and he called such rings Colorings. Beck proved many other results about Colorings. For example, he showed that if $R$ is a Coloring, then $Ass(R)$ is finite [137, Theorem 4.3], and for $P \in Ass(R)$, either $R_P$ is a field or $P$ is a maximal ideal of $R$ [137, Theorem 4.4]. He also investigated the stability of the family of Colorings and showed that a finite direct product of Colorings is a Coloring [137, Theorem 5.5], a localization of a Coloring is a Coloring [137, Theorem 5.8], and certain factor rings of Colorings are Colorings [137, Theorems 5.2, 5.4, and 5.6]. Moreover, Beck determined all the finite commutative rings $R$ with $\chi(\Gamma_0(R)) \leq 3$ [137, Theorem 7.4(a)-(b)].

Let $\omega(G)$ be the clique number of a graph $G$. Then it is easy to see that $\omega(G) \leq \chi(G)$ for any graph $G$. In this connection, Beck conjectured that $\omega(\Gamma_0(R)) = \chi(\Gamma_0(R))$ for any Coloring $R$. In 1993, five years after Beck's paper appeared, D. D. Anderson and M. Naseer [60] provided a counterexample to Beck's conjecture (also see [145] and [188]). In addition, their paper contained several positive results. For example, they showed that a Noetherian ring $R$ is

**(a)** $\Gamma(\mathbb{Z}_{27})$, $\Gamma(\mathbb{Z}_9[x]/(3x, x^2-3))$, $\Gamma(\mathbb{Z}_9[x]/(3x, x^2-6))$, and $\Gamma(\mathbb{Z}_3[x]/(x^3))$

**(b)** $\Gamma(\mathbb{Z}_2 \times \mathbb{Z}_2 \times \mathbb{Z}_3)$

Figure 2.9: Zero-divisor graphs on 8 and 9 vertices

**(a)** $\Gamma(\mathbb{Z}_4 \times \mathbb{F}_4)$ and $\Gamma(\mathbb{Z}_2[x]/(x^2) \times \mathbb{F}_4)$

**(b)** $\Gamma(\mathbb{Z}_2 \times \mathbb{Z}_9)$ and $\Gamma(\mathbb{Z}_2 \times \mathbb{Z}_3[x]/(x^2))$

Figure 2.10: Zero-divisor graphs on 9 and 11 vertices

a Coloring if and only if it is a subring of a finite direct product of fields and a finite ring [60, Theorem 3.6] and that a ring of the form $A/\mathfrak{m}_1^{r_1} \cdots \mathfrak{m}_n^{r_n}$, where $A$ is a regular Noetherian ring, $\mathfrak{m}_1, \ldots, \mathfrak{m}_n$ are maximal ideals of $A$ with each $A/\mathfrak{m}_i$ finite, and $r_1, \ldots, r_n$ are positive integers, is a Coloring and determined its chromatic number [60, Corollary 3.3]. They also determined all finite commutative rings $R$ with $\chi(\Gamma_0(R)) = 4$. More properties on coloring of zero-divisor graphs are given in subsequent sections.

## 2.2 Basic Properties of Zero-divisor Graphs

Note that $\Gamma(R)$ may be infinite (i.e., a ring may have an infinite number of zero-divisors). But $\Gamma(R)$ is probably of most interest when it is finite, for then one can draw $\Gamma(R)$. We will state most results in as general a setting as possible, and then often specialize to the finite case. Recall that if $R$ is finite, then every prime ideal of $R$ is an annihilator ideal, and every nonunit of $R$ is nilpotent if and only if $R$ is local. Moreover, if $R$ is a finite local ring with maximal ideal $\mathfrak{m}$, then $char(R) = p^n$ for some prime $p$ and integer $n \geq 1$. Hence $\mathfrak{m}\,(= Z(R))$ is a $p$-group; so $|\Gamma(R)| = p^m - 1$ for some integer $m \geq 0$. The first result on $\Gamma(R)$ is that $Z(R)$ is finite if and only if either $R$ is finite or an integral domain (this result, with a different proof, and the fact that $|R| \leq |Z(R)|^2$ when $2 \leq |Z(R)| < \infty$ is due to Ganesan [200, Theorem 1]).

**Theorem 2.2.1.** ([83, Theorem 2.2]) *Let $R$ be a commutative ring. Then $\Gamma(R)$ is finite if and only if either $R$ is finite or an integral domain. In particular, if*

(a) $\Gamma(\mathbb{Z}_5 \times \mathbb{Z}_4)$ and $\Gamma(\mathbb{Z}_5 \times \mathbb{Z}_2[x]/(x^2))$ (b) $\Gamma(\mathbb{Z}_2 \times \mathbb{Z}_8)$, $\Gamma(\mathbb{Z}_2 \times \mathbb{Z}_2[x]/(x^3))$, and $\Gamma(\mathbb{Z}_2 \times \mathbb{Z}_4[x]/(2x, x^2 - 2))$

Figure 2.11: Zero-divisor graphs on 11 vertices

(a) $\Gamma(\mathbb{Z}_2 \times \mathbb{Z}_2[x,y]/(x,y)^2)$ and $\Gamma(\mathbb{Z}_2 \times \mathbb{Z}_4[x]/(2,x)^2)$ (b) $\Gamma(\mathbb{Z}_4 \times \mathbb{Z}_4)$, $\Gamma(\mathbb{Z}_4 \times \mathbb{Z}_2[x]/(x^2))$, and $\Gamma(\mathbb{Z}_2[x]/(x^2) \times \mathbb{Z}_2[x]/(x^2))$

Figure 2.12: Zero-divisor graphs on 11 vertices

$1 \leq |\Gamma(R)| < \infty$, *then $R$ is finite and not a field.*

*Proof.* The proof of the "if" part is trivial. On the other hand, assume that $\Gamma(R)$ is finite and nonempty, i.e., $Z(R)^*$ is finite. Then there are $0 \neq x, y \in R$ with $xy = 0$. Let $I = Ann(x)$. Then $I \subseteq Z(R)$ is finite and $ry \in I$ for all $r \in R$. If $R$ is infinite, then there is an $i \in I$ with $J = \{r \in R \mid ry = i\}$ infinite. For any $r, s \in J$, $(r - s)y = 0$, and so $Ann(y) \subseteq Z(R)$ is infinite, a contradiction. Thus $R$ must be finite.

For an alternate proof, let $0 \neq x \in Z(R)$. Then the $R$-module homomorphism $f : R \longrightarrow R$, given by $f(r) = rx$, has $Ker(f) = Ann(x) \subseteq Z(R)$ and $im(f) = xR \subseteq Z(R)$. Thus $|R| = |Ann(x)||xR| \leq |Z(R)|^2$. □

We next discuss when $\Gamma(R)$ has a vertex adjacent to every other vertex (i.e., when $\Gamma(R)$ has a spanning tree which is a star graph). This characterization is the key concept for determining when $\Gamma(R)$ is either a complete graph or a star graph.

**Theorem 2.2.2.** ([83, Theorem 2.5]) *Let $R$ be a commutative ring. Then there is a vertex of $\Gamma(R)$ which is adjacent to every other vertex of $\Gamma(R)$ if and only if either $R \cong \mathbb{Z}_2 \times A$, where $A$ is an integral domain, or $Z(R)$ is an annihilator ideal (and hence is prime).*

*Proof.* ($\Rightarrow$) Suppose that $Z(R)$ is not an annihilator ideal and $0 \neq a \in Z(R)$ is adjacent to every other vertex. Now $a \notin Ann(a) = I$, for otherwise $Z(R) = I$

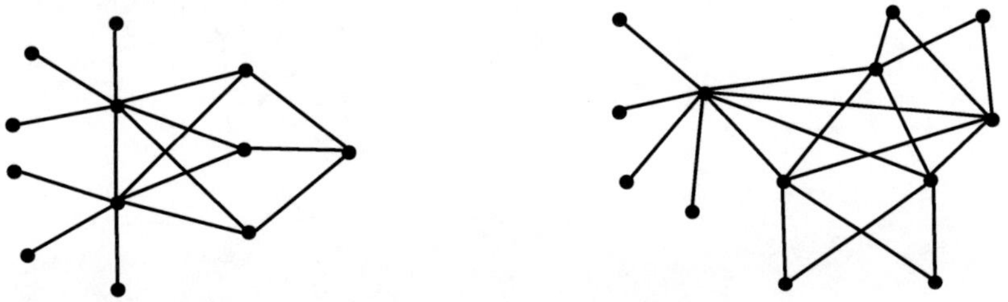

(a) $\Gamma(\mathbb{Z}_2 \times \mathbb{Z}_2 \times \mathbb{F}_4)$ (b) $\Gamma(\mathbb{Z}_2 \times \mathbb{Z}_3 \times \mathbb{Z}_3)$

Figure 2.13: Zero-divisor graphs on 12 and 13 vertices

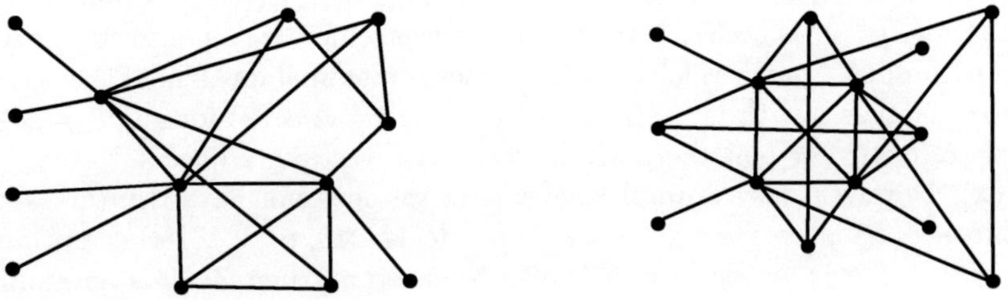

(a) $\Gamma(\mathbb{Z}_2 \times \mathbb{Z}_2 \times \mathbb{Z}_4)$ and $\Gamma(\mathbb{Z}_2 \times \mathbb{Z}_2 \times \mathbb{Z}_2[x]/(x^2))$ (b) $\Gamma(\mathbb{Z}_2 \times \mathbb{Z}_2 \times \mathbb{Z}_2 \times \mathbb{Z}_2)$

Figure 2.14: Zero-divisor graphs on 13 and 14 vertices

would be an annihilator ideal. Thus $I$ is maximal among annihilator ideals, and hence is prime [228, Theorem 6]. If $a^2 \neq a$, then $a^3 = a^2a = 0$, and hence $a \in I$, a contradiction. Thus $a^2 = a$; so $R = Ra \oplus R(1-a)$. Hence we may assume that $R = R_1 \times R_2$ with $(1,0)$ adjacent to every other vertex. For any $1 \neq c \in R_1, (c,0)$ is a nonzero zero-divisor; so $(c,0) = (c,0)(1,0) = (0,0)$, a contradiction unless $c = 0$. Hence $R_1 \cong \mathbb{Z}_2$. If $R_2$ is not an integral domain, then there is a $0 \neq b \in Z(R_2)$. Then $(1,b)$ is a zero-divisor of $R$ which is not adjacent to $(0,1)$, a contradiction. Thus $R_2$ must be an integral domain. Note that if $Z(R)$ is an annihilator ideal, then it is certainly maximal among annihilator ideals, and hence is prime.

($\Leftarrow$) If $R \cong \mathbb{Z}_2 \times A$ for an integral domain $A$, then $(1,0)$ is adjacent to every other vertex. If $Z(R) = Ann(x)$ for some $0 \neq x \in R$, then $x$ is adjacent to every other vertex. □

Example 2.1.3($c$) shows that both cases in Theorem 2.2.2 may occur for the same graph. However, if $R$ is reduced and $\Gamma(R)$ has a vertex adjacent to every other vertex, then $R$ must have the form $\mathbb{Z}_2 \times A$ for an integral domain $A$.

Note that the proof of Theorem 2.2.2 shows that if a vertex $x$ of $\Gamma(R)$ is adjacent to every other vertex, then either $x$ is idempotent with $Rx = \{0, x\}$ a

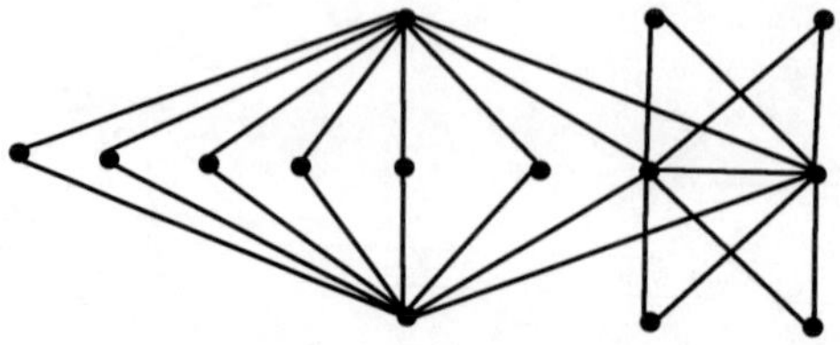

$\Gamma(\mathbb{Z}_3 \times \mathbb{Z}_9)$ and $\Gamma(\mathbb{Z}_3 \times \mathbb{Z}_3[x]/(x^2))$

Figure 2.15: Zero-divisor graphs on 14 vertices

prime ideal of $R$, or $Z(R) = Ann(x)$. In fact, if $Z(R)$ is an annihilator ideal, then $Ann(Z(R))^*$ is precisely the set of vertices which are adjacent to every other vertex. Suppose that $R$ is local with (nonzero) maximal ideal $\mathfrak{m}$. If there is a least positive integer $k$ with $\mathfrak{m}^k = 0$, then $Z(R) = \mathfrak{m} = Ann(x)$ for any $0 \neq x \in \mathfrak{m}^{k-1}$ (so if $|R/\mathfrak{m}| \geq 3$, then there are at least two vertices adjacent to every other vertex). However, easy examples show that we need not have $Ann(\mathfrak{m}) = \mathfrak{m}^{k-1}$.

Since $Z(R)$ is always a union of prime ideals [228, p. 3], $Z(R)$ is a prime ideal if and only if it is an ideal. If $R$ is also Noetherian, then $Z(R)$ is an annihilator ideal if and only if it is an (prime) ideal [228, Theorems 6 and 82]. Recall that $\{0\}$ is a primary ideal of $R$ if and only if $Z(R) = Nil(R)$. If $dim(R) = 0$, then $Z(R) = Nil(R)$ if and only if $Z(R)$ is a (prime) ideal of $R$; if $R$ is finite, this is equivalent to $R$ being local.

**Corollary 2.2.3.** ([83, Corollary 2.6]) *Let $R$ be a commutative Noetherian ring. Then there is a vertex of $\Gamma(R)$ which is adjacent to every other vertex if and only if either $R \cong \mathbb{Z}_2 \times A$, where $A$ is an (Noetherian) integral domain, or $Z(R)$ is an (prime) ideal of $R$. If in addition, $dim(R) = 0$, then either $R \cong \mathbb{Z}_2 \times A$, where $A$ is a field, or $\{0\}$ is a primary ideal of $R$ (i.e., $Z(R) = Nil(R)$).*

The Noetherian hypothesis is needed in Corollary 2.2.3. For, let $\{x_\beta\}$ be an infinite family of indeterminates. Then $R = \mathbb{Z}_2[\{x_\beta\}]/(\{x_\beta^2\})$ is a nonNoetherian zero-dimensional local ring with maximal ideal $Z(R) = Nil(R) = (\{x_\beta\})/(\{x_\beta^2\})$, but $Z(R)$ is not an annihilator ideal. We next specialize to the case when $R$ is finite.

**Corollary 2.2.4.** ([83, Corollary 2.7]) *Let $R$ be a finite commutative ring. Then there is a vertex of $\Gamma(R)$ which is adjacent to every other vertex if and only if either $R \cong \mathbb{Z}_2 \times F$, where $F$ is a finite field, or $R$ is local. Moreover, for some prime $p$ and integer $n \geq 1, |\Gamma(R)| = |F| = p^n$ if $R \cong \mathbb{Z}_2 \times F$ and $|\Gamma(R)| = p^n - 1$ if $R$ is local.*

The next result determines when $\Gamma(R)$ is a complete graph. By definition, $\Gamma(R)$ is complete if and only if $xy = 0$ for all distinct $x, y \in Z(R)$. Except for the case when $R \cong \mathbb{Z}_2 \times \mathbb{Z}_2$, the next theorem shows that we must also have $x^2 = 0$

for all $x \in Z(R)$ when $\Gamma(R)$ is complete. So, except for that one case, nilpotent elements are detected by complete graphs.

**Theorem 2.2.5.** ([83, Theorem 2.8]) *Let $R$ be a commutative ring. Then $\Gamma(R)$ is complete if and only if either $R \cong \mathbb{Z}_2 \times \mathbb{Z}_2$ or $xy = 0$ for every $x, y \in Z(R)$.*

*Proof.* ($\Leftarrow$) By definition.
($\Rightarrow$) Suppose that $\Gamma(R)$ is complete. Then trivially $xy = 0$ for all distinct $x, y \in Z(R)$. Suppose there is an $x \in Z(R)$ with $x^2 \neq 0$. We show that $x^2 = x$. If not, then $x^3 = x^2x = 0$. Hence $x^2(x + x^2) = 0$ with $x^2 \neq 0$; so $x + x^2 \in Z(R)$. If $x + x^2 = x$, then $x^2 = 0$, a contradiction. Thus $x + x^2 \neq x$; so $x^2 = x^2 + x^3 = x(x + x^2) = 0$ since $\Gamma(R)$ is complete, again a contradiction. Hence $x^2 = x$. As in the proof of Theorem 2.2.2, we have $R \cong \mathbb{Z}_2 \times A$, and necessarily $A \cong \mathbb{Z}_2$. □

**Corollary 2.2.6.** ([83, Corollary 2.9]) *Let $R$ be a commutative ring. For $x, y \in Z(R)$, define $x \sim y$ if $xy = 0$ or $x = y$, and define $x \sim^* y$ if $xy = 0$.*

(a) *The relation $\sim$ is transitive (equivalently, an equivalence relation) if and only if $\Gamma(R)$ is complete;*

(b) *The relation $\sim^*$ is transitive (equivalently, an equivalence relation) if and only if $\Gamma(R)$ is complete and $R \ncong \mathbb{Z}_2 \times \mathbb{Z}_2$.*

Suppose that $R$ is a finite commutative ring and not isomorphic to $\mathbb{Z}_2 \times \mathbb{Z}_2$. By Corollary 2.2.4 and Theorem 2.2.5, $\Gamma(R)$ is complete if and only if $R$ is local with maximal ideal $\mathfrak{m}$ and $\mathfrak{m}^2 = 0$. However, this does not hold if $R$ is not finite. For example, if $Z(R)^2 = 0$, then by McCoy's Theorem, also $Z(R[x])^2 = 0$. Hence $\Gamma(\mathbb{Z}_4[x])$ is an infinite complete graph with vertex set $2\mathbb{Z}_4[x] \setminus \{0\}$, but $\mathbb{Z}_4[x]$ is not local. Suppose that $\Gamma(R)$ is complete and $R \ncong \mathbb{Z}_2 \times \mathbb{Z}_2$; then by Theorem 2.2.5, $Z(R)$ is an ideal of $R$ with $Z(R)^2 = 0$. Hence $Z(R) = Nil(R)$ is a prime ideal of $R$.

**Theorem 2.2.7.** ([83, Theorem 2.10]) *Let $R$ be a finite commutative ring. If $\Gamma(R)$ is complete, then either $R \cong \mathbb{Z}_2 \times \mathbb{Z}_2$ or $R$ is local with $char(R) = p$ or $p^2$, and $|\Gamma(R)| = p^n - 1$, where $p$ is prime and $n \geq 1$.*

*Proof.* For a field $F$, the graph $\Gamma(\mathbb{Z}_2 \times F)$ is not complete unless $F = \mathbb{Z}_2$. So assume that $R \ncong \mathbb{Z}_2 \times \mathbb{Z}_2$. By Corollary 2.2.4, $R$ must be local with maximal ideal $\mathfrak{m}$. Hence $char(R) = p^m$ for some prime $p$ and integer $m \geq 1$. If $m \geq 3$, then $R$ would have nonadjacent zero-divisors, a contradiction. Hence $char(R) = p$ or $p^2$. Since $\mathfrak{m}$ is a $p$-group, $|\Gamma(R)| = p^n - 1$ for some prime $p$ and integer $n \geq 1$. □

In the following example, we show conversely that for every prime $p$ and integer $n \geq 1$, there is a ring $R$ such that $\Gamma(R)$ is complete and $|\Gamma(R)| = p^n - 1$. Thus, $K_m$ may be realized as $\Gamma(R)$ if and only if $m = p^n - 1$ for some prime $p$ and integer $n \geq 1$. Example 2.2.8 also shows that any graph $\Gamma$ is a subgraph of a zero-divisor graph $\Gamma(R)$ for some (quasilocal) commutative ring $R$.

**Example 2.2.8.** ([83, Example 2.11]) $(a)$ Let $p$ be prime, $X = \{x_\alpha\}_{\alpha\in\mathfrak{B}}$ a family of indeterminates, $A = \mathbb{Z}_p[x]$, and $I = (\{x_\alpha x_\beta \mid \alpha, \beta \in \mathfrak{B}\})$. Then $R = A/I$ is a zero-dimensional quasilocal ring with maximal ideal $\mathfrak{m} = (x)/I$ and $\mathfrak{m}^2 = 0$. Thus $\Gamma(R)$ is complete. Moreover, $|\Gamma(R)| = p^n - 1$ if $|\mathfrak{B}| = n$ and $|\Gamma(R)| = |\mathfrak{B}|$ if $\mathfrak{B}$ is infinite. Also, for any field $F$, the zero-dimensional local ring $S = F[x]/(x^2)$ has a complete zero-divisor graph with $|\Gamma(S)| = |F| - 1$.

$(b)$ Let $p$ be prime and $R = \mathbb{Z}_{p^2}$. Then $\Gamma(R)$ is complete with $|\Gamma(R)| = p - 1$. Note that $\mathbb{Z}_{p^2}$ and $\mathbb{Z}_p[x]/(x^2)$ are not isomorphic as rings, but have the same zero-divisor graph, namely, $K_{p-1}$. In particular, none of $\mathbb{Z}_2 \times \mathbb{Z}_2, \mathbb{Z}_9$, or $\mathbb{Z}_3[x]/(x^2)$ are isomorphic as rings, but all have the same complete zero-divisor graph $K_2$.

Hence, for a finite commutative ring $R$, $\Gamma(R)$ is a complete graph if and only if either $R \cong \mathbb{Z}_2 \times \mathbb{Z}_2$ or $R$ is a local ring with maximal ideal $\mathfrak{m}$ such that $\mathfrak{m}^2 = 0$.

We next consider when $\Gamma(R)$ has exactly one vertex which is adjacent to every other vertex.

**Lemma 2.2.9.** ([83, Lemma 2.12]) *Let $R$ be a finite commutative ring. If $\Gamma(R)$ has exactly one vertex adjacent to every other vertex and no other adjacent vertices, then either $R \cong \mathbb{Z}_2 \times F$, where $F$ is a finite field with $|F| \geq 3$, or $R$ is local with maximal ideal $\mathfrak{m}$ satisfying $R/\mathfrak{m} \cong \mathbb{Z}_2, \mathfrak{m}^3 = 0$, and $|\mathfrak{m}^2| \leq 2$. Thus $|\Gamma(R)|$ is either $p^n$ or $2^n - 1$ for some prime $p$ and integer $n \geq 1$.*

*Proof.* If $R \ncong \mathbb{Z}_2 \times F$, then $R$ is local with maximal ideal $\mathfrak{m}$ by Corollary 2.2.4. Thus $\mathfrak{m} = Ann(x)$ for a unique $x \in \mathfrak{m}$. Let $k$ be the least positive integer with $\mathfrak{m}^k = 0$. Then $\mathfrak{m} = Ann(y)$ for any $0 \neq y \in \mathfrak{m}^{k-1}$. Hence $\mathfrak{m}^{k-1} = \{0, x\}$, and thus $|\mathfrak{m}^{k-1}/\mathfrak{m}^k| = 2$ yields $R/\mathfrak{m} \cong \mathbb{Z}_2$. If $k \geq 4$, then $|\mathfrak{m}^{k-2}| \geq 4$. Hence for any distinct $a, b \in \mathfrak{m}^{k-2} \setminus \mathfrak{m}^{k-1}, ab \in \mathfrak{m}^{2k-4} \subseteq \mathfrak{m}^k$. Thus $ab = 0$, a contradiction. Hence $\mathfrak{m}^3 = 0$, and thus $|\mathfrak{m}^2| \leq 2$. (Note that we may have $\mathfrak{m}^2 = 0$.) □

In Remark 2.1.4$(c)$, we observed that $\Gamma(\mathbb{Z}_2 \times F)$ is a star graph of order $|F|$ when $F$ is a finite field. We next use Lemma 2.2.9 to show that, except for zero-divisor graphs of small order (see Example 2.1.3$(a)$), this is the only way a finite star graph arises as $\Gamma(R)$.

**Theorem 2.2.10.** ([83, Theorem 2.13]) *Let $R$ be a finite commutative ring with $|\Gamma(R)| \geq 4$. Then $\Gamma(R)$ is a star graph if and only if $R \cong \mathbb{Z}_2 \times F$, where $F$ is a finite field. In particular, if $\Gamma(R)$ is a star graph, then $|\Gamma(R)| = p^n$ for some prime $p$ and integer $n \geq 1$. Conversely, every star graph of order $p^n$ can be realized as $\Gamma(R)$.*

*Proof.* Suppose that $\Gamma(R)$ is a star graph and $R \ncong \mathbb{Z}_2 \times F$. By Corollary 2.2.4 and Lemma 2.2.9, we may assume that $(R, \mathfrak{m})$ is local with $|\mathfrak{m}| = 2^k$ for some $k \geq 3$ (since $\mathfrak{m}$ is a 2-group and $|\Gamma(R)| \geq 4$) and $|\mathfrak{m}^2| = 2$. Let $\mathfrak{m} = Ann(x)$ and choose distinct $a, b, c, d \in \mathfrak{m}^* \setminus \{x\}$. Then $ab = ac = ad = x$ since $\mathfrak{m}^2 = \{0, x\}$, and there are no other zero-divisor relations. Hence $a(b - c) = a(b - d) = 0$. Note that $Ann(a) = \{0, x\}(a \notin Ann(a)$ since $Ann(a)$ has even order), and thus

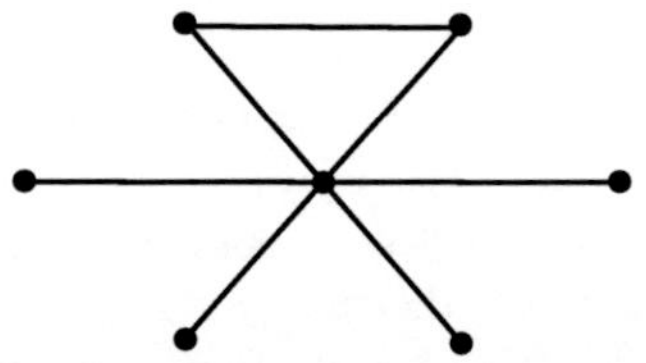

Figure 2.16: $\Gamma(\mathbb{Z}_2[x]/(x^4))$

$b - c = b - d = x$. Hence $c = d$, a contradiction. The "in particular" statement is clear and the "conversely" statement follows from the remarks preceding the theorem. □

For a complete characterization of star zero-divisor graphs, see Remark 2.3.5.

**Example 2.2.11.** ([83, Example 2.14]) For every integer $n \geq 1$, consider the finite local ring $R_n = \mathbb{Z}_2[x]/(x^{n+1})$. Then $x^n$ is the only vertex adjacent to every other vertex. However, for $n \geq 3, \Gamma(R_n)$ is not a star graph since the vertices $x^{n-1}+x^n$ and $x^{n-1}$ are also adjacent. Note that $|\Gamma(R_n)| = 2^n-1$. The zero-divisor graph $\Gamma(R_3) = \Gamma(\mathbb{Z}_2[x]/(x^4))$ is given in Figure 2.16.

In 2004, Akbari and Mohammadian [38] characterized all regular graphs which can be the zero-divisor graph of a commutative ring. First of all, they proved the following theorem that any infinite zero-divisor graph has a vertex with infinite degree.

**Theorem 2.2.12.** ([38, Theorem 6]) *If $R$ is a commutative ring such that $R$ is not an integral domain and every vertex of $\Gamma(R)$ has finite degree, then $R$ is finite.*

**Theorem 2.2.13.** ([38, Theorem 7]) *Let $R$ be a finite commutative ring. If $\Gamma(R)$ is a regular graph, then it is either a complete graph or a complete bipartite graph.*

*Proof.* Assume that $\Gamma(R)$ is a regular graph of degree $r$. First, we assume that $R = R_1 \times R_2$ is a decomposable ring. Since the degree of $(1,0)$ is $|R_2| - 1$ and the degree of $(0,1)$ is $|R_1| - 1$, we have $|R_1| = |R_2| = r+1$. We show that $R_1$ is a field. If not, then there are two nonzero elements $a$ and $b$ in $R_1$ such that $ab = 0$. But $(\{0\} \times R_2) \cup \{(b,1)\} \subseteq Ann((a,0))$, and it follows that $deg((a,0)) \geq r+1$, a contradiction. Similarly, $R_2$ must be a field. So in this case, $\Gamma(R) \cong K_{r,r}$. Now, suppose that $R$ is an indecomposable ring. Then $R$ is a local ring and $Z(R)$ is a nilpotent ideal. Thus $Ann(Z(R)) \neq 0$, and since $\Gamma(R)$ is a regular graph, we conclude that $\Gamma(R)$ is a complete graph. □

## 2.3 Girth of Zero-divisor Graphs

In this section, we discuss the distance concept in zero-divisor graphs of commutative rings, and hence the girth of $\Gamma(R)$. Let us start with some definitions which are related to the distance concept. A graph $G$ is *connected* if there is a path between any two distinct vertices. For distinct vertices $x$ and $y$ of $G$, let $d(x,y)$ be the length of a shortest path from $x$ to $y$ ($d(x,y) = \infty$ if there is no such path). The *diameter* of $G$ is $diam(G) = \sup\{d(x,y) \mid x \text{ and } y \text{ are distinct vertices of } G\}$. The *girth* of $G$, denoted by $gr(G)$, is defined as the length of a shortest cycle in $G$ ($gr(G) = \infty$ if $G$ contains no cycles).

We discuss the girth of the zero-divisor graph of a commutative ring and its extension rings. Note that if $G$ contains a cycle, then $gr(G) \leq 2 \cdot diam(G) + 1$ [185, Proposition 1.3.2]. The next result shows that the zero-divisor graph of a commutative ring is connected and has girth at most seven.

**Theorem 2.3.1.** ([83, Theorem 2.3]) *Let $R$ be a commutative ring. Then $\Gamma(R)$ is connected and $gr(\Gamma(R)) \leq 7$ whenever $\Gamma(R)$ contains a cycle.*

*Proof.* Let $x, y \in Z(R)^*$ be distinct. If $xy = 0$, then $d(x,y) = 1$. So assume that $xy$ is nonzero. If $x^2 = y^2 = 0$, then $x - xy - y$ is a path of length 2; thus $d(x,y) = 2$. If $x^2 = 0$ and $y^2 \neq 0$, then there is a $b \in Z(R)^* \setminus \{x,y\}$ with $by = 0$. If $bx = 0$, then $x - b - y$ is a path of length 2. If $bx \neq 0$, then $x - bx - y$ is a path of length 2. In either case, $d(x,y) = 2$. A similar argument holds if $y^2 = 0$ and $x^2 \neq 0$. Thus we may assume that $xy$, $x^2$, and $y^2$ are all nonzero. Hence there are $a, b \in Z(R)^* \setminus \{x,y\}$ with $ax = by = 0$. If $a = b$, then $x - a - y$ is a path of length 2. Thus we may assume that $a \neq b$. If $ab = 0$, then $x - a - b - y$ is a path of length 3, and hence $d(x,y) \leq 3$. If $ab \neq 0$, then $x - ab - y$ is a path of length 2; thus $d(x,y) \leq 2$. Hence $d(x,y) \leq 3$, and thus $diam(\Gamma(R)) \leq 3$. Since $gr(G) \leq 2diam(G) + 1$, we have $gr(\Gamma(R)) \leq 7$.

For a simpler proof that $d(x,y) \leq 3$, suppose that $xy \neq 0$. Then there are $z, w \in Z(R)^*$ with $xz = wy = 0$. If $zw \neq 0$, then $x - zw - y$ is a path of length 2; so $d(x,y) = 2$. If $zw = 0$, then $x - z - w - y$ is a path of length at most 3 (we could have $x = z$). Thus $d(x,y) \leq 3$. □

The next result shows that $gr(\Gamma(R)) \leq 4$ when $R$ is a commutative Artinian ring. Example 2.1.3 gives several rings $R$ with $gr(\Gamma(R)) = 3, 4$, or $\infty$.

**Theorem 2.3.2.** ([83, Theorem 2.4]) *Let $R$ be a commutative Artinian ring (in particular, $R$ could be a finite commutative ring). If $\Gamma(R)$ contains a cycle, then $gr(\Gamma(R)) \leq 4$.*

*Proof.* Suppose that $\Gamma(R)$ contains a cycle. By [105, Theorem 8.7], $R$ is a finite direct product of Artinian local rings. First, suppose that $R$ is local with nonzero maximal ideal $\mathfrak{m}$. Then $\mathfrak{m} = Ann(x)$ for some $x \in \mathfrak{m}^*$ [228, Theorem 82]. If there are distinct $y, z \in \mathfrak{m}^* \setminus \{x\}$ with $yz = 0$, then $y - x - z - y$ is a triangle. Otherwise, $\Gamma(R)$ contains no cycles, a contradiction. In this case, $gr(\Gamma(R)) = 3$.

Next, suppose that $R = R_1 \times R_2$. If both $|R_1| \geq 3$ and $|R_2| \geq 3$, then we may choose $a_i \in R_i \setminus \{0,1\}$. Then $(1,0)-(0,1)-(a_1,0)-(0,a_1)-(1,0)$ is a square. So in this case, $gr(\Gamma(R)) \leq 4$. Thus we may assume that $R_1 = \mathbb{Z}_2$. If $|Z(R_2)| \leq 2$, then $\Gamma(R)$ contains no cycles, a contradiction. Hence we must have $|Z(R_2)| \geq 3$. Since $\Gamma(R)$ is connected, there are distinct $x, y \in Z(R_2)^*$ with $xy = 0$. Thus $(0,x) - (1,0) - (0,y) - (0,x)$ is a triangle. Hence in this case, $gr(\Gamma(R)) = 3$. Thus in all cases, $gr(\Gamma(R)) \leq 4$. □

The proof of the above theorem shows that a finite commutative ring $R$ has $gr(\Gamma(R)) = 4$ if and only if either $R \cong F \times K$ for finite fields $F$ and $K$ with $|F|, |K| \geq 3$ (cf. Remark 2.1.4(c)), or $R \cong F \times A$ for $F$ a finite field with $|F| \geq 3$ and $A$ a finite ring with $|Z(A)| = 2$ (so either $A \cong \mathbb{Z}_4$ or $\mathbb{Z}_2[x]/(x^2)$). Also, one may easily show that $gr(\Gamma(R)) = \infty$ if and only if either $|\Gamma(R)| \leq 2$, $|\Gamma(R)| = 3$ and $\Gamma(R)$ is not complete, or $R \cong \mathbb{Z}_2 \times A$ for A a finite field or a finite ring with $|Z(A)| = 2$ (cf. Theorem 2.2.10).

Anderson and Livingston [83] conjectured that $gr(\Gamma(R)) \in \{3, 4, \infty\}$ for any commutative ring $R$. Theorem 2.3.2 proved this if the ring was Artinian. The conjecture was proven independently by Mulay [295] and DeMeyer and Schneider [178]. Additionally, Axtell, Coykendall, and Stickles [119] provided a simple proof. Their proof is given below. (A similar simple proof is given in [433, Proposition 6].)

**Theorem 2.3.3.** ([119, Theorem 2.2], [295, Theorem 1.4], [178, Theorem 1.6]) *Let $R$ be a commutative ring (not necessarily with identity). If $\Gamma(R)$ contains a cycle, then $gr(\Gamma(R)) \leq 4$.*

*Proof.* If $\Gamma(R)$ contains a cycle $x_0 - x_1 - \cdots - x_n - x_0$ with $n \geq 4$ and $x_i x_j = 0$ for some $i$ and $j$, where $j > i+1$ and either $0 \leq i < j \leq n-1$ or $1 \leq i < j \leq n$, then we can eliminate $x_k$, where $i < k < j$, to obtain a shorter cycle. So, assume $\Gamma(R)$ contains a cycle $x_0 - x_1 - \cdots - x_n - x_0$ with $n \geq 4$ and $x_i x_j \neq 0$ for any $i$ and $j$ under the above conditions. Then we can form the cycle $x_0 - x_1 x_{n-1} - x_n - x_0$ of length 3.

So assume, without loss of generality, that $x_1 x_{n-1} = x_0$. This necessarily forces $x_0^2 = 0$. If there is a $y \in R$ such that $x_0 y \neq 0$ or $x_0$, then either $x_0 - x_1 - x_0 y - x_0$ is a cycle of length 3 (if $x_0 y \neq x_1$), or $x_0 - x_n - x_0 y - x_0$ is a cycle of length 3 (if $x_0 y = x_1$). Finally, if $(x_0) = \{0, x_0\}$, then $(x_0) \subsetneq (x_1)$, yielding a $y \in R$ such that $x_1 y \notin \{0, x_0, x_1\}$. Now, $x_1 y \neq x_2$ since $x_0 x_2 \neq 0$. Then $x_0 - x_1 - x_2 - x_1 y - x_0$ is a cycle of length 4. □

The upper bound in the previous theorem can be achieved by the ring $\mathbb{Z}_3 \times \mathbb{Z}_3$ (cf. Remark 2.1.4(c)). Thus $gr(\Gamma(R)) = 3, 4$, or $\infty$, and so we have the following theorem.

**Theorem 2.3.4.** ([178, Theorem 1.6], [75, Theorem 2.3]) *Let $R$ be a commutative ring which is not an integral domain. Then exactly one of the following holds:*

(a) $\Gamma(R)$ *has a cycle of length* 3 *or* 4 *(i.e.,* $gr(\Gamma(R)) \leq 4$*);*

(b) $\Gamma(R)$ *is a star graph; or*

(c) $R \cong \mathbb{Z}_2 \times \mathbb{Z}_4$ *or* $R \cong \mathbb{Z}_2 \times \mathbb{Z}_2[x]/(x^2)$.

**Remark 2.3.5.** ([75, Remark 2.4(a)]) One can explicitly list the rings which occur in part (b) of the Theorem 2.3.4. The star graph $K_{1,r}$ may be realized as $\Gamma(R)$ if and only if $r = p^n - 1$ for some prime $p$ and integer $n \geq 1$, i.e., if and only if the order of the star graph is a prime power (cf. Theorem 2.2.10). See Example 2.1.3(a) for star graphs $\Gamma(R)$ with $|\Gamma(R)| \leq 3$. A finite star graph $\Gamma(R)$ with $|\Gamma(R)| \geq 4$ can only be realized by $R \cong \mathbb{Z}_2 \times F$ for $F$ a finite field (cf. Theorem 2.2.10). For infinite star graphs $\Gamma(R)$, either $R \cong \mathbb{Z}_2 \times D$ for $D$ an infinite integral domain or $R$ is a ring with a $0 \neq x \in R$ such that $Z(R) = Ann(x)$, $Nil(R) = \{0, x\}$, and $R/Nil(R)$ is an infinite integral domain (for example: $R = \mathbb{Z}[x]/(2x, x^2)$).

Hence $\Gamma(R)$ is a star graph if and only if either $R \cong A$, where

$$A \in \{\mathbb{Z}_9, \mathbb{Z}_3[x]/(x^2), \mathbb{Z}_8, \mathbb{Z}_2[x]/(x^3), \mathbb{Z}_4[x]/(x^2 - 2, 2x)\}; R \cong \mathbb{Z}_2 \times D,$$

where $D$ is an integral domain; or there is a $0 \neq x \in R$ such that $Z(R) = Ann(x)$, $Nil(R) = \{0, x\}$, and $R/Nil(R)$ is an infinite integral domain.

**Remark 2.3.6.** If $\Gamma(R)$ is finite and $gr(\Gamma(R)) = 4$ (i.e., $\Gamma(R)$ contains a rectangle, but no triangles), then either $R \cong F_1 \times F_2$, and thus $\Gamma(R)$ is complete bipartite or $R \cong F_1 \times A$, where $F_i$ are finite fields with $|F_i| \geq 3$ and $A$ is isomorphic to either $\mathbb{Z}_4$ or $\mathbb{Z}_2[x]/(x^2)$ (by the paragraph following Theorem 2.3.2).

Now we characterize when $\Gamma(R)$ has girth $\geq 4$ in terms of the zero-divisor graph of the total quotient ring $T(R)$ and $\Gamma(R)$. Recall that the total quotient ring of $R$ is $T(R) = R_S$, where $S = R \setminus Z(R)$. First, we start with a lemma which gives a relationship between the girth of $\Gamma(T(R))$ and $\Gamma(R)$. (This also follows since $\Gamma(T(R)) \cong \Gamma(R)$ by [81, Theorem 2.2].)

**Lemma 2.3.7.** ([296, Theorem 1.1], [85, Lemma 2.1]) *Let $R$ be a commutative ring with total quotient ring $T(R)$. Then $gr(\Gamma(T(R))) = gr(\Gamma(R))$.*

*Proof.* Since $\Gamma(R)$ is a subgraph of $\Gamma(T(R))$, $gr(\Gamma(T(R))) \leq gr(\Gamma(R))$. Suppose that $gr(\Gamma(T(R))) = 3$. Then there are distinct $0 \neq q_1, q_2, q_3 \in T(R)$ such that $q_1q_2 = q_2q_3 = q_3q_1 = 0$. Let each $q_i = a_i/t$ with $a_i \in R$ and $t \in R \setminus Z(R)$. Then $a_1, a_2, a_3$ are distinct elements in $R$ with $a_1a_2 = a_2a_3 = a_3a_1 = 0$. Thus $a_1 - a_2 - a_3 - a_1$ is a triangle in $\Gamma(R)$; so $gr(\Gamma(R)) = 3$. Similarly, $gr(\Gamma(R)) = 4$ if $gr(\Gamma(T(R))) = 4$. The result for the girth now follows since the girth of a zero-divisor graph is either 3, 4, or $\infty$. □

We next use the above concepts and results to determine when $gr(\Gamma(R)) = 4$. We have two cases, depending on whether or not $R$ has any nonzero nilpotent elements. Let $\overline{K}_{m,3}$ be the graph formed by joining $G_1 = K_{m,3}(= A \cup B$ with $|A| = m$ and $|B| = 3)$ to the star graph $G_2 = K_{1,m}$ by identifying the center of $G_2$ and a point of $B$. For a similar characterization, see [296, Theorem 2.3].

**Theorem 2.3.8.** ([85, Theorem 2.2]) *The following statements are equivalent for a reduced commutative ring $R$:*

(a) $gr(\Gamma(R)) = 4$;

(b) $T(R) = F_1 \times F_2$, *where each* $F_i$ *is a field with* $|F_i| \geq 3$;

(c) $\Gamma(R) = K_{m,n}$ *with* $m, n \geq 2$.

**Theorem 2.3.9.** ([85, Theorem 2.3]) *The following statements are equivalent for a commutative ring $R$ with $Nil(R)$ nonzero:*

(a) $gr(\Gamma(R)) = 4$;

(b) $R \cong D \times B$, *where $D$ is an integral domain with $|D| \geq 3$ and $B = \mathbb{Z}_4$ or $\mathbb{Z}_2[x]/(x_2)$ (thus $T(R) \cong T(D) \times B$);*

(c) $\Gamma(R) = \overline{K}_{m,3}$ *with* $m \geq 2$.

Next, let us see when $gr(\Gamma(R)) = \infty$. Again, there are two cases, depending on whether or not $R$ is reduced. For another characterization, see [296, Theorem 2.5, Remarks 1]. Since $gr(\Gamma(R)) = 3, 4$, or $\infty$, it is also implicitly characterized when $gr(\Gamma(R)) = 3$.

**Theorem 2.3.10.** ([85, Theorem 2.4]) *The following statements are equivalent for a reduced commutative ring $R$:*

(a) $\Gamma(R)$ *is nonempty with* $gr(\Gamma(R)) = \infty$;

(b) $T(R) = \mathbb{Z}_2 \times K$, *where $K$ is a field;*

(c) $\Gamma(R) = K_{1,n}$ *for some* $n \geq 1$.

**Theorem 2.3.11.** ([85, Theorem 2.5]) *The following statements are equivalent for a commutative ring $R$ with $Nil(R)$ nonzero:*

(a) $gr(\Gamma(R)) = \infty$;

(b) $R \cong B$ *or* $R \cong \mathbb{Z}_2 \times B$, *where* $B = \mathbb{Z}_4$ *or* $\mathbb{Z}_2[x]/(x^2)$, *or $\Gamma(R)$ is a star graph;*

(c) $\Gamma(R)$ *is a singleton,* $\overline{K}_{1,3}$, *or* $K_{1,n}$ *for some* $n \geq 1$.

**Remark 2.3.12.** ([85, Remark 2.6]) (a) If $R$ is finite, then $T(R) = R$; so Theorems 2.3.8 - 2.3.11 characterize which finite rings $R$ have $gr(\Gamma(R)) = 4$ or $\infty$. For a finite ring $R$, the only case not explicitly handled above is when $R$ is not reduced and $\Gamma(R)$ is a star graph (i.e., the third case in Theorem 2.3.11(b)). In this case, necessarily $|\Gamma(R)| \leq 3$ (cf. Theorem 2.2.10); if $\Gamma(R) = K_{1,1}$, then $R \cong \mathbb{Z}_9$, or $\mathbb{Z}_3[x]/(x^2)$, and if $\Gamma(R) = K_{1,2}$, then $R \cong \mathbb{Z}_8$, $\mathbb{Z}_2[x]/(x^3)$, or $\mathbb{Z}_4[x]/(2x, x^2 - 2)$. Thus the finite graphs $K_{m,n}$, $K_{1,n}$, and $\overline{K}_{m,3}$ can be realized as a zero-divisor graph $\Gamma(R)$ if and only if $m = p^i - 1$ and $n = q^j - 1$ for some primes $p, q$ and integers $i, j \geq 1$. If $\Gamma(R)$ is an infinite star graph, then either $R \cong \mathbb{Z}_2 \times D$ for an integral domain $D$, or $Nil(R) = \{0, x\}$, where $Z(R) = Ann(x)$. Moreover, if $\Gamma(R) = K_{m,n}$ with $m, n \geq 2$, then $R$ is a subdirect sum of two integral domains, each with more than 2 elements.

(b) Let $T(R) = R_1 \times R_2$. If there are distinct $0 \neq x, y \in R_1$ with $xy = 0$, then $gr(\Gamma(R)) = 3$ since $(x, 0) - (y, 0) - (0, 1) - (x, 0)$ is a triangle. In particular, if $T(R) = \prod_{\alpha \in A} K_\alpha$ is a direct product of fields (for example, if $R$ is a reduced Noetherian ring), then $gr(\Gamma(R)) = 3$ if and only if $|A| \geq 3$.

(c) Note that if $gr(\Gamma(R)) = 4$, then $|Nil(R)| \leq 2$. Also, if $gr(\Gamma(R)) = \infty$, then $|Nil(R)| \leq 2$, except for the $|\Gamma(R)| \leq 3$ case mentioned above in (a), where $|Nil(R)| = 3$ or 4. In particular, $|Nil(R)| = 3$ when $R \cong \mathbb{Z}_9$ or $\mathbb{Z}_3[x]/(x^2)$, and $|Nil(R)| = 4$ when $R \cong \mathbb{Z}_8$, $\mathbb{Z}_2[x]/(x^3)$, or $\mathbb{Z}_4[x]/(2x, x^2 - 2)$.

We next give some results by Axtell, Stickles, and Warfel [123] which give the girth of the zero-divisor graph for the direct product of two commutative rings (not necessarily with identity).

**Theorem 2.3.13.** ([123, Theorem 4.1]) *Let $R_1$ and $R_2$ be nontrivial commutative rings (not necessarily with identity). Then $gr(\Gamma(R_1 \times R_2)) = 3$ if and only if one (or both) of the following hold:*

(i) *$|Z(R_1)^*| \geq 2$ or $|Z(R_2)^*| \geq 2$;*

(ii) *Both $R_1$ and $R_2$ contain a nonzero nilpotent element.*

**Theorem 2.3.14.** ([123, Theorem 4.2]) *Let $R_1$ and $R_2$ be commutative rings (not necessarily with identity). Then $gr(\Gamma(R_1 \times R_2)) = 4$ if and only if both of the following hold:*

(i) *$|R_1|, |R_2| \geq 3$;*

(ii) *Without loss of generality, $R_1$ is an integral domain and $|Z(R_2)^*| \leq 1$.*

**Theorem 2.3.15.** ([123, Corollary 4.3]) *Let $R_1$ and $R_2$ be commutative rings (not necessarily with identity). Then $gr(\Gamma(R_1 \times R_2)) = \infty$ if and only if*

(i) *without loss of generality, $R_1 \cong \mathbb{Z}_2$ and $|Z(R_2)^*| \leq 1$, or*

(ii) *without loss of generality, $R_1$ is an integral domain, and $|R_2| = 2$ with $R_2^2 = 0$.*

We end this section with a brief discussion of the girth of ring extensions. Let $A \subseteq B$ be an extension of commutative rings with identity. Then $\Gamma(A)$ is an induced subgraph of $\Gamma(B)$. It may happen that $\Gamma(A) = \Gamma(B)$ for $A \subsetneq B$ (this happens if and only if $A$ is a pullback of a finite local ring [75, Theorem 4.3]). It is clear that $gr(\Gamma(B)) \leq gr(\Gamma(A))$. Moreover, for every $m, n \in \{3, 4, \infty\}$ with $m \leq n$, there is an extension $A \subsetneq B$ of reduced finite commutative rings with $gr(\Gamma(B)) = m$ and $gr(\Gamma(A)) = n$ [64, Example 2.1]. More generally, if we fix the ring $A$, then there is a proper extension ring $B$ with $gr(\Gamma(B)) = m$ and $gr(\Gamma(A)) = n$ as long as $A$ is reduced, but not in general when $gr(\Gamma(A)) = \infty$ [64, Example 2.2]. However, for a fixed commutative ring $B$, there need not be a subring $A$ of $B$ with $gr(\Gamma(B)) = m$ and $gr(\Gamma(A)) = n$ [64, Example 2.3].

### 2.3.1 Rings of Polynomials and Power Series

In this section, we are interested in the girth of zero-divisor graphs under extensions to polynomial and power series rings. First of all, observe that $\Gamma(R)$ is a subgraph of both $\Gamma(R[x])$ and $\Gamma(R[[x]])$. Therefore $gr(\Gamma(R[x])) \leq gr(\Gamma(R))$ and $gr(\Gamma(R[[x]])) \leq gr(\Gamma(R))$. It is well known that a polynomial $f(x)$ is a zero-divisor in $R[x]$ if and only if there is a $0 \neq r \in R$ with $rf(x) = 0$ (McCoy's Theorem) [285]. The following is due to Fields [196]; it will be used later.

**Lemma 2.3.16.** ([119, Lemma 3.1]) *Let $R$ be a commutative ring, and let $f(x) = \sum_{i=0}^{\infty} f_i x^i \in R[[x]]$. If for some nonnegative integer $t$, $f_t$ is regular in $R$, while $f_i$ is nilpotent for $0 \leq i \leq t-1$, then $f(x)$ is regular in $R[[x]]$.*

The first result in this regard deals with the completeness of $\Gamma(R)$, $\Gamma(R[x])$, and $\Gamma(R[[x]])$.

**Theorem 2.3.17.** ([119, Theorem 3.2]) *If $R \ncong \mathbb{Z}_2 \times \mathbb{Z}_2$, then the following are equivalent:*

(a) *$\Gamma(R[[x]])$ is complete;*

(b) *$\Gamma(R[x])$ is complete;*

(c) *$\Gamma(R)$ is complete.*

*Proof.* (a) $\Rightarrow$ (b) $\Rightarrow$ (c) is obvious since $\Gamma(R)$ is a subgraph of $\Gamma(R[x])$ and $\Gamma(R[x])$ is a subgraph of $\Gamma(R[[x]])$.

(c) $\Rightarrow$ (a) Since $\Gamma(R)$ is complete and $R \ncong \mathbb{Z}_2 \times \mathbb{Z}_2$, we have $xy = 0$ for every $x, y \in Z(R)^*$. Therefore $(Z(R))^2 = 0$. Let $f(x), g(x) \in Z(R[[x]])^*$. By Lemma 2.3.16, no coefficient of $f$ or $g$ can be regular since all nonregular coefficients of $f$ and $g$ are nilpotent. Thus all coefficients of $f$ and $g$ are zero-divisors in $R$. Since $\Gamma(R)$ complete, $f(x)g(x) = 0$, and hence $\Gamma(R[[x]])$ is also complete. □

Consider the ring $R = \mathbb{Z}_2 \times \mathbb{Z}_2$. Then clearly $\Gamma(R)$ is complete. However, $\Gamma(R[x])$ is not complete. For, if $a = (1,0) + (1,0)x$ and $b = (1,0) + (1,0)x^2$, then $ab \neq 0$. Thus $\Gamma(R[x])$ is not complete, and hence $\Gamma(R[[x]])$ is not complete.

Now let us see the answer for the following question: If $f$ and $g$ are distinct, nonconstant polynomials in $R[x]$ with $fg = 0$, is there a cycle involving $f$ and $g$ that has at least one vertex contained in $\Gamma(R)$? Proofs of these results are omitted and can be found in the references.

**Proposition 2.3.18.** ([119, Proposition 4.1]) *Let $R$ be a commutative ring (not necessarily with identity). If $f, g \in Z(R[x])^*$ are distinct, nonconstant polynomials with $fg = 0$, then there are $a, b \in Z(R)^*$ such that $a - b - g - f - a$ is a cycle in $\Gamma(R[x])$, or $b - f - g - b$ is a cycle in $\Gamma(R[x])$.*

Observe that if $f \in Z(R[x])^*$ with $f$ nonconstant, then we can always find a nonconstant polynomial $g \in R[x]$ such that $f \neq g$ and $fg = 0$ by letting $g(x) = ax^{1+deg(f)}$, where $af = 0$. Also, note that there is always some nonconstant $f \in Z(R[x])^*$ as long as $Z(R)^* \neq \emptyset$. Thus we get the following result.

**Corollary 2.3.19.** ([119, Corollary 4.2]) *If $f \in Z(R[x])^*$ is a nonconstant polynomial, then there is a cycle of length 3 or 4 in $\Gamma(R[x])$ with $f$ as one vertex and some $a \in Z(R)^*$ as another.*

If $Z(R)^* \neq \emptyset$, then $gr(\Gamma(R[x])), gr(\Gamma(R[[x]])) < \infty$. Also, $\Gamma(R[x])$ always has cycles if $R$ contains nonzero zero-divisors, and there is no upper bound on the length of cycles in $\Gamma(R[x])$. Proposition 2.3.18 is not true in general for $\Gamma(R[[x]])$. For example, consider the zero-divisor graph of the commutative ring

$$R = F[y, \{x_i\}_{i=0}^{\infty}]/(\{x_0 y\} \cup \{x_i - x_{i+1}y\}_{i=0}^{\infty}),$$

where $F$ is a field (see Example 2.4.14 ([119, Example 3.3 and p. 2049])). However, Proposition 2.3.18 will hold for $\Gamma(R[[x]])$ if $R$ is Noetherian. Also, in the proof of Proposition 2.3.18, we see that there could be a cycle of length 3 in $\Gamma(R[x])$. This is not always the case. If $R = \mathbb{Z} \times \mathbb{Z}$, then $gr(\Gamma(R)) = gr(\Gamma(R[x])) = 4$. It is not true in general that for every ring $R$ we have $gr(\Gamma(R)) = gr(\Gamma(R[x]))$, since if $R = \mathbb{Z}_4$, then $gr(\Gamma(R)) = \infty$, but $gr(\Gamma(R[x])) = 3$. Things improve if $\Gamma(R)$ contains a cycle.

The following result gives a sufficient condition for $gr(\Gamma(R)) = gr(\Gamma(R[x])) = gr(\Gamma(R[[x]]))$.

**Theorem 2.3.20.** ([119, Theorem 4.3]) *Let $R$ be a commutative ring (not necessarily with identity). Then $gr(\Gamma(R)) \geq gr(\Gamma(R[x])) = gr(\Gamma(R[[x]]))$. In addition, if $R$ is reduced and $\Gamma(R)$ contains a cycle, then $gr(\Gamma(R)) = gr(\Gamma(R[x])) = gr(\Gamma(R[[x]]))$.*

**Corollary 2.3.21.** ([119, Corollary 4.4]) *If $R$ is a reduced commutative ring and $gr(\Gamma(R[x])) = 3$, then $gr(\Gamma(R)) = 3$.*

**Example 2.3.22.** ([119, Example 4.5]) If $R \cong \mathbb{Z}_6$, then $R$ has no nonzero nilpotent elements and $\Gamma(R)$ contains no cycles. Hence $gr(\Gamma(\mathbb{Z}_6[x])) = 4$.

Observe that Example 2.3.22 shows that if $R$ is reduced and $gr(\Gamma(R[x])) = 4$, then we cannot say that $gr(\Gamma(R)) = 4$. It is not true in general that if $\Gamma(R)$ contains a cycle, then $gr(\Gamma(R)) = gr(\Gamma(R[x]))$. Let $R = \mathbb{Z} \times \mathbb{Z}_4$. Then since

$(0,2)-(0,2)x-(0,2)x^2-(0,2)$ is a cycle of length three, we have $gr(\Gamma(R[x]))=3$. It is straightforward to verify that $gr(\Gamma(R))=4$.

The study of the girth of the zero-divisor graph of a polynomial ring or power series ring was continued by Anderson and Mulay [85]. As a first step, the following result shows that $\Gamma(R[x])$ and $\Gamma(R[[x]])$ always contain a cycle.

**Lemma 2.3.23.** ([85, Lemma 3.1]) *Let $R$ be a commutative ring which is not an integral domain. Then $gr(\Gamma(R[x]))$ and $gr(\Gamma(R[[x]]))$ are either 3 or 4. If $R$ is not reduced, then both zero-divisor graphs have girth 3.*

The following theorem completely determines $gr(\Gamma(R[x]))$ and $gr(\Gamma(R[[x]]))$ in terms of $gr(\Gamma(R))$.

**Theorem 2.3.24.** ([85, Theorem 3.2]) *Let $R$ be a commutative ring.*

(a) *Suppose that $\Gamma(R)$ is nonempty with $gr(\Gamma(R))=\infty$;*
  (i) *If $R$ is reduced, then $gr(\Gamma(R[x]))=gr(\Gamma(R[[x]]))=4$;*
  (ii) *If $R$ is not reduced, then $gr(\Gamma(R[x]))=gr(\Gamma(R[[x]]))=3$;*

(b) *If $gr(\Gamma(R))=3$, then $gr(\Gamma(R[x]))=gr(\Gamma(R[[x]]))=3$;*

(c) *Suppose that $gr(\Gamma(R))=4$;*
  (i) *If $R$ is reduced, then $gr(\Gamma(R[x]))=gr(\Gamma(R[[x]]))=4$;*
  (ii) *If $R$ is not reduced, then $gr(\Gamma(R[x]))=gr(\Gamma(R[[x]]))=3$.*

### 2.3.2 Idealization

In this section, we are interested in the girth of the zero-divisor graph of a ring formed by idealization. First, we briefly recall the idealization construction. Let $R$ be a commutative ring (with identity) and $M$ an (unitary) $R$-module. The *idealization* of $R$ and $M$, denoted by $R(+)M$, is defined to be the ring $R\times M$ with addition and multiplication given by $(a,m)+(b,n)=(a+b,m+n)$ and $(a,m)(b,n)=(ab,an+bm)$, respectively. For more on idealizations, see [63] and [219]. The zero-divisor graph of a somewhat similar construction, the amalgamated duplication of a commutative ring $R$ along an ideal $I$, i.e., the subring $R\bowtie I=\{(r,r+i)\mid r\in R, i\in I\}$ of $R\times R$, has been studied in [278].

**Remark 2.3.25.** (a) It is known that $Z(R(+)M)=(Z(R)\cup Z(M))(+)M$ [219, Theorem 25.3], where $Z(M)=\{r\in R\mid rm=0 \text{ for some } 0\neq m\in M\}$. We will always assume that $M$ is nonzero, and hence $R(+)M$ is not reduced since $\{0\}(+)M\subseteq Nil(R(+)M)$.

(b) If $a\in Z(R)^*$, then $(a,m)\in Z(R(+)M)^*$ for every $m\in M$. To see this, consider $b\in Z(R)^*$ with $ab=0$. If $bM=0$, then $(a,m)(b,0)=0$. If $bM\neq 0$, then there is an $n\in M$ such that $bn\neq 0$. Hence $(a,m)(0,bn)=0$.

The following result is a special case of Theorem 25.3 [219].

**Proposition 2.3.26.** ([120, Proposition 1.1]) *Let $R$ be a commutative ring and let $M$ be an $R$-module. Then* $Z(R(+)M)^* = \{(0,m) \mid m \in M^*\} \cup \{(a,n) \mid a \in R^*,\ n \in M,$ *and* $am = 0$ *for some* $m \in M^*\} \cup \{(a,n) \mid a \in Z(R)^*,\ n \in M\}$.

**Lemma 2.3.27.** ([120, Theorem 2.8(i)]) *Let $R$ be a commutative ring and let $M$ be an $R$-module. If $|M| \geq 4$, then $gr(\Gamma(R(+)M)) = 3$.*

*Proof.* Let $m_1$, $m_2$, and $m_3$ be distinct nonzero elements of $M$. Then $(0,m_1) - (0,m_2) - (0,m_3) - (0,m_1)$ is a cycle of length 3, and hence $g(\Gamma(R(+)M)) = 3$. □

In view of Lemma 2.3.27, we only need to consider when the module has two or three elements. First we look at when $M \cong \mathbb{Z}_3$ and consider $R(+)\mathbb{Z}_3$. In most cases, the girth of $\Gamma(R(+)\mathbb{Z}_3)$ is three. One item worthy of note is that if $R$ has more than three elements, then there is always a $0 \neq r \in R$ such that $r\mathbb{Z}_3 = 0$. To see this, assume $r\mathbb{Z}_3 \neq 0$ for all $r \in R^*$. Then there are distinct $r_1, r_2 \in R^*$ such that $r_1 1 = r_2 1$ and hence $(r_1 - r_2)1 = 0$, where $r_1 - r_2 \neq 0$, a contradiction. Also, since the module is unitary, the ring cannot have fewer than three elements.

**Theorem 2.3.28.** ([120, Theorem 2.1]) *Let $R$ a commutative ring with identity and $M \cong \mathbb{Z}_3$ an $R$-module. Then*

(a) *$gr(\Gamma(R(+)\mathbb{Z}_3)) = 3$ if and only if $Ann(\mathbb{Z}_3) \neq 0$;*

(b) *$gr(\Gamma(R(+)\mathbb{Z}_3)) = \infty$ if and only if $Ann(\mathbb{Z}_3) = 0$. This occurs precisely when $R \cong \mathbb{Z}_3$.*

The above result classifies the girth of $R(+)\mathbb{Z}_3$, and it is somewhat surprising that the girth is never 4. We now consider the situation when $M \cong \mathbb{Z}_2$. We shall state the classification when the girth of $\Gamma(R(+)\mathbb{Z}_2)$ is 3 and when it is infinite. We begin with the girth 3 case.

**Theorem 2.3.29.** ([120, Theorem 2.2]) *The girth of $\Gamma(R(+)\mathbb{Z}_2)$ is three if and only if one of the following holds:*

(a) *$gr(\Gamma(R)) = 3$;*

(b) *There is an $r \in R^*$ such that $r^2 = 0$;*

(c) *There are distinct $a, b \in Z(R)^*$ such that $ab = 0 = a\mathbb{Z}_2 = b\mathbb{Z}_2$.*

We will now provide necessary and sufficient conditions for ensuring that the girth of $R(+)\mathbb{Z}_2$ is infinite.

**Proposition 2.3.30.** ([120, Proposition 2.5]) *Let $R \cong \mathbb{Z}_2 \times F$, where $F$ is a field and $|F| \geq 3$. Then $gr(\Gamma(R(+)\mathbb{Z}_2)) = 4$.*

**Lemma 2.3.31.** ([120, Lemma 2.6]) *If $diam(\Gamma(R)) = 3$, then $gr(\Gamma(R(+)\mathbb{Z}_2))$ is finite.*

**Theorem 2.3.32.** ([120, Theorem 2.7]) *$gr(\Gamma(R(+)\mathbb{Z}_2)) = \infty$ if and only if $R \cong \mathbb{Z}_2 \times \mathbb{Z}_2$ or $R$ is an integral domain.*

The following theorem summarizes the results on the girth of $\Gamma(R(+)M)$.

**Theorem 2.3.33.** ([120, Theorem 2.8]) *Let $R$ be a commutative ring and $M$ an $R$-module.*

(a) *$gr(\Gamma(R(+)M)) = 3$ if and only if exactly one of the following holds:*
  - (i) $|M| \geq 4$;
  - (ii) *$M \cong \mathbb{Z}_3$ and $Ann(M) \neq 0$; or*
  - (iii) *$M \cong \mathbb{Z}_2$ and one of the following holds:*
    - (1) $gr(\Gamma(R)) = 3$;
    - (2) *there is a $0 \neq r \in R$ such that $r^2 = 0$, or*
    - (3) *there are distinct $a, b \in Z(R)^*$ such that $ab = 0 = aM = bM$;*

(b) *$gr(\Gamma(R(+)M)) = \infty$ if and only if exactly one of the following holds:*
  - (1) *$M \cong \mathbb{Z}_3$ and $Ann(M) = 0$ (and $R \cong \mathbb{Z}_3$), or*
  - (2) *$M \cong \mathbb{Z}_2$ and either $R \cong \mathbb{Z}_2 \times \mathbb{Z}_2$ or $R$ is an integral domain.*

Axtell et al. [120] observed that the only case in which $gr(\Gamma(R(+)M)) = 4$ is when $M \cong \mathbb{Z}_2$ and $R$ does not meet any of the above conditions. For example, by Proposition 2.3.30, $gr(\Gamma(R(+)\mathbb{Z}_2)) = 4$ when $R \cong \mathbb{Z}_2 \times \mathbb{Z}_3$. The study on the girth of the zero-divisor graph of a idealization ring was continued by Anderson and Mulay [85]. In the next theorem, they determined precisely when $gr(\Gamma(R(+)M)) = 4$.

**Theorem 2.3.34.** ([85, Theorem 3.4]) *Let $R$ be a commutative ring and $M$ a nonzero $R$-module. Then $gr(\Gamma(R(+)M)) = 4$ if and only if $R \cong D \times \mathbb{Z}_2$, where $D$ is an integral domain with $|D| \geq 3$ and $M \cong \mathbb{Z}_2$ (so $gr(\Gamma(R)) = \infty$).*

**Remark 2.3.35.** ([85, Remark 3.6]) Let $R$ be an infinite commutative ring. As mentioned earlier, if $R$ is reduced, then $\Gamma(R)$ is a star graph if and only if $R \cong D \times \mathbb{Z}_2$ for some infinite integral domain $D$. An idealization $R$ has $\Gamma(R)$ a star graph if and only if $R \cong D(+)\mathbb{Z}_2$ for some infinite integral domain $D$. However, not all infinite commutative rings with $Nil(R)$ nonzero and $\Gamma(R)$ a star graph are idealizations. For example, let $R = \mathbb{Z}_2[x, y, z]/I$, where $I = fN$ for $f = xz + y^2$ and $N = (x, y, z)$.

## 2.3.3 Rings with Linearly Ordered Primes

We next discuss the girth of $\Gamma(R)$ when the prime ideals of $R$ contained in $Z(R)$ are linearly ordered. These are precisely the commutative rings $R$ such that the prime ideals of the total quotient ring $T(R)$ are linearly ordered, and include chained rings, divided rings, rings with $Z(R) = Nil(R)$, and zero-dimensional local rings. This study was first initiated by Anderson and Badawi [68]. They first handled the case where $Z(R) = Nil(R)$. In this case, $gr(\Gamma(R)) \neq 4$, and we can explicitly say when the girth is either 3 or $\infty$.

**Theorem 2.3.36.** ([68, Theorem 2.11]) *Let $R$ be a commutative ring with $Z(R) = Nil(R) \neq 0$. Then exactly one of the following four cases must occur:*

(a) $|Z(R)^*| = 1$. *In this case, $R$ is isomorphic to $\mathbb{Z}_4$ or $\mathbb{Z}_2[x]/(x^2)$, and $gr(\Gamma(R)) = \infty$;*

(b) $|Z(R)^*| = 2$. *In this case, $R$ is isomorphic to $\mathbb{Z}_9$ or $\mathbb{Z}_3[x]/(x^2)$, and $gr(\Gamma(R)) = \infty$;*

(c) $|Z(R)^*| = 3$. *If $R$ is isomorphic to $\mathbb{Z}_8$, $\mathbb{Z}_2[x]/(x^3)$, or $\mathbb{Z}_4[x]/(2x, x^2 - 2)$, then $gr(\Gamma(R)) = \infty$;*
*Otherwise, $R$ is isomorphic to $\mathbb{Z}_2[x, y]/(x, y)^2$, $\mathbb{Z}_4[x]/(2, x)^2$, $\mathbb{Z}_4[x]/(x^2 + x + 1)$, or $\mathbb{F}_4[x]/(x^2)$; and in this case, $gr(\Gamma(R)) = 3$;*

(d) $|Z(R)^*| \geq 4$. *In this case, $gr(\Gamma(R)) = 3$.*

Note that in Theorem 2.3.36, $gr(\Gamma(R)) = \infty$ if and only if $\Gamma(R)$ is a finite star graph. We next handle the $Nil(R) \subsetneq Z(R)$ case when $Nil(R)$ is a prime ideal of $R$. If $Nil(R) \subsetneq Z(R)$ and $Nil(R)$ is a prime ideal of $R$, then $dim(R) \geq 1$ and $\Gamma(R)$ must be infinite. For in this case, $R$ is not an integral domain, and thus if $\Gamma(R)$ is finite, then $R$ must also be finite, contradicting $dim(R) \geq 1$. The next theorem, and Theorem 2.3.36, completely characterize $gr(\Gamma(R))$ in terms of $|Nil(R)^*|$ when the prime ideals of $R$ contained in $Z(R)$ are linearly ordered. In particular, we have $gr(\Gamma(R)) = 3$ or $\infty$, with $gr(\Gamma(R)) = \infty$ if and only if $\Gamma(R)$ is a star graph.

**Theorem 2.3.37.** ([68, Theorem 2.12]) *Let $R$ be a commutative ring such that $Nil(R)$ is a prime ideal of $R$ and $Nil(R) \subsetneq Z(R)$. In particular, this holds when the prime ideals of $R$ contained in $Z(R)$ are linearly ordered and $Nil(R) \subsetneq Z(R)$. Then $gr(\Gamma(R)) = 3$ or $\infty$. Moreover, $gr(\Gamma(R)) = \infty$ if and only if $|Nil(R)^*| = 1$; and in this case, $\Gamma(R)$ is an infinite star graph.*

The hypothesis that $Nil(R)$ is a prime ideal of $R$ is needed in Theorem 2.3.37. For example, let $R = \mathbb{Z}_3 \times \mathbb{Z}_3$. Then $Nil(R) \subsetneq Z(R)$, $Nil(R)$ is not a prime ideal of $R$, and $gr(\Gamma(R)) = 4$.

## 2.4 Diameter of Zero-divisor Graphs

In this section, we are interested in the diameter of zero-divisor graphs. Recall that for distinct vertices $x$ and $y$ of a graph $G$, $d(x, y)$ denotes the length of a shortest path from $x$ to $y$ ($d(x, y) = \infty$ if there is no such path). The *diameter* of $G$ is $diam(G) = \sup\{d(x,y) \mid x \text{ and } y \text{ are distinct vertices of } G\}$. The first theorem in this section shows that zero-divisor graphs have exceedingly small ($\leq 3$) diameter. Thus, for distinct $x, y \in Z(R)^*$, either $xy = 0, xz = zy = 0$ for some $z \in Z(R)^* \setminus \{x, y\}$, or $xz_1 = z_1z_2 = z_2y = 0$ for some distinct $z_1, z_2 \in Z(R)^* \setminus \{x, y\}$. More specifically, we are interested in the equality of diameters of zero-divisor graphs of rings and their products.

From the proof of Theorem 2.3.1, we have the following.

**Theorem 2.4.1.** ([83, Theorem 2.3]) *Let $R$ is a commutative ring. Then $diam(\Gamma(R)) \leq 3$.*

**Example 2.4.2.** Example 2.1.3(a) gives several rings $R$ with $diam(\Gamma(R)) = 0, 1$, or 2. In $R = \mathbb{Z}_2 \times \mathbb{Z}_4$, the path $(0,1) - (1,0) - (0,2) - (1,2)$ shows that $diam(\Gamma(R)) = 3$. Thus $diam(\Gamma(R)) = 0$, 1, 2, or 3. Note that $\Gamma(R)$ has diameter zero if and only if $R \cong \mathbb{Z}_4$ or $\mathbb{Z}_2[x]/(x^2)$ (Example 2.1.3(c)).

The next result gives the relationship between the diameter of the zero-divisor graph of the total quotient ring $T(R)$ and the ring $R$. (This also follows since $\Gamma(T(R)) \cong \Gamma(R)$ by [81, Theorem 2.2].) As an immediate consequence, $diam(\Gamma(B)) = diam(\Gamma(R))$ for any overring $B$ of $R$.

**Theorem 2.4.3.** ([296, Theorem 1.1], [85, Lemma 2.1]) *Let $R$ be a commutative ring with total quotient ring $T(R)$. Then $diam(\Gamma(T(R))) = diam(\Gamma(R))$.*

*Proof.* Clearly $diam(\Gamma(T(R))) = 1$ if and only if $diam(\Gamma(R)) = 1$. Suppose that $diam(\Gamma(T(R))) = 2$. Then $diam(\Gamma(R)) \geq 2$. Let $a, b \in Z(R)^*$ with $a \neq b$ and $ab \neq 0$. Then $aq = 0 = bq$ for some $q \in Z(T)^* \setminus \{a, b\}$. Let $q = c/t$ with $c \in R$ and $t \in R \setminus Z(R)$. Then $ac = 0 = bc$. Thus $d(a, b) = 2$, and hence $diam(\Gamma(R)) = 2$. A similar argument shows that $diam(\Gamma(T(R))) = 2$ if $diam(\Gamma(R)) = 2$. The result for the diameter now follows since the diameter of a zero-divisor graph is at most 3 as given in Theorem 2.4.1. □

Let $A \subseteq B$ be an extension of commutative rings with identity. The relationship between $diam(\Gamma(A))$ and $diam(\Gamma(B))$ when $B$ is not an overring of $A$ is not clear because although $Z(A) \subseteq Z(B)$, it need not be the case that $Z(A) = Z(B) \cap A$. In fact, for $m, n \in \{0, 1, 2, 3\}$, there is an extension of commutative rings $A \subsetneq B$ with $diam(\Gamma(A)) = m$ and $diam(\Gamma(B)) = n$ unless $(m, n) \in \{(0,0), (1,0), (2,0), (2,1), (3,0), (3,2)\}$ [64, Proposition 3.2]. Thus, $diam(\Gamma(A)) \leq diam(\Gamma(B))$ unless $diam(\Gamma(A)) = 3$ and $diam(\Gamma(B)) = 2$. Specific examples with $diam(\Gamma(A)) = 3$ and $diam(\Gamma(B)) = 2$ are given in Example 2.4.57 and [64, Example 3.7]. The next theorem gives conditions for this to happen; an example of a reduced quasilocal commutative ring $(A, \mathfrak{m})$ with $diam(\Gamma(A)) = 3$ (and necessarily $Z(A) \subseteq \mathfrak{m}$) is given in [64, Example 3.13].

**Theorem 2.4.4.** (a) ([64, Theorem 3.8]) *Let $A$ be a commutative ring with $diam(\Gamma(A)) = 3$. Then there is a commutative extension ring $B$ of $A$ with*

*$diam(\Gamma(B)) = 2$ if and only if $Z(A) \subseteq \mathfrak{m}$ for some maximal ideal $\mathfrak{m}$ of $A$. Moreover, if $A$ is reduced, then $B$ can also be chosen to be reduced.*

(b) ([64, Corollary 3.12]) *Let $A \subseteq B$ be an extension of commutative rings with $diam(\Gamma(A)) = 0$. Then $diam(\Gamma(A)) \leq diam(\Gamma(B))$. In particular, this happens if $A$ is Artinian or a finite commutative ring.*

It is easy to check that $diam(\Gamma(\mathbb{Z}_2 \times \mathbb{Z}_2)) = 1$ and $diam(\Gamma(R_1 \times R_2)) = 2$ if $R_1$ and $R_2$ are integral domains, not both $\mathbb{Z}_2$. In 2006, Axtel, Stickles, and Warfel [123] determined the diameter of $\Gamma(R_1 \times R_2)$ for commutative rings $R_1$ and $R_2$, possibly without identity. The list of cases is rather long; so we will only state the case for commutative rings with identity.

**Theorem 2.4.5.** ([123, Corollaries 3.2 and 3.10]) *Let $R_1$ and $R_2$ be commutative rings with identity that are not both integral domains. Then $diam(\Gamma(R_1 \times R_2)) = 3$.*

In 2007, Atani and Kohan [113] established a set of theorems that describes the diameter of the zero-divisor graph for a finite direct product $R_1 \times \cdots \times R_n$ in terms of the diameters of the zero-divisor graphs of the rings $R_i$ (not necessarily with identity), where $1 \leq i \leq n$ and $n > 2$. The results in this regard are stated below.

**Theorem 2.4.6.** ([113, Theorem 2.2]) *Let $R_1, \ldots, R_n$ be commutative rings (not necessarily with 1) such that $diam(\Gamma(R_1)) = \cdots = diam(\Gamma(R_n)) = 1$, and let $R = R_1 \times \cdots \times R_n$ $(n > 2)$. Then the following hold:*

(a) *$diam(\Gamma(R)) = 1$ if and only if $R_i = Z(R_i)$ for every $i = 1, \ldots, n$;*

(b) *$diam(\Gamma(R)) = 2$ if and only if $R_i = Z(R_i)$ and $R_j \neq Z(R_j)$ for some $i, j \in \{1, \ldots, n\}$;*

(c) *$diam(\Gamma(R)) = 3$ if and only if $R_i \neq Z(R_i)$ for every $i = 1, \ldots, n$.*

**Theorem 2.4.7.** ([113, Theorem 2.4]) *Let $R_1, \ldots, R_n$ be commutative rings (not necessarily with 1) such that $diam(\Gamma(R_1)) = \cdots = diam(\Gamma(R_n)) = 2$, and let $R = R_1 \times \cdots \times R_n$ $(n > 2)$. Then the following hold:*

(a) *$diam(\Gamma(R)) \neq 1$;*

(b) *$diam(\Gamma(R)) = 2$ if and only if $R_i = Z(R_i)$ for some $i = 1, \ldots, n$;*

(c) *$diam(\Gamma(R)) = 3$ if and only if $R_i \neq Z(R_i)$ for every $i = 1, \ldots, n$.*

**Theorem 2.4.8.** ([113, Theorem 2.5]) *Let $R_1, \ldots, R_n$ be commutative rings (not necessarily with 1) such that $diam(\Gamma(R_1)) = \cdots = diam(\Gamma(R_n)) = 3$, and let $R = R_1 \times \cdots \times R_n$ $(n > 2)$. Then $diam(\Gamma(R)) = 3$.*

**Theorem 2.4.9.** ([113, Theorem 2.6]) *Let $R_1, \ldots, R_n$ be commutative rings (not necessarily with 1) such that $diam(\Gamma(R_i)) = 1$ and $diam(\Gamma(R_j)) = 2$ for some $i, j \in \{1, \ldots, n\}$ and there is no $k \in \{1, \ldots, n\}$ with $diam(\Gamma(R_k)) = 3$, and let $R = R_1 \times \cdots \times R_n$ $(n > 2)$. Then the following hold:*

(a) $diam(\Gamma(R)) \neq 1$;

(b) $diam(\Gamma(R)) = 2$ *if and only if* $R_i = Z(R_i)$ *for some* $i = 1, \ldots, n$;

(c) $diam(\Gamma(R)) = 3$ *if and only if* $R_i \neq Z(R_i)$ *for every* $i = 1, \ldots, n$.

**Theorem 2.4.10.** ([113, Theorem 2.7]) *Let* $R_1, \ldots, R_n$ *be commutative rings (not necessarily with* 1*) such that* $diam(\Gamma(R_i)) = 1$ *and* $diam(\Gamma(R_j)) = 3$ *for some* $i, j \in \{1, \ldots, n\}$ *and there is no* $k \in \{1, \ldots, n\}$ *with* $diam(\Gamma(R_k)) = 2$, *and let* $R = R_1 \times \cdots \times R_n$ $(n > 2)$. *Then the following hold:*

(a) $diam(\Gamma(R)) \neq 1$;

(b) $diam(\Gamma(R)) = 2$ *if and only if* $diam(\Gamma(R_i)) = 1$ *and* $R_i = Z(R_i)$ *for some* $i = 1, \ldots, n$;

(c) $diam(\Gamma(R)) = 3$ *if and only if there is no* $i \in \{1, \ldots, n\}$ *with* $diam(\Gamma(R_i)) = 1$ *and* $R_i \neq Z(R_i)$.

**Theorem 2.4.11.** ([113, Theorem 2.8]) *Let* $R_1, \ldots, R_n$ *be commutative rings (not necessarily with* 1*) such that* $diam(\Gamma(R_i)) = 2$ *and* $diam(\Gamma(R_j)) = 3$ *for some* $i, j \in \{1, \ldots, n\}$ *and there is no* $k \in \{1, \ldots, n\}$ *with* $diam(\Gamma(R_k)) = 1$, *and let* $R = R_1 \times \cdots \times R_n$ $(n > 2)$. *Then the following hold:*

(a) $diam(\Gamma(R)) \neq 1$;

(b) $diam(\Gamma(R)) = 2$ *if and only if* $diam(\Gamma(R_i)) = 2$ *and* $R_i = Z(R_i)$ *for some* $i = 1, \ldots, n$;

(c) $diam(\Gamma(R)) = 3$ *if and only if there is no* $i \in \{1, \ldots, n\}$ *with* $diam(\Gamma(R_i)) = 2$ *and* $R_i \neq Z(R_i)$.

**Theorem 2.4.12.** ([113, Theorem 2.9]) *Let* $R_1, \ldots, R_n$ *be commutative rings (not necessarily with* 1*) such that* $diam(\Gamma(R_i)) = 1$, $diam(\Gamma(R_j)) = 2$, *and* $diam(\Gamma(R_k)) = 3$ *for some* $i, j, k \in \{1, \ldots, n\}$, *and let* $R = R_1 \times \cdots \times R_n$ $(n > 2)$. *Then the following hold:*

(a) $diam(\Gamma(R)) \neq 1$;

(b) $diam(\Gamma(R)) = 2$ *if and only if* $diam(\Gamma(R_i)) \leq 2$ *and* $R_i = Z(R_i)$ *for some* $i = 1, \ldots, n$;

(c) $diam(\Gamma(R)) = 3$ *if and only if there is no* $i \in \{1, \ldots, n\}$ *with* $diam(\Gamma(R_i)) \leq 2$ *and* $R_i \neq Z(R_i)$.

**Corollary 2.4.13.** ([113, Corollary 2.10]) *Let* $R_1, \ldots, R_n$ *be commutative rings with identity and* $R = R_1 \times \cdots \times R_n$ $(n > 2)$. *Then* $diam(\Gamma(R)) = 3$.

### 2.4.1 Rings of Polynomials and Power Series

Next, we study the diameter of the zero-divisor graph of extension rings such as polynomial and power series rings. As mentioned earlier, $\Gamma(R)$ is a subgraph of both $\Gamma(R[x])$ and $\Gamma(R[[x]])$; so $diam(\Gamma(R)) \leq diam(\Gamma(R[x]))$ and $diam(\Gamma(R)) \leq diam(\Gamma(R[[x]]))$. First of all, observe that if $R = \mathbb{Z}_2 \times \mathbb{Z}_2$, then $diam(\Gamma(R[x])) = diam(\Gamma(R[[x]])) = 2$. The example given below exhibits a ring $R$ with $diam(\Gamma(R)) < diam(\Gamma(R[[x]]))$, and it also shows that $diam(\Gamma(R[x])) = 2$ need not imply $diam(\Gamma(R[[x]])) = 2$.

**Example 2.4.14.** Let $F$ be a field and

$$R = F[y, \{x_i\}_{i=0}^{\infty}]/(\{x_0 y\} \cup \{x_i - x_{i+1} y\}_{i=0}^{\infty}).$$

Then $diam(\Gamma(R)) = diam(\Gamma(R[x])) = 2$ and $diam(\Gamma(R[[x]])) = 3$. For a proof of these facts, see [119, Example 3.3].

**Theorem 2.4.15.** ([119, Proposition 3.4]) *Let $R$ be a commutative ring with $diam(\Gamma(R)) \leq 2$, and let $Z(R) = \cup_{i\in\Lambda} I_i$, where each $I_i$ is a (maximal) prime ideal. If there is an element in $Z(R)$ that is contained in a unique maximal $I_i$, then $|\Lambda| \leq 2$.*

*Proof.* We choose our element $p_1 \in I_1$ such that $p_1$ is contained in none of the other maximal primes in $Z(R)$. Assume that we have at least two other maximal primes which we denote as $I_2$ and $I_3$. Using [228, Theorem 81], we can choose $p_2 \in I_2 \setminus (I_1 \cup I_3)$. Note that the ideal $(p_1, p_2)$ is contained in no maximal prime ideal in $Z(R)$; hence there must be a regular element in $(p_1, p_2)$. We write $ap_1 + bp_2 = r$ with $r$ a regular element. Note that if there is a mutual annihilator of $p_1$ and $p_2$, then this element must annihilate $r$, and hence is 0 since $r$ is regular. Since the diameter of $\Gamma(R)$ is 2, this means that $p_1 p_2 = 0 \in I_3$. But since $p_1 \notin I_3$, this means that $p_2 \in I_3$, contradicting the choice of $p_2$. □

**Corollary 2.4.16.** ([119, Corollary 3.5]) *Let $R$ be a commutative ring with $diam(\Gamma(R)) \leq 2$, and let $Z(R) = \cup_{i\in\Lambda} I_i$, where each $I_i$ is a prime ideal. If $\Lambda$ is a finite set (in particular, if $R$ is Noetherian), then $|\Lambda| \leq 2$.*

*Proof.* We first remark that [228, Theorem 80] shows that any Noetherian ring satisfies the hypothesis of the corollary. We will apply Theorem 2.4.15 after showing that in $Z(R)$ there is an element that is contained in a unique maximal prime. Suppose that $Z(R) = I_1 \cup I_2 \cup \cdots \cup I_n$, and suppose that every element of $I_1$ is contained in one of the other maximal primes. This implies that $I_1 \setminus (I_2 \cup \cdots \cup I_n) = \emptyset$. Equivalently, we have $I_1 \subseteq I_2 \cup \cdots \cup I_n$. But by [228, Theorem 81], this implies that $I_1 \subseteq I_i$ for some $2 \leq i \leq n$, which contradicts the maximality of $I_1$. □

**Theorem 2.4.17.** ([119, Proposition 3.6]) *Let $R$ be a commutative ring such that $diam(\Gamma(R)) = 2$ and $Z(R) = I_1 \cup I_2$ with $I_1$ and $I_2$ distinct maximal primes in $Z(R)$. Then $I_1 \cap I_2 = \{0\}$ (in particular, for all $p_1 \in I_1$ and $p_2 \in I_2$, $p_1 p_2 = 0$).*

*Proof.* We begin by verifying some annihilator relations of elements of $Z(R)$. Suppose first that $p_1 \in I_1 \setminus I_2$ and $p_2 \in I_2 \setminus I_1$. Since $(p_1, p_2)$ is not contained in either $I_1$ or $I_2$, there must be a regular element in $(p_1, p_2)$. Hence there is a linear combination of $p_1$ and $p_2$ that is regular, and hence $p_1$ and $p_2$ can possess no mutual annihilator. Since $diam(\Gamma(R)) = 2$, we have that $p_1p_2 = 0$.

Now suppose that $p_1 \in I_1 \setminus I_2$ and $p_2 \in I_2$. If $p_2$ is not in $I_1$, then $p_1p_2 = 0$. So we can assume that $p_2 \in I_2 \cap I_1$, and we select $y \in I_2 \setminus I_1$. Note that $p_2 + y \in I_2 \setminus I_1$, and hence $p_1(p_2 + y) = 0 = p_1p_2$.

We have shown that any element of $I_1 \setminus I_2$ annihilates $I_2$ (and of course, every element of $I_2 \setminus I_1$ annihilates $I_1$). Suppose $0 \neq x \in I_1 \cap I_2$. Note that the annihilator of $x$ necessarily contains all of $I_1 \setminus I_2$ and $I_2 \setminus I_1$. So the annihilator of $(x) \subseteq Z(R)$ is contained in neither $I_1$ nor $I_2$, a contradiction. □

**Theorem 2.4.18.** ([119, Proposition 3.7]) *Let $R$ be a commutative ring such that $diam(\Gamma(R)) = 2$. If $Z(R) = I_1 \cup I_2$ is the union of precisely two maximal primes in $Z(R)$, then $diam(R[x]) = 2$.*

*Proof.* Let $f(x), g(x)$ be zero-divisors (with the degree of $f(x)$ greater than 0). It is easy to see that it is necessary for $f(x)$ to be contained in $I_1[x]$ or $I_2[x]$ (indeed, if $a \in I_1 \setminus I_2$ and $b \in I_2 \setminus I_1$ are two coefficients of $f(x)$, then $a$ and $b$ must have a mutual annihilator that is necessarily in $I_1 \cap I_2 = \{0\}$ by the previous result). There are two natural cases. If $f(x) \in I_1[x]$ and $g(x) \in I_2[x]$, then Theorem 2.4.17 shows that $f(x)g(x) = 0$. On the other hand, if $f(x), g(x) \in I_1[x]$, then any element of $I_2$ suffices as a mutual annihilator. So $diam(R[x]) = 2$. □

**Theorem 2.4.19.** ([119, Proposition 3.8]) *Let $R$ be a commutative Noetherian ring such that $diam(\Gamma(R)) = 2$. If $Z(R) = I$ is a prime ideal, then $diam(R[x]) = 2$.*

*Proof.* By [228, Theorem 82], $I$ is annihilated by a single element (say $z$). Suppose that $f(x), g(x)$ are zero-divisors (with degree of $f(x)$ greater than 0). If $f(x)g(x) = 0$, we are done. If $f(x)g(x) \neq 0$, then neither $f(x)$ nor $g(x)$ is $z$, and so $z$ is a mutual annihilator of $f(x)$ and $g(x)$. □

**Theorem 2.4.20.** ([119, Theorem 3.9]) *Let $R$ be a commutative Noetherian ring with $diam(\Gamma(R)) = 2$. Then $diam(\Gamma(R[[x]])) = 2$.*

*Proof.* By Corollary 2.4.16, either $Z(R) = I_1 \cup I_2$ is the union of precisely two maximal primes in $Z(R)$ or $Z(R) = I$ is a prime ideal. By [196, Theorem 5], if $R$ is Noetherian, then $f(x) \in Z(R[[x]])^*$ if and only if $rf(x) = 0$ for some $0 \neq r \in R$. Thus, every coefficient of $f$ is a zero-divisor of $R$.

Assume $Z(R) = I_1 \cup I_2$ is the union of precisely two maximal primes in $Z(R)$. If $f(x) = \sum f_i x^i \in Z(R[[x]])^*$, then by the proof of Theorem 2.4.18, we have that, without loss of generality, $f_i \in I_1$ for all $i$. Hence every element of $I_2$ will annihilate $f(x)$ by Theorem 2.4.17. Thus $diam(\Gamma(R[[x]])) = 2$.

If $Z(R) = I$ is a prime ideal, then by [228, Theorem 82], $I$ is annihilated by a single element (say $z$). The result follows in much the same way as the proof of Theorem 2.4.19. □

The following theorem summarizes these facts.

**Theorem 2.4.21.** ([119, Theorem 3.10]) *Let $R$ be a commutative ring such that $diam(\Gamma(R)) = 2$ and $Z(R)$ is a finite union of primes. Then $Z(R)$ is necessarily the union of at most two primes. If $Z(R)$ is the union of two primes, then $diam(\Gamma(R[x])) = 2$; and additionally, if $R$ is Noetherian, then $diam(\Gamma(R[x])) = diam(\Gamma(R[[x]])) = 2$.*

**Proposition 2.4.22.** ([119, Theorem 3.11]) *Let $R \not\cong \mathbb{Z}_2 \times \mathbb{Z}_2$ be a commutative Noetherian ring with nontrivial zero-divisors. The following conditions are equivalent:*

(a) *$diam(\Gamma(R)) = 2$;*

(b) *$diam(\Gamma(R[x])) = 2$;*

(c) *$diam(\Gamma(R[x_1, x_2, \ldots, x_n])) = 2$ for all $n > 0$;*

(d) *$diam(\Gamma(R[[x]])) = 2$;*

(e) *$Z(R)$ is either the union of two prime ideals with intersection $\{0\}$, or $Z(R)$ is prime and $(Z(R))^2 \neq 0$.*

**Proposition 2.4.23.** ([119, Proposition 3.12]) *Let $R$ be a commutative ring. If $(Z(R))^{n-1} \neq 0$ and $(Z(R))^n = 0$ for some integer $n > 2$ , then $diam(\Gamma(R[[x]])) = diam(\Gamma(R[x])) = diam(\Gamma(R)) = 2$.*

*Proof.* By hypothesis and [83, Theorem 2.8], $\Gamma(R)$ is not complete. Hence, there are distinct $a, b \in Z(R)$ with $ab \neq 0$. Since $(Z(R))^{n-1} \neq 0$, there are $c_1, c_2, \ldots, c_{n-1} \in Z(R)$ with $c = \prod_{i=1}^{n-1} c_i \neq 0$. However, $ac = 0 = bc$, and so $diam(\Gamma(R)) = 2$. Let $f(x), g(x) \in Z(R[[x]])^*$. By Lemma 2.3.16, we have that all the coefficients of $f(x)$ and $g(x)$ are elements of $Z(R)$. Thus either $f - g$ or $f - c - g$. Hence $diam(\Gamma(R[[x]])) = 2$. □

The study of the diameter of $\Gamma(R)$, $\Gamma(R[x])$, and $\Gamma(R[[x]])$ was continued by Lucas [269]. He was mainly interested in characterizing the diameter of $\Gamma(R)$, $\Gamma(R[x])$, and $\Gamma(R[[x]])$ strictly in terms of properties of the ring $R$.

The following theorem provides a sufficient condition for $\Gamma(R)$ to have diameter 3 when $R$ is a reduced ring.

**Theorem 2.4.24.** ([269, Theorem 2.1]) *Let $R$ be a reduced commutative ring. If $R$ has more than two minimal primes and there are $0 \neq a, b \in Z(R)$ such that $(a, b)$ has no nonzero annihilator, then $diam(\Gamma(R)) = 3$.*

**Theorem 2.4.25.** ([269, Theorem 2.2]) *Let $R$ be a reduced commutative ring. If $Z(R)$ is not an ideal of $R$, then $diam(\Gamma(R)) \leq 2$ if and only if $R$ has exactly two minimal primes.*

If $R$ is a reduced commutative ring with exactly two minimal primes, then $Z(R)$ cannot be an ideal. Thus the initial assumption that $Z(R)$ is not an ideal is not needed to show $diam(\Gamma(R)) = 2$ when $R$ has exactly two minimal primes. On the other hand, the reduced ring $R$ in Example 2.4.40 has infinitely many minimal primes, but $diam(\Gamma(R)) = 2$, so necessarily, $Z(R)$ is an ideal (which must be prime). Thus the assumption that $Z(R)$ is not an ideal does play a significant role in establishing the converse in Theorem 2.4.25. On the other hand, it is not entirely necessary. A slight modification in the construction used in Example 2.4.40 will yield a reduced ring whose set of zero-divisors forms an ideal, yet the diameter of the zero-divisor graph is three (see [64, Example 3.13] for details).

The next result is the nonreduced version of Theorem 2.4.24. Note that here no assumption is made that $R$ has more than two minimal primes.

**Theorem 2.4.26.** ([269, Theorem 2.4]) *Let $R$ be a nonreduced commutative ring. If there is a pair of zero-divisors $a, b \in Z(R)$ such that $(0 : (a, b)) = (0)$, then $diam(\Gamma(R)) = 3$.*

**Corollary 2.4.27.** ([269, Corollary 2.5]) *If $R$ is a nonreduced commutative ring such that $Z(R)$ is not an ideal, then $diam(\Gamma(R)) = 3$.*

The next result characterizes the diameter of $\Gamma(R)$ in terms of the ideals of $R$.

**Theorem 2.4.28.** ([269, Theorem 2.6]) *Let $R$ be a commutative ring. Then*

(a) *$diam(\Gamma(R)) = 0$ if and only if $R$ is (nonreduced and) isomorphic to either $\mathbb{Z}_4$ or $\mathbb{Z}_2[x]/(x^2)$;*

(b) *$diam(\Gamma(R)) = 1$ if and only if $xy = 0$ for every distinct pair of zero-divisors and $R$ has at least two nonzero zero-divisors;*

(c) *$diam(\Gamma(R)) = 2$ if and only if either $R$ is reduced with exactly two minimal primes and at least three nonzero zero-divisors, or $Z(R)$ is an ideal whose square is not $(0)$ and every pair of distinct zero-divisors has a nonzero annihilator;*

(d) *$diam(\Gamma(R)) = 3$ if and only if there are zero-divisors $a \neq b$ such that $(0 : (a, b)) = (0)$ and either $R$ is a reduced ring with more than two minimal primes or $R$ is nonreduced.*

The following characterization of when $diam(\Gamma(R))) \leq 2$ was given by Mulay in [296].

**Theorem 2.4.29.** ([296, Theorem 1.2]) *Let $R$ be a commutative ring. If $diam(\Gamma(R)) \leq 2$, then exactly one of the following holds:*

(i) *$Z(R)$ is a prime ideal of $R$;*

(ii) *R is a subring of a product of two integral domains.*

*If* (ii) *holds for (a nondomain) R, then diam*$(\Gamma(R)) \leq 2$. *If* (i) *holds for (a nondomain R) and R is Noetherian, then diam*$(\Gamma(R)) \leq 2$.

A ring $R$ is said to be a *McCoy ring* if every finitely generated ideal contained in $Z(R)$ has a nonzero annihilator. Note that the polynomial ring $R[x]$ is a McCoy ring.

**Theorem 2.4.30.** ([269, Theorem 3.3]) *For a commutative ring R,* $Z(R[x])$ *is an ideal of* $R[x]$ *if and only if R is a McCoy ring such that* $Z(R)$ *is an ideal.*

The above result shows that it is enough to characterize the diameter of $\Gamma(R[x])$ based on the ideals of $R$.

**Theorem 2.4.31.** ([269, Theorem 3.4]) *Let R be a commutative ring. Then*

(a) *diam*$(\Gamma(R[x])) \geq 1$;

(b) *diam*$(\Gamma(R[x])) = 1$ *if and only if R is a nonreduced ring such that* $Z(R)^2 = 0$;

(c) *diam*$(\Gamma(R[x])) = 2$ *if and only if either R is a reduced ring with exactly two minimal primes, or R is a McCoy ring and* $Z(R)$ *is an ideal with* $Z(R)^2 \neq 0$;

(d) *diam*$(\Gamma(R[x])) = 3$ *if and only if R is not a reduced ring with exactly two minimal primes and either R is not a McCoy ring or* $Z(R)$ *is not an ideal.*

If the total quotient ring of $R$ is von Neumann regular, then the only primes that contain only zero-divisors are the minimal primes. Hence $Z(R)$ is an ideal only in the trivial case that $R$ is an integral domain, a case we have assumed does not happen.

**Corollary 2.4.32.** ([269, Corollary 3.5]) *Let R be a reduced ring that is not isomorphic to* $\mathbb{Z}_2 \times \mathbb{Z}_2$. *If the total quotient ring of R is von Neumann regular, then diam*$(\Gamma(R)) =$ *diam*$(\Gamma(R[x]))$. *Moreover, diam*$(\Gamma(R)) = 3$ *if and only if R has more than two minimal primes.*

As mentioned earlier, it is clear that *diam*$(\Gamma(R)) \leq$ *diam*$(\Gamma(R[x]))$. In the following theorem, we include all of the cases where *diam*$(\Gamma(R)) =$ *diam*$(\Gamma(R[x]))$.

**Theorem 2.4.33.** ([269, Theorem 3.6]) *Let R be a commutative ring. The following cases describe all possibilities for the pair diam*$(\Gamma(R))$ *and diam*$(\Gamma(R[x]))$:

(a) *diam*$(\Gamma(R)) = 0$ *and diam*$(\Gamma(R[x])) = 1$ *if and only if R is isomorphic to either* $\mathbb{Z}_4$ *or* $\mathbb{Z}_2[x]/(x^2)$;

(b) *diam*$(\Gamma(R)) =$ *diam*$(\Gamma(R[x])) = 1$ *if and only if R is a nonreduced ring with more than one nonzero zero-divisor such that* $Z(R)^2 = 0$;

(c) *diam*$(\Gamma(R)) = 1$ *and diam*$(\Gamma(R[x])) = 2$ *if and only if* $R$ *is isomorphic to* $\mathbb{Z}_2 \times \mathbb{Z}_2$;

(d) *diam*$(\Gamma(R)) =$ *diam*$(\Gamma(R[x])) = 2$ *if and only if either* $R$ *is a reduced ring with exactly two minimal primes and* $R$ *has more than two nonzero zero-divisors, or* $Z(R)$ *is an ideal with* $Z(R)^2 \neq 0$ *and* $R$ *is a McCoy ring;*

(e) *diam*$(\Gamma(R)) = 2$ *and diam*$(\Gamma(R[x])) = 3$ *if and only if* $Z(R)$ *is an ideal,* $R$ *is not a McCoy ring, but every pair of zero-divisors of* $R$ *has a nonzero annihilator;*

(f) *diam*$(\Gamma(R)) =$ *diam*$(\Gamma(R[x])) = 3$ *if and only if* $R$ *is not a reduced ring with exactly two minimal primes and there is a pair of zero-divisors* $a$ *and* $b$ *such that* $(0 : (a, b)) = 0$.

For power series, Lucas [269] has completely characterized *diam*$(\Gamma(R[[x]]))$ in terms of the ideals of $R$ when $R$ is reduced. For nonreduced rings, he provides a few special cases.

For the power series ring $R[[x]]$ in a single indeterminate, a Gilmer et al. [207] result says that if $f = \sum f_i x^i$ and $g = \sum g_j x^j$ are power series for which $fg = 0$, then the product $c(f)c(g)$ is contained in the nilradical of $R$. Thus, if $R$ is reduced, we have $fg = 0$ if and only if $f_i g_j = 0$ for every pair of coefficients $f_i$ and $g_j$. As we will see, this result is key in determining *diam*$(\Gamma(R[[x]])$ when $R$ is a reduced ring.

**Theorem 2.4.34.** ([269, Theorem 4.3]) *Let* $R$ *be a reduced commutative ring with exactly two minimal primes. Then diam*$(\Gamma(R[[x]])) = 2$.

Recall that *diam*$(\Gamma(R)) \leq$ *diam*$(\Gamma(R[[x]]))$. Thus, if *diam*$(\Gamma(R)) = 3$, then also *diam*$(\Gamma(R[[x]])) = 3$. This observation and Theorem 2.4.25 give the following.

**Theorem 2.4.35.** ([269, Theorem 4.4]) *Let* $R$ *be a reduced commutative ring with more than two minimal primes. If* $Z(R)$ *is not an ideal, then diam*$(\Gamma(R[[x]])) = 3$.

We know that if there is a two-generated ideal contained in $Z(R)$ with no nonzero annihilator, then *diam*$(\Gamma(R)) = 3$. Also, we know that if $Z(R)$ is an ideal and $R$ is not a McCoy ring, then there are polynomials $f(x), g(x) \in Z(R[x]) = Z(R)[x]$ with $d(f(x), g(x)) = 3$. Viewed in $R[[x]]$, we will still have $d(f(x), g(x)) = 3$ when $R$ is reduced since $f(x)$ and $g(x)$ have no common annihilator in $R$. The next result characterizes when *diam*$(\Gamma(R[[x]])) = 2$ for a reduced ring such that $Z(R)$ is an ideal.

**Theorem 2.4.36.** ([269, Theorem 4.5]) *Let* $R$ *be a reduced commutative ring such that* $Z(R)$ *is an ideal of* $R$. *Then diam*$(\Gamma(R[[x]])) = 2$ *if and only if* $I + J$ *has a nonzero annihilator whenever* $I$ *and* $J$ *are both countably generated with nonzero annihilators.*

Note that if $I + J$ has a nonzero annihilator whenever $I$ and $J$ are countably generated ideals with nonzero annihilators, then $R$ will be a McCoy ring and $Z(R)$ will be an ideal.

**Corollary 2.4.37.** ([269, Corollary 4.6]) *Let $R$ be a reduced commutative ring. If $diam(\Gamma(R[[x]])) = 2$, then $R$ is a McCoy ring and $diam(\Gamma(R[x])) = 2$.*

A commutative ring $R$ is said to be *countably McCoy* if every ideal $I \subseteq Z(R)$ that is countably generated has a nonzero annihilator.

**Corollary 2.4.38.** ([269, Corollary 4.7,4.8]) *Let $R$ be a commutative reduced ring.*

(a) *If $Z(R)$ is an ideal of $R$ and $R$ is countably McCoy, then $diam(\Gamma(R[[x]])) = 2$;*

(b) *$diam(\Gamma(R[[x]])) = 2$ if and only if either $R$ has exactly two minimal primes or $R$ is a McCoy ring where $Z(R)$ is an ideal and $I+J$ has a nonzero annihilator whenever $I$ and $J$ are countably generated ideals of $R$ with nonzero annihilators.*

In the following theorem, we summarize all of the various results concerning the diameters of the graphs $\Gamma(R)$, $\Gamma(R[x])$, and $\Gamma(R[[x]])$ for $R$ a reduced commutative ring.

**Theorem 2.4.39.** ([269, Theorem 4.9]) *Let $R$ be a reduced commutative ring that is not an integral domain. Then*

$$1 \leq diam(\Gamma(R)) \leq diam(\Gamma(R[x])) \leq diam(\Gamma(R[[x]])) \leq 3.$$

*Moreover, here are all possible sequences for these dimensions:*

(a) *$diam(\Gamma(R)) = 1$ and $diam(\Gamma(R[x])) = diam(\Gamma(R[[x]])) = 2$ if and only if $R$ is isomorphic to $\mathbb{Z}_2 \times \mathbb{Z}_2$;*

(b) *$diam(\Gamma(R)) = diam(\Gamma(R[x])) = diam(\Gamma(R[[x]])) = 2$ if and only if either $R$ has exactly two minimal primes and is not isomorphic to $\mathbb{Z}_2 \times \mathbb{Z}_2$, or for every pair of countably generated ideals $I$ and $J$ with nonzero annihilators, the sum $I+J$ has a nonzero annihilator (and $R$ is a McCoy ring with $Z(R)$ an ideal);*

(c) *$diam(\Gamma(R)) = diam(\Gamma(R[x])) = 2$ and $diam(\Gamma(R[[x]])) = 3$ if and only if $R$ is a McCoy ring with $Z(R)$ an ideal, but there are countably generated ideals $I$ and $J$ with nonzero annihilators such that $I + J$ does not have a nonzero annihilator;*

(d) *$diam(\Gamma(R)) = 2$ and $diam(\Gamma(R[x])) = diam(\Gamma(R[[x]])) = 3$ if and only if $Z(R)$ is an ideal and every two generated ideal contained in $Z(R)$ has a nonzero annihilator, but $R$ is not a McCoy ring;*

(e) *$diam(\Gamma(R)) = diam(\Gamma(R[x])) = diam(\Gamma(R[[x]])) = 3$ if and only if $R$ has more than two minimal primes and there is a pair of zero-divisors $a$ and $b$ such that $(a, b)$ does not have a nonzero annihilator.*

In Example 2.4.14, Axtell et al. provided a commutative ring $R$ with $diam(\Gamma(R)) = diam(\Gamma(R[x])) = 2$ and $diam(\Gamma(R[[x]])) = 3$, and they left open the existence of a reduced ring with the same sequence of diameters. Then Lucas [269] constructed several examples regarding the various possible diameters of $\Gamma(R)$, $\Gamma(R[x])$, and $\Gamma(R[[x]])$.

In the following first two examples, he gave examples of both reduced (Example 2.4.40) and nonreduced (Example 2.4.41) commutative rings $R$, where $diam(\Gamma(R)) = 2$ and $diam(\Gamma(R[x])) = diam(\Gamma(R[[x]])) = 3$. Both of these examples are based on the domain $D = F[w, y, z]_M$, where $F$ is a field, $w, y$, and $z$ are algebraically independent indeterminates, and $M = (w, y, z)F[w, y, z]$. The constructions are similar to ones employed in [270] to construct rings where each two-generated ideal containing only zero-divisors has a nonzero annihilator, but some three-generated ideal (containing only zero-divisors) does not. For the nonreduced ring, $R$ is built using idealization. For the reduced ring, he used a variation on the so-called "$A + B$" construction (cf. for example, [270]).

**Example 2.4.40.** ([269, Example 5.1]) Let $Q$ be the maximal ideal of $D = F[w, y, z]_{(w,y,z)}$, and let $\mathcal{P}$ denote the set of height-two primes of $D$. For every $P_\alpha \in \mathcal{P}$, let $Q_\alpha = Q/P_\alpha$. Let $\mathcal{I} = \mathcal{A} \times \mathbb{N}$, where $\mathcal{A}$ is an index set for $\mathcal{P}$, and let $B = \sum Q_i$, where $Q_i = Q_\alpha$ for every $i = (\alpha, n) \in \mathcal{I}$. Finally, let $R = D + B$ be the ring constructed from $D \times B$ by setting $(r, a) + (s, b) = (r + s, a + b)$ and $(r, a)(s, b) = (rs, rb + sa + ab)$.

(i) $Z(R) = Q + B$ is a prime ideal of $R$;

(ii) Every two-generated ideal contained in $Z(R)$ has a nonzero annihilator;

(iii) $R$ is not a McCoy ring;

(iv) $diam(\Gamma(R)) = 2$ and $diam(\Gamma(R[x])) = diam(\Gamma(R[[x]])) = 3$.

A similar construction based on using the height-one primes for the set $\mathcal{P}$ will yield a reduced ring $R$ with $Z(R)$ an ideal and $diam(\Gamma(R)) = 3$ instead of 2 (see [64, Example 3.13] for details).

**Example 2.4.41.** ([269, Example 5.2]) As above, let $\mathcal{P}$ denote the height-two primes of $D = F[w, y, z]_{(w,y,z)}$, and let $Q$ be the maximal ideal of $D$, but now let $B = \sum K_\alpha$, where $K_\alpha = qf(D/P_\alpha)$ for every $P_\alpha \in \mathcal{P}$. Let $R = D(+)B$ be the idealization of $B$ over $D$.

(i) $R$ is a local ring with maximal ideal $Q(+)B = Z(R)$;

(ii) Every two-generated ideal contained in $Z(R)$ has a nonzero annihilator, but $R$ is not a McCoy ring;

(iii) $diam(\Gamma(R)) = 2$, but $diam(\Gamma(R[x])) = diam(\Gamma(R[[x]])) = 3$.

In the following example, Lucas constructed a reduced ring $R$ with $\text{diam}(\Gamma(R)) = \text{diam}(\Gamma(R[x])) = 2$ and $\text{diam}(\Gamma(R[[x]])) = 3$.

Let the domain $D = F[\mathcal{X}]_{(\mathcal{X})}$, where $F$ is a field and $\mathcal{X} = \{x_n\}$ is a countably infinite set of indeterminates. Let $M(= (\mathcal{X})D)$ denote the maximal ideal of $D$. Next, let $\mathcal{P}$ denote the primes of $D$ that are generated by finite subsets of $\mathcal{X}$. The set $\mathcal{P}$ includes $P_0 = (0)$, the prime generated by the empty subset of $\mathcal{X}$. Also for $n \geq 1$, let $P_n = (x_1, x_2, \ldots, x_n)D$. Note that given a prime $P_\alpha \in \mathcal{P}$, there is an integer $n$ such that $P_\alpha \subseteq P_k$ for every $k \geq n$. For every $P_\alpha \in \mathcal{P}$, let $Q_\alpha = M/P_\alpha$.

**Example 2.4.42.** ([269, Example 5.3]) For the domain $D$ above, let $\mathcal{O} = \{P_n \mid n \geq 0\}$ and let $C = \sum Q_n$. Let $R = D + C$ be the ring formed from the product $D \times C$ by setting $(r, b) + (s, c) = (r + s, b + c)$ and $(r, b)(s, c) = (rs, rc + sb + bc)$. Then $R$ is a reduced McCoy ring such that $Z(R)$ is an ideal; so $diam(\Gamma(R)) = diam(\Gamma(R[x])) = 2$. However, $diam(\Gamma(R[[x]])) = 3$ since $Z(R[[x]])$ is not an ideal.

The following is an example constructed by Lucas which provides a reduced McCoy ring $R$ in which $Z(R)$ is an ideal and the sum of any two countably generated ideals with nonzero annihilators will have a nonzero annihilator, but where some countably generated ideal containing only zero-divisors will not have a nonzero annihilator. Thus, each of the graphs $\Gamma(R)$, $\Gamma(R[x])$, and $\Gamma(R[[x]])$ will have diameter 2 and $R$ will have infinitely many minimal primes, but $R$ will not be countably McCoy.

Let $D = F[\mathcal{X}]_{(\mathcal{X})}$, where $\mathcal{X} = \{x_n\}$ and $\mathcal{P}$ is the set of primes generated by finite subsets of $\mathcal{X}$. Let $\mathcal{N} = \{P_\alpha\}$ denote the set of nonzero primes of $\mathcal{P}$. Let $\mathcal{A}$ be an index set for $\mathcal{N}$ and let $\mathcal{I} = \mathcal{A} \times \mathbb{R}^+$ (with $\mathbb{R}^+$ the positive reals). Let $E = \prod D_i$, where $D_i = D_\alpha = D/P_\alpha$ for every $i = (\alpha, r)$, also let $Q_i = Q_\alpha = M/P_\alpha$. We let $\mathcal{I}^0 = \{i = (\alpha, r) \in \mathcal{I} \mid r \leq 1\}$, and for every $i = (\alpha, r) \in \mathcal{I}^o$, we let $\overline{r} = \{t \in \mathbb{R}^+ \mid t - r \in \mathbb{Z}\}$ and $\overline{i} = \{j \in \mathcal{I} \mid j = (\alpha, t) \text{ for some } t \in \overline{r}\}$.

For every nonempty finite subset $X$ of $\mathcal{X}$ and every positive integer $n$, let $X_{(n)} = \{x_{nk} \mid x_k \in X\}$. Obviously, for a pair of finite sets $X$ and $Y$, $X \subseteq Y$ if and only if $X_{(n)} \subseteq Y_{(n)}$ for every (equivalently, some) $n$. Note that $X_{(1)} = X$. For a nonzero polynomial $b \in F[\mathcal{X}]$ with constant term 0, there is a minimal finite subset $X \subseteq \mathcal{X}$ such that $b(X) \in F[X]$. For such a $b = b(X)$ and minimal set $X$, if $X \subseteq Y$, then we also have $b(Y) \in F[Y]$. Moreover, we also have $b(X_{(n)}) = b(Y_{(n)})$ for every $n$ using the substitution of $x_{nk}$ for every $x_k$ in $X$ (and $Y$). For a fixed $i = (\alpha, r) \in \mathcal{I}^o$, let $b(i)$ denote the element of $E$, where $(b(i))_j = 0$ if $j \notin \overline{i}$ and $(b(i))_j = (b(X_{(n+1)}))_j$ if $j = (\alpha, r + n) \in \overline{i}$. Since the constant term of $b$ is 0, $b(i) \in \prod Q_i$. Note that for a fixed $i \in \mathcal{I}^o$ and two such polynomials $b(X)$ and $d(X), b(i)d(i) = (bd)(i)$. Let $B \subseteq E$ be the $D$-subalgebra of $E$ consisting of the finite sums of the form $d = s_1 b_1(i_1) + s_2 b_2(i_2) + \cdots + s_m b_m(i_m)$ with every $s_k \in D$, every $b_k \in \mathcal{X}F[\mathcal{X}]$, and every $i_k \in \mathcal{I}^o$. Since every sum is finite, there is a finite subset $X$ such that every $s_k$ is in $F[X]_{(X)}$ and every $b_k \in F[X]$. Let $R = D + B$

be the ring formed from the product $D \times B$ using the same definition for addition and multiplication as that used in the previous example.

**Example 2.4.43.** ([269, Example 5.4]) Let $D$ and $R = D+B$ be the commutative rings defined in the previous paragraph. The ring $R$ has the following properties:

(i) $R$ is a reduced McCoy ring;

(ii) $Z(R) = M + B$ is an ideal of $R$;

(iii) $R$ is not countably McCoy, but if $I$ and $J$ are countably generated ideals with nonzero annihilators, then $I + J$ has a nonzero annihilator;

(iv) $diam(\Gamma(R)) = diam(\Gamma(R[x])) = diam(\Gamma(R[[x]])) = 2$.

**Example 2.4.44.** ([269, Example 5.5]) Let $D = F[\mathcal{Y}, \mathcal{Z}]$, where $F$ is a field of characteristic 2 and $\mathcal{Y} = \{y_n\}_{n=0}^{\infty}$ and $\mathcal{Z} = \{z_n\}_{n=0}^{\infty}$ are disjoint countably infinite sets of indeterminates. Let $R = D/A$, where

$$A = (y_0 z_0, \{y_0 y_n\}, \{y_n^2\}, \{z_0 z_n\}, \{z_n^2\})D.$$

(i) $R$ is a zero-dimensional commutative ring with a single maximal (minimal) prime, $M = (\mathcal{Y}, \mathcal{Z})R$;

(ii) For every $0 \neq r \in M$, $r^2 = 0$;

(iii) For every pair of zero-divisors $r$ and $s$, either $rs = 0$ or $rs$ is a nonzero annihilator of the ideal $(r, s)$;

(iv) $R$ is a McCoy ring, $Z(R)$ is an ideal, and $diam(\Gamma(R)) = diam(\Gamma(R[x])) = 2$;

(v) $Z(R[[x]]) = M[[x]]$ is an ideal of $R[[x]]$. Moreover, $f^2 = 0$ for every $f \in M[[x]]$;

(vi) $diam(\Gamma(R[[x]])) = 2$ even though there are countably generated ideals $I$ and $J$ each with a nonzero annihilator such that $I + J$ has no nonzero annihilator.

**Example 2.4.45.** ([269, Example 5.6]) Let $R = \mathbb{Z}(+)\mathbb{Z}_p$, the idealization of the integers modulo some prime $p$, and let $S = \mathbb{Z}(+)\mathbb{Z}(p^{\infty})$.

(i) $Z(R) = p\mathbb{Z}(+)\mathbb{Z}_p$ is a maximal ideal of $R$;

(ii) $Z(S) = p\mathbb{Z}(+)\mathbb{Z}(p^{\infty})$ is a maximal ideal of $S$;

(iii) Both $Z(R)$ and $Z(S)$ have nonzero annihilators. Also, $Z(R)Nil(R) = 0$, but $Z(S)Nil(S) \neq 0$;

(iv) Both $R$ and $S$ are McCoy rings;

(v) $diam(\Gamma(R)) = diam(\Gamma(R[x])) = diam(\Gamma(R[[x]])) = 2$;

(vi) $diam(\Gamma(S)) = diam(\Gamma(S[x])) = 2$, but $diam(\Gamma(S[[x]])) = 3$.

As a $\mathbb{Z}$-module, $\mathbb{Z}(p^\infty)$ is divisible, i.e., for every $0 \neq r \in \mathbb{Z}$ and every $0 \neq h \in \mathbb{Z}(p^\infty)$, there is an element $k \in \mathbb{Z}(p^\infty)$ such that $rk = h$. In an analogous fashion, an ideal $I$ of a ring $R$ is said to be *divided* if it is comparable with every principal ideal.

**Theorem 2.4.46.** ([269, Theorem 5.7]) *Let $R$ be a nonreduced commutative ring with prime nilradical $Nil(R)$. If $Nil(R)$ is divided and $Z(R)$ properly contains $Nil(R)$, then for every $r \in Z(R) \setminus Nil(R)$, every power series with constant term $r$ is a zero-divisor of $R[[x]]$.*

**Theorem 2.4.47.** ([269, Theorem 5.8]) *Let $R$ be a nonreduced commutative ring with prime nilradical $Nil(R)$. If $Nil(R)$ is divided and $Z(R)$ properly contains $Nil(R)$, then $diam(\Gamma(R[[x]])) = 3$.*

The next result provides a condition which is sufficient to give $diam(\Gamma(R[[x]])) = 2$ when $R$ is not reduced with nonnilpotent zero-divisors.

**Theorem 2.4.48.** ([269, Theorem 5.10]) *Let $R$ be a nonreduced commutative ring such that $Z(R)$ is not the nilradical of $R$. If $Z(R)$ has a nonzero annihilator, then $Z(R)$ is an ideal of $R$, $R$ is a McCoy ring, $diam(\Gamma(R)) = diam(\Gamma(R[x])) = 2$, and $Z(R)[[x]] \subseteq Z(R[[x]])$. Moreover, if $Z(R)Nil(R) = 0$, then $diam(\Gamma(R[[x]])) = 2$.*

The study of the diameter of the zero-divisor graph of a polynomial ring or power series ring was continued by Anderson and Mulay in [85]. The next theorem handles the case when $T(R)$ is von Neumann regular

**Theorem 2.4.49.** ([85, Theorem 3.3]) *Let $R$ be a (reduced) commutative ring with $T(R)$ von Neumann regular.*

*If $R \cong \mathbb{Z}_2 \times \mathbb{Z}_2$, then $1 = diam(\Gamma(R)) < diam(\Gamma(R[x])) = diam(\Gamma(R[[x]])) = 2$. Otherwise, $diam(\Gamma(R)) = diam(\Gamma(R[x])) = diam(\Gamma(R[[x]]))$. (It is 2 if $R$ has exactly 2 minimal prime ideals, and 3 if $R$ has at least 3 minimal prime ideals.)*

Some of the results from this section on diameter and from Section 2.3.1 on girth have been extended to skew generalized power series rings by Moussavi and Paykan in [294]. These rings include (skew) polynomial rings, (skew) power series rings, (skew) monoid rings, and generalized power series rings. Hurwitz series rings have been studied in [10] and [307].

### 2.4.2 Idealization

For a commutative ring $R$ and $R$-module $M$, the *idealization* of $M$ is the ring $R(+)M$ built from the product $R \times M$ by setting $(r, b) + (s, c) = (r + s, b + c)$ and $(r, b)(s, c) = (rs, rc + sb)$. In this section, we study the diameter of the zero-divisor graph of a ring formed by idealization. For some basic results on idealization, one may refer to Remark 2.3.25 and Proposition 2.3.26. In studying the zero-divisor graph of an idealization, it quickly becomes clear that the diameter of the ring's graph need not be preserved. It is also clear that if $diam(\Gamma(R)) > 1$, then $diam(\Gamma(R(+)M)) > 1$. However, it is possible to have a ring $R$ such that $\Gamma(R)$ is complete and $diam(\Gamma(R(+)M)) > 1$, as well as all other possible combinations between the diameter of $R$ and that of $R(+)M$.

We begin with some examples illustrating ways in which the diameter of $\Gamma(R)$ is not preserved upon moving to $R(+)M$.

**Example 2.4.50.** ([120, Example 3.1]) Consider $\mathbb{Z}_9(+)\mathbb{Z}_9$ under the usual module operation. Clearly $diam(\Gamma(\mathbb{Z}_9)) = 1$. Observe that if $r \in Reg(\mathbb{Z}_9)$, then $rm = 0$ if and only if $m = 0$. Thus for $r \in Reg(\mathbb{Z}_9)$, we have $(r, m) \notin Z(\mathbb{Z}_9(+)\mathbb{Z}_9)^*$. Now $(0, 3)(r, m) = 0$ for any $(r, m) \in Z(\mathbb{Z}_9(+)\mathbb{Z}_9)$ and $(3, 1)(3, 3) \neq 0$. Hence $diam(\Gamma(\mathbb{Z}_9(+)\mathbb{Z}_9)) = 2$.

**Example 2.4.51.** ([120, Example 3.2]) Let $R = \mathbb{Z}[X]/(X^2) = \mathbb{Z}[x]$. Then $diam(\Gamma(R)) = 1$ since $Z(R)^* = \{ax \mid a \in \mathbb{Z}^*\}$. Let $M = \mathbb{Z}_6$, and define $(a + bx)m = am$. Consider $(2, 1), (3, 1) \in Z(R(+)M)^*$. Clearly $Ann((2, 1)) = \{(0, 3), (0, 0)\}$, while $Ann((3, 1)) = \{(0, 2), (0, 0), (0, 4)\}$. Hence $d((2, 1), (3, 1)) = 3$, and so $diam(\Gamma(R(+)M)) = 3$. We can also consider $\mathbb{Z}_4$ as an $R$-module under the same operation, in which case $diam(\Gamma(R(+)\mathbb{Z}_4)) = 2$.

We now discuss necessary and sufficient conditions on $R$ and $M$ to ensure that $\Gamma(R(+)M)$ is complete. Notice in the previous two examples $(Z(R))^2 = 0$, yet $(Z(R(+)M))^2 \neq 0$. So, simply requiring $\Gamma(R)$ to be complete will not guarantee that $\Gamma(R(+)M)$ will be complete.

We next state the three properties that will be considered numerous times in this section.

**Remark 2.4.52.** Let $M$ be a $R$-Module. Then

(a) $(Z(R))^2 = 0$;

(b) For every $r \in Reg(R)$, $rm \neq 0$ for all $m \in M^*$;

(c) If $r \in Z(R)^*$, then $rM = 0$.

**Theorem 2.4.53.** ([120, Theorem 3.3]) *Let $\Gamma(R) \neq \emptyset$. Then $\Gamma(R(+)M)$ is complete if and only if $R(+)M$ satisfies the following three properties:*

(a) $(Z(R))^2 = 0$;

(b) *For every $r \in Reg(R)$, $rm \neq 0$ for all $m \in M^*$;*

(c) *If $r \in Z(R)^*$, then $rM = 0$.*

Note that in Example 2.4.50, the ring $\mathbb{Z}_9$ does not satisfy condition $(c)$, and in Example 2.4.51, the ring $\mathbb{Z}[x]/(x^2)$ does not satisfy condition $(b)$.

**Corollary 2.4.54.** ([120, Corollary 3.4]) *$\Gamma(R(+)M)$ is complete if and only if $(Z(R(+)M))^2 = 0$.*

Also, note that any integral domain $D$ satisfies condition $(a)$ and vacuously satisfies condition $(c)$. Hence $\Gamma(D(+)M)$ is complete if and only if $D$ satisfies condition $(b)$.

We now focus on when $diam(\Gamma(R)) \geq 1$. In this case, it is clear that $diam(\Gamma(R(+)M)) > 1$. However, it is possible to form idealizations with zero-divisor graphs having diameters of 2 and 3 from rings whose zero-divisor graphs have diameter 2, as well as rings whose zero-divisor graphs have diameter 3, as the following examples illustrate.

**Example 2.4.55.** ([120, Example 3.5]) Consider $\mathbb{Z}_8(+)\mathbb{Z}_2$. It is easy to verify that $diam(\Gamma(\mathbb{Z}_8)) = diam(\Gamma(\mathbb{Z}_8(+)\mathbb{Z}_2)) = 2$.

**Example 2.4.56.** ([120, Example 3.6]) Let $R = \mathbb{Z}_6$ and $M = \mathbb{Z}_6$, and consider $R(+)M$ under the usual module operation. It is easy to see that $diam(\Gamma(R)) = 2$ and $Z(R(+)M)^* = \{(0,m) \mid m \in M^*\} \cup \{(a,n) \mid a \in Z(R)^*,\ n \in M\}$. However, $diam(\Gamma(R(+)M)) = 3$ since $Ann((2,1)) = \{(0,3),(0,0)\}$, but $(3,1)(0,3) = (0,3) \neq (0,0)$.

**Example 2.4.57.** ([120, Example 3.7]) Let $F$ be a field,

$$S = \mathbb{F}[y, \{x_i\}_{i=0}^{\infty}]/(\{x_0 y\} \cup \{x_i - x_{i+1}y\}_{i=0}^{\infty}),$$

and $R = S[[w]]$. Then by Example 2.4.14, $diam(\Gamma(R)) = 3$. Let $f(w) = f_0 + f_1 w + f_2 w^2 + \cdots$, where $f_i \in S$. Make $F$ an $R$-module with the operation $f \cdot k = f_0(0) \cdot k$, where $f_0(0)$ is $f_0$ evaluated when all indeterminates appearing in $f_0$ are 0. Note that $(f,k) \in Z(R(+)K)^*$ if and only if $f_0(0) = 0$. So $(f,k)(0,\ell) = (0,0)$ for every $(f,k) \in Z(R(+)K)^*$. Hence $diam(\Gamma(R(+)K)) = 2$.

**Example 2.4.58.** ([120, Example 3.8]) Let $R = \mathbb{Z}_6 \times \mathbb{Z}_6$. Then $diam(\Gamma(R)) = 3$. Make $\mathbb{Z}_6$ an $R$-module with the operation $(a,b) \cdot m = am$. Observe that $Ann(((2,3),1)) = \{((0,a),b) \mid a = 0,2,4 \text{ and } b = 0,3\}$ and $((3,1),1)((0,a),b) \neq ((0,0),0)$ for $a = 0,2,4$ and $b = 0,3$ unless $a = 0$ and $b = 0$. Hence $diam(\Gamma(R(+)M)) = 3$.

If necessary and sufficient conditions for ensuring $diam(\Gamma(R(+)M)) = 2$ could be found, then the classification of the diameter of an idealization would be complete. It may prove easier to classify the diameter 3 case instead.

Below, we provide two results that ensure $diam(\Gamma(R(+)M)) = 2$ based on properties (a), (b), and (c) in Remark 2.4.52.

**Theorem 2.4.59.** ([120, Theorem 3.9]) *If $R(+)M$ satisfies properties* (b) *and* (c), *but not property* (a), *then $diam(\Gamma(R(+)M)) = 2$.*

**Theorem 2.4.60.** ([120, Theorem 3.10]) *If $R(+)M$ satisfies properties* (a) *and* (b), *but not property* (c), *then $diam(\Gamma(R(+)M)) = 2$.*

It is also worth briefly mentioning the diameter behavior of the idealization of nondomains $R$ with $diam(\Gamma(R)) = 0$, namely $R \cong \mathbb{Z}_4$ and $R \cong \mathbb{Z}_2[x]/(x^2)$.

If $R \cong \mathbb{Z}_4$, then any $R$-module $M$ has the property that $2M = 0$ or $4M = 0$. It is easy to show using Theorem 2.4.53 that $diam(\Gamma(\mathbb{Z}_4(+)M)) = 1$ if and only if $2M = 0$. Thus, using Theorem 2.4.60, it is also clear that $diam(\Gamma(\mathbb{Z}_4(+)M)) = 2$ if and only if $2M \neq 0$.

Now, consider $R \cong \mathbb{Z}_2[x]/(x^2)$. The diameter of an idealization of $R$ is less clear, though 1 and 2 are again the only possibilities. Clearly $R$ satisfies property (a). Also, note that every $R$-module $M$ has the property that $2M = 0$, and so $m = -m$ for every $m \in M$. Observe that the only regular elements of $R$ are 1 and $x + 1$. Clearly, for every $m \in M^*$, we have $1m \neq 0$. If $(x + 1)m = 0$ for some $m \in M^*$, then $xm = m$. Hence $x(xm) = xm$, and since $x^2m = 0m = 0$, we obtain $xm = m = 0$, a contradiction. Thus $R(+)M$ also satisfies property (b). Then by Theorems 2.4.53 and 2.4.60, we have that $diam(\Gamma(R(+)M)) = 1$ or 2 for $R \cong \mathbb{Z}_2[x]/(x^2)$. For example, if $M = \mathbb{Z}_2[x]/(x^2)$ using ring multiplication as the module operation, then $R(+)M$ does not satisfy property (c), and hence the diameter is 2. If $M = \mathbb{Z}_2$ with module operation defined by $(ax + b)m = bm$, then $R(+)M$ does satisfy property (c), and hence the diameter is 1.

By Proposition 2.3.26, it is clear that $diam(\Gamma(R(+)M)) = 0$ if and only if $R(+)M = \mathbb{Z}_2(+)\mathbb{Z}_2$ (note that $\mathbb{Z}_2(+)\mathbb{Z}_2 \cong \mathbb{Z}_2[x]/(x^2)$). Example 2.4.50 serves as both an example of Theorem 2.4.60 and a counterexample to the converse of Theorem 2.4.59. Likewise, Example 2.4.55 serves as both an example of Theorem 2.4.59 and a counterexample to the converse of Theorem 2.4.60. In addition, Example 2.4.51 shows that if properties (a) and (c) hold, but not (b), then diameters of 2 and 3 are possible.

The study of the diameter of the zero-divisor graph of idealizations was continued by Anderson and Mulay in [85]. First of all, they determined when $diam(\Gamma(R(+)M)) \leq 1$. An equivalent characterization is given in Theorem 2.4.53 and Corollary 2.4.54. (The conditions (b) and (c) are equivalent to $Z(M) \subset Z(R)$ and $Z(R) \subseteq Ann(M)$, respectively.)

**Theorem 2.4.61.** ([85, Theorem 3.7]) *The following statements are equivalent for a commutative ring $R$ and a nonzero $R$-module $M$:*

(a) *$\Gamma(R(+)M)$ is a complete graph;*

(b) *$diam(\Gamma(R(+)M)) \leq 1$;*

(c) *$Z(R) = Z(M) = Ann(M)$ and $Z(R)^2 = 0$;*

(d) *$Z(R(+)M)^2 = 0$.*

The next theorem characterizes when $diam(\Gamma(R(+)M)) \leq 2$. This answers an open question from [120]. Combining this with Theorem 2.4.61, thus, in some sense, characterizes when $diam(\Gamma(R(+)M)) = 3$.

**Theorem 2.4.62.** ([85, Theorem 3.9]) *Let $R$ be a commutative ring and $M$ a nonzero $R$-module. Then $diam(\Gamma(R(+)M)) \leq 2$ if and only for every $x, y \in Z(R) \cup Z(M)$, either* (i) *there is a $0 \neq z \in Ann(M)$ such that $xz = yz = 0$ or* (ii) *there is a $0 \neq m \in M$ such that $xm = ym = 0$.*

**Remark 2.4.63.** ([85, Remark 3.10]) (1) Note that either $Z(R) \subseteq Z(M)$ or $Z(M) \subseteq Z(R)$ if $diam(\Gamma(R(+)M)) \leq 2$. For suppose that $Z(R) \not\subseteq Z(M)$ and $Z(M) \not\subseteq Z(R)$. Choose $x \in Z(R) \setminus Z(M)$ and $y \in Z(M) \setminus Z(R)$. Then $x, y \in Z(R) \cup Z(M)$ do not satisfy either $(i)$ or $(ii)$ of Theorem 2.4.62.

(2) In Theorems 2.4.59 and 2.4.60, two sufficient conditions are given to have $diam(\Gamma(R(+)M)) = 2$. Since the conditions $(b)$ and $(c)$ are equivalent to $Z(M) \subseteq Z(R)$ and $Z(R) \subseteq Ann(M)$, respectively, Theorem 2.4.59 translates to $diam(\Gamma(R(+)M)) = 2$ if $Z(R) = Z(M) = Ann(M)$ and $Z(R)^2 \neq 0$ (this implies $(ii)$ of our Theorem 2.4.62), and Theorem 2.4.60 translates to $diam(\Gamma(R(+)M)) = 2$ if $Z(R) = Z(M)$, $Z(R) \not\subseteq Ann(M)$, and $Z(R)^2 = 0$ (note that $Z(R) \subseteq Z(M)$ if $Z(R)^2 = 0$), which implies $(i)$ of Theorem 2.4.62.

### 2.4.3 Rings with Linearly Ordered Primes

In this section, we investigate the diameter of the zero-divisor graph of a commutative ring $R$ when the prime ideals of $R$ contained in $Z(R)$ are linearly ordered. These are precisely the rings $R$ such that the prime ideals of the total quotient ring $T(R)$ are linearly ordered, and include chained rings, divided rings, PVRs, $\varphi$-PVRs, rings with $Z(R) = Nil(R)$, and zero-dimensional quasilocal rings. This is due to Anderson and Badawi [68].

When $Z(R) = Nil(R)$, it is easy to explicitly describe the diameter of $\Gamma(R)$; and moreover, $diam(\Gamma(R)) \leq 2$ in this case. We record this as our first theorem (cf. Theorem 2.4.28). Note that in this case, $Nil(R)$ is the unique minimal prime ideal of $R$ and is the only prime ideal of $R$ contained in $Z(R)$; so this is the simplest case where the prime ideals of $R$ contained in $Z(R)$ are linearly ordered.

**Theorem 2.4.64.** ([68, Theorem 2.2]) *Let $R$ be a commutative ring with $Z(R) = Nil(R) \neq 0$. Then exactly one of the following three cases must occur:*

(a) $|Z(R)^*| = 1$. *In this case, $R$ is isomorphic to $\mathbb{Z}_4$ or $\mathbb{Z}_2[x]/(x^2)$, and $diam(\Gamma(R)) = 0$;*

(b) $|Z(R)^*| \geq 2$ *and $Z(R)^2 = 0$. In this case, $\Gamma(R)$ is a complete graph, and $diam(\Gamma(R)) = 1$;*

(c) $Z(R)^2 \neq 0$. *In this case, $diam(\Gamma(R)) = 2$.*

Thus, when studying the diameter of the zero-divisor graph of a ring $R$, the interesting case is when $Nil(R) \subsetneq Z(R)$. We next give several results. Note that in Lemma 2.4.66, we need only assume that $x \in Z(R) \setminus N(R)$, where $N(R) = \{x \in R \mid x^2 = 0\}$.

**Lemma 2.4.65.** ([68, Lemma 2.3]) *Let $R$ be a commutative ring with $x \in Nil(R)^*$ and $y \in Z(R)^*$. Then $d(x, y) \leq 2$ in $\Gamma(R)$.*

**Lemma 2.4.66.** ([68, Lemma 2.4]) *Let $R$ be a commutative ring with $x \in Z(R) \setminus Nil(R)$ and $y \in Z(R)^*$ such that $x \mid zy^n$ for some integer $n \geq 1$ and $z \in R \setminus Z(R)$. Then $d(x, y) \leq 2$ in $\Gamma(R)$.*

By [126, Theorem 1], the prime ideals of $R$ are linearly ordered if and only if the radical ideals of $R$ are linearly ordered, if and only if for every $x, y \in R$, there is an integer $n = n(x, y) \geq 1$ such that either $x \mid y^n$ or $y \mid x^n$. This result easily extends to the prime ideals of $R$ contained in $Z(R)$.

**Theorem 2.4.67.** ([68, Theorem 2.5]) *Let $R$ be a commutative ring.*

(a) *The prime ideals of $R$ contained in $Z(R)$ are linearly ordered if and only if for every $x, y \in Z(R)$, there is an integer $n = n(x, y) \geq 1$ and an element $z \in R \setminus Z(R)$ such that either $x \mid zy^n$ or $y \mid zx^n$;*

(b) *The radical ideals of $R$ contained in $Z(R)$ are linearly ordered if and only if for every $x, y \in Z(R)$, there is an integer $n = n(x, y) \geq 1$ such that either $x \mid y^n$ or $y \mid x^n$;*

(c) *If the prime ideals of $R$ contained in $Z(R)$ are linearly ordered, then $Nil(R)$ and $Z(R)$ are prime ideals of $R$.*

The prime ideals of $R$ contained in $Z(R)$ are linearly ordered if and only if the prime ideals of $T(R)$ are linearly ordered. Moreover, $\Gamma(R) \cong \Gamma(T(R))$ by [81, Theorem 2.2]. Thus we can often reduce to the case where the prime ideals of $R$ are linearly ordered. Note that a reduced ring $R$ with its prime ideals contained in $Z(R)$ linearly ordered is an integral domain. Also, observe that a nonreduced ring $R$ has $\Gamma(R)$ complete if and only if $Z(R)^2 = 0$ (by Theorem 2.2.5), i.e., if $xy = 0$ for every $x, y \in Z(R)$ with $x \neq y$, then $x^2 = 0$ for every $x \in Z(R)$. So if $R$ is a nonreduced ring with $Z(R)^2 = 0$, then $0 \neq N(R) = Nil(R) = Z(R)$ and $diam(\Gamma(R)) \leq 1$, with equality when $|Z(R)^*| \geq 2$.

**Theorem 2.4.68.** ([68, Theorem 2.7]) *Let $R$ be a commutative ring with $Z(R)^2 \neq 0$ such that the prime ideals of $R$ contained in $Z(R)$ are linearly ordered. Then $diam(\Gamma(R)) = 2$.*

## 2.5 Radius and Center of Zero-divisor Graphs

In this section, we are interested in the radius and center of the zero-divisor graph of a commutative ring. These results are due to Redmond [324]. For any vertex $v$ of a connected graph $G$, the *eccentricity* of $v$, denoted by $e(v)$, is the maximum of the distances from $v$ to the other vertices of $G$. The set of vertices with minimum eccentricity is called the *center* of the graph, and this minimum eccentricity value is the *radius* of $G$. That is, a vertex is central in $G$ if its greatest distance from any other vertex is as small as possible. A graph $G$ is called *self-centered* if the center of $G$ is itself (i.e., $e(v) = k$ for every $v \in V(G)$). A well-known graph theory result says that for any connected graph with radius $r$ and diameter $d$, we have $r \leq d \leq 2r$.

Note that any graph $G$ with radius 1 necessarily has at least one vertex adjacent to all other vertices of $G$. Therefore by Corollaries 2.2.3 and 2.2.4, we have the following result.

**Theorem 2.5.1.** ([324, Corollary 2.2]) *Let $R$ be a commutative Noetherian ring with identity. The radius of $\Gamma(R)$ is 0 if and only if either $R \cong \mathbb{Z}_4$ or $R \cong \mathbb{Z}_2[x]/(x^2)$. The radius of $\Gamma(R)$ is 1 if and only if either $R \cong \mathbb{Z}_2 \times A$, where $A$ is an integral domain, or $Z(R)$ is an ideal of $R$. If, in addition, $R$ is finite, then the radius of $\Gamma(R)$ is 1 if and only if either $R \cong \mathbb{Z}_2 \times F$, where $F$ is a finite field or $R$ is local.*

**Theorem 2.5.2.** ([324, Theorem 2.3]) *Let $R$ be a commutative Noetherian ring with identity that is not an integral domain. Then the radius of $\Gamma(R)$ is at most 2.*

*Proof.* By Theorem 2.5.1, assume that $Z(R)$ is not an ideal.

Suppose that $R$ is a reduced ring. Let $Z(R) = \cup_{i=1}^{n} P_i$, where every $P_i$ is a minimal prime ideal of $R$. Since $Z(R)$ is not an ideal of $R$, $n \geq 2$. Further, for $j = 1, \ldots, n$, one can choose a $0 \neq y_j \in \cap\{P_i \mid i = 1, \ldots, j-1, j+1, \ldots, n\}$. Let $x \in Z(R)$. Then $x \in P_m$ for some $1 \leq m \leq n$. Note that $xy_m \in \cap_{i=1}^{n} P_i = Nil(R) = \{0\}$, and so $xy_m = 0$. For $j \neq m$, $y_m y_j = 0$ by the same argument. Thus, the distance from any $y_j$ to $x$ is 1 (if $j = m$) or 2 (if $j \neq m$). Hence, the radius of $\Gamma(R)$ is at most 2.

Now, suppose that $R$ is not reduced (that is, $Nil(R) \neq 0$). Let $Z(R) = \cup_{i=1}^{N} P_i$, where every $P_i$ is a minimal prime ideal of $R$. By [228, Theorem 86], for every $i = 1, \ldots, N$, there is a $0 \neq a_i \in R$ such that $P_i = Ann(a_i)$. By our hypothesis, one can choose a $0 \neq v \in Nil(R) \subseteq \cap_{i=1}^{N} P_i = \cap_{i=1}^{N} Ann(a_i)$. Then $va_i = 0$ for every $i = 1, \ldots, N$. Let $x \in Z(R)$. Then $x \in P_j$ for some index $j$. Thus, either $x - v$ (if $xv = 0$) or $x - a_j - v$ (if $xv \neq 0$) is a path in $\Gamma(R)$. Hence, the eccentricity of $v$ is at most 2, implying that the radius of $\Gamma(R)$ is also at most 2. □

Using the above results, one has the following ring-theoretic result.

**Corollary 2.5.3.** ([324, Corollary 2.4]) *Let $R$ be a commutative Noetherian ring. If $R$ is not an integral domain, then there is a $0 \neq x \in R$ such that either $xy = 0$ or $Ann(x) \cap Ann(y) \neq \{0\}$ for all $y \in Z(R)$.*

Now that it has been established that the radius of the zero-divisor graph of a commutative ring is either 0, 1, or 2, identifying the center of the graph is the next goal. As one might expect from the previous results, several cases are needed.

If the radius of the zero-divisor graph is zero, then the graph consists of a single vertex. Thus the center must be this lone vertex. If the radius of the zero-divisor graph is 1, then those elements in the center are precisely those elements of eccentricity 1.

**Lemma 2.5.4.** ([324, Lemma 3.1]) *Let $(R, \mathfrak{m})$ be a commutative Artinian local ring that is not an integral domain. If $x$ is an element of the center of $\Gamma(R)$, then $x^2 = 0$.*

*Proof.* Since $R$ is Artinian, there is a least integer $k \geq 2$ such that $\mathfrak{m}^k = 0$. Since every nonzero element of $\mathfrak{m}^{k-1}$ has eccentricity 1, the radius of $\Gamma(R)$ is 1. Assume $x$ is in the center of $\Gamma(R)$ and $x^2 \neq 0$. Note that $x^2 \neq x$ since $x$ cannot be idempotent. Then $x^2$ is another vertex of $\Gamma(R)$. Since $e(x) = 1$, $x^3 = x(x^2) = 0$. Then $x + x^2 \neq 0$ (since otherwise $x^2 = (-x^2)^2 = x^4 = 0$) and $x + x^2 \neq x$. Thus $x + x^2$ is also a vertex of $\Gamma(R)$ with $0 = x(x + x^2) = x^2 + x^3 = x^2$, which is a contradiction. □

We next consider the case when $R$ is a nonlocal commutative Artinian ring; so $n + m \geq 2$ is the next theorem.

**Theorem 2.5.5.** ([324, Theorem 3.6]) *Let $n$ and $m$ be nonnegative integers with $n + m \geq 2$. Let $R = R_1 \times \cdots \times R_n \times F_1 \times \cdots \times F_m$, where every $R_i$ is a commutative Artinian local ring that is not a field and every $F_i$ is a field. For every $j = 1, \ldots, m$, define the ideal $I_j = \{0\} \times \cdots \times \{0\} \times F_j \times \{0\} \times \cdots \times \{0\}$. Then the center of $\Gamma(R)$ is $J(R) \cup \left(\bigcup_{j=1}^m I_j\right) \setminus \{(0, \ldots, 0)\}$.*

**Corollary 2.5.6.** ([324, Corollary 3.7])

(a) *Let $n \geq 2$ be an integer. Let $R = R_1 \times \cdots \times R_n$, where every $R_i$ is a commutative Artinian local ring that is not a field. Then the center of $\Gamma(R)$ is $J(R) \setminus \{(0, \ldots, 0)\}$;*

(b) *Let $m \geq 2$ be an integer. Let $R = F_1 \times \cdots \times F_m$, where every $F_i$ is a field. Furthermore, suppose that $R$ is not isomorphic to $\mathbb{Z}_2 \times A$ for any integral domain $A$. For every $j = 1, \ldots, m$, define the ideal $I_j = \{0\} \times \cdots \times \{0\} \times F_j \times \{0\} \times \cdots \times \{0\}$. Then the center of $\Gamma(R)$ is $\left(\bigcup_{j=1}^m I_j\right) \setminus \{(0, \ldots, 0)\}$.*

Let $R$ be a commutative Artinian ring that is not local. Then $R \cong R_1 \times \cdots \times R_n \times F_1 \times \cdots \times F_m$, where every $R_i$ is a commutative Artinian local ring that is not a field, every $F_i$ is a field, and $n$ and $m$ are nonnegative integers such that $n + m \geq 2$. (We will call this isomorphic image of $R$ the Artinian decomposition of $R$). The ideal $I_j = \{0\} \times \cdots \times \{0\} \times F_j \times \{0\} \times \cdots \times \{0\}$ is the annihilator of the maximal ideal $N_j = R_1 \times \cdots \times F_{j-1} \times \{0\} \times F_{j+1} \times \cdots \times F_m$. Note that $N_j^2 = N_j$. That is, $I_j$ is the isomorphic image of some $Ann(N_i)$, where $N_i$ is a maximal ideal of $R$ such that $N_i^2 = N_i$. Now, a characterization of the center of the zero-divisor graph of a ring can be given in terms of the Jacobson radical and the annihilators of maximal ideals for any commutative Artinian ring with identity.

**Theorem 2.5.7.** ([324, Theorem 3.8]) *Let $R$ be a commutative Artinian ring that is not an integral domain. Then the center of $\Gamma(R)$ is the set of nonzero elements in one of the following sets:*

(a) *$Ann(\mathfrak{m})$ if $R$ is local with maximal ideal $\mathfrak{m}$;*

(b) *$Ann(\{0\} \times A) = \{(1,0)\}$ if $R \cong \mathbb{Z}_2 \times A$, where $A$ is an integral domain with $A \ncong \mathbb{Z}_2$;*

(c) *$J(R) \cup (\bigcup_{i=1}^{n} Ann(\mathfrak{m}_i))$ otherwise, where $\mathfrak{m}_1, \ldots, \mathfrak{m}_n$ are the maximal ideals of $R$ such that $\mathfrak{m}_i^2 = \mathfrak{m}_i$.*

**Corollary 2.5.8.** ([324, Corollary 3.9]) *Let $R$ be a commutative Artinian ring that is not local and not an integral domain. Then any nonzero nilpotent element of $\Gamma(R)$ is contained in the center of $\Gamma(R)$.*

Note that in the local case, nilpotent elements of $R$ need not be in the center of $\Gamma(R)$. For example, the center of $\Gamma(\mathbb{Z}_{16})$ is $\{8\}$, but all the vertices of $\Gamma(R)$ are nilpotent elements.

Now that the radius and center of the zero-divisor graph have been determined, the diameter can also be determined. Further in any graph, the radius and diameter are equal if and only if the center is the entire vertex set.

**Theorem 2.5.9.** ([324, Theorem 3.10]) *Let $R$ be a commutative Artinian ring that is not an integral domain. Then the following are true:*

(a) *The radius of $\Gamma(R)$ is zero if and only if $diam(\Gamma(R)) = 0$, if and only if $R \cong \mathbb{Z}_4$ or $R \cong \mathbb{Z}_2[x]/(x^2)$;*

(b) *If the radius of $\Gamma(R)$ is 1, then $diam(\Gamma(R)) = 1$ if and only if $\Gamma(R)$ is complete. Otherwise, $diam(\Gamma(R)) = 2$;*

(c) *If the radius of $\Gamma(R)$ is 2, then $diam(\Gamma(R)) = 2$ if and only if $R \cong F_1 \times F_2$, where $F_1$ and $F_2$ are both fields not isomorphic to $\mathbb{Z}_2$. Otherwise, $diam(\Gamma(R)) = 3$.*

Thus, in all the remaining cases, the center is not the entire vertex set of $\Gamma(R)$. Therefore, the diameter is strictly larger than the radius, implying that $diam(\Gamma(R)) = 3$.

For any vertex $v$ of a connected graph $G$, the *status* of $v$, denoted by $s(v)$, is the sum of the distances from $v$ to the other vertices of $G$. That is, $s(v) = \sum\{d(v,u) \mid u \in V(G)\}$. The set of vertices with minimal status is called the *median* of the graph.

The status is most meaningful in a finite graph. Although both the center and the median relate to the topic of centrality in a graph, they need not coincide.

One can easily construct examples where the center is a proper subset of the median, or the median is a proper subset of the center. However, in the case of zero-divisor graphs of finite commutative rings, the following theorem gives a relationship between the center and median.

**Theorem 2.5.10.** ([324, Theorem 4.1]) *Let $R$ be a finite commutative ring that is not an integral domain. Then the median and center of $\Gamma(R)$ are equal if the radius of $\Gamma(R)$ is at most 1, and the median is a subset of the center if the radius is 2.*

**Corollary 2.5.11.** ([324, Corollary 4.2]) *Let $R$ be a finite commutative ring that is not an integral domain. If the radius of $\Gamma(R)$ is 2, then the center equals the median if and only if $R$ is isomorphic to a direct product of a finite number of copies of a single finite field (i.e., $R \cong F^d$ for some finite field $F$ and some integer $d \geq 2$).*

To illustrate Theorem 2.5.10 and Corollary 2.5.11, here are several examples. In the reduced case, if $R = \mathbb{Z}_3 \times \mathbb{Z}_3$, both the center and the median of $\Gamma(R)$ are the entire vertex set. If $R = \mathbb{Z}_2 \times \mathbb{Z}_2 \times \mathbb{Z}_2$, then the center and median of $\Gamma(R)$ are both the set $\{(1,0,0),(0,1,0),(0,0,1)\}$. However, if $R = \mathbb{Z}_3 \times \mathbb{Z}_5$, the center of $\Gamma(R)$ is the entire vertex set and the median is $\{(1,0),(2,0)\}$. In the nonreduced case, if $R = \mathbb{Z}_{12}$, then the center of $\Gamma(R)$ is $\{4,6,8\}$ and the median is $\{6\}$. If $R = \mathbb{Z}_4 \times \mathbb{Z}_2$, then the center of $\Gamma(R)$ is $\{(0,1),(2,0)\}$ and the median is $\{(0,1)\}$. Note that the last two examples show that in some cases, the elements of the median are nilpotent.

Zero-divisor graphs with radius one have also been studied by J. LaGrange in [247].

## 2.6 Rings without Identity

It is customary to assume that all rings have an identity as we have done in this book. However, we have stated some results for rings without identity, and for other results the proofs do not use an identity. In this section, we discuss zero-divisor graphs for commutative rings without identity. This has been systematically studied by D. F. Anderson and D. Weber [87] in 2018. The following fundamental results hold for all commutative rings, with or without identity; they have all been given in earlier sections.

**Theorem 2.6.1.** ([87, Theorem 2.2]) *Let $R$ be a commutative ring, not necessarily with identity. Then*

(a) *$\Gamma(R)$ is connected with $diam(\Gamma(R)) \in \{0,1,2,3\}$;*

(b) *$gr(\Gamma(R)) \in \{3,4,\infty\}$;*

(c) *$\Gamma(R)$ is complete if and only if $R \cong \mathbb{Z}_2 \times \mathbb{Z}_2$ or $Z(R)^2 = \{0\}$;*

(d) *If* $Z(R) \neq \{0\}$, *then* $\Gamma(R)$ *is finite if and only if* $R$ *is finite.*

When studying commutative rings $R$ without identity, there are two distinct natural cases, either $R = Z(R)$ or $Z(R) \subsetneq R$. The first case holds when $R$ is finite, and either case may occur when $R$ is infinite. Things are particularly nice for finite commutative rings $R$ without identity since $|V(\Gamma(R))| = |R| - 1$. In the second case, $\Gamma(R)$ is isomorphic to the zero-divisor graph of a commutative ring with identity, namely $\Gamma(R_S)$, where $S = R \setminus Z(R)$ [87, Theorem 4.2].

Commutative rings without identity arise naturally from commutative rings with identity. First, let $A$ be a commutative ring with identity. If $Z(A)$ is an (necessarily prime) ideal of $R$, then $R = Z(A)$ is a commutative ring without identity, $R = Z(R) = Z(A)$, and $\Gamma(R) = \Gamma(A)$. This would be the case when $A$ is a zero-dimensional (e.g., finite) local ring. Secondly, let $A$ be a commutative ring with identity and $x \in A \setminus (Z(A) \cup U(A))$. Then $R = xA$ is a commutative ring without identity and $\Gamma(R) \cong \Gamma(A)$ via $x \leftrightarrow xa$. In this case, $A$ and hence $R$, is necessarily infinite. For a third example, let $G$ be an abelian group. Then $G$ becomes a commutative ring without identity when endowed with the trivial multiplication, $xy = 0$ for every $x, y \in G$. We will denote this ring by $G^0$.

The following results hold for finite commutative rings without identity.

**Theorem 2.6.2.** ([87, Theorems 2.3 and 3.7]) *Let $R$ be a finite commutative ring without identity. Then*

(a) $R = Z(R)$;

(b) ([62, Lemma 4.4]) *Either* $R = Nil(R)$ *or* $R = R_1 \times R_2$, *where* $R_1$ *has an identity and* $R_2 = Nil(R_2) \neq \{0\}$;

(c) $\Gamma(R)$ *is connected with* $diam(\Gamma(R)) \in \{0, 1, 2\}$;

(d) $gr(\Gamma(R)) \in \{3, \infty\}$.

The following results hold for commutative rings with $R = Z(R)$.

**Theorem 2.6.3.** ([87, Theorems 3.1 and 3.2]) *Let $R$ be a commutative ring with* $R = Z(R)$.

(a) $\Gamma(R)$ *is complete if and only if* $xy = 0$ *for every* $x, y \in R$;

(b) $gr(\Gamma(R)) \in \{3, \infty\}$;

(c) *If* $diam(\Gamma(R)) = 3$, *then* $gr(\Gamma(R)) = 3$;

(d) *If* $gr(\Gamma(R)) = \infty$, *then* $\Gamma(R)$ *is either* $K_1, K_2 = K_{1,1}, K_{1,2}, K_{1,2p^n-2}$, *or* $K_{1,m}$, *where $p$ is prime, $n \geq 1$ is an integer, and $m$ is an infinite cardinal number.*

The following examples illustrate certain differences for zero-divisor graphs of commutative rings without identity.

**Example 2.6.4.** ([87, Example 2.5]) (a) There are many graphs that can only be realized as the zero-divisor graph of commutative rings with identity or commutative rings without identity, but not both. For example, the complete bipartite graph $K_{2,2}$ can only be realized as the zero-divisor graph of a commutative ring with identity (let $R = \mathbb{Z}_3 \times \mathbb{Z}_3$), and the complete graph $K_5$ can only be realized by a commutative ring without identity (let $R = \mathbb{Z}_6^0$). Moreover, 5 is the fewest number of vertices for which there are graphs that can be realized as zero-divisor graphs only by a commutative ring with identity and other graphs that can be realized as zero-divisor graphs only by a commutative ring without identity.

(b) Let $n$ be a positive integer. Then $\Gamma(R) = K_n$ for a commutative ring with identity if and only if $n = p^k - 1$ for a prime $p$ and integer $k \geq 1$ by Theorem 2.2.7. However, for every positive integer $n$, $\Gamma(\mathbb{Z}_{n+1}^0) = K_n$.

(c) Let $m, n$ be positive integers. Then $\Gamma(R) = K_{m,n}$ for a commutative ring with identity if and only if $m = p^i - 1$ and $n = q^j - 1$ for primes $p, q$ and positive integers $i, j$ by Remark 2.3.12(a). However, for a finite commutative ring without identity, $\Gamma(R)$ is complete bipartite if and only if it is either $K_1, K_{1,1}, K_{1,2}$, or $K_{1,2p^n-2}$ for a prime $p$ and positive integer $n$. For example, $K_{1,3}$ can only be realized by $\mathbb{F}_4 \times \mathbb{Z}_2$, $K_{1,14}$ by $\mathbb{F}_8 \times \mathbb{Z}_2^0$, and $K_{1,2}$ by either $\mathbb{Z}_3 \times \mathbb{Z}_2$ or $\mathbb{Z}_2 \times \mathbb{Z}_2^0$.

(d) For every positive integer $n$, then is a commutative ring $R_n$ without identity such that $|V(\Gamma(R_n))| = n$, namely, $R_n = \mathbb{Z}_{n+1}^0$ has $\Gamma(R_n) = K_n$. However, for commutative rings with identity, there need not be a zero-divisor graph with $n$ vertices. For example, there are no commutative rings with 1210, 3342, or 5466 zero-divisors, and thus there are no zero-divisor graphs with 1209, 3341, or 5465 vertices [327].

(e) There are no finite commutative rings $R$ without identity and either $gr(\Gamma(R)) = 4$ or $diam(\Gamma(R)) = 3$. However, for finite rings with identity, $gr(\Gamma(R)) = 4$ and $diam(\Gamma(R)) = 3$ for $R = \mathbb{Z}_{12}$.

In Section 2.1, we gave Redmond's classification of all commutative rings with identity that have zero-divisor graphs with 14 or fewer vertices. In [87, Section 6], a similar classification is given for commutative rings without identity. In addition, tables are given that list all possible values for $(diam(\Gamma(R)), gr(\Gamma(R)))$ for six different types of commutative rings: finite and infinite rings with identity, where every element is either a zero-divisor or a unit; rings with identity with a nonunit regular element; finite and infinite rings without identity, where every element is a zero-divisor; and rings without identity with a regular element. For example, for finite commutative rings with identity (resp., without identity), the set of realizable pairs is $\{(0,\infty), (1,3), (1,\infty), (2,3), (2,4), (2,\infty), (3,3), (3,4), (3,\infty)\}$ (resp., $\{(0,\infty), (1,3), (1,\infty), (2,3), (2,\infty)\}$). For additional results on zero-divisor graphs for commutative rings without identity, see [87], [226], and [227].

# Chapter 3

# Properties of Zero-divisor Graphs

In this chapter, we deal with some graph-theoretical properties of the zero-divisor graph of a commutative ring such as colorings, connectedness, bipartite nature, isomorphisms, and automorphisms.

## 3.1 Coloring of Zero-divisor Graphs

One of the best known, and most studied, areas in graph theory is coloring. Dividing a given set into subsets is a fundamental procedure in mathematics. When the set is associated with a graph in some manner, then we are dealing with graph colorings. Graph colorings is a subject of great interest, largely because of its diverse theoretical results, its unsolved problems, and its numerous applications. For more on graph coloring, see [161].

In this section, we are interested in properties related to coloring of the zero-divisor graph of commutative rings. Using this, we also deal with the clique number of zero-divisor graphs. We start the section with vertex colorings of zero-divisor graphs.

### 3.1.1 Vertex Coloring

The problems in graph colorings that have received the most attention involve coloring the vertices of a graph. Furthermore, the problems in vertex colorings that have been studied most often are those referred to as proper vertex colorings. We begin with these.

A proper *k-vertex coloring* of a graph $G$ is an assignment of $k$ colors $\{1, \ldots, k\}$ to the vertices of $G$ such that no two adjacent vertices have the same color. The *vertex chromatic number* $\chi(G)$ of a graph $G$ is the minimum $k$ for which $G$ has a $k$-vertex coloring. Recall that a complete subgraph of a graph $G$ is called a *clique* of $G$. The *clique number* $\omega(G)$ is the greatest integer $r \geq 1$ such that $K_r$ is a clique of $G$. If $K_r \subseteq G$ for every integer $r \geq 1$, we set $\omega(G) = \infty$.



D. F. Anderson et al., *Graphs from Rings*, https://doi.org/10.1007/978-3-030-88410-9_3

As mentioned earlier, Beck [137] was mostly concerned with the coloring of $\Gamma_0(R)$. Actually, Beck made a conjecture that the clique number and chromatic number of $\Gamma_0(R)$ are equal. This conjecture is called Beck's Conjecture. A commutative ring $R$ is called a *Coloring* provided that $\chi(\Gamma_0(R)) < \infty$. Further, a graph $G$ is called *weakly perfect* if $\chi(G) = \omega(G)$.

The following is trivial.

**Proposition 3.1.1.** ([137, Propositions 2.1 and 2.2]) *Let $R$ be a commutative ring. Then the following hold:*

(a) $\chi(\Gamma_0(R)) = 1$ *if and only if $R$ is the zero ring;*

(b) $\chi(\Gamma_0(R)) = 2$ *if and only if $R$ is an integral domain, $R \cong \mathbb{Z}_4$, or $R \cong \mathbb{Z}_2[x]/(x^2)$.*

In view of the definitions of $\Gamma_0(R)$ and $\Gamma(R)$, one notes that $\chi(\Gamma_0(R)) = \chi(\Gamma(R)) + 1$. In the development of this section, we extensively use this property without mention. Note that $\omega(\Gamma(R)) = \omega(\Gamma_0(R)) - 1$, and hence many results from [137] and [60] for $\Gamma_0(R)$ can be directly translated to $\Gamma(R)$.

The clique number $\omega(\Gamma_0(R))$ was extensively studied in [137] and [60] because of its relationship to the chromatic number $\chi(\Gamma_0(R))$. By definition, it is clear that $\omega(\Gamma_0(R)) \leq \chi(\Gamma_0(R))$. Beck [137] showed that for $n = 2, 3$, or $4$, $\chi(\Gamma_0(R)) = n$ if and only if $\omega(\Gamma_0(R)) = n$. Further, Beck [137] proved that if $\chi(\Gamma_0(R)) = 5$, then $\omega(\Gamma_0(R)) = 5$; and thus conjectured that $\chi(\Gamma_0(R)) = \omega(\Gamma_0(R))$ for all rings $R$ with $\omega(\Gamma_0(R))$ finite. However, an example was given in [60, Theorem 2.1] of a finite local ring $R$ with maximal ideal $\mathfrak{m}$ satisfying $\mathfrak{m}^3 = \{0\}$ for which $\omega(\Gamma_0(R)) = 5$, but $\chi(\Gamma_0(R)) = 6$.

First of all, we present a characterization of rings with finite chromatic number. An element $x$ of $R$ is said to be *finite* if the ideal $Rx$ is finite.

**Lemma 3.1.2.** ([137, Lemma 3.1]) *Suppose that a commutative ring $R$ has an infinite number of finite elements. Then $\Gamma_0(R)$ contains an infinite clique.*

**Lemma 3.1.3.** ([137, Lemma 3.2]) *Let $I$ be a finite ideal of a commutative ring $R$. Then $\Gamma_0(R)$ contains an infinite clique if and only if $R/I$ has an infinite clique.*

**Lemma 3.1.4.** ([137, Lemma 3.3]) *If a commutative ring $R$ contains a nilpotent element which is not finite, then $\Gamma_0(R)$ contains an infinite clique.*

**Lemma 3.1.5.** ([137, Lemma 3.4]) *Let $R$ be a commutative ring. If $Nil(R)$ is infinite, then $\Gamma_0(R)$ has an infinite clique.*

**Lemma 3.1.6.** ([137, Lemma 3.5]) *Let $R$ be a reduced commutative ring which does not contain an infinite clique. Then $R$ has ACC on ideals of the form $Ann(x)$.*

A commutative ring $R$ is called *chromatic* if $\chi(\Gamma_0(R)) = \omega(\Gamma_0(R)) < \infty$.

**Theorem 3.1.7.** ([137, Theorem 3.7]) *For a reduced commutative ring $R$, the following conditions are equivalent:*

(a) $\chi(\Gamma_0(R))$ *is finite (i.e., $R$ is a Coloring);*

(b) $\omega(\Gamma_0(R))$ *is finite;*

(c) *The zero ideal of $R$ is a finite intersection of prime ideals;*

(d) $\Gamma_0(R)$ *does not contain an infinite clique.*

**Theorem 3.1.8.** ([137, Theorem 3.8]) *Let $R$ be a nonzero reduced commutative ring. If $\chi(\Gamma_0(R))$ is finite, then $R$ has only a finite number of minimal prime ideals. If $n$ is this number, then $\chi(\Gamma_0(R)) = n + 1$.*

The following theorem is one of the main results proved by Beck.

**Theorem 3.1.9.** ([137, Theorem 3.9]) *The following conditions are equivalent for a commutative ring $R$:*

(a) $\chi(\Gamma_0(R))$ *is finite (i.e., $R$ is a Coloring);*

(b) $\omega(\Gamma_0(R))$ *is finite;*

(c) $Nil(R)$ *is finite and is a finite intersection of prime ideals;*

(d) $\Gamma_0(R)$ *does not contain an infinite clique.*

Beck [137] proved many other results about Colorings. For example, he showed that if $R$ is a Coloring, then $Ass(R)$ (the set of associated prime ideals of $R$) is finite [137, Theorem 4.3], and for $P \in Ass(R)$, either $R_P$ is a field or $P$ is a maximal ideal [137, Theorem 4.4]. He also investigated the stability of the family of Colorings and showed that a finite direct product of Colorings is a Coloring [137, Theorem 5.5], a localization of a Coloring is a Coloring [137, Theorem 5.8], and certain factor rings of Colorings are Colorings [137, Theorems 5.2, 5.4, and 5.6]. Results concerning all these facts are presented below.

**Theorem 3.1.10.** ([137, Theorem 6.13]) *Let $R$ be a commutative ring with $\chi(\Gamma_0(R)) < \infty$. If $R$ is a finite direct product of reduced rings and principal ideal rings, then $\chi(\Gamma_0(I)) = \omega(\Gamma_0(I))$ for every $I \subseteq R$.*

**Theorem 3.1.11.** ([137, Theorem 7.3]) *Let $R$ be a commutative ring with $\chi(\Gamma_0(R)) < \infty$. For a positive integer $n \leq 4$, $\chi(\Gamma_0(R)) = n$ if and only if $\omega(\Gamma_0(R)) = n$. Moreover, $\chi(\Gamma_0(R)) = 5$ implies $\omega(\Gamma_0(R)) = 5$.*

Moreover, Beck determined all finite commutative rings $R$ with $\chi(\Gamma_0(R)) \leq 3$.

**Theorem 3.1.12.** ([137, p. 226]) *The following hold:*

(a) $\chi(\Gamma_0(R)) = 1$ *if and only if $R$ is the zero ring;*

(b) $\chi(\Gamma_0(R)) = 2$ *if and only if* $R$ *is isomorphic to* $F$, $\mathbb{Z}_4$, *or* $\mathbb{Z}_2[x]/(x^2)$, *where* $F$ *is a finite field;*

(c) $\chi(\Gamma_0(R)) = 3$ *if and only if* $R$ *is isomorphic to* $F \times K$, $F \times \mathbb{Z}_4$, $F \times \mathbb{Z}_2[x]/(x^2)$, $\mathbb{Z}_8$, $\mathbb{Z}_9$, $\mathbb{Z}_3[x]/(x^2)$, $\mathbb{Z}_2[x]/(x^3)$, *or* $\mathbb{Z}_4[x]/(2x, x^2 - 2)$, *where* $F$ *and* $K$ *are finite fields.*

Beck did not find any commutative ring $R$ with $\chi(\Gamma_0(R)) > \omega(\Gamma_0(R))$. Due to the lack of such counterexamples, he made the following conjecture.

**Conjecture 3.1.13.** *For a commutative ring* $R$, $\chi(\Gamma_0(R)) = \omega(\Gamma_0(R))$ *whenever* $\omega(\Gamma_0(R)) < \infty$.

After five years, D. D. Anderson and M. Naseer [60] provided a counterexample to Beck's conjecture.

**Example 3.1.14.** ([60, Theorem 2.1]) Let

$$R = \mathbb{Z}_4[x, y, z]/(x^2 - 2, y^2 - 2, z^2, 2x, 2y, 2z, xy, xz, yz - 2).$$

Here, $\omega(\Gamma_0(R)) = 5$, but $\chi(\Gamma_0(R)) = 6$. This is a local ring $(R, \mathfrak{m})$ with 32 elements, $R/\mathfrak{m} \cong \mathbb{Z}_2$, $\mathfrak{m}^2 \neq 0$, but $\mathfrak{m}^3 = 0$. This counterexample is important in several senses. First, it has the smallest possible clique or chromatic number for a counterexample of Beck's conjecture. Second, it is minimal in the sense that a Coloring $S$ with $Nil(S)^2 = 0$ has $\chi(\Gamma_0(S)) = \omega(\Gamma_0(S)) = |Nil(S)| + 1$ by [60, Theorem 3.1]. Finally, it has the smallest number of elements possible since if $S$ is a finite commutative ring with $|S| \leq 31$, then $\chi(\Gamma_0(S)) = \omega(\Gamma_0(S))$.

Having given a counterexample to Beck's conjecture, Anderson and Naseer [60] proved several positive results. Some of them are listed below.

**Theorem 3.1.15.** ([60, Theorem 3.1]) *Let* $R$ *be a commutative ring,* $Nil(R)$ *be the nilradical of* $R$, $\epsilon(R)$ *be the number of primes with* $R_P$ *a field, and* $\chi(\Gamma_0(R)) < \infty$.

(a) *If* $Nil(R)^2 = 0$, *then* $\chi(\Gamma_0(R)) = \omega(\Gamma_0(R)) = |Nil(R)| + \epsilon(R)$;

(b) *If* $Nil(R)^2 \neq 0$, *then* $\chi(\Gamma_0(R)) \geq \omega(\Gamma_0(R)) \geq |Nil(R) \cap (0 : Nil(R))| + \epsilon(R) + 1$. *If* $\omega(\Gamma_0(R)) = |Nil(R) \cap (0 : Nil(R))| + \epsilon(R) + 1$, *then* $\chi(\Gamma_0(R)) = \omega(\Gamma_0(R))$.

The following theorem shows that the commutative ring $R/\mathfrak{m}_1^{r_1} \cdots \mathfrak{m}_n^{r_n}$, where $R$ is a regular Noetherian ring, $\mathfrak{m}_1, \ldots, \mathfrak{m}_n$ are maximal ideals of $R$ with every $R/\mathfrak{m}_i$ finite, and $r_1, \ldots, r_n$ are positive integers, is a Coloring.

For a nilpotent ideal $I$ of a ring $R$, the *index of nilpotency* of $I$ is the smallest positive integer $n$ such that $I^n = 0$.

**Theorem 3.1.16.** ([60, Theorem 3.2]) *Let* $R \cong R_1 \times \cdots \times R_k \times R_{k+1} \times \cdots \times R_{k+r}$, *where every* $R_i$ *is a commutative Coloring. Let* $N_i$ *be the nilradical of* $R_i$ *(so* $N_i$ *is finite). Suppose that the index of nilpotency of* $N_i$ *is* $2n_i$ *for* $i = 1, \ldots, k$ *and* $2m_i - 1$ *for* $i = k+1, \ldots, k+r$ $(n_i, m_i \geq 1)$. *Then*

$$|N_1^{n_1}| \cdots |N_k^{n_k}||N_{k+1}^{m_{k+1}}| \cdots |N_{k+r}^{m_{k+r}}| + r \leq \omega(\Gamma_0(R)) \leq \chi(\Gamma_0(R)).$$

*Furthermore, suppose that for every* $i$, *the following conditions hold:*

(i) *For* $1 \leq i \leq k$, $x, y \in R_i$ *with* $xy = 0$ *implies* $x \in N_i^{n_i}$ *or* $y \in N_i^{n_i}$, *and if* $x \notin N_i^{n_i+1}$, *then* $y \in N_i^{n_i}$;

(ii) *For* $k+1 \leq i \leq k+r$, $x, y \in R_i$ *with* $xy = 0$ *implies* $x \in N_i^{m_i}$ *or* $y \in N_i^{m_i}$. *Then*

$$\chi(\Gamma_0(R)) = \omega(\Gamma_0(R)) = |N_1^{n_1}| \cdots |N_k^{n_k}||N_{k+1}^{m_{k+1}}| \cdots |N_{k+r}^{m_{k+r}}| + r.$$

If $R$ is a subring of a finite direct product of integral domains and finite rings, then $R$ is a Coloring. For Noetherian rings, the converse is true.

**Theorem 3.1.17.** ([60, Theorem 3.6]) *A commutative Noetherian ring* $R$ *is a Coloring if and only if it is a subring of a finite direct product of fields and a finite ring.*

The following is the characterization for all finite commutative rings $R$ for which $\chi(\Gamma_0(R)) = 4$.

**Theorem 3.1.18.** ([60, Theorem 4.4]) *Let* $R$ *be a finite commutative ring with* $\chi(\Gamma_0(R)) = 4$. *Then* $R$ *is isomorphic to one of the following* 31 *types of rings:*

$$\mathbb{Z}_4 \times \mathbb{Z}_4,\ \mathbb{Z}_4 \times \mathbb{Z}_2[x]/(x^2),\ \mathbb{Z}_2[x]/(x^2) \times \mathbb{Z}_2[x]/(x^2),\ F_1 \times F_2 \times F_3,$$
$$F_1 \times F_2 \times \mathbb{Z}_4,\ F_1 \times F_2 \times \mathbb{Z}_2[x]/(x^2),\ F \times \mathbb{Z}_8,\ F \times \mathbb{Z}_9,\ F \times \mathbb{Z}_3[x]/(x^2),$$
$$F \times \mathbb{Z}_2[x]/(x^3),\ F \times \frac{\mathbb{Z}_4[x]}{(2x, x^2-2)},\ \mathbb{Z}_{16},\ \frac{\mathbb{Z}_2[x]}{(x^4)},\ \frac{\mathbb{Z}_4[x]}{(2x, x^3-2)},\ \mathbb{Z}_4[x]/(x^2 -$$
$$2),\ \frac{\mathbb{Z}_4[x]}{(x^2+2x+2)},\ \frac{\mathbb{F}_{2^2}[x]}{(x^2)},\ \frac{\mathbb{Z}_4[x]}{(x^2+x+1)},\ \frac{\mathbb{Z}_2[x,y]}{(x,y)^2},\ \frac{\mathbb{Z}_4[x]}{(2x)^2},\ \mathbb{Z}_{27},\ \frac{\mathbb{Z}_3[x]}{(x^3)},$$
$$\frac{\mathbb{Z}_9[x]}{(x^2-3, 3x)},\ \frac{\mathbb{Z}_9[x]}{(x^2-6, 3x)},\ \frac{\mathbb{Z}_2[x,y]}{(x^2, y^2-xy)},\ \frac{\mathbb{Z}_2[x,y]}{(x^2, y^2)},\ \frac{\mathbb{Z}_8[x]}{(2x-4, x^2)},\ \frac{\mathbb{Z}_4[x]}{(x^2)},$$
$$\frac{\mathbb{Z}_4[x]}{(x^2-2x)},\ \frac{\mathbb{Z}_4[x,y]}{(x^2, xy-2, y^2, 2x, 2y)},\ \frac{\mathbb{Z}_4[x,y]}{(x^2, xy-2, y^2-xy, 2x, 2y)},$$

*where* $F, F_i$ *are finite fields and* $\mathbb{F}_{p^n}$ *is the finite field with* $p^n$ *elements.*

The study of non-chromatic rings $R$ (i.e., $\chi(\Gamma_0(R)) > \omega(\Gamma_0(R))$) was continued by Bhatwadekar, Dumaldar, and Sharma in [145]. They gave a large class of non-chromatic rings.

**Theorem 3.1.19.** ([145, Theorem 1.1]) *Let $R$ be a finite commutative local ring with maximal ideal $\mathfrak{m}$ and residue field $F$ with $|F| = q$. Assume that $\mathfrak{m}^3 = 0$, $dim_F \mathfrak{m}^2 = 1$, $dim_F \mathfrak{m}/\mathfrak{m}^2 = 3$, and $Ann(\mathfrak{m}) = \mathfrak{m}^2$. Then for any maximal clique $C \subseteq \mathfrak{m}$ in $R$, $|C| = q^2 + 1$ or $q + 3$. Moreover, $\chi(\Gamma_0(R)) > \omega(\Gamma_0(R))$.*

**Example 3.1.20.** ([145, Remark, p. 486]) Let $R_1 = R \times \mathbb{Z}$, where $R$ is a finite commutative local ring which satisfies the conditions of Theorem 3.1.19. Then $\omega(\Gamma_0(R_1)) = \omega(\Gamma_0(R)) + 1$ and $\chi(\Gamma_0(R_1)) \geq \chi(\Gamma_0(R)) + \chi(\Gamma_0(\mathbb{Z})) - 1 = \chi(\Gamma_0(R)) + 1$. Therefore, $R_1$ is an infinite ring with $\chi(\Gamma_0(R_1)) > \omega(\Gamma_0(R_1))$.

**Theorem 3.1.21.** ([145, Theorem 3.1]) *Let $R$ be a commutative local ring with maximal ideal $\mathfrak{m}$ and residue field $F$. Assume that $\mathfrak{m}^3 = 0$, $dim_F \mathfrak{m}^2 = 1$, $dim_F \mathfrak{m}/\mathfrak{m}^2 = 3$, and $Ann(\mathfrak{m}) = \mathfrak{m}$. If $R$ has prime residue field $\mathbb{F}_p$ with $p \geq 3$, then $p^2 + 2 \leq \chi(\Gamma_0(R)) \leq p^2 + \frac{p+1}{2}$. Moreover, $\chi(\Gamma_0(R)) = p^2 + 2$ for $p = 2, 3, 5$.*

Bhatwadekar et al. [145] did not provide any example of a commutative local ring $R$ with prime residue field $\mathbb{F}_p$ and $\chi(\Gamma_0(R)) \neq p^2 + 2$. But Dumaldar and Sharma [188] continued the study of non-chromatic rings, and they gave the least number $p$ such that $\chi(\Gamma_0(R)) \neq p^2 + 2$. The corresponding result is given below.

**Theorem 3.1.22.** ([188, Theorem 2.2]) *Let $(R, \mathfrak{m})$ be a finite commutative local ring with prime residue field $\mathbb{F}_p$, where $p = 7$. Assume that $\mathfrak{m}^3 = 0$, $dim_{\mathbb{F}_p} \mathfrak{m}^2 = 1$, $dim_{\mathbb{F}_p} \mathfrak{m}/\mathfrak{m}^2 = 3$, and $Ann(\mathfrak{m}) = \mathfrak{m}^2$. Then $\chi(\Gamma_0(R)) = p^2 + 3$.*

Thereafter, Wang [421] obtained some sufficient conditions for commutative rings $R$ to satisfy $\chi(\Gamma_0(R)) = \omega(\Gamma_0(R))$; the relevant result is given below.

**Lemma 3.1.23.** ([421, Lemma 3.1]) *Let $R \cong \mathbb{Z}_{p^r}$, where $p$ is a prime number and $r$ is a positive integer. Then $R$ is a chromatic ring with*

$$\chi(\Gamma_0(R)) = \omega(\Gamma_0(R)) = \begin{cases} p^{\frac{r}{2}} & \text{if } r \text{ is even;} \\ p^{\frac{r-1}{2}} + 1 & \text{if } r \text{ is odd.} \end{cases}$$

**Example 3.1.24.** ([421, Example 3.3]) Note that $\chi(\Gamma_0(\mathbb{Z}_8 \times \mathbb{Z}_{16})) = \omega(\Gamma_0(\mathbb{Z}_8 \times \mathbb{Z}_{16})) = 9$.

In the backdrop of this example, the results given below deal with product rings.

**Lemma 3.1.25.** ([421, Lemma 3.4]) *Let $R_1$ and $R_2$ be commutative rings and $R \cong R_1 \times R_2$. Then the following hold:*

(a) $\chi(\Gamma_0(R)) \geq \chi(\Gamma_0(R_1)) + \chi(\Gamma_0(R_2)) - 1$;

(b) *If $R_2$ is reduced, then* $\chi(\Gamma_0(R)) = \chi(\Gamma_0(R_1)) + \chi(\Gamma_0(R_2)) - 1$.

**Lemma 3.1.26.** ([421, Lemma 3.5]) *Let $R_1$ and $R_2$ be commutative rings and $R \cong R_1 \times R_2$. Then the following hold:*

(a) $\omega(\Gamma_0(R)) \geq \omega(\Gamma_0(R_1)) + \omega(\Gamma_0(R_2)) - 1$;

(b) *If $R_2$ is reduced, then* $\omega(\Gamma_0(R)) = \omega(\Gamma_0(R_1)) + \omega(\Gamma_0(R_2)) - 1$.

**Corollary 3.1.27.** ([421, Corollary 3.6]) *Let $R \cong R_1 \times \cdots \times R_k$, where every $R_i$ is an integral domain. Then $\chi(\Gamma_0(R)) = \omega(\Gamma_0(R)) = k+1$.*

**Theorem 3.1.28.** ([421, Theorem 3.7]) *Let $R \cong R_1 \times \cdots \times R_k$, where every $R_i$ is a Coloring. Suppose that there is a finite set $S_i$ of $\Gamma_0(R_i)$ for every $i$ such that the following hold:*

(a) *$S_i$ is a maximal clique of $\Gamma_0(R_i)$;*

(b) *If $a, b \in \Gamma_0(R_i) \setminus S_i$, then $a^2 \neq 0$ and $ab \neq 0$.*

*Let $N_i = \{a \in S_i \mid a^2 = 0\} \subseteq S_i$ and $n_i = |N_i|$ for every $i$. Then $\chi(\Gamma_0(R)) = \omega(\Gamma_0(R)) = \prod_{i=1}^{k} n_i + \sum_{i=1}^{k}(s_i - n_i)$, where $s_i = |S_i|$.*

**Corollary 3.1.29.** ([421, Corollaries 3.8 and 3.9]) *If $R \cong \underbrace{\mathbb{Z} \times \cdots \times \mathbb{Z}}_{\ell \text{ times}} \times \mathbb{Z}_{p_1^{r_1}} \times \cdots \times \mathbb{Z}_{p_k^{r_k}}$, then $\chi(\Gamma_0(R)) = \omega(\Gamma_0(R)) = \prod_{i=1}^{k} p_i^{\lfloor r_i/2 \rfloor} + t + \ell$, where $\lfloor n \rfloor$ is the largest integer less than or equal to $n$ and $t = |\{i \mid r_i \text{ is odd}\}|$. In particular, $\mathbb{Z}_{p_1^{r_1}} \times \cdots \times \mathbb{Z}_{p_k^{r_k}}$ is chromatic and so is $\mathbb{Z}_n$.*

Note that Theorem 3.1.16 provides some sufficient conditions for the product of Colorings to be chromatic. However, Theorem 3.1.16 does not imply Theorem 3.1.28. For example, if $R \cong \mathbb{Z} \times \cdots \times \mathbb{Z}$, then Theorem 3.1.28 can be used to obtain that $R$ is chromatic. However, Theorem 3.1.16 cannot.

Wang [421] studied several properties of coloring of the complement graph $C\Gamma(R)$, where $R \cong \mathbb{Z}_{p_1^{r_1}} \times \cdots \times \mathbb{Z}_{p_k^{r_k}}$.

**Lemma 3.1.30.** ([421, Lemma 4.2]) *If $R \cong \mathbb{Z}_{p^r}$, where $p$ is a prime number and $r \geq 2$ is a positive integer, then $\chi(C\Gamma(R)) = \omega(C\Gamma(R)) = p^{r-1} - p^{r-\frac{r}{2}}$ if $r$ is even and $\chi(C\Gamma(R)) = \omega(C\Gamma(R)) = p^{r-1} - p^{\frac{r-1}{2}}$ if $r$ is odd.*

**Theorem 3.1.31.** ([421, Theorem 4.4]) *If $R \cong \mathbb{Z}_{p_1^{r_1}} \times \cdots \times \mathbb{Z}_{p_k^{r_k}}$, where $p_i$ is a prime number for every $i$, then $\chi(C\Gamma(R)) = \omega(C\Gamma(R))$.*

**Corollary 3.1.32.** ([421, Corollary 4.5]) *If $R \cong F_1 \times \cdots \times F_k$, where $F_i$ is a finite field for every $i$, then $\chi(C\Gamma(R)) = \omega(C\Gamma(R))$.*

Work on Hedetniemi's Conjecture regarding the chromatic number of product graphs has been attempted for the class of zero-divisor graphs by Córdova et al. in [168]. More specifically, Hedetniemi conjectured that for all graphs $G$ and $H$, the chromatic number $\chi(G \times H) = \min\{\chi(G), \chi(H)\}$. Recall that $\chi(\Gamma(\mathbb{Z}_n)) = \omega(\Gamma(\mathbb{Z}_n))$. Using this, he first proved the conjecture for $\Gamma(\mathbb{Z}_n)$.

**Theorem 3.1.33.** ([168, Theorem 4.1]) *Let $n$ be a positive integer. Then*

$$\chi(\Gamma(\mathbb{Z}_n) \times \Gamma(\mathbb{Z}_m)) = \min\{\chi(\Gamma(\mathbb{Z}_n)), \chi(\Gamma(\mathbb{Z}_m))\}.$$

The following result shows that the Hedetniemi's Conjecture is true for all zero-divisor graphs $\Gamma(R) \times \Gamma(R')$, where $\chi(\Gamma(R)) = \omega(\Gamma(R))$ and $\chi(\Gamma(R')) = \omega(\Gamma(R'))$. The corresponding result is given below.

**Theorem 3.1.34.** ([168, Theorem 4.4]) *Let $R$ and $R'$ be commutative rings such that $\chi(\Gamma(R)) = \omega(\Gamma(R))$ and $\chi(\Gamma(R')) = \omega(\Gamma(R'))$. Then $\chi(\Gamma(R) \times \Gamma(R')) = \min\{\chi(\Gamma(R)), \chi(\Gamma(R'))\}$.*

Next, we discuss properties of coloring the finite ring $\mathbb{Z}_n$. As mentioned earlier, Beck showed that the chromatic number of $\Gamma(\mathbb{Z}_n)$ is equal to its clique number. In continuation, Duane [187] calculated the chromatic number of the zero-divisor graph of $\mathbb{Z}_n$. If $p$ is a prime number, then $\Gamma(\mathbb{Z}_p)$ is empty. Suppose that $p$ and $q$ are two distinct primes; then $\Gamma(\mathbb{Z}_{pq})$ is a bipartite graph, and so $\chi(\Gamma(\mathbb{Z}_{pq})) = 2$. Some of the remaining cases are dealt with in the following results.

**Theorem 3.1.35.** ([187, Corollary 4.2]) *If $p$ and $q$ are distinct primes, then $\chi(\Gamma(\mathbb{Z}_{pq^2})) = q$.*

*Proof.* If $a$ and $b$ are zero-divisors in $\mathbb{Z}_{pq^2}$ such that $pq$ divides both $a$ and $b$, then $a$ and $b$ are adjacent in $\Gamma(\mathbb{Z}_{pq^2})$. Thus all such zero-divisors form a clique in $\Gamma(\mathbb{Z}_{pq^2})$. There are $\frac{pq^2}{pq} - 1 = q - 1$ such zero-divisors, not including $pq^2 = 0$ itself; so the clique contains $q-1$ vertices and can be colored with $k_1, k_2, \ldots, k_{q-1}$. Thus $\chi(\Gamma(\mathbb{Z}_{pq^2})) \geq q - 1$.

Now, if $c$ is any other zero-divisor, either $p \nmid c$ or $q \nmid c$, because otherwise $c$ would be in the clique. If $q \nmid c$, then $c$ is not adjacent to any vertex in the clique; so we can color it with $k_1$. If $q \mid c$, then $c$ is adjacent to every member of the clique; so we must color it with $k_q$. We can color all such zero-divisors with $k_q$, because if we have $c, d$ not in the clique and $q \mid c, d$, then $p \nmid c, d$ by the fact that $c$ and $d$ are outside the clique. Thus $c$ and $d$ cannot be adjacent, and so can be colored with the same color. Thus we have a proper coloring of $\Gamma(\mathbb{Z}_{pq^2})$ with exactly $q$ colors, and so $\chi(\Gamma(\mathbb{Z}_{pq^2})) \leq q$. Hence $\chi(\Gamma(\mathbb{Z}_{pq^2}))$ is either $q - 1$ or $q$.

Suppose $\chi(\Gamma(\mathbb{Z}_{pq^2})) = q - 1$. Then we can color every vertex with a color used in the clique consisting of elements divisible by $pq$. Suppose $a$ is a member of the clique. Then $pq^2 \mid qa$, so $q$ is adjacent to all such $a$. Thus there is no color in the clique with which we can color $q$, so we have a contradiction. Thus $\chi(\Gamma(\mathbb{Z}_{pq^2})) = q$. □

**Theorem 3.1.36.** ([187, Theorem 4.3]) *If $p$ and $q$ are distinct primes, then $\chi(\Gamma(\mathbb{Z}_{p^2q^2})) = pq - 1$.*

*Proof.* As in Theorem 3.1.35, we observe that for all zero-divisors $a, b$ such that $pq \mid a, b$, that $a$ and $b$ are adjacent in the zero-divisor graph of $\mathbb{Z}_{p^2q^2}$. These vertices form a clique on $\frac{p^2q^2}{pq} - 1 = pq - 1$ vertices, and so $\chi(\Gamma(\mathbb{Z}_{p^2q^2})) \geq pq - 1$. For every other zero-divisor $c$, we see that either $p \nmid c$ or $q \nmid c$. Regardless which case we have, we see that $c$ is not adjacent to the zero-divisor $pq$. In the former case, we see that $p^2 \nmid pqc$, and in the latter, $q^2 \nmid pqc$. Either way, $p^2q^2 \nmid pqc$; so $c$ and $pq$ are not adjacent in $\Gamma(\mathbb{Z}_{p^2q^2})$. Thus we can use the color already assigned to the vertex $pq$ to color all other vertices in the zero-divisor graph, and we have a proper coloring with $pq - 1$ colors. Thus $\chi(\Gamma(\mathbb{Z}_{p^2q^2})) \leq pq - 1$. Hence $\chi(\Gamma(\mathbb{Z}_{p^2q^2})) = pq - 1$. □

**Theorem 3.1.37.** ([187, Theorem 4.5]) *If $p_1, p_2, p_3$, and $p_4$ are distinct primes, then $\chi(\Gamma(\mathbb{Z}_{p_1p_2p_3p_4})) = 4$.*

*Proof.* For every zero-divisor $a$, we must have that at least one of $p_1, p_2, p_3$, or $p_4$ does not divide $a$ since if all of these primes divide $a$, then $a = p_1p_2p_3p_4 = 0$. Thus if $p_i \nmid a$, where $i$ is minimal, we color $a$ with $k_i$. Using this scheme, if $a$ and $b$ have the same color, say $k_j$, then $j \nmid a$ or $b$; so $j \nmid ab$, and thus $ab \neq 0$. Hence $a$ and $b$ cannot be adjacent. Therefore, we have a proper coloring with four colors; so $\chi(\Gamma(\mathbb{Z}_{p_1p_2p_3p_4})) \leq 4$. Furthermore, the elements $p_1p_2p_3, p_2p_3p_4, p_1p_3p_4$, and $p_1p_2p_4$ form a clique on four vertices. Thus $\chi(\Gamma(\mathbb{Z}_{p_1p_2p_3p_4})) \geq 4$ so that $\chi(\Gamma(\mathbb{Z}_{p_1p_2p_3p_4})) = 4$. □

**Theorem 3.1.38.** ([187, Corollary 4.8]) *If $n$ is a positive integer and $p$ is a prime number, then*

$$\chi(\Gamma(\mathbb{Z}_{p^n})) = \begin{cases} p^{\frac{n-1}{2}} & \text{if } n \text{ is odd;} \\ p^{\frac{n}{2}} - 1 & \text{if } n \text{ is even.} \end{cases}$$

*Proof.* If we have two zero-divisors, $u$ and $v$, such that $p^{\lceil n/2 \rceil} \mid u, v$, then $u$ and $v$ must be adjacent. There are $(p^n)/(p^{\lceil n/2 \rceil}) - 1 = p^{\lfloor n/2 \rfloor} - 1$ such zero-divisors; thus we have a clique on those vertices on which we can use the colors $k_1, k_2, \ldots, k_{p^{\lfloor n/2 \rfloor}-1}$. If $n$ is odd, then all multiples of $p^{\lfloor n/2 \rfloor}$ less than $p^{\lfloor n/2 \rfloor + 1}$ will not be included in the clique since they will not be adjacent to each other, but they will be adjacent to every vertex in the clique. Because none of these vertices can be adjacent to one another, we can color all of them with $k_{p^{\lfloor n/2 \rfloor}}$. If $a$ is any vertex not colored thus far, we know that $ap^{\lceil n/2 \rceil} \neq 0$; so we can use the color for $p^{\lceil n/2 \rceil}$ to color all such $a$. Thus we have a proper coloring with $p^{\lfloor n/2 \rfloor}$ colors. If $n$ is odd, then clearly $\chi(\Gamma(\mathbb{Z}_{p^n})) = p^{\frac{n-1}{2}}$. If $n$ is even, no vertex outside of the clique defined above can be attached to $p^{n/2}$; so we can use the color for this vertex to color everything else. Thus we require $p^{\frac{n}{2}} - 1$ colors for a proper coloring. Therefore $\chi(\Gamma(\mathbb{Z}_{p^n})) = p^{\frac{n}{2}} - 1$. □

Some attempts were made by Duane [187], Skowronek-Kaziów [350]), and Tengxia Ju et al. [413]) to find the chromatic number of $\Gamma(\mathbb{Z}_n)$ for every $n \geq 2$, but there are errors in certain results. These errors were corrected by Yanzhao and Qijiao in [438], and they gave the following corrected formula for calculating the chromatic number of $\Gamma(\mathbb{Z}_n)$.

**Theorem 3.1.39.** ([438, Theorem 2]) *Let $n = p_1^{\alpha_1} p_2^{\alpha_2} \cdots p_m^{\alpha_m}$ be the prime power factorization of $n$, where $p_1 < p_2 < \cdots < p_m$ are distinct primes and $\alpha_i \geq 1$, $m \geq 1$. If $n$ is square-free, then the clique number of $\Gamma(\mathbb{Z}_n)$ is $m$. If $\alpha_i$ are even integers for every $1 \leq i \leq m$, then the clique number is $p_1^{\alpha_1/2} p_2^{\alpha_2/2} \cdots p_m^{\alpha_m/2} - 1$. Otherwise, the clique number is $p_1^{\delta_1} \ldots p_s^{\delta_s} q_1^{(\beta_1-1)/2} \cdots q_t^{(\beta_t-1)/2} + t - 1$, where $\delta_i$ is even for $i = 1, \ldots, s$ and $\beta_i$ is odd for $i = 1, \ldots, t$.*

The work on coloring the zero-divisor graph of $\mathbb{Z}_n$ was continued by Endean, Henry, and Manlove in [192]. They concentrated on the perfectness of $\Gamma(\mathbb{Z}_n)$. A graph $G$ is said to be *perfect* if for every subgraph $H$ of $G$, $\chi(H) = \omega(H)$. In most cases, $\mathbb{Z}_n$ does not yield a perfect zero-divisor graph. However, the authors identified two cases in which the zero-divisor graphs are perfect.

**Theorem 3.1.40.** ([192, Theorem 1.2]) *The graph $\Gamma(\mathbb{Z}_{p^n})$, where $p$ is prime, is perfect.*

**Theorem 3.1.41.** ([192, Theorem 1.3]) *The graph $\Gamma(\mathbb{Z}_{p_1 p_2})$, where $p_1$ and $p_2$ are distinct primes, is perfect.*

They not only proved that $\Gamma(\mathbb{Z}_n)$ is perfect for these cases, but they also proved that these are the only cases.

**Theorem 3.1.42.** ([192, Theorem 1.3]) *$\Gamma(\mathbb{Z}_n)$ is perfect if and only if $n = p^k$ for some prime $p$ or $n = p_1 p_2$ for some distinct primes $p_1$ and $p_2$.*

### 3.1.2 Clique Number

We start this section with the clique number of certain zero-divisor graphs.

**Example 3.1.43.** ([75, Example 3.1]) (a) Let $R$ be a commutative ring. Note that $\omega(\Gamma(R)) = 1$ if and only if $|\Gamma(R)| = 1$, and hence $R \cong \mathbb{Z}_4$ or $R \cong \mathbb{Z}_2[x]/(x^2)$. And $\omega(\Gamma(R)) = 2$ if and only if $|\Gamma(R)| \geq 2$ and $\Gamma(R)$ contains no triangles (such finite rings $R$ are classified in Theorem 3.1.44 below). Thus $\omega(\Gamma(R)) \geq 3$ if and only if $\Gamma(R)$ contains a triangle.

(b) Let $R = F_1 \times F_2$, where $F_1$ and $F_2$ are finite fields with $|F_1| = m \geq 3$ and $|F_2| = n \geq 3$. Then $\Gamma(R)$ is the complete bipartite graph $K_{m-1,n-1}$. Note that $\Gamma(R)$ has no triangles, $\omega(\Gamma(R)) = 2$, and $gr(\Gamma(R)) = 4$.

(c) Let $R$ be a finite local ring with maximal ideal $\mathfrak{m}$ and $\mathfrak{m}^2 = \{0\}$. Then $\Gamma(R)$ is complete and $\omega(\Gamma(R)) = |\Gamma(R)| = |\mathfrak{m}^*|$. Also, the number of complete subgraphs of $\Gamma(R)$ of order $r$ is $\binom{n}{r}$, where $n = |\mathfrak{m}^*|$.

The following theorem characterizes all commutative rings $R$ for which $\omega(\Gamma(R)) = 2$ or 3, and consequently for $\omega(\Gamma_0(R)) = 3$ or 4 (cf. [137] and [60]). The following theorem is a consequence of Theorem 3.1.12 and Theorem 3.1.18.

**Theorem 3.1.44.** ([75, Theorem 3.2]) *Let $R$ be a finite commutative ring, and let $F_i$ be a finite field. Then*

(a) $\omega(\Gamma(R)) = 2$ *if and only if $R$ is isomorphic to one of the following* 8 *types of rings:* $F_1 \times F_2$, $F \times \mathbb{Z}_4$, $F \times \frac{\mathbb{Z}_2[x]}{(x^2)}$, $\mathbb{Z}_8, \mathbb{Z}_9, \frac{\mathbb{Z}_3[x]}{(x^2)}, \frac{\mathbb{Z}_2[x]}{(x^3)}, \frac{\mathbb{Z}_4[x]}{(2x, x^2-2)}$;

(b) $\omega(\Gamma(R)) = 3$ *if and only if $R$ is isomorphic to one of the following* 31 *types of rings:*

$$\mathbb{Z}_4\times\mathbb{Z}_4,\ \mathbb{Z}_4\times\frac{\mathbb{Z}_2[x]}{(x^2)},\ \frac{\mathbb{Z}_2[x]}{(x^2)}\times\frac{\mathbb{Z}_2[x]}{(x^2)},\ F_1\times F_2\times F_3,\ F_1\times F_2\times\mathbb{Z}_4,\ F_1\times F_2\times\frac{\mathbb{Z}_2[x]}{(x^2)},$$

$$F\times\mathbb{Z}_8,\ F\times\mathbb{Z}_9,\ F\times\frac{\mathbb{Z}_3[x]}{(x^2)},\ F\times\frac{\mathbb{Z}_2[x]}{(x^3)},\ F\times\frac{\mathbb{Z}_4[x]}{(2x,\ x^2-2)},\ \mathbb{Z}_{16},\ \frac{\mathbb{Z}_2[x]}{(x^4)},$$

$$\frac{\mathbb{Z}_4[x]}{(2x, x^3-2)}, \frac{\mathbb{Z}_4[x]}{(x^2-2)}, \frac{\mathbb{Z}_4[x]}{(x^2+2x+2)}, \frac{\mathbb{F}_4[x]}{(x^2)}, \frac{\mathbb{Z}_4[x]}{(x^2+x+1)}, \frac{\mathbb{Z}_2[x,y]}{(x,y)^2},$$

$$\frac{\mathbb{Z}_4[x]}{(2,x)^2}, \mathbb{Z}_{27}, \frac{\mathbb{Z}_3[x]}{(x^3)}, \frac{\mathbb{Z}_9[x]}{(3x, x^2-3)}, \frac{\mathbb{Z}_9[x]}{(3x, x^2-6)}, \frac{\mathbb{Z}_2[x,y]}{(x^2, y^2-xy)},$$

$$\frac{\mathbb{Z}_2[x,y]}{(x^2,y^2)}, \frac{\mathbb{Z}_8[x]}{(2x-4, x^2)}, \frac{\mathbb{Z}_4[x]}{(x^2)}, \frac{\mathbb{Z}_4[x]}{(x^2-2x)},$$

$$\frac{\mathbb{Z}_4[x,y]}{(x^2, xy-2, y^2, 2x, 2y)}, \frac{\mathbb{Z}_4[x,y]}{(x^2, xy-2, x^2-xy, 2x, 2y)}.$$

The following characterization of when $\omega(\Gamma(R)) < \infty$ is from [75]. As usual, $Nil(R)$ is the ideal of nilpotent elements of $R$ and $Min(R)$ is the set of minimal prime ideals of $R$. We have the following theorem from Theorem 3.1.9.

**Theorem 3.1.45.** ([75, Theorem 3.3]) *Let $R$ be a commutative ring.*

(a) $\omega(\Gamma(R)) = \infty$ *if and only if $\Gamma(R)$ has an infinite clique;*

(b) $\omega(\Gamma(R)) < \infty$ *if and only if $|Nil(R)| < \infty$ and $Nil(R)$ is a finite intersection of prime ideals (i.e., $|Min(R)| < \infty$).*

Bounds for $\omega(\Gamma(R))$ can be given using either idempotent or nilpotent elements of $R$. We first give a lower bound for $\omega(\Gamma(R))$ using the index of nilpotence (cf. [60, Theorem 3.2]).

**Theorem 3.1.46.** ([75, Theorem 3.4]) *Let $R$ be a commutative ring and $0 \neq x \in Nil(R)$, and let $n$ be the least positive integer such that $x^n = 0$.*

(a) *If* $n = 2k$, *then* $\omega(\Gamma(R)) \geq 2^k - 1$;

(b) *If* $n = 2k + 1$, *then* $\omega(\Gamma(R)) \geq 2^k$.

*Proof.* (a) Let $S = \{\alpha_1 x^k + \cdots + \alpha_k x^{2k-1} \mid \text{each } a_i = 0 \text{ or } 1\}^* \subseteq R$. Note that $|S| = 2^k - 1$ (if $\alpha_1 x^k + \cdots + \alpha_k x^{2k-1} = \beta_1 x^k + \cdots + \beta_k x^{2k-1}$ in $S$ with some $\alpha_i \neq \beta_i$, then $x^{n-1} = 0$, a contradiction) and $yz = 0$ for distinct $y, z \in S$. Thus $S$ is a clique; so $\omega(\Gamma(R)) \geq 2^k - 1$.

(b) In this case, let $S = \{\alpha_1 x^{k+1} + \cdots + \alpha_k x^{2k} \mid \text{every } a_i = 0 \text{ or } 1\}^* \cup \{x^k\}$. Then $|S| = 2^k$ and $S$ is a clique; so $\omega(\Gamma(R)) \geq 2^k$. □

**Corollary 3.1.47.** ([75, Corollary 3.5]) *Let $R$ be a commutative ring. If there is no bound on the index of nilpotence of elements of $R$, then $\omega(\Gamma(R)) = \infty$.*

We next investigate the above two invariants $\omega(\Gamma(R))$ and $\chi(\Gamma(R))$ for reduced commutative rings $R$. Note that $\omega(\Gamma(R)) \geq 2$ when $R$ is reduced and not an integral domain.

**Theorem 3.1.48.** ([75, Theorem 3.7]) *Let $R_1, \ldots, R_n$ be integral domains ($n \geq 2$), and let $R = R_1 \times \cdots \times R_n$. Then $\omega(\Gamma(R)) = n$.*

*Proof.* Clearly $\omega(\Gamma(R)) \geq n$. The $n = 2$ case is clear; so let $n \geq 3$, $X = \{x_1, \ldots, x_m\}$ be a complete subgraph of $\Gamma(R)$ and every $x_i = (x_{i1}, \ldots, x_{in})$. We may assume that $x_{11} \neq 0$, and thus $x_{21} = \cdots = x_{m1} = 0$. Hence we may consider $X \setminus \{x_1\}$ as a complete subgraph of $\Gamma(R_2 \times \cdots \times R_n)$. By induction, we have $m - 1 \leq n - 1$, and thus $m \leq n$. □

**Theorem 3.1.49.** ([75, Theorem 3.8]) *Let $R$ be a reduced commutative ring which is not an integral domain. Then $\omega(\Gamma(R)) = |min(R)|$.*

*Proof.* Since $R$ is reduced, by Theorem 3.1.45(b), $\omega(\Gamma(R))$ is finite if and only if $min(R)$ is finite. Let $P_1, \ldots P_n$ be the minimal prime ideals of $R$. Then $P_1 \cap \cdots \cap P_n = Nil(R) = \{0\}$ since $R$ is reduced, and thus $R \subseteq T = R/P_1 \times \cdots \times R/P_n$. Hence $\omega(\Gamma(R)) \leq \omega(\Gamma(T)) = n$ by Theorem 3.1.48. For every $1 \leq i \leq n$, we may choose a $0 \neq x_i \in P_1 \cap \cdots \cap P_{i-1} \cap P_{i+1} \cap \cdots \cap P_n$. Then, for each $i \neq j$, $x_i x_j \in Nil(R) = \{0\}$. Thus $\{x_1, \ldots, x_n\}$ is a complete subgraph of $\Gamma(R)$. Hence $\omega(\Gamma(R)) = n$. Conversely, if $\omega(\Gamma(R)) = n (\geq 2)$, then the above proof shows that $R$ has exactly $n$ minimal prime ideals. □

For a finite graph $G$ and $n \geq 1$ an integer, we define $\lambda(G, n)$ to be the number of complete subgraphs (cliques) of $G$ of order $n$. Note that $\lambda(G, 1)$ is the number of vertices of $G$, $\lambda(G, 2)$ is the number of edges of $G$, $\lambda(G, 3)$ is the number of triangles in $G$, and $\lambda(G, n) = 0$ for every $n \geq \omega(G) + 1$.

In Example 3.1.43(c), we observed that if $R$ is a finite local ring with maximal ideal $\mathfrak{m}$ satisfying $\mathfrak{m}^2 = 0$, then $\lambda(\Gamma, r) = \binom{n}{r}$, where $n = |\mathfrak{m}^*|$.

We next compute $\lambda(\Gamma(R), n)$ when $R$ is a finite reduced ring, i.e., a direct product of finitely many finite fields. Note that if some field $F_i$ is infinite, then $\lambda(\Gamma(F_1 \times \cdots \times F_m), n) = \infty$ for all $1 \leq n \leq m$.

**Theorem 3.1.50.** ([75, Theorem 3.11]) *Let* $F_1, \ldots, F_m$ *be finite fields* ($m \geq 2$), *and let* $\Gamma_m = \Gamma(F_1 \times \cdots \times F_m)$.

(a) $\lambda(\Gamma_m, 1) = |F_1| \cdots |F_m| - (|F_1| - 1) \cdots (|F_m| - 1) - 1$;

(b) $\lambda(\Gamma_m, n) = \frac{1}{n!} \sum\limits_{i=0}^{n} (-1)^i \binom{n}{i} [\prod\limits_{j=1}^{m} [(n-i)(|F_j| - 1) + 1]$ *for every integer* $n \geq 2$.

**Corollary 3.1.51.** ([75, Corollary 3.12]) *Let* $F$ *be a finite field, and let* $\Gamma_m = \Gamma(F^m)(m \geq 2)$.

(a) $\lambda(\Gamma_m, 1) = |F|^m - (|F| - 1)^m - 1$;

(b) $\lambda(\Gamma_m, n) = \frac{1}{n!} \sum_{i=0}^{n} (-1)^i \binom{n}{i} [(n-i)(|F| - 1) + 1]^m$ *for every integer* $n \geq 2$.

*Proof.* (a) This is clear.

(b) We first prove the case in Corollary 3.1.51 when every $F_i = F$. Let $|F| = \alpha$. Let $C_{n,m} = \lambda(\Gamma_m, n)$, i.e., the number of cliques in $\Gamma(F^m)$ of order $n$, and let $D_{n,m} = (1/n!) \sum\limits_{i=0}^{n} (-1)^i \binom{n}{i} [(n-i)(\alpha - 1) + 1]^m$ for integers $n \geq 2$ and $m \geq 1$. We first show that $C_{2,m} = D_{2,m}$ for every $m \geq 2$. Clearly $C_{n,1} = 0$ for every $n \geq 2$ since $\Gamma(F) = \emptyset$. We then show that $D_{n,1} = 0$ for every $n \geq 2$. Finally, we show that $C_{n,m}$ and $D_{n,m}$ each satisfy the same recursion relation, and hence $C_{n,m} = D_{n,m}$ for every $n \geq 2$ and $m \geq 1$. This proves (b).

First, observe that $C_{2,m+1} = (2\alpha - 1)C_{2,m} + (\alpha - 1)(\alpha^m - 1)$ for all $m \geq 2$ (the first term is the number of edges where each vertex has a nonzero entry in at least one of the first $m$ slots, and the second term is the number of edges where one vertex has all zeros in the first $m$ slots). Note that $D_{2,m} = (1/2)[(2\alpha - 1)^m - 2\alpha^m + 1]$, and thus one easily verifies that $D_{2,m+1} = (2\alpha - 1)D_{2,m} + (\alpha - 1)(\alpha^m - 1)$ for all $m \geq 2$. Since $C_{2,2} = (|K| - 1)^2 = D_{2,2}$, we have $C_{2,m} = D_{2,m}$ for all $m \geq 2$.

Clearly $C_{n,1} = 0$ for all $n \geq 2$. We also have $D_{n,1} = (1/n!) \sum_{i=0}^{n} (-1)^i \binom{n}{i} [(n - i)(\alpha - 1) + 1] = (\alpha - 1/n!) \sum_{i=0}^{n} (-1)^i \binom{n}{i} (n - i) + (1/n!) \sum_{i=0}^{n} ( 1)^i \binom{n}{i} = ((\alpha - 1)/(n-1)!) \sum_{i=0}^{n-1} (-1)^i \binom{n-1}{i} + (1/n!) \sum_{i=0}^{n} (-1)^i \binom{n}{i} = ((\alpha - 1)/(n-1)!)(1-1)^{n-1} + (1/n!)(1-1)^n = 0 + 0 = 0$ for all $n \geq 2$.

Finally, let $m \geq 2$. Then one can easily check that $C_{n,m+1} = (n(\alpha - 1) + 1)C_{n,m} + (\alpha - 1)C_{n-1,m}$ for all $n \geq 3$ (the first term is the number of cliques where each vertex has a nonzero entry in at least one of the first $m$ slots, and the second term is the number of cliques where one vertex has all zeros in the first $m$ slots). We show that also $D_{n,m+1} = (n(\alpha - 1) + 1)D_{n,m} + (\alpha - 1)D_{n-1,m}$ for $n \geq 3$. This follows from the observation that $(n(\alpha - 1) + 1)(-1)^i \binom{n}{i} [(n - i)(\alpha - 1) + 1]^m + n(\alpha - 1)(-1)^{i-1} \binom{n-1}{i-1} [(n - i)(\alpha - 1) + 1]^m = (-1)^i \binom{n}{i} [n(\alpha - 1) + 1 - n(\alpha - 1)i/n][(n - i)(\alpha - 1) + 1]^m = (-1)^i \binom{n}{i} [(n - i)(\alpha - 1) + 1][(n - i)(\alpha - 1) + 1]^m = (-1)^i \binom{n}{i} [(n - i)(\alpha - 1) + 1]^{m+1}$. Thus $C_{n,m} = D_{n,m}$ for all $m \geq 2$ and $n \geq 1$.

The proof of the general case in Theorem 3.1.50 is similar to that above, but it is notationally much messier. One uses the recursion relations $C_{2,m+1} =$

$(2|F_{m+1}|-1)C_{2,m}+(|F_{m+1}|-1)(|F_1|\cdots|F_m|-1)$ and $C_{n,m+1}=(n(|F_{m+1}|-1)+1)C_{n,m}+(|F_{m+1}|-1)C_{n-1,m}$. □

**Example 3.1.52.** ([75, Example 3.13]) (a) For $F$ a finite field, the number of edges in $\Gamma_m=\Gamma(F^m)$ is given by

$$\lambda(\Gamma_m,2)=\frac{1}{2}[(2|F|-1)^m-2|F|^m+1]$$

and the number of triangles in $\Gamma(F^m)$ is

$$\lambda(\Gamma_m,3)=\frac{1}{6}[(3|F|-2)^m-3(2|F|-1)^m+3|F|^m-1].$$

(b) When $F=\mathbb{Z}_2$, we have

$$\lambda(\Gamma(\mathbb{Z}_2^m),1)=2^m-2,$$

$$\lambda(\Gamma(\mathbb{Z}_2^m),2)=\frac{1}{2}[3^m-2^{m+1}+1],$$

$$\lambda(\Gamma(\mathbb{Z}_2^m),3)=\frac{1}{6}[4^m-3^{m+1}+3\cdot 2^m-1],$$

and

$$\lambda(\Gamma(\mathbb{Z}_2^m),n)=\frac{1}{n!}\sum_{i=0}^{n}(-1)^i\binom{n}{i}(n-i+1)^m.$$

Next, we compute a formula for $\lambda(\Gamma(R,n)$ when $R$ is the finite local ring $\mathbb{Z}_{p^m}$ for $p$ prime and $m\geq 2$. Recall that $|\Gamma(\mathbb{Z}_{p^m})|=p^{m-1}-1$ and $\omega(\Gamma(\mathbb{Z}_{p^m}))=p^{\frac{m}{2}}-1$ when $m$ is even and $p^{\frac{m-1}{2}}$ when $m$ is odd.

**Theorem 3.1.53.** ([75, Theorem 3.14]) *Let $p$ be prime, $m\geq 2$, and $n\geq 1$.*

(a) *If $m$ is even, then* $\lambda(\Gamma(\mathbb{Z}_{p^m}),n)=\binom{p^{\frac{m}{2}}-1}{n}+\sum_{i=1}^{\frac{m-2}{2}}\binom{p^i-1}{n-1}p^{m-i-1}(p-1)$;

(a) *If $m$ is odd, then* $\lambda(\Gamma(\mathbb{Z}_{p^m}),n)=\binom{p^{\frac{m-1}{2}}-1}{n}+\sum_{i=1}^{\frac{m-1}{2}}\binom{p^i-1}{n-1}p^{m-i-1}(p-1)$.

### 3.1.3 Edge Coloring

Now, we discuss the edge coloring of the zero-divisor graph of a commutative ring. A *$k$-edge coloring* of a graph $G$ is an assignment of $k$ colors $\{1,\ldots,k\}$ to the edges of $G$ such that no two adjacent edges have the same color. The *edge chromatic number* $\chi'(G)$ of a graph $G$ is the minimum $k$ for which $G$ has a $k$-edge coloring.

If $G$ is a graph, clearly in any edge coloring of $G$, the edges incident with one vertex should be colored with different colors. This observation implies that $\chi'(G) \geq \Delta(G)$. An important theorem due to Vizing is the following.

**Theorem 3.1.54.** ([161, Theorem 10.2]) ***Vizing's Theorem.*** *If $G$ is a simple graph, then either $\chi'(G) = \Delta(G)$ or $\chi'(G) = \Delta(G) + 1$.*

**Example 3.1.55.** ([161]) For the complete graph $K_n$,

$$\chi'(K_n) = \begin{cases} \Delta(K_n) & \text{if } n \text{ is even;} \\ \Delta(K_n) + 1 & \text{if } n \text{ is odd.} \end{cases}$$

A graph $G$ is said to be *critical* if $G$ is connected, $\chi'(G) = \Delta(G) + 1$, and $\chi'(G \setminus \{e\}) < \chi'(G)$ for any edge $e$ of $G$. The following are essential for future sections.

**Theorem 3.1.56.** ***Vizing's Adjacency Lemma.*** *If $G$ is a critical graph, then $G$ has at least $\Delta(G) - \delta(G) + 2$ vertices of maximum degree.*

**Theorem 3.1.57.** ([161, Theorem 10.8]) ***König's Theorem.*** *For any bipartite graph $G$, we have $\chi'(G) = \Delta(G)$.*

Akbari and Mohammadian [38] studied the edge chromatic number of zero-divisor graphs.

**Remark 3.1.58.** ([38, Remark 1]) We note that if $G$ is a graph and $\chi'(G) = \Delta(G)+1$, then there exists a subgraph of $G$, say $G_1$, such that $\chi'(G_1) = \Delta(G)+1$ and for any edge $e$ of $G_1$ we have $\chi'(G \setminus \{e\}) = \Delta(G)$. Clearly $G_1$ has a connected subgraph, say $H$, such that $\chi'(H) = \Delta(G) + 1$. The graph $H$ is a critical graph with maximum degree $\Delta(G)$. If $x$ is a vertex of $H$ with degree $\Delta(G)$, then by Vizing's Adjacency Lemma, $H$ has at least $\Delta(G) - deg_H(v) + 2$ vertices of degree $\Delta(G)$ for any vertex $v$ which is adjacent to $x$. Therefore if $G$ is a graph such that for every vertex $u$ of maximum degree there exists an edge $uv$ such that $\Delta(G) - deg(v) + 2$ is more than the number of vertices with maximum degree in $G$, then by the above argument and Vizing's Theorem, we have $\chi'(G) = \Delta(G)$.

**Remark 3.1.59.** ([38, Remark 2]) Assume that $R = R_1 \times \cdots \times R_n$ is a finite decomposable commutative ring. We note that if $x = (x_1, \ldots, x_n)$ has maximum degree in $\Gamma(R)$, then $x$ has exactly one nonzero component, say $x_1$. Suppose that $R_1$ is a local ring. We consider two cases: If $R_1$ is a field, then $\Delta(\Gamma(R)) = deg(x) = |R_2| \cdots |R_n| - 1$; If $R_1$ is not a field, then we have $x_1 \in Ann(Z(R_1))^*$ and $\Delta(\Gamma(R)) = deg(x) = |Z(R_1)||R_2| \cdots |R_n| - 2$.

**Theorem 3.1.60.** ([38, Theorem 3]) *If $R$ is a finite commutative ring, then $\chi'(\Gamma(R)) = \Delta(\Gamma(R))$, unless $\Gamma(R)$ is a complete graph of odd order.*

*Proof.* Let $R$ be a finite commutative ring.

**Case : 1** Suppose $R$ is a local ring. If $R$ is a field, then $\Gamma(R)$ is the empty graph, and so there is nothing to prove. Assume that $R$ is not a field. Note that if $R$ is an Artinian local ring, then the Jacobson radical of $R$ equals $Z(R)$. Thus $Z(R)$ is a nilpotent ideal, and this implies that if $R$ is not a field, then $Ann(Z(R)) \neq \{0\}$. Moreover, each element of $Ann(Z(R))^*$ is adjacent to each other vertex of $\Gamma(R)$. If $\Gamma(R)$ is a complete graph, then by Example 3.1.55, we are done. Thus suppose that $\Gamma(R)$ is not a complete graph, and so $Ann(Z(R)) \neq Z(R)$. If $x \in Z(R) \setminus Ann(Z(R))$, then there is an element $a \in Z(R)$ such that $ax \neq 0$. This implies that $x$ is adjacent to no vertices of $a + Ann(Z(R))$. Therefore $deg(x) \leq |Z(R)^*| - |Ann(Z(R))|$. Hence $\Delta(\Gamma(R)) - d(x) + 2 \geq |Ann(Z(R))| + 1$. Clearly, $Ann(Z(R))^*$ is the set of all vertices of maximum degree in $\Gamma(R)$. So, by Remark 3.1.58, we have $\chi'(\Gamma(R)) = \Delta(\Gamma(R))$.

**Case : 2** Assume that $R$ is not a local ring. It is well known that every commutative Artinian ring is isomorphic to a direct product of finitely many local rings. Suppose that $R = R_1 \times \cdots \times R_n$, where $n \geq 2$ and each $R_i$ is a local ring. By Remark 3.1.59, without loss of generality suppose that the nonzero components of the vertices with maximum degree in $\Gamma(R)$ occur in $R_1, \ldots, R_k$. First, we claim that all of the rings $R_1, \ldots, R_k$ are fields or none of them are fields. Working towards a contradiction, suppose that $R_1$ is a field and $R_2$ is not a field. Now, every vertex with maximum degree in $R_1 \times \{0\} \times \cdots \times \{0\}$ has degree $|R_2| \cdots |R_n| - 1$ and each vertex with maximum degree in $\{0\} \times R_2 \times \{0\} \times \cdots \times \{0\}$ has degree $|R_1||Z(R_2)||R_3| \cdots |R_n| - 2$. Thus we have $|Z(R_2)||R_3| \cdots |R_n|(|R_1| - |R_2/Z(R_2)|) = 1$, a contradiction. Therefore by Remark 3.1.59, for any $i$, $1 \leq i \leq k$, $\Delta(\Gamma(R)) = |R_1| \cdots |R_{i-1}||Z(R_i)||R_{i+1}| \cdots |R_n| - \epsilon$, where $\epsilon = 1$ or 2. Hence, we have $|R_1/Z(R_1)| = \cdots = |R_k/Z(R_k)|$. Moreover, since for each $j$, $k+1 \leq j \leq n$, the degree of any vertex in $\{0\} \times \cdots \times \{0\} \times R_j \times \{0\} \times \cdots \times \{0\}$ is less than $\Delta(\Gamma(R))$, we have

$$|R_j/Z(R_j)| \geq |R_1/Z(R_1)|. \tag{3.1}$$

For any $t$, $1 \leq t \leq n$, suppose that $e_t$ is the element whose $t^{th}$ component is one and other components are zero. First, suppose that the rings $R_1, \ldots, R_k$ are not fields. Then $\Gamma(R)$ has $\sum_{t=1}^{k} |Ann(Z(R_t))^*|$ vertices of maximum degree. Clearly, every vertex of maximum degree in $\Gamma(R)$ is adjacent to at least one of the $e_t$'s. Now for any $i$, $1 \leq i \leq n$, we have

$$\begin{aligned}
\Delta(\Gamma(R)) - deg(e_i) + 2 \geq & (|R_1| \cdots |R_{i-1}||Z(R_i)||R_{i+1}| \cdots |R_n| - 2) \\
& - (|R_1| \cdots |R_{i-1}||R_{i+1}| \cdots |R_n| + 2) \\
= & |R_1| \cdots |R_{i-1}|(|Z(R_i)| - 1)|R_{i+1}| \cdots |R_n| + 1 \\
> & \sum_{t=1}^{k} |Ann(Z(R_t))^*|.
\end{aligned}$$

Hence by Remark 3.1.58, we conclude that $\chi'(\Gamma(R)) = \Delta(\Gamma(R))$. Next, suppose that the rings $R_1, \ldots, R_k$ are fields. Then $\Gamma(R)$ has $\sum_{t=1}^{k} |R_t^*|$ vertices of maximum degree. If $n > 2$, then every vertex of maximum degree in $\Gamma(R)$ is adjacent to $1-e_t$, for some $t$, $1 \leq t \leq k$. Note that in this case $|R_1| = \cdots = |R_k|$, and if we set $|R_1| = a$, then by (3.1) we have $|R_j| \geq a$ for any $j$, $j > k$. Now since $a^{n-1} - a + 2 > n(a-1)$, for any $i$, $1 \leq i \leq k$, we have

$$\begin{aligned}\Delta(\Gamma(R)) - deg(1-e_i) + 2 =& (|R_1| \cdots |R_{i-1}||R_{i+1}| \cdots |R_n| - 1) - (|R_i| - 1) + 2 \\ \geq& a^{n-1} - a + 2 > n(a-1) > \sum_{t=1}^{k} |R_t^*|.\end{aligned}$$

Thus by Remark 3.1.58, we conclude that $\chi'(\Gamma(R)) = \Delta(\Gamma(R))$. So assume that $n = 2$. If $k = 1$ and $R_2$ is not a field, then by (3.1) we have $|R_2| \geq 2|R_1|$. Since in this case any vertex of maximum degree in $\Gamma(R)$ is adjacent to $e_2$ and $\Delta(\Gamma(R)) - deg(e_2) + 2 = (|R_2| - 1) - (|R_1| - 1) + 2 > |R_1^*|$, by Remark 3.1.58, we obtain $\chi'(\Gamma(R)) = \Delta(\Gamma(R))$. If either $k = 1$ and $R_2$ is a field or $k = 2$, then $\Gamma(R)$ is a complete bipartite graph. Hence, by König's Theorem, we have $\chi'(\Gamma(R)) = \Delta(\Gamma(R))$. □

Hence the edge chromatic number of $\Gamma(R)$ is equal to its maximum degree.

## 3.2 Connectivity of Zero-divisor Graphs

Let $v$ be a vertex and $e$ an edge of a graph $G$. If $G - v$ has more components than $G$, then $v$ is a *cut-vertex* of $G$; while if $G - e$ has more components than $G$, then $e$ is a *bridge* of $G$. In particular, if $v$ is a cut-vertex of a connected graph $G$, then $G - v$ is disconnected; and if $e$ is a bridge of a connected graph $G$, then $G - e$ is disconnected, necessarily a graph with exactly two components. A nontrivial connected graph having no cut-vertices is called *nonseparable*. An important theorem in graph theory says that "every nontrivial connected graph contains at least two vertices that are not cut-vertices". The vertex-connectivity of a graph is the size of a smallest subset of vertices whose removal renders the graph disconnected or leaves a single vertex, while the edge-connectivity is the size of a smallest subset of edges whose removal renders the graph disconnected.

A *vertex cut* is a subset $S \subseteq V(G)$ such that $G - S$ is disconnected, and a *disconnecting set* of edges of $G$ is a subset $T \subseteq E(G)$ such that the graph $G - T$ is disconnected. An *edge cut* is a disconnecting set of edges which is minimal. The *vertex-connectivity* of $G$, denoted by $\kappa(G)$, is the size of smallest set $S$ such that $S$ is a vertex cut or $G - S$ has only one vertex. Similarly, the *edge-connectivity* of $G$, denoted by $\lambda(G)$, is the size of the smallest edge cut. Note that for any graph $G$, $\kappa(G) \leq \lambda(G) \leq \delta(G)$, where $\delta(G)$ denotes the minimum vertex degree in $G$.

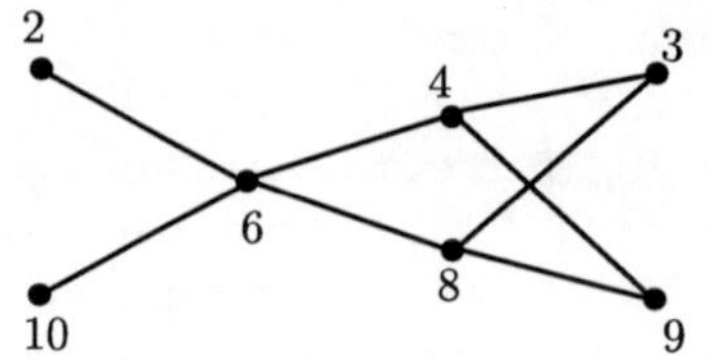


Figure 3.1: $\Gamma(\mathbb{Z}_{12})$

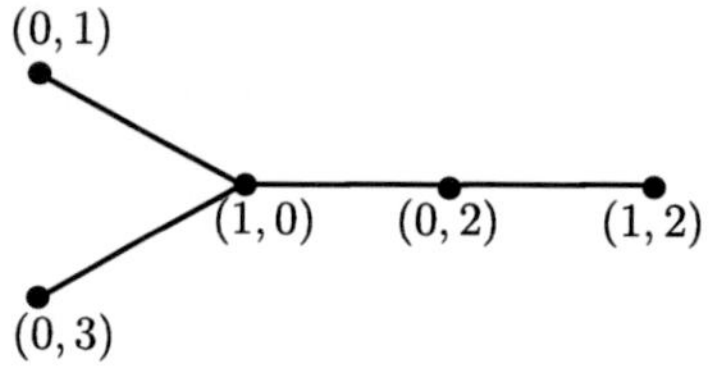


Figure 3.2: $\Gamma(\mathbb{Z}_2 \times \mathbb{Z}_4)$

The study of cut-vertices in a zero-divisor graph began in [122]. Actually, the authors were analyzing the realizability of graphs as zero-divisor graphs of commutative rings with identity through endpoint and cut-vertex analysis. Throughout this section, we allow loops in zero-divisor graphs.

**Example 3.2.1.** ([169, Example 1.3]) Consider the zero-divisor graph of $\mathbb{Z}_{12}$ shown in Figure 3.1. In $\Gamma(\mathbb{Z}_{12})$, 6 is a cut-vertex isolating 2 and 10. In addition, $\{4, 8\}$ is a cut-set isolating 3 and 9.

The following few theorems were proved by Axtell et al. in [122].

**Theorem 3.2.2.** ([122, Theorem 4.2]) *Let $R$ be a commutative ring with identity. If $\Gamma(R)$ is partitioned into two subgraphs $X$ and $Y$ with cut-vertex "$a$" such that $X \setminus \{a\}$ is a complete subgraph, then $I = V(X) \cup \{0\}$ is an ideal.*

The converse of Theorem 3.2.2 is false. For example, let $R = \mathbb{Z}_2 \times \mathbb{Z}_4$. Then $(1, 0)$ is a cut-vertex. The set $\{(0, 0), (1, 0), (0, 2), (1, 2)\}$ is an ideal of $R$. However, the corresponding subgraph is not complete (see Figure 3.2).

**Theorem 3.2.3.** ([122, Theorem 4.3]) *Let $R$ be a commutative ring with identity such that $\Gamma(R)$ is partitioned into two subgraphs $X$ and $Y$ with cut-vertex "$a$" and $|V(X)| > 2$. If $X$ is a complete subgraph, then every vertex of $X$ is looped (i.e., $b^2 = 0$ for every $b \in V(X)$).*

**Theorem 3.2.4.** ([122, Theorem 4.4]) *Let $R$ be a commutative ring with identity. If $a$ is a cut-vertex of $\Gamma(R)$, then $\{0, a\}$ is an ideal of $R$.*

In continuation of this, the problem of identifying cut-sets and cut-vertices in the zero-divisor graph of $\prod_{i=1}^{n} \mathbb{Z}_{n_i}$ was examined by Coté et al. in [169, 170].

Initially, they examined the cut-vertices in the zero-divisor graph of $\mathbb{Z}_n$, and then generalized the same to direct products.

**Theorem 3.2.5.** ([169, Theorem 2.1], [170, Lemma 2.2]) *Let $n \geq 6$. Then an element $a$ is a cut-vertex of $\Gamma(\mathbb{Z}_n)$ if and only if $2a = n$.*

**Theorem 3.2.6.** ([169, Theorem 2.2]) *Let $\mathbb{Z}_n \times \mathbb{Z}_m \ncong \mathbb{Z}_2 \times \mathbb{Z}_2$. Then $(a, 0)$ is a cut-vertex of $\Gamma(\mathbb{Z}_n \times \mathbb{Z}_m)$ if and only if $2a = n$.*

**Theorem 3.2.7.** ([169, Theorem 2.3], [170, Theorem 2.3]) *Let $R = \mathbb{Z}_{n_1} \times \cdots \times \mathbb{Z}_{n_i} \times \cdots \times \mathbb{Z}_{n_m}$ for $m \geq 3$. Then $(0, \ldots, a_i, \ldots, 0)$ is a cut-vertex of $\Gamma(R)$ if and only if $2a_i = n_i$.*

Note that when $n = p$ is prime, the ring $\mathbb{Z}_n$ is a field, and so $\Gamma(\mathbb{Z}_n)$ is empty. When $n = 2p$ with $p$ prime and $p > 2$, $\Gamma(\mathbb{Z}_n)$ is a star graph with center $p$, where the only cut-set is $A = Ann(2) \setminus \{0\} = \{p\}$. Notice that $Ann(p) \setminus \{0\} = V(\Gamma(\mathbb{Z}_n)) \setminus \{p\}$. When $n = p^2$, $\Gamma(\mathbb{Z}_n)$ is a complete graph; hence there are no cut-sets.

**Theorem 3.2.8.** ([169, Theorem 3.1], [170, Theorem 2.6]) *Let $n$ be a positive integer such that $n \neq p$, $2p$, $p^2$ for any prime $p$. A set $A$ is a cut-set of $\Gamma(\mathbb{Z}_n)$ if and only if $A = Ann(p) \setminus \{0\}$ for some prime $p$ which divides $n$.*

Note that for a prime number $p$ which divides $n$, $|Ann(p)| = p$ in $\mathbb{Z}_n$. Therefore the size of any cut-set in $\Gamma(\mathbb{Z}_n)$ is known.

**Theorem 3.2.9.** ([169, Theorem 3.7], [170, Theorem 2.9]) *Let $p$ be a prime such that $p \mid n_i \in \mathbb{Z}_{n_i}$. A subset $A$ of $V(\Gamma(\prod_{i=1}^{n} \mathbb{Z}_{n_i}))$ with $m \geq 2$ is a cut-set if and only if $A = \{(0, \ldots, 0, a_{i_1}, 0, \ldots, 0), \ldots, (0, \ldots, 0, a_{i_k}, 0, \ldots, 0)\}$, where $a_{i_j} \in \mathbb{Z}_{n_i}$ and $\{a_{i_1}, a_{i_2}, \ldots, a_{i_k}\} = Ann(p) \setminus \{0\} \in \mathbb{Z}_{n_i}$.*

The following result concerns cut-sets in the zero-divisor graph of a commutative ring. Recall that for any zero-divisor graph, the maximum distance between any two vertices is 3. However, the next result tells us that any two vertices connected to some vertex in the cut-set also have to be connected to a common vertex in the cut-set.

**Theorem 3.2.10.** ([170, Theorem 2.10]) *Let $A = \{a_1, a_2, \ldots, a_n\}$ be a cut-set of $\Gamma(R)$ for some commutative ring $R$, and let $X$ and $Y$ be the subgraphs associated with the cut-set $A$. Denote $\overline{X} = \{x_1, x_2, \ldots, x_m\}$ as the set of all $x_i \in X \setminus A$ such that $x_i - a_j$ is a path for some $a_j \in A$, and denote $\overline{Y} = \{y_1, y_2, \ldots y_k\}$ as the set of all $y_j \in Y \setminus A$ such that $y_j - a_g$ is a path for some $a_g \in A$. Then the minimal path in $\Gamma(R)$ between any $x_i \in \overline{X}$ and $y_j \in \overline{Y}$ has length 2.*

The following two theorems show that for $R \cong R_1 \times \cdots \times R_n$, where each $R_i$ is any finite commutative ring, all cut-vertices and vertices in cut-sets are of the form $(0, \ldots, 0, a_i, 0, \ldots, 0)$.

**Theorem 3.2.11.** ([169, Theorem 2.4], [170, Lemma 2.1]) *Let $R \cong R_1 \times \cdots \times R_n$ where $n \geq 2$. If $a$ is a cut-vertex of $\Gamma(R)$, then there exists some $i$, $1 \leq i \leq n$, such that $a = (0, \ldots, 0, a_i, 0, \ldots, 0)$.*

**Theorem 3.2.12.** ([169, Theorem 3.6]) *Let $R \cong R_1 \times \cdots \times R_n$ with $n \geq 2$. If $A$ is a cut-set of $\Gamma(R)$, then there exists some $i$, $1 \leq i \leq n$, such that $a = (0, \ldots, 0, a_i, 0, \ldots, 0)$ for every $a \in A$.*

Next, we discuss the cut-vertices and cut-sets of the zero-divisor graph of a finite commutative ring. Recall that any finite commutative ring $R$ with identity can be written as $R \cong R_1 \times \cdots \times R_n \times F_1 \times \cdots F_m$, where each $R_i$ is local and each $F_j$ is a field.

**Theorem 3.2.13.** ([170, Theorem 3.3]) *Let $R$ be a nonlocal commutative ring not isomorphic to $\mathbb{Z}_2 \times F$, where $F$ is a field and write $R$ as $R \cong R_1 \times \cdots \times R_n \times F_1 \times \cdots \times F_m$, where each $R_i$ is a local ring and each $F_j$ is a field. Then*

(a) *$A = \{(0, \ldots, 0, r_i, 0, \ldots, 0) \mid r_i \in Ann(r) \setminus \{0\}\}$ is a cut-set in $\Gamma(R)$, where $Ann(r)$ is a minimal annihilator ideal;*

(b) *$A = \{(0, \ldots, 0, f_j, 0, \ldots, 0) \mid f_j \in F_j \setminus \{0\}\}$ is a cut-set in $\Gamma(R)$.*

*Moreover, every cut-set in $\Gamma(R)$ has the form of (a) or (b).*

Note that $\Gamma(\mathbb{Z}_2 \times \mathbb{Z}_2)$ has only two vertices, and therefore cannot have a cut-set. Also, $\Gamma(\mathbb{Z}_2 \times F)$, where $F \ncong \mathbb{Z}_2$, yields a star graph where $\{(1,0)\}$ is the only cut-set because the only other vertices in $\Gamma(\mathbb{Z}_2 \times F)$ are $\{(0, f) \mid f \in F\}$. So by removing that set, one is left with the single vertex, namely, $\{(1,0)\}$. The next theorem is a generalization of Theorem 3.2.4.

**Theorem 3.2.14.** ([170, Theorem 3.4]) *Let $A$ be a cut-set of $\Gamma(R)$, where $R$ is a nonlocal ring. Then $A \cup \{0\}$ is an ideal of $R$.*

**Theorem 3.2.15.** ([170, Theorem 3.5]) *In a finite local ring $R$, let $\mathfrak{m} = Z(R) = (q_1, q_2, \ldots, q_n)$. If a vertex $x$ is a cut-vertex of $\Gamma(R)$, then $\cap_{k=1}^{n} Ann(q_k) \setminus \{0\} = \{a\}$.*

The study of cut-vertices was continued by Axtell et al. in [118]. Actually, they mainly concentrated on understanding when annihilator ideals corresponding to cut-vertices in $\Gamma(R)$ are properly maximal. Further, they partially classified finite local rings having zero-divisor graphs with cut-vertices. The following are some of the results proved in this regard.

**Lemma 3.2.16.** ([118, Proposition 2.2]) *Let $R$ be a commutative ring with identity. If $x$ is a cut-vertex of $\Gamma(R)$, then $r + s \in Z(R)$ for all $r, s \in R \setminus Z(R)$.*

The converse of Lemma 3.2.16 is false since $Ann(2)$ is properly maximal in $\mathbb{Z}_4$, but 2 is not a cut-vertex of $\Gamma(\mathbb{Z}_4)$. Note that if $R$ is a Noetherian ring with the property that $Ann(a) \neq Ann(b)$ whenever $a \neq b$, then $\Gamma(R)$ has only finitely many cut-vertices.

**Lemma 3.2.17.** ([118, Lemma 3.3]) *Let $R$ be a finite commutative ring, $R \ncong \mathbb{Z}_2 \times \mathbb{Z}_2$, and $a \in Z(R)^*$. If $|Z(R)^*| \geq 2$ and $a$ is not a cut-vertex of $\Gamma(R)$, then $(a, 0)$ is not a cut-vertex of $\Gamma(R \times S)$ for any ring $S$.*

The hypothesis that $|Z(R)^*| \geq 2$ is necessary, since 2 is not a cut-vertex of $\Gamma(\mathbb{Z}_4)$, but $(0, 2)$ is a cut vertex of $\Gamma(\mathbb{Z}_2 \times \mathbb{Z}_4)$.

**Lemma 3.2.18.** ([118, Lemma 3.5]) *Let $R_1$ and $R_2$ be finite commutative rings such that $\Gamma(R_1)$ is split into two subgraphs $X$ and $Y$ via $a_1$. Then $(a_1, 0)$ is a cut vertex of $\Gamma(R_1 \times R_2)$ if and only if one of $X$ or $Y$ contains at least one isolated vertex when $a_1$ and its incident edges are removed.*

**Theorem 3.2.19.** ([118, Theorem 3.6]) *Let $R = R_1 \times R_2 \times \cdots \times R_n$ with each $R_i$ a finite local ring and $n \geq 2$. If $R \ncong \mathbb{Z}_2 \times \mathbb{Z}_2$, then a vertex $a = (a_1, 0, \ldots, 0) \in R$ is a cut-vertex of $\Gamma(R)$ if and only if one of the following holds:*

(a) *$a_1 = 1$ with $R_1 \cong \mathbb{Z}_2$;*

(b) *$R_1 \cong \mathbb{Z}_4$ or $\mathbb{Z}_2[x]/(x^2)$ with $a_1 = 2$ or $x + (x^2)$;*

(c) *$a_1$ is a cut-vertex in $\Gamma(R_1)$ that isolates at least one vertex when removed.*

Theorem 3.2.19 also shows that cut-vertices of zero-divisor graphs of direct products always isolate vertices. If $R_1 \cong \mathbb{Z}_2$ and $R_2$ is not isomorphic to $\mathbb{Z}_2$, then the cut-vertex $(1, 0)$ of $\Gamma(R_1 \times R_2)$ isolates $(0, 1)$. Therefore we have the following result.

**Corollary 3.2.20.** ([118, Corollary 3.7]) *Let $R$ be a finite commutative ring. If $\Gamma(R)$ contains a cut-vertex that does not isolate any vertices, then $R$ is a local ring.*

The converse of Corollary 3.2.20 is false. In $\Gamma(\mathbb{Z}_8)$, 4 is the only cut-vertex, and it isolates 2. At this point, note that the classification of the cut-vertices of graphs of decomposable rings has reduced to classifying cut-vertices of graphs of local rings.

**Theorem 3.2.21.** ([118, Theorem 4.1, Corollary 4.2]) *Let $R$ be a finite local ring with maximal ideal $\mathfrak{m}$ and $k$ the least positive integer such that $\mathfrak{m}^k = 0$. If $x$ is a cut-vertex of $\Gamma(R)$, then $\mathfrak{m}^{k-1} = \{x, 0\}$. In particular, $x$ is a cut-vertex of $\Gamma(R)$ if and only if $x$ is the unique cut-vertex of $\Gamma(R)$.*

**Theorem 3.2.22.** ([118, Theorem 4.4, Corollary 4.5]) *Let $R$ be a finite local ring with maximal ideal $\mathfrak{m}$. If $x$ is a cut-vertex of $\Gamma(R)$, then $|R| = 2^{n+1}$ and $|\mathfrak{m}| = 2^n$ for some $n \in \mathbb{N}$. That is, if $x$ is a cut-vertex of $\Gamma(R)$, then $R$ has characteristic $2^m$.*

Recall that a principal ideal ring (PIR) is a ring in which all ideals are principal, and a special principal ideal ring (SPIR) is a PIR with a unique prime ideal, and this prime ideal is nilpotent. Clearly, an SPIR is also local. In the SPIR case, the converse of Theorem 3.2.21 is true.

**Theorem 3.2.23.** ([118, Theorem 4.6]) *Let $R$ be a finite SPIR with maximal ideal $\mathfrak{m} = (a)$. Let $x = a^{k-1}$, where $k \geq 3$ is the minimal positive integer such that $a^k = 0$. The following are equivalent:*

(a) *$x$ is a cut-vertex of $\Gamma(R)$;*

(b) *$\mathfrak{m}^{k-1} = \{0, x\}$;*

(c) *$R/\mathfrak{m} \cong \mathbb{Z}_2$;*

(d) *$ua^{k-1} = a^{k-1}$ for all $u \in U(R)$;*

(e) *$Ann(x)$ is properly maximal.*

As per the properties in the ring $\mathbb{Z}_4$, $k \geq 3$ is necessary in the statement of Theorem 3.2.23.

**Corollary 3.2.24.** ([118, Corollary 4.7]) *Let $R$ be a finite SPIR and let $k \geq 3$. Then $\Gamma(R)$ has a cut-vertex if and only if $R \cong \dfrac{\mathbb{Z}_{2^n}[x]}{(g(x), 2^{n-1}x^t)}$, where $t = k - (n - 1)s > 0$ and $g(x)$ is an Eisenstein polynomial of degree $s$ over $\mathbb{Z}_{2^n}$.*

In [45], the authors investigated the size of minimal vertex cuts and edge cuts in $\Gamma(R)$. The following theorem gives the vertex-connectivity and edge-connectivity of $\Gamma(R)$ in the case of a nonlocal commutative ring $R$.

**Theorem 3.2.25.** ([45, Theorem 3.1]) *Let $R$ be a nonlocal finite commutative ring. Then $\kappa(\Gamma(R)) = \lambda(\Gamma(R)) = \delta(\Gamma(R))$.*

Now assume that $R$ is a local ring with maximal ideal $\mathfrak{m}$. Let $r$ be the smallest positive integer such that $\mathfrak{m}^r = 0$. As mentioned earlier, $\Gamma(R)$ is the empty graph if $r = 1$, and $\Gamma(R)$ is a complete graph if $r = 2$. So in both cases, $\kappa(\Gamma(R)) = \lambda(\Gamma(R)) = \delta(\Gamma(R)) = |\mathfrak{m}| - 2$. Thus assume that $r \geq 3$. Since $\mathfrak{m}^{r-1} \subseteq Ann(\mathfrak{m})$, it follows immediately that $Ann(\mathfrak{m}) \setminus \{0\}$ is nonempty, and also that $\Gamma(R)$ is not a complete graph. Viewed as a subset of $V(\Gamma(R))$, $Ann(\mathfrak{m}) \setminus \{0\}$ is a dominating set in $\Gamma(R)$. Clearly any vertex cut must contain $Ann(\mathfrak{m}) \setminus \{0\}$. Thus, we have the following elementary bounds:

$$|Ann(\mathfrak{m})| - 1 \leq \kappa(\Gamma(R)) \leq \lambda(\Gamma(R)) \leq \delta(\Gamma(R)).$$

**Theorem 3.2.26.** ([45, Proposition 3.2]) *Let $R$ be a local finite commutative ring. If there exists $x \in \mathfrak{m}$ such that $Ann(x) = Ann(\mathfrak{m})$, then $\kappa(\Gamma(R)) = \lambda(\Gamma(R)) = \delta(\Gamma(R)) = |Ann(\mathfrak{m})| - 1$.*

**Corollary 3.2.27.** ([45, Corollary 3.3]) *Let $p$ be a prime number and $n \geq 3$. Then $\kappa(\Gamma(\mathbb{Z}_{p^n})) = \lambda(\Gamma(\mathbb{Z}_{p^n})) = p - 1$.*

Recall that a vertex $x \in V(G)$ is called *dominant* if $N(x) = V(G) \setminus \{x\}$.

**Theorem 3.2.28.** ([45, Proposition 3.4]) *Let $G$ be a graph with a dominant vertex. Then $\lambda(G) = \delta(G)$. In particular, if $R$ is a local finite commutative ring with $\mathfrak{m}^2 \neq 0$, then $\lambda(\Gamma(R)) = \delta(\Gamma(R))$.*

It is natural to look how tight the bounds $|Ann(\mathfrak{m})| - 1 \leq \kappa(\Gamma(R)) \leq \delta(\Gamma(R))$ are. The following result says that the lower bound is not met for certain nonlocal rings.

**Theorem 3.2.29.** ([45, Proposition 3.6]) *Let $G$ be a graph with $r \geq 4$. If $R$ be a local ring with $Ann(x) \neq Ann(\mathfrak{m})$ for all $x \in \mathfrak{m}$, then $\kappa(\Gamma(R)) > |Ann(\mathfrak{m})| - 1$.*

The hypothesis $r \geq 4$ in Theorem 3.2.29 is necessary. For example, consider $R = \mathbb{F}_2[x, y]/(x^2, y^2)$. By abuse of notation, we will use elements of $\mathbb{F}_2[x, y]$ to describe the cosets they represent in $R$. Then $\mathfrak{m} = (x, y)$ has 8 elements and $m^2 = Ann(\mathfrak{m}) = \{0, xy\}$. Thus, $\Gamma(R)$ has 7 vertices, with $xy$ a dominant vertex. Moreover, $\Gamma(R) - \{xy\}$ is a graph on 6 vertices with 3 connected components $\{x, x+xy\}$, $\{y, y+xy\}$, and $\{x+y, x+y+xy\}$; so $\kappa(\Gamma(R)) = |Ann(\mathfrak{m})| - 1 = 1$. Also, note that for any $t \in R$, $Ann(t)$ contains $(t)$; since $(t)$ has at least 4 elements for any $t \neq 0$, there is no way for the equality $Ann(t) = Ann(\mathfrak{m})$ to hold for any $t \in V(\Gamma(R))$. Hence, $Ann(t) \neq Ann(\mathfrak{m})$ for all $t \in \mathfrak{m}$.

The following example shows that both the bounds in the inequality $|Ann(\mathfrak{m})| - 1 \leq \kappa(\Gamma(R)) \leq \delta(\Gamma(R))$ are quite loose.

**Example 3.2.30.** ([45, Proposition 3.7]) Let $F$ be a field of order $f = 2^s$ and $R = F[x, y, z]/(x^2, y^2, z^2)$. Then $|Ann(\mathfrak{m})| - 1 = f - 1$, $\kappa(\Gamma(R)) = f^3 - 1$, and $\delta(\Gamma(R)) = f^4 - 2$.

In Example 3.2.30, $\kappa(\Gamma(R))$ is roughly $\frac{1}{|F|}\delta(\Gamma(R))$; so by taking $F$ to be arbitrarily large, we see that there is no hope for a general upper bound on $\kappa(\Gamma(R))$ which is linear in $\delta(\Gamma(R))$. In fact, in this family, $\kappa(\Gamma(R))$ is roughly $\delta(\Gamma(R))^{\frac{3}{4}}$. It is natural, then, to ask for the maximum value of $n$, $0 < n \leq \frac{3}{4}$, such that $\kappa(\Gamma(R))$ can be bounded below by a function of order $\delta(\Gamma(R))^n$. One such bound is given below.

**Theorem 3.2.31.** ([45, Proposition 3.8]) *Let $R$ be a finite commutative ring. Then $\kappa(\Gamma(R)) \geq (\frac{\delta(\Gamma(R))}{2})^{\frac{1}{3}} - \frac{1}{\sqrt{3}}$.*

A continuation of the examination of cut-vertices in the zero-divisor graphs of commutative rings was done by Redmond in [328]. Actually, Redmond continued the work on cut-vertices by Axtell et al. in [118] (see Lemma 3.2.16 to Corollary 3.2.24). Recall that the *center* of a graph $G$ is set of all vertices whose eccentricity is equal to the radius. The next proposition establishes a relationship between the cut-vertices and the center of a zero-divisor graph.

**Proposition 3.2.32.** ([328, Proposition 1]) *Let $R$ be a commutative ring with 1. If $a$ is a cut-vertex of $\Gamma(R)$, then $a$ is in the center of $\Gamma(R)$.*

**Proposition 3.2.33.** ([328, Proposition 2]) *Let $R$ be a commutative ring with 1. Suppose $a$ is a cut-vertex of $\Gamma(R)$ dividing the graph into disjoint components $X$ and $Y$ when $a$ is removed. Then either $d(a, x) = 1$ for all $x \in X$ or $d(a, y) = 1$ for all $y \in Y$.*

The following theorem shows that, except in a finite number of known cases, cut-vertices occur in zero-divisor graphs only when the graph has degree-one vertices.

**Theorem 3.2.34.** ([328, Theorem 3]) *Let $R$ be a finite commutative ring with 1 such that $\Gamma(R)$ has at least 3 vertices. Then $\Gamma(R)$ has a cut vertex if and only if either (a) $\Gamma(R)$ has a vertex of degree one or (b) $R$ is isomorphic to one of the following seven rings:*

$$\frac{\mathbb{Z}_4[x,y]}{(x^2, y^2, xy-2, 2x, 2y)}, \frac{\mathbb{Z}_2[x,y]}{(x^2, y^2)}, \frac{\mathbb{Z}_4[x]}{(x^2)}, \frac{\mathbb{Z}_4[x]}{(x^2+2x)}, \frac{\mathbb{Z}_8[x]}{(x^2+4, 2x)},$$
$$\frac{\mathbb{Z}_2[x,y]}{(x^2, y^2-xy)}, \frac{\mathbb{Z}_4[x,y]}{(x^2, y^2-xy, xy-2, 2x, 2y)}.$$

In light of Theorem 3.2.34, the study has shifted to the degree-one vertices in zero-divisor graphs.

**Theorem 3.2.35.** ([328, Lemma 4]) *Let $R$ be a finite commutative ring with 1. Then $\Gamma(R)$ has a vertex of degree one if and only if either (a) there is some $x \in R$ such that $|Ann(x)| = 2$ or (b) $R$ is isomorphic to $\mathbb{Z}_9$ or $\mathbb{Z}_3[x]/(x^2)$.*

**Corollary 3.2.36.** ([328, Corollary 5]) *Let $R$ be a finite commutative ring with 1. Then $\Gamma(R)$ has a cut-vertex if and only if either (a) there is some $x \in R$ such that $|Ann(x)| = 2$ or (b) $R$ is isomorphic to one of the following seven rings:*

$$\frac{\mathbb{Z}_4[x,y]}{(x^2, y^2, xy-2, 2x, 2y)}, \frac{\mathbb{Z}_2[x,y]}{(x^2, y^2)}, \frac{\mathbb{Z}_4[x]}{(x^2)}, \frac{\mathbb{Z}_4[x]}{(x^2+2x)}, \frac{\mathbb{Z}_8[x]}{(x^2+4, 2x)},$$
$$\frac{\mathbb{Z}_2[x,y]}{(x^2, y^2-xy)}, \frac{\mathbb{Z}_4[x,y]}{(x^2, y^2-xy, xy-2, 2x, 2y)}.$$

**Theorem 3.2.37.** ([328, Corollary 6]) *Let $R$ be a finite commutative ring with 1. Then the cut-vertices of $\Gamma(R)$ form a complete subgraph of $\Gamma(R)$.*

**Corollary 3.2.38.** ([328, Corollary 7]) *Let $R$ be a finite commutative ring with 1. If $\Gamma(R)$ has a degree-one vertex, then either the characteristic of $R$ is even or $R$ is isomorphic to $\mathbb{Z}_9$ or $\mathbb{Z}_3[x]/(x^2)$.*

The converse of Corollary 3.2.38 is false. For example, $\Gamma(\mathbb{F}_4 \times \mathbb{F}_4)$ is the complete bipartite graph $K_{3,3}$ and $\Gamma(\mathbb{F}_8[x]/(x^2))$ is the complete graph on 7 vertices.

**Corollary 3.2.39.** ([328, Corollary 8]) *Let $R$ be a finite commutative ring with 1. If $n$ is the number of cut-vertices of $\Gamma(R)$, then $0 \leq n \leq \frac{|V(\Gamma(R))|}{2}$.*

Define $R_k = \prod_{i=1}^{k} \mathbb{Z}_2$ for $k \geq 2$. Then the ratio of cut-vertices of $\Gamma(R_k)$ to all vertices of $\Gamma(R_k)$ is $\frac{k}{2^k-2}$. In general, any finite commutative ring with 1 can be written as a product of local rings, say $R \cong R_1 \times \cdots \times R_n$ for some $n \geq 1$. The graph $\Gamma(R)$ has a cut-vertex if and only if some $R_i$ is local and $\Gamma(R_i)$ has a cut-vertex or if $n \geq 2$ with some $R_i$ isomorphic to $\mathbb{Z}_2$ , $\mathbb{Z}_4$, or $\mathbb{Z}_2[x]/(x^2)$. Thus, the number of cut-vertices of $R$ is at most $n$. However, the number of zero-divisors of $R$ is $|R_1| \cdot |R_2| \cdots |R_n| - |R_1 - Z(R_1)| \cdot |R_2 - Z(R_2)| \cdots |R_n - Z(R_n)| \geq 2^n - 1$. So, the ratio of cut-vertices to all vertices in $\Gamma(R)$ is at most $\frac{n}{2^n-2}$. Hence, not every rational number in the interval $[0, 0.5]$ will be the ratio of cut-vertices to all vertices in some zero-divisor graph of a commutative ring with 1. However, the sequence of rings $R_k$ shows that 0 is a limit point of the set of all such possible ratios.

A more interesting question is the ratio of degree-one vertices to all vertices. Clearly, the ratio must be between 0 and 1, with the ratio 1 for $\Gamma(\mathbb{Z}_2 \times \mathbb{Z}_2)$, $\frac{1}{2}$ for $\Gamma(\mathbb{Z}_2 \times \mathbb{Z}_2 \times \mathbb{Z}_2)$, and 0 for any graph without degree-one vertices, such as $\Gamma(\mathbb{Z}_{25})$. Using the sequence of rings $R_k$, 0 is a limit point of this set of ratios. Letting $\mathbb{F}_p$ denote the field with $p$ elements, the ratio of degree-one vertices to all vertices in $\Gamma(\mathbb{Z}_2 \times \mathbb{F}_p)$ is $\frac{p-1}{p}$, showing 1 is a limit point. The ratio for $\Gamma(\mathbb{Z}_4 \times \mathbb{F}_p)$ is $\frac{p-1}{2p-3}$, showing $\frac{1}{2}$ is a limit point. It is a routine exercise to find an infinite number of other limit points. Keeping these facts in mind, Redmond proposed the following open question.

**Open Question**. For any rational number $q \in [0, 1]$, does there exist a finite commutative ring with 1 such that $q$ is equal to the ratio of the number of degree-one vertices of $\Gamma(R)$ to the total number of vertices of $\Gamma(R)$? If not, is the set of all rational numbers that do represent the ratio of degree-one vertices to all vertices in a zero-divisor graph a dense subset of $[0, 1]$?

## 3.3 Isomorphisms and Automorphisms

A very natural question when studying zero-divisor graphs is whether they are unique; that is, is $\Gamma(R) \cong \Gamma(S)$ if and only if $R \cong S$? Clearly, if $R \cong S$, then we have $\Gamma(R) \cong \Gamma(S)$. But Example 2.1.3 shows that non-isomorphic rings may have isomorphic zero-divisor graphs. Specifically, the zero-divisor graphs of $\mathbb{Z}_2 \times \mathbb{Z}_2$ and $\mathbb{Z}_9$ are isomorphic, yet the two rings are clearly not isomorphic. This question has a positive answer when the rings are finite direct products of finite fields.

Such work was first initiated by Anderson, Frazier, and Lauve [75]. The next theorem shows that for finite reduced rings which are not fields, the ring $R$ is determined by its zero-divisor graph $\Gamma(R)$.

**Theorem 3.3.1.** ([75, Theorem 4.1]) *Let $R$ and $S$ be finite reduced commutative rings which are not fields. Then $\Gamma(R) \cong \Gamma(S)$ as graphs if and only if $R \cong S$ as rings.*

The"finite" hypothesis is needed in Theorem 3.3.1. Let $R = \mathbb{Z}_2 \times \mathbb{Z}$ and $S = \mathbb{Z}_2 \times \mathbb{Q}$. Then the infinite star graphs $\Gamma(R)$ and $\Gamma(S)$ are isomorphic since $|\mathbb{Z}| = |\mathbb{Q}|$, but $R$ and $S$ are not isomorphic since $S$ is a total quotient ring while $R$ is not. Note that Theorem 3.3.1 is a remarkable result which says that for certain rings, the behavior of the zero-divisors uniquely determines the entire ring. A generalization of Theorem 3.3.1 was given by R. Levy and J. Shapiro [260].

**Theorem 3.3.2.** ([260, Corollary 2.4]) *Let $R = \prod_{i \in I} F_i$ and $S = \prod_{j \in J} F'_j$, where $F_i$ and $F'_j$ are finite fields for all $i \in I$ and $j \in J$. If $\Gamma(R) \cong \Gamma(S)$, then $R \cong S$.*

In fact, Theorem 3.3.1 and Theorem 3.3.2 have been generalized by S. Akbari and A. Mohammadian [38] as shown in Theorem 3.3.4. The authors of [38] first proved the following result.

**Theorem 3.3.3.** ([38, Theorem 4]) *Let $R_1, \ldots, R_n$ and $S_1, \ldots, S_m$ be finite local rings for $n, m \geq 2$. Then $\Gamma(R_1 \times \cdots \times R_n) \cong \Gamma(S_1 \times \cdots \times S_m)$ if and only if $n = m$ and there exists a permutation $\varphi$ of $\{1, \ldots, n\}$ such that for any $i \leq i \leq n$, $|R_i| = |S_{\varphi(i)}|$ and $\Gamma(R_i) \cong \Gamma(S_{\varphi(i)})$.*

In what follows, the authors of [38] generalize both Theorem 3.3.1 and Theorem 3.3.2. Indeed, we show that if one of the two rings is reduced, then the assertion remains true.

**Theorem 3.3.4.** ([38, Theorem 5]) *Let $R$ be a finite reduced ring and $S$ be a ring such that $S$ is not an integral domain. If $\Gamma(R) \cong \Gamma(S)$, then $R \cong S$, unless $R \cong \mathbb{Z}_2 \times \mathbb{Z}_2$, $\mathbb{Z}_6$ and $S$ is a local ring.*

In the next theorem, D. F. Anderson, R. Levy, and J. Shapiro [81] give another generalization of both Theorem 3.3.1 and Theorem 3.3.2.

**Theorem 3.3.5.** ([81, Theorem 2.1]) *Let $\{R_i\}_{i \in I}$, ($|I| \geq 2$) and $\{S_j\}_{j \in J}$ be two families of integral domains, and let $R = \prod_{i \in I} R_i$ and $S = \prod_{j \in J} S_j$. Then $\Gamma(R) \cong \Gamma(S)$ if and only if there is a bijection $\varphi : I \to J$ such that $|R_i| = |S_{\varphi(i)}|$ for each $i \in I$. In particular, if $\Gamma(R) \cong \Gamma(S)$ and each $R_i$ is a finite field, then each $S_j$ is also a finite field and $R_i \cong S_{\varphi(i)}$ for each $i \in I$, and thus $R \cong S$.*

In [81], they also showed that $\Gamma(T(R))$ and $\Gamma(R)$ are isomorphic graphs, where $T(R)$ is the total quotient ring of $R$. Recall that $T(R)$ is the quotient ring of $R$ with respect to the multiplicatively closed subset $R \setminus Z(R)$ of $R$. The next theorem shows that $\Gamma(T(R))$ and $\Gamma(R)$ are isomorphic by showing that there is a bijection between equivalence classes of $\Gamma(T(R))$ and $\Gamma(R)$ such that the corresponding equivalence classes have the same cardinality.

**Theorem 3.3.6.** ([81, Theorem 2.2]) *Let $R$ be a commutative ring with total quotient ring $T(R)$. Then the graphs $\Gamma(T(R))$ and $\Gamma(R)$ are isomorphic.*

**Corollary 3.3.7.** ([81, Corollary 2.3]) *Let $R$ and $S$ be commutative rings with isomorphic total quotient rings. Then $\Gamma(R)$ and $\Gamma(S)$ are isomorphic graphs. In particular, $\Gamma(R) \cong \Gamma(S)$ when $R$ and $S$ have the same total quotient ring.*

**Corollary 3.3.8.** ([81, Corollaries 2.4 and 2.5]) *Let $R$ and $S$ be reduced commutative Noetherian rings which are not integral domains. Then $\Gamma(R) \cong \Gamma(S)$ if and only if there is a bijection $\varphi : Min(R) \to Min(S)$ such that $|R/P| = |S/\varphi(P)|$ for each $P \in Min(R)$, where $Min(R)$ is the set of minimal prime ideals of $R$. In particular, if $Min(R) = \{P_1, \ldots, P_n\}$, then $\Gamma(R) \cong \Gamma(F_1 \times \cdots \times F_n)$, where every $F_i = T(R/P_i)$ is a field.*

J. LaGrange ([244], [248], and [250]) has investigated the zero-divisor graph of the complete ring of quotients $Q(R)$ of $R$ (see [255] and [244] for the definition of $Q(R)$). In particular, he gave an example of a ring $R$ for which $\Gamma(R) \ncong \Gamma(Q(R))$ ([244]). Let $R = \{N \subseteq \mathbb{N} \mid |N| < \infty \text{ or } |\mathbb{N} \setminus N| < \infty\}$, where $\mathbb{N}$ is the set of natural numbers. It is well known that $R$ is a Boolean ring with addition defined as "symmetric difference" and multiplication defined as "intersection". Then $\Gamma(R) \ncong \Gamma(Q(R))$.

Levy and Shapiro [260] presented a number of examples of von Neumann regular rings that have isomorphic zero-divisor graphs and yet are not isomorphic as rings. For instance, if $R$ and $S$ are the infinite product of finite fields, none of which is the field with two elements, then their zero-divisor graphs will always be isomorphic. Another example: let $\mathbb{F}_9$ be the field with 9 elements and $R$ be the product of $\mathbb{F}_9$ with itself countably many times. Let $S$ be the subring of $R$ consisting of all $a \in R$ such that all but finitely many of the coordinates of $a$ are in $\mathbb{Z}_3$. Then $S$ is a von Neumann regular ring and $\Gamma(R) \cong \Gamma(S)$, whereas $R \ncong S$.

Note that the two von Neumann regular rings may have isomorphic zero-divisor graphs, but are not isomorphic, and every Boolean ring is a von Neumann regular ring. However, a Boolean ring is determined by its zero-divisor graph (see Theorem 3.3.10). In general, the von Neumann regular ring $R$ is quite different from the ring $B(R)$, where $B(R)$ denotes the Boolean algebra of $R$. (Note that while multiplication in $B(R)$ is just multiplication in $R$, the addition in $B(R)$ is given by $e + f - 2ef$ for idempotents $e$ and $f$ of $R$.) For example, if $R$ is any direct product of fields indexed by a set $I$, then $B(R)$ is the direct product of copies of $\mathbb{Z}_2$ indexed by $I$. The following theorem was proved by Anderson et al. [81] regarding the ring $B(R)$.

**Theorem 3.3.9.** ([81, Theorem 4.1]) *Let $R$ and $S$ be commutative von Neumann regular rings. Then $\Gamma(R)$ and $\Gamma(S)$ are isomorphic if and only if there exists a Boolean algebra isomorphism between $B(R)$ and $B(S)$ such that $|[e]| = |[\varphi(e)]|$ for all $1 \neq e \in B(R)$.*

In the above theorem, if one of the rings is Boolean, then we can say even more. The following theorem due to LaGrange [244] says that a Boolean ring is determined by its zero-divisor graph.

**Theorem 3.3.10.** ([244, Theorem 4.1]) *Let $R$ be a commutative ring with nonzero zero-divisors, not isomorphic to $\mathbb{Z}_9$ or $\mathbb{Z}_3[x]/(x^2)$. If $S$ is a Boolean ring such that $\Gamma(R) \cong \Gamma(S)$, then $R \cong S$. In particular, if $R$ and $S$ are Boolean rings, then $\Gamma(R) \cong \Gamma(S)$ if and only if $R \cong S$.*

In the second half of this section, we study $Aut(\Gamma(R))$, the group of graph-automorphisms of the zero-divisor graph $\Gamma(R)$. Recall that a *graph-automorphism* $f$ of a graph $G$ is a bijection $f : G \to G$ which preserves adjacency as well as non-adjacency. The set $Aut(G)$ of all graph-automorphisms of $G$ forms a group under the usual composition of functions. If $|G| = n$, then in the obvious way $Aut(G)$ is isomorphic to a subgroup of $S_n$, and clearly $Aut(K_n) \cong S_n$. In fact, for a graph $G$ of order $n$, $Aut(G) \cong S_n$ if and only if $G = K_n$.

Let $Aut(R)$ be the group of ring-automorphisms of $R$. Note that each ring-automorphism of $R$ induces a graph-automorphism of $\Gamma(R)$; we thus have a natural group-homomorphism $\varphi : Aut(R) \to Aut(\Gamma(R))$ given by just restricting each $f \in Aut(R)$ to $Z(R)^*$. In general, $Aut(\Gamma(R))$ is much larger than $Aut(R)$; for example, each $f \in Aut(\mathbb{Z}_n)$ is trivial.

An interesting question is to see how the structure of $\Gamma(R)$ determines the structure of $R$. It is first showed that if $R$ is a finite ring which is not a field, then $\varphi : Aut(R) \to Aut(\Gamma(R))$ is injective, i.e., an $f \in Aut(R)$ is completely determined by its action on $Z(R)$.

**Theorem 3.3.11.** ([83, Theorem 3.1]) *Let $R$ be a finite commutative ring which is not a field and let $f \in Aut(R)$. If $f(x) = x$ for all $x \in Z(R)$, then $f = 1_R$. Thus $\varphi : Aut(R) \to Aut(\Gamma(R))$ is a monomorphism.*

*Proof.* First suppose that $R$ has distinct maximal ideals $\mathfrak{m}_1$ and $\mathfrak{m}_2$. Then $\mathfrak{m}_1, \mathfrak{m}_2 \subseteq Z(R)$. Since $R = \mathfrak{m}_1 + \mathfrak{m}_2$, we must have $f = 1_R$. Thus we may assume that $R$ is local with nonzero maximal ideal $\mathfrak{m}(= Z(R))$. Let $char(R) = p^n$ and $|R/\mathfrak{m}| = p^m$ for some $p$ and integers $m, n \geq 1$. We show that $f(u) = u$ for each $u \in U(R)$. Observe that for any $m \in \mathfrak{m}$, $um = f(um) = f(u)m$, and hence $(f(u) - u)m = 0$. Thus $f(u) = u + x$ for some $x \in \mathfrak{m}$ (note that $x$ may depend on $u$) with $x\mathfrak{m} = 0$. Since $|R/\mathfrak{m}| = p^m$, we have $u^{p^m-1} = 1 + y$ for some $y \in \mathfrak{m}$, and hence $f$ fixes $u^{p^m-1}$. Thus $u^{p^m-1} = f(u^{p^m-1}) = f(u)^{p^m-1} = (u + x)^{p^m-1} = u^{p^m-1} + (p^m - 1)(u^{p^m-2})x$ since $x\mathfrak{m} = 0$, and hence $(p^m - 1)(u^{p^m-2})x = 0$. Since $(p^m - 1)$ and $(u^{p^m-2})$ are units in $R$, we have $x = 0$. Thus $f(u) = u$ for all $u \in U(R)$, and hence $f = 1_R$. □

It was also proved that $Aut(\Gamma(\mathbb{Z}_n))$ is a (finite) direct product of symmetric groups, and thus $\Gamma(\mathbb{Z}_n)$ is highly symmetrical. The proof of this fact gives the explicit construction of $Aut(\Gamma(\mathbb{Z}_n))$. Specific examples are given in parts (c) and (d) of Example 3.3.14.

Recall that for a vertex $x$ of a zero-divisor graph $\Gamma(R)$, $deg(x) = |Ann(x) \setminus \{0, x\}|$. Now specialize to $\Gamma(\mathbb{Z}_n)$ and establish notation. Let $n \geq 4$ be a nonprime integer, and let $X = \{d \in \mathbb{Z} \mid 1 < d < n \text{ and } d \mid n\}$. For each $d \in X$, let $V_d = \{\overline{x} \in \mathbb{Z}_n \mid 1 < x < n \text{ and } gcd(x,n) = d\}(= U(\mathbb{Z}_n)\overline{d} \subseteq Z(\mathbb{Z}_n)^*)$ and $n_d = |V_d|$. Note that $\Gamma(\mathbb{Z}_n)(= Z(\mathbb{Z}_n)^*)$ is the disjoint union of the $V_d$'s and that for each $x \in V_d$, the subgroup $Ann(x) = Ann(\overline{d})$ has order $d$ in $\mathbb{Z}_n$ (i.e., for any $x, y \in V_d$ and $a \in \mathbb{Z}_n$, $ax = 0$ if and only if $ay = 0$). In particular, $deg(x) = deg(\overline{d})$ for all $x \in V_d$. Clearly, $deg(\overline{d}) = d - 1$ if $n \nmid d^2$ (i.e., $\overline{d} \notin Ann(\overline{d})$ in $\mathbb{Z}_n$), and $deg(\overline{d}) = d - 2$ if $n \mid d^2$ (i.e., $\overline{d} \in Ann(\overline{d})$ in $\mathbb{Z}_n$). The key fact is that for $d, e \in X$, we have $deg(\overline{d}) = deg(\overline{e})$ if and only if $d = e$, and thus two vertices of $\Gamma(\mathbb{Z}_n)$ have the same degree if and only if they are in the same $V_d$. For suppose $deg(\overline{d}) = deg(\overline{e})$ for some $d, e \in X$ with $d < e$. Then necessarily $d-1 = e-2$ and hence $e = d+1$. But, then $n \mid (d+1)^2$ and $d(d+1) \mid n$ (since $d \mid n, (d+1) \mid n$, and $gcd(d, d+1) = 1$) yield that $d \mid (d+1)$, a contradiction.

**Theorem 3.3.12.** ([83, Theorem 3.2]) *Let $n \geq 4$ be a nonprime integer. Then $Aut(\Gamma(\mathbb{Z}_n))$ is a (finite) direct product of symmetric groups. Specifically, $Aut(\Gamma(\mathbb{Z}_n)) \cong \prod\{S_{n_d} \mid d \in X\}$, where $X = \{d \in \mathbb{Z} \mid 1 < d < n \text{ and } d \mid n\}$ and $n_d = |\{x \in \mathbb{Z} \mid 1 < x < n \text{ and } gcd(x,n) = d\}|$ $(= |V_d|)$.*

*Proof.* Use the same notation as above. Since two vertices of $\Gamma(\mathbb{Z}_n)$ have the same degree if and only if they are in the same $V_d$ for each $f \in Aut(\Gamma(\mathbb{Z}_n))$ and $d \in X$. Define $\varphi : Aut(\Gamma(\mathbb{Z}_n)) \to \prod\{S_n \mid d \in X\}$ by $\varphi(f) = (f|_{V_d})$, with $f|_{V_d}$ viewed in the natural way as an element of $S_{n_d}$. By the above comments, $\varphi$ is surjective; it suffices to show that for each fixed $d \in X$ and permutation $\alpha$ of $V_d$, there is an $f \in Aut(\Gamma(\mathbb{Z}_n))$ with $f|_{V_d} = \alpha$ and $f|_{V_{d'}} = 1_{V_{d'}}$ for all $d' \neq d$ in $X$. This follows since for any $x, y \in V_d$ and $a \in \mathbb{Z}_n$, $ax = 0$ if and only if $ay = 0$. □

It is well known that any finite group $G$ may be realized as $Aut(\Gamma)$ for some graph $\Gamma$. It would be interesting to know exactly which groups may be realized as $Aut(\Gamma(R))$ or $Aut(\Gamma(\mathbb{Z}_n))$. The next corollary gives a partial answer.

**Corollary 3.3.13.** ([83, Corollary 3.3]) *Let $n \geq 4$ be a nonprime integer. Then*

(a) *$Aut(\Gamma(\mathbb{Z}_n))$ is trivial if and only if $n = 4$;*

(b) *$Aut(\Gamma(\mathbb{Z}_n))$ is abelian if and only if $n = 4, 6, 8, 9$, or $12$.*

*In particular, $Aut(\Gamma(\mathbb{Z}_n)) \cong \mathbb{Z}_2$ when $n = 6, 8$, or $9$, and $Aut(\Gamma(\mathbb{Z}_{12})) \cong \mathbb{Z}_2 \times \mathbb{Z}_2 \times \mathbb{Z}_2$.*

*Proof.* The computations of $Aut(\Gamma(\mathbb{Z}_n))$ for $n = 4, 6, 8$, and 9 follow easily from Theorem 3.3.12 or Example 2.1.3; for $n = 12$, see Example 3.3.14. We show that $Aut(\Gamma(\mathbb{Z}_n))$ is nonabelian for all other nonprime integers $n \geq 4$. It is sufficient to show that in each case, some $|V_d| \geq 3$, and hence $Aut(\Gamma(\mathbb{Z}_n))$ has a direct factor of $S_m$ for some $m \geq 3$.

If $n = 2^r$ with $r \geq 4$, then $\{\overline{2}, \overline{6}, \overline{10}\} \subseteq V_2$. If $n = 3^r$ with $r \geq 3$, then $\{\overline{3}, \overline{6}, \overline{15}\} \subseteq V_3$. If $n = p^r$ with $p \geq 5$ prime, then $\{\overline{p}, \overline{2p}, \overline{3p}\} \subseteq V_p$. Let $n = p_1^{r_1} \cdots p_k^{r_k}$ be its prime factorization. If $p_1 \geq 3$ and $k \geq 2$, then $\{\overline{p_1}, \overline{2p_1}, \overline{4p_1}\} \subseteq$

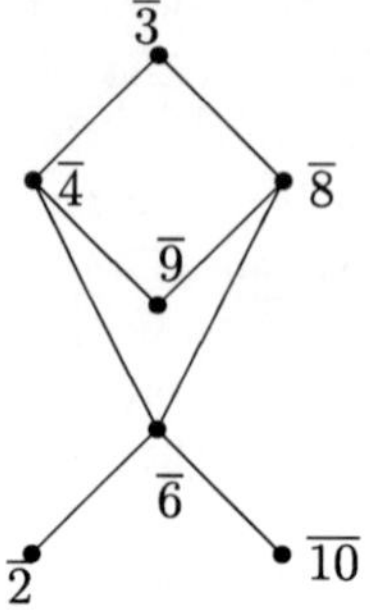


Figure 3.3: $\Gamma(\mathbb{Z}_{12})$

$V_{p_1}$. Thus we may assume that $p_1 = 2$ with $r = r_1 \geq 1$ and $k \geq 2$. If $n > 2^r \cdot 3$ then $\{\bar{2}^r, \bar{2}^{r+1}, \bar{2}^{r+2}\} \subseteq V_{2^r}$. Hence we may assume that $n = 2^r \cdot 3$. If $r \geq 3$, then $\{\bar{2}, \bar{10}, \bar{14}\} \subseteq V_2$. Thus $Aut(\Gamma(\mathbb{Z}_n))$ is abelian only if $n = 4, 8, 9$, or $12$. $\square$

Some specific examples of $Aut(\Gamma(R))$ are given in the following example.

**Example 3.3.14.** ([83, Example 3.4]) (a) The zero-divisor graphs of $\mathbb{F}_4[x]/(x^2)$ and $\mathbb{Z}_3 \times \mathbb{Z}_3$ are a triangle and a square, respectively. Thus $Aut(\Gamma(\mathbb{F}_4[x]/(x^2))) \cong S_3$ and $Aut(\Gamma(\mathbb{Z}_3 \times \mathbb{Z}_3)) \cong D_4 (\cong S_2 \wr S_2)$, where $\wr$ denotes the wreath product).

(b) Let $F$ and $K$ be finite fields with distinct orders $m$ and $n$, respectively. Then $\Gamma(F \times K)$ is the complete bipartite graph $K_{m-1,n-1}$. Hence $Aut(\Gamma(F \times K)) \cong S_{m-1} \times S_{n-1}$ has order $(m-1)!(n-1)!$. Also $\Gamma(F \times F)$ is the complete bipartite graph $K_{m-1,m-1}$. It is also known that $Aut(K_{m-1,m-1}) \cong S_{m-1} \wr S_2$, which has order $2((m-1)!)^2$.

(c) We illustrate the proof of Theorem 3.3.12 by computing $Aut(\Gamma(\mathbb{Z}_{12}))$. We have $X = \{2, 3, 4, 6\}$, and thus $V_2 = \{\bar{2}, \overline{10}\}$, $V_3 = \{\bar{3}, \bar{9}\}$, $V_4 = \{\bar{4}, \bar{8}\}$, and $V_6 = \{\bar{6}\}$. Hence $Aut(\Gamma(\mathbb{Z}_{12})) \cong S_2 \times S_2 \times S_2 \times S_1 \cong \mathbb{Z}_2 \times \mathbb{Z}_2 \times \mathbb{Z}_2$. This is also evident from the obvious symmetries of the zero-divisor graph of $\mathbb{Z}_{12}$ given in Figure 3.3.

(d) For a more complicated example, let $R = \mathbb{Z}_{1225} (\cong \mathbb{Z}_{25} \times \mathbb{Z}_{49})$ with $|\Gamma(R)| = 384$. Then $X = \{5, 7, 25, 35, 49, 175, 245\}$, and one may calculate that $|V_5| = 168$, $|V_7| = 120$, $|V_{25}| = 42$, $|V_{35}| = 24$, $|V_{49}| = 20$, $|V_{175}| = 6$, and $|V_{245}| = 4$. Thus $Aut(\Gamma(R)) \cong S_{168} \times S_{120} \times S_{42} \times S_{24} \times S_{20} \times S_6 \times S_4$ has order $168! \cdot 120! \cdot 42! \cdot 24! \cdot 6! \cdot 4!$, which is approximately $6.192877 \times 10^{598}$.

In 2002, two papers by Mulay [295] and DeMeyer and Schneider [178] gave a more detailed study of automorphisms of $\Gamma(R)$. It is natural to expect that there is a relationship between the graph-automorphisms and the ring-automorphisms. The objective of Mulay's [295] work was to exhibit the extent of such relationship. We need the following notation throughout this section.

Let $J$ be an ideal of $R$ such that $J^2 = 0$, and let $B$ be a subring of $R$ containing $J$. Define $Der(R/B, J)$ to be the set of derivations $h : R \to J$ with $h(B) = 0$. Let $i$ denote the identity automorphism of $R$. Define

$$Id + Der(R/B, J) = \{\theta : R \to R \mid \theta = i + h \text{ for some } h \in Der(R/B, J)\}.$$

Define $G(R, B, J)$ to be the set of automorphisms $g$ of $R$ such that $g(b) = b$ for all $b \in B$ and $g$ induces the identity automorphism on $R/J$. Clearly, $G(R, B, J)$ is a subgroup of $Aut(R)$.

Let $\pi(R)$ denote the prime subring of $R$ and let $X$ be an indeterminate over $\pi(R)$. Let $M$ be the ideal generated by $Z(R)$. An element $r \in R$ is called *special* provided there is a polynomial $f(X) \in \pi(R)[X]$ such that $f(r)$ is in $M$ and $f'(r)$ is a non-zero-divisor of $R$ (as usual $f'(X)$ is the derivative of $f(X)$ with respect to $X$). Note that every element of $M$ is special. The subring obtained by adjoining the special elements to $\pi(R)$ is denoted by $\bar{\omega}(R)$.

Define $\mathfrak{J}(R) = \{x \in Z(R) \mid Z(R) \subseteq (0 : x)\}$. Note that $\mathfrak{J}(R) \subseteq \bar{\omega}(R)$ and $\mathfrak{J}(R) \subseteq (0 : x)$ for all $x \in Z(R)$.

**Theorem 3.3.15.** ([295, Theorem 3.1]) *$G(R, B, J) = Id + Der(R/B, J)$ and the group $G(R, B, J)$ is isomorphic to the additive group $Der(R/B, J)$. In particular $G(R, B, J)$ is abelian.*

**Theorem 3.3.16.** ([295, Theorem 3.2]) *Let $R$ be a commutative ring and $M = (Z(R))$. Then the following hold:*

(a) *The set $\mathfrak{J}(R)$ is an ideal with $\mathfrak{J}(R)^2 = 0$. Further, if $\mathfrak{J}(R) \neq 0$, then for an $x \in \mathfrak{J}(R)^*$ we have $M = (0 : x)$ and $M$ is a prime ideal with $M = Z(R)$;*

(b) *Assume that either $M = R$ or $M$ is the unique maximal ideal of $R$ and $R/M$ is algebraic over a finite field. Then we have $\bar{\omega}(R) = R$;*

(c) *Assume $Z(R)^* \neq \emptyset$ and let $\sigma \in Aut(R)$. Then $\sigma(z) = z$ for all $z \in Z(R)$ if and only if $\sigma$ is in $G(R, \omega(R), \mathfrak{J}(R))$.*

Note that there is a natural homomorphism from $Aut(R)$ to $Aut(\Gamma(R))$. Denote the kernel of this map by $ker(R)$. It consists of those ring-automorphisms which map each zero-divisor to itself. For finite rings $R$, it follows that $ker(R)$ is trivial as noted in Theorem 3.3.11. The following theorem determines the kernel $ker(R)$.

**Theorem 3.3.17.** ([295, Theorem 3.3]) *The kernel of the natural homomorphism from $Aut(R)$ to $Aut(\Gamma(R))$ is*

$$ker(R) = G(R, \bar{\omega}(R), \mathfrak{J}(R)) = Id + Der(R/\bar{\omega}(R), \mathfrak{J}(R)).$$

From the above characterization of $ker(R)$, it is evident that if $R$ has no nonzero nilpotents, then the group $ker(R)$ is trivial.

**Theorem 3.3.18.** ([295, Theorem 3.4]) *If $ker(R)$ is nontrivial, then $ker(R)$ is not a finitely generated group.*

Then Mulay [295] introduced an equivalence relation on the vertices of the zero-divisor graph and observed basic properties of the set of equivalence classes. Using this, he proved the following main theorem.

**Theorem 3.3.19.** ([295, Theorem 3.11]) *Let $R$ be a Noetherian ring. Then*

(a) *$Aut(\Gamma(R))$ is finite if and only if $R$ is finite;*

(b) *$Aut(\Gamma(R))$ is trivial if and only if $R$ is a binary local ring of cardinality 4;*

(c) *$Aut(\Gamma(R))$ is simple if and only if $|R| \leq 8$;*

(d) *If $Aut(\Gamma(R))$ is solvable, then $|R|$ is either $25, 27$, or $2^r 3^s 5^t$ with $t \leq 1, s + t \leq 2$, and $rst = 0$. Moreover, each residue field has at most 5 elements. There are only finitely many such rings if and only if there are only finitely many finite local rings $A$ of cardinality $2^m$ with solvable $Aut(\Gamma(A))$;*

(e) *If $Aut(\Gamma(R))$ is abelian, then $|R|$ is either 9 or $2^r 3^s$ with $s \leq 1$ and each residue field has at most 3 elements. There are only finitely many such rings if and only if there are only finitely many finite local rings $A$ with residue field $\mathbb{F}_2$ and solvable $Aut(\Gamma(A))$.*

Thus $Aut(\Gamma(R))$ is trivial if and only if there is exactly one nonzero zero-divisor in $R$. In this regard, finally Mulay [295] proposed an open question.

**Open Question:** Are there only finitely many finite rings $R$ for which $Aut(\Gamma(R))$ is solvable (resp., abelian)?

The study of $Aut(\Gamma(R))$ was continued by DeMeyer and Schneider [178]. Let $Aut_{Z(R)}(R)$ be the group of ring-automorphisms of $R$ fixing $Z(R)$ pointwise. Assume that $Z(R)^*$ is nonempty. Restriction induces an exact sequence $1 \to Aut_{Z(R)}(R) \to Aut(R) \to Aut(\Gamma(R))$. If $R$ is finite, then by Theorem 3.3.11, $Aut_{Z(R)}(R) = \{1\}$; so the restriction map is one-to-one in this case.

**Theorem 3.3.20.** ([178, Lemma 2.1]) *Let $R$ be a commutative ring with at least two zero-divisors. Let $\sigma \in Aut(R)$ and assume $\sigma \neq 1$, but $\sigma|_{Z(R)} = 1$. Then*

(a) *$Z(R) = Ann_R(x)$ for some $x \in Z(R)$, and $Z(R)$ is a prime ideal of $R$;*

(b) *If $J = \cap_{y \in Z(R)} Ann_R(y)$, then $J \neq (0)$ and $J^2 = (0)$;*

(c) *$\Gamma(R)$ contains a spanning tree which is a star graph;*

(d) *The induced automorphism $\bar{\sigma} = 1$ on $R/J$ and $R/Z(R)$.*

The following result characterizes the abelian nature of the group $Aut_{Z(R)}(R)$.

**Theorem 3.3.21.** ([178, Corollary 2.3]) *Let $R$ be a commutative ring. If $R$ has at least two zero-divisors, then $Aut_{Z(R)}(R)$ is an abelian group.*

**Theorem 3.3.22.** ([178, Corollary 2.7]) *Let $R$ be a commutative ring. If $\Gamma(R)$ does not contain a cycle, then $Aut_{Z(R)}(R) = \{1\}$.*

The next theorem illustrates the group of ring-automorphisms of the polynomial ring $R[x]$ fixing $Z(R[x])$ pointwise.

**Theorem 3.3.23.** ([178, Corollary 2.9]) *Let $R$ be a commutative ring such that $\Gamma(R)$ contains a spanning tree which is a star graph, and let $x$ be an indeterminate. Then $Aut_{Z(R[x])}(R[x]) \neq \{1\}$.*

The next result illustrates the group of ring-automorphisms of the localization $R_{Z(R)}$ fixing $Z(R_{Z(R)})$ pointwise.

**Theorem 3.3.24.** ([178, Theorem 2.14]) *Assume that $P = Z(R)$ is a proper ideal of $R$ and $R_P$ is the localization of $R$ at $P$. Then $Aut_{Z(R_P)}(R_P) = \{1\}$ implies $Aut_{Z(R)}(R) = \{1\}$. Conversely, if $R/Z(R)$ is finitely generated as an algebra over its prime subring, then $Aut_{Z(R)}(R) = \{1\}$ implies $Aut_{Z(R_P)}(R_P) = \{1\}$. Moreover, $Aut_{Z(R_P)}(R_P) = \{1\}$ if and only if the quotient field $R/P$ is algebraic over its prime subfield.*

## 3.4 Bipartite Zero-divisor Graphs

In this section, we realize the zero-divisor graph as some special types of graphs. In a later part of the section, we discuss properties about the zero-divisor graph of the ring of Gaussian integers. Two of the most elementary forms that a connected graph can take are being complete or complete bipartite. In this section, we study the following question:

"For which positive integers $r$ does there exist a ring $R$ such that $\Gamma(R)$ is a complete $r$-partite graph?"

First of all, note that the complete graph $K_n$ is a complete $n$-partite graph. Recall that the completeness of the zero-divisor graph of a commutative ring was discussed by Anderson and Livingston [83], see Theorems 2.2.5 and 2.2.7.

**Theorem 3.4.1.** ([83, Theorems 2.8, 2.10])

(a) *Let $R$ be a commutative ring. Then $\Gamma(R)$ is complete if and only if either $R \cong \mathbb{Z}_2 \times \mathbb{Z}_2$ or $xy = 0$ for all $x, y \in Z(R)$;*

(b) *Let $R$ be a finite commutative ring. If $\Gamma(R)$ is complete, then either $R \cong \mathbb{Z}_2 \times \mathbb{Z}_2$ or $R$ is local with $char(R) = p$ or $p^2$, and $|\Gamma(R)| = p^n - 1$, where $p$ is prime and $n \geq 1$.*

Conversely, for each prime $p$ and integer $n \geq 1$, the ring $R = \mathbb{F}_{p^n}[x]/(x^2)$ has $\Gamma(R)$ complete with $|\Gamma(R)| = p^n - 1$ (see Example 2.2.8). Characterizations of complete zero-divisor graphs was continued by D. C. Lu and T. S. Wu in [267]. We call a local ring $R$ *$Z$-local* if $Z(R)^2 = 0$, i.e., $xy = 0$ for all $x, y \in Z(R)$. By the above, $\Gamma(R)$ is a finite complete graph with $|Z(R)| \geq 4$ if and only if $R$ is finite and $Z$-local. The following result gives some structure results about finite $Z$-local rings.

**Theorem 3.4.2.** ([267, Theorem 3.2]) *If $R$ is a finite $Z$-local ring (i.e., $\Gamma(R)$ is a finite complete graph with $|Z(R)| \geq 4$), then the following statements hold:*

(a) *$R/J(R)$ is isomorphic to $\mathbb{F}_{p^n}$ for some prime $p$ and positive integer $n$, where $\mathbb{F}_{p^n}$ is the field with $p^n$ elements;*

(b) *$J(R)$ is an $R/J(R)$-algebra, and in particular $|J(R)| = |R/J(R)|^m$ for some positive integer $m$;*

(c) *If $char(R/J(R)) = p$, then $pa = 0$ for any $a \in J(R)$;*

(d) *$char(R) = p$ or $char(R) = p^2$ for some prime $p$;*

(e) *Assume that $|R/J(R)| = p^m$. Then there exists a $\beta \in R$ such that $\beta^{p^m-1} = 1$ and $R/J(R) = \{0, 1, \bar{\beta}, \bar{\beta}^2, \ldots, \bar{\beta}^{p^m-2}\}$. Moreover, if further $char(R) = p$, then $A = \{0, 1, \beta, \beta^2, \ldots, \beta^{p^m-2}\}$ is a field as a subring of $R$.*

**Corollary 3.4.3.** ([267, Corollary 3.3,3.4])

(a) *Let $R$ be a finite ring with $char(R) = p$. Then $R$ is a finite $Z$-local ring if and only if $R \cong \dfrac{F[x_1, x_2, \ldots, x_n]}{(\{x_i x_j \mid 1 \leq i, j \leq n\})}$ for some finite field $F$ and positive integer $n$;*

(b) *Let $R$ be a finite ring with $char(R) = p^2$ and $R/J(R) = \mathbb{Z}_p$. Then $R \cong \dfrac{\mathbb{Z}_{p^2}[x_1, x_2, \ldots, x_{n-1}]}{(\{px_i, x_i x_j \mid 1 \leq i, j \leq n-1\})}$ for some positive integer $n$.*

Let $G$ be an infinite complete graph, and let $S$ be a set having the same cardinality as $V(G)$. Then for any field $F$, the ring $F[\{x_i \mid i \in S\}]/(\{x_i x_j \mid i, j \in S\})$ is a *corresponding ring of the graph $G$*. Therefore, every infinite complete graph has infinitely many corresponding rings. The following result determines all the corresponding rings with characteristic $p$ for any complete graph $K_n$, and in particular, gives all the corresponding rings of $K_n$ if $n+1 = p^q$ for some primes $p, q$.

**Theorem 3.4.4.** ([267, Theorem 3.6])

(a) *If $n + 1 = p^q$ for some primes $p, q$, then $K_n$ has exactly four corresponding rings;*

(b) *If $n+1 = p^m$ for some prime $p$, then $K_n$ has at least $d(m)+2$ corresponding rings, where $d(m)$ is the number of positive divisors of $m$.*

More generally, for each infinite cardinal number $\alpha$, let $F$ be a field with $|F| = \alpha$ and $R = F[x]/(x^2)$. Then $\Gamma(R)$ is complete and $|\Gamma(R)| = \alpha$. However, for infinite rings, $\Gamma(R)$ may be complete when $R$ is not quasilocal. For example, $R = \mathbb{Z}_4[x]$ is not quasilocal, but $\Gamma(R)$ is complete since $Z(R) = 2R$.

The study of complete bipartite zero-divisor graphs has been done by DeMeyer and Schneider [178]. Note that for integral domains $R_1$ and $R_2$, $\Gamma(R_1 \times R_2) = K_{m,n}$, where $m = |R_1| - 1$ and $n = |R_2| - 1$. The converse holds for finite rings except when the graph is either $K_{1,1}$ or $K_{1,2}$.

**Theorem 3.4.5.** ([83, Theorem 2.13], [178, Theorem 1.14]) *Let $R$ be a finite commutative ring. Then*

(a) $\Gamma(R)$ *is a star graph (i.e.,* $\Gamma(R) = K_{1,|\Gamma(R)|-1}$*) if and only if* $R \cong \mathbb{Z}_2 \times F$, *where $F$ is a finite field or $R$ is isomorphic to* $\mathbb{Z}_9$, $\dfrac{\mathbb{Z}_3[x]}{(x^2)}$, $\mathbb{Z}_8$, $\dfrac{\mathbb{Z}_2[x]}{(x^3)}$, *or* $\dfrac{\mathbb{Z}_4[x]}{(x^2-2,2x)}$;

(b) $\Gamma(R)$ *is complete bipartite if and only if* $R \cong F_1 \times F_2$, *where $F_1$ and $F_2$ are finite fields or $R$ is isomorphic to* $\mathbb{Z}_9$, $\dfrac{\mathbb{Z}_3[x]}{(x^2)}$, $\mathbb{Z}_8$, $\dfrac{\mathbb{Z}_2[x]}{(x^3)}$, *or* $\dfrac{\mathbb{Z}_4[x]}{(x^2-2,2x)}$.

The study of $r$-partite zero-divisor graphs of a commutative ring was discussed by Akbari, Maimani, and Yassemi in [37]. First note that for any prime number $p$ and any positive integer $n$, there exists a finite ring $R$ whose zero-divisor graph $\Gamma(R)$ is a complete $p^n$-partite graph. For example, if $\mathbb{F}_{p^n}$ is a finite field with $p^n$ elements, then $R = \dfrac{\mathbb{F}_{p^n}[x,y]}{(xy, y^2 - x)}$ is the desired ring.

The following theorem answers the question "For which positive integers $r$ does there exist a ring $R$ such that $\Gamma(R)$ is a complete $r$-partite graph?" in terms of $r = 2$ (i.e., bipartite graph).

**Theorem 3.4.6.** ([37, Theorem 2.4], [178, Theorem 1.14]) *Let $R$ be a reduced commutative ring. Then $\Gamma(R)$ is bipartite if and only if there exist two distinct prime ideals $I_1$ and $I_2$ of $R$ such that $I_1 \cap I_2 = \{0\}$. In addition, if $\Gamma(R)$ is bipartite, then it is a complete bipartite graph.*

The above theorem is true if one replaces the condition "reduced" by "the minimum degree of $\Gamma(R)$ is at least two".

Hence, hereafter in this section, assume that $r \geq 3$ and $V_1, \ldots, V_r$ are the $r$ parts of the complete $r$-partite graph $\Gamma(R)$.

**Theorem 3.4.7.** ([37, Theorem 3.1]) *Let $R$ be a commutative ring. If $\Gamma(R)$ is a complete $r$-partite graph with $r \geq 3$, then at most one part has more than one vertex. If $V_i$ is one of the parts such that $V_i = \{x\}$, then $x^2 = 0$. Further,*

*$Z(R) \in Max(R) \cap Ass(R)$, where $Max(R)$ denotes the set of maximal ideals of $R$ and $Ass(R)$ denotes the set of associated prime ideals of $R$.*

**Theorem 3.4.8.** ([37, Theorem 3.2]) *Let $R$ be a finite commutative ring, and let $\Gamma(R)$ be a complete $r$-partite graph with $r \geq 3$. Then the following statements hold:*

(a) *$R$ is a local ring;*

(b) *If $|V_r| \geq 2$, then there exists a prime integer $p$ and positive integers $t$ and $k$ such that $r = p^t$ and $|R| = p^k$;*

(c) *If $|V_r| \geq 2$, then for any $z \in V_r$, we have $z^2 \neq 0$ and $z^3 = 0$;*

(d) *For any $x \in V_r$ and for any $y \in Z(R)$, $xy \notin V_r$.*

**Theorem 3.4.9.** ([37, Lemma 3.3]) *Let $R$ be a finite commutative ring, $r \geq 2$, and $\Gamma(R)$ be a complete $r$-partite graph. Then $|Z(R)| \leq r^2$.*

**Theorem 3.4.10.** ([37, Theorem 3.4]) *Let $R$ be a commutative ring. If there is a prime integer $p \geq 3$ such that $\Gamma(R)$ is a finite complete $p$-partite graph, then $|Z(R)| = p^2$, $|R| = p^3$, and $R$ is isomorphic to one of the rings: $\mathbb{Z}_{p^3}$, $\dfrac{\mathbb{Z}_p[x, y]}{(xy, y^2 - x)}$, $\dfrac{\mathbb{Z}_{p^2}[x]}{(px, x^2 - ps)}$, where $1 \leq s < p$.*

**Theorem 3.4.11.** ([37, Theorem 3.5]) *Let $R$ be an infinite commutative ring and let $\Gamma(R)$ be a complete $r$-partite graph with $r \geq 3$. Then $Nil(R) = \bigcup_{i=1}^{r-1} V_i \cup \{0\}$ is a prime ideal and $|Nil(R)| = r = p^t$, where $p$ is a prime integer. In addition, for any $x \in V_r$, $Rx \subseteq V_r \cup \{0\}$.*

**Theorem 3.4.12.** ([37, Theorem 3.6]) *Let $R$ be a commutative ring such that the zero-divisor graph $\Gamma(R)$ is complete $r$-partite with $r \geq 3$. Then the following hold:*

(a) *If $R$ is Artinian, then $R$ is a finite local ring;*

(b) *If $R$ is Noetherian, then $R$ is a subring of a ring $F \times S$, where $F$ is a field and $S$ is a finite ring.*

There are infinite Noetherian rings whose zero-divisor graphs are complete $r$-partite. For example, the ring $R = \dfrac{\mathbb{Z}_p[x_1, \ldots, x_n]}{(x_1, \ldots, x_m)(x_1, \ldots, x_n)}$, where $m < n$. If $f(x_1, \ldots, x_n) \in \mathbb{Z}_p[x_1, \ldots, x_n]$ such that $f(0, \ldots, 0) = 0$, then $f + I$ is a zero-divisor, where $I = (x_1, \ldots, x_m)(x_1, \ldots, x_n)$. It is clear that if $g(x_1, \ldots, x_m) \in \mathbb{Z}_p[x_1, \ldots, x_m] \setminus I$ with $g(0, \ldots, 0) = 0$, then $Ann(g + I) = Z(R)$. The cardinality of the set $\{g + I \mid Ann(g + I) = Z(R)\}$ is $p^m - 1$. The elements of the set $V_r$ are $h + I$, where $h \in \mathbb{Z}_p[x_1, \ldots, x_n]$, $h(0, \ldots, 0) = 0$, and $h$ has at least a monomial as

$$x_{m+1}^{i_{m+1}} x_{m+2}^{i_{m+2}} \cdots x_n^{i_n}$$

such that for every $t$, $i_t \geq 0$ and $\sum i_t \neq 0$. If $f \in \mathbb{Z}_p[x_n]$ and $f(0) = 0$, then it is clear that $f + I \in V_r$, and hence $V_r$ is infinite. Thus $R$ is an infinite ring and $r = p^m$.

In [187], Anna Duane continued the work on this topic. The author actually concentrated on the $r$-partite structure of algebraically simple rings, namely $\mathbb{Z}_n$. Note that $\Gamma(\mathbb{Z}_{p^2}) = K_{p-1}$.

The following result exhibits that for all primes $p$, there is at least one complete $p$-partite graph which is realizable as a zero-divisor graph.

**Theorem 3.4.13.** ([187, Theorem 3.2]) *The graph $\Gamma(\mathbb{Z}_{p^3})$ is complete $p$-partite, where $p$ is prime.*

One can see in the available literature that complete bipartite structures are not only interesting when they comprise the entire graph, but also when they are merely a piece of the whole. In the following two results, one can get the underlying $p$-partite structures in $\mathbb{Z}_n$ for certain values of $n$.

**Theorem 3.4.14.** ([187, Theorem 3.3]) *The graph $\Gamma(\mathbb{Z}_{p^5})$, where $p$ is prime, has an induced complete $p$-partite subgraph.*

**Theorem 3.4.15.** ([187, Theorem 3.4]) *The graph $\Gamma(\mathbb{Z}_{p^q})$, where $p$ and $q$ are prime and $q$ is odd, has $\frac{q-1}{2}$ induced complete $p$-partite subgraphs.*

One of the limitations of Theorems 3.4.14 and 3.4.15 is that they deal with induced subgraphs. In order to form any nontrivial induced subgraph, vertices must be removed from the original graph. In the context of zero-divisor graphs, the removal of a vertex is tantamount to removing a zero-divisor within the ring. Therefore, the algebraic repercussions of these results in and of themselves might be minimal. One direction for further research would be to extend the above results to deal with subgraphs in which only edges are removed, as these graphical structures have proved much more useful in finding algebraic connections than the induced subgraphs used above.

## 3.5 Zero-divisor Graphs of Gaussian Integers

Several authors have studied the zero-divisor graph of $\mathbb{Z}_n = \{0, 1, \ldots, n-1\}$, the set of all residue classes modulo $n$. Such work was first discussed by Phillips, Rogers, Tolliver, and Worek in [309]. Recall that two elements $a$ and $b$ are said to be *associates* if there exists a unit $u$ such that $a = ub$. Note that if two elements of the vertex set of $\Gamma(\mathbb{Z}_n)$ are associates, then they have the same set of neighbors. Because of this, one can partition the elements of $V(\Gamma(\mathbb{Z}_n))$ in such a way that two vertices are in the same class if and only if they are associates. This partition divides the elements of $V(\Gamma(\mathbb{Z}_n))$ into associate classes. Let $A_v$ represent the associate class of $v$; that is, $A_v$ consists of all the associates of $v$. In the result given below, the authors are interested in the center of $\Gamma(\mathbb{Z}_n)$.

**Theorem 3.5.1.** ([309, Propositions 3.1, 3.2, 3.3]) *Let $n$ be a positive integer.*

(a) *If $n = 2p$ for a prime $p$, then the center of $\Gamma(\mathbb{Z}_n)$ is $p$;*

(b) *If $n = p^\alpha$ for a prime $p$, then the center of $\Gamma(\mathbb{Z}_n)$ is $A_{p^{\alpha-1}}$;*

(c) *Let $n = p_1^{\alpha_1} \cdots p_m^{\alpha_m}$ for distinct primes $p_1 < \cdots < p_m$, where $m > 2$, or if $m = 2$, then $n \neq 2p_2$. Also, let $v_i = n/p_i$ and $B = \{x \in \mathbb{Z}_n \mid p_i \mid x$ for all $i\}$. Then the center of $\Gamma(\mathbb{Z}_n)$ is the subgraph induced by $(\cup_{i=1}^m A_{v_i}) \cup B)$.*

Moreover, the authors showed that an associate class of a zero-divisor in $\mathbb{Z}_n$ is either a clique or an independent set in the graph $\Gamma(\mathbb{Z}_n)$. The corresponding result says that "every nonindependent associate class is a clique". The authors also characterized when an associate class is a clique or an independent set. The result in this regard says that "Let $n = p_1^{\alpha_1} \cdots p_m^{\alpha_m}$, where $p_1 < \cdots < p_m$ are distinct primes. Then an associate class $A_{p_1^{\beta_1} \cdots p_m^{\beta_m}}$ is a clique of $\Gamma(\mathbb{Z}_n)$ if and only if $\beta_1 \geq \frac{\alpha_1}{2} \geq \cdots \geq \beta_m \geq \frac{\alpha_m}{2}$".

Then, this work was continued by Córdova, Gholston, Hauser in [168]. In this work, the authors gave a formula to calculate the number of elements in an associate class, i.e., $|A_r| = \varphi(\frac{n}{gcd(n,r)})$. The authors are interested in the core of $\Gamma(\mathbb{Z}_n)$. A graph $G$ is a *core* if any homomorphism from $G$ to itself is an automorphism. Also, a subgraph $H$ of $G$ is called a core of $G$ if $H$ is a core itself and there is a homomorphism from $G$ to $H$. Every graph has at least one core, and in cases where a graph has more than one core, all of the cores are isomorphic.

**Theorem 3.5.2.** ([168, Theorem 5.1]) *The core of the graph $\Gamma(\mathbb{Z}_n)$ is the maximal clique in $\Gamma(\mathbb{Z}_n)$. In particular, $\Gamma(\mathbb{Z}_n)$ is a core if and only if $n = p^2$ for some prime $p$.*

Using the above result, the authors generalized the result to a general commutative ring.

**Theorem 3.5.3.** ([168, Theorem 5.3]) *Let $R$ be a commutative ring such that $\chi(\Gamma(R)) = \omega(\Gamma(R))$. Then the core of $\Gamma(R)$ is the maximal clique in $\Gamma(R)$.*

The study of zero-divisor graphs of the quotient ring $Q = \mathbb{Z}_p[x]/(x^n)$, where $p$ is prime and $n \geq 2$, was initiated by Endean, Henry, and Manlove in [192]. The elements of this ring are congruence classes of polynomials modulo the principal ideal $(x^n)$. For sake of simplifying notation, $q(x)$ will denote the congruence class of polynomials congruent to $q(x) \bmod (x^n)$. Thus an element in $Q$ is of the form $q(x) = q_0 + q_1x + q_2x^2 + \cdots + q_{n-1}x^{n-1}$, where $q_i \in \mathbb{Z}_p$ for each $i$. Note that an element $a(x) = a_0 + a_1x + \cdots + a_{n-1}x^{n-1} \in Z(Q)$, then $a_0 \equiv 0 \mod p$ and $|Z(Q)| = p^{n-1} - 1$.

**Theorem 3.5.4.** ([192, Theorem 2.3]) *The clique number of* $\Gamma(Q)$ *is* $p^{\frac{n}{2}} - 1$ *if* $n$ *even and* $p^{\frac{n-1}{2}}$ *if* $n$ *odd.*

**Corollary 3.5.5.** ([192, Corollary 2.4]) *The chromatic number of* $\Gamma(Q)$ *is equal to the clique number of* $\Gamma(Q)$.

Note that $\Gamma(Q)$ is similar to $\Gamma(\mathbb{Z}_{p^n})$, having the same clique number and chromatic number. In fact, it is proved that the graphs are isomorphic. To prove this, the authors used the following lemma.

**Lemma 3.5.6.** ([265, Lemma 2.5]) *Let* $b$ *be an integer greater than* 1. *Then every integer* $a > 0$ *can be uniquely represented in the form* $a = c_n b^n + c_{n-1} b^{n-1} + \cdots + c_1 b + c_0$ *with* $c_n \neq 0$, $n \geq 0$, *and* $0 \geq c_i < b$ *for* $i = 0, 1, 2, \ldots, n$.

**Theorem 3.5.7.** ([192, Theorem 2.6]) *If* $p$ *is prime, then* $\Gamma(\mathbb{Z}_{p^n}) \cong \Gamma(Q)$.

The zero-divisor graph of $\mathbb{Z}_{p^n}(x)$ was studied by AbuGhneim, AbdAlJawad, and Al-Ezeh in [12] and [13]. In [12], the goal of the authors was to find the clique number of the graph $\Gamma(\mathbb{Z}_{p^n}(x))$, where $p$ is a prime number. Let $\mathbb{Z}_{p^n}(x) = \{a + bx \mid a, b \in \mathbb{Z}_{p^n}$ and $x^2 = 0\}$. Note that the ring $\mathbb{Z}_{p^n}(x)$ can also be described using Nagata's principle of idealization. In general, for a commutative ring $R$ and an $R$-module $M$, the idealization of $M$ is the ring $R(+)M$ formed from $R \times M$ using $(r, a) + (s, b) = (r + s, a + b)$ and $(r, a)(s, b) = (rs, rb + sa)$. Hence $\mathbb{Z}_{p^n}(x) \cong \mathbb{Z}_{p^n}(+)\mathbb{Z}_{p^n}$.

Recall that if $R$ be a commutative ring with identity, then $a + bx$ is a zero-divisor in $R(x)$ if and only if $a$ is a zero-divisor in $R$. One can find the zero-divisors of $\mathbb{Z}_{p^n}$ by taking the numbers $1, 2, \ldots, p^{n-1} - 1$, and then multiplying these numbers by $p$. One can divide the zero-divisors into $n - 1$ sets according to how many factors of $p$ each divisor has. Further, one can categorize the zero-divisors of $\mathbb{Z}_{p^n}(x)$ into three types. In the first type, one has $n - 1$ sets, and these sets are $S_p, S_{p^2}, \ldots, S_{p^{n-1}}$, where $S_{p^i} = \{sp^i \mid gcd(s, p) = 1\}$. The Euler phi-function gives the sizes of the $S_{p^i}$'s, and one gets $|S_{p^i}| = p^{n-i} - p^{n-(i+1)}$ for $1 \leq i \leq n - 1$. The second type consists of $n$ sets, and these sets are $E_{p^0}, E_{p^1}, \ldots, E_{p^{n-1}}$, where $E_{p^i} = \{bx \mid b \in S_{p^i}\}$. Note that $S_{p^0}$ is the set of units in $\mathbb{Z}_{p^n}$. We have $|E_{p^i}| = |S_{p^i}| = p^{n-i} - p^{n-(i+1)}$ for $1 \leq i \leq n-1$ and $|E_{p^0}| = p^n - p^{n-1}$. The third type consists of $n(n-1)$ sets. These sets are $S_{p^i,p^j}$, $i \in \{1, 2, \ldots, n-1\}$ and $j \in \{1, 2, \ldots, n-1\}$, where $S_{p^i,p^j} = \{sp^i + tp^j x \mid gcd(s, p) = 1$ and $gcd(t, p) = 1\}$. Again, the Euler phi-function gives the sizes of the $S_{p^i,p^j}$'s, and one gets $|S_{p^i,p^j}| = (p^{n-i} - p^{n-(i+1)})(p^{n-j} - p^{n-(j+1)})$ for $1 \leq i \leq n - 1$ and $0 \leq j \leq n - 1$.

The following few results will be used in calculating the clique number of $\Gamma(\mathbb{Z}_{p^n}(x))$, where $p$ is an odd prime.

**Lemma 3.5.8.** ([12, Lemma 2]) *Let* $p$ *be an odd prime. Suppose that* $y_1 \in S_{p^i,p^j}$, $y_2 \in S_{p^\ell,p^m}$, *and* $y_3 \in S_{p^u,p^v}$, *with* $i + \ell \geq n$, $i + u \geq n$, $\ell + u \geq n$, *and* $i + m = j + \ell = i + v = u + j = \ell + v = u + m \leq n - 1$. *Moreover, assume* *that $y_1, y_2$, and $y_3$ are distinct elements and $S_{p^i,p^j}$, $S_{p^\ell,p^m}$, and $S_{p^u,p^v}$ are not necessarily distinct sets. Then $\{y_1, y_2, y_3\}$ cannot form a complete subgraph of $\Gamma(\mathbb{Z}_{p^n}(x))$.*

This lemma tells us that we cannot have a complete subgraph of order greater than two in $S_{p^i,p^j} \cup S_{p^\ell,p^m} \cup S_{p^u,p^v} \subseteq \Gamma(\mathbb{Z}_{p^n}(x))$, with $i+\ell \geq n$, $i+u \geq n$, $\ell+u \geq n$, and $i+m = j+\ell = i+v = u+j = \ell+v = u+m \leq n-1$.

Let $p \geq 3$ be a prime and $n > 1$ be an integer. Let

$$L_m = \left( \bigcup_{i=\lceil \frac{n}{2} \rceil + m}^{n-1} \bigcup_{j=\lfloor \frac{n}{2} \rfloor - m}^{n-1} S_{p^i,p^j} \right) \cup \left( \bigcup_{k=\lceil \frac{n}{2} \rceil + m}^{n-1} S_{p^k} \right) \cup \left( \bigcup_{\ell=\lfloor \frac{n}{2} \rfloor - m}^{n-1} E_{p^\ell} \right)$$

with $m \in \{0, 1, \ldots, \lfloor \frac{n}{2} \rfloor - 1\}$ and $N = \bigcup_{\ell=0}^{n-1} E_{p^\ell}$. Observe that each of the $L_m$'s is a complete subgraph of $\Gamma(\mathbb{Z}_{p^n}(x))$ and $N$ is also a complete subgraph of $\Gamma(\mathbb{Z}_{p^n}(x))$.

The authors presented a sequence of results that will be used in showing that the $L_m$'s and $N$ have maximum cardinality among all complete subgraphs of $\Gamma(\mathbb{Z}_{p^n}(x))$.

**Lemma 3.5.9.** ([12, Lemma 4]) *Suppose that $H$ is a complete subgraph of $\Gamma(\mathbb{Z}_{p^n}(x))$ that contains an element of $\bigcup_{k=1}^{\lceil \frac{n}{2} \rceil - 1} S_{p^k}$, where $p$ is an odd prime and $n > 1$ is an integer. Then $|H| < p^n - 1$. Moreover, $H$ cannot have the maximum cardinality among all complete subgraphs of $\Gamma(\mathbb{Z}_{p^n}(x))$.*

**Lemma 3.5.10.** ([12, Lemma 5]) *Suppose that $H$ is a complete subgraph of $\Gamma(\mathbb{Z}_{p^n}(x))$ that contains an element of $\bigcup_{i=1}^{\lceil \frac{n}{2} \rceil - 1} \bigcup_{j=0}^{n-1} S_{p^i,p^j}$, where $p$ is an odd prime and $n > 1$ is an integer. Then $|H| < p^n - 1$. Moreover, $H$ cannot have the maximum cardinality among all complete subgraphs of $\Gamma(\mathbb{Z}_{p^n}(x))$.*

**Lemma 3.5.11.** ([12, Lemma 6]) *Suppose that $H$ is a complete subgraph of $\Gamma(\mathbb{Z}_{p^n}(x))$ ($p$ is an odd prime and $n > 1$ is an integer) that contains an element of the form $z = ap^i + bp^j x$ with $gcd(a,p) = gcd(b,p) = 1$, $i \geq \lceil \frac{n}{2} \rceil$, and $i + j < n$. Then $|H| < p^n - 1$. Moreover, $H$ cannot have the maximum cardinality among all complete subgraphs of $\Gamma(\mathbb{Z}_{p^n}(x))$.*

**Lemma 3.5.12.** ([12, Lemma 7]) *Let $p$ be an odd prime and $n > 1$ be an integer. If $M = N \cup \{y\}$, where $y \notin N$, or $M = L_m \cup \{y\}$, where $y \notin L_m$ and $m \in \{0, 1, \ldots, \lfloor \frac{n}{2} \rfloor - 1\}$, then $M$ is not a complete subgraph of $\Gamma(\mathbb{Z}_{p^n}(x))$.*

The following result gives the complete subgraphs in $\Gamma(\mathbb{Z}_{p^n}(x))$ that achieve the clique number.

**Theorem 3.5.13.** ([12, Theorem 1]) *If $p$ is an odd prime and $n > 1$ is an integer, then every one of the complete subgraphs $L_m$, where $m \in \{0, 1, \ldots, \lfloor \frac{n}{2} \rfloor - 1\}$, and $N$ has the maximum cardinality among all complete subgraphs of $\Gamma(\mathbb{Z}_{p^n}(x))$.*

Thus, to find the clique number of $\Gamma(\mathbb{Z}_{p^n}(x))$, we want to calculate the sizes of the $L_m$'s and $N$. One can view the elements of $L_m$ in the plane $\mathbb{Z}_{p^n} \times \mathbb{Z}_{p^n}$. The entire set consists of the nonzero points of the form $(ap^s, bp^t)$, where $s = \lceil \frac{n}{2} \rceil + m$, $0 \leq a < p^{n-s}$, $t = \lfloor \frac{n}{2} \rfloor - m$, and $0 \leq b < p^{n-t}$, (but not both $a$ and $b$ are zero). Adding $(0, 0)$ gives a rectangular grid. Thus $|L_m| = p^{2n-(s+t)} - 1 = p^n - 1$. For $N$, it is clear that $N$ corresponds to the nonzero elements of $\mathbb{Z}_{p^n}$, and hence $|N| = p^n - 1$. Thus the clique number of $\Gamma(\mathbb{Z}_{p^n}(x))$ is $p^n - 1$. Indeed, Theorem 3.5.13 shows that the only complete subgraphs with maximum cardinality are $N$ and $L_m$, where $m \in \{0, 1, \ldots, \lfloor \frac{n}{2} \rfloor - 1\}$.

The study of the zero-divisor graph of $\mathbb{Z}_{p^n}(x)$ was continued by AbuGhneim, AbdAlJawad, and Al-Ezeh in [13]. In [13], the goal of the authors was to compute the independence number of $\Gamma(\mathbb{Z}_{p^n}(x))$, where $p$ is a prime number. Recall that a subset of the vertex set in a graph $G$ is said to be *independent* if no two vertices in the set are adjacent. The *independence number*, denoted by $\beta(G)$, is the maximum cardinality of an independent set in $G$.

Suppose that $I$ is an independent set with maximum cardinality in $\Gamma(\mathbb{Z}_{p^n}(x))$. The authors of [13] presented a series of results that determine the elements of the set $I$.

**Lemma 3.5.14.** ([13, Lemma 3.1]) *Suppose that $I$ is an independent set with maximum cardinality in $\Gamma(\mathbb{Z}_{p^n}(x))$. Then $I$ contains exactly one element of the set $\bigcup_{i=0}^{n-1} E_{p^i}$. Moreover, $I$ can be chosen so that this element is in $E_{p^0}$.*

**Lemma 3.5.15.** ([13, Lemma 3.2]) *Suppose that $I$ is an independent set with maximum cardinality in $\Gamma(\mathbb{Z}_{p^n}(x))$, where $n \geq 3$. Then $I$ cannot contain any element of the set $\bigcup_{i=\lfloor \frac{n}{2} \rfloor+1}^{n-1} S_{p^i}$.*

**Lemma 3.5.16.** ([13, Lemma 3.3]) *Suppose that $I$ is an independent set with maximum cardinality in $\Gamma(\mathbb{Z}_{p^n}(x))$, where $n \geq 3$. Then $I$ contains no elements of the set $\bigcup_{i=\lfloor \frac{n}{2} \rfloor+1}^{n-1} \bigcup_{j=\lfloor \frac{n}{2} \rfloor+1}^{n-1} S_{p^i,p^j}$.*

**Lemma 3.5.17.** ([13, Lemma 3.4]) *Suppose that $I$ is an independent set with maximum cardinality in $\Gamma(\mathbb{Z}_{p^n}(x))$, where $n \geq 3$. Then* $\bigcup_{i=1}^{\lfloor \frac{n-1}{2} \rfloor} S_{p^i}$ *is a subset of $I$.*

**Lemma 3.5.18.** ([13, Lemma 3.5]) *Suppose that $I$ is an independent set with maximum cardinality in $\Gamma(\mathbb{Z}_{p^n}(x))$, where $n \geq 3$. Then* $I \cap \bigcup_{i=\lfloor \frac{n}{2} \rfloor+1}^{n-1} \bigcup_{j=i-\lfloor \frac{n-1}{2} \rfloor}^{n-1} S_{p^i,p^j}$ $= \emptyset$ *and* $\bigcup_{i=1}^{\lfloor \frac{n-1}{2} \rfloor} \bigcup_{j=0}^{n-1} S_{p^i,p^j}$ *is a subset of $I$.*

**Lemma 3.5.19.** ([13, Corollary 3.1]) *Suppose that $I$ is an independent set with maximum cardinality in $\Gamma(\mathbb{Z}_{p^n}(x))$, where $n \geq 3$. Then $I$ contains exactly*

$$p^{2n-\lfloor \frac{n-1}{2} \rfloor -1}\left(p^{\lfloor \frac{n-1}{2} \rfloor} - 1\right) + 1$$

*elements from the set*

$$\bigcup_{i=1}^{\lfloor \frac{n-1}{2} \rfloor} S_{p^i} \cup \bigcup_{i=1}^{\lfloor \frac{n-1}{2} \rfloor} \bigcup_{j=0}^{n-1} S_{p^i,p^j} \cup \bigcup_{i=\lfloor \frac{n}{2} \rfloor+1}^{n-1} S_{p^i} \cup \bigcup_{i=\lfloor \frac{n}{2} \rfloor+1}^{n-1} \bigcup_{j=i-\lfloor \frac{n-1}{2} \rfloor}^{n-1} S_{p^i,p^j} \cup \bigcup_{i=0}^{n-1} E_{p^i}.$$

**Lemma 3.5.20.** ([13, Lemma 3.6]) *If $0 \leq j \leq n_1$, then the induced subgraph on $S_{p^{2n_1+2+j},p^j}$ is isomorphic to the disjoint union of $\dfrac{p^{2n_1+1-2j} - p^{2n_1-2j}}{2}$ copies of the complete bipartite graph $K_{p^{4n_1+2}(p-1),\ p^{4n_1+2}(p-1)}$. For $n_1 + 1 \leq j \leq 2n_1$, the induced subgraph on $S_{p^{2n_1+2+j},\ p^j}$ is complete. Moreover, $I$ contains at most $\dfrac{p^{4n_1+2}(p-1)(p^{2n_1+2}-1)}{2(p+1)}$ elements of $\bigcup_{j=0}^{n_1} S_{p^{2n_1+2+j},\ p^j}$.*

The following five theorems give the exact value of the independence number of $\Gamma(\mathbb{Z}_{p^n}(x))$.

**Theorem 3.5.21.** ([13, Theorem 3.1]) *Suppose that $n = 4n_1 + 3$, where $n_1$ is a positive integer. Then the independence number of $\Gamma(\mathbb{Z}_{p^n}(x))$ is equal to*

$$\frac{(1-p)p^{4n_1+3}}{2(p+1)}\left(\frac{n_1+1}{p} + n_1 - \frac{(1-p^{2n_1+1})p}{1-p}\right) + p^{6n_1+4}(p^{2n_1+1} - 1) + n_1 + 1.$$

**Theorem 3.5.22.** ([13, Theorem 3.2]) *Suppose that* $n = 4n_1 + 1$, *where* $n_1$ *is a positive integer. Then the independence number of* $\Gamma(\mathbb{Z}_{p^n}(x))$ *is equal to*

$$\frac{(p-1)p^{4n_1+1}}{2(p+1)}\left(\frac{(p^{2n_1}-1)p}{p-1} - (n_1 + \frac{n_1}{p})\right) + n_1 + p^{6n_1+1}(p^{2n_1}-1) + 1.$$

**Theorem 3.5.23.** ([13, Theorem 3.3]) *Suppose that* $n = 4n_1 + 2$, *where* $n_1$ *is a positive integer. Then the independence number of* $\Gamma(\mathbb{Z}_{p^n}(x))$ *is equal to*

$$\frac{(p-1)p^{4n_1+2}}{2(p+1)}\left(\frac{-(n_1+1)}{p-n_1} + \frac{(p^{2n_1+1}-1)p}{p-1}\right) + n_1 + p^{6n_1+3}(p^{2n_1-1}) + 1 +$$

$$\frac{p^{4n_1+1}(p^{2n_1}-1)(p-1)}{2} + 1.$$

**Theorem 3.5.24.** ([13, Theorem 3.4]) *Suppose that* $n = 4n_1$ *where* $n_1$ *is a positive integer. Then the independence number of* $\Gamma(\mathbb{Z}_{p^n}(x))$ *is equal to*

$$\frac{(p-1)p^{4n_1+2}}{2(p+1)}\left(\frac{-n_1}{p-n_1} + \frac{(p^{2n_1}-1)p}{p-1}\right) + n_1 + p^{6n_1}(p^{2n_1-1}-1) + 1 +$$

$$\frac{p^{4n_1-1}(p^{2n_1-1}-1)(p-1)}{2} + 1.$$

**Theorem 3.5.25.** ([13, Theorem 3.5]) *The independence number of* $\Gamma(\mathbb{Z}_{p^n}(x))$, *where* $n = 1, 2, 3$ *is* $1, \frac{(p^2-p)(p-1)}{2} + 1, (p^2-p)p^3 + \frac{(p^3-p^2)(p-1)}{2} + 1$, *respectively.*

The zero-divisor graph of the Gaussian integers modulo $n$ was studied by Osba et al. in [306, 305]. The set of all complex numbers $a + bi$, where $a$ and $b$ are integers, forms a Euclidean domain with the usual complex number operations and Euclidean norm $N(a + bi) = a^2 + b^2$. This domain will be denoted by $\mathbb{Z}[i]$ and will be called the *ring of Gaussian integers*. It is clear that $a + bi$ is a unit in $\mathbb{Z}[i]$ if and only if $N(a + bi) = 1$, which implies that the only units in $\mathbb{Z}[i]$ are $1, -1, i$, and $-i$. The prime integers are $a = 2$, $a \equiv 1 \ (mod\ 4)$, and $a \equiv 3 \ (mod\ 4)$. A prime element in $\mathbb{Z}[i]$ will be called a *Gaussian prime*. The Gaussian primes can be described as follows:

(1) $1 + i$ and $1 - i$ are Gaussian primes;

(2) If $q$ is a prime integer with $q \equiv 3 \ (mod\ 4)$, then $q$ is a Gaussian prime;

(3) If $p$ is a prime integer with $p \equiv 1 \ (mod\ 4)$ and $p = a^2 + b^2$ for some integers $a$ and $b$, then $a + bi$ and $a - bi$ are Gaussian primes.

Moreover, if $\alpha$ is a Gaussian prime, then its complex conjugate and its associates are also Gaussian primes. The quotient ring $\mathbb{Z}[i]/(n) \cong \mathbb{Z}_n[i] = \{a + bi \mid a, b \in \mathbb{Z}_n\}$ is called the *ring of Gaussian integers modulo n*. Note that $\mathbb{Z}_n[i]$ is a principal ideal ring.

Note that 2 is not a Gaussian prime, since $2 = (1 + i)(1 - i)$; however, $2 = -i(1 + i)^2$, so $\mathbb{Z}_2[i]$ is isomorphic to the local ring $\mathbb{Z}[i]/(1 + i)^2$ with its only maximal ideal $\{0, 1 + i\}$. Also $1 - i = -i(1 + i)$, and so they generate the same maximal ideal. Thus $V(\Gamma(\mathbb{Z}_2[i])) = \{1 + i\}$, which implies that $\Gamma(\mathbb{Z}_2[i])$ is the empty graph.

Let $n$ be an integer greater than 1. Then $2^n = (-i)^n(1+i)^{2n}$, and so $\mathbb{Z}_{2^n}[i] \cong \mathbb{Z}[i]/(1 + i)^{2n}$. Hence $\mathbb{Z}_{2^n}[i]$ is local with its only maximal ideal $\mathfrak{m} = (1 + i) = \{a + bi \mid a \text{ and } b \text{ are both even or odd}\}$. Since $\mathbb{Z}_{2^n}[i]$ is a local ring, $Z(\mathbb{Z}_{2^n}[i])$ is an annihilator ideal, and so there exists a vertex adjacent to every vertex in $\Gamma(\mathbb{Z}_{2^n}[i])$. Also note that $|V(\Gamma(\mathbb{Z}_{2^n}[i]))| = 2^{2n-1} - 1$ for all $n \geq 1$.

**Theorem 3.5.26.** ([306, Theorem 4,5]) *For $n > 1$, $diam(\Gamma(\mathbb{Z}_{2^n}[i])) = 2$ and $gr(\Gamma(\mathbb{Z}_{2^n}[i])) = 3$.*

For the remainder of this section, when we write $q \equiv 3 \ (mod\ 4)$ or $p \equiv 1 \ (mod\ 4)$, we assume that $p$ and $q$ are prime integers. If $q \equiv 3 \ (mod\ 4)$, then $q$ is a Gaussian prime, and so $\mathbb{Z}_q[i]$ is a splitting field for the polynomial $g(x) = x^2+1$ over the field $\mathbb{Z}_q$. Thus $\mathbb{Z}_q[i]$ is a field. So in this case, $\mathbb{Z}_q[i]$ has no nonzero zero-divisors. If $n > 1$, then $\mathbb{Z}_{q^n}[i]$ is a local ring with maximal ideal $(q)$. Note that $|V(\Gamma(\mathbb{Z}_{q^n}[i]))| = q^{2n-2} - 1$ for all $n > 1$.

**Theorem 3.5.27.** ([306, Theorem 7,8]) *For $q \equiv 3 \ (mod\ 4)$ and $n > 1$, $gr(\Gamma(\mathbb{Z}_{q^n}[i])) = 3$ and*

$$diam(\Gamma(\mathbb{Z}_{q^n}[i])) = \begin{cases} 1 & \text{if } n = 2; \\ 2 & \text{if } n > 2. \end{cases}$$

Let $p \equiv 1 \ (mod\ 4)$. Then there is an integer $c$ such that $c^2 \equiv -1 \ (mod\ p)$, and there also exist $a, b \in \mathbb{Z}$ such that $p = a^2 + b^2 = (a + bi)(a - bi)$. Moreover, $a + bi$ and $a - bi$ are Gaussian primes in $\mathbb{Z}[i]$. Thus the ideals $(a + bi)$ and $(a - bi)$ are the only maximal ideals in $\mathbb{Z}[i]$ containing $p$ since $\mathbb{Z}[i]$ is a unique factorization domain. Hence we have $\mathbb{Z}_p[i] \cong \dfrac{\mathbb{Z}[i]}{(a + bi)} \times \dfrac{\mathbb{Z}[i]}{(a - bi)}$. Thus the ideals $(a + bi)$ and $(a - bi)$ are the only maximal ideals in $\mathbb{Z}_p[i]$. Note that $|V(\Gamma(\mathbb{Z}_p[i]))| = 2p - 2$ and $\Gamma(\mathbb{Z}_p[i]) = K_{p-1,p-1}$.

Let $p \equiv 1 \ (mod\ 4)$, $n > 1$, and $p = a^2 + b^2$. Then $p^n = (a^2 + b^2)^n = (a + bi)^n(a - bi)^n$, and therefore the ideals $(a + bi)$ and $(a - bi)$ are the only maximal ideals in $\mathbb{Z}[i]$ containing $p^n$. Hence we have $\mathbb{Z}_p[i] \cong \dfrac{\mathbb{Z}[i]}{(a + bi)^n} \times \dfrac{\mathbb{Z}[i]}{(a - bi)^n}$. Note that $|V(\Gamma(\mathbb{Z}_{p^n}[i]))| = 2p^{2n-1} - p^{2n-2} - 1$ for all $n > 1$.

**Theorem 3.5.28.** ([306, Theorems 10, 11]) *Let* $p \equiv 1 \ (mod\ 4)$ *and* $n > 1$. *Then* $diam(\Gamma(\mathbb{Z}_{p^n}[i])) = 3$ *and* $gr(\Gamma(\mathbb{Z}_{p^n}[i])) = 3$.

Let $n = 2^k \times \prod_{j=1}^{m} q_j^{\alpha_j} \times \prod_{s=1}^{\ell} p_s^{\beta_s}$, where $q_j \equiv 3 \ (mod\ 4)$ for $j = 1, 2, \ldots, m$ and $p_s \equiv 1 \ (mod\ 4)$ for $s = 1, 2, \ldots, \ell$. Then

$$|V(\Gamma(\mathbb{Z}_n[i]))| = n - \left(2^{2k-1} \times \prod_{j=1}^{m}(q_j^{2\alpha_j} - q_j^{2\alpha_j - 2}) \times \prod_{s=1}^{\ell}(p_s^{\beta_s} - p_s^{\beta_s - 1})^2\right) - 1.$$

**Theorem 3.5.29.** ([306, Theorem 13]) *Let* $n$ *be a positive integer greater than* 1, $q \equiv 3 \ (mod\ 4)$, *and* $p \equiv 1 \ (mod\ 4)$. *Then*

(a) $diam(\Gamma(\mathbb{Z}_n[i])) = 1$ *if and only if* $n = q^2$;

(b) $diam(\Gamma(\mathbb{Z}_n[i])) = 2$ *if and only if* $n = p$ *or* $n = 2^m$ *with* $m \geq 2$ *or* $n = q^m$ *with* $m \geq 3$;

(c) $diam(\Gamma(\mathbb{Z}_n[i])) = 3$ *if and only if* $m = p^m$ *with* $m \geq 2$ *or* $n$ *is divisible by at least two distinct primes.*

**Theorem 3.5.30.** ([306, Theorem 14]) *Let* $n$ *be a positive integer greater than* 1 *and* $q_i \equiv 3 \ (mod\ 4)$ *for* $i = 1, 2$. *Then*

$$gr(\Gamma(\mathbb{Z}_n[i])) = \begin{cases} 4 & if \ \ n = 2q_1 \ \ or \ \ n = q_1 \cdot q_2; \\ 3 & otherwise. \end{cases}$$

Next, the authors of [306] were interested in characterizing all values of $n$ for which $\Gamma(\mathbb{Z}_n[i])$ is either complete, complete bipartite, or a star graph.

**Theorem 3.5.31.** ([306, Theorem 15]) *The graph* $\Gamma(\mathbb{Z}_n[i])$ *is complete if and only if* $n = q^2$, *where* $q \equiv 3 \ (mod\ 4)$.

**Theorem 3.5.32.** ([306, Theorem 17]) *Let* $R = R_1 \times R_2$. *Then* $\Gamma(R)$ *is a complete bipartite graph if and only if* $R_1$ *and* $R_2$ *are integral domains. In particular, the graph* $\Gamma(\mathbb{Z}_n[i])$ *is complete bipartite if and only if* $n = p$ *or* $n = q_1q_2$, *where* $q_i \equiv 3 \ (mod\ 4)$ *for* $i = 1, 2$ *and* $p \equiv 1 \ (mod\ 4)$.

**Theorem 3.5.33.** ([306, Theorem 18]) *For each* $n$, $\Gamma(\mathbb{Z}_n[i])$ *is never a star graph.*

Osba et al. [306] completely determined the regular and planar zero-divisor graphs in the class of $\mathbb{Z}_n[i]$.

**Theorem 3.5.34.** ([306, Theorem 25]) *The graph* $\Gamma(\mathbb{Z}_n[i])$ *is regular if and only if* $n = 2$, $n = p$, *or* $n = q^2$, *where* $q \equiv 3 \ (mod\ 4)$ *and* $p \equiv 1 \ (mod\ 4)$.

**Theorem 3.5.35.** ([306, Theorem 22]) *The graph* $\Gamma(\mathbb{Z}_n[i])$ *is planar if and only if* $n = 2$ *or* 4.

In [305], Osba, Al-Addasi, and Al-Khamaiseh continued the study of the zero-divisor graph of the Gaussian integers modulo $n$. In the following result, the authors determine all values of $n$ for which $\Gamma(\mathbb{Z}_n[i])$ is bipartite.

**Theorem 3.5.36.** ([305, Theorem 17]) *For any integer $n > 2$, the graph $\Gamma(\mathbb{Z}_n[i])$ is bipartite if and only if $n = p$, $n = 2q_1$, or $n = q_1q_2$, where $q_i \equiv 3 \pmod 4$ for $i = 1, 2$ and $p \equiv 1 \pmod 4$.*

The following result generalizes Theorem 3.5.33 to a general tree.

**Theorem 3.5.37.** ([305, Corollary 18]) *For any integer $n > 2$, the graph $\Gamma(\mathbb{Z}_n[i])$ is not a tree.*

Then, Osba et al. [305] were interested in the radius of $\Gamma(\mathbb{Z}_n[i])$. Recall that the *eccentricity* of a vertex $v$ of a connected graph $G$ is the distance between $v$ and a vertex farthest from $v$. The minimum eccentricity among the vertices of $G$ is its *radius*, which is denoted by $rad(G)$.

**Theorem 3.5.38.** ([305, Corollaries 12,13 and 14]) *For any integers $n > 1$, $m > 1$ with $n \neq q$ for any $q \equiv 3 \pmod 4$,*

$$rad(\Gamma(\mathbb{Z}_n[i])) = \begin{cases} 0 & \text{if } n = 2\,; \\ 1 & \text{if } n = 2^m \quad \text{or } n = q^m\,; \\ 2 & \text{otherwise.} \end{cases}$$

The coloring of $\Gamma(\mathbb{Z}_n[i])$ was also discussed by Osba et al.[305]. The following theorem gives the general formula for the chromatic number of $\Gamma(\mathbb{Z}_n[i])$.

**Theorem 3.5.39.** ([305, Corollary 26]) *Let $n = 2^\ell \times (\prod_{k=1}^{r} q_k^{m_k} \times \prod_{k=r+1}^{t} q_k^{m_k}) \times (\prod_{k=1}^{z} p_k^{n_k} \times \prod_{k=z+1}^{c} p_k^{n_k})$, where $q_k \equiv 3 \pmod 4$ for $k = 1, 2, \ldots, t$ and $p_k \equiv 1 \pmod 4$ for $k = 1, 2, \ldots, c$, $m_k$ is odd for all $k \leq r$ while $m_k$ is even otherwise, $n_k$ is odd for all $k \leq z$ while $n_k$ is even otherwise, and let $s = 2^\ell \times (\prod_{k=1}^{t} q_k^{2\lfloor \frac{m_k}{2} \rfloor} \times \prod_{k=1}^{c} p_k^{2\lfloor \frac{m_k}{2} \rfloor}$. Then $\chi(\Gamma(\mathbb{Z}_n[i])) = s + r + 2z - 1$.*

The complement of the zero-divisor graph of $\mathbb{Z}_n$ was studied by Brickel in [153]. Even though the author did not determine a formula for the coloring of the complement of $\Gamma(\mathbb{Z}_n)$; he proved that the chromatic number of $C\Gamma(\mathbb{Z}_n)$ equals the clique number of $C\Gamma(\mathbb{Z}_n)$ for the cases $n = p^e$ and $n = p_1p_2p_3$, where $p$ and $p_1 < p_2 < p_3$ are primes. Then the coloring of $C\Gamma(\mathbb{Z}_n)$ was continued by Córdova, Gholston, and Hauser in [168].

**Theorem 3.5.40.** ([153], [168, Propositions 6.3 and 6.4]) *Let $n = p^e$ or $p_1p_2p_3$, where $p$ and $p_1 < p_2 < p_3$ are primes. Then $\chi(C\Gamma(\mathbb{Z}_n)) = \omega(C\Gamma(\mathbb{Z}_n))$.*

**Theorem 3.5.41.** ([168, Propositions 6.5, 6.6, and 6.8]) *Let* $n = p_1p_2$, $p_1^2p_2$, *or* $p_1^2p_2^2$, *where* $p_1$ *and* $p_2$ *are distinct primes. Then* $\chi(C\Gamma(\mathbb{Z}_n)) = \omega(C\Gamma(\mathbb{Z}_n))$.

The above result was extended to all finite reduced commutative rings by Sarika Devhare et al. in [182, Corollary 4.1]. In fact, they proved that both the zero-divisor graph and its complement of reduced commutative rings are weakly perfect. Their result is given below.

**Theorem 3.5.42.** ([182, Corollary 4.1]) *Let* $R$ *be a reduced commutative ring such that* $Z(R) \neq \{0\}$. *If* $\omega(C\Gamma(R)) < \infty$, *then* $\Gamma(R)$ *and* $C\Gamma(R)$ *are weakly perfect.*

The properties of the complement of $\Gamma(\mathbb{Z}_n)$ were discussed by Phillips, Rogers, Tolliver, and Worek in [309]. Let $A_v$ represent the associate class of $v$; that is, $A_v$ consists of all the associates of $v$. The authors were first interested in the center of $C\Gamma(\mathbb{Z}_n)$.

**Theorem 3.5.43.** ([309, Propositions 3.5 and 3.6]) *Let* $n$ *be a positive integer and* $p_1, \ldots, p_m$ *distinct primes.*

(a) *Let* $n = p_1^{\alpha_1} \cdots p_m^{\alpha_m}$, *where* $m \geq 3$. *Then* $C\Gamma(\mathbb{Z}_n)$ *is self-centered;*

(b) *Let* $n = p_1^{\alpha_1}p_2^{\alpha_2}$, *where* $\alpha_1 > 1$ *or* $\alpha_2 > 1$, *and let* $v_i = n/p_i$ *for* $i = 1, 2$. *Then the center of* $C\Gamma(\mathbb{Z}_n)$ *is the subgraph induced by* $V(C\Gamma(\mathbb{Z}_n)) \setminus (A_{v_1} \cup A_{v_2})$.

Next, Phillips et al. [309] determine when $C\Gamma(\mathbb{Z}_{n_1} \times \cdots \times \mathbb{Z}_{n_r})$ is planar, where $n_1, \ldots, n_r \in \mathbb{N}$.

**Theorem 3.5.44.** ([309, Theorem 4.1]) $C\Gamma(\mathbb{Z}_{n_1} \times \cdots \times \mathbb{Z}_{n_r})$ *is planar if and only if* $\mathbb{Z}_{n_1} \times \cdots \times \mathbb{Z}_{n_r}$ *is isomorphic to* $\mathbb{Z}_{p^2}$, $\mathbb{Z}_2 \times \mathbb{Z}_2$, $\mathbb{Z}_2 \times \mathbb{Z}_3$, $\mathbb{Z}_8$, $\mathbb{Z}_2 \times \mathbb{Z}_4$, $\mathbb{Z}_2 \times \mathbb{Z}_2 \times \mathbb{Z}_2$, $\mathbb{Z}_3 \times \mathbb{Z}_3$, $\mathbb{Z}_2 \times \mathbb{Z}_5$, $\mathbb{Z}_4 \times \mathbb{Z}_3$, *or* $\mathbb{Z}_3 \times \mathbb{Z}_5$ *for some prime* $p$.

Then the authors of [309] looked at vertices of minimum degree in $C\Gamma(\mathbb{Z}_n)$ and the connectivity of $C\Gamma(\mathbb{Z}_n)$.

**Theorem 3.5.45.** ([309, Theorems 6.9 and 7.1]) *Let* $n = p_1^{\alpha_1} \cdots p_m^{\alpha_m}$ *be the prime decomposition of* $n$ *with* $p_1 < \cdots < p_m$. *Then a vertex of minimum degree in* $C\Gamma(\mathbb{Z}_n)$ *is* $n/p_1$. *Moreover,* $\delta(C\Gamma(\mathbb{Z}_n)) = p_1^{\alpha_1-1}(p_1-1)p_2^{\alpha_2} \cdots p_m^{\alpha_m} - p_1^{\alpha_1-1}(p_1-1)p_2^{\alpha_2-1}(p_2-1) \cdots p_m^{\alpha_m-1}(p_m-1)$.

**Theorem 3.5.46.** ([309, Theorems 7.2, 7.3]) *Let* $n = p_1^{\alpha_1}p_2^{\alpha_2}$, *where* $p_1, p_2$ *are primes with* $p_1 < p_2$, *and* $v_1 = n/p_1$. *Then*

$$\kappa(C\Gamma(\mathbb{Z}_n)) = \begin{cases} \delta(C\Gamma(\mathbb{Z}_n)) & \textit{if } \alpha_1 > 1; \\ \delta(C\Gamma(\mathbb{Z}_n)) - |A_{v_1}| + 1 & \textit{if } \alpha_1 = 1 \textit{ and } \alpha_2 > 1. \end{cases}$$

Finally, Phillips et al. [309] proposed the following conjecture.

**Conjecture 3.5.47.** *Let $n = p_1^{\alpha_1} \cdots p_m^{\alpha_m}$, where the $p_i$'s are primes with $p_1 < \cdots < p_m$, and $v_1 = n/p_1$. Then $\kappa(C\Gamma(\mathbb{Z}_n)) = \delta(C\Gamma(\mathbb{Z}_n)) - |A_{v_1}| + 1$ when $\alpha_1 = 1$, and $\kappa(C\Gamma(\mathbb{Z}_n)) = \delta(C\Gamma(\mathbb{Z}_n))$ when $\alpha_1 > 1$.*

In [304], Osba studied the complement of the zero-divisor graph of Gaussian integers modulo $n$. First of all, basic properties of $C\Gamma(\mathbb{Z}_{t^n}[i])$ when $t = 2$, $t \equiv 3\ (mod\ 4)$, or $t \equiv 1\ (mod\ 4)$ were studied. Note that $C\Gamma(\mathbb{Z}_2[i]) = K_1$, $C\Gamma(\mathbb{Z}_q[i]) = K_0$, $C\Gamma(\mathbb{Z}_{q^2}[i]) = (q^2-1)K_1$, and $C\Gamma(\mathbb{Z}_p[i]) = K_{p-1} \cup K_{p-1}$, where $q \equiv 3\ (mod\ 4)$ and $p \equiv 1\ (mod\ 4)$.

**Theorem 3.5.48.** ([304, Theorem 1]) *Let $n$ be a positive integer, $q \equiv 3\ (mod\ 4)$, and $p \equiv 1\ (mod\ 4)$. Then*

(a) *$diam(C\Gamma(\mathbb{Z}_{2^n}[i])) = 2$ and $gr(C\Gamma(\mathbb{Z}_{2^n}[i])) = 3$ for all $n > 1$;*

(b) *$diam(C\Gamma(\mathbb{Z}_{q^n}[i])) = 2$ and $gr(C\Gamma(\mathbb{Z}_{q^n}[i])) = 3$ for all $n > 2$;*

(c) *$diam(C\Gamma(\mathbb{Z}_{p^n}[i])) = 3$ for all $n > 1$.*

Next, the connectedness and girth of $C\Gamma(R)$ were discussed. Note that for a finite commutative ring $R$, $\Gamma(R)$ has a vertex adjacent to every other vertex if and only if either $R \cong \mathbb{Z}_2 \times F$, where $F$ is a finite field or $R$ is local. So, for these rings, $C\Gamma(R)$ cannot be connected. Also, if $R = R_1 \times R_2$, where $R_1$ and $R_2$ are integral domains, then $\Gamma(R) = K_{|R_1|-1,|R_2|-1}$, and so $C\Gamma(R)$ is a disconnected graph. The following results show that if $R$ is not of these two categories, then $C\Gamma(R)$ is connected.

**Theorem 3.5.49.** ([304, Theorem 2]) *Let $R$ be a commutative ring that is a product of two rings with at least one of them not an integral domain. Then $C\Gamma(R)$ is connected with $diam(C\Gamma(R)) \leq 3$.*

**Theorem 3.5.50.** ([304, Theorem 3]) *Let $n > 1$, $q \equiv 3\ (mod\ 4)$, and $p \equiv 1\ (mod\ 4)$. If $m \neq p$, $m \neq 2^n$, $m \neq q^n$, and $m$ is not a product of two distinct primes, then $C\Gamma(\mathbb{Z}_m[i])$ is connected with $diam(C\Gamma(\mathbb{Z}_m[i])) \leq 3$.*

**Theorem 3.5.51.** ([304, Theorem 5]) *Let $n$ be an integer such that $n \neq 2$, $n \neq q$, and $n \neq q^2$, where $q \equiv 3\ (mod\ 4)$. Then $gr(C\Gamma(\mathbb{Z}_n[i])) = 3$.*

Finally, the authors of [304] investigated the cases in which $C\Gamma(\mathbb{Z}_n[i])$ is planar.

**Theorem 3.5.52.** ([304, Theorem 9]) *$C\Gamma(\mathbb{Z}_n[i])$ is planar if and only if $n = 2,\ 5$, or $q^2$, where $q \equiv 3\ (mod\ 4)$.*

The following characterizations of Eulerian and Hamiltonian complements for finite reduced rings were proved by Devhare et al. in [184].

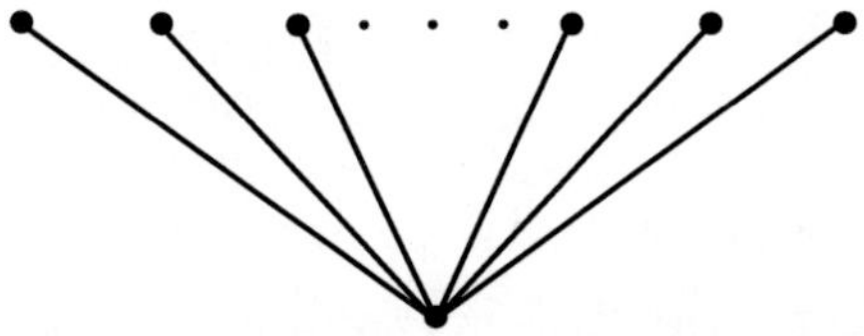

Figure 3.4: Cycle free zero-divisor graph

**Theorem 3.5.53.** ([184, Corollary 3.5]) *If $R$ is a finite reduced commutative ring, then $C\Gamma(R)$ is Eulerian if and only if $R$ is of characteristic 2 and has at least three prime ideals.*

**Theorem 3.5.54.** ([184, Corollary 4.10]) *If $R$ is a finite reduced commutative ring with at least three prime ideals, then $C\Gamma(R)$ is Hamiltonian.*

## 3.6 Cycles in Zero-divisor Graphs

Note that cycles play an important role in analyzing the structural properties of a graph. In view of this, in this section, we discuss the presence of cycles in zero-divisor graphs. As a consequence, important concepts like Eulerian and Hamiltonian graphs are also discussed.

### 3.6.1 Trees and Complements

An *acyclic* graph, one not containing any cycles, is called a *forest*. A connected forest is called a *tree*. (Thus, a forest is a graph whose components are trees.) The vertices of degree one in a tree are its *leaves*. Every nontrivial tree has at least two leaves, for example, the ends of a path. This little fact often comes in handy, especially in induction proofs about trees: if we remove a leaf from a tree, what remains is still a tree.

To begin with, one may ask for a characterization of those zero-divisor graphs which are acyclic, i.e., free of cycles. The answer, is simple: the graph must be either as in Figure 3.4 or as in Figure 3.2. The graph in Figure 3.4 is called a fan (or star graph) and is allowed to be a single point. Moreover, if the zero-divisor graph of $R$ is isomorphic to the graph in Figure 3.2, then $R = S\times\mathbb{F}_2$, where $S$ is a ring of cardinality 4 with a nonzero nilpotent and $\mathbb{F}_2$ is the field with 2 elements. For finite rings $R$, this is Theorem 2.2.10. A description of rings $R$ whose zero-divisor-graphs are as in Figure 3.4 is provided in the following theorem. A *binary local ring* is a finite local principal ideal ring with residue field $\mathbb{F}_2$. The following few results are due to Mulay [295].

**Theorem 3.6.1.** ([295, Theorem 2.1]) *Let* $R_1 = \mathbb{Z}_4 \times \mathbb{F}_2$, $R_2 = \mathbb{F}_2[x]/(x^2) \times \mathbb{F}_2$, $R_3 = \mathbb{Z}_9$, *and* $R_4 = \mathbb{F}_3[x]/(x^2)$. *Assume that* $R$ *is not a binary local ring of cardinality* 8, $R \ncong R_i$ *for* $i = 1, 2, 3, 4$, *and* $Z(R)^* \neq \emptyset$. *Then* $\Gamma(R)$ *is acyclic if and only if there is a* $y \in Z(R)^*$ *such that* $Z(R) \subseteq Ann(y) \cup \{y\}$, $Ann(x) = \{y, 0\}$ *for all* $x \in Z(R)^* \setminus \{y\}$, *and* $Nil(R) \subseteq \{y, 0\}$.

Let $V_n(R) = \{a \in R \setminus \{0\} \mid a \in V(C_n)$ for some cycle $C_n$ of $\Gamma(R)\}$, $V(R) = \cup_n V_n(R)$, and $P_n(R) = \{(a_1, \ldots, a_n) \mid a_i \in R^*$ are distinct and $a_i a_{i+1} = 0$ for $1 \leq i < n\}$. In the following theorem, it is proved that the connected components of $V(R) \setminus V_3(R)$ are either singleton sets or complete bipartite graphs.

**Theorem 3.6.2.** ([295, Theorem 2.2]) *Let* $B_1 = \mathbb{Z}_4$ *and* $B_2 = \mathbb{F}_2[x]/(x^2)$.

(a) *Assume* $(w, x, y, z) \in P_4(R)$ *is such that* $Ann(x) \cap Ann(y) \subseteq \{x, y, 0\}$. *Then,* $Ann(x) \cap Ann(y) \neq 0$ *if and only if* $R \cong B_i \times T$ *for some ring* $T$ *and some* $i = 1, 2$;

(b) *Assume* $x \in V(R) \setminus V_3(R)$ *and* $R$ *is not (isomorphic to a ring) of type* $B_i \times T$. *Then* $x$ *is not nilpotent and for* $a, b \in Ann(x)^*$, *we have* $Ann(a) = Ann(b)$;

(c) *Assume* $R$ *is not (isomorphic to a ring) of type* $B_i \times T$, $(x, y) \in P_2(S)$, *where* $S = V(R) \setminus V_3(R)$, *and let* $K = Ann(x)^* \cup Ann(y)^*$. *Then* $K, K \cap S$ *are 2-partite sets and* $K \cap S$ *is a connected component of* $S$.

**Theorem 3.6.3.** ([295, Theorem 2.3]) *If* $R$ *is not isomorphic to a ring of type* $B_i \times T$ *for some ring* $T$, *then* $Nil(R) \cap Z(R) \cap V(R) \subseteq V_3(R)$.

In 2002, DeMeyer and Schneider [178] provided a few more results regarding the acyclic nature of the zero-divisor graph of a commutative ring.

**Lemma 3.6.4.** ([178, Lemma 1.4]) *If* $R \cong \mathbb{Z}_2 \times S$, *then* $\Gamma(R)$ *does not contain a cycle if and only if* $S$ *is an integral domain,* $S \cong \mathbb{Z}_4$, *or* $S \cong \mathbb{Z}_2[x]/(x^2)$.

**Lemma 3.6.5.** ([178, Lemma 1.5]) *If* $R$ *is a commutative ring and* $\Gamma(R)$ *contains a path of length three, then exactly one of the following holds:*

(a) $\Gamma(R)$ *contains a cycle of length* $\leq 4$;

(b) $R \cong \mathbb{Z}_2 \times \mathbb{Z}_4$ *or* $R \cong \mathbb{Z}_2 \times \mathbb{Z}_2[x]/(x^2)$.

**Theorem 3.6.6.** ([178, Theorem 1.6]) *Let* $R$ *be a commutative ring containing at least one nonzero zero-divisor. Then exactly one of the following holds:*

(a) $\Gamma(R)$ *contains a cycle of length* $\leq 4$. *i.e.,* $gr(\Gamma(R)) \leq 4$;

(b) $\Gamma(R)$ *is a star graph;*

(c) $R \cong \mathbb{Z}_2 \times \mathbb{Z}_4$ *or* $R \cong \mathbb{Z}_2 \times \mathbb{Z}_2[x]/(x^2)$.

The following theorem gives a complete list of the finite rings whose zero-divisor graphs do not contain a cycle. Recall that, by Theorem 2.2.10, if $R$ is a finite commutative ring with $|\Gamma(R)| \geq 4$, then $\Gamma(R)$ is a star graph if and only if $R \cong \mathbb{Z}_2 \times F$, where $F$ is a finite field with at least five elements.

**Theorem 3.6.7.** ([178, Theorem 1.7]) *Let $R$ be a finite commutative ring. Then $\Gamma(R)$ does not contain a cycle if and only if $R$ is isomorphic to one of the rings in the following list:*

(a) $\mathbb{Z}_4$ *or* $\mathbb{Z}_2[x]/(x^2)$;

(b) $\mathbb{Z}_8$, $\mathbb{Z}_2[x]/(x^3)$, *or* $\mathbb{Z}_4[x]/(x^2-2,2x)$;

(c) $\mathbb{Z}_9$ *or* $\mathbb{Z}_3[x]/(x^2)$;

(d) $\mathbb{Z}_2 \times \mathbb{Z}_4$ *or* $\mathbb{Z}_2 \times \mathbb{Z}_2[x]/(x^2)$;

(e) $F$ *or* $\mathbb{Z}_2 \times F$ *for any finite field* $F$.

**Theorem 3.6.8.** ([178, Theorem 1.12] (also see Theorem 2.2.2)) *Let $R$ be a commutative ring with more than nine elements and at least two zero-divisors. Then the following are equivalent:*

(a) $\Gamma(R)$ *does not contain a cycle;*

(b) $\Gamma(R)$ *is a star graph;*

(c) *either $R \cong \mathbb{Z}_2 \times D$ for some integral domain $D$ or $I = \{0, x\}$ is the radical of $R$, $R/I$ is an integral domain, and $Ann_R(x) = Z(R)$.*

Recall that a connected graph having no cut-vertices is called *nonseparable* (i.e., the graph remains connected after any vertex is removed).

**Theorem 3.6.9.** ([178, Theorem 1.13]) *Let $R$ be a commutative ring with $|Z(R)^*| \geq 3$ and with the property that $\{0, x\}$ is not an ideal for any $0 \neq x \in R$. Then any two distinct zero-divisors in $Z(R)^*$ are contained in a cyclic of length less than or equal to* 6. *Thus $\Gamma(R)$ is nonseparable.*

A graph $G$ is called *triangulated* if each vertex of $G$ is a vertex of a triangle, and $G$ is called *hyper-triangulated* if each edge of $G$ is an edge of a triangle. A subset $S$ of $V(G)$ is said to be *hyper-triangulated* if each edge of the subgraph induced by $S$ in $G$ is an edge of a triangle. The work on hyper-triangulated graphs for $\Gamma(R)$ is due to Mulay [295]. In the following result, it is exhibited how the presence of nonzero nilpotents affects the cycle structure. For instance, it is proved in Theorem 3.6.10(b) that if every zero-divisor is nilpotent, then in most cases the zero-divisor graph is hyper-triangulated.

**Theorem 3.6.10.** ([295, Theorem 2.4]) *Let $B_1 = \mathbb{Z}_4$, $B_2 = \mathbb{F}_2[x]/(x^2)$, and $N(R) = Nil(R) \cap Z(R)^*$.*

(a) *If $N(R)$ is not hyper-triangulated, then $R$ has a distinguished nilpotent and $N(R)$ is not triangulated;*

(b) *If $N(R) = Z(R)^*$ and $R$ is not isomorphic to a binary local ring, then $Z(R)^*$ is hyper-triangulated. On the other hand, if $R$ is isomorphic to a binary local ring, then $N(R) \setminus V(R)$ is nonempty;*

(c) *Assume* $Z(R)^*$ *has at least three elements. Then every path of length* 3 *is a cycle if and only if* $Z(R)$ *generates a prime ideal* $M$ *with* $M^2 = 0$.

The *circumference* of a graph $G$, denoted by $c(G)$, is the length of a longest cycle in $G$. If $G$ has no cycles, we write $c(G) = 0$; if $G$ has cycles of arbitrarily long length, we write $c(G) = \infty$. Note that $gr(G) \leq c(G)$ and $\omega(G) \leq c(G)$ if $G$ contains a cycle. For a subring $T$ of a commutative ring $R$, we always have $c(\Gamma(T)) \leq c(\Gamma(R))$.

The circumference of the zero-divisor graph of a commutative ring was discussed by Anderson et al. in [75]. Some additional (unpublished) results on $c(\Gamma(F^m))$ for a finite field $F$ were given by N. Smith in [354, Chapter 4].

**Example 3.6.11.** Let $R = F_1 \times F_2$, where $F_1$ and $F_2$ are finite fields with $|F_1| = m \geq 3$ and $|F_2| = n \geq 3$. Then $c(\Gamma(R)) = 2(\min\{m, n\} - 1)$.

**Theorem 3.6.12.** ([75, Theorem 3.9]) *Let* $D_1, \ldots, D_n$ *be integral domains* ($n \geq 2$), *and let* $R = D_1 \times \cdots \times D_n$. *Then* $c(\Gamma(R)) = \infty$ *if and only if at least two of the* $D_i$*'s are infinite.*

Anderson et al. [75] proposed the following open question.

**Question:** Determine $c(\Gamma(R))$ when $R$ is a direct product of finitely many finite fields. In concrete terms, let $R = F_1 \times \cdots \times F_m$, where the $F_i$'s are finite fields and $m \geq 2$. What is the largest integer $n \geq 2$ such that there is an $m \times n$ matrix $A = [a_{ij}]$ with each $a_{ij} \in F_i$ satisfying (*i*) $A$ has distinct nonzero columns and (*ii*) if $a_{ij} \neq 0$, then $a_{ij+1} = 0$ when $j < n$ and $a_{ij} = 0$ when $j = n$?

The study of the complemented or uniquely complemented nature of $\Gamma(R)$ was initiated by Anderson, Levy, and Shapiro [81]. Recall that two distinct vertices $a$ and $b$ of a graph $G$ is said to be *orthogonal*, written $a \perp b$, if $a$ and $b$ are adjacent and there is no vertex $c$ of $G$ which is adjacent to both $a$ and $b$, i.e., the edge $a - b$ is not part of any triangle in $G$. We say that $G$ is *complemented* if for each vertex $a$ of $G$, there is a vertex $b$ of $G$ (called a *complement* of $a$) such that $a \perp b$, and that $G$ is *uniquely complemented* if $G$ is complemented and whenever $a \perp b$ and $a \perp c$, then $N(b) = N(c)$, where $N(x)$ denotes open neighborhood of $x \in V(G)$. For $a, b \in Z(R)^*$, we have $N(a) = N(b)$ in $\Gamma(R)$ if and only if $Ann_R(a) \setminus \{a\} = Ann_R(b) \setminus \{b\}$.

Some of the results of Anderson et al. [81] are listed below. First of all, the authors proved that $R$ and its total quotient ring $T(R)$ have isomorphic zero-divisor graphs.

**Theorem 3.6.13.** ([81, Theorem 2.2]) *Let* $R$ *be a commutative ring with total quotient ring* $T(R)$. *Then the graphs* $\Gamma(T(R))$ *and* $\Gamma(R)$ *are isomorphic.*

**Lemma 3.6.14.** ([81, Lemma 3.3]) *Let* $R$ *be a commutative ring and* $a, b \in Z(R)^*$. *Then the following statements are equivalent:*

(a) $a \perp b$, $a^2 \neq 0$, *and* $b^2 \neq 0$;

(b) $ab = 0$ *and* $a + b$ *is a regular element of* $R$.

**Lemma 3.6.15.** ([81, Lemma 3.4]) *Let* $R$ *be a reduced commutative ring and* $a, b, c \in Z(R)^*$. *If* $a \perp b$ *and* $a \perp c$, *then* $N(b) = N(c)$. *Thus* $\Gamma(R)$ *is uniquely complemented if and only if* $\Gamma(R)$ *is complemented.*

Clearly star graphs are uniquely complemented. Suppose that $R$ is von Neumann regular. Then, for each $a \in Z(R)^*$, we have $a = ue$, where $u \in U(R)$ and $e \in R$ is idempotent. Clearly $ue \perp (1 - e)$. Thus a von Neumann regular ring $R$ has $\Gamma(R)$ complemented. In fact, by Lemma 3.6.15, $\Gamma(R)$ is uniquely complemented since a von Neumann regular ring $R$ is reduced. Hence by Theorem 3.6.13, $\Gamma(R) \cong \Gamma(T(R))$ is uniquely complemented if $T(R)$ is von Neumann regular. The next theorem shows that the converse holds when $R$ is a reduced commutative ring.

**Theorem 3.6.16.** ([81, Theorem 3.5]) *The following statements are equivalent for a reduced commutative ring* $R$:

(a) $T(R)$ *is von Neumann regular;*

(b) $\Gamma(R)$ *is uniquely complemented;*

(c) $\Gamma(R)$ *is complemented.*

*Moreover, a nonempty* $\Gamma(R)$ *is a star graph if and only if* $R \cong D \times \mathbb{Z}_2$ *for some integral domain* $D$.

Next, consider the case when $R$ has nonzero nilpotent elements. In this case, the following theorem shows that $\Gamma(R)$ is uniquely complemented if and only if it is a star graph. Moreover, either $\Gamma(R)$ has one or two edges, or $\Gamma(R)$ is an infinite star graph with center $x$, where $Nil(R) = \{0, x\}$.

**Theorem 3.6.17.** ([81, Theorem 3.9]) *Let* $R$ *be a commutative ring with* $Nil(R)$ *nonzero. If* $\Gamma(R)$ *is uniquely complemented, then either* $\Gamma(R)$ *is a star graph with at most two edges or* $\Gamma(R)$ *is an infinite star graph with center* $x$, *where* $Nil(R) = \{0, x\}$.

Combining Theorem 3.6.16 and Theorem 3.6.17, one can get the following result.

**Theorem 3.6.18.** ([81, Corollary 3.10]) *Let* $R$ *be a commutative ring. Then* $\Gamma(R)$ *is uniquely complemented if and only if either* $T(R)$ *is von Neumann regular or* $\Gamma(R)$ *is a star graph.*

Note that if $\Gamma(R)$ is uniquely complemented, then the total quotient ring of $R/Nil(R)$ is von Neumann regular. In fact, $R/Nil(R)$ is an integral domain when $Nil(R)$ is nonzero. However, the converse is false even for finite rings. For an infinite example, let $T$ be any von Neumann regular ring and $R = T[x]/(x^2)$. Then $Nil(R) = Z(R) = (x)/(x^2)$ and $R/Nil(R) = T$ is von Neumann regular. But $\Gamma(R)$ is the complete graph on $|T| - 1$ vertices, which is uniquely complemented if and only if $T = \mathbb{Z}_3$.

The next theorem characterizes those commutative rings $R$ for which $\Gamma(R)$ is complemented, but not uniquely complemented.

**Lemma 3.6.19.** ([81, Lemma 3.13]) *Let $R = A \times B$, where $A$ is a reduced commutative ring and $B$ is a commutative ring with $Nil(B)$ nonzero. Then $\Gamma(R)$ is complemented, but not uniquely complemented, if and only if $A$ is an integral domain and $B$ is isomorphic to either $\mathbb{Z}_4$ or $\mathbb{Z}_2[x]/(x^2)$.*

**Theorem 3.6.20.** ([81, Theorem 3.14]) *Let $R$ be a commutative ring. Then $\Gamma(R)$ is complemented, but not uniquely complemented, if and only if $R$ is isomorphic to $D \times B$, where $D$ is an integral domain and $B$ is $\mathbb{Z}_4$ or $\mathbb{Z}_2[x]/(x^2)$.*

**Corollary 3.6.21.** ([81, Corollary 3.15,3.16])

(a) *Let $R$ be a commutative ring such that $\Gamma(R)$ is complemented. Then $T(R/Nil(R))$ is von Neumann regular. Moreover, if $Nil(R) \neq \{0\}$, then $\Gamma(R/Nil(R))$ is either the empty graph or a star graph;*

(b) *Let $R$ be a finite commutative ring which is not a field. Then $\Gamma(R)$ is complemented, but not uniquely complemented, if and only if $R$ is isomorphic to $F \times B$, where $F$ is a finite field and $B$ is $\mathbb{Z}_4$ or $\mathbb{Z}_2[x]/(x^2)$.*

LaGrange [244] continued the investigation of complemented zero-divisor graphs. Actually, he investigated the zero-divisor structure of Boolean rings, as well as ring-theoretic properties that arise when a ring has a specifically classified zero-divisor graph.

In what follows, we see that if $R$ is Boolean, then the atoms of $B(R)$ are precisely the elements of $V(\Gamma(R))$ that are adjacent to an end vertex. Recall that an *atom* in a Boolean algebra $B$ is an element $0 \neq a \in B$ such that $0 \neq b \in B$ with $b \leq a$ implies $b = a$.

**Theorem 3.6.22.** ([244, Theorem 2.2]) *Let $R \not\cong \mathbb{Z}_2$ be a Boolean ring. Then the atoms of $B(R)$ are precisely the elements of $V(\Gamma(R))$ that are adjacent to an end vertex.*

Now, we see that every vertex of the zero-divisor graph of a Boolean ring (not isomorphic to $\mathbb{Z}_2$) has a unique complement. By excluding the rings $\mathbb{Z}_9$ and $\mathbb{Z}_3[x]/(x^2)$, we see that the converse is also true.

**Theorem 3.6.23.** ([244, Lemma 2.3, Theorem 2.5])

(a) *Let $R \ncong \mathbb{Z}_2$ be a Boolean ring. Then every element of $V(\Gamma(R))$ has a unique complement;*

(b) *A commutative ring $R$ is Boolean if and only if either $R \cong \mathbb{Z}_2$, or $\Gamma(R)$ is not the empty graph, $R \notin \{\mathbb{Z}_9, \mathbb{Z}_3[x]/(x^2)\}$, and $\Gamma(R)$ has the property that every vertex has a unique complement. In particular, if $|V(\Gamma(R))| \geq 3$, then $R$ is Boolean if and only if every vertex of $\Gamma(R)$ has a unique complement.*

Next, we are interested in answering the question "which rings have the property that every vertex of $\Gamma(R)$ is either an end or is adjacent to an end?". Of course, all star graphs have this property.

By Remark 2.3.5, we have $\Gamma(R)$ is a star graph if and only if either $R \cong A$, where $A \in \{\mathbb{Z}_9, \frac{\mathbb{Z}_3[x]}{(x^2)}, \mathbb{Z}_8, \frac{\mathbb{Z}_2[x]}{(x^3)}, \frac{\mathbb{Z}_4[x]}{(x^2-2, 2x)}\}$; $R \cong \mathbb{Z}_2 \times D$, where $D$ is an integral domain; or there exists a $0 \neq x \in R$ such that $Z(R) = Ann(x)$, $Nil(R) = \{0, x\}$, and $R/Nil(R)$ is an infinite integral domain. We will see that only two other graphs that have this property are realizable as zero-divisor graphs.

**Theorem 3.6.24.** ([244, Theorem 4.3]) *Let $R$ be a commutative ring with the property that every vertex of $\Gamma(R)$ is either an end or is adjacent to an end. Then exactly one of the following holds:*

(a) *$R \cong A$, where $A \in \{\mathbb{Z}_2 \times \mathbb{Z}_4, \mathbb{Z}_2 \times \mathbb{Z}_2[x]/(x^2)\}$;*

(b) *$\Gamma(R)$ is a star graph;*

(c) *$R \cong \mathbb{Z}_2 \times \mathbb{Z}_2 \times \mathbb{Z}_2$.*

Note that if $a, b \in V(\Gamma(R))$ are non-end-vertices with $a \perp b$, then every vertex in $V(\Gamma(R)) \setminus \{a, b\}$ is an end-vertex. For, if $c$ is a non-end-vertex distinct from $a$ and $b$, then $ac = 0 = bc$, a contradiction. So $a$ and $b$ are vertices that are not contained in any cycle, and hence $R$ is isomorphic to either $\mathbb{Z}_2 \times \mathbb{Z}_4$ or $\mathbb{Z}_2 \times \mathbb{Z}_2[x]/(x^2)$. Thus a ring that satisfies the hypothesis of Theorem 3.6.24, but does not satisfy (a), has the property that $a \perp b$ implies either $a \in E(b)$ or $b \in E(a)$, where $E(a) = \{x \in V(\Gamma(R)) \setminus \{a\} \mid x \text{ is a end-vertex which is adjacent to a}\}$. Therefore, since the zero-divisor graph of any ring satisfying the hypothesis of Theorem 3.6.24 is complemented, Theorem 3.6.24 will be established upon proving the following result.

**Theorem 3.6.25.** ([244, Theorem 4.4]) *Let $R$ be a commutative ring such that $\Gamma(R)$ is complemented, and $a \perp b$ if and only if either $a \in E(b)$ or $b \in E(a)$. Then either $R \cong \mathbb{Z}_2 \times \mathbb{Z}_2 \times \mathbb{Z}_2$ or $\Gamma(R)$ is a star graph.*

**Corollary 3.6.26.** ([244, Corollary 4.5]) *Let $R \ncong \mathbb{Z}_2 \times \mathbb{Z}_2 \times \mathbb{Z}_2$ be a commutative ring with nonzero zero-divisors. Then $\Gamma(R)$ has a vertex of a cycle that is not adjacent to an end-vertex if and only if $\Gamma(R)$ contains a cycle.*

### 3.6.2 Eulerian and Hamiltonian Graphs

The Königsberg Bridge Problem and Hamilton's game give rise to two concepts in graph theory named after Euler and Hamilton. These concepts are the main subjects of this section.

First, let us see the Hamiltonian nature of the zero-divisor graph. Recall that a *Hamiltonian cycle* of $G$ is a cycle that contains every vertex of $G$. A graph is *Hamiltonian* if it contains a Hamiltonian cycle. The name "Hamiltonian cycle" arises from the fact that Sir William Hamilton investigated their existence in the dodecahedron graph. One of the major unsolved problems of graph theory is to obtain simple characterizations for Hamiltonian graphs. Most existing theorems have the form "if $G$ has enough edges, then $G$ is Hamiltonian". Probably, the most celebrated of these is the result of Dirac, see [432].

The study of Hamiltonian zero-divisor graphs of a commutative ring was initiated by Akbari and Mohammadian [38]. Actually, they determined a family of commutative rings whose zero-divisor graphs are Hamiltonian.

**Theorem 3.6.27.** ([38, Theorem 8]) *Let $R$ be a finite commutative decomposable ring (i.e., $R$ can be written as $R_1 \times R_2$, where $R_1$ and $R_2$ are rings). If $\Gamma(R)$ is a Hamiltonian graph, then $\Gamma(R) \cong K_{n,n}$ for some natural number $n$.*

For a characterization of complete bipartite zero-divisor graphs of a commutative ring, see Theorem 3.4.5. By Theorem 3.4.5(b), we have $\Gamma(R)$ is complete bipartite if and only if $R \cong F_1 \times F_2$, where $F_1$ and $F_2$ are finite fields, or $R \cong \mathbb{Z}_9, \dfrac{\mathbb{Z}_3[x]}{(x^2)}, \mathbb{Z}_8, \dfrac{\mathbb{Z}_2[x]}{(x^3)}$, or $\dfrac{\mathbb{Z}_4[x]}{(x^2-2,2x)}$. But $\Gamma(R)$ is not Hamiltonian whenever $R \cong \mathbb{Z}_9, \dfrac{\mathbb{Z}_3[x]}{(x^2)}, \mathbb{Z}_8, \dfrac{\mathbb{Z}_2[x]}{(x^3)}$, or $\dfrac{\mathbb{Z}_4[x]}{(x^2-2,2x)}$. Hence, if a finite commutative decomposable ring with identity has a Hamiltonian zero-divisor graph, then the ring is isomorphic to $F \times F$, where $F$ is a finite field with more than two elements.

**Theorem 3.6.28.** ([38, Theorem 9]) *Let $R$ be a finite principal ideal ring. If $\Gamma(R)$ is a Hamiltonian graph, then it is either a complete graph or a complete bipartite graph.*

**Corollary 3.6.29.** ([38, Corollary 1]) *The graph $\Gamma(\mathbb{Z}_n)$ is a Hamiltonian graph if and only if $n = p^2$, where $p$ is a prime greater than 3, and in this case, $\Gamma(\mathbb{Z}_n) \cong K_{p-1}$.*

Consider the local ring $R = \dfrac{\mathbb{Z}_3[x,y]}{(x^2+xy, y^2+2xy)}$. Then $\Gamma(R)$ is a Hamiltonian graph which is neither a complete graph nor a complete bipartite graph.

Theorem 3.6.27 was generalized to the case of noncommutative rings by Mal'tsev and Kuz'mina [282], and their result is given below.

**Theorem 3.6.30.** ([282, Theorem 1]) *Let $R$ be a finite decomposable ring with identity. Then $\Gamma(R)$ is Hamiltonian if and only if $R \cong \mathbb{F}_q \times \mathbb{F}_q$ with $q \neq 2$, where $\mathbb{F}_q$ is the field with $q$ elements.*

Let $Z\langle X\rangle = Z\langle x_1, x_2, \ldots\rangle$ be the free associative ring in countably many variables $X = \{x_1, x_2, \ldots\}$ and $f(x_1, \ldots, x_d) \in Z\langle X\rangle$. Let $\mathfrak{M}$ be a variety of rings. Denote by $T(\mathfrak{M})$ the set of all polynomials $f \in Z\langle X\rangle$ such that $f = 0$ is an identity on all rings in $\mathfrak{M}$. The set $T(\mathfrak{M})$ forms an ideal of the ring $Z\langle X\rangle$, which is called an *ideal of identities* of $\mathfrak{M}$.

**Corollary 3.6.31.** ([282, Corollary 1]) *Let $\mathfrak{M}$ be a nontrivial ring variety in which every finite ring has a Hamiltonian zero-divisor graph. Then $\mathfrak{M}$ does not contain a field.*

**Theorem 3.6.32.** ([282, Proposition 1]) *Let $\mathfrak{M}$ be a nontrivial ring variety in which every finite ring has a Hamiltonian zero-divisor graph. Then $x^n \in T(\mathfrak{M})$ and $mx \in T(\mathfrak{M})$, where $m, n \geq 2$.*

**Theorem 3.6.33.** ([282, Proposition 2]) *Let $\mathfrak{M}$ be a nontrivial ring variety in which every finite ring has a Hamiltonian zero-divisor graph. Then $exp(\mathfrak{M})$ is not divisible by* 3.

**Theorem 3.6.34.** ([282, Proposition 3]) *Let $\mathfrak{M}$ be a nontrivial ring variety in which every finite ring has a Hamiltonian zero-divisor graph. Then $x^2 \in T(\mathfrak{M})$.*

**Theorem 3.6.35.** ([282, Theorem 2]) *For an arbitrary variety $\mathfrak{M}$ of associative rings, the following conditions are equivalent:*

(a) *Every ring in $\mathfrak{M}$ has a Hamiltonian zero-divisor graph;*

(b) *Every finite ring in $\mathfrak{M}$ has a Hamiltonian zero-divisor graph;*

(c) *$xy, mx \in T(\mathfrak{M})$, where $m$ is a natural number not divisible by* 3.

Hamiltonian zero-divisor graphs of the ring of Gaussian integers modulo $n$ have been investigated by Osba, Al-Addasi, and Al-Khamaiseh [305].

**Theorem 3.6.36.** ([305, Theorems 5, 6, and 8])

(a) *For each $m \geq 1$, the graph $\Gamma(\mathbb{Z}_{2^m}[i])$ is not Hamiltonian;*

(b) *Let $p$ be a prime integer such that $p \equiv 1 (mod\ 4)$. Then the graph $\Gamma(\mathbb{Z}_{p^m}[i])$ is Hamiltonian if and only if $m = 1$;*

(c) *Let $q$ be a prime integer such that $q \equiv 3 (mod\ 4)$. Then the graph $\Gamma(\mathbb{Z}_{q^m}[i])$ is Hamiltonian if and only if $m = 2$.*

**Theorem 3.6.37.** ([305, Corollary 11]) *The graph $\Gamma(\mathbb{Z}_n[i])$ is Hamiltonian if and only if $n = p$ or $n = q^2$, where $p$ and $q$ are prime integers such that $p \equiv 1 (mod\ 4)$ and $q \equiv 3 (mod\ 4)$.*

Next, we discuss Eulerian zero-divisor graphs. A graph is called *Eulerian* if it contains a closed trail containing every edge. A well-known theorem of Euler and Hierholzer states that a graph $G$ is Eulerian if and only if every vertex of $G$ has even degree. Eulerian zero-divisor graphs were studied by Córdova, Gholston, and Hauser [168]. The authors were interested in the Eulerian property for $\Gamma(\mathbb{Z}_n)$. The following result characterizes when the zero-divisor graph of $\mathbb{Z}_n$ is Eulerian.

**Theorem 3.6.38.** ([168, Theorem 3.1]) *The graph $\Gamma(\mathbb{Z}_n)$ is Eulerian if and only if either (i) $n$ is odd and square-free or (ii) $n = 4$.*

A ring $R$ is called *decomposable* if $R$ is a direct sum of two nonzero ideals of $R$, i.e., $R = I \oplus J$ (direct sum of ideals) for some nonzero ideals $I, J$ of $R$. The characterization of finite rings (not necessarily commutative) with Eulerian zero-divisor graphs was done by Kuz'mina [240]. First, she described all nilpotent rings with Eulerian zero-divisor graphs.

**Theorem 3.6.39.** ([240, Proposition 2.2]) *Let $R$ be a finite nonzero nilpotent ring. Then the graph $\Gamma(R)$ is Eulerian if and only if the order of $R$ is even and $x^2 = 0$ for every $x \in R$.*

Next, all nonnilpotent finite rings without identity that have Eulerian zero-divisor graphs are described.

**Theorem 3.6.40.** ([240, Proposition 3.1]) *Let $R$ be a finite nonnilpotent ring without identity element. Then the graph $\Gamma(R)$ is Eulerian if and only if the ring $R$ satisfies one of following conditions:*

(a) *$R \cong S \oplus N$, where $S$ is a local ring such that $|S| = 2^\alpha, \alpha \geq 1$, and $s^2 = 0$ for every $s \in J(S)$, and $N$ is a nonzero nilpotent ring of odd order such that $x^2 = 0$ for every $x \in N$;*

(b) *$R \cong S \oplus N$, where $S$ is an indecomposable ring without one-sided identity such that $|S| = 2^\alpha, \alpha \geq 2$, and $s^2 = 0$ for every $s \in J(S)$, the quotient ring $S/J(S)$ is a field, and $N$ is a nonzero nilpotent ring of odd order such that $x^2 = 0$ for every $x \in N$ (it is a possible that $N = (0)$).*

Finally, the following theorem describes all finite rings that have Eulerian zero-divisor graphs.

**Theorem 3.6.41.** ([240, Theorem 1.1]) *Let $R$ be a finite ring ($R$ is not a field). The graph $\Gamma(R)$ is Eulerian if and only if $R$ satisfies one of the following conditions:*

(a) *$R \cong \oplus_{i=1}^{k} \mathbb{F}_{p_i^{\alpha_i}}$, where $p_i \neq 2$ for every $i$, $k \geq 2$, and $\mathbb{F}_q$ is the finite field with $q$ elements;*

(b) *$R$ is a nilpotent ring of even order and $x^2 = 0$ for every $x \in R$;*

(c) $R \cong S \oplus N$, *where* $S$ *is a local ring such that* $|S| = 2^\alpha, \alpha \geq 1$, *and* $s^2 = 0$ *for every* $s \in J(S)$, *and* $N$ *is a nonzero nilpotent ring of odd order such that* $x^2 = 0$ *for every* $x \in N$ *(it is a possible that* $N = (0)$*);*

(d) $R \cong S \oplus N$, *where* $S$ *is an in decomposable ring such that* $|S| = 2^\alpha, \alpha \geq 2$, *and* $s^2 = 0$ *for every* $s \in J(S)$, *every one-sided identity of* $S$ *is a two-sided identity of* $S$, *the quotient ring* $S/J(S)$ *is a field, and* $N$ *is a nonzero nilpotent ring of odd order such that* $x^2 = 0$ *for every* $x \in N$ *(it is a possible that* $N = (0)$*).*

For a characterization of Eulerian directed zero-divisor graphs, see [40].

# Chapter 4

# Genus of Zero-divisor Graphs

In this chapter, we study topological concepts like the genus of zero-divisor graphs. The prime objective of topological graph theory is to draw a graph on a surface so that no two edges cross, an intuitive geometric problem that can be enriched by specifying symmetries or combinatorial side-conditions. Graphs on surfaces form a natural link between discrete and continuous mathematics. They enable us to understand both graphs and surfaces better.

There are two kinds of closed surfaces, orientable and nonorientable. The sphere, the torus, the double torus, the triple torus, and so on, are orientable. They are commonly denoted $S_0, S_1, S_2, S_3, \ldots$. If we add $k$ handles to $S_0$, we obtain the surface $S_k$, which we refer to as the *orientable surface of genus* $k$. If we add $k$ crosscaps to $S_0$, we get the *nonorientable surface of genus* $k$. This surface will be denoted by $\tilde{S}_k$. The surfaces $S_1$, $S_2$, $\tilde{S}_1$, and $\tilde{S}_2$ are also called the torus, the double torus, the projective plane, and the Klein bottle, respectively.

## 4.1 Genus of Graphs

If a surface is orientable, its "genus" is defined to be the number of handles one must add to the sphere to obtain its homeomorphism type. Let $S_k$ denote the sphere with $k$ handles, where $k$ is a nonnegative integer, that is, $S_k$ is an oriented surface with $k$ handles. The *genus of a graph* $G$, denoted by $g(G)$, is the smallest integer $n$ such that the graph can be embedded in $S_n$. Intuitively, $G$ is embedded in a surface if it can be drawn in the surface so that its edges intersect only at their common vertices. We say that a graph $G$ is *planar* if $g(G) = 0$, and *toroidal* if $g(G) = 1$. Thus a planar graph $G$ has an embedding in the plane. Note that if $H$ is a subgraph of a graph $G$, then $g(H) \leq g(G)$. For details on the notion of embedding of a graph in a surface, see White [429, Chapter 6].

A *minor* of a graph $G$ is the graph obtained from $G$ by contracting edges in $G$ or deleting edges and isolated vertices in $G$. A classical theorem due to Wagner [419] states that a graph $G$ is planar if and only if $G$ does not have $K_5$ or



D. F. Anderson et al., *Graphs from Rings*, https://doi.org/10.1007/978-3-030-88410-9_4

$K_{3,3}$ as a minor. It is well known that if $G'$ is a minor of $G$, then $g(G') \leq g(G)$. For $xy \in E(G)$, we denote the contracted edge by the vertex $[x, y]$. Also, if $H$ is a subgraph of $G$ and $H'$ is a minor of $H$, then we call $H'$ a minor subgraph of $G$.

In this section, we collect certain bounds on the genus of a graph.

**Lemma 4.1.1.** ([429, Theorems 6.37, 6.38, 6.39, and 6.42]) *Let $m, n \geq 3$ be integers. Then*

(a) $g(K_n) = \left\lceil \frac{(n-3)(n-4)}{12} \right\rceil$;

(b) $g(K_{m,n}) = \left\lceil \frac{(m-2)(n-2)}{4} \right\rceil$;

(c) $g(K_{mn,n,n}) = \frac{(mn-2)(n-1)}{2}$;

(d) $g(K_{n,n,n,n}) = (n-1)^2$ *for* $n \neq 3$, *and* $g(K_{3,3,3,3}) = 5$.

**Corollary 4.1.2.** *If $G$ is a graph with $n$ vertices, then* $g(G) \leq \left\lceil \frac{(n-3)(n-4)}{12} \right\rceil$.

**Lemma 4.1.3.** ([330, Theorem 1.1]) *For all integers $n \not\equiv 5$ or $9 \pmod{12}$ with $n \geq 2$, $g(K_n \square K_2) = \left\lceil \frac{(n-2)(n-3)}{6} \right\rceil$. If $n = 5$, then $g(K_n \square K_2) = 2$.*

**Lemma 4.1.4.** ([429, Euler's Formula]) *If $G$ is a finite connected graph with $n$ vertices, $m$ edges, and genus $g$, then $n - m + f = 2 - 2g$, where $f$ is the number of faces created when $G$ is minimally embedded on a surface of genus $g$.*

**Lemma 4.1.5.** ([430, Proposition 2.1]) *If $G$ is a graph with $n$ vertices and genus $g$, then*

$$\delta(G) \leq 6 + \frac{12g - 12}{n}.$$

*In particular, if $G$ is a graph of genus 1, then $\delta(G) \leq 6$ with equality if and only if $G$ is a triangulation of the torus and 6-regular.*

**Lemma 4.1.6.** ([89, Lemma 2.1]) *If $G$ is a graph with $n$ vertices, $m$ edges, girth $gr$, and genus $g$, then*

$$\frac{m(gr-2)}{2gr} - \frac{n}{2} + 1 \leq g.$$

**Lemma 4.1.7.** ([292, p. 114]) *If $G = G_1 \cup G_2$, where $G_1, G_2$ are edge-disjoint and $G_1 \cap G_2$ contains $k$ vertices, then*

$$g(G) \leq g(G_1) + g(G_2) + k - 1.$$

**Lemma 4.1.8.** ([428, Theorem 2]) *Let $G_1$ and $G_2$ be two graphs, and let $n_i$ be the number of vertices of $G_i$. Then*

$$max\{n_1 g(G_2) + g(G_1), n_2 g(G_1) + g(G_2)\} \leq g(G_1 \square G_2).$$

It is well known that every nonorientable compact surface is a connected sum of a finite number of copies of projective planes. The number of copies of projective spaces is called the *crosscap number* of this surface. For example, *projective space* is of crosscap one, and the *Klein bottle* is of crosscap two. A nonplanar graph is said to be *projective* if it can be embedded in the projective plane. According to the crosscap formula, the nonplanar graphs $K_5$, $K_6$, $K_{3,3}$, and $K_{3,4}$ are all projective.

Let $j$ be a nonnegative integer, and let $\tilde{S}_j$ denote the surface obtained by attaching $j$ crosscaps to the sphere. Then $\tilde{S}_j$ is a nonorientable surface of genus $j$. The *nonorientable genus of a graph* $G$, denoted by $\tilde{g}(G)$, is the minimum integer $n$ such that the graph can be embedded in $\tilde{S}_n$. Euler's Formula for nonorientable surfaces says that if $G$ is a graph with $n$ vertices, $m$ edges, and nonorientable genus $\tilde{g}$, then $n - m + r = 2 - \tilde{g}$, where $r$ is the number of regions created when $G$ is embedded on a nonorientable surface of genus $\tilde{g}$.

The following result gives the crosscap formula for a complete graph and a complete bipartite graph.

**Lemma 4.1.9.** ([292, Theorems 4.4.5, 4.4.6, and 4.4.7]) *Let $m, n \geq 3$ be integers. Then*

(i) $$\tilde{g}(K_n) = \begin{cases} \left\lceil \frac{(n-3)(n-4)}{6} \right\rceil & \text{if } n \geq 3 \text{ and } n \neq 7; \\ 3 & \text{if } n = 7. \end{cases}$$

*In particular,* $\tilde{g}(K_n) = 1$ *if* $n = 5, 6$;

(ii) $\tilde{g}(K_{m,n}) = \left\lceil \frac{(m-2)(n-2)}{2} \right\rceil$. *In particular,* $\tilde{g}(K_{3,3}) = \tilde{g}(K_{3,4}) = 1$.

## 4.2 Planar Zero-divisor Graphs

The study of planar embeddings has a long and rich history that intertwines with chromatic graph theory, algorithmic analysis, enumeration, and much more. There are numerous examples showing the importance of planar graphs in discrete mathematics. One beautiful example is Steinitz' Theorem that the graphs (or 1-skeletons) of 3-dimensional polytopes are precisely the 3-connected planar graphs. Planar graphs and their duals form an important link between graphs and matroids.

Graph drawing has appeared as a lively area in computer science due to its applications in almost all branches of science and technology. Many researchers have concentrated their attention on drawing planar graphs for the following reasons:

- drawings of planar graphs have no edge crossings and look nice;
- drawings of planar graphs have practical applications in VLSI floor planning and routing, architectural floor planning, displaying RNA structures in bio informatics, etc.; and
- algorithms for drawing planar graphs can be successfully used for drawing a nonplanar graph by transforming it into a similar planar graph.

A graph is *planar* if it can be drawn in the plane in such a way that no edges intersect, except of course at a common end-vertex. The famous Kuratowski's Theorem says that a graph $G$ is planar if and only if it contains no subdivisions homeomorphic to either $K_5$ or $K_{3,3}$.

Recall that a finite commutative ring is the direct product of a finite number of finite local rings. If $\Gamma(R)$ is planar, then $R$ has at most 3 local ring factors since otherwise it would contain a $K_{3,3}$. The cases for 2 or 3 local ring factors are then handled separately. For a finite local ring $(R, \mathfrak{m})$, $\mathfrak{m}^n = 0$ for some positive integer $n$. If $n \geq 5$, then $\Gamma(R)$ would contain a $K_5$, and thus not be planar. The first work in this direction was given in [75], where they asked which finite commutative rings $R$ have $\Gamma(R)$ planar and gave the following partial answer.

**Theorem 4.2.1.** ([75, Theorem 5.1])

(a) *Let $R = \mathbb{Z}_n$, where $n \geq 2$ is not prime. Then $\Gamma(R)$ is planar if and only if $n \in \{8, 12, 16, 18, 25, 27\} \cup \{2p, 3p \mid p \text{ is prime}\}$;*

(b) *Let $R = \mathbb{Z}_{n_1} \times \cdots \times \mathbb{Z}_{n_r}$, where $r \geq 2$ and $2 \leq n_1 \leq \cdots \leq n_r$. Then $\Gamma(R)$ is planar if and only if $R$ is one of $\mathbb{Z}_2 \times \mathbb{Z}_4$, $\mathbb{Z}_2 \times \mathbb{Z}_6$, $\mathbb{Z}_2 \times \mathbb{Z}_8$, $\mathbb{Z}_2 \times \mathbb{Z}_9$, $\mathbb{Z}_2 \times \mathbb{Z}_p$, $\mathbb{Z}_3 \times \mathbb{Z}_4$, $\mathbb{Z}_3 \times \mathbb{Z}_9$, $\mathbb{Z}_3 \times \mathbb{Z}_q$, $\mathbb{Z}_2 \times \mathbb{Z}_2 \times \mathbb{Z}_2$, $\mathbb{Z}_2 \times \mathbb{Z}_2 \times \mathbb{Z}_3$, where $p \geq 2$ and $q \geq 3$ are primes.*

**Theorem 4.2.2.** ([75, Theorem 5.2]) *Let $R_{n,m} = \mathbb{Z}_n[x]/(x^m)$, where $m, n \geq 2$.*

(a) *$\Gamma(R_{n,2})$ is planar if and only if $n \leq 5$;*

(b) *$\Gamma(R_{n,3})$ is planar if and only if $n \leq 3$;*

(c) *$\Gamma(R_{n,4})$ is planar if and only if $n = 2$;*

(d) *$\Gamma(R_{n,m})$ is never planar when $m \geq 5$.*

After establishing the above two results, Anderson et al. [75], considered the problem of which finite rings in general determine a planar zero-divisor graph. Also, if $\Gamma(R)$ is planar, then necessarily $\omega(\Gamma(R)) \leq 4$. Note that Theorem 3.1.44 identified the 31 non-isomorphic rings $R$ for which $\omega(\Gamma(R)) = 3$. Thus it was needed to determine the rings $R$ with $\omega(\Gamma(R)) = 4$ for which $\Gamma(R)$ is planar. This was partially answered by Akbari, Maimani, and Yassemi [37], who were able to refine the question to local rings of cardinality at most thirty-two.

**Theorem 4.2.3.** ([37, Theorems 1.2 and 1.4]) *Let $R$ be a finite commutative local ring with maximal ideal $\mathfrak{m}$. Then $\Gamma(R)$ is nonplanar if one of the following holds:*

(a) $|R/\mathfrak{m}| \geq 4$ *and* $|R| \geq 26$;

(b) $|R/\mathfrak{m}| = 3$ *and* $|R| \geq 28$;

(c) $|R/\mathfrak{m}| = 2$ *and* $|R| \geq 33$.

Thus, if $R$ is a finite commutative local ring that is not a field and contains at least thirty-three elements, then $\Gamma(R)$ is not planar. Also note that $\Gamma(\mathbb{Z}_{27})$ is planar. In view of these, Akbari, Maimani, and Yassemi [37] posed the question: "Is it true that, for any local ring $R$ of cardinality 32, which is not a field, $\Gamma(R)$ is nonplanar?". They also established that a finite commutative local ring (not a field) with more than 27 elements (or more than 9 zero-divisors) cannot have a planar zero-divisor graph. Akbari et al. [37] also proved the following result regarding the planarity of zero-divisor graphs.

**Theorem 4.2.4.** ([37, Theorem 1.6]) *Let $R$ be local commutative Artinian ring such that $\Gamma(R)$ is a planar graph. Then $R$ is quasi-Frobenius, $R \cong \mathbb{Z}_4[x]/(2,x)^2$, or $R \cong \mathbb{Z}_2[x,y]/(x,y)^2$.*

However, at the same time, N. Smith [355] independently provided a complete answer, as well as a classification, of the finite commutative rings $R$ for which $\Gamma(R)$ is planar, listing forty-four isomorphism classes in all. The question posed by Akbari, Maimani, and Yassemi [37] has also been answered independently by Wang [420], and Belshoff and Chapman [142]. Belshoff and Chapman also gave all finite commutative local rings with planar zero-divisor graphs using somewhat different techniques than [355].

First, we discuss answers for the question raised by Akbari, Maimani, and Yassemi [37] (also see Corollary 4.2.15).

**Theorem 4.2.5.** ([355, Corollary 3.8], [420, Theorem 3.2], [142, Proposition 5]) *If $R$ is a commutative local ring of cardinality* 32 *that is not a field, then $\Gamma(R)$ is not planar.*

*Proof.* Since $R$ is not a field and $|R| = 32 = 2^5$, $|R/\mathfrak{m}| = 2, 4, 8, 16$. If $|R/\mathfrak{m}| = 8$ or 16, then $\mathfrak{m} \neq 0$ and $|\mathfrak{m}| \leq 4$, so that $\mathfrak{m}/\mathfrak{m}^2$ is a nonzero vector space over the residue field $R/\mathfrak{m}$. It follows that $|\mathfrak{m}|/|\mathfrak{m}^2| = |R/\mathfrak{m}|^{dim_{R/\mathfrak{m}}\mathfrak{m}/\mathfrak{m}^2} \geq 8^1 = 8$, which is a contradiction to $|\mathfrak{m}| \leq 4$. Suppose now that $|R/\mathfrak{m}| = 4$. Then $|\mathfrak{m}| = 8$ and $\mathfrak{m}/\mathfrak{m}^2$ is a nonzero vector space over the residue field $R/\mathfrak{m}$. If $dim_{R/\mathfrak{m}}\mathfrak{m}/\mathfrak{m}^2 \geq 2$, then $|\mathfrak{m}|/|\mathfrak{m}^2| \geq 16$, which is a contradiction to $|\mathfrak{m}| = 8$. Therefore $dim_{R/\mathfrak{m}}\mathfrak{m}/\mathfrak{m}^2 = 1$, and then $|\mathfrak{m}|/|\mathfrak{m}^2| = 4$. It follows that $|\mathfrak{m}^2| = 2$. However, $\mathfrak{m}^2/\mathfrak{m}^3$ is a nonzero vector space over the residue field $R/\mathfrak{m}$; so $|\mathfrak{m}^2|/|\mathfrak{m}^3| \geq 4$, which contradicts the fact that $|\mathfrak{m}^2| = 2$. Consequently, we conclude that $R/\mathfrak{m}| = 2$.

We proceed with the proof by considering the dimension of $\mathfrak{m}/\mathfrak{m}^2$ over the field $R/\mathfrak{m}$. Note that $|\mathfrak{m}| = 16$; so $dim_{R/\mathfrak{m}}\mathfrak{m}/\mathfrak{m}^2 \leq 4$.

**Case 1.** $dim_{R/\mathfrak{m}}\mathfrak{m}/\mathfrak{m}^2 = 4$. In this case, $\mathfrak{m}^2 = 0$ and there are 15 nonzero elements in $\mathfrak{m}$. Thus, $\Gamma(R) \cong K_{15}$ and $\Gamma(R)$ is not planar by Lemma 4.1.1.

**Case 2.** $dim_{R/\mathfrak{m}}\mathfrak{m}/\mathfrak{m}^2 = 3$. In this case, $|\mathfrak{m}^2| = 2$ and $\mathfrak{m} = (x, y, z)$ for some elements $x, y, z \in \mathfrak{m}$. Since $|\mathfrak{m}^2| = 2$ and $\mathfrak{m}^2/\mathfrak{m}^3$ is a nonzero vector space over the residue field $R/\mathfrak{m}$, $dim_{R/\mathfrak{m}}\mathfrak{m}^2/\mathfrak{m}^3 = 1$ and $\mathfrak{m}^3 = 0$. Therefore, at least one of the elements of the set $\{x^2, y^2, z^2, xy, xz, yz\}$ is nonzero. If $x^2 \neq 0$, then $\{x^2\}$ is a basis for the vector space $\mathfrak{m}^2/\mathfrak{m}^3$; so there are elements $\alpha$ and $\beta$ of $R$ such that $xy - \alpha x^2 = xz - \beta x^2 = 0$. Therefore, we may replace $y$ by $y - \alpha x$, $z$ by $z - \beta x$, and assume that $xy = xz = 0$. Let $u_1 = x$, $u_2 = x^2$, $u_3 = x + x^2$, $v_1 = y$, $v_2 = z$, and $v_3 = y + z$; then $u_i \cdot v_j = 0$ for all $i, j$; so $K_{3,3} \subseteq \Gamma(R)$. It follows that $\Gamma(R)$ is not planar by Lemma 4.1.1. The cases where $y^2 \neq 0$ and $z^2 \neq 0$ are similar. Thus, we may assume that $u^2 = 0$ for any generator $u$ of $\mathfrak{m}$ and $xy \neq 0$. Since $\{xy\}$ is a basis for the vector space $\mathfrak{m}^2/\mathfrak{m}^3$, there is an element $\alpha$ of $R$ such that $xz - \alpha xy = 0$. Therefore, we may replace $z$ by $z - \alpha y$ and assume that $xz = 0$. (Note that $z - \alpha y$ is a generator of $\mathfrak{m}$.) Let $u_1 = xy$, $u_2 = x + xy$, $u_3 = z + xy$, $v_1 = x$, $v_2 = z$, and $v_3 = x + z$. Then $u_i \cdot v_j = 0$ for all $i, j$; so $K_{3,3} \subseteq \Gamma(R)$. It follows that $\Gamma(R)$ is not planar by Lemma 4.1.1.

**Case 3.** $dim_{R/\mathfrak{m}}\mathfrak{m}/\mathfrak{m}^2 = 2$. In this case, let $\mathfrak{m} = (x, y)$ for some $x, y \in \mathfrak{m}$. Since $|\mathfrak{m}^2| = 4$, either $|\mathfrak{m}^3| = 2$ or $\mathfrak{m}^3 = 0$.

Assume that $|\mathfrak{m}^3| = 2$. Then $\mathfrak{m}^4 = 0$ and $dim_{R/\mathfrak{m}}\mathfrak{m}^2/\mathfrak{m}^3 = dim_{R/\mathfrak{m}}\mathfrak{m}^3/\mathfrak{m}^4 = 1$. If $x^2, y^2 \in \mathfrak{m}^3$, then $\mathfrak{m}^3 = (x^3, x^2y, xy^2, y^3) \subseteq \mathfrak{m}^4$, a contradiction. Therefore, we may assume that $x^2 \notin \mathfrak{m}^3$. Hence $\{x^2\}$ is a basis for the vector space $\mathfrak{m}^2/\mathfrak{m}^3$. As before, we may further assume that $xy \in \mathfrak{m}^3$. If $y^2 \notin \mathfrak{m}^3$, then there is a unit $\lambda$ such that $x^2 - \lambda y^2 \in \mathfrak{m}^3$; so $y^3 = x^3 = 0$. It follows that $\mathfrak{m}^3 = 0$, a contradiction. Therefore, $y^2 \in \mathfrak{m}^3$ and $\{x^3\}$ is a basis for the vector space $\mathfrak{m}^3/\mathfrak{m}^4$. Let $u_1 = x^2$, $u_2 = x^3$, $u_3 = x^2 + x^3$, $v_1 = y, v_2 = y + x^2$, and $v_3 = y + x^3$. Then $u_i \cdot v_j = 0$ for all $i, j$; so $K_{3,3} \subseteq \Gamma(R)$. It follows that $\Gamma(R)$ is not planar by Lemma 4.1.1. Assume that $\mathfrak{m}^3 = 0$. Then there are 12 elements in $\mathfrak{m}$ which are not in $\mathfrak{m}^2$, and there are 3 elements in $\mathfrak{m}^2$ which are not zero. Since $\mathfrak{m}^3 = 0$, $K_{3,12} \subseteq \Gamma(R)$; so $\Gamma(R)$ is not planar by Lemma 4.1.1.

**Case 4.** $dim_{R/\mathfrak{m}}\mathfrak{m}/\mathfrak{m}^2 = 1$. In this case, $\mathfrak{m}$ is principal, and therefore $\mathfrak{m}^i$ is principal for every $i$. Thus $|\mathfrak{m}^2| = 8$, $|\mathfrak{m}^3| = 4$, $|\mathfrak{m}^4| = 2$, and $\mathfrak{m}^5 = 0$. Moreover, there is an element $x \in \mathfrak{m}$ such that $\mathfrak{m}^i = (x^i)$ for $i = 1, 2, 3, 4$, and $x^5 = 0$. Let $u_1 = x^3$, $u_2 = x^4$, $u_3 = x^3 + x^4$, $v_1 = x^2$, $v_2 = x^2 + x^3$, and $v_3 = x^2 + x^4$. Then $u_i \cdot v_j = 0$ for all $i, j$; so $K_{3,3} \subseteq \Gamma(R)$. It follows that $\Gamma(R)$ is not planar. □

Next, we are interested in the characterization of all commutative local rings with planar zero-divisor graph. This was first done by Smith in [355]. Let $(R, \mathfrak{m})$ be a finite commutative local ring. Then it is well known that there is an integer $n \geq 2$ such that $\mathfrak{m}^n = 0$ and $\mathfrak{m}^{n-1} \neq 0$. Thus $\mathfrak{m}^i/\mathfrak{m}^{i+1}$ is an $R/\mathfrak{m}$-vector space for every $1 \leq i \leq n-1$, and hence $|\mathfrak{m}^{n-1}| = |R/\mathfrak{m}|^d$, where $d = dim_{R/\mathfrak{m}}\mathfrak{m}^{n-1}$.

**Lemma 4.2.6.** ([355, Proposition 3.1]) *Let $(R, \mathfrak{m})$ be a finite commutative local ring with an integer $n \geq 2$ such that $\mathfrak{m}^n = 0$ and $\mathfrak{m}^{n-1} \neq 0$. If $\Gamma(R)$ is planar, then $n \leq 5$.*

**Lemma 4.2.7.** ([355, Lemma 3.2]) *If $(R, \mathfrak{m})$ is a finite commutative local ring with $0 = \mathfrak{m}^5 \subsetneq \mathfrak{m}^4 \subsetneq \cdots \subsetneq \mathfrak{m}$, then $\Gamma(R)$ is not planar.*

**Lemma 4.2.8.** ([355, Corollary 3.4]) *Let $(R, \mathfrak{m})$ be a finite commutative local ring with $\omega(\Gamma(R)) = 4$ and $\mathfrak{m}^2 = 0$. Then $R$ is isomorphic to either $\mathbb{Z}_{25}$ or $\mathbb{Z}_5[x]/(x^2)$, and so $\Gamma(R)$ is planar.*

**Lemma 4.2.9.** ([355, Lemma 3.5]) *Let $(R, \mathfrak{m})$ be a finite commutative local ring with $0 = \mathfrak{m}^4 \subsetneq \mathfrak{m}^3 \subsetneq \mathfrak{m}^2 \subsetneq \mathfrak{m}$. If $\Gamma(R)$ is planar, then $\omega(\Gamma(R)) = 3$. Thus $R$ is isomorphic to one of the following 5 rings: $\mathbb{Z}_{16}$, $\mathbb{Z}_2[x]/(x^4)$, $\mathbb{Z}_4[x]/(2x, x^3 - 2)$, $\mathbb{Z}_4[x]/(x^2 - 2)$, or $\mathbb{Z}_4[x]/(x^2 + 2x + 2)$.*

**Lemma 4.2.10.** ([355, Lemma 3.6]) *Let $(R, \mathfrak{m})$ be a finite commutative local ring with $0 = \mathfrak{m}^3 \subsetneq \mathfrak{m}^2 \subsetneq \mathfrak{m}$ and $\omega(\Gamma(R)) \geq 3$. If $\Gamma(R)$ is planar, then $\omega(\Gamma(R)) = 3$ and $R$ is isomorphic to one of the following 11 rings:*

$$\mathbb{Z}_{27}, \ \frac{\mathbb{Z}_3[x]}{(x^3)}, \ \frac{\mathbb{Z}_9[x]}{(x^2-3, 3x)}, \ \frac{\mathbb{Z}_9[x]}{(x^2-6, 3x)}, \ \frac{\mathbb{Z}_2[x,y]}{(x^2, y^2-xy)}, \ \frac{\mathbb{Z}_2[x,y]}{(x^2, y^2)}, \ \frac{\mathbb{Z}_8[x]}{(2x-4, x^2)},$$

$$\frac{\mathbb{Z}_4[x]}{(x^2)}, \ \frac{\mathbb{Z}_4[x]}{(x^2-2x)}, \ \frac{\mathbb{Z}_4[x,y]}{(x^2, xy-2, y^2-xy, 2x, 2y)}, \ \frac{\mathbb{Z}_4[x,y]}{(x^2, xy-2, y^2, 2x, 2y)}.$$

**Theorem 4.2.11.** ([355]) *Let $R$ be a finite commutative local ring which is not a field. Then $\Gamma(R)$ is a planar graph if and only if $R$ is isomorphic to one of the following 29 rings:*

$$\mathbb{Z}_4, \ \frac{\mathbb{Z}_2[x]}{(x^2)}, \ \mathbb{Z}_9, \ \frac{\mathbb{Z}_3[x]}{(x^2)}, \ \mathbb{Z}_8, \ \frac{\mathbb{Z}_2[x]}{(x^3)}, \ \frac{\mathbb{Z}_4[x]}{(2x, x^2-2)}, \ \mathbb{Z}_{16}, \ \frac{\mathbb{Z}_2[x]}{(x^4)}, \ \frac{\mathbb{Z}_4[x]}{(2x, x^3-2)},$$
$$\frac{\mathbb{Z}_4[x]}{(x^2-2)}, \ \frac{\mathbb{Z}_4[x]}{(x^2+2x+2)}, \ \frac{\mathbb{F}_4[x]}{(x^2)}, \ \frac{\mathbb{Z}_4[x]}{(x^2+x+1)}, \ \frac{\mathbb{Z}_2[x,y]}{(x,y)^2}, \ \frac{\mathbb{Z}_4[x]}{(2,x)^2}, \ \mathbb{Z}_{27}, \ \frac{\mathbb{Z}_3[x]}{(x^3)},$$
$$\frac{\mathbb{Z}_9[x]}{(x^2-3, 3x)}, \ \frac{\mathbb{Z}_9[x]}{(x^2-6, 3x)}, \ \frac{\mathbb{Z}_2[x,y]}{(x^2, y^2-xy)}, \ \frac{\mathbb{Z}_2[x,y]}{(x^2, y^2)}, \ \frac{\mathbb{Z}_8[x]}{(2x-4, x^2)}, \ \frac{\mathbb{Z}_4[x]}{(x^2)}, \ \frac{\mathbb{Z}_4[x]}{(x^2-2x)},$$
$$\frac{\mathbb{Z}_4[x,y]}{(x^2, xy-2, y^2-xy, 2x, 2y)}, \ \frac{\mathbb{Z}_4[x,y]}{(x^2, xy-2, y^2, 2x, 2y)}, \ \mathbb{Z}_{25}, \ \frac{\mathbb{Z}_5[x]}{(x^2)}.$$

As a continuation of this, Smith [355] characterized all commutative nonlocal rings with planar zero-divisor graph.

**Lemma 4.2.12.** ([355, Proposition 2.5]) *Let $R_1, \ldots, R_n$ be finite commutative rings, $R$ a finite commutative ring which is not an integral domain, and $D$ an integral domain. Then*

(a) $\omega(\Gamma(R_1 \times \cdots \times R_n)) \geq \omega(\Gamma(R_1)) + \cdots + \omega(\Gamma(R_n))$;
(b) $\omega(\Gamma(R \times D)) = \omega(\Gamma(R)) + 1$.

**Theorem 4.2.13.** ([355, Theorem 2.12]) *Let $R$ be a finite commutative nonlocal ring and $F$ be a finite field. Then $\Gamma(R)$ is a planar graph if and only if $R$ is isomorphic to one of the following* 15 *rings:*

$$\mathbb{Z}_2 \times F,\ \mathbb{Z}_3 \times F,\ \mathbb{Z}_2 \times \mathbb{Z}_4,\ \mathbb{Z}_2 \times \frac{\mathbb{Z}_2[x]}{(x^2)}, \mathbb{Z}_3 \times \mathbb{Z}_4,\ \mathbb{Z}_3 \times \frac{\mathbb{Z}_2[x]}{(x^2)},\ \mathbb{Z}_2 \times \mathbb{Z}_2 \times \mathbb{Z}_2,$$
$$\mathbb{Z}_2 \times \mathbb{Z}_2 \times \mathbb{Z}_3, \mathbb{Z}_2 \times \mathbb{Z}_8,\ \mathbb{Z}_2 \times \frac{\mathbb{Z}_2[x]}{(x^3)},\ \mathbb{Z}_2 \times \frac{\mathbb{Z}_4[x]}{(2x, x^2-2)}, \mathbb{Z}_2 \times \mathbb{Z}_9, \mathbb{Z}_2 \times \frac{\mathbb{Z}_3[x]}{(x^2)},$$
$$\mathbb{Z}_3 \times \mathbb{Z}_9,\ \mathbb{Z}_3 \times \frac{\mathbb{Z}_3[x]}{(x^2)}.$$

From Lemma 4.2.12 and Theorem 4.2.13, we get the following result.

**Theorem 4.2.14.** ([355, Theorem 3.7]) *Let $R$ be a finite commutative ring (not a field) and $F$ be a finite field. Then $\Gamma(R)$ is a planar graph if and only if $R$ is isomorphic to one of the following* 44 *types of rings:*

$$\mathbb{Z}_2 \times F,\ \mathbb{Z}_3 \times F,\ \mathbb{Z}_2 \times \mathbb{Z}_4,\ \mathbb{Z}_2 \times \frac{\mathbb{Z}_2[x]}{(x^2)}, \mathbb{Z}_3 \times \mathbb{Z}_4,\ \mathbb{Z}_3 \times \frac{\mathbb{Z}_2[x]}{(x^2)},\ \mathbb{Z}_2 \times \mathbb{Z}_2 \times \mathbb{Z}_2,$$
$$\mathbb{Z}_2 \times \mathbb{Z}_2 \times \mathbb{Z}_3, \mathbb{Z}_2 \times \mathbb{Z}_8,\ \mathbb{Z}_2 \times \frac{\mathbb{Z}_2[x]}{(x^3)},\ \mathbb{Z}_2 \times \frac{\mathbb{Z}_4[x]}{(2x, x^2-2)}, \mathbb{Z}_2 \times \mathbb{Z}_9,\ \mathbb{Z}_2 \times \frac{\mathbb{Z}_3[x]}{(x^2)},$$
$$\mathbb{Z}_3 \times \mathbb{Z}_9,\ \mathbb{Z}_3 \times \frac{\mathbb{Z}_3[x]}{(x^2)}, \mathbb{Z}_4,\ \frac{\mathbb{Z}_2[x]}{(x^2)},\ \mathbb{Z}_9,\ \frac{\mathbb{Z}_3[x]}{(x^2)},\ \mathbb{Z}_8,\ \frac{\mathbb{Z}_2[x]}{(x^3)},\ \frac{\mathbb{Z}_4[x]}{(2x, x^2-2)},\ \mathbb{Z}_{16},\ \frac{\mathbb{Z}_2[x]}{(x^4)},$$
$$\frac{\mathbb{Z}_4[x]}{(2x, x^3-2)},\ \frac{\mathbb{Z}_4[x]}{(x^2-2)},\ \frac{\mathbb{Z}_4[x]}{(x^2+2x+2)},\ \frac{\mathbb{F}_4[x]}{(x^2)},\ \frac{\mathbb{Z}_4[x]}{(x^2+x+1)},\ \frac{\mathbb{Z}_2[x,y]}{(x,y)^2},\ \frac{\mathbb{Z}_4[x]}{(2,x)^2},\ \mathbb{Z}_{27},$$
$$\frac{\mathbb{Z}_3[x]}{(x^3)},\ \frac{\mathbb{Z}_9[x]}{(x^2-3, 3x)},\ \frac{\mathbb{Z}_9[x]}{(x^2-6, 3x)},\ \frac{\mathbb{Z}_2[x,y]}{(x^2, y^2-xy)},\ \frac{\mathbb{Z}_2[x,y]}{(x^2, y^2)},\ \frac{\mathbb{Z}_8[x]}{(2x-4, x^2)},\ \frac{\mathbb{Z}_4[x]}{(x^2)},$$
$$\frac{\mathbb{Z}_4[x]}{(x^2-2x)},\ \frac{\mathbb{Z}_4[x,y]}{(x^2, xy-2, y^2-xy, 2x, 2y)},\ \frac{\mathbb{Z}_4[x,y]}{(x^2, xy-2, y^2, 2x, 2y)},\ \mathbb{Z}_{25},\ \frac{\mathbb{Z}_5[x]}{(x^2)}.$$

**Corollary 4.2.15.** ([355, Corollary 3.8]) *Let $R$ be a finite commutative local ring (not a field) with either $|R| \geq 28$ or $|Z(R)| \geq 10$. Then $\Gamma(R)$ is not planar.*

Note that the above corollary is the best possible since $R = \mathbb{Z}_{27}$ is a local ring with $|R| = 27$, $|Z(R)| = 9$, and planar zero-divisor graph. Smith [356] has also characterized the infinite planar zero-divisor graphs.

Hereafter, in this section we will denote the cardinality of the real numbers by $c$, and $\aleph_0$ will denote the cardinality of a countably infinite set.

**Proposition 4.2.16.** ([356, Proposition 2.1]) *Let $D$ be an integral domain with $|D| \leq c$. Then $\Gamma(D \times \mathbb{Z}_2)$ and $\Gamma(D \times \mathbb{Z}_3)$ are planar graphs.*

In the case where $R$ decomposes (nontrivially) as the product $R \cong \prod_{\alpha\in A} R_\alpha$, we will see that the rings described in the above proposition are the only infinite rings whose zero-divisor graph is planar. Note that if $R \cong S \times T$, then $\Gamma(R)$ contains a $K_{|S|-1,|T|-1}$. Further, it is easy to see that if $R \cong \prod_{\alpha\in A} R_\alpha$ and $\Gamma(R)$ is planar, then $|A| \leq 3$; for $|A| \geq 4$ would force $\Gamma(R)$ to contain a $K_{3,3}$. Thus, if $R \cong \prod_{\alpha\in A} R_\alpha$, we may assume that $A$ is finite and that one of the factors is an infinite ring. This, together with the previous comments, forces $|A| = 2$, i.e., $R \cong R_1 \times R_2$, where without loss of generality, $R_1$ is infinite and $R_2$ is either $\mathbb{Z}_2$ or $\mathbb{Z}_3$.

In the case where $R \cong R_1 \times \mathbb{Z}_3$, we can almost immediately see that if $R$ is infinite and $\Gamma(R)$ is planar, then $R_1$ must be a domain.

**Lemma 4.2.17.** ([356, Lemma 2.2]) *Let $R$ be a commutative ring. If $\Gamma(R \times \mathbb{Z}_3)$ is planar, then $\Gamma(R)$ contains no vertices of degree 3 or more.*

**Proposition 4.2.18.** ([356, Proposition 2.3]) *Let $R$ be an infinite commutative ring. If $\Gamma(R \times \mathbb{Z}_3)$ is planar, then $R$ is an integral domain with $|R| \leq c$.*

**Lemma 4.2.19.** ([356, Lemma 2.4]) *Let $R$ be an infinite commutative ring with $\Gamma(R)$ planar. Then $deg(v) = 1,\ 2,$ or $\infty$ for every $v \in V(\Gamma(R))$.*

**Proposition 4.2.20.** ([356, Proposition 2.5]) *Let $R$ be an infinite commutative ring. If $\Gamma(R \times \mathbb{Z}_2)$ is planar, then $R$ is an integral domain with $|R| \leq c$.*

**Theorem 4.2.21.** ([356, Theorem 2.6]) *Suppose that $R$ is an infinite commutative ring and $R \cong \prod_{\alpha\in A} R_\alpha$ with $|A| \geq 2$. If $\Gamma(R)$ is planar, then $|A| = 2$ and further, $R \cong R_1 \times \mathbb{Z}_2$ or $R \cong R_1 \times \mathbb{Z}_3$, where $R_1$ is an infinite integral domain with $|R_1| \leq c$.*

On the basis of the above result, one might hope to prove that if $R$ is infinite and $\Gamma(R)$ is planar, then $R$ is isomorphic to one of the two types of rings listed in the above theorem. Unfortunately, it is not possible to make such a strong statement. For example, if $R = \mathbb{Z}[x]/(2x, x^2)$, then $\Gamma(R)$ is a star graph with a countably infinite number of vertices. Thus, in this case, $\Gamma(R)$ is planar

**Proposition 4.2.22.** ([356, Proposition 2.9]) *Let $R$ be an infinite commutative ring with $\Gamma(R)$ planar, and suppose that $\Gamma(R)$ contains a vertex $v$ of degree two. If $R$ is reduced, then $R \cong D \times \mathbb{Z}_3$, where $D$ is an infinite integral domain with $|D| \leq c$.*

In the above proposition, $\Gamma(R) = K_{2,|D|}$. If we drop the requirement that $R$ be reduced, we can generate examples so that $\Gamma(R)$ is planar and contains vertices of degree two, but $\Gamma(R)$ is not complete bipartite. Let $R = \mathbb{Z}[x]/(x^2, 3x)$. Then every vertex of $\Gamma(R)$ has degree two (or has infinite degree) and $\Gamma(R)$ is planar. Further, note that $\Gamma(R)$ has the form of a $K_{2,\aleph_0}$ with one additional edge.

**Proposition 4.2.23.** ([356, Proposition 2.11]) *Let $R$ be an infinite commutative ring. If $\Gamma(R)$ is planar, then $\Gamma(R)$ does not contain vertices of both degree one and degree two.*

**Proposition 4.2.24.** ([356, Proposition 2.12]) *Let $R$ be an infinite commutative ring with $\Gamma(R)$ planar. If $x$ and $y$ are two distinct vertices of degree 2 in $\Gamma(R)$, then $(0 : x) = (0 : y)$.*

**Theorem 4.2.25.** ([356, Theorem 2.14]) *Let $R$ be an infinite commutative ring such that $\Gamma(R)$ is planar and contains a vertex $x$ of degree two. Then $\Gamma(R)$ is graph-isomorphic to either $K_{2,\alpha}$ (where $\aleph_0 \leq \alpha \leq c$) or the graph obtained from such a $K_{2,\alpha}$ by adding an edge joining its two vertices of infinite degree.*

The previous theorem allows us to generate examples of planar graphs which cannot be realized as $\Gamma(R)$ for any infinite commutative ring $R$. For example, let $G = K_{2,\alpha}$, where $\alpha \leq c$. Let $a_1$ and $a_2$ denote the vertices of infinite degree in $G$. Denote by $H_1$ the graph formed by taking $G$ and adding an edge joining $a_1$ and $a_2$. Let $H_2$ denote another copy of $H_1$, and denote the vertices of infinite degree in $H_2$ by $b_1$ and $b_2$. Let $H$ be the graph formed from $H_1$ and $H_2$ by adding edges between $a_1$ and $b_1$, and between $a_2$ with $b_2$. Then $H$ is a planar graph, for $H_1$ and $H_2$ are both planar graphs. Also, $H$ has diameter 3 and girth 4, but by the previous theorem this graph cannot be realized as $\Gamma(R)$ for any (infinite) commutative ring $R$.

It is possible to construct planar graphs where every vertex has infinite degree. To see how this is done, let $G_1$ denote a copy of $K_{1,\aleph_0}$. Note that $G_1$ is a planar graph. Let $x$ be the vertex of infinite degree in $G_1$ and let $x_1, x_2, \ldots$ denote the vertices of degree 1 of this graph. Then, let $G_2$ be the graph obtained by taking $G_1$ and attaching a (graph-isomorphic) copy of $G_1$ at each of the vertices of degree 1. Inductively, we then define $G_{n+1}$ to be the graph obtained by taking $G_n$ and attaching an isomorphic copy of $G_1$ at each of the vertices of degree 1 in $G_n$. Then, let $G = \cup_{i=1}^{\infty} G_i$. It is easy to check that this graph is planar. However, this graph cannot be realized as $\Gamma(R)$ for any commutative ring $R$. For in the graph in question, given any positive integer $n$, there exists a pair of points at distance $n$. Since for any commutative ring $R$, we have $diam(\Gamma(R)) \leq 3$, the graph $G$ cannot be $\Gamma(R)$ for any commutative ring $R$. This example motivates the following proposition.

**Proposition 4.2.26.** ([356, Proposition 2.17]) *Let $R$ be an infinite commutative ring. If every vertex of $\Gamma(R)$ has infinite degree, then $\Gamma(R)$ is not planar.*

The next result shows that in the case $R$ is an infinite commutative ring, if $\Gamma(R)$ contains a vertex of degree one and $\Gamma(R)$ is planar, then $\Gamma(R)$ must be a star graph.

**Theorem 4.2.27.** ([356, Theorem 2.19]) *Let $R$ be an infinite commutative ring with $\Gamma(R)$ planar. Then $\Gamma(R)$ is graph-isomorphic to either a star graph, a $K_{2,\alpha}$,* *where $\alpha \leq c$, or the graph obtained by taking such a $K_{2,\alpha}$ and adding an edge between the two vertices of infinite degree.*

Note that if $\Gamma(R)$ is to be planar, we must have $|V(\Gamma(R))| \leq c$. Note that if $F$ is a field of cardinality $\alpha$, where $\alpha$ is an infinite cardinal, then $\Gamma(F \times \mathbb{Z}_2)$ and $\Gamma(F \times \mathbb{Z}_3)$ are a $K_{1,\alpha}$ and a $K_{2,\alpha}$ respectively, and each of these is a graph on $\alpha$ vertices. Also, if $R = \mathbb{Z}[x]/(x^2, 3x)$, then $\Gamma(R)$ has the form of a $K_{2,\beta}$ with one additional edge. Note that this is a graph on $\aleph_0$ vertices.

Let $R = \dfrac{\mathbb{Z}[x_0, \{x_\gamma \mid \gamma \in S\}]}{(x_0^2, 3x_0, \{x_0 x_\gamma \mid \gamma \in S\})}$, where $S$ is a nonempty set with $|S| \leq c$ and $0 \notin S$. Then $\Gamma(R)$ has the form of a $K_{2,\beta}$ with one additional edge. This is a graph on $\max\{\aleph_0, |S|\}$ vertices. Thus every graph as listed in Theorem 4.2.27 can be realized as $\Gamma(R)$ for an appropriate infinite commutative ring $R$.

Finally, Smith, posed a question: "Is it possible to explicitly classify those infinite commutative rings $R$ with $\Gamma(R)$ isomorphic to a star graph, a $K_{2,\alpha}$, or the graph obtained from adding an edge between the two vertices of infinite degree in such a $K_{2,\alpha}$?".

If we add edges to $K_{n,\alpha}$ by joining all distinct vertices within the vertex subset of cardinality $n$, then the resulted graph is denoted by $K_{[n],\alpha}$. Therefore, at present we have for an infinite commutative ring $R$ (not a an integral domain), $\Gamma(R)$ is planar if and only if $\Gamma(R)$ is isomorphic to either $K_{1,\alpha}$, $K_{2,\alpha}$, or $K_{[2],\alpha}$, with $|\mathbb{Z}| \leq |\alpha| \leq |\mathbb{R}|$.

The open question of Smith, finding a ring-theoretic characterization of an infinite ring $R$ such that $\Gamma(R)$ is isomorphic to either $K_{1,\alpha}$, $K_{2,\alpha}$, or $K_{[2],\alpha}$, with $|\mathbb{Z}| \leq |\alpha| \leq |\mathbb{R}|$, was studied by Yongwei Yao in [441]. In fact, Y. Yao studied the ring structure of $R$ such that $\Gamma(R)$ is $K_{p-1,\alpha}$ or $K_{[p-1],\alpha}$ for any prime $p$, which gives a characterization of all infinite commutative rings $R$ such that $\Gamma(R)$ is planar. Note that the structure of finite commutative rings $R$ with bipartite $\Gamma(R)$ was discussed in Section 3.4.

Hereafter, throughout this section, we assume that $\Gamma(R) \cong K_{p-1,\alpha}$ or $\Gamma(R) \cong K_{[p-1],\alpha}$, where $p$ is a prime number (note that $K_{1,\alpha} = K_{[1],\alpha}$). In either case, there is a bipartition of the vertex set $V = Z(R)^*$, say $V = U \cup W$, a disjoint union, such that $|U| = p - 1$, while $|W| = \alpha$. Let $\mathfrak{a} = U \cup \{0\}$, which is a subset of $R$ and $\mathfrak{i} = (0 :_R \mathfrak{a}) = \{r \in R \mid ru = 0, \ \forall\, u \in \mathfrak{a}\}$, an ideal of $R$. These notations and assumptions will be kept throughout this section.

Let $p$ be prime and $n \in \mathbb{Z}$ be such that $p \nmid n$. We say $n$ is a quadratic residue modulo $p$ if $n \equiv k^2$ $(mod\ p)$ for some $k \in \mathbb{Z}$. Otherwise, we say $n$ is a quadratic nonresidue modulo $p$.

**Lemma 4.2.28.** ([441, Lemma 1.1]) *Let $R$ be a ring such that $\Gamma(R) \cong K_{p-1,\alpha}$ or $\Gamma(R) \cong K_{[p-1],\alpha}$. Moreover, assume $\alpha \geq 2$.*

(a) *We have $W \subseteq \mathfrak{i}$ and the subset $\mathfrak{a}$ is an ideal of $R$. In fact, we have $\mathfrak{a} = (0 :_R W) = (0 :_R w)$ and $ww' \neq 0$ for every $w, w' \in W$;*

(b) *Thus* $\mathfrak{a} \cong \mathbb{Z}_p$ *as* $\mathbb{Z}$*-modules. Hence,* $\mathfrak{a}$ *is generated by* $u$ *over* $\mathbb{Z}$ *(and therefore over* $R$*) for every* $u \in U$*. Consequently, we have* $\mathfrak{i} = (0 :_R u)$ *for every* $u \in U$*;*

(c) *We have* $\mathfrak{a} \cong R/\mathfrak{i}$ *as* $R$*-modules and* $R/\mathfrak{i} \cong \mathbb{Z}_p$ *as rings, which implies that* $\mathfrak{i}$ *is a maximal ideal of* $R$*. In particular,* $p \in \mathfrak{i}$*;*

(d) *If* $\mathfrak{a} \not\subseteq \mathfrak{i}$*, then* $\Gamma(R) \cong K_{p-1,\alpha}$*. If* $\mathfrak{a} \subseteq \mathfrak{i}$*, then* $\Gamma(R) \cong K_{[p-1],\alpha}$*;*

(e) $\mathfrak{a} \subseteq \mathfrak{i}$ *if and only if* $\mathfrak{a}^2 = 0$*, if and only if* $uu' = 0$ *for some* $u,\ u' \in U$*. In this case, we have a disjoint union* $\mathfrak{i} = \{0\} \cup U \cup W$*, which forces* $\mathfrak{a} \subsetneq \mathfrak{i}$*.*

**Proposition 4.2.29.** ([441, Proposition 1.2]) *Assume* $\alpha \geq 2$*. If* $\mathfrak{a} \not\subseteq \mathfrak{i}$*, then* $\mathfrak{a} \cap \mathfrak{i} = 0$*,* $\mathfrak{a} \in Spec(R)$*, and* $R \cong R_1 \times \mathbb{Z}_p$ *with* $R_1$ *an integral domain such that* $|R_1| = \alpha + 1$*. Conversely, if* $R \cong R_1 \times \mathbb{Z}_p$ *for any integral domain* $R_1$*, we have* $\Gamma(R) \cong K_{p-1,|R_1|-1}$*.*

Now that we are done with the case where $\mathfrak{a} \not\subseteq \mathfrak{i}$, we next study the case where $\mathfrak{a} \subseteq \mathfrak{i}$. In light of Lemma 4.2.28(a), we see that it cannot actually happen that $\mathfrak{a} = \mathfrak{i}$ under our assumption on $\Gamma(R)$ and that $\alpha \geq 2$. Thus it remains to study the case where $\mathfrak{a} \subsetneq \mathfrak{i}$ only.

**Proposition 4.2.30.** ([441, Proposition 1.3]) *Assume that* $\alpha \geq 2$*. If* $\mathfrak{a} \subsetneq \mathfrak{i}$ *and* $\mathfrak{a} \in Spec(R)$*, then* $R$ *can be characterized as an infinite ring with a prime ideal* $\mathfrak{a}$ *such that* $\mathfrak{a}^2 = 0$ *and* $|\mathfrak{a}| = p$*. In this case, we have* $dim(R) \geq 1$ *and* $\mathfrak{a} = Nil(R)$*, where* $Nil(R)$ *denotes the nilradical of* $R$*.*

*Conversely, if* $R$ *is an infinite ring such that the nilradical* $Nil(R)$ *is a prime ideal and* $|Nil(R)| = p$*, then* $\Gamma(R) \cong K_{p-1,|R|}$*.*

**Proposition 4.2.31.** ([441, Proposition 1.5]) *Assume that* $\alpha \geq 2$*,* $\mathfrak{a} \subsetneq \mathfrak{i}$*, and* $\mathfrak{a} \in Spec(R)$*. Then the following statements hold:*

(a) *We have* $w_1 w_2 \in U$ *for all* $w_1, w_2 \in W$*. Hence we have* $\mathfrak{i}^2 \subseteq \mathfrak{a}$ *and* $\mathfrak{i}^3 = 0$*;*

(b) *Fix any* $x \in W$*. Then* $px^2 = 0$ *and every element* $r \in \mathfrak{i}$ *can be written as* $r = n_2 x^2 + n_1 x$ *with unique* $n_i \in \mathbb{Z}$ *such that* $0 \leq n_i \leq p-1$*. Thus,* $|\mathfrak{i}| = p^2$*;*

(c) *Consequently, any element* $s \in R$ *can be written as* $s = n_2 x^2 + n_1 x + n_0$ *with unique* $n_i \in \mathbb{Z}$ *such that* $0 \leq n_i \leq p-1$*. Thus* $R$ *is finite with* $|R| = p^3$*, and as an algebra,* $R$ *is generated by* $x$ *over* $\mathbb{Z}$*.*

**Proposition 4.2.32.** ([441, Proposition 1.6]) *Assume that* $\alpha \geq 2$*,* $\mathfrak{a} \subsetneq \mathfrak{i}$*, and* $\mathfrak{a} \notin Spec(R)$*. Then* $\Gamma(R) \cong K_{[p-1],p(p-1)}$*, and exactly one of the following will happen:*

(a) *If* $p \notin \mathfrak{a}$*, then* $R \cong \mathbb{Z}_{p^3}$*;*

(b) *If* $0 = p \in R$*, then* $R \cong \mathbb{Z}[x]/(p, x^3) \cong \mathbb{Z}_p[x]/(x^3)$*;*

(c) *If* $0 \neq p \in \mathfrak{a}$ *and* $x^2 = p \in R$ *for some* $x \in W$*, then* $R \cong \mathbb{Z}[x]/(px, x^2 - p) \cong \mathbb{Z}_{p^2}[x]/(px, x^2 - p)$*;*

(d) *If* $0 \neq p \in \mathfrak{a}$ *and* $w^2 \neq p \in R$ *for any* $w \in W$, *then* $R \cong \mathbb{Z}_{p^2}[x]/(px, x^2 - np)$, *where* $n$ *is the least positive integer that is a quadratic nonresidue modulo* $p$. *This case will never happen when* $p = 2$.

We next concentrate on what happens when $\alpha = 1$, i.e., $\Gamma(R) \cong K_{p-1,1}$ or $\Gamma(R) \cong K_{[p-1],1}$. In fact, we consider cases where $\Gamma(R) \cong K_{n,1}$ and $\Gamma(R) \cong K_{[n],1}$ for any integer $n \geq 1$.

**Remark 4.2.33.** ([441, Remark 1.7]) Let $R$ be a commutative ring such that $\Gamma(R) \cong K_{n,1}$ or $\Gamma(R) \cong K_{[n],1}$ for any integer $n \geq 1$. As $\Gamma(R)$ is finite, we see that $R$ is a finite ring.

(1) Assume that $\Gamma(R) \cong K_{n,1}$ and $n \geq 2$, that is, $\Gamma(R) \cong K_{1,n}$ and $n \geq 2$. By the above propositions, we see that either $R \cong \mathbb{F}_{n+1} \times \mathbb{Z}_2$, where $\mathbb{F}_{n+1}$ is the field of cardinality $n+1$ (in case Proposition 4.2.29 applies and $n+1$ is a prime power), or else $R \cong \mathbb{Z}_8$, $\mathbb{Z}_2[x]/(x^3)$, or $\mathbb{Z}_4[x]/(2x, x^2-2)$ (in case Proposition 4.2.32 applies and $n = 2$).

(2) Assume that $\Gamma(R) \cong K_{1,1}$ or $\Gamma(R) \cong K_{[n],1}$ for some $n \geq 1$. Then $\Gamma(R)$ is just a complete graph on 2 or more vertices.

In the next theorem, Yao [441] characterized the structure of an infinite commutative ring $R$ when $\Gamma(R)$ is planar.

**Theorem 4.2.34.** ([441, Theorem 2.1]) *Let* $R$ *be a commutative ring that is not an integral domain. Then the following statements are equivalent:*

(a) $R$ *is an infinite ring such that* $\Gamma(R)$ *is planar;*

(b) $|\mathbb{Z}| \leq |R| \leq |\mathbb{R}|$, *and moreover,* $R$ *falls into exactly one of the two categories;*
   (i) $R \cong R_1 \times \mathbb{Z}_p$ *with* $R_1$ *an integral domain and* $p = 2$ *or* $p = 3$;
   (ii) *The nilradical of* $R$ *satisfies* $Nil(R) \in Spec(R)$ *and* $|Nil(R)| = p$ *with* $p = 2$ *or* $p = 3$. *If this is the case, then* $Nil(R)^2 = 0$;

(c) $|\mathbb{Z}| \leq |R| \leq |\mathbb{R}|$ *and there exists* $\mathfrak{a} \in Spec(R)$ *such that* $|\mathfrak{a}| = p$ *with* $p = 2$ *or* $p = 3$;

(d) $R \cong T/I$, *where* $T = \mathbb{Z}[\{X_\lambda \mid \lambda \in \mathbb{R}\}]$, $I$ *is an ideal of* $T$ *such that* $I \subsetneq \mathfrak{p}$ *and* $\mathfrak{p}/I \cong T/\mathfrak{n}$ *as* $T$*-modules for some* $\mathfrak{p} \in Spec(T)$ *satisfying* $|T/\mathfrak{p}| = \infty$, *and* $\mathfrak{n} = (p, \{X_\lambda \mid \lambda \in \mathbb{R}\})$ *with* $p = 2$ *or* $p = 3$;

(e) $R \cong T/I$, *where* $T = \mathbb{Z}[\{X_\lambda \mid \lambda \in \Lambda\}]$ *(for some index set* $\Lambda$*) and* $I$ *is an ideal of* $T$ *such that* $I \subsetneq \mathfrak{p}$ *for some* $\mathfrak{p} \in Spec(T)$ *satisfying* $|\mathbb{Z}| \leq |T/\mathfrak{p}| \leq |\mathbb{R}|$ *and* $|\mathfrak{p}/I| = p$ *with* $p = 2$ *or* $p = 3$.

Theorem 4.2.34(d) provides a way to construct all infinite commutative rings $R$ (up to isomorphism) such that $\Gamma(R)$ is planar. Form the polynomial ring $T = \mathbb{Z}\{[X_\lambda \mid \lambda \in \mathbb{R}\}]$. Let $\mathfrak{n} = (p, \{X_\lambda \mid \lambda \in \Lambda\})$. Choose any nonzero prime ideal $\mathfrak{p} \neq 0$ of $T$ such that $|T/\mathfrak{p}| = \infty$. We claim that $\mathfrak{p}\mathfrak{n} \subsetneq \mathfrak{p}$, and hence $\mathfrak{p}/(\mathfrak{p}\mathfrak{n})$ is a nonzero vector space over $T/\mathfrak{n}$. To show the claim, suppose to the contrary that $\mathfrak{p}\mathfrak{n} = \mathfrak{p}$. Then we would have $\mathfrak{p} = \mathfrak{p}\mathfrak{n}^i$ for every $i \geq 1$, and hence $\mathfrak{p} \subseteq \cap_{i=1}^{\infty}\mathfrak{n}^i = 0$, a contradiction. To see why $\cap_{i=1}^{\infty}\mathfrak{n}^i = 0$, consider any $0 \neq f \in T$. Choose any nonzero term of $f$, say of monomial degree $d$ and with coefficient $n = p^m a$ such that $p \nmid a$. Then we see that $f \notin \mathfrak{n}^{m+d+1}$. Thus, there exists an ideal $I$ of $T$ such that $\mathfrak{p}\mathfrak{n} \subseteq I \subsetneq \mathfrak{p}$ and $\mathfrak{p}/I \cong T/\mathfrak{n}$ as $T$-modules. Let $R = T/I$ and let $\mathfrak{m} = \mathfrak{n}/I$ and $\mathfrak{b} = \mathfrak{p}/I$, which are ideals of $R$. Then $\mathfrak{b}$ is a prime ideal of $R$ such that $|\mathfrak{b}| = |\mathfrak{p}/I| = p$. By Theorem 4.2.34(d), we see that $R$ is an infinite commutative ring and $\Gamma(R)$ is planar.

Also, Theorem 4.2.34(b) implies that the set of all infinite rings with planar zero-divisor graphs contains the set of rings of the form $R_1 \times \mathbb{Z}_p$ or $R_2(+)\mathbb{Z}_p$, where each $R_i$ is an integral domain with $|\mathbb{Z}| \leq |R_i| \leq |\mathbb{R}|$, $p = 2$ or $3$, and $R_2$ maps onto $\mathbb{Z}_p$. However, the latter set is a proper subset of the former.

More generally, the following theorem is a characterization of the structure of infinite rings $R$ with $\Gamma(R)$ graph-isomorphic to $K_{p-1,\alpha}$ or $K_{[p-1],\alpha}$.

**Theorem 4.2.35.** ([441, Theorem 2.3]) *Let $R$ be a commutative ring that is not an integral domain and $p$ be a prime number. Then the following statements are equivalent:*

(a) *$R$ is an infinite ring and $\Gamma(R)$ is graph-isomorphic to $K_{p-1,\alpha}$ or $K_{[p-1],\alpha}$;*

(b) *$R$ is an infinite ring that falls into exactly one of the two categories:*

  (i) *$R \cong R_1 \times \mathbb{Z}_p$ with $R_1$ an integral domain;*

  (ii) *The nilradical of $R$ satisfies $Nil(R) \in Spec(R)$ and $|Nil(R)| = p$;*

(c) *$|R| = \infty$ and there exists $\mathfrak{a} \in Spec(R)$ such that $|\mathfrak{a}| = p$;*

(d) *$R \cong T/I$, where $T = \mathbb{Z}[\{X_\lambda \mid \lambda \in \Lambda\}]$ (for some index set $\Lambda$), $I$ is an ideal of $T$ such that $I \subsetneq \mathfrak{p}$ and $\mathfrak{p}/I \cong T/\mathfrak{n}$ as $T$-modules for some $\mathfrak{p} \in Spec(T)$ satisfying $|T/\mathfrak{p}| = \infty$ and $\mathfrak{n} = (p, \{X_\lambda \mid \lambda \in \Lambda\})$;*

(e) *$R \cong T/I$, where $T = \mathbb{Z}[\{X_\lambda \mid \lambda \in \Lambda\}]$ (for some index set $\Lambda$) and $I$ is an ideal of $T$ such that $I \subsetneq \mathfrak{p}$ for some $\mathfrak{p} \in Spec(T)$ satisfying $|T/\mathfrak{p}| = \infty$ and $|\mathfrak{p}/I| = p$.*

## 4.3 Zero-divisor Graphs with Genus One

Let $S_k$ be an oriented surface with $k$ handles, where $k$ is a nonnegative integer. Recall that the orientable *genus* of $G$, denoted by $g(G)$, is the minimum integer $n$ such that $G$ can be embedded in $S_n$. Intuitively, $G$ is embedded in a surface if it can be drawn in the surface so that its edges intersect only at their common vertices. A graph $G$ is called *toroidal* if $g(G) = 1$. Note that if $H$ is a subgraph of $G$, then $g(H) \leq g(G)$. For details on the notion of embedding of graphs in a surface, we refer to White [429].

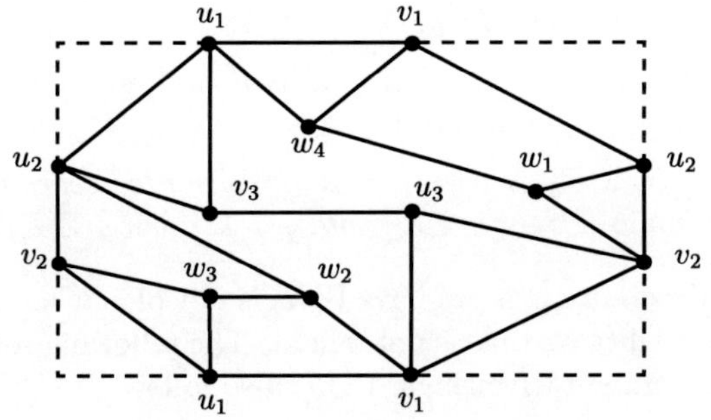


Figure 4.1: The graph $\tilde{G}$

A *minor* of $G$ is a graph obtained from $G$ by contracting edges in $G$ or deleting edges and isolated vertices in $G$. A classical theorem due to Wagner states that a graph $G$ is planar if and only if $G$ does not have $K_5$ or $K_{3,3}$ as a minor. It is well known that if $G'$ is a minor of $G$, then $g(G') \le g(G)$. For $xy \in E(G)$, we denote the contracted edge by the vertex $[x, y]$. Also, if $H$ is a subgraph of $G$ and $H'$ is a minor of $H$, then we call $H'$ a minor subgraph of $G$.

In this section, we are interested in the commutative rings that have genus one zero-divisor graphs. This study was first initiated by Hsin-Ju Wang in [420]. Actually, he found all rings of the form $\mathbb{Z}_{p_1^{\alpha_1}} \times \cdots \times \mathbb{Z}_{p_n^{\alpha_n}}$ and $\mathbb{Z}_n[x]/(x^m)$ that have genus one zero-divisor graph.

Let $G = (V, E)$ be a graph. If $V' = \{v \in V(G) \mid deg(v) = 1\}$, then the subgraph induced by $G \setminus V'$ is called the *reduction* of $G$ and is denoted by $\tilde{G}$. Note that $g(\tilde{G}) = g(G)$.

Let us determine the orientable genus one zero-divisor graphs, that is, the toroidal zero-divisor graphs. We start with an example of a commutative nonlocal ring which has a toroidal zero-divisor graph.

**Example 4.3.1.** If $R = \mathbb{Z}_2 \times \mathbb{Z}_2 \times \mathbb{Z}_2 \times \mathbb{Z}_2$, then $g(\Gamma(R)) = 1$. For, let $u_1 = (1, 0, 0, 0)$, $u_2 = (0, 1, 0, 0)$, $u_3 = (1, 1, 0, 0)$, $v_1 = (0, 0, 1, 0)$, $v_2 = (0, 0, 0, 1)$, $v_3 = (0, 0, 1, 1)$, $w_1 = (1, 0, 1, 0)$, $w_2 = (1, 0, 0, 1)$, $w_3 = (0, 1, 1, 0)$, $w_4 = (0, 1, 0, 1)$, $w_5 = (1, 1, 1, 0)$, $w_6 = (1, 1, 0, 1)$, $w_7 = (1, 0, 1, 1)$, and $w_8 = (0, 1, 1, 1)$. Since $u_i \cdot v_j = 0$ for every $i$, $j$, we have $K_{3,3}$ is a subgraph of $\Gamma(R)$, and so $g(\Gamma(R)) \ge 1$. However, we can draw $\tilde{G} = \Gamma(R) \setminus \{w_5, w_6, w_7, w_8\}$ on the surface of a torus; see Figure 4.1. Therefore $g(\Gamma(R)) = 1$.

**Lemma 4.3.2.** ([420, Lemma 3.1]) *Let $R$ be a finite commutative ring which is not a field.*

(i) *If $R$ has at least 5 maximal ideals, then $g(\Gamma(R)) \ge 2$;*

(ii) *Let* $R = R_1 \times R_2 \times R_3 \times R_4$*, where* $R_i$ *is a local ring for every* $i$*. If* $|R_j| \geq 4$ *for some* $j \in \{1, 2, 3, 4\}$*, then* $g(\Gamma(R)) \geq 2$*;*

(iii) *Let* $R = R_1 \times R_2 \times R_3$*, where* $R_i$ *is a local ring for every* $i$*. If* $|R_j| \geq 8$ *for some* $j \in \{1, 2, 3\}$*, then* $g(\Gamma(R)) \geq 2$*;*

(iv) *Let* $R = R_1 \times R_2 \times R_3$*, where* $R_i$ *is a local ring for every* $i$*. If* $|R_j| \geq 4$ *and* $|R_k| \geq 4$ *for some* $j, k \in \{1, 2, 3\}$ *with* $j \neq k$*, then* $g(\Gamma(R)) \geq 2$*.*

Note that by Theorem 4.2.5, we have $\Gamma(R)$ is not planar for any commutative local ring $R$ of cardinality 32 that is not a field. The following result improves the result for all finite commutative rings. (This also follows from Theorem 4.2.13.)

**Theorem 4.3.3.** ([420, Theorem 3.4]) *Let* $R$ *be a commutative ring and* $|R| = 32$*. Then* $\Gamma(R)$ *is planar if and only if* $R \cong \mathbb{Z}_2 \times \mathbb{F}_{16}$*, where* $\mathbb{F}_{16}$ *is a field with* 16 *elements.*

The next theorem characterizes all rings of the form $\mathbb{Z}_n$ whose $\Gamma(R)$ has genus at most one.

**Theorem 4.3.4.** ([420, Theorem 3.5]) *Let* $R$ *be a homomorphic image of* $\mathbb{Z}_n$ *and* $p$ *be a prime number. Then* $g(\Gamma(R)) \leq 1$ *if and only if* $R$ *is one of* $\mathbb{Z}_p$, $\mathbb{Z}_4$, $\mathbb{Z}_9$, $\mathbb{Z}_{25}$, $\mathbb{Z}_{49}$, $\mathbb{Z}_8$, $\mathbb{Z}_{27}$, $\mathbb{Z}_{16}$, $\mathbb{Z}_{32}$, $\mathbb{Z}_2 \times \mathbb{Z}_p$, $\mathbb{Z}_3 \times \mathbb{Z}_p$, $\mathbb{Z}_5 \times \mathbb{Z}_5$, $\mathbb{Z}_2 \times \mathbb{Z}_4$, $\mathbb{Z}_3 \times \mathbb{Z}_4$, $\mathbb{Z}_2 \times \mathbb{Z}_9$, $\mathbb{Z}_3 \times \mathbb{Z}_9$, $\mathbb{Z}_4 \times \mathbb{Z}_5$, $\mathbb{Z}_4 \times \mathbb{Z}_7$, $\mathbb{Z}_4 \times \mathbb{Z}_4$, $\mathbb{Z}_2 \times \mathbb{Z}_8$, $\mathbb{Z}_3 \times \mathbb{Z}_8$, $\mathbb{Z}_2 \times \mathbb{Z}_2 \times \mathbb{Z}_2$, $\mathbb{Z}_2 \times \mathbb{Z}_2 \times \mathbb{Z}_3$, $\mathbb{Z}_2 \times \mathbb{Z}_2 \times \mathbb{Z}_5$, $\mathbb{Z}_2 \times \mathbb{Z}_2 \times \mathbb{Z}_7$, $\mathbb{Z}_2 \times \mathbb{Z}_3 \times \mathbb{Z}_3$, $\mathbb{Z}_3 \times \mathbb{Z}_3 \times \mathbb{Z}_3$, $\mathbb{Z}_2 \times \mathbb{Z}_2 \times \mathbb{Z}_4$, $\mathbb{Z}_2 \times \mathbb{Z}_2 \times \mathbb{Z}_2 \times \mathbb{Z}_2$.

The following result generalizes Theorem 4.2.3.

**Theorem 4.3.5.** ([420, Theorem 3.6]) *If* $(R, \mathfrak{m})$ *is a finite commutative local ring and* $g(\Gamma(R)) \leq 1$*, then the following hold:*

(i) $|R/\mathfrak{m}| \leq 8$;

(ii) *If* $|R/\mathfrak{m}| = 8$*, then* $\mathfrak{m}^2 = 0$ *and* $|R| = 64$;

(iii) *If* $|R/\mathfrak{m}| = 7$*, then* $\mathfrak{m}^2 = 0$ *and* $|R| = 49$;

(iv) *If* $|R/\mathfrak{m}| = 5$*, then* $\mathfrak{m}^2 = 0$ *and* $|R| = 25$;

(v) *If* $|R/\mathfrak{m}| = 4$*, then* $\mathfrak{m}^2 = 0$ *and* $|R| = 16$;

(vi) *If* $|R/\mathfrak{m}| = 3$*, then* $\mathfrak{m}^3 = 0$ *and* $|R| \leq 27$;

(vii) *If* $|R/\mathfrak{m}| = 2$*, then* $\mathfrak{m}^5 = 0$ *and* $|R| \leq 32$.

**Corollary 4.3.6.** ([420, Corollary 3.8]) *Let* $(R, \mathfrak{m})$ *be a finite commutative local ring which is not a field. Then* $g(\Gamma(R)) = 1$ *if one of the following holds:*

(i) $|R| = 64$ *and* $|R/\mathfrak{m}| = 8$;

(ii) $|R| = 49$ *and* $|R/\mathfrak{m}| = 7$.

**Lemma 4.3.7.** ([420, Lemma 3.9]) *Let $p > q$ be prime numbers and $R = \mathbb{Z}_{pq}[x]/(x^m)$, where $m \geq 2$. Then $g(\Gamma(R)) \geq 2$.*

**Lemma 4.3.8.** ([420, Lemma 3.10]) *Let $p$ be a prime number and $m \geq 3$. For $R = \mathbb{Z}_{p^2}[x]/(x^m)$, $g(\Gamma(R)) \geq 2$.*

**Theorem 4.3.9.** ([420, Theorem 3.11]) *Let $R_{m,n} = \mathbb{Z}_n[x]/(x^m)$, where $m, n \geq 2$. Then*

(i) $g(\Gamma(R_{2,n})) \leq 1$ *if and only if* $n = 2, 3, 4, 5, 7$;

(ii) $g(\Gamma(R_{3,n})) \leq 1$ *if and only if* $n = 2, 3$;

(iii) $g(\Gamma(R_{4,n})) \leq 1$ *if and only if* $n = 2$;

(iv) $g(\Gamma(R_{5,n})) \leq 1$ *if and only if* $n = 2$;

(v) $g(\Gamma(R_{m,n})) \geq 2$ *whenever* $m \geq 6$.

After the work of Wang [420], Cameron Wickham [430] and Hung-Jen Chiang-Hsieh et al. [164] independently characterized all the finite commutative rings whose zero-divisor graph has genus one. The following few observations are due to Wickham.

**Lemma 4.3.10.** ([430, Proposition 2.5]) *Let $(R, \mathfrak{m})$ be a finite commutative local ring with cardinality $p^k$ for some prime $p$ (and $k > 1$). Assume that $g(\Gamma(R)) \leq 1$.*

(i) *If $p$ is odd, then $p = 3, 5, 7$ and $|R| \leq p^3$;*

(ii) *If $p = 2$ and $|R/\mathfrak{m}| = 2$, then $|R| \leq 32$.*

**Lemma 4.3.11.** ([430, Theorem 2.6]) *Let $R$ be a finite commutative local ring with maximal ideal $\mathfrak{m} \neq 0$. Then there is a prime power $q$ and integers $s > r$ such that $|R| = q^s$, $|\mathfrak{m}| = q^r$, and $|R/\mathfrak{m}| = q$. Let $t$ be the smallest positive integer such that $\mathfrak{m}^{t+1} = 0$. Then*

(i) *$\Gamma(R)$ contains a copy of $K_v$, where $v = |\mathfrak{m}|^{(t+1)/2}$;*

(ii) *$\Gamma(R)$ contains a copy of $K_{q-1}$;*

(iii) *If $r > 1$, then $\Gamma(R)$ contains a copy of $K_{q-1,q^r-q}$.*

In order to characterize all finite commutative rings with genus one zero-divisor graphs, Hung-Jen Chiang-Hsieh et al. [164] considered several special rings.

**Lemma 4.3.12.** ([164, Proposition 2.1]) *Let $(R, \mathfrak{m})$ be a commutative local ring with $|R| = 32$. If $|R/\mathfrak{m}| = 2$ and $|\mathfrak{m}^2| = 4$, then $g(\Gamma(R)) \geq 2$.*

**Lemma 4.3.13.** ([164, Proposition 2.2]) *Let $(R, \mathfrak{m})$ be a commutative local ring with $|R| = 32$. If $|R/\mathfrak{m}| = 2$ and $|\mathfrak{m}^2| = 2$, then $g(\Gamma(R)) \geq 2$.*

**Lemma 4.3.14.** ([164, Proposition 2.11]) *Let* $(R, \mathfrak{m})$ *be a commutative local ring with* $|\mathfrak{m}| = 4$ *and* $\mathfrak{m}^2 = 0$. *Then* $g(\Gamma(\mathbb{Z}_2 \times R)) = 1$.

**Lemma 4.3.15.** ([164, Lemma 3.1]) *Let* $(R, \mathfrak{m})$ *be a commutative local ring with* $|R| = q^2$ *and* $|R/\mathfrak{m}| = q = p^d$ *for some prime* $p$ *and* $d \in \mathbb{N}$. *If* $char(R) = p^2$, *then* $R \cong \mathbb{Z}_{p^2}[x]/(f(x))$, *where* $f$ *is a monic polynomial of* $\mathbb{Z}_{p^2}[x]$ *of degree* $d$ *such that the image of* $f$ *in* $\mathbb{Z}_p[x]$ *is irreducible.*

**Lemma 4.3.16.** ([164, Corollary 3.2]) *Let* $(R, \mathfrak{m})$ *be a commutative local ring with* $|R| = 64$ *and* $|R/\mathfrak{m}| = 8$. *If* $char(R) = 4$, *then* $R \cong \mathbb{Z}_4[x]/(x^3 + x + 1)$.

**Lemma 4.3.17.** ([164, Corollary 3.3]) *Let* $(R, \mathfrak{m})$ *be a commutative local ring with* $|R| = 16$ *and* $|R/\mathfrak{m}| = 4$. *If* $char(R) = 4$, *then* $R \cong \mathbb{Z}_4[x]/(x^3 + x + 1)$.

The following theorem completely characterizes all finite local commutative rings whose zero-divisor graphs have genus one.

**Theorem 4.3.18.** ([164, Theorem 3.5.2], [430, Theorem 4.1]) *Let* $R$ *be a finite commutative local ring which is not a field. Then* $g(\Gamma(R)) = 1$ *if and only if* $R$ *is isomorphic to one of the following* 17 *rings:*

$$\mathbb{Z}_{32},\ \mathbb{Z}_{49},\ \frac{\mathbb{Z}_2[x]}{(x^5)},\ \frac{\mathbb{F}_8[x]}{(x^2)},\ \frac{\mathbb{Z}_2[x,y]}{(x^3, xy, y^2)},\ \frac{\mathbb{Z}_2[x,y,z]}{(x,y,z)^2},\ \frac{\mathbb{Z}_4[x]}{(x^3+x+1)},\ \frac{\mathbb{Z}_4[x]}{(x^3-2, x^5)},$$
$$\frac{\mathbb{Z}_4[x]}{(x^4-2, x^5)},\ \frac{\mathbb{Z}_4[x,y]}{(x^3, x^2-2, xy, y^2)},\ \frac{\mathbb{Z}_4[x]}{(x^3, 2x)},\ \frac{\mathbb{Z}_4[x,y]}{(2x, 2y, x^2, xy, y^2)},\ \frac{\mathbb{Z}_7[x]}{(x^2)},$$
$$\frac{\mathbb{Z}_8[x]}{(x^2, 2x)},\ \frac{\mathbb{Z}_8[x]}{(x^2-2, x^5)},\ \frac{\mathbb{Z}_8[x]}{(x^2+2x-2, x^5)},\ \frac{\mathbb{Z}_8[x]}{(x^2-2x+2, x^5)}.$$

After characterizing all finite local commutative rings $R$ with $g(\Gamma(R)) = 1$, Wickham [430] proposed two open questions regarding the genus of zero-divisor graphs of local rings.

Let $q$ be a prime power and define two numbers $g(q)$ and $G(q)$ as follows:

$$g(q) = min\{g(\Gamma(R)) \mid R \text{ is a local ring, not a field, with } |R| = q\}$$

and

$$G(q) = max\{g(\Gamma(R)) \mid R \text{ is a local ring, not a field, with } |R| = q\}.$$

Let $p$ be a prime. Note that the only commutative local rings of order $p^2$ are $\mathbb{Z}_{p^2}$ and $\mathbb{A}_{p^2} = \mathbb{F}_p[x]/(x^2)$. Since $\Gamma(\mathbb{Z}_{p^2}) = \Gamma(\mathbb{A}_{p^2}) = K_{p-1}$, then $g(p^2) = G(p^2)$. Also, it is easy to see that $g(8) = G(8) = 0$. Also, Wickham showed that $0 = g(27) < G(27) = 2$ and $g(2^i) < G(2^i)$ for $i = 4, 5, 6$. Therefore, he proposed the following question.

**Question**. Let $p$ be a prime. Is $g(p^i) < G(p^i)$ for every $i \geq 3$ and $p$ odd, and $g(2^i) < G(2^i)$ for every $i \geq 4$?

Recall that a finite local ring $R$ with maximal ideal $\mathfrak{m}$ is called *Gorenstein* if $dim_{R/\mathfrak{m}}\, Ann(\mathfrak{m}) = 1$. Let $q = 16, 32, 64$, or $27$ and $R$ be a commutative local ring with $|R| = q$. Then Wickham showed that $R$ is Gorenstein whenever $g(q) < G(q)$ and $g(\Gamma(R)) = g(q)$. Thus, he proposed another question as follows:

**Question**. Let $q$ be a power of a prime and $R$ a commutative local ring with $|R| = q$. If $g(q) < G(q)$ and $g(\Gamma(R)) = g(q)$, is $R$ Gorenstein?

The following results provide a complete list of nonlocal commutative rings whose zero-divisor graphs have genus one.

**Theorem 4.3.19.** ([164, Theorem 3.6.2], [430, Theorem 3.1]) *Let $R$ be a finite commutative ring which is not local. Then $g(\Gamma(R)) = 1$ if and only if $R$ is isomorphic to one of the following* 29 *rings:*

$$\mathbb{F}_4\times\mathbb{F}_4,\ \mathbb{F}_4\times\mathbb{Z}_5,\ \mathbb{F}_4\times\mathbb{Z}_7,\ \mathbb{Z}_5\times\mathbb{Z}_5,\ \mathbb{Z}_2\times\frac{\mathbb{F}_4[x]}{(x^2)},\ \mathbb{Z}_2\times\frac{\mathbb{Z}_4[x]}{(x^2+x+1)},\ \mathbb{Z}_2\times\frac{\mathbb{Z}_2[x,y]}{(x^2,xy,y^2)},$$
$$\mathbb{Z}_2\times\frac{\mathbb{Z}_4[x]}{(2x,x^2)},\ \mathbb{Z}_3\times\frac{\mathbb{Z}_2[x]}{(x^3)},\ \mathbb{Z}_3\times\frac{\mathbb{Z}_4[x]}{(x^2-2,x^3)},\ \mathbb{Z}_3\times\mathbb{Z}_8,\ \mathbb{Z}_4\times\mathbb{F}_4,\ \mathbb{F}_4\times\frac{\mathbb{Z}_2[x]}{(x^2)},\ \mathbb{Z}_4\times\mathbb{Z}_4,$$
$$\mathbb{Z}_4\times\frac{\mathbb{Z}_2[x]}{(x^2)},\ \frac{\mathbb{Z}_2[x]}{(x^2)}\times\frac{\mathbb{Z}_2[x]}{(x^2)},\ \mathbb{Z}_4\times\mathbb{Z}_5,\ \mathbb{Z}_5\times\frac{\mathbb{Z}_2[x]}{(x^2)},\ \mathbb{Z}_4\times\mathbb{Z}_7,\ \mathbb{Z}_7\times\frac{\mathbb{Z}_2[x]}{(x^2)},$$
$$\mathbb{Z}_2\times\mathbb{Z}_3\times\mathbb{Z}_3,\ \mathbb{Z}_3\times\mathbb{Z}_3\times\mathbb{Z}_3,\ \mathbb{Z}_2\times\mathbb{Z}_3\times\mathbb{F}_4,\ \mathbb{Z}_2\times\mathbb{Z}_2\times\mathbb{Z}_5,\ \mathbb{Z}_2\times\mathbb{Z}_2\times\mathbb{Z}_7,$$
$$\mathbb{Z}_2\times\mathbb{Z}_2\times\mathbb{F}_4,\ \mathbb{Z}_2\times\mathbb{Z}_2\times\mathbb{Z}_4,\ \mathbb{Z}_2\times\mathbb{Z}_2\times\frac{\mathbb{Z}_2[x]}{(x^2)},\ \mathbb{Z}_2\times\mathbb{Z}_2\times\mathbb{Z}_2\times\mathbb{Z}_2,$$

*where $\mathbb{F}_n$ is the field with $n$ elements.*

## 4.4 Zero-divisor Graphs with Genus Two

Now, we discuss the genus two zero-divisor graphs. This work was first initiated by Bloomfield and Wickham in [147]. Actually, they enumerated all finite commutative local rings whose zero-divisor graphs have genus two.

**Theorem 4.4.1.** ([147, Theorem 1]) *Let $R$ be a finite commutative local ring. Then $g(\Gamma(R)) = 2$ if and only if $R$ is isomorphic to one of the following* 14 *rings:*

$$\mathbb{Z}_{81},\ \frac{\mathbb{F}_3[x]}{(x^4)},\ \frac{\mathbb{Z}_9[x]}{(3x,x^3-3)},\ \frac{\mathbb{Z}_9[x]}{(x^2-3)},\ \frac{\mathbb{Z}_9[x]}{(x^2-6)},\ \frac{\mathbb{Z}_9[x,y]}{(x,y)^2},\ \frac{\mathbb{Z}_9[x]}{(3x,x^2)},\ \frac{\mathbb{F}_9[x]}{(x^2)},$$
$$\frac{\mathbb{Z}_9[x]}{(x^2+1)},\ \frac{\mathbb{F}_2[x,y]}{(x^4,xy,y^2-x^3)},\ \frac{\mathbb{Z}_4[x,y]}{(x^2-2,xy,y^2-2x,2y)},\ \frac{\mathbb{Z}_4[x,y]}{(x^3-2,xy,2x,y^2-2)},$$
$$\frac{\mathbb{Z}_8[x]}{(2x,x^3-4)},\ \frac{\mathbb{Z}_{16}[x]}{(2x,x^2-8)}.$$

Now we return to the questions raised by Wickham [430]. By Theorem 4.3.18, there are no local rings of cardinality 81 that have genus one. Thus, by Theorem 4.4.1, we have $g(81) = 2$. Since $\Gamma(\frac{\mathbb{Z}_3[x,y,z]}{(x,y,z)^2}) = K_{26}$, $G(81) = 43$. Hence $g(81) < G(81)$. Also note that the local rings in Theorem 4.4.1 with cardinality 81 are in fact Gorenstein.

Then in [217], Huadong Su and Pailing Li determined when $g(\Gamma(\mathbb{Z}_n)) = 2$ or 3.

**Theorem 4.4.2.** ([217, Theorem 9]) *The following are true:*

(a) $g(\Gamma(\mathbb{Z}_n)) = 2$ *if and only if* $n \in \{30, 35, 36, 44, 50, 81\}$;

(b) $g(\Gamma(\mathbb{Z}_n)) = 3$ *if and only if* $n \in \{42, 45, 52, 54, 64\}$.

The problem of characterizing nonlocal finite commutative rings with genus two zero-divisor graph was open for quite sometime. The following example provides a lower bound for the genus of the zero-divisor graph of nonlocal rings.

**Example 4.4.3.** ([420, Example 2.5]) If $R = \mathbb{Z}_2 \times \mathbb{Z}_{16}$, then $g(\Gamma(R)) \geq 2$. Let $G = \Gamma(\mathbb{Z}_2 \times \mathbb{Z}_{16})$. Let $u_1 = (1,0)$, $u_2 = (1,8)$, $u_3 = (0,8)$, $v_1 = (0,2)$, $v_2 = (0,4)$, $v_3 = (0,6)$, $v_4 = (0,10)$, $v_5 = (0,12)$, $v_6 = (0,14)$, $w_1 = (1,4)$, $w_2 = (1,12)$, $w_3 = (1,2)$, $w_4 = (1,6)$, $w_5 = (1,10)$, $w_6 = (1,14)$, $w_7 = (0,1)$, $w_8 = (0,3)$, $w_9 = (0,5)$, $w_{10} = (0,7)$, $w_{11} = (0,9)$, $w_{12} = (0,11)$, $w_{13} = (0,13)$, and $w_{14} = (0,15)$. Then $\tilde{G} = G \setminus \{w_3, \ldots, w_{14}\}$.

Let $G' = \tilde{G} \setminus \{u_1u_3, u_2u_3, v_2v_5\}$. Then $G' \setminus \{w_1, w_2\} \cong K_{3,6}$. Suppose that $g(G) = 1$. Then $g(G') = 1$ and $G'$ has 13 faces as $|V(G'))| = 11$ and $|E(G')| = 24$. Fix a representation of $G'$ (on the surface of a torus), and let $\{F_1, \ldots, F_{13}\}$ be the set of faces of $G'$ corresponding to this representation. Let $\{F'_1, \ldots, F'_n\}$ be the set of faces of $G' \setminus \{w_1, w_2\}$ obtained by deleting $w_1$, $w_2$, and all edges incident with $w_1, w_2$ from the representation corresponding to $\{F_1, \ldots, F_{13}\}$.

Notice that $G' \setminus \{w_1, w_2\} \cong K_{3,6}$. From the fact that $n - m + f = 2 - 2g$, $K_{3,6}$ has 9 faces each with 4 boundary edges, and hence $n = 9$. Moreover, for every $i$, each boundary of $F'_i$ cannot have consecutive repetitions of a single edge. Therefore, in $K_{3,6}$, the only way to have a closed walk of length 4 without consecutive repetitions of a single edge is to have a 4-cycle. Thus in $K_{3,6}$, all face boundaries are 4-cycles. That is, the boundary of each $F'_i$ is a 4-cycle.

Now $\{F_1, \ldots, F_{13}\}$ can be recovered by inserting $w_1$, $w_2$, and all edges incident with $w_1$, $w_2$ into the representation corresponding to $\{F'_1, \ldots, F'_9\}$. Therefore, there are faces, say $F'_{t_1}$ and $F'_{t_2}$, $1 \leq t_i \leq 9$, such that if we insert $w_i$ and all edges incident with $w_i$ into $F'_{t_i}$ for every $i$, we are able to obtain the set of faces $\{F_1, \ldots, F_{13}\}$. From the above, the boundary of $F'_i$ is a cycle of length 4 for every $i$. Moreover, $t_1 = t_2$ since $w_iv_2 = w_iv_5 = w_iu_3 = 0$ for $i = 1, 2$. We may assume $t_1 = 1$. Let $e_1 = w_1v_2$, $e_2 = w_1v_5$, $e_3 = w_1u_3$ be the three edges incident with $w_1$, and let $e_4 = w_2v_2$, $e_5 = w_2v_5$, $e_6 = w_2u_3$ be the three edges incident with

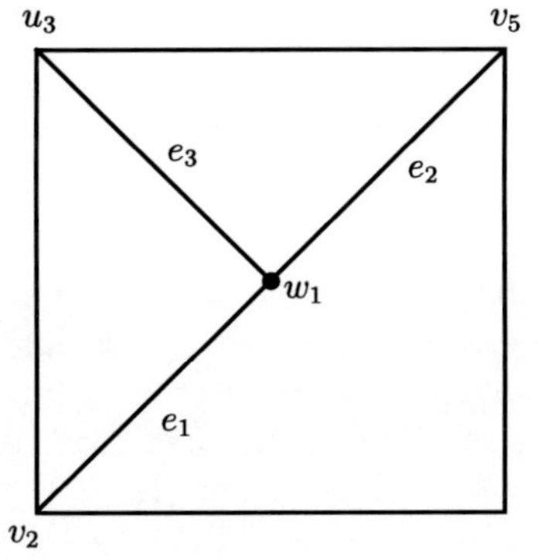


Figure 4.2: The graph $F_1'$

$w_2$. We can obtain Figure 4.2 by inserting $w_1$ and $e_1$, $e_2$, $e_3$ into $F_1'$. However, it is easy to see from Figure 4.2 that we cannot insert $w_2$ and $e_4$, $e_5$, $e_6$ into $F_1'$ without crossings, a contradiction. Therefore, we conclude that $g(G) \geq 2$.

In 2019, Asir and Mano [98] completely determined all nonlocal commutative rings whose zero-divisor graphs have genus two. In what follows, the graph $\widetilde{G}$ denotes the induced subgraph of $G$ induced by the vertices $G - V'$, where $V' = \{v \in V \mid deg(v) \leq 2\}$, and we call this graph the *reduction* of $G$. It is clear that $g(\widetilde{G}) = g(G)$ for any graph $G$.

The following two results are useful tools to determine the genus of zero-divisor graphs of nonlocal rings.

**Proposition 4.4.4.** ([430, Theorem 2.2]) *Let $R$ and $S$ be (finite) commutative rings such that $\Gamma(R)$ is (isomorphic to) a subgraph of $\Gamma(S)$ and $|R^* \setminus Z(R)| \leq |S^* \setminus Z(S)|$. If $T$ is a commutative ring, then $\Gamma(R \times T)$ is (isomorphic to) a subgraph of $\Gamma(S \times T)$.*

**Remark 4.4.5.** ([164, Remark 2.12]) Let $\mathbb{F}_q$ denote the finite field with $q$ elements and $\chi : \mathbb{F}_q \to \mathbb{Z}_q$ be any bijective map such that $\chi(0) = 0$. Let $R$ be a commutative ring and define a map $\varphi : R \times \mathbb{F}_q \to R \times \mathbb{Z}_q$ such that $\varphi((a,b)) = (a, \chi(b))$. Then $\varphi$ induces an embedding from $\Gamma(R \times \mathbb{F}_q)$ to $\Gamma(R \times \mathbb{Z}_q)$. Therefore $g(\Gamma(R \times \mathbb{F}_q)) \leq g(\Gamma(R \times \mathbb{Z}_q))$.

The following lemma says that if the genus of the zero-divisor graph of a local ring is less than or equal to 2, then the number of elements in its residue field is bounded by 9.

**Lemma 4.4.6.** ([98, Lemma 1]) *Let $R$ be a commutative local ring with maximal ideal $\mathfrak{m}$. If $g(\Gamma(R)) \leq 2$, then $|R/\mathfrak{m}| \leq 9$. Further,*

(i) *If* $|R/\mathfrak{m}| = 9, 8, 7, 5, 4$, *then* $\mathfrak{m}^2 = 0$;

(ii) *If* $|R/\mathfrak{m}| = 3$, *then* $\mathfrak{m}^4 = 0$;

(iii) *If* $|R/\mathfrak{m}| = 2$, *then* $\mathfrak{m}^5 = 0$.

*Proof.* Assume that $g(\Gamma(R)) \leq 2$. Let $R$ be a local ring with maximal ideal $\mathfrak{m}$. Then there is a prime power $q$ and integers $s > r$ such that $|R| = q^s$, $|\mathfrak{m}| = q^r$, and $|R/\mathfrak{m}| = q$. Let $t$ be the largest positive integer such that $\mathfrak{m}^t \neq 0$; so $\mathfrak{m}^{t+1} = 0$.

If $|R/\mathfrak{m}| > 9$, then $|\mathfrak{m}^t - \mathfrak{m}^{t+1}| \geq 10$. Since $\mathfrak{m}^t \cdot \mathfrak{m}^t = 0$, $xy = 0$ for any $x, y \in \mathfrak{m}^t$. Therefore $K_{10}$ is a subgraph of $\Gamma(R)$. Then $g(\Gamma(R)) \geq 3$, a contradiction. Thus $|R/\mathfrak{m}| \leq 9$. If $|R/\mathfrak{m}| \geq 4$ with $t \geq 2$, then $|\mathfrak{m}^t - \mathfrak{m}^{t+1}| \geq 3$ and $|\mathfrak{m}^{t-1} - \mathfrak{m}^t| \geq 12$. Since $\mathfrak{m}^{t-1} \cdot \mathfrak{m}^t = 0$, $K_{3,12}$ is a subgraph of $\Gamma(R)$ which has genus 3. Hence we conclude that $4 \leq |R/\mathfrak{m}| \leq 9$ with $\mathfrak{m}^2 = 0$.

If $|R/\mathfrak{m}| = 3$ with $t \geq 4$, then $|\mathfrak{m}^{t-1} - \mathfrak{m}^t| \geq 6$ and $|\mathfrak{m}^{t-2} - \mathfrak{m}^{t-1}| \geq 18$. Since $\mathfrak{m}^{t-1} \cdot \mathfrak{m}^{t-2} \subseteq \mathfrak{m}^{t+1} = 0$, $K_{6,18}$ is a subgraph of $\Gamma(R)$, a contradiction. Thus $t \leq 3$.

If $|R/\mathfrak{m}| = 2$ with $t \geq 5$, then $|\mathfrak{m}^{t-3} - \mathfrak{m}^{t-1}| \geq 12$ and $|\mathfrak{m}^{t-1} - \mathfrak{m}^{t+1}| \geq 3$. Since $\mathfrak{m}^{t-3} \cdot \mathfrak{m}^{t-1} = 0$, $K_{3,12}$ is a subgraph of $\Gamma(R)$, a contradiction. Thus $t \leq 4$. □

The graph given in Figure 4.3 plays a vital role in characterizing nonlocal commutative rings with genus two zero-divisor graphs. So we find its genus bound in the following result.

**Lemma 4.4.7.** ([98, Lemma 2]) For the graph $H$ as shown in Figure 4.3, we have $g(H) \geq 3$.

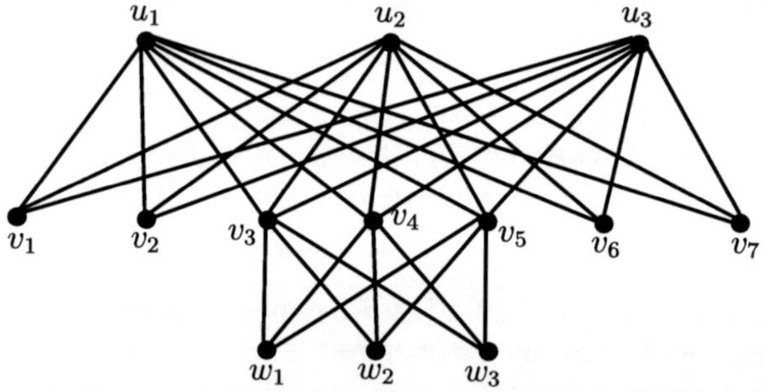


Figure 4.3: Graph $H$

*Proof.* Consider the subgraph $H' = H - \{w_1, w_2, w_3\}$. Then $H' \cong K_{3,7}$; so $g(H') = 2$. Note that $H'$ has 10 vertices, 21 edges, and 9 faces on $S_2$. Let $e$, $f'$, and $f'_i$ be the number of edges, the number of faces, and the number of $i$-gons in $H'$ respectively. Since $H'$ is a complete bipartite graph with parts of size 3 and 7, $H'$ contains either rectangular or hexagonal faces. Then $f' = f'_4 + f'_6$ and $2e = 4f'_4 + 6f'_6$. By solving these two equations, we get that $f'_4 = 6$ and $f'_6 = 3$.

Fix a representation of $H'$, and let $F'_1, F'_2, \ldots, F'_9$ be the faces of $H'$ corresponding to this representation.

Now we are trying to insert $w_i$ for $i = 1, 2, 3$ and all edges incident with $w_i$, say $e_i = w_i v_3, e_{i+3} = w_i v_4, e_{i+6} = w_i v_5$, into the representation corresponding to $F'_1, F'_2, \ldots, F'_9$. Let $X = \{v_3, v_4, v_5\}$. Note that it is possible to have exactly one hexagonal face (say $F'_9$) in the embedding of $H'$ in $S_2$ containing the three vertices of $X$ and the other faces of $H'$ contain at the maximum two vertices from $X$. Now insert $w_1$, $e_1$, $e_4$, and $e_7$ into $F'_9$ without any crossings as shown in Figure 4.4. After inserting $w_1$, no face contains all three vertices of $X$; so one cannot insert $w_2, w_3$, and its edges in $S_2$. Thus $g(H) \geq 3$. □

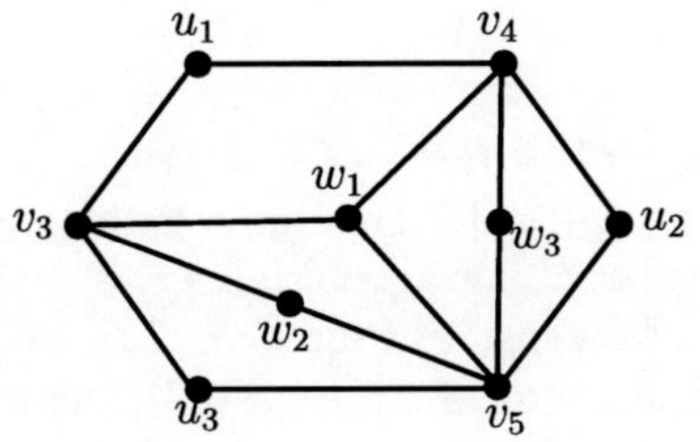


Figure 4.4: $F'_9$

**Example 4.4.8.** ([98, Example 1]) Let $R \cong \mathbb{Z}_2 \times \mathbb{Z}_2 \times \mathbb{Z}_9$, $\mathbb{Z}_3 \times \mathbb{Z}_3 \times \mathbb{Z}_4$, or $\mathbb{Z}_2 \times \mathbb{Z}_4[x]/(x^2)$. Then $g(\Gamma(R)) \geq 3$.

*Proof.* (a) Let $R \cong \mathbb{Z}_2 \times \mathbb{Z}_2 \times \mathbb{Z}_9$. Then the vertices of $\Gamma(R)$ are $u_1 = (1,0,0), u_2 = (0,1,0), u_3 = (1,1,0), v_1 = (0,0,1), v_2 = (0,0,2), v_3 = (0,0,3), v_4 = (0,0,4), v_5 = (0,0,5), v_6 = (0,0,6), v_7 = (0,0,7), v_8 = (0,0,8), w_1 = (0,1,3), w_2 = (1,0,3), w_3 = (0,1,6)$, and $w_4 = (1,0,6)$.

Let us take the subgraph $G' = \Gamma(R) - \{w_1, w_2, w_3, w_4\}$. Then $G' \cong K_{3,8}$, and a similar technique used for $K_{3,7}$ in Lemma 4.4.7 gives us that any embedding of $G'$ in $S_2$ contains 9 rectangular faces and 2 hexagonal faces.

Suppose that $g(\Gamma(R)) = 2$. Insert $w_i$ for $i = 1, 2, 3, 4$ and all edges incident with $w_i$ into the representation of $G'$ in $S_2$. Since $e_1 = w_1 w_2, e_2 = w_2 w_3, e_3 = w_3 w_4, e_4 = w_4 w_1, e_{i+4} = w_i v_3, e_{i+8} = w_i v_6 \in E(\Gamma(R)) \setminus E(G')$ for $i = 1, 2, 3, 4$, all the $w_i$'s should be inserted into a single face of $G'$ which contains both $v_3$ and $v_6$. As seen in Figure 4.5(a), one cannot draw all the edges $e_5, e_6, \ldots, e_{12}$ without crossings, a contradiction. Thus $g(\Gamma(R)) \geq 3$.

(b) Let $R \cong \mathbb{Z}_3 \times \mathbb{Z}_3 \times \mathbb{Z}_4$. Then the vertices of $\Gamma(R)$ are $u_1 = (0,0,1), u_2 = (0,0,2), u_3 = (0,0,3), v_1 = (1,0,0), v_2 = (2,0,0), v_3 = (0,1,0), v_4 = (0,2,0), v_5 = (1,1,0), v_6 = (1,2,0), v_7 = (2,1,0), v_8 = (2,2,0), w_1 = (0,1,2), w_2 = (1,0,2), w_3 = (0,2,2)$, and $w_4 = (2,0,2)$.

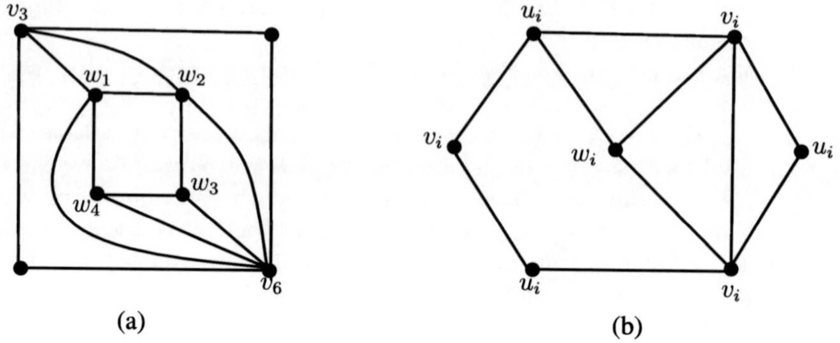


Figure 4.5:

Let us take the subgraph $G = \Gamma(R) - \{e_1 = v_1v_3, e_2 = v_1v_4, e_3 = v_2v_3, e_4 = v_2v_4\}$ and $G' = G - \{w_1, w_2, w_3, w_4\}$. Then any embedding of $G' = K_{3,8}$ in $S_2$ contains 9 rectangular faces and 2 hexagonal faces.

Suppose that $g(G) = 2$. Now insert $w_i$ for $i = 1, 2, 3, 4$ in any embedding of $G'$ in $S_2$. Since $w_1w_2, w_2w_3, w_3w_4, w_4w_1 \in E(G) \setminus E(G')$, the $w_i$'s should be inserted into a single face of $G'$, say $F'$. Further, $w_iv_1,\ w_iv_2,\ w_jv_3,\ w_jv_4 \in E(G) \setminus E(G')$ for $i = 1, 2$ and $j = 3, 4$; so $F'$ needs to contain $v_1, v_2, v_3, v_4$. But even the largest face in $G'$ is a hexagon which contains only three $v_\ell$'s for $\ell = 1, \ldots, 8$, a contradiction. Hence $g(\Gamma(R)) \geq g(G) \geq 3$.

(c) Let $R \cong \mathbb{Z}_2 \times \mathbb{Z}_4[x]/(x^2)$. Then the vertices of $\Gamma(R)$ are $u_1 = (0, 2x), u_2 = (1, 0), u_3 = (1, 2x), v_1 = (0, 2), v_2 = (0, 2 + 2x), v_3 = (0, x), v_4 = (0, 3x), v_5 = (0, 2 + x), v_6 = (0, 2 + 3x), w_1 = (1, 2), w_2 = (1, 2 + 2x), w_3 = (1, x), w_4 = (1, 3x), w_5 = (1, 2 + x)$, and $w_6 = (1, 2 + 3x)$. Consider the subgraph $G'' = \langle\{u_1, u_2, u_3, v_1, v_2, v_3, v_4, v_5, v_6\}\rangle - \{e_1 = u_1u_2, e_2 = u_1u_3, e_3 = v_1v_2, e_4 = v_3v_4, e_5 = v_5v_6\}$ of $\Gamma(R)$. Then $G'' \cong K_{3,6}$, and so one can embed $G''$ in $S_2$. Further, by Euler's Formula, any embedding of $G''$ in $S_2$ consists of 3 rectangular and 4 hexagonal faces. For simplicity, we denote this embedding of $G''$ in $S_2$ by $E''$.

Suppose $g(\Gamma(R)) = 2$. Then we have to insert $w_i$'s for $i = 1, \ldots, 6$ with its incident edges and $e_j$'s for $j = 1, \ldots, 5$ in $E''$ without edge crossing. Since the degree of $u_1$ is six in $G''$, $u_1$ belongs to at most six faces of $E''$. Note that $N(w_1) = N(w_2) = \{u_1, v_1, v_2\}$, $N(w_3) = N(w_4) = \{u_1, v_3, v_4\}$, $N(w_5) = N(w_6) = \{u_1, v_5, v_6\}$ and all the faces in $E''$ contain a maximum of three $v_k$'s $(1 \leq k \leq 6)$. So no two $w_i$'s are inserted in a single face of $E''$. Further, since $N(w_1) = N(w_2)$, we should have two distinct faces in $E''$ containing both $v_1$ and $v_2$. Similarly, out of the remaining four faces containing $u_1$, two of the faces should have $v_3, v_4$, and the other two faces should have $v_5, v_6$ in common. But such an embedding of $G''$ in $S_2$ is not possible. Thus $g(\Gamma(\mathbb{Z}_2 \times \mathbb{Z}_4[x]/(x^2))) \geq 3$. □

The following result provides the list of all nonlocal commutative rings with genus two zero-divisor graphs.

**Theorem 4.4.9.** ([98, Theorem 4], [99, Theorem 1]) *Let $R$ be a nonlocal commutative ring. Then the genus of the zero-divisor graph is 2 if and only if $R$ is isomorphic to one of the following* 31 *rings:*

$\mathbb{F}_4 \times \mathbb{F}_8$, $\mathbb{F}_4 \times \mathbb{F}_9$, $\mathbb{F}_4 \times \mathbb{F}_{11}$, $\mathbb{Z}_5 \times \mathbb{Z}_7$, $\mathbb{Z}_4 \times \mathbb{Z}_8$, $\mathbb{Z}_2[x]/(x^2) \times \mathbb{Z}_8$, $\mathbb{Z}_4 \times \mathbb{Z}_2[x]/(x^3)$, $\mathbb{Z}_2[x]/(x^2) \times \mathbb{Z}_2[x]/(x^3)$, $\mathbb{Z}_2[x]/(x^2) \times \mathbb{Z}_9$, $\mathbb{Z}_4 \times \mathbb{Z}_3[x]/(x^2)$, $\mathbb{Z}_2[x]/(x^2) \times \mathbb{Z}_3[x]/(x^2)$, $\mathbb{Z}_4 \times \mathbb{Z}_9$, $\mathbb{Z}_2 \times \mathbb{Z}_{25}$, $\mathbb{Z}_2 \times \mathbb{Z}_5[x]/(x^2)$, $\mathbb{Z}_2 \times \mathbb{Z}_{16}$, $\mathbb{Z}_2 \times \mathbb{Z}_2[x]/(x^4)$, $\mathbb{Z}_2 \times \mathbb{Z}_4[x]/(x^2-2, x^4)$, $\mathbb{Z}_2 \times \mathbb{Z}_4[x]/(x^3-2, x^4)$, $\mathbb{Z}_2 \times \mathbb{Z}_4[x]/(x^3+x^2-2, x^4)$, $\mathbb{F}_4 \times \mathbb{Z}_8$, $\mathbb{F}_4 \times \mathbb{Z}_2[x]/(x^3)$, $\mathbb{F}_4 \times \mathbb{Z}_9$, $\mathbb{F}_4 \times \mathbb{Z}_3[x]/(x^2)$, $\mathbb{Z}_2 \times \mathbb{Z}_2 \times \mathbb{F}_8$, $\mathbb{Z}_2 \times \mathbb{Z}_2 \times \mathbb{F}_9$, $\mathbb{Z}_2 \times \mathbb{Z}_2 \times \mathbb{Z}_{11}$, $\mathbb{Z}_2 \times \mathbb{Z}_3 \times \mathbb{Z}_4$, $\mathbb{Z}_2 \times \mathbb{Z}_3 \times \mathbb{Z}_2[x]/(x^2)$, $\mathbb{Z}_2 \times \mathbb{Z}_3 \times \mathbb{Z}_5$, $\mathbb{Z}_3 \times \mathbb{Z}_3 \times \mathbb{F}_4$, $\mathbb{Z}_2 \times \mathbb{Z}_2 \times \mathbb{Z}_2 \times \mathbb{Z}_3$.

*Proof.* Assume that $g(\Gamma(R)) = 2$.

Let $R$ be a nonlocal ring. Then $R \cong R_1 \times R_2 \times \cdots \times R_n$, where each $R_i$ is local and $n \geq 2$. Suppose that $n \geq 5$.

Let $V_1 = \{(1,0,\ldots,0), (0,1,0,\ldots,0), (1,1,0,\ldots,0), [(1,0,0,1,0,\ldots,0), (0,1,1,0,\ldots,0), (1,0,0,0,1,0,\ldots,0)]\}$ and $V_2 = \{(0,0,1,0,\ldots,0), (0,0,0,1,0, \ldots,0), (0,0,0,0,1,0,\ldots,0), (0,0,1,1,0,\ldots,0), (0,0,1,0,1,0,\ldots,0), (0,0,0,1,1, 0,\ldots,0), (0,0,1,1,1,0,\ldots,0)\}$, where the vertex $[x,y,z]$ denotes the contracted path $x-y-z$. Then it is easy to check that the subgraph induced by $V_1 \cup V_2$ in $\Gamma(R)$ contains $K_{4,7} - \{\text{an edge}\}$. Let us take $G' = K_{4,7} - \{\text{an edge}\}$ as a subgraph of $\Gamma(R)$. Note that the girth $gr(G') = 4$, the number of edges $e = 27$, and any genus two embedding of $G'$ has $f = 14$ faces, which contradicts the inequality $2e \geq gr(G') \cdot f$. Hence $n \leq 4$.

**Case 1:** Let $n = 4$. Assume that $|R_1| \leq |R_2| \leq |R_3| \leq |R_4|$.

If $|R_4| \geq 5$, then every vertex in $\{0\} \times \{0\} \times \{0\} \times R_4$ is adjacent to all the vertices of $R_1 \times R_2 \times R_3 \times \{0\}$; so $K_{4,7}$ is a subgraph of $\Gamma(R)$. Since $g(K_{4,7}) = 3$, we have $g(\Gamma(R)) \geq 3$. Therefore $|R_4| \leq 4$.

(a) Suppose $|R_4| = 4$. Then $R_4 \cong \mathbb{F}_4$, $\mathbb{Z}_4$, or $\mathbb{Z}_2[x]/(x^2)$. Let $R \cong \mathbb{Z}_2 \times \mathbb{Z}_2 \times \mathbb{Z}_2 \times \mathbb{F}_4$, and let $H'$ be the subgraph induced by $\{(0,0,0,1), (0,0,0,\omega), (0,0,0,\omega^2), (0,0,0,\omega^3), (1,0,0,0), (0,1,0,0), (0,0,1,0), (1,1,0,0), (1,0,1,0), (0,1,1,0), (0,0,1,1), (0,0,1,\omega), (0,0,1,\omega^2)\}$. One can easily verify that the graph in Figure 4.3 is a subgraph of $H'$, and so $g(\Gamma(R)) \geq g(H') \geq 3$. Further, by Remark 4.4.5, we have $\Gamma(\mathbb{Z}_2 \times \mathbb{Z}_2 \times \mathbb{Z}_2 \times \mathbb{F}_4) \subseteq \Gamma(\mathbb{Z}_2 \times \mathbb{Z}_2 \times \mathbb{Z}_2 \times R_4)$ whenever $R_4 \cong \mathbb{Z}_4$ or $\mathbb{Z}_2[x]/(x^2)$; so $g(\Gamma(\mathbb{Z}_2 \times \mathbb{Z}_2 \times \mathbb{Z}_2 \times R_4)) \geq 3$.

(b) Suppose $|R_4| = 3$; that is, $R_4 \cong \mathbb{Z}_3$. Let $R \cong \mathbb{Z}_2 \times \mathbb{Z}_2 \times \mathbb{Z}_2 \times \mathbb{Z}_3 \cong \mathbb{Z}_2 \times \mathbb{Z}_2 \times \mathbb{Z}_6$. Now, by Figure 4.6, we get $g(\Gamma(\mathbb{Z}_2 \times \mathbb{Z}_2 \times \mathbb{Z}_6)) = 2$. Implies that $g(\Gamma(\mathbb{Z}_2 \times \mathbb{Z}_2 \times \mathbb{Z}_2 \times \mathbb{Z}_3)) = 2$.

If $R \cong \mathbb{Z}_2 \times \mathbb{Z}_2 \times \mathbb{Z}_3 \times \mathbb{Z}_3$, then the subsets $V_1 = \{(0,0,2,0), [(0,0,1,0), (1,1,0,0)], [(0,1,0,0), (1,0,1,0)], [(1,0,0,0), (0,1,1,0)]\}$ and $V_2 = \{(0,0,0,1), (0,0,0,2), (0,0,1,1), (0,1,0,1), (0,1,0,2), (1,0,0,1), (1,0,0,2)\}$ form a subgraph $K_{4,7} - \{\text{an edge}\}$ in $\Gamma(R)$, and so $g(\Gamma(R)) \geq 3$.

If $R \cong \mathbb{Z}_2 \times \mathbb{Z}_3 \times \mathbb{Z}_3 \times \mathbb{Z}_3$ or $R \cong \mathbb{Z}_3 \times \mathbb{Z}_3 \times \mathbb{Z}_3 \times \mathbb{Z}_3$, then $\Gamma(R) \supseteq \Gamma(\mathbb{Z}_2 \times \mathbb{Z}_2 \times \mathbb{Z}_3 \times \mathbb{Z}_3)$; so $g(\Gamma(R)) \geq 3$.

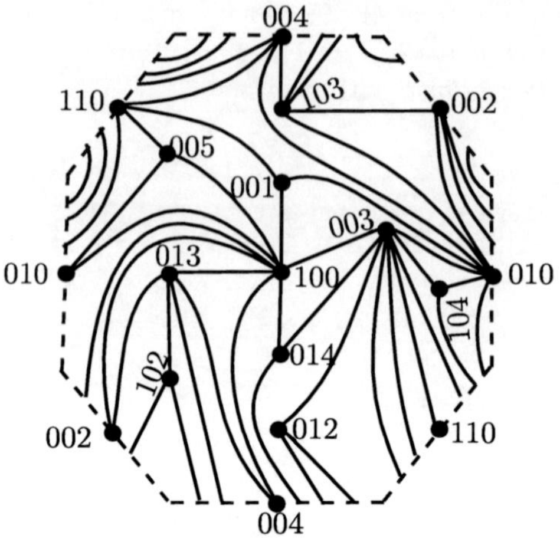


Figure 4.6: The embedding of $\Gamma(\mathbb{Z}_2 \times \mathbb{Z}_2 \times \mathbb{Z}_6)$ in $S_2$.

(c) If $|R_4| = 2$, then $R \cong \mathbb{Z}_2 \times \mathbb{Z}_2 \times \mathbb{Z}_2 \times \mathbb{Z}_2$, and by Theorem 4.3.19 we have $\Gamma(R)$ is toroidal.

Hence, in this case $g(\Gamma(R)) = 2$ if and only if $R$ is isomorphic to $\mathbb{Z}_2 \times \mathbb{Z}_2 \times \mathbb{Z}_2 \times \mathbb{Z}_3$.

**Case 2:** Let $n = 3$ and assume that $|R_1| \leq |R_2| \leq |R_3|$.

If $|R_3| \geq 12$, then let $X = \{0\} \times \{0\} \times R_3^*$ and $Y = (R_1 \times R_2 \times \{0\})^*$; each vertex in $X$ is adjacent to every other vertex in the set $Y$. Therefore the subgraph $K_{3,11}$ is contained in $\Gamma(R)$, which means $g(\Gamma(R)) \geq 3$. Thus $|R_i| \leq 11$. Also, if $|R_2| \geq 3$ and $|R_3| \geq 7$, then the subsets $(\{0\} \times \{0\} \times R_3^*)$ and $(R_1 \times R_2 \times \{0\})^*$ form a subgraph $K_{5,6}$ in $\Gamma(R)$, and so $g(\Gamma(R)) \geq 3$. Moreover, if $|R_2| \geq 4$ and $|R_3| \geq 5$, then the subsets $(\{0\} \times \{0\} \times R_3^*)$ and $(R_1 \times R_2 \times \{0\})^*$ form a subgraph $K_{4,9}$ in $\Gamma(R)$ so that $g(\Gamma(R)) \geq 3$. Further, if $|R_1| \geq 3$ and $|R_3| \geq 5$, then the subsets $(\{0\} \times \{0\} \times R_3^*)$ and $(R_1 \times R_2 \times \{0\})^*$ form a subgraph $K_{4,8}$ in $\Gamma(R)$, and so $g(\Gamma(R)) \geq 3$. This leaves the following rings to check.

$$\mathbb{Z}_2 \times \mathbb{Z}_2 \times \mathbb{Z}_2, \mathbb{Z}_2 \times \mathbb{Z}_2 \times \mathbb{Z}_3, \mathbb{Z}_2 \times \mathbb{Z}_2 \times \mathbb{Z}_4, \mathbb{Z}_2 \times \mathbb{Z}_2 \times \mathbb{Z}_2[x]/(x^2), \mathbb{Z}_2 \times \mathbb{Z}_2 \times \mathbb{F}_4,$$
$$\mathbb{Z}_2 \times \mathbb{Z}_2 \times \mathbb{Z}_5, \mathbb{Z}_2 \times \mathbb{Z}_2 \times \mathbb{Z}_7, \mathbb{Z}_2 \times \mathbb{Z}_2 \times \mathbb{Z}_8, \mathbb{Z}_2 \times \mathbb{Z}_2 \times \mathbb{Z}_2[x]/(x^3), \mathbb{Z}_2 \times \mathbb{Z}_2 \times \mathbb{F}_8,$$
$$\mathbb{Z}_2 \times \mathbb{Z}_2 \times \mathbb{Z}_9, \mathbb{Z}_2 \times \mathbb{Z}_2 \times \mathbb{Z}_3[x]/(x^2), \mathbb{Z}_2 \times \mathbb{Z}_2 \times \mathbb{F}_9, \mathbb{Z}_2 \times \mathbb{Z}_2 \times \mathbb{Z}_{11}, \mathbb{Z}_2 \times \mathbb{Z}_3 \times \mathbb{Z}_3,$$
$$\mathbb{Z}_2 \times \mathbb{Z}_3 \times \mathbb{Z}_4, \mathbb{Z}_2 \times \mathbb{Z}_3 \times \mathbb{Z}_2[x]/(x^2), \mathbb{Z}_2 \times \mathbb{Z}_3 \times \mathbb{F}_4, \mathbb{Z}_2 \times \mathbb{Z}_4 \times \mathbb{Z}_2[x]/(x^2),$$
$$\mathbb{Z}_2 \times \mathbb{Z}_4 \times \mathbb{Z}_4, \mathbb{Z}_2 \times \mathbb{Z}_4 \times \mathbb{F}_4, \mathbb{Z}_2 \times \mathbb{Z}_2[x]/(x^2) \times \mathbb{F}_4, \mathbb{Z}_2 \times \mathbb{Z}_2[x]/(x^2) \times \mathbb{Z}_2[x]/(x^2),$$
$$\mathbb{Z}_2 \times \mathbb{F}_4 \times \mathbb{F}_4, \mathbb{Z}_2 \times \mathbb{Z}_3 \times \mathbb{Z}_5, \mathbb{Z}_3 \times \mathbb{Z}_3 \times \mathbb{Z}_3, \mathbb{Z}_3 \times \mathbb{Z}_3 \times \mathbb{Z}_4, \mathbb{Z}_3 \times \mathbb{Z}_3 \times \mathbb{Z}_2[x]/(x^2),$$
$$\mathbb{Z}_3 \times \mathbb{Z}_3 \times \mathbb{F}_4, \mathbb{Z}_3 \times \mathbb{Z}_4 \times \mathbb{Z}_4, \mathbb{Z}_3 \times \mathbb{F}_4 \times \mathbb{Z}_4, \mathbb{Z}_3 \times \mathbb{Z}_4 \times \mathbb{Z}_2[x]/(x^2),$$
$$\mathbb{Z}_3 \times \mathbb{Z}_2[x]/(x^2) \times \mathbb{Z}_2[x]/(x^2), \mathbb{Z}_3 \times \mathbb{F}_4 \times \mathbb{F}_4.$$

Note that by Theorem 4.2.13 and Theorem 4.3.19, the zero-divisor graph of the rings $\mathbb{Z}_2 \times \mathbb{Z}_2 \times \mathbb{Z}_2$, $\mathbb{Z}_2 \times \mathbb{Z}_2 \times \mathbb{Z}_3$, and $\mathbb{Z}_2 \times \mathbb{Z}_2 \times \mathbb{F}_4$ are planar, and the rings $\mathbb{Z}_2 \times \mathbb{Z}_2 \times \mathbb{Z}_4, \mathbb{Z}_2 \times \mathbb{Z}_2 \times \mathbb{Z}_2[x]/(x^2)$, $\mathbb{Z}_2 \times \mathbb{Z}_2 \times \mathbb{Z}_5$, $\mathbb{Z}_2 \times \mathbb{Z}_2 \times \mathbb{Z}_7$, $\mathbb{Z}_2 \times \mathbb{Z}_3 \times \mathbb{Z}_3$, $\mathbb{Z}_2 \times \mathbb{Z}_3 \times \mathbb{F}_4$, and $\mathbb{Z}_3 \times \mathbb{Z}_3 \times \mathbb{Z}_3$ are toroidal. Therefore all the remaining rings in the list have genus at least 2.

Let $R \cong \mathbb{Z}_2 \times \mathbb{Z}_2 \times \mathbb{F}_8$. Then a $S_2$-embedding of $\Gamma(\mathbb{Z}_2 \times \mathbb{Z}_2 \times \mathbb{F}_8)$ is given in Figure 4.7 so that $g(\Gamma(\mathbb{Z}_2 \times \mathbb{Z}_2 \times \mathbb{F}_8)) = 2$.

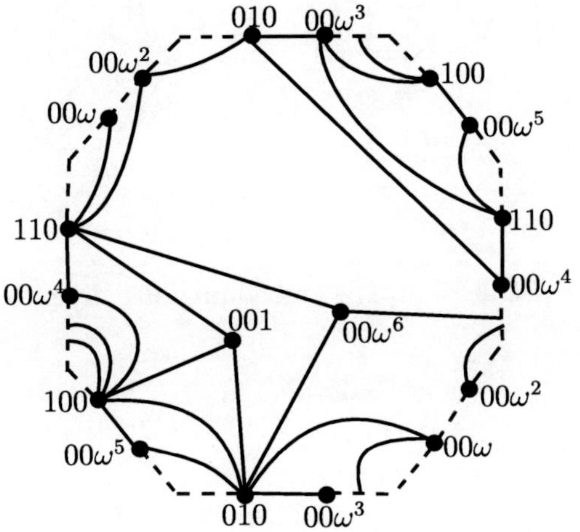


Figure 4.7: The embedding of $\widetilde{\Gamma(\mathbb{Z}_2 \times \mathbb{Z}_2 \times \mathbb{F}_8)}$ in $S_2$.

Consider the ring $R = \mathbb{Z}_2 \times \mathbb{Z}_2 \times \mathbb{Z}_8$. In this case, $\Gamma(R)$ contains $K_{3,7}$ with bipartite sets $X = \{(1,0,0), (1,1,0), (0,1,0)\}$ and $Y = \{(0,0,1), (0,0,2), (0,0,3), (0,0,4), (0,0,5), (0,0,6), (0,0,7)\}$. Note that $g(K_{3,7}) = 2$ and there are 6 rectangular faces and 3 hexagonal faces in any $S_2$-embedding of $K_{3,7}$. Now, the vertices $(0,1,4)$ and $(1,0,4)$ would have to be embedded in a face of $K_{3,7}$. Since $((0,1,4),(1,0,4)) \in E(\Gamma(R))$, we observe that they are attached into a single face of $S_2$-embedding of $K_{3,7}$. Further, in $\Gamma(R)$, the neighbor of $(0,1,4)$ contains the vertices $(0,0,2), (0,0,4), (0,0,6), (1,0,0)$ and the neighbor of $(1,0,4)$ contains the vertices $(0,0,2), (0,0,4), (0,0,6), (0,1,0)$. But any hexagonal faces does not adopt such 9 edges together. Therefore $g(\Gamma(\mathbb{Z}_2 \times \mathbb{Z}_2 \times \mathbb{Z}_8)) \geq 3$ and thus $g(\Gamma(\mathbb{Z}_2 \times \mathbb{Z}_2 \times \mathbb{Z}_2[x]/(x^3))) \geq 3$.

If $R \cong \mathbb{Z}_2 \times \mathbb{Z}_2 \times \mathbb{Z}_9$ or $\mathbb{Z}_2 \times \mathbb{Z}_2 \times \mathbb{Z}_3[x]/(x^2)$, then Example 4.4.8(a) and the fact that $\Gamma(\mathbb{Z}_2 \times \mathbb{Z}_2 \times \mathbb{Z}_9) \cong \Gamma(\mathbb{Z}_2 \times \mathbb{Z}_2 \times \mathbb{Z}_3[x]/(x^2))$ gives $g(\Gamma(R)) \geq 3$.

If $R \cong \mathbb{Z}_2 \times \mathbb{Z}_2 \times \mathbb{Z}_{11}$, then $\widetilde{\Gamma(R)} = \Gamma(R) - \{(1,0,i),\ (0,1,i) \mid i \in \mathbb{Z}_{11}^*\}$ and the embedding of $\widetilde{\Gamma(R)}$ is given in Figure 4.8(b). Thus $g(\Gamma(R)) = 2$. Further, since $\Gamma(\mathbb{Z}_2 \times \mathbb{Z}_2 \times \mathbb{F}_9) \subseteq \Gamma(R)$, we have $g(\Gamma(\mathbb{Z}_2 \times \mathbb{Z}_2 \times \mathbb{F}_9)) = 2$.

If $R \cong \mathbb{Z}_2 \times \mathbb{Z}_3 \times \mathbb{Z}_4$, then by Figure 4.9(a), the reduction graph $\widetilde{\Gamma(R)}$ can

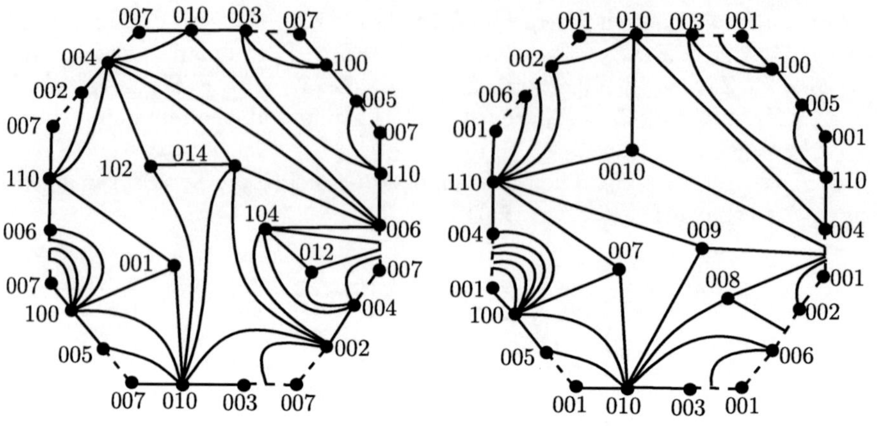


(a) $R = \mathbb{Z}_2 \times \mathbb{Z}_2 \times \mathbb{Z}_8$ (b) $R = \mathbb{Z}_2 \times \mathbb{Z}_2 \times \mathbb{Z}_{11}$

Figure 4.8: The embedding of $\widetilde{\Gamma(R)}$ in $S_2$

be embedded in $S_2$; so $g(\Gamma(\mathbb{Z}_2 \times \mathbb{Z}_3 \times \mathbb{Z}_4)) = 2$. Moreover, $\Gamma(\mathbb{Z}_2 \times \mathbb{Z}_3 \times \mathbb{Z}_4) \cong \Gamma(\mathbb{Z}_2 \times \mathbb{Z}_3 \times \mathbb{Z}_2[x]/(x^2))$; so $g(\Gamma(\mathbb{Z}_2 \times \mathbb{Z}_3 \times \mathbb{Z}_2[x]/(x^2))) = 2$.

If $R \cong \mathbb{Z}_2 \times \mathbb{Z}_3 \times \mathbb{Z}_5$, then the reduction zero-divisor graph $\widetilde{\Gamma(R)}$ can be embedded in $S_2$ as in Figure 4.9(b). Therefore $g(\Gamma(\mathbb{Z}_2 \times \mathbb{Z}_3 \times \mathbb{Z}_5)) = 2$.

If $R \cong \mathbb{Z}_2 \times \mathbb{F}_4 \times \mathbb{F}_4$, then the vertices $u_1 = (1,0,0), u_2 = (0,1,0), u_3 = (0,\omega,0), u_4 = (0,\omega^2,0), u_5 = (1,1,0), u_6 = (1,\omega,0), u_7 = (1,\omega^2,0), v_1 = (0,0,1), v_2 = (0,0,\omega), v_3 = (0,0,\omega^2), w_1 = (1,0,1), w_2 = (1,0,\omega), w_3 = (1,0,\omega^2)$ induce a subgraph $H$ of $\Gamma(R)$ the same as the graph in Figure 4.3, and so $g(\Gamma(R)) \geq 3$. Further, by Remark 4.4.5, $g(\Gamma(\mathbb{Z}_2 \times \mathbb{F}_4 \times \mathbb{Z}_4)) \geq 3$; so $g(\Gamma(\mathbb{Z}_2 \times \mathbb{F}_4 \times \mathbb{Z}_2[x]/(x^2))) \geq 3$. Moreover, if $T \cong \mathbb{Z}_4 \times \mathbb{Z}_4$, $\mathbb{Z}_4 \times \mathbb{Z}_2[x]/(x^2)$, or $\mathbb{Z}_2[x]/(x^2) \times \mathbb{Z}_2[x]/(x^2)$, then $\Gamma(\mathbb{Z}_2 \times T)$ contains $\Gamma(\mathbb{Z}_2 \times \mathbb{F}_4 \times \mathbb{F}_4)$ as a subgraph; this implies that $g(\Gamma(\mathbb{Z}_2 \times T)) \geq 3$.

Note that by Example 4.4.8(b), we have $g(\Gamma(\mathbb{Z}_3 \times \mathbb{Z}_3 \times \mathbb{Z}_4)) \geq 3$, and so $g(\Gamma(\mathbb{Z}_3 \times \mathbb{Z}_3 \times \mathbb{Z}_2[x]/(x^2))) \geq 3$. Also, $\widetilde{\Gamma(\mathbb{Z}_3 \times \mathbb{Z}_3 \times \mathbb{F}_4)}$ can be embedded in $S_2$ as shown in Figure 4.10. Thus $g(\Gamma(\mathbb{Z}_3 \times \mathbb{Z}_3 \times \mathbb{F}_4)) = 2$. Moreover, for $S \cong \mathbb{F}_4 \times \mathbb{F}_4$, $\mathbb{Z}_4 \times \mathbb{F}_4$, $\mathbb{Z}_4 \times \mathbb{Z}_4$, $\mathbb{Z}_4 \times \mathbb{Z}_2[x]/(x^2)$, or $\mathbb{Z}_2[x]/(x^2) \times \mathbb{Z}_2[x]/(x^2)$, the graph $\Gamma(\mathbb{Z}_3 \times S)$ contains $\Gamma(\mathbb{Z}_2 \times \mathbb{F}_4 \times \mathbb{F}_4)$ as a subgraph; so $g(\Gamma(\mathbb{Z}_3 \times S)) \geq g(\Gamma(\mathbb{Z}_2 \times \mathbb{F}_4 \times \mathbb{F}_4)) \geq 3$.

Hence, in this case $g(\Gamma(R)) = 2$ if and only if $R$ is isomorphic to one of

$$\mathbb{Z}_2 \times \mathbb{Z}_2 \times \mathbb{Z}_8, \mathbb{Z}_2 \times \mathbb{Z}_2 \times \mathbb{Z}_2[x]/(x^3), \mathbb{Z}_2 \times \mathbb{Z}_2 \times \mathbb{F}_8, \mathbb{Z}_2 \times \mathbb{Z}_2 \times \mathbb{F}_9, \mathbb{Z}_2 \times \mathbb{Z}_2 \times \mathbb{Z}_{11},$$
$$\mathbb{Z}_2 \times \mathbb{Z}_3 \times \mathbb{Z}_4, \mathbb{Z}_2 \times \mathbb{Z}_3 \times \mathbb{Z}_2[x]/(x^2), \mathbb{Z}_2 \times \mathbb{Z}_3 \times \mathbb{Z}_5, \mathbb{Z}_3 \times \mathbb{Z}_3 \times \mathbb{F}_4.$$

**Case 3:** Let $n = 2$ and $|R_1| \leq |R_2|$.

If both $R_1$ and $R_2$ are fields, then $\Gamma(R) \cong K_{|R_1|-1,|R_2|-1}$. Therefore $g(\Gamma(R)) =$

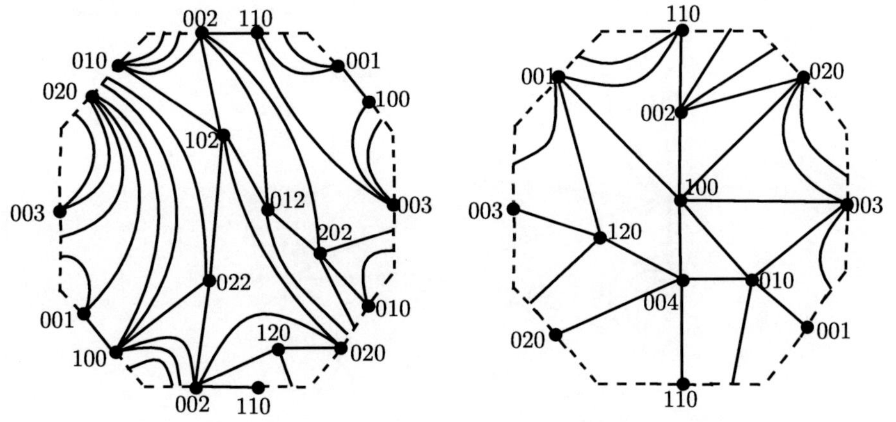


(a) $R = \mathbb{Z}_2 \times \mathbb{Z}_3 \times \mathbb{Z}_4$ (b) $R = \mathbb{Z}_2 \times \mathbb{Z}_3 \times \mathbb{Z}_5$

Figure 4.9: The embedding of $\widetilde{\Gamma(R)}$ in $S_2$

2 if and only if $R$ is isomorphic to one of the following rings

$$\mathbb{F}_4 \times \mathbb{F}_8, \mathbb{F}_4 \times \mathbb{F}_9, \mathbb{F}_4 \times \mathbb{F}_{11}, \mathbb{Z}_5 \times \mathbb{Z}_7.$$

**Case 3.1:** Suppose that exactly one of $R_1$ or $R_2$ is a field.

Without loss of generality, assume that $R_1$ is a field. Let $\mathfrak{m}_2$ be the maximal ideal of $R_2$. If $|R_1| \geq 5$, then $|R_2| \geq 8$ and the sets $\{0\} \times R_2^*$ and $R_1^* \times \{0\}$ induce a subgraph $K_{4,7}$ in $\Gamma(R)$; so $g(\Gamma(R)) \geq 3$. Thus $|R_1| \leq 4$.

(a) Suppose $R_1 \cong \mathbb{Z}_2$. If $|R_2/\mathfrak{m}_2| \geq 6$, then take $u_i = (0, a_i)$, $v_i = (1, a_i)$, and $v_6 = (1, 0)$ for $i = 1, \ldots, 5$ and $a_i \in \mathfrak{m}_2$. Here $u_i v_j = 0$ for all $i, j$ ($1 \leq i \leq 5$ and $1 \leq j \leq 6$); so $K_{5,6}$ is a subgraph of $\Gamma(R)$, and therefore $g(\Gamma(R)) \geq 3$. Thus $|R_2/\mathfrak{m}_2| \leq 5$.

If $|R_2/\mathfrak{m}_2| = 5$, then by Lemma 4.4.6, $\mathfrak{m}_2^2 = 0$. Thus $R \cong \mathbb{Z}_{25}$ or $\mathbb{Z}_5[x]/(x^2)$. By Figure 4.11(a), $\widetilde{\Gamma(R)} = \Gamma(R) - (\{0\} \times U(\mathbb{Z}_{25}))$ can be embedded in $S_2$. Here $U(R)$ denotes the set of all unit elements of $R$. Therefore $g(\Gamma(\mathbb{Z}_2 \times \mathbb{Z}_{25})) = 2$. Since $\Gamma(\mathbb{Z}_2 \times \mathbb{Z}_{25}) \cong \Gamma(\mathbb{Z}_2 \times \mathbb{Z}_5[x]/(x^2))$, we have $g(\Gamma(\mathbb{Z}_2 \times \mathbb{Z}_5[x]/(x^2))) = 2$.

If $|R_2/\mathfrak{m}_2| = 4$, then by Lemma 4.4.6, $\mathfrak{m}_2^2 = 0$, and so $|R_2| = 16$. Therefore $R_2 \cong \mathbb{F}_4[x]/(x^2)$ or $\mathbb{Z}_4[x]/(x^2 + x + 1)$, but by Theorem 4.3.19, we get $\Gamma(\mathbb{Z}_2 \times \mathbb{F}_4[x]/(x^2))$ and $\Gamma(\mathbb{Z}_2 \times \mathbb{Z}_4[x]/(x^2 + x + 1))$ are toroidal.

Let $|R_2/\mathfrak{m}_2| = 3$, and let $|R_2| = 3^t$ for some $t \in \mathbb{N}$. Then by Lemma 4.4.6, $\mathfrak{m}_2^4 = 0$. If $t \geq 4$, then $|\mathfrak{m}_2^2| \geq 9$. Let $u_i = (1, a_i)$ and $v_i = (0, a_i)$ for $0 \neq a_i \in \mathfrak{m}_2^2$. Since $\mathfrak{m}_2^2 \cdot \mathfrak{m}_2^2 = 0$, $K_{8,8}$ is a subgraph of $\Gamma(R)$. Thus $|R_2| \leq 27$. Moreover, if $|R_2| \leq 9$, then by Theorem 4.2.13, $\Gamma(R)$ is planar. Therefore it remains to verify the rings $R_2$ of order 27; so $R_2 \cong \mathbb{Z}_{27}, \mathbb{Z}_3[x]/(x^3), \mathbb{Z}_9[x]/(x^2 - 3, x^3)$, or $\mathbb{Z}_9[x]/(x^2 + 3, x^3)$. If $R \cong \mathbb{Z}_2 \times \mathbb{Z}_{27}$, then the sets $\{(0,3), (0,6), (0,12), (0,15), (0,21), (0,24)\}$ and

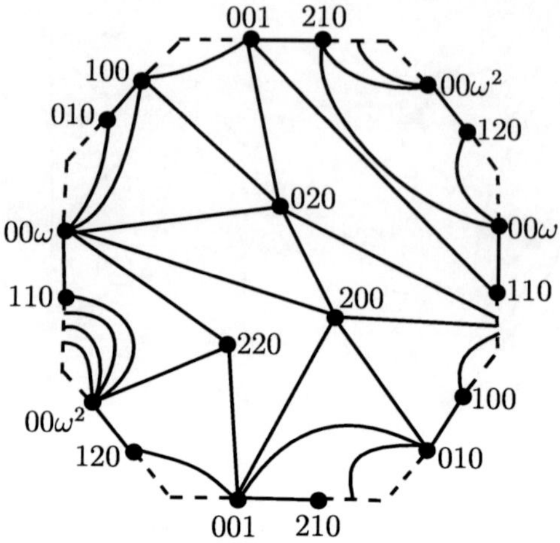


Figure 4.10: The embedding of $\Gamma(\mathbb{Z}_3 \times \mathbb{Z}_3 \times \mathbb{F}_4)$ in $S_2$

$\{(1,0),(1,9),(1,18),(0,9),(0,18)\}$ form a subgraph $K_{5,6}$ in $\Gamma(R)$, and so $g(\Gamma(\mathbb{Z}_2 \times \mathbb{Z}_{27})) \geq 3$. Further, $\Gamma(\mathbb{Z}_{27}) \cong \Gamma(S)$ whenever $S \cong \mathbb{Z}_3[x]/(x^3), \mathbb{Z}_9[x]/(x^2-3,x^3)$, or $\mathbb{Z}_9[x]/(x^2+3,x^3)$, and so by Proposition 4.4.4, $g(\Gamma(\mathbb{Z}_2 \times S)) \geq 3$.

Let $|R_2/\mathfrak{m}_2| = 2$, and let $|R_2| = 2^t$ for some $t \in \mathbb{N}$. Note that by Lemma 4.4.6, $\mathfrak{m}_2^5 = 0$. If $t \geq 5$, then $|\mathfrak{m}_2^3| \geq 4$ and $|\mathfrak{m}_2^2 - \mathfrak{m}_2^3| \geq 4$. For $i = 1,2,3$ and $j = 4,\ldots,7$, let $u_i = (1,a_i)$, $v_i = (0,a_i)$, and $v_j = (0,b_j)$, where $0 \neq a_i \in \mathfrak{m}_2^3$ and $b_j \in \mathfrak{m}_2^2 - \mathfrak{m}_2^3$. Let $u_4 = (1,0)$. Then $u_iv_j = 0$ for all $i = 1,\ldots,4$ and $j = 1,\ldots,7$; so $K_{4,7}$ is a subgraph of $\Gamma(R)$. Therefore $|R_2| \leq 16$. Also, by Theorems 4.2.13 and 4.3.19, the zero-divisor graph of all rings $R_2$ of order less than 16 is either planar or toroidal. Thus $|R_2| = 16$. There are 18 local rings of order 16 with cardinality of the residue field as 2. We partition these 18 local rings into six sets and they are given below:

$$
\begin{aligned}
Q_1 &\cong \mathbb{Z}_{16}, \mathbb{Z}_4[x]/(x^2), \mathbb{Z}_2[x,y]/(x^3,xy,y^2);\\
Q_2 &\cong \mathbb{Z}_2[x]/(x^4), \mathbb{Z}_4[x]/(x^2-2,x^4), \mathbb{Z}_4[x]/(x^3-2,x^4), \mathbb{Z}_4[x]/(x^3+x^2-2,x^4);\\
Q_3 &\cong \mathbb{Z}_4[x,y]/(x^2,y^2,xy-2), \mathbb{Z}_2[x,y]/(x^2,y^2);\\
Q_4 &\cong \mathbb{Z}_2[x,y]/(x^3,xy,y^2-x^2), \mathbb{Z}_4[x]/(x^3,x^2-2x), \mathbb{Z}_8[x]/(x^2-4,2x),\\
&\quad\ \mathbb{Z}_4[x,y]/(x^3,x^2-2,xy,y^2-2);\\
Q_5 &\cong \mathbb{Z}_4[x,y]/(2x,2y,x^2,xy,y^2), \mathbb{Z}_2[x,y,z]/(x,y,z)^2;\\
Q_6 &\cong \mathbb{Z}_4[x]/(x^3,2x), \mathbb{Z}_4[x,y]/(x^3,x^2-2,xy,y^2), \mathbb{Z}_8[x]/(x^2,2x).
\end{aligned}
$$

First consider the ring $\mathbb{Z}_{16}$ for $R_2$. The graph $\widetilde{\Gamma(R)} = \Gamma(R) - \{(0,1),(0,3),(0,5),(0,7),(0,9),(0,11),(0,13),(0,15),(1,2),(1,6),(1,10),(1,14)\}$ can be embedded in $S_2$ as shown in Figure 4.11(b). So $g(\Gamma(\mathbb{Z}_2 \times \mathbb{Z}_{16})) = 2$. Clearly

$\Gamma(\mathbb{Z}_{16}) \cong \Gamma(Q_2)$, and therefore $g(\Gamma(\mathbb{Z}_2 \times Q_2)) = 2$.

By Example 4.4.8(c), $g(\Gamma(\mathbb{Z}_2 \times \mathbb{Z}_4[x]/(x^2))) \geq 3$. Note that $\Gamma(\mathbb{Z}_4[x]/(x^2))$ is isomorphic to $\Gamma(Q_3)$; so $g(\Gamma(\mathbb{Z}_2 \times Q_3)) \geq 3$. Also, by [164, Example 2.6], we have $\Gamma(Q_3)$ is contained in $\Gamma(Q_4)$. Further $\Gamma(Q_5) = K_7$ (see [164, Table 2]), and so $\Gamma(Q_5)$ contains $\Gamma(Q_3)$ as a subgraph. Thus $g(\Gamma(\mathbb{Z}_2 \times Q_k)) \geq 3$ for $k = 4, 5$.

Suppose that $R_2 \cong \mathbb{Z}_2[x,y]/(x^3, xy, y^2)$. Then $Z(R_2)^* = \{x, y, x^2, y + x^2, x + x^2, x + y, x + y + x^2\}$. Let $u_1 = (0, x), u_2 = (0, x + x^2), u_3 = (0, y + x^2), u_4 = (0, x+y), u_5 = (0, x+y+x^2), v_1 = (0, y), v_2 = (1, y), v_3 = (0, x^2), v_4 = (1, x^2), v_5 = (1, y + x^2)$. Clearly $u_i v_j = 0$ for all $i, j \in \{1, \ldots, 5\}$; so $K_{5,5}$ is contained in $\Gamma(R)$, and therefore $g(\Gamma(R)) \geq 3$. Note that $\Gamma(\mathbb{Z}_2[x,y]/(x^3, xy, y^2))$ is isomorphic to $\Gamma(Q_6)$; so $g(\Gamma(\mathbb{Z}_2 \times Q_6)) \geq 3$.

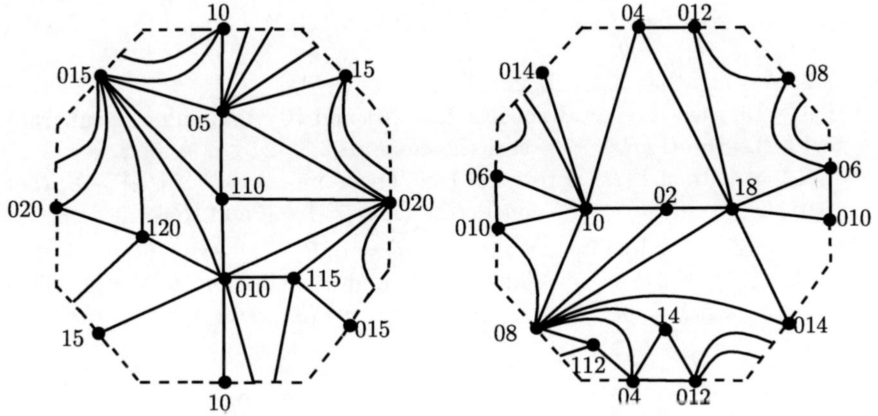


(a) $R = \mathbb{Z}_2 \times \mathbb{Z}_{25}$ (b) $R = \mathbb{Z}_2 \times \mathbb{Z}_{16}$

Figure 4.11: The embedding of $\widetilde{\Gamma(R)}$ in $S_2$

(b) Suppose $R_1 \cong \mathbb{Z}_3$. If $|R_2/\mathfrak{m}_2| \geq 5$, then $|\mathfrak{m}_2| \geq 5$, and by Lemma 4.4.6, $\mathfrak{m}_2^2 = 0$. Therefore every vertex in the set $\{(0, a_i) \mid 0 \neq a_i \in \mathfrak{m}_2^2\}$ is adjacent to all the vertices of the set $\{(1, a_i), (2, a_i) \mid 0 \neq a_i \in \mathfrak{m}_2^2\}$; so $K_{4,8}$ is a subgraph of $\Gamma(R)$. Thus $|R_2/\mathfrak{m}_2| \leq 4$. If $|R_2/\mathfrak{m}_2| = 4$, then by Lemma 4.4.6, $\mathfrak{m}_2^2 = 0$ and $|\mathfrak{m}_2| = 4$. But there is no ring $R_2$ of order 16 such that $|R_2/\mathfrak{m}_2| = 4$. Further, the fact that $\Gamma(\mathbb{Z}_2 \times R_2) \subseteq \Gamma(\mathbb{Z}_3 \times R_2)$ and a similar argument to part (a) gives us that no local ring $R_2$ with $|R_2/\mathfrak{m}_2| = 3$ satisfies $g(\Gamma(\mathbb{Z}_3 \times R_2)) = 2$. Thus $|R_2/\mathfrak{m}_2| = 2$. As in part(a), we have

$$R_2 \cong \mathbb{Z}_{16}, \mathbb{Z}_2[x]/(x^4), \mathbb{Z}_4[x]/(x^2 - 2, x^4), \mathbb{Z}_4[x]/(x^3 - 2, x^4), \mathbb{Z}_4[x]/(x^3 + x^2 - 2, x^4).$$

But the sets $\{(1,0), (2,0), (1,8), (2,8)\}$ and $\{(0,2), (0,4), (0,6), (0,8), (0,10), (0,12), (0,14)\}$ form a subgraph $K_{4,7}$ in $\Gamma(\mathbb{Z}_3 \times \mathbb{Z}_{16})$, and so $g(\Gamma(\mathbb{Z}_3 \times \mathbb{Z}_{16}) \geq 3$. Therefore $g(\Gamma(\mathbb{Z}_3 \times R_2)) \geq 3$.

(c) Let $|R_1| = 4$; so $R_1 \cong \mathbb{F}_4$. If $|R_2/\mathfrak{m}_2| \geq 4$, then by Lemma 4.4.6, $\mathfrak{m}_2^2 = 0$. So every vertex of the form $\{0\} \times \mathfrak{m}_2^*$ is adjacent to all the vertices of the form $R_1^* \times \mathfrak{m}_2$, and therefore $K_{3,12}$ is a subgraph of $\Gamma(R)$. Thus $|R_2/\mathfrak{m}_2| \leq 3$.

If $|R_2/\mathfrak{m}_2| = 3$, then $\mathfrak{m}_2^4 = 0$. If $|\mathfrak{m}_2^2| \geq 9$, then all the vertices of the form $(0, a_i)$ are adjacent to $(1, a_i)$ for all $0 \neq a_i \in \mathfrak{m}_2^2$; so $K_{8,8} \subseteq \Gamma(R)$. Therefore $|R_2| \leq 27$. Also, if $|R_2| = 27$, then by part (b), $g(\Gamma(\mathbb{F}_4 \times R_2)) \geq 3$ because $\Gamma(\mathbb{Z}_3 \times R_2) \subseteq \Gamma(\mathbb{F}_4 \times R_2)$. Further, if $|R_2| = 3$, then by Theorem 4.2.13, $\Gamma(R)$ is planar. So $|R_2| = 9$. Then by Remark 4.4.5, $g(\Gamma(\mathbb{F}_4 \times \mathbb{Z}_9)) \leq g(\Gamma(\mathbb{Z}_4 \times \mathbb{Z}_9)) = 2$ (see Figure 4.12(b)). Thus $g(\Gamma(\mathbb{F}_4 \times \mathbb{Z}_9)) = g(\Gamma(\mathbb{F}_4 \times \mathbb{Z}_3[x]/(x^2))) = 2$.

If $|R_2/\mathfrak{m}_2| = 2$, then the fact $\Gamma(\mathbb{Z}_2 \times R_2) \subseteq \Gamma(\mathbb{F}_4 \times R_2)$ and part $(a)$ implies that $|R_2| \leq 16$. If $|R_2| \leq 4$, then by Theorem 4.2.13, $\Gamma(R)$ is planar. If $|R_2| = 16$, then by part (b), $\Gamma(\mathbb{F}_4 \times R_2) \geq 3$. Therefore $|R_2| = 8$. Note that $g(\Gamma(\mathbb{F}_4 \times \mathbb{Z}_8)) \leq g(\Gamma(\mathbb{Z}_4 \times \mathbb{Z}_8)) = 2$ (see Case 3.2 and Figure 4.12(a)). So $g(\Gamma(\mathbb{F}_4 \times \mathbb{Z}_8)) = g(\Gamma(\mathbb{F}_4 \times \mathbb{Z}_2[x]/(x^2))) = 2$.

**Case 3.2:** Assume that both $R_1$ and $R_2$ are not fields.

If $|R_2| \geq 16$, then the vertex subsets $R_1^* \times \{0\}$ and $\{0\} \times R_2^*$ induce a subgraph $K_{3,15}$ in $\Gamma(R)$. So $g(\Gamma(R)) \geq 3$, and therefore $|R_i| = 4, 8$, or $9$ for $i = 1, 2$. If $|R_1| = |R_2| = 4$, then $\Gamma(R)$ is toroidal (see Theorem 4.3.19). If $|R_1| \geq 8$, then $\langle (R_1^* \times \{0\}) \cup (\{0\} \times R_2^*) \rangle = K_{7,7}$ in $\Gamma(R)$. Thus $|R_1| = 4$ and $|R_2| = 8$ or $9$.

If $R \cong \mathbb{Z}_4 \times \mathbb{Z}_8$, then by Figure 4.12(a), $\widetilde{\Gamma(R)} = \Gamma(R) - \{(1,2),(1,6),(3,2),(3,6),(2,1),(2,3),(1,4),(3,4)\}$ is embedded in $S_2$, and so $g(\Gamma(\mathbb{Z}_4 \times \mathbb{Z}_8)) = 2$. Thus $g(\Gamma(\mathbb{Z}_2[x]/(x^2) \times \mathbb{Z}_8)) = g(\Gamma(\mathbb{Z}_4 \times \mathbb{Z}_2[x]/(x^3))) = g(\Gamma(\mathbb{Z}_2[x]/(x^2) \times \mathbb{Z}_2[x]/(x^3))) = 2$.

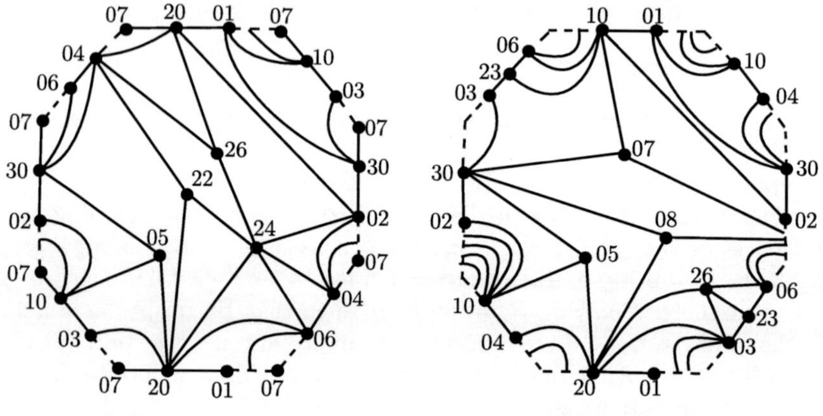


(a) $R = \mathbb{Z}_4 \times \mathbb{Z}_8$ (b) $R = \mathbb{Z}_4 \times \mathbb{Z}_9$

Figure 4.12: The embedding of $\widetilde{\Gamma(R)}$ in $S_2$

If $R \cong \mathbb{Z}_4 \times \mathbb{Z}_9$, then by Figure 4.12(b), $g(\Gamma(R)) = g(\widetilde{\Gamma(R)}) = 2$, where $\widetilde{\Gamma(R)} =$

$\Gamma(R) - (0,1),(0,2),(0,4),(0,5),(0,7),(0,8),(1,3),(1,6),(3,3),(3,6)\}$. Hence $g(\Gamma(\mathbb{Z}_2[x]/(x^2)\times\mathbb{Z}_9)) = g(\Gamma(\mathbb{Z}_4\times\mathbb{Z}_3[x]/(x^2))) = g(\Gamma(\mathbb{Z}_2[x]/(x^2)\times\mathbb{Z}_3[x]/(x^2))) = 2$. □

## 4.5 Zero-divisor Graphs with Large Genus

Graphs on higher surfaces are useful. Consider, for example, the following seemingly simple question of characterizing those graphs that do not contain two disjoint cycles of odd length. This problem was solved by Lovasz using Seymour's deep characterization of regular matroids.

In this section, we discuss the existence of higher order genus in the zero-divisor graph of a commutative ring. In [431], Wickham initiated such studies. The main result of his paper showed that for every positive integer $g$, there are only finitely many finite commutative rings with genus $g$. Thereafter, Aliniaeifard and Behboodi [48] gave a short and direct proof for this result. The corresponding result is given below.

**Theorem 4.5.1.** ([431, Theorem 2], [48, Theorem 2.2]) *For every positive integer $g$, there are only finitely many finite commutative rings of genus $g$.*

Aliniaeifard et al. [48] also showed that if $R$ is a commutative ring such that the zero-divisor graph of $R$ has finite genus $g$, then either $g = 0$ or $R$ is a finite ring. That is, there is no infinite commutative ring $R$ with $0 < g(\Gamma(R)) < \infty$.

**Theorem 4.5.2.** ([48, Theorem 3.6]) *Let $R$ be an infinite commutative ring. Then either $g(\Gamma(R)) = 0$ or $g(\Gamma(R)) = \infty$.*

By the use of this theorem, the following statement is a generalization of Theorem 4.5.1 to arbitrary (not necessarily finite) commutative rings.

**Corollary 4.5.3.** ([48, Corollary 3.7]) *For a fixed positive integer $g$, there are only finitely many isomorphism classes of commutative rings whose zero-divisor graph has genus $g$.*

## 4.6 Crosscap of Zero-divisor Graphs

The "Möbius band" is a surface that is neither closed nor orientable. The "Klein bottle", another nonorientable closed surface, can be obtained by cutting two holes in a sphere and then closing each of them off with a Mobius band. Note that every closed, connected, nonorientable surface can be obtained by cutting holes in a sphere and then closing them off with Möbius bands.

Let $j$ be a nonnegative integer, and let $\tilde{S}_j$ denote the surface obtained by attaching $j$ crosscaps to the sphere. So $\tilde{S}_j$ is a nonorientable surface of genus

$j$. Recall that the *nonorientable genus of a graph G*, denoted by $\tilde{g}(G)$, is the minimal integer $n$ such that the graph can be embedded in $\tilde{S}_n$.

The embedding of a zero-divisor graph in nonorientable compact surfaces was first studied by Hung-Jen Chiang-Hsieh [165]. First of all, he determined all finite commutative local rings whose zero-divisor graphs are of crosscap one. The following lemmas are used to characterize all finite commutative local rings $R$ with $\tilde{g}(\Gamma(R)) = 1$.

**Lemma 4.6.1.** ([165, Proposition 2.2]) *Let $(R, \mathfrak{m})$ be a finite commutative local ring such that $\mathfrak{m}^3 \neq 0$. If $dim_{R/\mathfrak{m}}\mathfrak{m}/\mathfrak{m}^2 \geq 2$, then $\tilde{g}(\Gamma(R)) \geq 2$.*

**Lemma 4.6.2.** ([165, Corollary 2.3]) *Let $(R, \mathfrak{m})$ be a finite commutative local ring. If $\mathfrak{m}^3 \neq 0$ and $\tilde{g}(\Gamma(R)) \leq 1$, then $\mathfrak{m}$ is a principal ideal.*

**Lemma 4.6.3.** ([165, Corollary 2.4]) *Let $(R, \mathfrak{m})$ be a finite commutative local ring such that $|\mathfrak{m}^2| \geq 3$. If $dim_{R/\mathfrak{m}}\mathfrak{m}/\mathfrak{m}^2 \geq 2$, then $\tilde{g}(\Gamma(R)) \geq 2$.*

**Lemma 4.6.4.** ([165, Proposition 2.5]) *Let $(R, \mathfrak{m})$ be a finite commutative local ring. If $dim_{R/\mathfrak{m}}\mathfrak{m}/\mathfrak{m}^2 \geq 2$, then $\tilde{g}(\Gamma(R)) \geq 2$.*

**Theorem 4.6.5.** ([165, Theorem 2.8]) *Let $(R, \mathfrak{m})$ be a finite commutative local ring which is not a field. Then $\tilde{g}(\Gamma(R)) = 1$ if and only if $R$ is isomorphic to one of the following* 13 *rings.*

$$\mathbb{Z}_{49}, \frac{\mathbb{Z}_7[x]}{(x^2)}, \mathbb{Z}_{32}, \frac{\mathbb{Z}_2[x]}{(x^5)}, \frac{\mathbb{Z}_4[x]}{(x^3-2, x^5)}, \frac{\mathbb{Z}_4[x]}{(x^4-2, x^5)}, \frac{\mathbb{Z}_8[x]}{(x^2-2, x^5)},$$
$$\frac{\mathbb{Z}_8[x]}{(x^2-2x+2, x^5)}, \frac{\mathbb{Z}_8[x]}{(x^2+2x-2, x^5)}, \frac{\mathbb{Z}_2[x,y]}{(x^3, xy, y^2)}, \frac{\mathbb{Z}_4[x]}{(x^3, 2x)},$$
$$\frac{\mathbb{Z}_4[x,y]}{(x^3, x^2-2, xy, y^2)}, \frac{\mathbb{Z}_8[x]}{(x^2, 2x)}.$$

Next, we are interested in the classification of all finite nonlocal commutative rings $R$ with $\tilde{g}(\Gamma(R)) = 1$.

**Lemma 4.6.6.** ([165, Proposition 3.3]) *Let $S$ and $R$ be two finite commutative rings such that $(R, \mathfrak{m})$ is local. If $\mathfrak{m}^2 \neq 0$ and $dim_{R/\mathfrak{m}}\mathfrak{m}/\mathfrak{m}^2 \geq 2$, then $\tilde{g}(\Gamma(S \times R)) \geq 2$.*

**Lemma 4.6.7.** ([165, Proposition 3.4]) *Let $(R, \mathfrak{m})$ be a finite commutative local ring such that $|\mathfrak{m}| = p^n$ and $\mathfrak{m}^2 = 0$. Then $\tilde{g}(K_{p^n-1,p^n}) \leq \tilde{g}(\Gamma(\mathbb{Z}_2 \times R)) \leq \tilde{g}(\Gamma(\mathbb{Z}_{p^{2n+1}}))$.*

**Lemma 4.6.8.** ([165, Proposition 3.6]) *Let $B_n = \mathbb{Z}_2 \times \cdots \times \mathbb{Z}_2$ be a finite Boolean ring with $n$ components, where $n \geq 5$. Then $\tilde{g}(\Gamma(B_n)) \geq \left\lceil \frac{3^n - 2^{n+3} + 4n + 25}{6} \right\rceil$.*

**Lemma 4.6.9.** ([165, Corollary 3.7]) *Let $R$ be a finite commutative ring with $|Spec(R)| \geq 5$. Then $\tilde{g}(\Gamma(R)) \geq 6$.*

**Theorem 4.6.10.** ([165, Theorem 3.8]) *Let $R$ be a finite commutative ring with $|Spec(R)| = 2$. Then $\tilde{g}(\Gamma(R)) = 1$ if and only if $R$ is isomorphic to one of the following* 16 *rings:*

$$\mathbb{F}_4 \times \mathbb{F}_4,\ \mathbb{F}_4 \times \mathbb{Z}_5,\ \mathbb{Z}_2 \times \frac{\mathbb{F}_4[x]}{(x^2)},\ \mathbb{Z}_2 \times \frac{\mathbb{Z}_4[x]}{(x^2+x+1)},\ \mathbb{Z}_2 \times \frac{\mathbb{Z}_2[x,y]}{(x^2,xy,y^2)},$$
$$\mathbb{Z}_2 \times \frac{\mathbb{Z}_4[x]}{(x^2,2x)},\ \mathbb{Z}_3 \times \mathbb{Z}_8,\ \mathbb{Z}_3 \times \frac{\mathbb{Z}_2[x]}{(x^3)},\ \mathbb{Z}_3 \times \frac{\mathbb{Z}_4[x]}{(x^3,x^2-2)},\ \mathbb{Z}_4 \times \mathbb{Z}_5,\ \mathbb{Z}_5 \times \frac{\mathbb{Z}_2[x]}{(x^2)},$$
$$\mathbb{F}_4 \times \mathbb{Z}_4,\ \mathbb{F}_4 \times \frac{\mathbb{Z}_2[x]}{(x^2)},\ \mathbb{Z}_4 \times \mathbb{Z}_4,\ \mathbb{Z}_4 \times \frac{\mathbb{Z}_2[x]}{(x^2)},\ \frac{\mathbb{Z}_2[x]}{(x^2)} \times \frac{\mathbb{Z}_2[x]}{(x^2)}.$$

**Theorem 4.6.11.** ([165, Theorem 3.10]) *Let $R$ be a finite commutative ring with $|Spec(R)| = 3$. Then $\tilde{g}(\Gamma(R)) = 1$ if and only if $R$ is isomorphic to one of the following* 6 *rings: $\mathbb{Z}_2 \times \mathbb{Z}_2 \times \mathbb{F}_4$, $\mathbb{Z}_2 \times \mathbb{Z}_2 \times \mathbb{Z}_4$, $\mathbb{Z}_2 \times \mathbb{Z}_2 \times \mathbb{Z}_2[x]/(x^2)$, $\mathbb{Z}_2 \times \mathbb{Z}_2 \times \mathbb{Z}_5, \mathbb{Z}_2 \times \mathbb{Z}_3 \times \mathbb{Z}_3$, $\mathbb{Z}_3 \times \mathbb{Z}_3 \times \mathbb{Z}_3$.*

**Theorem 4.6.12.** ([165, Theorem 3.11]) *Let $R$ be a finite commutative ring such that $|Spec(R)| = 4$. Then $\tilde{g}(\Gamma(R)) = 1$ if and only if $R$ is isomorphic to $\mathbb{Z}_2 \times \mathbb{Z}_2 \times \mathbb{Z}_2 \times \mathbb{Z}_2$.*

We now list all commutative ring such that $\Gamma(R)$ is a projective graph.

**Proposition 4.6.13.** ([165, Theorems 2.8, 3.8, 3.10, and 3.11]) *Let $R$ be a commutative ring such that $\tilde{g}(\Gamma(R)) = 1$. Then the following hold:*

(a) *If $R$ is a local ring which is not a field, then $R$ is isomorphic to one of the following* 13 *rings:*

$$\mathbb{Z}_{49}, \mathbb{Z}_7[x]/(x^2), \mathbb{Z}_{32}, \mathbb{Z}_2[x]/(x^5), \mathbb{Z}_4[x]/(x^3-2,x^5), \mathbb{Z}_4[x]/(x^4-2,x^5),$$
$$\mathbb{Z}_8[x]/(x^2-2,x^5), \mathbb{Z}_8[x]/(x^2-2x+2,x^5), \mathbb{Z}_8[x]/(x^2+2x-2,x^5),$$
$$\mathbb{Z}_2[x,y]/(x^3,xy,y^2), \mathbb{Z}_4[x]/(x^3,2x), \mathbb{Z}_4[x,y]/(x^3,x^2-2,xy,y^2),$$
$$\mathbb{Z}_8[x]/(x^2,2x);$$

(b) *If $R$ is not a local ring, then $R$ is isomorphic to the one of the following* 23 *rings:*

$$\mathbb{Z}_2 \times \mathbb{F}_4[x]/(x^2), \mathbb{Z}_2 \times \mathbb{Z}_4[x]/(x^2+x+1), \mathbb{Z}_2 \times \mathbb{Z}_2[x,y]/(x^2,xy,y^2),$$
$$\mathbb{Z}_2 \times \mathbb{Z}_4[x]/(2x,x^2), \mathbb{Z}_3 \times \mathbb{Z}_8, \mathbb{Z}_3 \times \mathbb{Z}_2[x]/(x^3), \mathbb{Z}_3 \times \mathbb{Z}_4[x]/(x^2-2,x^3),$$
$$\mathbb{Z}_4 \times \mathbb{Z}_5, \mathbb{Z}_2[x]/(x^2) \times \mathbb{Z}_5, \mathbb{F}_4 \times \mathbb{F}_4, \mathbb{F}_4 \times \mathbb{Z}_4, \mathbb{F}_4 \times \mathbb{Z}_2[x]/(x^2), \mathbb{Z}_4 \times \mathbb{Z}_4,$$
$$\mathbb{Z}_4 \times \mathbb{Z}_2[x]/(x^2), \mathbb{Z}_2[x]/(x^2) \times \mathbb{Z}_2[x]/(x^2), \mathbb{F}_4 \times \mathbb{Z}_5, \mathbb{Z}_2 \times \mathbb{Z}_2 \times \mathbb{F}_4,$$
$$\mathbb{Z}_2 \times \mathbb{Z}_2 \times \mathbb{Z}_4, \mathbb{Z}_2 \times \mathbb{Z}_2 \times \mathbb{Z}_2[x]/(x^2), \mathbb{Z}_2 \times \mathbb{Z}_2 \times \mathbb{Z}_5, \mathbb{Z}_2 \times \mathbb{Z}_3 \times \mathbb{Z}_3,$$
$$\mathbb{Z}_3 \times \mathbb{Z}_3 \times \mathbb{Z}_3, \mathbb{Z}_2 \times \mathbb{Z}_2 \times \mathbb{Z}_2 \times \mathbb{Z}_2.$$

When comparing the list of finite commutative rings with toroidal zero-divisor graphs and the ones with projective zero-divisor graphs, one finds that if a zero-divisor graph is projective, then it must be toroidal, i.e., $\tilde{g}(\Gamma(R)) \geq g(\Gamma(R))$ when $g(\Gamma(R)) = 1$. However, since a zero-divisor graph usually contains some complete subgraphs or complete bi-partite subgraphs, Hsieh [165] thought that the above inequality holds not merely by accident. Hence, he proposed the following conjecture.

**Conjecture 4.6.14.** ([165, Conjecture 3.12]) *$\tilde{g}(\Gamma(R)) \geq g(\Gamma(R))$ for every finite commutative ring $R$.*

A slight modification of the proof of Theorem 4.5.1 will prove the following result.

**Theorem 4.6.15.** ([431, Theorem 3]) *For every positive integer $\tilde{g}$, there are only finitely many finite commutative rings of nonorientable genus $\tilde{g}$.*

In 2019, Asir and Mano [95] enumerated all isomorphism classes of commutative rings with identity whose zero-divisor graph has crosscap two. We discuss it in detail. It is well known that every finite commutative ring is a direct product of local rings, and so we start our discussion with the local case.

The first result gives a bound for the size of the residue field $|R/\mathfrak{m}|$, which helps us to exclude some cases of higher crosscap of zero-divisor graphs.

**Lemma 4.6.16.** ([165, Lemma 3.1]) *Let $(R, \mathfrak{m})$ be a finite local ring. If $\tilde{g}(\Gamma(R)) \leq 2$, then $|R/\mathfrak{m}| \leq 7$; and the following hold:*

(a) *If $|R/\mathfrak{m}| = 4, 5,$ or $7$, then $\mathfrak{m}^2 = 0$;*

(b) *If $|R/\mathfrak{m}| = 3$, then $\mathfrak{m}^3 = 0$;*

(c) *If $|R/\mathfrak{m}| = 2$, then $\mathfrak{m}^5 = 0$.*

*Proof.* From the proof of [165, Proposition 2.1], the contradictory assumption of the cases in the statement gives us that $K_7$, $K_8$, or $K_{3,12}$ is a subgraph of $\Gamma(R)$. Thus $\tilde{g}(\Gamma(R)) \geq 3$. □

Let us start by examining the crosscap number of the zero-divisor graph of several rings.

**Example 4.6.17.** ([165, Example 3.5]) Let $R \cong \mathbb{Z}_2 \times \mathbb{Z}_{16}$, $\mathbb{Z}_3 \times \mathbb{Z}_{16}$, $\mathbb{Z}_2 \times \mathbb{Z}_3 \times \mathbb{Z}_4$, or $\mathbb{Z}_2 \times \mathbb{Z}_3 \times \mathbb{Z}_2[x]/(x^2)$. Then $\tilde{g}(\Gamma(R)) \geq 3$.

*Proof.* (a) Let $G = \Gamma(\mathbb{Z}_2 \times \mathbb{Z}_{16})$ and $u_1 = (1,0), u_2 = (1,8), u_3 = (0,8), v_1 = (0,2), v_2 = (0,4), v_3 = (0,6), v_4 = (0,10), v_5 = (0,12), v_6 = (0,14), w_1 = (1,4), w_2 = (1,12), w_3 = (1,2), w_4 = (1,6), w_5 = (1,10), w_6 = (1,14), w_7 = (0,1), w_8 = (0,3), w_9 = (0,5), w_{10} = (0,7), w_{11} = (0,9), w_{12} = (0,11), w_{13} = (0,13), w_{14} =$

$(0, 15)$. Let $\tilde{G} = G - \{w_3, w_4, w_5, w_6, w_7, w_8, w_9, w_{10}, w_{11}, w_{12}, w_{13}, w_{14}\}$. Since the vertices $w_3, w_4, w_5, w_6, w_7, w_8, w_9, w_{10}, w_{11}, w_{12}, w_{13}$, and $w_{14}$ have degree one, $\tilde{g}(\tilde{G}) = \tilde{g}(G)$.

Suppose that $\tilde{g}(\tilde{G}) = 2$. Then by Euler's Formula, there are 16 faces when drawing $\tilde{G}$ on $N_2$. Fix a representation of $\tilde{G}$, and let $F_1, \ldots, F_{16}$ be the faces of $\tilde{G}$ corresponding to this representation. Let $G' = \tilde{G} - \{w_1, w_2\} - \{u_1u_3, u_2u_3, v_2v_5\}$. Clearly $G' \cong K_{3,6}$; so $\tilde{g}(G') = 2$ and $G'$ has 9 faces. Write the faces of $G'$ as $F'_1, \ldots, F'_9$. Here for every $i$, each boundary of $F'_i$ cannot have consecutive repetitions of a single edge. Therefore, in $K_{3,6}$, the only way to have a closed walk of length 4 without consecutive repetitions of a single edge is a 4-cycle. Thus the boundary of each $F'_i$ is a 4-cycle.

Now the faces $\{F_1, \ldots, F_{16}\}$ can be recovered by inserting $w_1, w_2$, and all the edges incident with $w_1, w_2$ and $u_1u_3, u_2u_3, v_2v_5$ into the representation corresponding to $F'_1, \ldots, F'_9$. Since $w_iv_2, w_iv_5, w_iu_3, u_3v_2, u_3v_5 \in E(\tilde{G})$ for $i = 1, 2$, $w_1$ and $w_2$ must be in the neighboring faces of each other, as shown in Figure 4.13(a). Note that the two unlabeled vertices in Figure 4.13(a) are adjacent to both $v_2$ and $v_5$ in $G'$, and so they must be $u_1$ and $u_2$. Further, $u_3$ is adjacent to $u_1$ and $u_2$, so that one cannot draw it without crossing, a contradiction. Therefore, we may conclude that $\tilde{g}(G) \geq 3$.

(b) Consider $R \cong \mathbb{Z}_3 \times \mathbb{Z}_{16}$. Then $\Gamma(\mathbb{Z}_2 \times \mathbb{Z}_{16})$ is a subgraph of $\Gamma(\mathbb{Z}_3 \times \mathbb{Z}_{16})$, and so $\tilde{g}(\Gamma(R)) \geq 3$.

(c) Let $G = \Gamma(\mathbb{Z}_2 \times \mathbb{Z}_3 \times \mathbb{Z}_4)$ and $u_1 = (0, 0, 1), u_2 = (0, 0, 2), u_3 = (0, 0, 3)$, $v_1 = (1, 0, 0), v_2 = (0, 1, 0), v_3 = (0, 2, 0), v_4 = (1, 1, 0), v_5 = (1, 2, 0), w_1 = (1, 0, 2), w_2 = (0, 1, 2), w_3 = (0, 2, 2), w_4 = (1, 0, 1), w_5 = (1, 0, 3), w_6 = (0, 1, 1)$, $w_7 = (0, 1, 3), w_8 = (0, 2, 1), w_9 = (0, 2, 3), w_{10} = (1, 2, 2), w_{11} = (1, 1, 2)$.

Here the vertices $w_6, w_7, w_8, w_9, w_{10}$, and $w_{11}$ have degree one, and so $\tilde{g}(\tilde{G}) = \tilde{g}(G)$, where $\tilde{G} = G - \{w_6, w_7, w_8, w_9, w_{10}, w_{11}\}$. Let $G' = \tilde{G} - \{w_4, w_5\} - \{v_1v_2, v_1v_3\}$ and $G'' = G' - \{w_1, w_2, w_3\}$. Suppose that $\tilde{g}(\tilde{G}) = 2$. Then $\tilde{g}(G') = 2$, and by Euler's Formula there are 13 faces when drawing $G'$ on $N_2$. Fix a representation of $G'$, and let $F'_1, \ldots, F'_{13}$ be the faces of $G'$ corresponding to this representation. Note that $G'' \cong K_{3,5}$, which has 7 faces in any $N_2$ representation. Let $F''_1, \ldots, F''_7$ be the faces of $G''$. Here, for every $i$, each boundary of $F''_i$ cannot have consecutive repetitions of a single edge. Therefore in $K_{3,5}$, the only way to have a closed walk of length 6 without consecutive repetition of a single edge is to have a 6-cycle. So all face boundaries are 4-cycles, along with one 6-cycle. Let us assume that the boundary $F''_7$ has length 6.

Then $\{F'_1, \ldots, F'_{13}\}$ can be recovered by inserting $w_1, w_2, w_3$, and all edges incident with $w_1, w_2, w_3$ into the representation corresponding to $F''_1, \ldots, F''_7$. Let $F''_{t_i}$ denote the face of $G''$ into which $w_i$ is inserted during the recovering process from $G''$ to $G'$. Note that $w_1w_2, w_1w_3 \in E(G')$, and so $w_1, w_2, w_3$ should be inserted into the same face. Further, $w_1v_2, w_1v_3, w_2v_1 \in E(G')$; so $t_i = 7$ for $i = 1, 2, 3$, as $F''_7$ is the only face whose boundary is a 6-cycle. Let $e_1 = w_1v_2, e_2 = w_1v_3, e_3 = w_1u_2, e_4 = w_1w_2, e_5 = w_2u_2, e_6 = w_2v_1, e_7 = w_3u_2, e_8 = w_3v_1$, and

$e_9 = w_3w_1$. After inserting $w_1, w_2$, and $e_1, \ldots, e_6$ into $F_7''$, we obtain Figure 4.13(b) as below. However, from Figure 4.13(b), one cannot insert $w_3$ together with the edges $e_7, e_8, e_9$ into $F_7''$ without crossing, a contradiction. Hence $\tilde{g}(\Gamma(R)) \geq 3$.

Further, it is easy to see that $\Gamma(\mathbb{Z}_2 \times \mathbb{Z}_3 \times \mathbb{Z}_4) \cong \Gamma(\mathbb{Z}_2 \times \mathbb{Z}_3 \times \mathbb{Z}_2[x]/(x^2))$, and thus $\tilde{g}(\Gamma(\mathbb{Z}_2 \times \mathbb{Z}_3 \times \mathbb{Z}_2[x]/(x^2))) \geq 3$. □

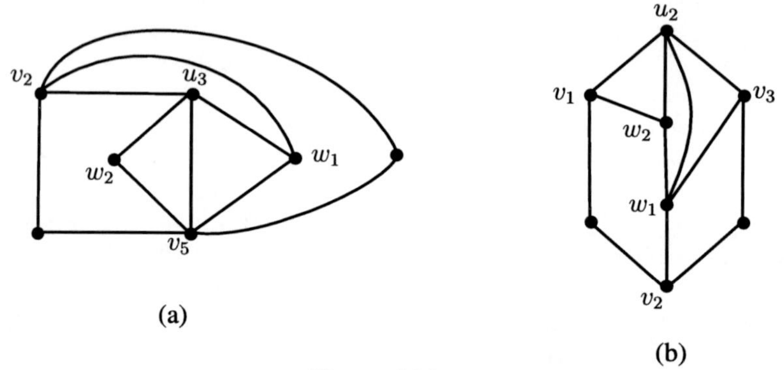


Figure 4.13:

Next, we characterize all commutative rings $R$ with $\tilde{g}(\Gamma(R)) = 2$. For proving this, we require the following five results.

**Proposition 4.6.18.** ([164, Proposition 2.4]) *Let $(R, \mathfrak{m})$ be a finite commutative local ring such that $|R| = p^n$, where $p$ is a prime and $n \in \mathbb{N}$. If $\mathfrak{m}^{n-1} \neq 0$, then $\mathfrak{m}$ is principal and $\Gamma(R) \cong \Gamma(\mathbb{Z}_{p^n})$. Moreover, for any ring $S$, one has that $\Gamma(S \times R) \cong \Gamma(S \times \mathbb{Z}_{p^n})$*

**Proposition 4.6.19.** ([165, Lemma 3.1]) *Let $R_1$ and $R_2$ be two commutative rings. Suppose that $|R_1| = n$ and that there are nonzero elements $x_1, \ldots, x_r \in R_2$ such that $x_ix_j = 0$ for $i \neq j$ and $x_1^2 = x_2^2 = \cdots = x_s^2 = 0$ for some $s \leq r$. Then $K_{r,l} \subseteq \Gamma(R_1 \times R_2)$, where $l = (s+1)(n-1)$.*

**Proposition 4.6.20.** ([165, Lemma 3.2]) *Let $S$ and $R$ be two finite commutative rings such that $(R, \mathfrak{m})$ is local and $|R/\mathfrak{m}| = q \geq 3$. Let $r$ be the largest integer such that $\mathfrak{m}^r \neq 0$. If $r \geq 2$, then $K_{c,d} \subseteq \Gamma(S \times R)$, where $c = |\mathfrak{m}| - |\mathfrak{m}^r|$, $d = |S| \cdot |\mathfrak{m}^r| - 1$. In particular, $\tilde{g}(\Gamma(S \times R)) \geq 6$.*

**Proposition 4.6.21.** ([165, Proposition 3.4]) *Let $(R, \mathfrak{m})$ be a commutative local ring such that $|\mathfrak{m}| = p^n$ and $\mathfrak{m}^2 = 0$. Then $\tilde{g}(K_{p^n-1,p^n}) \leq \tilde{g}(\Gamma(\mathbb{Z}_2 \times R)) \leq \tilde{g}(\Gamma(\mathbb{Z}_{p^{2n+1}}))$.*

Now we provide the proof of the crosscap two characterization.

**Theorem 4.6.22.** ([165, Theorem 3.11]) *Let $R$ be a finite commutative ring. Then $\tilde{g}(\Gamma(R)) = 2$ if and only if $R$ is isomorphic to one of the following 4 rings:*

$$\mathbb{F}_4 \times \mathbb{Z}_7, \mathbb{Z}_5 \times \mathbb{Z}_5, \mathbb{Z}_2 \times \mathbb{Z}_2 \times \mathbb{Z}_7, \mathbb{Z}_2 \times \mathbb{Z}_3 \times \mathbb{F}_4.$$

*Proof.* Assume that $\tilde{g}(\Gamma(R)) = 2$.

Let $R$ be a commutative local ring, and let $\mathfrak{m}$ be the maximal ideal of $R$. Then by Lemma 4.6.16, the number of elements in the residue field $R/\mathfrak{m}$ is bounded by 7. If $\mathfrak{m}^2 = 0$, then clearly $\Gamma(R)$ is complete. Note that the crosscap number of any complete graph is not two. Thus $\mathfrak{m}^2 \neq 0$, and so by Lemma 4.6.16, $|R/\mathfrak{m}|$ is either 3 or 2. If $|R/\mathfrak{m}| = 3$, then $\mathfrak{m}^3 = 0$ and $|R| = 3^t$ for some $t \in \mathbb{N}$. If $t \geq 4$, then $\Gamma(R)$ has $K_8$ or $K_{3,7}$ as a subgraph, a contradiction. Thus $t = 3$. Now apply $\mathfrak{m}^2 \neq 0$ in Proposition 4.6.18; we have $\Gamma(R) \cong \Gamma(\mathbb{Z}_{27})$, a planar graph. If $|R/\mathfrak{m}| = 2$, then from the discussion of 2.7 in [165], we have either $K_7$ is a subgraph of $\Gamma(R)$ or $\Gamma(R)$ is isomorphic to $K_{1,1,1,4}$. Here note that $\tilde{g}(K_{1,1,1,4}) = 1$ (see [165, Remark 1.6]) and $\tilde{g}(K_7) = 3$. Thus, no local ring $R$ satisfies $\tilde{g}(\Gamma(R)) = 2$. Hence $R$ must be nonlocal. Then it is well known that $R \cong R_1 \times \cdots \times R_n$, where each $R_i$ is a local ring and $n \geq 2$. Let $\mathfrak{m}_i$ be the maximal ideal of $R_i$ for $i = 1, \ldots, n$. Without loss of generality, we assume that $|R_1| \leq |R_2| \leq \cdots \leq |R_n|$.

Suppose $n \geq 5$. Let $V_1 = (\{0\} \times \{0\} \times R_3 \times \cdots \times R_n)^*$ and $V_2 = (R_1 \times R_2 \times \{0\} \times \cdots \times \{0\})^*$. Then every vertex in the set $V_1$ is adjacent to each vertex in the set $V_2$. Since $|V_1| \geq 7$ and $|V_2| \geq 3$, $K_{7,3}$ is a subgraph of $\Gamma(R)$, a contradiction to $\tilde{g}(\Gamma(R)) = 2$. Thus $n \leq 4$.

**Case 1:** Let $n = 4$. Note that every vertex in the set $V_3 = (\{0\} \times \{0\} \times R_3 \times R_4)^*$ is adjacent to each vertex in the set $V_4 = (R_1 \times R_2 \times \{0\} \times \{0\})^*$. Therefore, if $|R_4| \geq 4$ or $|R_3| \geq 3$, then $K_{3,7}$ or $K_{3,8}$ is a subgraph of $\Gamma(R)$ respectively, whose crosscap is more than two. Thus $R_i \cong \mathbb{Z}_2$ for $i = 1, 2, 3$, and $R_4 \cong \mathbb{Z}_2$ or $\mathbb{Z}_3$.

If $R \cong \mathbb{Z}_2 \times \mathbb{Z}_2 \times \mathbb{Z}_2 \times \mathbb{Z}_2$, then by Proposition 4.6.13, $\Gamma(R)$ is a projective graph.

Suppose $R \cong \mathbb{Z}_2 \times \mathbb{Z}_2 \times \mathbb{Z}_2 \times \mathbb{Z}_3$ and $H_1$ is a subgraph of $\Gamma(R)$ generated by the vertex sets $V_5, V_6$, and $V_7$, where $V_5 = \{a_1 = (0,0,1,1), a_2 = (0,0,1,2), b_1 = (0,1,0,1), b_2 = (0,1,0,2), c_1 = (1,0,0,1), c_2 = (1,0,0,2), d_1 = (0,0,0,1), d_2 = (0,0,0,2)\}$, $V_6 = \{u = (0,0,1,0), v = (0,1,0,0), w = (1,0,0,0)\}$, and $V_7 = \{x = (1,1,0,0), y = (1,0,1,0), z = (0,1,1,0)\}$. Note that $u$ is adjacent to all $b_i$'s, $c_i$'s, and $d_i$'s; $v$ is adjacent to all $a_i$'s, $c_i$'s, and $d_i$'s; and $w$ is adjacent to all $a_i$'s, $b_i$'s, and $d_i$'s in $H_1$. There are also edges from $u$ to $x$ and $x$ to the $a_i$'s; from $v$ to $y$ and $y$ to the $b_i$'s; and from $w$ to $z$ and $z$ to the $c_i$'s in $H_1$. Now contracting $u$ to $x$, $y$ to $v$, and $w$ to $z$ along the edges between them yields a subgraph $K_{3,8}$, a contradiction.

Hence, in this case, no commutative ring $R$ satisfies $\tilde{g}(\Gamma(R)) = 2$.

**Case 2:** Suppose $n = 3$. If $|R_3| \geq 8$, then every vertex in the set $(\{0\} \times \{0\} \times R_3)^*$ is adjacent to each vertex in the set $(R_1 \times R_2 \times \{0\})^*$, and so $\Gamma(R)$ would contain a subgraph $K_{7,3}$, a contradiction. Thus $|R_3| \leq 7$. Also if $|R_2| \geq 4$, then the vertex sets $(\{0\} \times \{0\} \times R_3)^*$ and $(R_1 \times R_2 \times \{0\})^*$ form a $K_{3,7}$ in $\Gamma(R)$, a contradiction. Thus $|R_i| \leq 3$ for $i = 1, 2$, and $|R_3| \leq 7$.

If $R \cong \mathbb{Z}_2 \times \mathbb{Z}_2 \times \mathbb{Z}_2$ or $\mathbb{Z}_2 \times \mathbb{Z}_2 \times \mathbb{Z}_3$, then by Proposition 4.2.13, $\Gamma(R)$ is planar. If $R \cong \mathbb{Z}_2 \times \mathbb{Z}_2 \times \mathbb{F}_4$, $\mathbb{Z}_2 \times \mathbb{Z}_2 \times \mathbb{Z}_4$, $\mathbb{Z}_2 \times \mathbb{Z}_2 \times \mathbb{Z}_2[x]/(x^2)$, $\mathbb{Z}_2 \times \mathbb{Z}_2 \times \mathbb{Z}_5$, $\mathbb{Z}_2 \times \mathbb{Z}_3 \times \mathbb{Z}_3$, or $\mathbb{Z}_3 \times \mathbb{Z}_3 \times \mathbb{Z}_3$, then by Proposition 4.6.13, $\Gamma(R)$ is a projective graph.

Suppose $R \cong \mathbb{Z}_2 \times \mathbb{Z}_2 \times \mathbb{Z}_7$. Then all vertices in the set $X = \{(0, 1, i), (1, 0, i) \mid i \in \mathbb{Z}_7^*\}$ have degree 1, and therefore $\tilde{g}(G) = \tilde{g}(\tilde{G}_1)$, where $\tilde{G}_1 = \Gamma(R) - X$. Now by Figure 4.14 (a), $\tilde{g}(\Gamma(\mathbb{Z}_2 \times \mathbb{Z}_2 \times \mathbb{Z}_7)) = 2$. Note that in any figure, an element $(x, y, z)$ of $R_1 \times R_2 \times R_3$ is simply denoted by $xyz$.

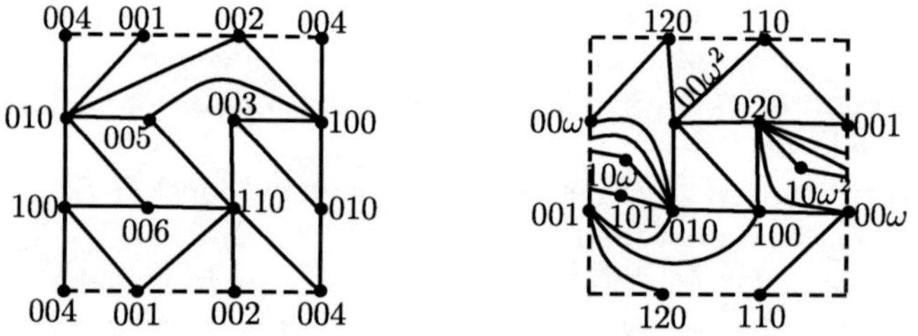


(a) Embedding of $\tilde{G}_1$ in $N_2$ (b) Embedding of $\tilde{G}_2$ in $N_2$

Figure 4.14:

If $R \cong \mathbb{Z}_2 \times \mathbb{Z}_3 \times \mathbb{F}_4$, then the vertices in the set $Y = \{(0, i, j) \mid i \in \mathbb{Z}_3^*$ and $j \in \mathbb{F}_4^*\}$ have degree 1. Let $\tilde{G}_2 = \Gamma(R) - Y$. Then by Figure 4.14(b), $\tilde{g}(\tilde{G}_2) = \tilde{g}(\Gamma(\mathbb{Z}_2 \times \mathbb{Z}_3 \times \mathbb{F}_4)) = 2$.

If $R \cong \mathbb{Z}_2 \times \mathbb{Z}_3 \times \mathbb{Z}_4$ or $\mathbb{Z}_2 \times \mathbb{Z}_3 \times \mathbb{Z}_2[x]/(x^2)$, then by Example 4.6.17(c), we have $\tilde{g}(\Gamma(R)) \geq 3$.

If $R \cong \mathbb{Z}_2 \times \mathbb{Z}_3 \times \mathbb{Z}_5$ or $\mathbb{Z}_2 \times \mathbb{Z}_3 \times \mathbb{Z}_7$, then it is easy to check that $K_{5,4}$ is a subgraph of $\Gamma(R)$, and thus $\tilde{g}(\Gamma(R)) \geq 3$.

Suppose that $R \cong \mathbb{Z}_3 \times \mathbb{Z}_3 \times S$ with $|S| \geq 4$. Then the vertex sets $(\{0\} \times \{0\} \times S)^*$ and $(\mathbb{Z}_3 \times \mathbb{Z}_3 \times \{0\})^*$ form a subgraph $K_{3,8}$ in $\Gamma(R)$. Thus $\tilde{g}(\Gamma(R)) \geq 3$.

Hence, in this case, $\tilde{g}(\Gamma(R)) = 2$ if and only if $R \cong \mathbb{Z}_2 \times \mathbb{Z}_2 \times \mathbb{Z}_7$ or $\mathbb{Z}_2 \times \mathbb{Z}_3 \times \mathbb{F}_4$.

**Case 3:** Let $n = 2$. If $R_1$ and $R_2$ are both fields, then $\Gamma(R) \cong K_{m,n}$, where $m = |R_1| - 1, n = |R_2| - 1$. So $\tilde{g}(\Gamma(R)) = 2$ if and only if $R \cong \mathbb{F}_4 \times \mathbb{Z}_7$ or $\mathbb{Z}_5 \times \mathbb{Z}_5$. Suppose that $R_1$ and $R_2$ are not both fields. If $|R_2| \geq 8$, then the subgraph induced by the set $(\{0\} \times R_2^*) \cup (R_1^* \times \{0\})$ is a $K_{7,3}$ in $\Gamma(R)$, a contradiction. Thus $R_i \cong \mathbb{Z}_4$ or $\mathbb{Z}_2[x]/(x^2)$ for $i = 1, 2$, which means $\Gamma(R)$ is a projective graph by Proposition 4.6.13.

Hence, assume that exactly one of the $R_i$ is a field. Let us assume $R_1$ is a field. If $|R_1| \geq 5$, then $|R_2| \geq 8$, and so $K_{4,7}$ is a subgraph of $\Gamma(R)$. Thus $|R_1| \leq 4$.

**Case 3.1** Let $|R_1| = 4$. That is $R_1 \cong \mathbb{F}_4$. If $|R_2| \geq 8$, then the subgraph $K_{3,7}$ is contained in $\Gamma(R)$. Therefore $|R_2| = 4$, which means $R_2 \cong \mathbb{Z}_4$ or $\mathbb{Z}_2[x]/(x^2)$. Thus by Proposition 4.6.13, $\Gamma(R)$ is a projective graph.

**Case 3.2** Let $|R_1| = 3$; so $R_1 \cong \mathbb{Z}_3$. Suppose $|R_2/\mathfrak{m}_2| \geq 4$. If $\mathfrak{m}_2^2 \neq 0$, then by Proposition 4.6.20, $\tilde{g}(\Gamma(\mathbb{Z}_3 \times R_2)) \geq 6$. Therefore $\mathfrak{m}_2^2 = 0$. Since $R_2$ is local and $|R_2/\mathfrak{m}_2| \geq 4$, $|\mathfrak{m}_2| \geq 4$. Now applying Proposition 4.6.19 with $n = |R_1| = 3$ and $r = s = |\mathfrak{m}_2^*| \geq 3$, we have $\tilde{g}(\Gamma(\mathbb{Z}_3 \times R_2)) \geq \tilde{g}(K_{3,8}) = 3$. Hence $|R_2/\mathfrak{m}_2| \leq 3$.

(a) Let $|R_2/\mathfrak{m}_2| = 3$. If $\mathfrak{m}_2^2 \neq 0$, then by Proposition 4.6.20, $\tilde{g}(\Gamma(\mathbb{Z}_2 \times R_2)) \geq 6$. Thus $\mathfrak{m}_2^2 = 0$. It is worth mentioning that if $|\mathfrak{m}_2| = p$ for any odd prime $p$ and $\mathfrak{m}_2^2 = 0$, then $|R_2| = p^2$. This implies that $R_2$ is isomorphic to $\mathbb{Z}_9$ or $\mathbb{Z}_3[x]/(x^2)$. But by Proposition 4.2.13, $\Gamma(\mathbb{Z}_3 \times \mathbb{Z}_9)$ and $\Gamma(\mathbb{Z}_3 \times \mathbb{Z}_3[x]/(x^2))$ are planar graphs.

(b) Let $|R_2/\mathfrak{m}_2| = 2$. Then $|R_2| = 2^k$ for some $k \in \mathbb{N}$. Let $k \geq 4$ and let $\mathfrak{m}_2 = ann(x)$ for some $x \in R_2$. Then every vertex in the set $\{(0, a) \mid a \in \mathfrak{m}_2^*\}$ is adjacent to each vertex in the set $\{(1, 0), (2, 0), (1, x), (2, x)\}$, and so $K_{7,4}$ is a subgraph of $\Gamma(R)$, a contradiction. If $k = 2$, then $R_2 \cong \mathbb{Z}_4$ or $\mathbb{Z}_2[x]/(x^2)$. But by Proposition 4.2.13, $\Gamma(\mathbb{Z}_3 \times R_2)$ is planar. Thus $k = 3$. Suppose $\mathfrak{m}_2^2 = 0$. Let $x_1, x_2, x_3$ be distinct nonzero elements in $\mathfrak{m}_2$, $u_i = (0, x_i), v_i = (1, x_i)$, $v_{i+3} = (2, x_i)$, for $i = 1, 2, 3$, and $v_7 = (1, 0)$. Note that $u_i \cdot v_j = 0$ for every $i, j$; so $K_{3,7} \subseteq \Gamma(R)$, a contradiction. Therefore $\mathfrak{m}_2^2 \neq 0$. There are three rings with $|R_2| = 8$ and $\mathfrak{m}_2^2 \neq 0$, namely $\mathbb{Z}_8$, $\mathbb{Z}_2[x]/(x^3)$, and $\mathbb{Z}_4[x]/(x^2 - 2, x^3)$. But by Proposition 4.6.13, $\Gamma(\mathbb{Z}_3 \times \mathbb{Z}_8)$, $\Gamma(\mathbb{Z}_3 \times \mathbb{Z}_2[x]/(x^3))$, and $\Gamma(\mathbb{Z}_3 \times \mathbb{Z}_4[x]/(x^2 - 2, x^3))$ are projective graphs.

**Case 3.3** Let $|R_1| = 2$; that is, $R_1 \cong \mathbb{Z}_2$. Applying Lemma 4.6.16 to the local ring $R_2$, we get that $\mathfrak{m}_2^2 = 0$ when $|R_2/\mathfrak{m}_2| \geq 5$. Then applying Proposition 4.6.21 with $|\mathfrak{m}_2| \geq 5$, we have $\tilde{g}(\Gamma(\mathbb{Z}_2 \times R_2)) \geq \tilde{g}(K_{4,5}) = 3$, a contradiction. Thus $|R_2/\mathfrak{m}_2| \leq 4$.

(a) Let $|R_2/\mathfrak{m}_2| = 4$. Then $|\mathfrak{m}_2| = 2^\ell$ for some $\ell \in \mathbb{N}$. Also, by Lemma 4.6.16, $\mathfrak{m}_2^2 = 0$. If $\ell \geq 3$, by Proposition 4.6.21, $\tilde{g}(\Gamma(\mathbb{Z}_2 \times R_2)) \geq 3$. Then the possibilities for $R_2$ are $\mathbb{F}_4[x]/(x^2)$ and $\mathbb{Z}_4[x]/(x^2 + x + 1)$. Therefore $R \cong \mathbb{Z}_2 \times \mathbb{F}_4[x]/(x^2)$ or $\mathbb{Z}_2 \times \mathbb{Z}_4[x]/(x^2 + x + 1)$, but by Proposition 4.6.13, $\Gamma(R)$ is a projective graph.

(b) Let $|R_2/\mathfrak{m}_2| = 3$. This implies $|\mathfrak{m}_2| = 3^\ell$ for some $\ell \in \mathbb{N}$. Note that if $\mathfrak{m}_2^2 \neq 0$, then by Proposition 4.6.20, $\tilde{g}(\Gamma(\mathbb{Z}_2 \times R_2)) \geq 6$, and so we only need to consider the case $\mathfrak{m}_2^2 = 0$. If $\ell \geq 2$, then by Proposition 4.6.21, $\tilde{g}(\Gamma(\mathbb{Z}_2 \times R_2)) \geq \tilde{g}(K_{8,9}) = 21$, a contradiction. Therefore $\ell = 1$. Again by Proposition 4.6.21, $\tilde{g}(\Gamma(\mathbb{Z}_2 \times R_2)) \leq \tilde{g}(\Gamma(\mathbb{Z}_{27})) = 0$ so that $\Gamma(\mathbb{Z}_2 \times R_2)$ is planar.

(c) Let $|R_2/\mathfrak{m}_2| = 2$. Then $|R_2| = 2^t$ and $|\mathfrak{m}_2| = 2^{t-1}$. If $t \geq 6$, then $K_{3,7}$ or $K_{3,8}$ is a subgraph of $\Gamma(R)$, a contradiction. So $t \leq 5$.

If $t = 2$, then $R_2 \cong \mathbb{Z}_4$ or $\mathbb{Z}_2[x]/(x^2)$, and by Proposition 4.2.13, $\Gamma(\mathbb{Z}_2 \times \mathbb{Z}_4)$ and $\Gamma(\mathbb{Z}_2 \times \mathbb{Z}_2[x]/(x^2))$ are planar.

Let $t = 3$. If $\mathfrak{m}_2^2 = 0$, then $R_2 \cong \mathbb{Z}_4[x]/(2x, x^2)$ or $\mathbb{Z}_2[x, y]/(x^2, xy, y^2)$. By Proposition 4.6.13, $\Gamma(\mathbb{Z}_2 \times \mathbb{Z}_4[x]/(2x, x^2))$ and $\Gamma(\mathbb{Z}_2 \times \mathbb{Z}_2[x, y]/(x^2, xy, y^2))$ are projective graphs. If $\mathfrak{m}_2^2 \neq 0$, then $R_2 \cong \mathbb{Z}_8$, $\mathbb{Z}_2[x]/(x^3)$, or $\mathbb{Z}_4[x]/(x^2 - 2, x^3)$. But $\Gamma(\mathbb{Z}_2 \times \mathbb{Z}_8)$, $\Gamma(\mathbb{Z}_2 \times \mathbb{Z}_2[x]/(x^3))$, and $\Gamma(\mathbb{Z}_2 \times \mathbb{Z}_4[x]/(x^2 - 2, x^3))$ are planar graphs (see Proposition 4.2.13).

Let $t = 4$. Then there are 18 local rings of order 16 with cardinality of the residue field as 2. We partition these 18 local rings into five sets, and they are given below:

$$\begin{aligned} P_1 &\cong \mathbb{Z}_{16}, \mathbb{Z}_2[x]/(x^4), \mathbb{Z}_4[x]/(x^2-2, x^4), \mathbb{Z}_4[x]/(x^3-2, x^4), \\ &\quad \mathbb{Z}_4[x]/(x^3+x^2-2, x^4); \\ P_2 &\cong \mathbb{Z}_2[x,y]/(x^3, xy, y^2-x^2), \mathbb{Z}_4[x]/(x^3, x^2-2x), \mathbb{Z}_8[x]/(x^2-4, 2x), \\ &\quad \mathbb{Z}_4[x,y]/(x^3, x^2-2, xy, y^2-2, y^3); \\ P_3 &\cong \mathbb{Z}_4[x,y]/(x^2, y^2, xy-2), \mathbb{Z}_4[x]/(x^2), \mathbb{Z}_2[x,y]/(x^2, y^2); \\ P_4 &\cong \mathbb{Z}_2[x,y]/(x^3, xy, y^2), \mathbb{Z}_4[x]/(x^3, 2x), \mathbb{Z}_4[x,y]/(x^3, x^2-2, xy, y^2), \\ &\quad \mathbb{Z}_8[x]/(x^2, 2x); \\ P_5 &\cong \mathbb{Z}_4[x,y]/(2x, 2y, x^2, xy, y^2), \mathbb{Z}_2[x,y,z]/(x,y,z)^2. \end{aligned}$$

Then it is not hard to check that $\Gamma(P_1)$ is isomorphic to $G_1$, $\Gamma(P_2)$ is isomorphic to $G_2$, and $\Gamma(P_3)$ is isomorphic to $G_3$, where $G_1, G_2$, and $G_3$ are given in Figure 4.15. Also, $\Gamma(P_4)$ is isomorphic to $K_{1,1,1,4}$ (see Example 2.9 of [164]) and $\Gamma(P_5)$ is isomorphic to $K_7$. Therefore $\Gamma(P_2)$, $\Gamma(P_3)$, $\Gamma(P_4)$, and $\Gamma(P_5)$ contain $\Gamma(P_1)$ as a subgraph. In particular, $\Gamma(\mathbb{Z}_{16})$ is a subgraph of $\Gamma(R_2)$ for any local ring $R_2$ of order 16. Now applying the fact $|\mathbb{Z}_{16}^* \setminus Z(\mathbb{Z}_{16})| \leq |R_2^* \setminus Z(R_2)|$ in Proposition 4.4.4, we get $\Gamma(\mathbb{Z}_2 \times \mathbb{Z}_{16})$ is a subgraph of $\Gamma(\mathbb{Z}_2 \times R_2)$. Since $\tilde{g}(\Gamma(\mathbb{Z}_2 \times \mathbb{Z}_{16})) \geq 3$ (see Example 4.6.17(a)), we have $\tilde{g}(\Gamma(\mathbb{Z}_2 \times R_2)) \geq 3$.

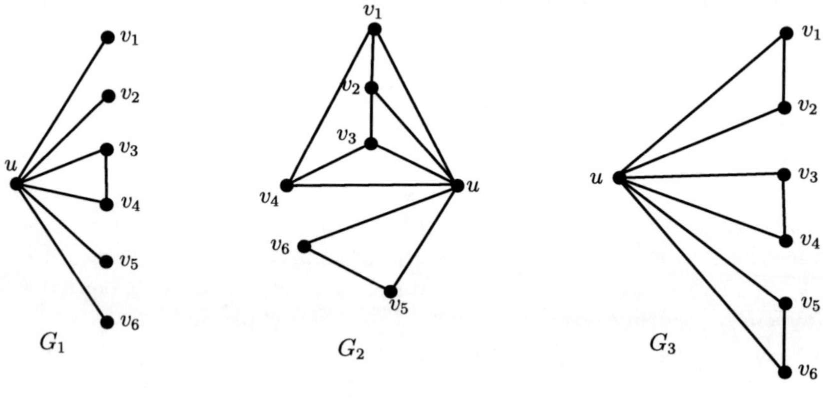


Figure 4.15:

Let $t = 5$. Then $R_2 \cong \mathbb{Z}_{32}, \mathbb{Z}_2[x]/(x^5), \mathbb{Z}_4[x]/(x^3-2, x^5), \mathbb{Z}_4[x]/(x^4-2, x^5)$, $\mathbb{Z}_8[x]/(x^2-2, x^5), \mathbb{Z}_8[x]/(x^2-2x+2, x^5)$, or $\mathbb{Z}_8[x]/(x^2+2x-2, x^5)$. Further, one can check that all the zero-divisor graphs of $R_2$ are the same and $\Gamma(\mathbb{Z}_{16})$ is a subgraph of $\Gamma(R_2)$. Since $|\mathbb{Z}_{16}^* \setminus Z(\mathbb{Z}_{16})| \leq |R_2^* \setminus Z(R_2)|$, applying Proposition 4.4.4, we get $\Gamma(\mathbb{Z}_2 \times \mathbb{Z}_{16})$ is a subgraph of $\Gamma(\mathbb{Z}_2 \times R_2)$, and so $\tilde{g}(\Gamma(\mathbb{Z}_2 \times R_2)) \geq 3$.

Hence, in this case, $\tilde{g}(\Gamma(R)) = 2$ if and only if $R$ is isomorphic to $\mathbb{F}_4 \times \mathbb{Z}_7$ or $\mathbb{Z}_5 \times \mathbb{Z}_5$. □

By comparing Theorem 4.3.19 and Theorem 4.6.22, the orientable genus of all four rings in Theorem 4.6.22 is one. So we have the following result, which answers a case of Conjecture 4.6.14.

**Theorem 4.6.23.** ([165, Theorem 3.12]) *Let $R$ be a commutative ring with the crosscap number of the zero-divisor graph $\Gamma(R)$ less than or equal to 2. Then $\tilde{g}(\Gamma(R)) \geq g(\Gamma(R))$.*

## 4.7 Embedding of Line Graphs

Motivated by the work on the genus of zero-divisor graphs, Chiang-Hsieh et al. [163] studied the minimal orientable and nonorientable embeddings of line graphs of zero-divisor graphs. Recall that the *line graph of a graph $G$*, denoted by $L(G)$, is the graph such that every vertex of $L(G)$ represents an edge of $G$, and two vertices of $L(G)$ are adjacent if and only if their corresponding edges in $G$ share a common vertex. The goal of Chiang-Hsieh et al. in [163] was to determine all finite commutative rings $R$ such that their line graphs $L(\Gamma(R))$ are of small genera or crosscaps up to two.

The following result provides an upper bound for the genera and crosscaps of line graphs.

**Lemma 4.7.1.** ([163, Proposition 2.4]) *Let $G$ be a graph. Then*

(a) $g(L(G)) \leq g(G) + \sum_{v\in V(G)} g(K_{deg(v)+1})$;

(b) $\tilde{g}(L(G)) \leq \tilde{g}(G) + \sum_{v\in V(G)} \tilde{g}(K_{deg(v)+1})$.

**Example 4.7.2.** ([163, Examples 2.12 and 2.14])

(i) $g(L(K_{3,3})) = 1$ and $\tilde{g}(L(K_{3,3})) = 1$;

(ii) $\tilde{g}(L(K_{2,4})) = 2$, $g(L(K_{3,4})) = 2$, $\tilde{g}(L(K_{3,4})) \geq 3$, and $g(L(K_{4,4})) \geq 3$.

Recall that a simple graph is said to be $k$-connected if the minimum number of vertices whose removal results in a disconnected or trivial graph is $k$.

**Lemma 4.7.3.** ([163, Lemma 2.13]) *Let $G$ be a connected graph and $G_1$, $G_2$ be two connected subgraphs of $G$. Suppose the following hold:*

(i) $V(G_1) \cap V(G_2) = \emptyset$;

(ii) *$G_1$ is a 3-connected planar graph;*

(iii) *Every vertex of* $G_1$ *is joined with* $G_2$.

*Then* $g(G) > g(G_2)$ *and* $\tilde{g}(G) > \tilde{g}(G_2)$.

**Lemma 4.7.4.** ([163, Lemma 2.15]) *The following are true:*

(i) $g(L(K_{1,1,n})) = g(L(K_{2,n+1}))$;

(ii) $\tilde{g}(L(K_{1,1,n})) = \tilde{g}(L(K_{2,n+1}))$.

**Theorem 4.7.5.** ([163, Theorem 3.1]) *Let* $(R, \mathfrak{m})$ *be a finite commutative local ring which is not a field. If either* $g(L(\Gamma(R))) \leq 2$ *or* $\tilde{g}(L(\Gamma(R))) \leq 2$, *then* $|\mathrm{R}/\mathfrak{m}| \leq 5$; *and the following hold:*

(i) *If* $|R/\mathfrak{m}| = 3, 4,$ *or* $5$, *then* $\mathfrak{m}^2 = 0$ *and* $|R| = |R/\mathfrak{m}|^2$;

(ii) *If* $|R/\mathfrak{m}| = 2$, *then* $\mathfrak{m}^4 = 0$ *and* $|R| \leq 16$.

The following theorem characterizes all finite commutative local rings $R$ for which $L(\Gamma(R))$ is planar.

**Theorem 4.7.6.** ([163, Theorem 3.5.1]) *Let* $R$ *be a finite commutative local ring which is not a field. Then* $L(\Gamma(R))$ *is planar if and only if* $R$ *is isomorphic to one of the following* 13 *rings:*

$$\mathbb{Z}_4,\ \mathbb{Z}_8,\ \mathbb{Z}_9,\ \mathbb{Z}_{25},\ \frac{\mathbb{Z}_3[x]}{(x^2)},\ \frac{\mathbb{Z}_2[x]}{(x^2)},\ \frac{\mathbb{Z}_2[x,y]}{(x^2,xy,y^2)},\ \frac{\mathbb{Z}_2[x]}{(x^3)},\ \frac{\mathbb{F}_4[x]}{(x^2)},$$
$$\frac{\mathbb{Z}_4[x]}{(x^2+x+1)},\ \frac{\mathbb{Z}_4[x]}{(x^2,2x)},\ \frac{\mathbb{Z}_4[x]}{(x^3,x^2-2)},\ \frac{\mathbb{Z}_5[x]}{(x^2)}.$$

**Theorem 4.7.7.** ([163, Theorem 3.5.2(a)]) *Let* $R$ *be a finite commutative local ring. Then* $g(L(\Gamma(R))) = 1$ *if and only if* $R$ *is isomorphic to one of the following* 12 *rings:*

$$\mathbb{Z}_{16},\ \frac{\mathbb{Z}_2[x]}{(x^4)},\ \frac{\mathbb{Z}_4[x]}{(x^2-2,x^4)},\ \frac{\mathbb{Z}_4[x]}{(x^3-2,x^4)},\ \frac{\mathbb{Z}_4[x]}{(x^3+x^2-2,x^4)},\ \frac{\mathbb{Z}_4[x]}{(x^3,x^2-2x)},$$
$$\frac{\mathbb{Z}_8[x]}{(x^2-4,2x)},\ \frac{\mathbb{Z}_2[x,y]}{(x^3,xy,y^2-x^2)},\ \frac{\mathbb{Z}_4[x]}{(x^2)},\ \frac{\mathbb{Z}_2[x,y]}{(x^3,x^2-2,xy,y^2-2,y^3)},$$
$$\frac{\mathbb{Z}_4[x,y]}{(x^2,xy-2,y^2)},\ \frac{\mathbb{Z}_2[x,y]}{(x^2,y^2)}.$$

**Theorem 4.7.8.** ([163, Theorem 3.5.2(b)]) *There is no finite commutative local ring* $R$ *with* $g(L(\Gamma(R))) = 2$.

**Theorem 4.7.9.** ([163, Theorem 3.5.3]) *Let* $R$ *be a finite commutative local ring. Then* $\tilde{g}(L(\Gamma(R))) = 1$ *if and only if* $R$ *is isomorphic to one of the following* 8 *rings:*

$$\mathbb{Z}_{16},\ \frac{\mathbb{Z}_2[x]}{(x^4)},\ \frac{\mathbb{Z}_4[x]}{(x^2-2,x^4)},\ \frac{\mathbb{Z}_4[x]}{(x^3-2,x^4)},\ \frac{\mathbb{Z}_4[x]}{(x^3+x^2-2,x^4)},\ \frac{\mathbb{Z}_4[x]}{(x^2)},$$
$$\frac{\mathbb{Z}_4[x,y]}{(x^2,xy-2,y^2)},\ \frac{\mathbb{Z}_2[x,y]}{(x^2,y^2)}.$$

**Theorem 4.7.10.** ([163, Theorem 3.5.3]) *Let $R$ be a finite commutative local ring. Then $\tilde{g}(L(\Gamma(R))) = 2$ if and only if $R$ is isomorphic to one of the following 4 rings:*

$$\frac{\mathbb{Z}_4[x]}{(x^3, x^2 - 2x)}, \ \frac{\mathbb{Z}_8[x]}{(x^2 - 4, 2x)}, \ \frac{\mathbb{Z}_2[x,y]}{(x^3, xy, y^2 - x^2)}, \ \frac{\mathbb{Z}_2[x,y]}{(x^3, x^2 - 2, xy, y^2 - 2, y^3)}.$$

After characterizing the finite commutative local rings $R$ with $g(L(\Gamma(R))) \leq 2$ and $\tilde{g}(L(\Gamma(R))) \leq 2$, Chiang-Hsieh et al. [163] classified all nonlocal finite commutative rings $R$ with $g(L(\Gamma(R))) \leq 2$ and $\tilde{g}(L(\Gamma(R))) \leq 2$.

**Example 4.7.11.** ([163, Example 4.3.3]) If $R = \mathbb{Z}_2 \times \mathbb{Z}_8$, then $g(L(\Gamma(R))) = 2$ and $\tilde{g}(L(\Gamma(R))) \geq 3$.

**Example 4.7.12.** ([163, Example 4.3.4]) If $R = \mathbb{Z}_2 \times \mathbb{Z}_9$, then $g(L(\Gamma(R))) \geq 3$ and $\tilde{g}(L(\Gamma(R))) \geq 5$.

**Example 4.7.13.** ([163, Example 4.3.5]) If $R = \mathbb{Z}_4 \times \mathbb{Z}_4$, then $g(L(\Gamma(R))) \geq 3$ and $\tilde{g}(L(\Gamma(R))) \geq 3$.

**Example 4.7.14.** ([163, Example 4.3.6]) If $R = \mathbb{Z}_4 \times \mathbb{Z}_5$, then $g(L(\Gamma(R))) \geq 3$ and $\tilde{g}(L(\Gamma(R))) \geq 4$.

**Example 4.7.15.** ([163, Example 4.3.7]) If $R = \mathbb{Z}_2 \times \mathbb{Z}_2[x,y]/(x^2, xy, y^2)$ or $R = \mathbb{Z}_2 \times \mathbb{Z}_3 \times \mathbb{Z}_3$, then $g(L(\Gamma(R))) \geq 3$ and $\tilde{g}(L(\Gamma(R))) \geq 3$.

**Example 4.7.16.** ([163, Example 4.3.8]) If $R = \mathbb{Z}_2 \times \mathbb{Z}_2 \times \mathbb{F}_4$, then $g(L(\Gamma(R))) \geq 3$ and $\tilde{g}(L(\Gamma(R))) \geq 5$.

**Lemma 4.7.17.** ([163, Lemma 4.4]) *Let $R$ be a nonlocal finite commutative ring. If $g(L(\Gamma(R))) \leq 2$ or $\tilde{g}(L(\Gamma(R))) \leq 2$, then $|Spec(R)| \leq 3$.*

**Theorem 4.7.18.** ([163, Theorem 4.7(a)]) *Let $R$ be a nonlocal finite commutative ring. Then $L(\Gamma(R))$ is planar if and only if $R$ is isomorphic to one of the following 11 rings:*

$$\mathbb{Z}_2 \times \mathbb{Z}_2,\ \mathbb{Z}_2 \times \mathbb{Z}_3,\ \mathbb{Z}_2 \times \mathbb{F}_4,\ \mathbb{Z}_2 \times \mathbb{Z}_4,\ \mathbb{Z}_2 \times \mathbb{Z}_2[x]/(x^2),\ \mathbb{Z}_2 \times \mathbb{Z}_5,$$
$$\mathbb{Z}_3 \times \mathbb{Z}_3,\ \mathbb{Z}_3 \times \mathbb{F}_4,\ \mathbb{Z}_3 \times \mathbb{Z}_4,\ \mathbb{Z}_3 \times \mathbb{Z}_2[x]/(x^2),\ \mathbb{Z}_2 \times \mathbb{Z}_2 \times \mathbb{Z}_2.$$

**Theorem 4.7.19.** ([163, Theorem 4.7(b)]) *Let $R$ be a nonlocal finite commutative ring. Then $g(L(\Gamma(R))) = 1$ if and only if $R$ is isomorphic to $\mathbb{Z}_2 \times \mathbb{Z}_7$, $\mathbb{Z}_2 \times \mathbb{F}_8$, $\mathbb{Z}_3 \times \mathbb{Z}_5$, or $\mathbb{F}_4 \times \mathbb{F}_4$.*

**Theorem 4.7.20.** ([163, Theorem 4.7(c)]) *Let $R$ be a nonlocal finite commutative ring. Then $g(L(\Gamma(R))) = 2$ if and only if $R$ is isomorphic to one of the following 9 rings:*

$$\mathbb{Z}_2 \times \mathbb{Z}_8,\ \mathbb{Z}_2 \times \mathbb{Z}_2[x]/(x^3),\ \mathbb{Z}_2 \times \mathbb{Z}_4[x]/(x^3, x^2 - 2),\ \mathbb{Z}_2 \times \mathbb{F}_9,\ \mathbb{Z}_3 \times \mathbb{Z}_7,\ \mathbb{Z}_4 \times \mathbb{F}_4,$$
$$\mathbb{F}_4 \times \mathbb{Z}_2[x]/(x^2),\ \mathbb{Z}_5 \times \mathbb{F}_4,\ \mathbb{Z}_2 \times \mathbb{Z}_2 \times \mathbb{Z}_3.$$

**Theorem 4.7.21.** ([163, Theorem 4.8]) *Let $R$ be a nonlocal finite commutative ring. Then $\tilde{g}(L(\Gamma(R))) = 1$ if and only if $R$ is isomorphic to $\mathbb{Z}_2 \times \mathbb{Z}_7$ or $\mathbb{F}_4 \times \mathbb{F}_4$.*

**Theorem 4.7.22.** ([163, Theorem 4.8(b)]) *Let $R$ be a nonlocal finite commutative ring. Then $\tilde{g}(L(\Gamma(R))) = 2$ if and only if $R$ is isomorphic to $\mathbb{Z}_4 \times \mathbb{F}_4$, $\mathbb{F}_4 \times \mathbb{Z}_2[x]/(x^2)$, $\mathbb{Z}_3 \times \mathbb{Z}_5$, or $\mathbb{Z}_2 \times \mathbb{Z}_2 \times \mathbb{Z}_3$.*

Note that the inequalities $g(\Gamma(R)) \leq \tilde{g}(\Gamma(R)) \leq g(L(\Gamma(R))) \leq \tilde{g}(L(\Gamma(R)))$ hold for all commutative rings $R$ with $\tilde{g}(L(\Gamma(R))) \leq 2$. Moreover, the above inequalities also hold when $\Gamma(R)$ is a complete graph or a complete bipartite graph. Hence, the conjecture of Hsieh [165] was restated as follows.

**Conjecture:** The inequalities $g(\Gamma(R)) \leq \tilde{g}(\Gamma(R)) \leq g(L(\Gamma(R))) \leq \tilde{g}(L(\Gamma(R)))$ hold for any finite commutative ring $R$.

# Chapter 5

# Generalized Zero-divisor Graphs

The zero-divisor graph of a commutative ring has been generalized by several authors. The two most notable generalizations are the ideal-based zero-divisor graph and annihilating-ideal graph of commutative rings. We first discuss the ideal-based zero-divisor graph of a commutative ring.

## 5.1 Ideal-based Zero-divisor Graphs

The focus of this section is on a generalization of the zero-divisor graph called the *ideal-based zero-divisor graph* for commutative rings. In 2001, Redmond [321, 323] gave the following definition as a generalization of the zero-divisor graph.

Let $R$ be a commutative ring with nonzero identity and $I$ an ideal of $R$. The *zero-divisor graph of $R$ with respect to $I$* is the undirected graph, denoted by $\Gamma_I(R)$, with vertices $\{x \in R \setminus I \mid xy \in I$ for some $y \in R \setminus I\}$, where distinct vertices $x$ and $y$ are adjacent if and only if $xy \in I$. It is clear that if $I = (0)$, then $\Gamma_I(R) = \Gamma(R)$.

Note that $\Gamma_I(R)$ is just the zero-divisor graph of the Rees semigroup $R_I$ of $R$ with respect to the ideal $I$ (see Section 6.2 for zero-divisor graphs of semigroups). In particular, $diam(\Gamma_I(R)) \in \{0, 1, 2, 3\}$ and $gr(\Gamma_I(R)) \in \{3, 4, \infty\}$ by Theorem 6.2.2.

**Example 5.1.1.** ($a$) Consider the ring $R = \mathbb{Z}_6 \times \mathbb{Z}_3$ and the ideal $I = \{0\} \times \mathbb{Z}_3$. Then the ideal-based zero-divisor graph $\Gamma_I(R)$ is given in Figure 5.1.

($b$) Consider the ring $S = \mathbb{Z}_{24}$ and the ideal $J = (8) = \{0, 8, 16\}$. Then the ideal-based zero-divisor graph $\Gamma_J(S)$ is given in Figure 5.2.

In general, $\Gamma_I(R)$ need not be isomorphic to $\Gamma(S)$ for any commutative ring $S$. For example, let $R = \mathbb{Z}_4 \times \mathbb{Z}_5$ and $I = \{0\} \times \mathbb{Z}_5$. Then $\Gamma_I(R) = K_5$. Thus $\Gamma_I(R) \ncong \Gamma(S)$ for any commutative ring $S$ by Theorem 2.2.7. To avoid trivialities, throughout this section, we will always assume that $I \neq R$ for an ideal $I$ of $R$.



D. F. Anderson et al., *Graphs from Rings*, https://doi.org/10.1007/978-3-030-88410-9_5

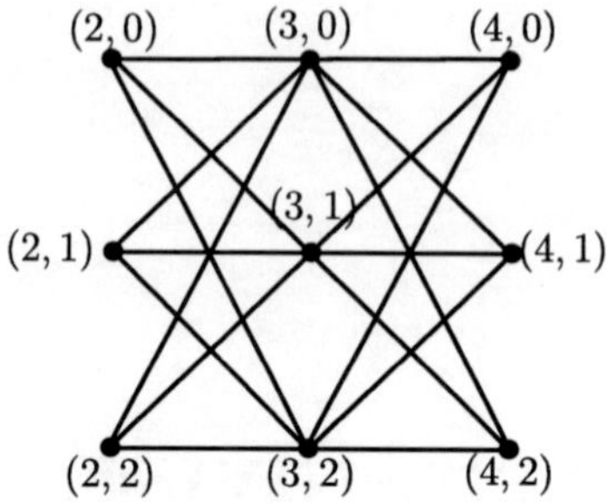


Figure 5.1: The graph $\Gamma_{\{0\}\times\mathbb{Z}_3}(\mathbb{Z}_6 \times \mathbb{Z}_3)$

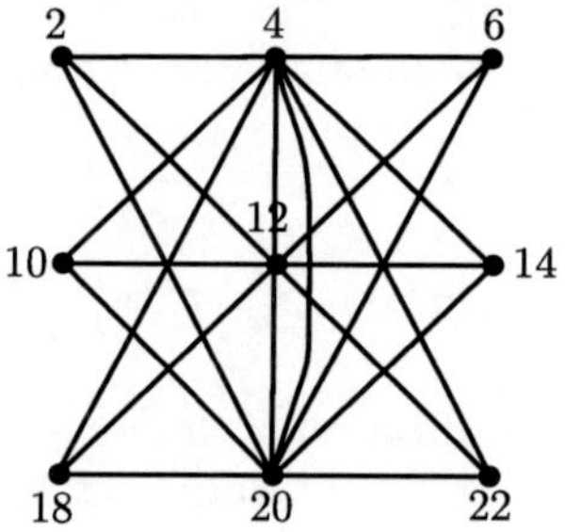


Figure 5.2: The graph $\Gamma_{(8)}(\mathbb{Z}_{24})$

First of all, Redmond [323] explored the relationship between the graphs $\Gamma_I(R)$ and $\Gamma(R/I)$. Note that $\Gamma_I(R) = \emptyset$ if and only if $I$ is a prime ideal of $R$, and hence $\Gamma_I(R) = \emptyset$ if and only if $\Gamma(R/I) = \emptyset$.

Let $R = \mathbb{Z}_6 \times \mathbb{Z}_3$, $I = \{0\} \times \mathbb{Z}_3$, $S = \mathbb{Z}_{24}$, and $J = \{0, 8, 16\}$. Then $\Gamma(R/I) \cong \Gamma(S/J)$, both being the graph on three vertices with two edges. But, by Example 5.1.1, $\Gamma_I(R) \ncong \Gamma_J(S)$. Thus we have an example where $\Gamma(R/I) \cong \Gamma(S/J)$, but $\Gamma_I(R) \ncong \Gamma_J(S)$. The next several results investigate the relationship between $\Gamma_I(R)$ and $\Gamma(R/I)$. Note that the natural epimorphism $\varphi : R \to R/I$, given by $\varphi(x) = x + I$, induces a surjective map $\bar{\varphi} : V(\Gamma_I(R)) \to V(\Gamma(R/I))$.

**Theorem 5.1.2.** ([323, Theorem 2.5]) *Let $I$ be an ideal of a commutative ring $R$, and let $x, y \in R \setminus I$. Then*

(a) *If $x + I$ is adjacent to $y + I$ in $\Gamma(R/I)$, then $x$ is adjacent to $y$ in $\Gamma_I(R)$;*

(b) *If $x$ is adjacent to $y$ in $\Gamma_I(R)$ and $x + I \neq y + I$, then $x + I$ is adjacent to $y + I$ in $\Gamma(R/I)$;*

(c) *If $x$ is adjacent to $y$ in $\Gamma_I(R)$ and $x + I = y + I$, then $x^2, y^2 \in I$.*

Therefore, for $x, y \in R \setminus I$, if $x + I$ is adjacent to $y + I$ in $\Gamma(R/I)$, then $x$ and $y$ are adjacent in $\Gamma_I(R)$; while the converse is true only when $x + I \neq y + I$. Moreover, if $x^2 \in I$, then all the distinct elements of $x + I$ are adjacent in $\Gamma_I(R)$.

**Lemma 5.1.3.** ([323, Corollary 2.7, Remark 2.8]) *Let $I$ be an ideal of a commutative ring $R$. Then $\Gamma_I(R)$ contains $|I|$ disjoint subgraphs isomorphic to $\Gamma(R/I)$. Moreover, if $\Gamma(R/I)$ is a graph on $N$ vertices, then $\Gamma_I(R)$ is a graph on $N \cdot |I|$ vertices.*

Thus $|V(\Gamma_I(R))| = |I||V(\Gamma(R/I))|$. At this point, remember that $\Gamma_I(R)$ is a graph on a finite number of vertices if and only if either $R$ is finite or $I$ is a prime ideal.

Clearly there is a strong relationship between $\Gamma(R/I)$ and $\Gamma_I(R)$. Let $I$ be an ideal of a commutative ring $R$. One can verify that the following method can be used to construct the graph $\Gamma_I(R)$. This construction is due to S. Redmond [323]. Let $\{a_\lambda\}_{\lambda \in \Lambda} \subseteq R$ be a set of coset representatives of the vertices of $\Gamma(R/I)$. For every $i \in I$, define a graph $G_i$ with vertices $\{a_\lambda + i \mid \lambda \in \Lambda\}$, where edges are defined by the relationship $a_\lambda + i$ is adjacent to $a_\beta + i$ in $G_i$ if and only if $a_\lambda + I$ is adjacent to $a_\beta + I$ in $G(R/I)$ (i.e., $a_\lambda \cdot a_\beta \in I$). Define the graph $G$ to have as its vertex set $V = \cup_{i \in I} G_i$. We define the edge set of $G$ to be: (1) all edges contained in $G_i$ for every $i \in I$, (2) for distinct $\lambda, \beta \in \Lambda$ and for any $i, j \in I$, $a_\lambda + i$ is adjacent to $a_\beta + j$ if and only if $a_\lambda + I$ is adjacent to $a_\beta + I$ in $\Gamma(R/I)$ (i.e., $a_\lambda \cdot a_\beta \in I$), (3) for $\lambda \in \Lambda$ and distinct $i, j \in I$, $a_\lambda + i$ is adjacent to $a_\lambda + j$ if and only if $a_\lambda^2 \in I$.

Using the notation as in the above construction, we call the subset $a_\lambda + I$ a *column* of $\Gamma_I(R)$. If $a_\lambda^2 \in I$, then we call $a_\lambda + I$ a *connected column* of $\Gamma_I(R)$.

The next result provides a relationship between the induced subgraph structure of $\Gamma_I(R)$ and $\Gamma(R/I)$.

**Theorem 5.1.4.** ([321, Proposition 12]) *Let $R \subseteq S$ be commutative rings and let $I$ be an ideal of $S$. Then $\Gamma(R/(I \cap R))$ is an induced subgraph of $\Gamma(S/I)$ and $\Gamma_{I \cap R}(R)$ is an induced subgraph of $\Gamma_I(S)$.*

Next, we study the relationship between the vertices of $\Gamma(R)$ and $\Gamma_I(R)$. In general, there is little relationship between the graphs $\Gamma(R)$ and $\Gamma_I(R)$. In this regard, it is a goal of many researchers to determine when $\Gamma_I(R) = \Gamma(R)$. We start by specifying when $V(\Gamma(R)) \subseteq V(\Gamma_I(R))$. The following several results are proved by Anderson and Shirinkam [86].

**Theorem 5.1.5.** ([86, Theorem 2.2]) *Let $I$ be a proper ideal of a commutative ring $R$. Then $V(\Gamma(R)) \subseteq V(\Gamma_I(R))$ if and only if either $I = \{0\}$ or $R$ is an integral domain. Moreover, if $I = \{0\}$, then $\Gamma_I(R) = \Gamma(R)$; and if $R$ is an integral domain, then $V(\Gamma(R)) = \emptyset$.*

The next result shows that $\Gamma_I(R) = \Gamma(R)$ in only the two trivial cases where either $I = \{0\}$ or both graphs are the empty graph.

**Theorem 5.1.6.** ([86, Theorem 2.3]) *Let $I$ be a proper ideal of a commutative ring $R$. Then the following statements are equivalent:*

(a) $\Gamma_I(R) = \Gamma(R)$;

(b) $V(\Gamma_I(R)) = V(\Gamma(R))$;

(c) *Either $I = \{0\}$ or $R$ is an integral domain with $I$ a prime ideal of $R$ (i.e., both graphs are the empty graph).*

In general, there is little relationship between $\Gamma_I(R)$ and $\Gamma(R)$. In fact, $V(\Gamma_I(R))$ and $V(\Gamma(R))$ need not be comparable. For example, let $R = \mathbb{Q}[x] \times \mathbb{Q}[x]$ and $I = x\mathbb{Q}[x] \times x\mathbb{Q}[x]$; so $I$ is a radical ideal of $R$ and $R/I = \mathbb{Q} \times \mathbb{Q}$. Then $V(\Gamma(R)) = Z(R)^* = (\mathbb{Q}[x]^* \times \{0\}) \cup (\{0\} \times \mathbb{Q}[x]^*)$, $V(\Gamma_I(R)) = Z(R)^* = (x\mathbb{Q}[x] \times (\mathbb{Q}[x] \setminus x\mathbb{Q}[x])) \cup ((\mathbb{Q}[x] \setminus x\mathbb{Q}[x]) \times x\mathbb{Q}[x])$ and $V(\Gamma(R/I)) = Z(R/I)^* = (\mathbb{Q}^* \times \{0\}) \cup (\{0\} \times \mathbb{Q}^*)$. Thus $(x, 1) \in V(\Gamma_I(R)) \setminus V(\Gamma(R))$ and $(x, 0) \in V(\Gamma(R)) \setminus V(\Gamma_I(R))$; so $V(\Gamma(R))$ and $V(\Gamma_I(R))$ are not comparable. One can refer to Example 5.2.16 for the smallest graph that can be realized as an ideal-based zero-divisor graph, but not as a zero-divisor graph, and vice versa.

Further, we may have $V(\Gamma_I(R)) \cap V(\Gamma(R)) = \emptyset$ for nonempty graphs $\Gamma_I(R)$ and $\Gamma(R)$. For example, let $R = \mathbb{Z}_4[x, y]$ and $I = (2, xy)$. Note that $Z(R) \cup U(R) \subsetneq R$ since $x$ is neither a zero-divisor nor a unit in $R$; also, $dim(R) = 2$. In fact, $Z(R) = 2R$; so $Z(R) \subsetneq I$. Thus $Z(R)^* \cap (R \setminus I) = \emptyset$; so $V(\Gamma_I(R)) \cap V(\Gamma(R)) = \emptyset$. Moreover, neither $\Gamma_I(R)$ nor $\Gamma(R)$ is the empty graph.

However, the next result shows that $V(\Gamma_I(R)) \subseteq Z(R)^* = V(\Gamma(R))$ when $R = Z(R) \cup U(R)$. Note that $R = Z(R) \cup U(R)$ if and only if $T(R) = R$. Recall that classes (not necessarily distinct) of commutative rings for which $R = Z(R) \cup U(R)$ include zero-dimensional rings, Artinian rings, and von Neumann regular rings.

**Theorem 5.1.7.** ([86, Theorem 2.4]) *Let $I$ be a proper ideal of a commutative ring $R$ with $R = Z(R) \cup U(R)$. Then $V(\Gamma_I(R)) \subseteq Z(R)^* = V(\Gamma(R))$. Moreover, if $V(\Gamma_I(R)) = V(\Gamma(R))$, then $I = \{0\}$ and $\Gamma_I(R) = \Gamma(R)$.*

Thus $|V(\Gamma_I(R))| \leq |V(\Gamma(R))|$, with equality if and only if $I = \{0\}$. Also Theorem 5.1.7 has the following converse.

**Theorem 5.1.8.** ([86, Theorem 2.6]) *Let $R$ be a commutative ring with $Z(R) \cup U(R) \subsetneq R$. Then $V(\Gamma_I(R)) \not\subseteq V(\Gamma(R))$ for some proper ideal $I$ of $R$.*

Hence, we have $V(\Gamma_I(R)) \subseteq V(\Gamma(R))$ for every proper ideal $I$ of $R$ if and only if $R = Z(R) \cup U(R)$. More generally, we have the following result.

**Theorem 5.1.9.** ([86, Theorem 2.8]) *Let $I \subseteq J$ be proper ideals of a commutative ring $R$. If $R/I = Z(R/I) \cup U(R/I)$, then $V(\Gamma_J(R)) \subseteq V(\Gamma_I(R))$.*

The next theorem is a generalization of several previous results on the zero-divisor graph of finite commutative rings.

**Theorem 5.1.10.** ([86, Theorem 2.10]) *Let $I$ be a proper ideal of a finite commutative ring $R$. Then the following statements are equivalent:*

(a) $\Gamma_I(R) = \Gamma(R)$;

(b) $\Gamma_I(R) \cong \Gamma(R)$;

(c) $|V(\Gamma_I(R))| = |V(\Gamma(R))|$;

(d) $V(\Gamma_I(R)) = V(\Gamma(R))$;

(e) $V(\Gamma(R)) \subseteq V(\Gamma_I(R))$;

(f) $I = \{0\}$.

Let $R$ be a finite commutative ring. Then $\Gamma_I(R) \cong \Gamma(R)$ if and only if $\Gamma_I(R) = \Gamma(R)$, if and only if $I = \{0\}$. Although $V(\Gamma_I(R)) \subseteq V(\Gamma(R))$, $\Gamma_I(R)$ need not be a subgraph of $\Gamma(R)$. Actually, both cases can happen. For example, let $R = \mathbb{Z}_4 \times \mathbb{Z}_2$ and $I = \{0\} \times \mathbb{Z}_2$. In this case, $V(\Gamma_I(R)) \subseteq V(\Gamma(R))$ and $\Gamma_I(R)$ is an induced subgraph of $\Gamma(R)$. But $\Gamma_J(R)$ is not a subgraph of $\Gamma(R)$ for $J = \{0,2\} \times \{0\}$ since $(1,0)$ and $(2,1)$ are adjacent in $\Gamma_J(R)$, but not in $\Gamma(R)$. However, if $x, y \in V(\Gamma_I(R))$ are adjacent in $\Gamma(R)$, then they are adjacent in $\Gamma_I(R)$. Thus, $\Gamma_I(R)$ is a subgraph of $\Gamma(R)$ if and only if $\Gamma_I(R)$ is an induced subgraph of $\Gamma(R)$.

So the natural question is "When is $\Gamma_I(R)$ a subgraph of $\Gamma(R)$?" To show that $\Gamma_I(R)$ is not a subgraph of $\Gamma(R)$, we only need to find distinct $x, y \in R \setminus I$ such that $0 \neq xy \in I$.

**Theorem 5.1.11.** ([86, Theorem 3.1]) *Let $R$ be a commutative quasilocal ring with maximal ideal $\mathfrak{m}$.*

(a) *If $\mathfrak{m}^2 = \{0\}$, then $\Gamma_I(R)$ is a (induced) subgraph of $\Gamma(R)$ for every proper ideal $I$ of $R$. Moreover, in this case, $\Gamma_I(R)$ is a complete subgraph of the complete graph $\Gamma(R)$;*

(b) *If $\mathfrak{m}^2 \neq \{0\}$, then there is a proper ideal $I$ of $R$ such that $\Gamma_I(R)$ is not a subgraph of $\Gamma(R)$.*

*In particular, $\Gamma_I(R)$ is a subgraph of $\Gamma(R)$ for every proper ideal $I$ of $R$ if and only if $\mathfrak{m}^2 = \{0\}$.*

We next see when $\Gamma_I(R)$ is a subgraph of $\Gamma(R)$ in the case when $R$ is not a quasilocal ring.

**Theorem 5.1.12.** ([86, Theorem 3.2]) *Let $R = R_1 \times R_2$, where $R_1$ and $R_2$ are commutative rings, and let $I$ be a proper ideal of $R$. Then $\Gamma_I(R)$ is a (induced) subgraph of $\Gamma(R)$ if and only if either of the following is true:*

(i) $I = \{0\}$;

(ii) *$I$ is a prime ideal of $R$;*

(iii) *$R \cong \mathbb{Z}_4 \times \mathbb{Z}_2$ or $R \cong \mathbb{Z}_2[x](x^2) \times \mathbb{Z}_2$ with $I = \{0\} \times \mathbb{Z}_2$.*

**Corollary 5.1.13.** ([86, Corollary 3.3]) *Let $R = R_1 \times \cdots \times R_n$, where every $R_i$ is a commutative ring and $n \geq 3$, and let $I$ be a proper ideal of $R$. Then $\Gamma_I(R)$ is a (induced) subgraph of $\Gamma(R)$ if and only if either $I = \{0\}$ or $I$ is a prime ideal of $R$. In particular, $\Gamma_I(R)$ is never a proper nonempty subgraph of $\Gamma(R)$.*

**Corollary 5.1.14.** ([86, Corollary 3.4]) *Let $R$ be a commutative Artinian ring (e.g., $R$ is finite) and $I$ a proper ideal of $R$. If $R$ is not a local ring, then $\Gamma_I(R)$ is a (induced) subgraph of $\Gamma(R)$ if and only if either* (i) *$I = \{0\}$,* (ii) *$I$ is a maximal ideal of $R$, or* (iii) *$R \cong \mathbb{Z}_4 \times \mathbb{Z}_2$ or $R \cong \mathbb{Z}_2[x]/(x^2) \times \mathbb{Z}_2$ with $I = \{0\} \times \mathbb{Z}_2$. In particular, if $R$ is a reduced commutative Artinian ring, then $\Gamma_I(R)$ is never a proper nonempty subgraph of $\Gamma(R)$.*

**Corollary 5.1.15.** ([86, Corollary 3.5]) *Let $R = F_1 \times \cdots \times F_n$, where every $F_i$ is a field and $n \geq 2$, and let $I$ be a proper ideal of $R$. If $I = \{0\}$, then $\Gamma_I(R) = \Gamma(R)$; if $I$ is a maximal ideal of $R$, then $V(\Gamma_I(R)) = \emptyset$; and otherwise, $\emptyset \neq V(\Gamma_I(R)) \subseteq V(\Gamma(R))$, but $\Gamma_I(R)$ is not a subgraph of $\Gamma(R)$.*

One of the main questions in the study of zero-divisor graphs is: "Let $R$ and $S$ be two commutative rings. If $\Gamma(R) \cong \Gamma(S)$, then does $R \cong S$?". See Theorems 3.3.1, 3.3.4, and 3.3.5 for well-known results on this question. Actually, by Theorem 3.3.1, for finite reduced commutative rings $R$ and $S$ that are not fields, we have $\Gamma(R) \cong \Gamma(S)$ if and only if $R \cong S$. Now, let $I$ be an ideal of $R$ and $J$ be an ideal of $S$. It is natural to ask the following question: "If $\Gamma_I(R) \cong \Gamma_J(S)$, then do we have $R/I \cong S/J$?". This question was focused on by Maimani, Pournaki, and Yassemi in [272]. Recall that for an ideal $I$ of $R$, $\sqrt{I} = \{r \in R \mid r^k \in I \text{ for some } k \in \mathbb{N}\}$.

**Theorem 5.1.16.** ([272, Theorem 2.1]) *Let $I$ be a finite ideal of a commutative ring $R$ such that $I = \sqrt{I}$. Then $\Gamma_I(R) \cong |I|^2\Gamma(R/I)$.*

**Theorem 5.1.17.** ([272, Theorem 2.2]) *Let $R$ and $S$ be commutative rings with finite radical ideals $I$ and $J$, respectively (i.e., $I = \sqrt{I}$ and $J = \sqrt{J}$). Then the following hold:*

(a) *If $|I| = |J|$ and $\Gamma(R/I) \cong \Gamma(S/J)$, then $\Gamma_I(R) \cong \Gamma_J(S)$;*

(b) *If $\Gamma_I(R) \cong \Gamma_J(S)$, then $\Gamma(R/I) \cong \Gamma(S/J)$.*

In [353, Remark 2.1], J. G. Smith Jr gave a counterexample to part (b) of the above theorem. Moreover, he conjectured ([353, Conjecture 2.1]) that part (b) is true if we add the additional hypothesis that $|I| = |J|$.

**Theorem 5.1.18.** ([272, Theorem 2.4]) *Let $I$ be a nonzero ideal of a commutative ring $R$ and $a \in \Gamma_I(R)$ be adjacent to every vertex of $\Gamma_I(R)$. Then $(I : a)$ is a maximal element of the set $\{(I : x) \mid x \in R\}$. Moreover, $(I : a)$ is a prime ideal.*

**Theorem 5.1.19.** ([272, Theorem 2.5]) *Let $I$ be an ideal of a commutative ring $R$, and let $S$ be a clique in $\Gamma_I(R)$ such that $x^2 = 0$ for every $x \in S$. Then $S \cup I$ is an ideal of $R$.*

Anderson et al. [86] investigated when $\Gamma_I(R) \cong \Gamma_J(S)$ for proper ideals $I$ and $J$ of commutative rings $R$ and $S$, respectively. The next theorem strengthens Theorem 5.1.17 when the rings $R$ and $S$ are both finite. However, Smith's counterexample ([353]) mentioned above also shows that the "$\Rightarrow$" implication of the next theorem is false.

**Theorem 5.1.20.** ([86, Theorem 5.3]) *Let $R$ and $S$ be finite commutative rings with nonmaximal radical ideals $I$ and $J$, respectively. Then $\Gamma_I(R) \cong \Gamma_J(S)$ if and only if $R/I \cong S/J$ and $|I| = |J|$.*

The next example shows that for nonmaximal radical ideals $I$ and $J$ of finite commutative rings $R$ and $S$, respectively, we may have both $\Gamma_I(R) \cong \Gamma_J(S)$ and $\Gamma(R/I) \cong \Gamma(S/J)$, but $R \ncong S$. We also give an example to show that we cannot replace the $R/I \cong S/J$ hypothesis with $|R/I| = |S/J|$ in Theorem 5.1.20.

**Example 5.1.21.** ([86, Example 5.4])

(a) Let $R = \mathbb{Z}_2 \times \mathbb{Z}_2 \times \mathbb{Z}_2 \times \mathbb{Z}_2$ and $I = \{0\} \times \{0\} \times \mathbb{Z}_2 \times \mathbb{Z}_2$, and $S = \mathbb{Z}_2 \times \mathbb{Z}_2 \times \mathbb{F}_4$ and $J = \{0\} \times \{0\} \times \mathbb{F}_4$. Then $|I| = |J| = 4$, and $I$ and $J$ are nonmaximal radical ideals of $R$ and $S$, respectively. Thus $R/I \cong S/J \cong \mathbb{Z}_2 \times \mathbb{Z}_2$, and hence $\Gamma_I(R) \cong \Gamma_J(S)$ by Theorem 5.1.17(a); however, $R \ncong S$.

(b) Let $R = \mathbb{Z}_2 \times \mathbb{Z}_2 \times \mathbb{Z}_2 \times \mathbb{Z}_2$ and $I = \mathbb{Z}_2 \times \{0\} \times \{0\} \times \{0\}$, and $S = \mathbb{Z}_2 \times \mathbb{Z}_2 \times \mathbb{F}_4$ and $J = \mathbb{Z}_2 \times \{0\} \times \{0\}$. Then $|I| = |J| = 2$, and $I$ and $J$ are nonmaximal radical ideals of $R$ and $S$, respectively, and $|R/I| = |S/J| = 8$. Note that $R/I \ncong S/J$, and so $\Gamma(R/I) \ncong \Gamma(S/J)$. Thus $\Gamma_I(R) \ncong \Gamma_J(S)$ by Theorem 5.1.17(b).

Given a proper ideal $I$ of a commutative ring $R$, one can ask if $\Gamma_I(R) \cong \Gamma(S)$ for some commutative ring $S$. We next show that if $I$ is a nonmaximal radical ideal of a finite commutative ring $R$, then this happens only in the trivial case where $I = \{0\}$ and $R \cong S$.

**Theorem 5.1.22.** ([86, Theorem 5.5]) *Let $I$ be a nonmaximal radical ideal of a finite commutative ring $R$. Then $\Gamma_I(R) \cong \Gamma(S)$ for a reduced finite commutative ring $S$ if and only if $I = \{0\}$ (and in this case, $S \cong R$).*

Consider the infinite ring $R = S = \mathbb{Q}[x] \times \mathbb{Q}[x]$ and $I = x\mathbb{Q}[x] \times x\mathbb{Q}[x]$. Then $I$ is a nonzero radical ideal of $R$ and $\Gamma_I(R) \cong \Gamma(S) = K_{\alpha,\alpha}$. So the hypothesis that $R$ and $S$ are finite is also needed in Theorem 5.1.22. The following example shows that we may have $\Gamma_I(R) \cong \Gamma(S)$ for finite commutative rings $R$ and $S$ with $I \neq \{0\}$; so the hypothesis that $R$ and $S$ are both reduced rings is needed in Theorem 5.1.22.

**Example 5.1.23.** ([86, Example 5.6]) Suppose that $n = p^k - 1$ for some prime $p$ and integer $k \geq 1$. Let $R_n = \mathbb{Z}_4 \times \mathbb{Z}_n$ and $I_n = \{0\} \times \mathbb{Z}_n$. Then $R_n$ is not reduced and $I_n$ is not a radical ideal of $R_n$. However, $\Gamma_{I_n}(R_n) = K_n = \Gamma(S_n)$ for the finite commutative local ring $S_n = \mathbb{F}_{p^k}[x]/(x^2)$ by Theorem 2.2.7 and Example 2.2.8. So $I_n \neq \{0\}$ (for $n \geq 2$) and $R_n \not\cong S_n$.

For $n = 2$, we also have $\Gamma_{I_2}(R_2) = K_2 = \Gamma(\mathbb{Z}_2 \times \mathbb{Z}_2)$. So $\Gamma_I(R) \cong \Gamma(S)$ with $S = \mathbb{Z}_2 \times \mathbb{Z}_2$ a finite reduced commutative ring and $I \neq \{0\}$.

We now turn to the investigation of which graphs on $n$ vertices can be realized as $\Gamma_I(R)$ for some commutative ring $R$ and ideal $I$ of $R$ for specific values of $n$. For all possible zero-divisor graphs $\Gamma(R)$ with $|V(\Gamma(R))| \leq 4$, see Example 2.1.3. Note that $\Gamma_I(R)$ can be a graph on one vertex only if $I = (0)$ and $\Gamma(R)$ is a graph on one vertex (for example, $R = \mathbb{Z}_4$). There is only one connected graph on two vertices, namely $K_2$. This graph can be realized as $\Gamma_I(R)$ for the rings $R = \mathbb{Z}_4 \times \mathbb{Z}_2$ with $I = \{0\} \times \mathbb{Z}_2$ and $R = \mathbb{Z}_9$ with $I = (0)$.

There are two connected graphs on three vertices, $K_3$ and $K_{1,2}$. The graph $K_3$ is realized if $R = \mathbb{Z}[x,y]/(x^2, xy, y^2)$ with $I = (0)$ and $K_{1,2}$ is realized if $R = \mathbb{Z}_6$ or $\mathbb{Z}_8$ with $I = (0)$. Further it is shown that $K_3$ can be realized with $I \neq (0)$, but $K_{1,2}$ cannot be realized by Theorem 5.1.22 since the girth of $K_{1,2}$ is infinite. The following theorem says that every finite complete graph can be realized as $\Gamma_I(R)$ for some commutative ring $R$ and an ideal $I \neq (0)$.

**Theorem 5.1.24.** ([323, Theorem 6.1]) *Let $n \geq 2$ be an integer. Then there is a commutative ring $R$ with a nonzero ideal $I$ such that $K_n \cong \Gamma_I(R)$.*

There are six connected graphs on 4 vertices. By Example 2.1.3, we have $K_4$, $K_{2,2}$, and $K_{1,3}$ are the only graphs realized as $\Gamma_I(R)$ with $I = (0)$. By Theorem 5.1.24, $K_4$ is realized as $\Gamma_I(R)$ with $I \neq (0)$. We will show that only $K_{2,2}$ and $K_4$ can be realized when $I \neq (0)$. For suppose that $R$ is a commutative ring with a nonzero ideal $I$ such that $\Gamma_I(R)$ has exactly four vertices. Then, by Lemma 5.1.3, $4 = N \cdot |I|$, where $N$ is the number of vertices of $\Gamma(R/I)$. If $|I| = 4$, then $\Gamma_I(R)$ is a single connected column, and therefore isomorphic to $K_4$. Hence, because $I \neq (0)$, we must have $|I| = 2$ and $\Gamma(R/I)$ is a graph on two vertices. Therefore $\Gamma_I(R)$ has either no or exactly two connected columns. However, if $\Gamma_I(R)$ has two connected columns, then $\Gamma_I(R)$ is isomorphic to $K_4$. If $\Gamma_I(R)$ has no connected columns, then $\Gamma_I(R)$ is isomorphic to $K_{2,2}$.

The following result gives a partial answer as to which graphs on $p$ vertices can be realized as $\Gamma_I(R)$ for some commutative ring $R$ and nonzero $I$ ideal of $R$, where $p$ is prime. It is an open question as to which graphs on $p$ vertices can be realized as $\Gamma(R)$ for some commutative ring $R$ for $p \geq 15$.

**Theorem 5.1.25.** ([323, Proposition 6.3]) *Let $p$ be a prime integer. Then the only graphs on $p$ vertices that can be realized as $\Gamma_I(R)$ for some commutative ring $R$ and ideal $I$ of $R$ are $K_p$ and those graphs realized as $\Gamma(S)$ for some commutative ring $S$. If we require that $I \neq (0)$, then the only graph that can be realized is $K_p$.*

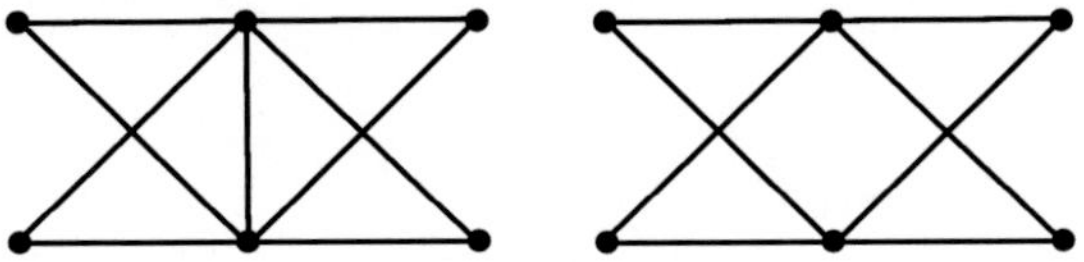

$\Gamma_{\{0\}\times\mathbb{Z}_2}(\mathbb{Z}_2[x]/(x^3)\times\mathbb{Z}_2)$ $\Gamma_{\{0\}\times\mathbb{Z}_2}(\mathbb{Z}_6\times\mathbb{Z}_2)$

Figure 5.3:

Next, we consider which graphs on 6 vertices can be realized as $\Gamma_I(R)$ for some commutative ring $R$ and ideal $I$ of $R$. If $I = (0)$, then a graph can be realized as $\Gamma_I(R)$ if and only if it can be realized as $\Gamma(R)$. If $I$ is nonzero, then either $\Gamma(R/I)$ is a graph on one, two, or three vertices. In the first case, the graph is isomorphic to $K_6$. If $\Gamma(R/I)$ is a graph on two vertices, then $\Gamma_I(R)$ has either no or exactly two connected columns. Therefore, our graph is isomorphic to either $K_{3,3}$ (for $R = \mathbb{Z}_2\times\mathbb{Z}_2\times\mathbb{Z}_3$ and $I = \{0\}\times\{0\}\times\mathbb{Z}_3$); or $K_6$ (for $R = \mathbb{Z}_9\times\mathbb{Z}_3$ and $I = \{0\}\times\mathbb{Z}_3$). If $\Gamma(R/I)\cong K_3$ and $|I| = 2$, then $\Gamma_I(R)\cong K_6$. Figure 5.3 gives the two remaining graphs on six vertices that are realized as $\Gamma_I(R)$ with $I\neq(0)$.

Now, we consider which graphs on nine vertices can be realized as $\Gamma_I(R)$ for some commutative ring $R$ and ideal $I$ of $R$. If $I = (0)$, then a graph can be realized as $\Gamma_I(R)$ if and only if it can be realized as $\Gamma(R)$. If $\Gamma(R/I)$ is a graph on one vertex, then $\Gamma_I(R)\cong K_9$. Otherwise, $\Gamma(R/I)$ is a graph on three vertices and $|I| = 3$. If $\Gamma(R/I)$ is complete, then again $\Gamma_I(R)\cong K_9$. If $\Gamma(R/I)$ is not complete, then the only possibilities are graphs isomorphic to Figure 5.1 or Figure 5.2.

Next, we analyze which graphs on eight vertices can be realized as $\Gamma_I(R)$ for some commutative ring $R$ and ideal $I$ of $R$. For $I = (0)$, a graph can be realized as $\Gamma_I(R)$ if and only if it can be realized as $\Gamma(R)$. If $I\neq(0)$, then $\Gamma(R/I)$ consists of a single, two, or four vertices. If $\Gamma(R/I)$ consists of a single vertex, then $\Gamma_I(R)\cong K_8$. If $\Gamma(R/I)$ consists of two vertices, then the graph has either no or exactly two connected columns. If $\Gamma_I(R)$ has two connected columns, then $\Gamma_I(R)\cong K_8$. For the other case, $\Gamma_I(R)\cong K_{4,4}$ (for $R = \mathbb{Z}_2\times\mathbb{Z}_2\times\mathbb{Z}_4$ and $I = \{0\}\times\{0\}\times\mathbb{Z}_4$). If $\Gamma(R/I)$ is a graph on four vertices, there are three possibilities: $K_{1,3}$, $K_4$, and $K_{2,2}$. If $\Gamma(R/I)\cong K_{1,3}$, then $\Gamma_I(R)$ is isomorphic to the graph in Figure 5.4. If $\Gamma(R/I)\cong K_4$, then $\Gamma_I(R)$ is isomorphic to $K_8$. If $\Gamma(R/I)\cong K_{2,2}$, then $\Gamma_I(R)$ is isomorphic to the graph in Figure 5.5.

Note that Example 2.1.3 shows the possible zero-divisor graphs with a maximum of 4 vertices. Up to isomorphism, each graph may be realized as $\Gamma(R)$ by precisely the following rings:

$$\mathbb{Z}_4,\ \frac{\mathbb{Z}_2[x]}{(x^2)},\ \mathbb{Z}_9,\ \mathbb{Z}_2\times\mathbb{Z}_2,\ \frac{\mathbb{Z}_3[x]}{(x^2)},\ \mathbb{Z}_6,\ \mathbb{Z}_8,\ \frac{\mathbb{Z}_2[x]}{(x^3)},\ \frac{\mathbb{Z}_4[x]}{(x^2-2,2x)},\ \frac{\mathbb{F}_4[x]}{(x^2)},$$
$$\frac{\mathbb{Z}_4[x]}{(x^2+x+1)},\ \frac{\mathbb{Z}_4[x]}{(x,2)^2},\ \frac{\mathbb{Z}_2[x,y]}{(x,y)^2},\ \mathbb{Z}_2\times\mathbb{F}_4,\ \mathbb{Z}_3\times\mathbb{Z}_3,\ \mathbb{Z}_{25},\ \frac{\mathbb{Z}_5[x]}{(x^2)}.$$

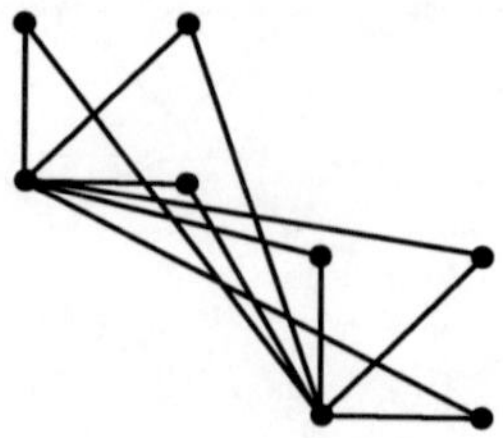

Figure 5.4: $\Gamma_I(R)$, where $R = \mathbb{Z}_2 \times \mathbb{F}_4 \times \mathbb{Z}_2$ and $I = \{0\} \times \{0\} \times \mathbb{Z}_2$.

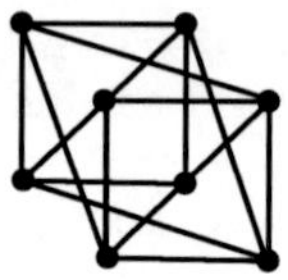

Figure 5.5: $\Gamma_I(R)$, where $R = \mathbb{Z}_2 \times \mathbb{Z}_3 \times \mathbb{Z}_2$ and $I = \{0\} \times \{0\} \times \mathbb{Z}_2$.

The next theorem gives all commutative rings $R$ for which the zero-divisor graphs $\Gamma(R)$ have 5 vertices (cf. the chart after Remark 2.1.4).

**Theorem 5.1.26.** ([323, Theorem 6.4]) *Let $R$ be a commutative ring. If $\Gamma(R)$ is a graph on 5 vertices, then $R$ is isomorphic to one of the following:* (a) $\mathbb{Z}_2 \times \mathbb{Z}_5$, (b) $\mathbb{Z}_2 \times \mathbb{Z}_4$, (c) $\mathbb{Z}_2 \times \mathbb{Z}_2[x]/(x^2)$, *or* (d) $\mathbb{F}_3 \times \mathbb{F}_4$.

Note that $\Gamma(\mathbb{Z}_2 \times \mathbb{Z}_5) \cong K_{1,4}$ and $\Gamma(\mathbb{F}_3 \times \mathbb{F}_4) \cong K_{2,3}$. Figure 5.6 is isomorphic to the zero-divisor graphs of $\mathbb{Z}_2 \times \mathbb{Z}_4$ and $\mathbb{Z}_2 \times \mathbb{Z}[x]/(x^2)$. By Theorem 5.1.24, $K_5$ is realizable as $\Gamma_I(R)$. Therefore, there are four graphs on five vertices which are realizable as $\Gamma_I(R)$. Also, by using the same method as in the cases for 6, 8, and 9 vertices, one can show that there are five graphs on exactly 10 vertices that are realized as $\Gamma_I(R)$ with $I \neq (0)$, six graphs on 15 vertices that can be realized as $\Gamma_I(R)$ with $I \neq (0)$, and four graphs on 25 vertices that can be realized as $\Gamma_I(R)$ with $I \neq (0)$.

In summary, the above work completely determined which graphs on $n$ vertices can be realized as $\Gamma_I(R)$ for some commutative ring $R$ and ideal $I$ of $R$ for $n = 1, 2, 3, 4, 5$. If we require $I$ to be nonzero, then we also have a complete determination for $n = 6, 8, 9, 10, 15, 25$, and any prime $n \geq 7$. If we knew which graphs on $n$ vertices could be realized as $\Gamma(R)$ for these values of $n$, then again

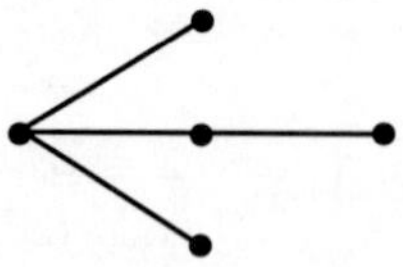

Figure 5.6: $\Gamma(R)$, where $R = \mathbb{Z}_2 \times \mathbb{Z}_8$ or $R = \mathbb{Z}_2 \times \mathbb{Z}_2[x]/(x^2)$.

we would have a complete determination of possibilities for $\Gamma_I(R)$ for any ideal $I$. At present, this is an open question.

Next, we discuss the diameter and girth of the zero-divisor graph with respect to ideals. First, we determine the diameter of $\Gamma_I(R)$.

**Theorem 5.1.27.** ([323, Theorem 2.4]) *Let $I$ be an ideal of a commutative ring $R$. Then $\Gamma_I(R)$ is connected with $diam(\Gamma_I(R)) \leq 3$. Furthermore, if $\Gamma_I(R)$ contains a cycle, then $gr(\Gamma_I(R)) \leq 7$.*

Darani [175] gave a few results regarding the diameter of the zero-divisor graph with respect to an ideal.

**Theorem 5.1.28.** ([175, Theorem 2.3]) *Let $I$ be an ideal of a commutative ring $R$ with $(V(\Gamma_I(R))\setminus I)^2 \not\subseteq I$ such that the prime ideals of $R$ contained in $(V(\Gamma_I(R))\setminus I)$ are linearly ordered. Then $diam(\Gamma_I(R)) = 2$.*

**Theorem 5.1.29.** ([175, Theorem 2.4]) *Let $I$ be an ideal of a commutative ring $R$ such that $R/I$ is a chained ring. Then $diam(\Gamma_I(R)) \leq 2$.*

In [65], Anderson, Atani, Kohan, and Sarvandi studied the structure of $\Gamma_I(R)$ when $R/I$ is a chained ring and gave the following result.

**Theorem 5.1.30.** ([65, Theorem 3.8]) *Let $I$ be an ideal of a commutative ring $R$ such that $R/I$ is a chained ring, $Z_I(R) = \{x \in R \mid xy \in I$ for some $y \in R\setminus I\}$, $Z_I(R)^* = Z_I(R) \setminus I$, and $N_I(R) = \{x \in R \mid x^2 \in I\}$. Then exactly one of the following three cases must occur:*

(1) $|Z_I(R)^*| \leq 1$. *In this case, $diam(\Gamma_I(R)) = 0$;*
(2) $|Z_I(R)^*| \geq 2$ *and* $N_I(R) = Z_I(R)$. *In this case, $diam(\Gamma_I(R)) = 1$;*
(3) $|Z_I(R)^*| \geq 2$ *and* $N_I(R) \subsetneq Z_I(R)$. *In this case, $diam(\Gamma_I(R)) = 2$.*

The following two results which describe the diameter of $\Gamma_I(R)$ were proved by Atani et al. [110]. The first one provides a relationship between $diam(\Gamma_I(R))$ and $diam(\Gamma(R/I))$.

**Theorem 5.1.31.** ([110, Corollary 2.1]) *Let $I$ be a nonzero ideal of a commutative ring $R$. Then*

(a) *$diam(\Gamma_I(R)) = 1$ if and only if $diam(\Gamma(R/I)) \leq 1$;*
(b) *$diam(\Gamma_I(R)) = 2$ if and only if $diam(\Gamma(R/I)) = 2$;*
(c) *$diam(\Gamma_I(R)) = 3$ if and only if $diam(\Gamma(R/I)) = 3$.*

**Theorem 5.1.32.** ([110, Theorem 2.1]) *Let $I$ be a nonzero ideal of a commutative ring $R$. Then*

(a) *$diam(\Gamma_I(R)) = 1$ if and only if either* (i) *$R/I \cong \mathbb{Z}_2 \times \mathbb{Z}_2$ or* (ii) *$V(\Gamma_I(R)) = \sqrt{I} \setminus I$, where $\sqrt{I}$ is the radical of $I$, i.e., $(\sqrt{I})^2 \subseteq I$;*

(b) *$diam(\Gamma_I(R)) = 2$ if and only if either* (i) $I = P_1 \cap P_2$ *for some distinct prime ideals $P_1$, $P_2$ of $R$ and $R/I \ncong \mathbb{Z}_2 \times \mathbb{Z}_2$, or* (ii) *the former conditions do not hold and every pair of distinct zero-divisors of $R/I$ has a nonzero annihilator.*

Brand, Jameson, Mcgowan, and Mckeel [152] entirely classified the diameter of ideal-based zero-divisor graphs of direct products of commutative rings (not necessarily with identity). Let $Z_I(R) = \{x \in R \mid \text{ there exists } y \in R \setminus I \text{ such that } xy \in I\}$ and $Z_I(R)^* = Z_I(R) \setminus I$.

**Theorem 5.1.33.** ([152, Theorem 3.3]) *Let $R_1$, $R_2$ be commutative rings and $I_1$, $I_2$ be their respective ideals such that $diam(\Gamma_{I_1}(R_1)) = diam(\Gamma_{I_2}(R_2)) = 0$. Then*

(a) *$diam(\Gamma_{I_1 \times I_2}(R_1 \times R_2)) = 0$ if and only if (without loss of generality) $I_1 = R_1$ and either $R_1 = \{0\}$ or $I_2$ is a prime ideal;*

(b) *$diam(\Gamma_{I_1 \times I_2}(R_1 \times R_2)) = 1$ if and only if (without loss of generality) $I_1 = \{0\}$ and either $Z_{I_1}(R_1)^* \neq \emptyset$ with $I_2 = R_2 \neq \{0\}$ or $R_1 \cong R_2$ with $|R_1| = 2$;*

(c) *$diam(\Gamma_{I_1 \times I_2}(R_1 \times R_2)) = 3$ if and only if $R_1 \neq Z_{I_1}(R_1)$, $R_2 \neq Z_{I_2}(R_2)$, and (without loss of generality) $|Z_{I_1}(R_1)^*| = 1$;*

(d) *$diam(\Gamma_{I_1 \times I_2}(R_1 \times R_2)) = 2$ otherwise.*

**Theorem 5.1.34.** ([152, Theorem 3.4]) *Let $R_1$ and $R_2$ be commutative rings with ideals $I_1$ and $I_2$, respectively. If $diam(\Gamma_{I_1}(R_1)) = 0$ and $diam(\Gamma_{I_2}(R_2)) = 1$, then*

(a) *$diam(\Gamma_{I_1 \times I_2}(R_1 \times R_2)) > 0$;*

(b) *$diam(\Gamma_{I_1 \times I_2}(R_1 \times R_2)) = 1$ if and only if $R_1^2 \subseteq I_1$ and $R_2^2 \subseteq I_2$ or $R_1 = I_1$;*

(c) *$diam(\Gamma_{I_1 \times I_2}(R_1 \times R_2)) = 2$ if and only if $R_1 \neq I_1$ and (without loss of generality) $R_1^2 \subseteq I_1$ and $R_2^2 \nsubseteq I_2$;*

(d) *$diam(\Gamma_{I_1 \times I_2}(R_1 \times R_2)) = 3$ if and only if $R_1^2 \nsubseteq I_1$ and $R_2^2 \nsubseteq I_2$.*

**Theorem 5.1.35.** ([152, Theorem 3.6]) *Let $R_1$ and $R_2$ be commutative rings with ideals $I_1$ and $I_2$, respectively, such that $diam(\Gamma_{I_1}(R_1)) = 0$ and $diam(\Gamma_{I_2}(R_2)) = 2$. Then*

(a) *$diam(\Gamma_{I_1 \times I_2}(R_1 \times R_2)) > 0$;*

(b) *$diam(\Gamma_{I_1 \times I_2}(R_1 \times R_2)) = 2$ if and only if $R_1 = Z_{I_1}(R_1)$ or $R_2 = Z_{I_2}(R_2)$;*

(c) *$diam(\Gamma_{I_1 \times I_2}(R_1 \times R_2)) = 3$ if and only if $R_1 \neq Z_{I_1}(R_1)$ and $R_2 \neq Z_{I_2}(R_2)$.*

**Theorem 5.1.36.** ([152, Theorem 3.8]) *Let $R_1$ and $R_2$ be commutative rings with ideals $I_1$ and $I_2$, respectively. If $diam(\Gamma_{I_1}(R_1)) = 0$ and $diam(\Gamma_{I_2}(R_2)) = 3$, then*

(a) *$diam(\Gamma_{I_1 \times I_2}(R_1 \times R_2)) > 1$;*

(b) *$diam(\Gamma_{I_1 \times I_2}(R_1 \times R_2)) = 2$ if and only if $R_1 = Z_{I_1}(R_1) \neq I_1$;*

(c) *$diam(\Gamma_{I_1 \times I_2}(R_1 \times R_2)) = 3$ if and only if $R_1 = I_1$ and $R_1 \neq Z_{I_1}(R_1)$.*

**Theorem 5.1.37.** ([152, Theorem 3.10]) *Let $R_1$ and $R_2$ be commutative rings with ideals $I_1$ and $I_2$, respectively, such that $diam(\Gamma_{I_1}(R_1)) = diam(\Gamma_{I_2}(R_2)) = 1$. Then*

(a) *$diam(\Gamma_{I_1 \times I_2}(R_1 \times R_2)) = 1$ if and only if $R_1^2 \subseteq I_1$ and $R_2^2 \subseteq I_2$;*

(b) *$diam(\Gamma_{I_1 \times I_2}(R_1 \times R_2)) = 2$ if and only if (without loss of generality) $R_1^2 \subseteq I_1$ and $R_2^2 \not\subseteq I_2$;*

(c) *$diam(\Gamma_{I_1 \times I_2}(R_1 \times R_2)) = 3$ if and only if $R_1^2 \not\subseteq I_1$ and $R_2^2 \not\subseteq I_2$.*

**Theorem 5.1.38.** ([152, Theorem 3.11]) *Let $I_1$ and $I_2$ be ideals of commutative rings $R_1$ and $R_2$, respectively, satisfying $diam(\Gamma_{I_1}(R_1)) = 1$ and $diam(\Gamma_{I_2}(R_2)) = 2$. Then*

(a) *$diam(\Gamma_{I_1 \times I_2}(R_1 \times R_2)) > 1$;*

(b) *$diam(\Gamma_{I_1 \times I_2}(R_1 \times R_2)) = 2$ if and only if $R_1 = Z_{I_1}(R_1)$ or $R_2 = Z_{I_2}(R_2)$;*

(c) *$diam(\Gamma_{I_1 \times I_2}(R_1 \times R_2)) = 3$ if and only if $R_1 \neq Z_{I_1}(R_1)$ and $R_2 \neq Z_{I_2}(R_2)$.*

**Theorem 5.1.39.** ([152, Theorem 3.12]) *Let $I_1$ and $I_2$ be ideals of commutative rings $R_1$ and $R_2$, respectively, satisfying $diam(\Gamma_{I_1}(R_1)) = 1$ and $diam(\Gamma_{I_2}(R_2)) = 3$. Then*

(a) *$diam(\Gamma_{I_1 \times I_2}(R_1 \times R_2)) > 1$;*

(b) *$diam(\Gamma_{I_1 \times I_2}(R_1 \times R_2)) = 2$ if and only if $R_1 = Z_{I_1}(R_1)$;*

(c) *$diam(\Gamma_{I_1 \times I_2}(R_1 \times R_2)) = 3$ if and only if $R_1 \neq Z_{I_1}(R_1)$.*

**Theorem 5.1.40.** ([152, Theorem 3.13]) *Let $R_1$ and $R_2$ be commutative rings with ideals $I_1$ and $I_2$, respectively, such that $diam(\Gamma_{I_1}(R_1)) = diam(\Gamma_{I_2}(R_2)) = 2$. Then*

(a) *$diam(\Gamma_{I_1 \times I_2}(R_1 \times R_2)) > 1$;*

(b) *$diam(\Gamma_{I_1 \times I_2}(R_1 \times R_2)) = 2$ if and only if $R_1 = Z_{I_1}(R_1)$ or $R_2 = Z_{I_2}(R_2)$;*

(c) *$diam(\Gamma_{I_1 \times I_2}(R_1 \times R_2)) = 3$ if and only if $R_1 \neq Z_{I_1}(R_1)$ and $R_2 \neq Z_{I_2}(R_2)$.*

**Theorem 5.1.41.** ([152, Theorem 3.14]) *Let $I_1$ and $I_2$ be ideals of commutative rings $R_1$ and $R_2$, respectively, satisfying $diam(\Gamma_{I_1}(R_1)) = 2$ and $diam(\Gamma_{I_2}(R_2)) = 3$. Then*

(a) *$diam(\Gamma_{I_1 \times I_2}(R_1 \times R_2)) > 1$;*

(b) *$diam(\Gamma_{I_1 \times I_2}(R_1 \times R_2)) = 2$ if and only if $R_1 = Z_{I_1}(R_1)$;*

(c) *$diam(\Gamma_{I_1 \times I_2}(R_1 \times R_2)) = 3$ if and only if $R_1 \neq Z_{I_1}(R_1)$.*

**Theorem 5.1.42.** ([152, Theorem 3.15]) *Let $R_1$ and $R_2$ be commutative rings with ideals $I_1$ and $I_2$, respectively, such that $diam(\Gamma_{I_1}(R_1)) = diam(\Gamma_{I_2}(R_2)) = 3$. Then $diam(\Gamma_{I_1 \times I_2}(R_1 \times R_2)) = 3$.*

Next, we consider the girth of $\Gamma_I(R)$. Recall that $gr(\Gamma(R)) \leq 4$ or $gr(\Gamma(R)) = \infty$ for any commutative ring $R$. The following result shows that given any ideal $I$ of a commutative ring $R$, the same result holds for $\Gamma_I(R)$.

**Lemma 5.1.43.** ([323, Lemma 5.1]) *Let $I$ be an ideal of a commutative ring $R$. Then $gr(\Gamma_I(R)) \leq gr(\Gamma(R/I))$. In particular, if $\Gamma(R/I)$ contains a cycle, then so does $\Gamma_I(R)$, and therefore $gr(\Gamma_I(R)) \leq gr(\Gamma(R/I)) \leq 4$.*

**Lemma 5.1.44.** ([323, Lemma 5.3]) *Let $I$ be an ideal of a commutative ring $R$. If $I \neq (0)$ and $\Gamma(R/I)$ has only one vertex, then*

$$gr(\Gamma_I(R)) = \begin{cases} 3 & \text{if } |I| \geq 3; \\ \infty & \text{if } |I| = 2. \end{cases}$$

**Theorem 5.1.45.** ([323, Theorem 5.5]) *Let $I$ be a nonzero ideal of a commutative ring $R$ that is not a prime ideal. Then*

$$gr(\Gamma_I(R)) = \begin{cases} \infty & \text{if } \Gamma(R/I) \text{ has only one vertex and } |I| = 2; \\ 4 & \text{if } gr(\Gamma(R/I)) > 3 \text{ and } \Gamma_I(R) \text{ has no} \\ & \text{connected columns;} \\ 3 & \text{otherwise.} \end{cases}$$

The girth of $\Gamma_I(R)$, when $I$ is an intersection of prime ideals, was investigated by Maimani et al. [272].

**Theorem 5.1.46.** ([272, Theorem 4.1]) *Let $P_1$ and $P_2$ be prime ideals of a commutative ring $R$ and $I = P_1 \cap P_2$. Then either $gr(\Gamma_I(R)) = 4$ or $R/I \cong \mathbb{Z}_2 \times \mathbb{Z}_2$.*

**Theorem 5.1.47.** ([272, Theorem 4.4]) *Let $I$ be an ideal of a commutative ring $R$. Suppose that either $I$ is a primary ideal of $R$ that is not prime and $|I| \geq 3$, or $|Ass(R/I)| \geq 3$. Then $gr(\Gamma_I(R)) = 3$.*

In the above theorem, one of the conditions "$|I| \geq 3$" or "$|Ass(R/I)| \geq 3$" is necessary. To see this, let $R = \mathbb{Z}_8$ and $I = \{0, 4\}$; then $|Ass(R/I)| = 1$ and $gr(\Gamma(R/I)) = \infty$.

**Theorem 5.1.48.** ([110, Theorem 3.1]) *Let $R$ be a commutative ring and $I$ be an ideal of $R$ which is not prime.*

(a) *Suppose that $R/I$ is a reduced ring. Then $gr(\Gamma_I(R)) = 4$ if and only if $I = P_1 \cap P_2$ for distinct prime ideals $P_1$, $P_2$ of $R$. In this case, $\Gamma_I(R)$ is a complete bipartite graph. Otherwise, $gr(\Gamma_I(R)) = 3$;*

(b) *Suppose that the ring $R/I$ is not reduced. Then $gr(\Gamma_I(R)) = 3$ or $gr(\Gamma_I(R)) = \infty$. The latter holds if and only if (up to isomorphism):*

(i) $R = \mathbb{Z}_2 \times \mathbb{Z}_4$ *and* $I = \mathbb{Z}_2 \times \{0\}$;

(ii) $R = \mathbb{Z}_2 \times \frac{\mathbb{Z}_2[x]}{(x^2)}$ *and* $I = \mathbb{Z}_2 \times \frac{\mathbb{Z}_2[x]}{(x^2)}$;

(iii) $R = \mathbb{Z}_8$ *and* $I = 4R$;

(iv) $R = \frac{\mathbb{Z}_4[x]}{(x^2-2,2x)}$ *and* $I = \frac{(x^2,2)}{(x^2-2,2x)}$;

(v) $R = \frac{\mathbb{Z}_4[x]}{(x^2,2x)}$ *and* $I = \frac{(x^2,2)}{(x^2,2x)}$ *or* $I = \frac{(x)}{(x^2,2x)}$ *or* $I = \frac{(2x,x-2)}{(x^2,2x)}$;

(vi) $R = \frac{\mathbb{Z}_2[x]}{(x^3)}$ *and* $I = \frac{(x^2)}{(x^3)}$;

(vii) $R = \frac{\mathbb{Z}_2[x,y]}{(x^2,y^2,xy)}$ *and* $I = \frac{(x,y^2,xy)}{(x^2,y^2,xy)}$.

The authors in [65] showed that $\mathrm{gr}(\Gamma_I(R)) \in \{3,\infty\}$ when $R/I$ is a chained ring.

**Theorem 5.1.49.** ([65, Theorem 3.9]) *Let $I$ be an ideal of a commutative ring $R$ such that $R/I$ is a chained ring, $Z_I(R) = \{x \in R \mid xy \in I$ for some $y \in R\setminus I\}$, $N_I(R) = \{x \in R \mid x^2 \in I\}$, and $N_I(R)^* = N_I(R) \setminus I$. Then exactly one of the following four cases must occur:*

(1) $|N_I(R)^*| \leq 1$. *In this case,* $gr(\Gamma_I(R)) = \infty$;

(2) $|N_I(R)^*| = 2$ *and* $N_I(R) = Z_I(R)$. *In this case,* $gr(\Gamma_I(R)) = \infty$;

(3) $|N_I(R)^*| = 2$ *and* $N_I(R) \subsetneq Z_I(R)$. *In this case,* $gr(\Gamma_I(R)) = 3$;

(4) $|N_I(R)^*| \geq 3$. *In this case,* $gr(\Gamma_I(R)) = 3$.

## 5.2 Properties of $\Gamma_I(R)$

In this section, we discuss some of the graph-theoretical properties of the ideal-based zero-divisor graph $\Gamma_I(R)$. First, we see when $\Gamma_I(R)$ is either a complete graph or a complete bipartite graph.

**Theorem 5.2.1.** ([86, Theorem 4.2]) *Let $I$ be a proper ideal of a commutative ring $R$. If $\Gamma_I(R)$ is a complete graph, then $\Gamma(R/I)$ is a complete graph.*

Next, we determine when $\Gamma_I(R)$ is a complete graph in the case where $I$ is a radical ideal of $R$.

**Theorem 5.2.2.** ([86, Theorem 4.3]) *Let $I$ be a radical ideal of a commutative ring $R$. Then $\Gamma_I(R)$ is a nonempty complete graph if and only if $R \cong \mathbb{Z}_2 \times \mathbb{Z}_2$ and $I = \{0\}$. Moreover, in this case, $\Gamma_I(R) = \Gamma(R) = K_2$.*

In some cases, $\Gamma_I(R)$ is a complete graph if $\Gamma(R)$ is a complete graph.

**Theorem 5.2.3.** ([86, Theorem 4.4]) *Let $I$ be a radical ideal of a commutative ring $R$. If $\Gamma(R)$ is a complete graph, then either $\Gamma_I(R)$ is a complete graph or $Z(R) \subseteq I$ (i.e., $V(\Gamma_I(R)) \cap V(\Gamma(R)) = \emptyset$). Moreover, if $\Gamma(R)$ is a complete graph and $Z(R) \not\subseteq I$, then either $\Gamma_I(R) = K_2$ or $\Gamma_I(R)$ is the empty graph.*

**Corollary 5.2.4.** ([86, Corollary 4.5]) *Let $I$ be a radical ideal of a commutative ring $R$ with $R = Z(R) \cup U(R)$. If $\Gamma(R)$ is a complete graph, then either $\Gamma_I(R) = K_2$ or $\Gamma_I(R)$ is the empty graph. In particular, this holds if $dim(R) = 0$.*

The hypothesis $R = Z(R) \cup U(R)$ is needed in Corollary 5.2.4. For example, let $R = \mathbb{Z}_4[x, y]$ and $I = (2, xy)$. Note that $Z(R) \cup U(R) \subsetneq R$ since $x$ is neither a zero-divisor nor a unit in $R$, and $I$ is radical ideal of $R$ since $I = (2, x) \cap (2, y)$ is the intersection of two prime ideals of $R$. Also, $Z(R) = 2R$; so $Z(R) \subsetneq I$. Clearly, $\Gamma(R)$ is a complete graph and $V(\Gamma_I(R)) \cap V(\Gamma(R)) = \emptyset$. Note that $\Gamma_I(R)$ is not a complete graph since $x, x^2 \in V(\Gamma_I(R))$ are not adjacent. In fact, $\Gamma_I(R)$ is the complete bipartite graph on disjoint vertex sets $A = (x) \setminus I$ and $B = (y) \setminus I$.

We next move to the case where $I$ is not a radical ideal of $R$.

**Theorem 5.2.5.** ([86, Theorem 4.7]) *Let $I$ be a proper nonradical ideal of a commutative ring $R$. Then $\Gamma_I(R)$ is a complete graph if and only if $\Gamma(R/I)$ is a complete graph. Moreover, if $\Gamma(R/I)$ is a complete graph on $\alpha$ vertices, then $\Gamma_I(R)$ is a complete graph on $\alpha|I|$ vertices.*

**Corollary 5.2.6.** ([86, Corollary 4.8]) *Let $I$ be a proper ideal of a finite commutative ring $R$. Then $\Gamma_I(R)$ is a nonempty complete graph if and only if either $R \cong \mathbb{Z}_2 \times \mathbb{Z}_2$ and $I = \{0\}$, or $I$ is contained in a unique maximal ideal $M$ of $R$ and $M^2 \subseteq I \subsetneq M$.*

Note that in Theorem 5.2.5, it is necessary to assume that $I$ is not a radical ideal of $R$. For example, let $R = \mathbb{Z}_2 \times \mathbb{Z}_2 \times \mathbb{Z}_2$ and $I = \{0\} \times \{0\} \times \mathbb{Z}_2$. Then $I$ is a radical ideal of $R$ and $R/I = \mathbb{Z}_2 \times \mathbb{Z}_2$. Thus $\Gamma(R/I)$ is a complete graph, but $\Gamma_I(R)$ is not a complete graph since the vertices $(1, 0, 0)$ and $(1, 0, 1)$ are not adjacent.

Recall that, by Theorem 2.2.7, for a finite commutative ring $R$, we have that $\Gamma(R)$ is a nonempty complete graph if and only if $\Gamma(R) = K_n$, where $n = p^k - 1$ for some prime $p$ and integer $k \geq 1$. Moreover, by Example 2.2.8(a), for every such $n$, there is a finite local commutative ring $R_n$ with $\Gamma(R_n) = K_n$, namely $R_n = \mathbb{F}_{p^k}[x]/(x^2)$. However, the situation is different for $\Gamma_I(R)$. By Theorem 5.1.24, we have the following result proved by Redmond [323] which says that every finite complete graph can be realized as $\Gamma_I(R)$ for some commutative ring $R$ and ideal $I \neq (0)$.

**Theorem 5.2.7.** ([323, Theorem 6.1]) *Let $n \geq 2$ be an integer. Then there is a commutative ring $R$ with a nonzero ideal $I$ such that $K_n \cong \Gamma_I(R)$.*

Thus $K_5$ can be realized as $\Gamma_I(R)$ (for example, let $R = \mathbb{Z}_4 \times \mathbb{Z}_5$ and $I = \{0\} \times \mathbb{Z}_5$), whereas $K_5$ cannot be realized as a zero-divisor graph. For any infinite cardinal $\alpha$, there is an integral domain $R_\alpha$ and a proper nonzero ideal $I_\alpha$ of $R_\alpha$ such that $\Gamma_{I_\alpha}(R_\alpha) = K_\alpha$. For example, let $R_\alpha = \mathbb{Z}[\{x_\beta\}]$ and $I_\alpha = (4, \{x_\beta\})$, where $\{x_\beta\}$ is a family of indeterminates with $|\{x_\beta\}| = \alpha$; so $|I_\alpha| = |R_\alpha| = \alpha$.

Then $R_\alpha/I_\alpha \cong \mathbb{Z}_4$; so $\Gamma(R_\alpha/I_\alpha) = K_1$. Thus $\Gamma_{I_\alpha}(R_\alpha)$ is a complete graph on $\alpha$ vertices by Theorem 5.2.5.

Theorem 5.2.7 shows that for every integer $n \geq 1$, there is a finite commutative ring $R$ and a proper ideal $I$ of $R$ such that $\Gamma_I(R)$ has exactly $n$ vertices (for example, let $R = \mathbb{Z}_4 \times \mathbb{Z}_n$ and $I = \{0\} \times \mathbb{Z}_n$, then $\Gamma_I(R) = K_n$). However, this is not true for $\Gamma(R)$. Indeed, recall that there is no commutative ring $R$ with $|V(\Gamma(R))| = 1209$, equivalently, there is no commutative ring $R$ with $|Z(R)| = 1210$ ([327]). Moreover, 1210 is the smallest such positive integer.

The complete $r$-partite structure of the ideal based zero-divisor graph $\Gamma_I(R)$ was discussed by Maimani et al. [272]. Recall that if $I$ is a prime ideal of $R$, then we have $\Gamma_I(R) = \emptyset$. The following theorem shows that if $I = P_1 \cap P_2$, where $P_1$ and $P_2$ are prime ideals of $R$, then $\Gamma_I(R)$ is a complete bipartite graph.

**Theorem 5.2.8.** ([272, Theorem 3.1]) *Let $I$ be a nonzero ideal of a commutative ring $R$. Then the following hold:*

(a) *If $P_1$ and $P_2$ are prime ideals of $R$ and $I = P_1 \cap P_2 \neq 0$, then $\Gamma_I(R)$ is a complete bipartite graph;*

(b) *If $I \neq 0$ is an ideal of $R$ for which $I = \sqrt{I}$, then $\Gamma_I(R)$ is a complete bipartite graph if and only if there exist prime ideals $P_1$ and $P_2$ of $R$ such that $I = P_1 \cap P_2$.*

Note that if we consider $R = \mathbb{Z}_8$ and $I = \{0, 4\}$, then it is easy to see that $\Gamma_I(R)$ is bipartite, but $I$ cannot be written as the intersection of two prime ideals. Therefore, the converse of Theorem 5.2.8(a) is not valid in general. Hence, the condition "$I = \sqrt{I}$" on the ideal $I$ is not superficial in Theorem 5.2.8(b).

**Theorem 5.2.9.** ([272, Theorem 3.2]) *Let $I$ be a nonzero proper ideal of a commutative ring $R$. If $\Gamma_I(R)$ is a complete $r$-partite graph, $r \geq 3$, then at most one of the parts has more than one vertex.*

The next result by Redmond [323] provides a necessary and sufficient condition for $\Gamma_I(R)$ to be a complete bipartite graph in terms of girth.

**Theorem 5.2.10.** ([323, Theorem 5.7]) *Let $I$ be a nonzero ideal of a commutative ring $R$. Then $\Gamma_I(R)$ is (complete) bipartite if and only if either* (a) $gr(\Gamma_I(R)) = \infty$ *or* (b) $gr(\Gamma_I(R)) = 4$ *and $\Gamma(R/I)$ is (complete) bipartite.*

Later, Anderson et al. [86] determined when $\Gamma_I(R)$ is a complete bipartite graph. First of all, the following theorem illustrates a relationship between the complete bipartite structure of the graphs $\Gamma_I(R)$ and $\Gamma(R/I)$.

**Theorem 5.2.11.** ([86, Theorem 4.10]) *Let $I$ be a proper ideal of a commutative ring $R$. If $\Gamma_I(R)$ is a complete bipartite graph and $|V(\Gamma_I(R))| \geq 3$, then $\Gamma(R/I)$ is a complete bipartite graph.*

The study of the complete bipartite problem for $\Gamma_I(R)$ breaks into the two cases as to whether or not $I$ is a radical ideal of $R$. First, we handle the case where $I$ is a radical ideal of $R$.

**Theorem 5.2.12.** ([86, Theorem 4.11]) *Let $I$ be a radical ideal of a commutative ring $R$. Then $\Gamma_I(R)$ is a complete bipartite graph if and only if $\Gamma(R/I)$ is a complete bipartite graph, if and only if $T(R/I) \cong F_1 \times F_2$ for fields $F_1$ and $F_2$. In particular, if $R/I = Z(R/I) \cup U(R/I)$ (e.g., $R/I$ is finite), then $\Gamma_I(R)$ is a complete bipartite graph if and only if $R/I \cong F_1 \times F_2$ for fields $F_1$ and $F_2$. Moreover, if $\Gamma(R/I) = K_{\alpha,\beta}$, then $\Gamma_I(R) = K_{\alpha|I|,\beta|I|}$.*

It is interesting to compare Theorems 5.2.12 and 5.2.2. Let $I$ be a radical ideal of a commutative ring $R$ such that $R/I \cong \mathbb{Z}_2 \times \mathbb{Z}_2$. Then $\Gamma_I(R) = K_{1,1} = K_2$. Thus $\Gamma_I(R) = K_{|I|,|I|}$ by Theorem 5.2.12; so $\Gamma_I(R)$ is a complete graph if and only if $I = \{0\}$. Hence Theorem 5.2.2 may be viewed as a special case of Theorem 5.2.12.

We finish with the case where $I$ is not a radical ideal of $R$.

**Theorem 5.2.13.** ([86, Theorem 4.12]) *Let $I$ be a proper nonradical ideal of a commutative ring $R$. Then $\Gamma_I(R)$ is a complete bipartite graph if and only if either $I = \{0\}$ and $\Gamma(R)(= \Gamma_I(R))$ is a star graph, or $|I| = 2$ and $\Gamma_I(R) = K_{1,1}$. In particular, $\Gamma_I(R)$ is a complete bipartite graph if and only if it is star graph.*

The following example lists the commutative rings $R$ with ideal $I$ such that $\Gamma_I(R)$ is a complete bipartite graph (or a star graph).

**Example 5.2.14.** ([86, Example 4.13])

(a) Let $I$ be a radical ideal of a finite commutative ring $R$ such that $\Gamma_I(R)$ is a complete bipartite graph. Then $T(R/I) = R/I$; so $R/I \cong F_1 \times F_2$ for finite fields $F_1$ and $F_2$ by Theorem 5.2.12, i.e., $R$ has exactly two maximal ideals $\mathfrak{m}_1$ and $\mathfrak{m}_2$ and $I = \mathfrak{m}_1 \cap \mathfrak{m}_2$, $F_1 \cong R/\mathfrak{m}_1$, and $F_2 \cong R/\mathfrak{m}_2$. In this case, $\Gamma(R/I) = K_{m,n}$, where $m = |F_1| - 1$ and $n = |F_2| - 1$; so $\Gamma_I(R) = K_{mj,nj}$, where $j = |I|$. Note that $\Gamma_I(R)$ is a star graph if and only if $I = \{0\}$ and (at least) one of the two fields is $\mathbb{Z}_2$.

(b) Let $R$ be a finite nonreduced commutative ring such that $\Gamma(R)$ is a complete bipartite graph (necessarily a star graph). Then $R \cong \mathbb{Z}_9$ or $\mathbb{Z}_3[x]/(x^2)$ for $\Gamma(R) = K_{1,1}$, and $R \cong \mathbb{Z}_8$, $\mathbb{Z}_2[x]/(x^3)$, or $\mathbb{Z}_4[x]/(x^2 - 2, 2x)$ for $\Gamma(R) = K_{1,2}$ (see Section 2.2). Thus, for a finite nonreduced commutative ring $R$ and $I = \{0\}$, $\Gamma_I(R)$ is a complete bipartite graph if and only if $R$ is isomorphic to one of the above five rings, and $\Gamma_I(R)$ is either $K_{1,1}$ or $K_{1,2}$.

Let $R$ be an infinite nonreduced commutative ring such that $\Gamma(R)$ is a complete bipartite graph (necessarily a star graph). Then either $R \cong \mathbb{Z}_2 \times D$

for an (infinite) integral domain $D$, or $Nil(R) = \{0, x\}$ is a prime ideal of $R$, where $Z(R) = ann_R(x)$ (see Theorem 3.6.8). Thus, for $R$ an infinite nonreduced commutative ring and $I = \{0\}$, $\Gamma_I(R)$ is a complete bipartite graph if and only if $R$ is isomorphic to one of these two types of rings, and $\Gamma_I(R)$ is an infinite star graph.

(c) Let $I$ be a proper, nonzero, nonradical ideal of a commutative ring $R$ such that $\Gamma(R)$ is a complete bipartite graph (necessarily a star graph). Then $|I| = 2$ and $|V(\Gamma_I(R))| = 2$ by Theorem 5.2.13; so $V(R/I) = 1$. Thus $R/I \cong \mathbb{Z}_4$ or $\mathbb{Z}_2[x]/(x^2)$; so $|R| = |R/I||I| = 8$.

If $R$ is not a local ring, then it is easily seen (since $R$ is a product of local rings) that $R$ is isomorphic to either (i) $\mathbb{Z}_4 \times \mathbb{Z}_2$ with $I = \{0\} \times \mathbb{Z}_2$ or (ii) $\mathbb{Z}_2[x]/(x^2) \times \mathbb{Z}_2$ with $I = \{0\} \times \mathbb{Z}_2$.

Next, suppose that $R$ is local. Then $R$ has maximal ideal $\mathfrak{m} = Z(R) \supsetneq I$ with $|\mathfrak{m}| = 4$; so $|V(\Gamma(R))| = 3$. Thus $R$ is isomorphic to either $\mathbb{Z}_8$, $\mathbb{Z}_2[x]/(x^3), \mathbb{Z}_4[x]/(x^2 - 2, 2x), \mathbb{Z}_4[x]/(x, 2)^2$, or $\mathbb{Z}_2[x, y]/(x, y)^2$; by Example 2.1.3(c) (note that the rings $\mathbb{Z}_6$, $\mathbb{F}_4[x]/(x^2)$, and $\mathbb{Z}_4[x]/(x^2 + x + 1)$ are eliminated from the list given in Example 2.1.3(c) since none has 8 elements). With a bit more work, one may then show that the following local rings with maximal ideal $\mathfrak{m}$ and ideal $I$ are cases where $\Gamma_I(R)$ is a complete bipartite graph: (iii) $R = \mathbb{Z}_8$ with $\mathfrak{m} = \{0, 2, 4, 6\}$ and $I = \{0, 4\}$; (iv) $R = \mathbb{Z}_2[X]/(X^3) = \mathbb{Z}_2[x]$ with $\mathfrak{m} = \{0, x, x^2, x^2 + x\}$ and $I = \{0, x^2\}$; (v) $R = \mathbb{Z}_4[X]/(X^2 - 2, 2X) = \mathbb{Z}_4[x]$ with $\mathfrak{m} = \{0, 2, x, x + 2\}$ and $I = \{0, 2\}$; (vi) $R = \mathbb{Z}_4[X]/(X, 2)^2 = \mathbb{Z}_4[x]$ with $\mathfrak{m} = \{0, 2, x, x+2\}$ and $I = \{0, 2\}$, $\{0, x\}$, or $\{0, x + 2\}$; and (vii) $R = \mathbb{Z}_2[X, Y]/(X, Y)^2 = \mathbb{Z}_2[x, y]$ with $M = \{0, x, y, x + y\}$ and $I = \{0, x\}$, $\{0, y\}$, or $\{0, x + y\}$.

In summary, for a commutative ring $R$ and $I$ a proper, nonzero, nonradical ideal of $R$, $\Gamma_I(R)$ is a complete bipartite graph if and only if $R$ is isomorphic to one of the above seven rings with given ideal $I$, and in every case, $\Gamma_I(R) = K_{1,1}$.

Recall that the complete bipartite graphs that can be realized as $\Gamma(R)$ for a finite commutative ring $R$ are precisely $K_{m,n}$, where $m = p_1^{k_1} - 1$, $n = p_2^{k_2} - 1$ for primes $p_1$, $p_2$ and positive integers $k_1$, $k_2$.

**Example 5.2.15.** ([86, Example 4.14]) Let $R = S \times F_1 \times F_2$ for $S$ a commutative ring and fields $F_1$ and $F_2$, and let $I = S \times \{0\} \times \{0\}$. Then $I$ is a radical ideal of $R$ with $R/I \cong F_1 \times F_2$. Thus $\Gamma(R/I) = K_{\alpha,\beta}$, where $\alpha = |F_1| - 1$ and $\beta = |F_2| - 1$, and hence $\Gamma_I(R) = K_{\alpha|S|,\beta|S|}$ by Theorem 5.2.12. Thus, the complete bipartite graphs that can be realized as $\Gamma_I(R)$ for $I$ a proper nonzero ideal of a finite commutative ring $R$ are $K_{1,1}$ and $K_{mj,nj}$, where $m = p_1^{k_1} - 1$, $n = p_2^{k_2} - 1$ for primes $p_1, p_2$ and integers $k_1, k_2 \geq 1$ and $j \geq 2$. This construction yields more examples of graphs that can be realized as ideal-based zero-divisor graphs, but not as zero-divisor graphs. Specifically, let $R = \mathbb{Z}_5 \times \mathbb{Z}_2 \times \mathbb{Z}_2$ and $I = \mathbb{Z}_5 \times \{0\} \times \{0\}$. Then $\Gamma_I(R) = K_{5,5}$. However there is no commutative ring $R$ with $\Gamma(R) = K_{5,5}$.

Recall that (paragraph following Theorem 5.2.7) $K_5$ can be realized as $\Gamma_I(R)$, whereas $K_5$ cannot be realized as a zero-divisor graph. The next example shows that $K_5$ is the smallest graph that can be realized as an ideal-based zero-divisor graph, but not as a zero-divisor graph.

**Example 5.2.16.** ([86, Example 5.1]) Every connected graph on at most 3 vertices can be realized as a zero-divisor graph, and the only connected graphs on 4 vertices that can be realized as zero-divisor graphs are $K_{1,3}$, $K_{2,2}$, and $K_4$ (Example 2.1.3(b)). Let $R$ be a commutative ring and $I$ a proper ideal of $R$ such that $|V(\Gamma_I(R))| = 4$. By Lemma 5.1.3, either (i) $|I| = 1$ and $|V(\Gamma(R/I)| = 4$, (ii) $|I| = 4$ and $|V(\Gamma(R/I))| = 1$, or (iii) $|I| = |V(\Gamma(R/I))| = 2$. In case (i), $I = \{0\}$; so $\Gamma_I(R) = \Gamma(R)$. In case (ii), $R/I$ is isomorphic to $\mathbb{Z}_4$ or $\mathbb{Z}_2[x]/(x^2)$; so $I$ is not a radical ideal of $R$. Thus $\Gamma_I(R) = K_4$ by Theorem 5.2.5. In case (iii), $\Gamma_I(R) = K_{1,1} = K_2$. If $I$ is a radical ideal of $R$, then $\Gamma_I(R) = K_{2,2}$ by Theorem 5.2.12; and if $I$ is not a radical ideal of $R$, then $\Gamma_I(R) = K_4$ by Theorem 5.2.5. Hence any graph on 4 vertices that can be realized as an ideal-based zero-divisor graph can also be realized as a zero-divisor graph. At the same time, $K_5$ can be realized as an ideal-based zero-divisor graph, but not as a zero-divisor graph. Thus $K_5$ is the smallest graph that can be realized as an ideal-based zero-divisor graph, but not as a zero-divisor graph. Moreover, by Proposition 5.1.25, $K_5$ is the only graph on 5 vertices that can be realized as an ideal-based zero-divisor graph, but not as a zero-divisor graph.

However, $K_{1,2}$ can be realized as a zero-divisor graph (note that $\Gamma(\mathbb{Z}_6) = K_{1,2}$), but not as $\Gamma_I(R)$ for any commutative ring $R$ and proper nonzero ideal $I$ of $R$ since the complete bipartite graphs that can be realized as $\Gamma_I(R)$ are $K_{1,1}$ and $K_{mj,nj}$, where $m = p_1^{k_1} - 1$, $n = p_2^{k_2} - 1$ for primes $p_1, p_2$ and integers $k_1, k_2 \geq 1$ and $j \geq 2$.

### 5.2.1 Clique Number of Ideal-based Zero-divisor Graphs

Next, we consider the clique number of $\Gamma_I(R)$. Note that $\omega(\Gamma(R/I)) \leq \omega(\Gamma_I(R))$ since $\Gamma(R/I)$ is isomorphic to a subgraph of $\Gamma_I(R/I)$. Note that for the ring $R = \mathbb{Z}_4 \times \mathbb{Z}_2$ and $I = \{0\} \times \mathbb{Z}_2$, $1 = \omega(\Gamma(R/I)) < \omega(\Gamma_I(R)) = 2$. Also, equality holds if $\omega(\Gamma(R/I)) = \infty$. The next result identifies other cases in which equality may hold.

**Theorem 5.2.17.** ([323, Theorem 4.5]) *Let $I$ be an ideal of a commutative ring $R$. If $\Gamma_I(R)$ has no connected columns, then $\omega(\Gamma(R/I)) = \omega(\Gamma_I(R))$.*

Work on the clique number of $\Gamma_I(R)$ was continued by Maimani [272].

**Theorem 5.2.18.** ([272, Theorem 4.2]) *Let $I$ be an ideal of a commutative ring $R$ such that $I = P_1 \cap \cdots \cap P_n$ and for every $1 \leq j \leq n$, $I \neq \bigcap_{1\leq j\leq n,\ i\neq j} P_i$, where the $P_i$'s are prime ideals of $R$. Then $\omega(\Gamma_I(R)) = n$.*

**Corollary 5.2.19.** ([272, Corollary 4.3]) *Let $R$ be a commutative ring. Then the following hold:*

(a) *If for any $P \in Min(R)$, $P$ is a finitely generated ideal, then $\omega(\Gamma_{Nil(R)}(R)) = |Min(R)|$ (which is always finite);*

(b) *If $R$ is semilocal and not local, then $\omega(\Gamma_{J(R)}(R)) = |Max(R)|$;*

(c) *If $n$ is a square-free integer, then $\omega(\Gamma_{n\mathbb{Z}}(\mathbb{Z})) = k$, where $k$ is the number of primes in the decomposition of $n$ into primes.*

Next, we are interested in the connectivity of $\Gamma_I(R)$. The following few results were proved by Redmond in [323].

**Theorem 5.2.20.** ([323, Theorem 3.2]) *If $I$ is a nonzero proper ideal of $R$, then $\Gamma_I(R)$ has no cut-points.*

**Theorem 5.2.21.** ([323, Theorem 3.3]) *Let $I$ be a nonzero proper ideal of a commutative ring $R$.*

(a) *If $\Gamma(R/I)$ is the graph on one vertex, then $\kappa(\Gamma_I(R)) = |I| - 1$;*

(b) *If $\Gamma(R/I)$ has at least two vertices, then $2 \leq \kappa(\Gamma_I(R)) \leq |I| \cdot \kappa(\Gamma(R/I))$;*

(c) $|I| - 1 \leq \kappa(\Gamma_I(R))$.

Therefore, if $I \neq (0)$ is a proper ideal, but not a prime ideal, then $|I| - 1 \leq \kappa(\Gamma_I(R)) \leq (|I| \cdot \kappa(\Gamma(R/I)))$. In particular, $\kappa(\Gamma_I(R)) = \infty$ if $I$ is infinite.

**Theorem 5.2.22.** ([323, Proposition 3.5]) *Let $I$ be an ideal of a commutative ring $R$. Then $\Gamma_I(R)$ has a bridge if and only if either* (i) *$\Gamma_I(R)$ is the graph on two vertices, or* (ii) *$I = (0)$ and $\Gamma(R)$ has a bridge.*

Next, we classify when the ideal-based zero-divisor graph of a commutative ring is complemented or uniquely complemented. This was done by J. Smith in [352]. One can refer to Section 3.6.1 for the definitions of complemented and uniquely complemented graphs. We consider the situation in two cases: either $I$ is a radical ideal of $R$ or $I$ is not a radical ideal of $R$.

**Theorem 5.2.23.** ([352, Proposition 2.1]) *Let $R$ be a commutative ring and $I$ a nonzero, nonradical ideal of $R$. If $|V(\Gamma(R/I))| = 2$, then $\Gamma_I(R)$ is not complemented.*

**Theorem 5.2.24.** ([352, Theorem 2.3]) *Let $R$ be a commutative ring and $I$ a nonradical ideal of $R$. Then $\Gamma_I(R)$ is complemented if and only $\Gamma_I(R) = K_2$.*

**Theorem 5.2.25.** ([352, Theorem 2.4]) *Let $R$ be a commutative ring and $I$ a nonzero, nonprime ideal of $R$. Then $\Gamma_I(R)$ is complemented and $|V(\Gamma(R/I))| = 2$ if and only if $\Gamma(R/I)$ is complemented and $\sqrt{I} = I$.*

**Corollary 5.2.26.** ([352, Corollary 2.5]) *Let $R$ be a commutative ring and $I$ a proper nonzero nonprime ideal of $R$. Then $\Gamma_I(R)$ is complemented if and only if exactly one of the following statements holds:*

(a) *$R/I \cong \mathbb{Z}_4$ or $R/I \cong \mathbb{Z}_2[x]/(x^2)$, and $|I| = 2$;*

(b) *$\Gamma(R/I)$ is complemented and $I$ is a radical ideal of $R$.*

**Corollary 5.2.27.** ([352, Corollary 2.6]) *Let $R$ be a commutative ring and $I$ a radical ideal of $R$. Then the following statements are equivalent:*

(a) *$\Gamma_I(R)$ is complemented;*

(b) *$\Gamma(R/I)$ is complemented;*

(c) *$\Gamma(R/I)$ is uniquely complemented;*

(d) *$T(R/I)$ is von Neumann regular.*

Based on the preceding results, one can see that when $I$ is a radical ideal, then $\Gamma_I(R)$ is uniquely complemented if and only $\Gamma_I(R)$ is complemented. Actually, the following statement proves the same.

**Theorem 5.2.28.** ([352, Theorem 2.9]) *Let $R$ be a commutative ring and $I$ a radical ideal of $R$. Then $\Gamma_I(R)$ is complemented if and only if $\Gamma_I(R)$ is uniquely complemented.*

**Theorem 5.2.29.** ([352, Theorem 2.10]) *Let $R$ be a commutative ring and $I$ a proper radical ideal of $R$. Then the following statements are equivalent:*

(a) *$\Gamma_I(R)$ is complemented;*

(b) *$\Gamma_I(R)$ is uniquely complemented;*

(c) *$\Gamma(R/I)$ is complemented;*

(d) *$\Gamma(R/I)$ is uniquely complemented;*

(e) *$T(R/I)$ is von Neumann regular.*

*Moreover, regardless if $I$ is a radical or nonradical ideal, $\Gamma_I(R)$ is complemented if and only if $\Gamma_I(R)$ is uniquely complemented.*

## 5.3 Genus of Ideal-based Zero-divisor Graphs

One simple question which one may ask is "What kind of rings can have $g(\Gamma_I(R)) = k$?". First, let us study the planarity of $\Gamma_I(R)$. In view of Theorem 5.1.26, it is clear that $\Gamma(R)$ is planar if $\Gamma(R)$ has 5 or fewer vertices. Since $\Gamma(\mathbb{Z}_{49}) \cong K_6$, we cannot improve on this value. Note that every star graph is planar. Therefore, by Theorem 2.2.10, one can produce arbitrarily large planar zero-divisor graphs. The next theorem, which provides a necessary and sufficient condition for the planarity of $\Gamma_I(R)$, was proved by Redmond [323].

**Theorem 5.3.1.** ([323, Theorem 7.2]) *Let $I$ be a proper, nonzero ideal of a commutative ring $R$ that is not a prime ideal. Then $\Gamma_I(R)$ is planar if and only if $\omega(\Gamma(R/I)) \leq 2$ (i.e., $\Gamma(R/I)$ has no cycles) and either* (a) $|I| = 2$ *or* (b) *$\Gamma(R/I)$ consists of a single vertex and $|I| \leq 4$.*

The above theorem shows that one can produce arbitrarily large planar ideal-based zero-divisor graphs. For example, let $F$ be a finite field and define $R = \mathbb{Z}_2 \times \mathbb{Z}_2 \times F$ and $I = \mathbb{Z}_2 \times \{0\} \times \{0\}$. Then $\Gamma(R/I)$ is a star graph on $|F|$ vertices and $\Gamma_I(R)$ is a planar graph on $2 \cdot |F|$ vertices. Also, Theorem 5.3.1 states that $\omega(G) \leq 2$ if and only if $G$ contains no cycles. However, this statement is wrong since $G = C_4$ (the cycle of length 4) satisfies $\omega(G) \leq 2$. Hence Theorem 5.3.1 should be restated as follows (a detailed proof along with the possible finite planar ideal-based zero-divisor graphs is given in [353]).

**Theorem 5.3.2.** (cf. [353, Theorem 6]) *Let $I$ be a proper, nonzero ideal of a commutative ring $R$ that is not a prime ideal. Then $\Gamma_I(R)$ is planar if and only if $\Gamma(R/I)$ contains no cycles and either* (a) $|I| = 2$ *or* (b) *$\Gamma(R/I)$ consists of a single vertex and $|I| \leq 4$.*

Next, we discuss the genus one ideal-based zero-divisor graph of a commutative ring, which was done by Wang in [422].

Let us start with a new notation. Let $G$ be a finite simple graph with vertex set $\{v_1, \ldots, v_m\}$ and edge set $E(G)$. Then $G^{(t)}$ is the finite simple graph with vertex set $\{v_{ij} \mid 1 \leq i \leq t,\ 1 \leq j \leq m\}$ and edge set $\{v_{ij}v_{kl} \mid j \neq l,\ v_jv_l \in E(G)\}$. Note that, if $G = K_n$ then $G^{(t)} = K_{t,\ldots,t}$ and if $G = K_{m,n}$ then $G^{(t)} = K_{mt,nt}$.

Now, we see some necessary and sufficient conditions for a finite commutative ring $R$ and a nonzero ideal $I$ of $R$ to have $g(\Gamma_I(R)) = 1$. First, we start with some observations regarding the genus of a zero-divisor graph.

**Lemma 5.3.3.** ([422, Lemma 4.6]) *Let $(R, \mathfrak{m})$ be a finite commutative local ring with $\mathfrak{m}^2 \neq 0$ and $|R/\mathfrak{m}| \geq 3$. Then $g(\Gamma(R)^{(2)}) \geq 2$.*

**Lemma 5.3.4.** ([422, Lemma 4.8]) *Let $R \cong \mathbb{Z}_2 \times S$, where $S$ is a finite local ring. Then the following hold:*

(a) *If $|\Gamma(S)| = 1$, then $\Gamma(R)$ is planar and contains no cycles;*

(b) *If* $|\Gamma(S)| \geq 2$, *then* $K_3$ *and* $K_{2,3}$ *are subgraphs of* $\Gamma(R)$.

**Lemma 5.3.5.** ([422, Lemma 4.9], cf. Example 2.1.3(c)) *Let* $(R, \mathfrak{m})$ *be a finite commutative local ring and* $\Gamma(R) = K_3$. *Then* $R \cong \mathbb{Z}_2[x,y]/(x^2, xy, y^2)$, $\mathbb{Z}_4[x]/(x^2, 2x)$, $\mathbb{F}_4[x]/(x^2)$, *or* $\mathbb{Z}_4[x]/(x^2+x+1)$. *In particular,* $\mathfrak{m}^2 = 0$.

**Theorem 5.3.6.** ([422, Lemma 4.10]) *Let* $R = R_1 \times \cdots \times R_n$, *where* $R_i$ *is a finite local ring for every* $i$ *and* $n \geq 3$. *If* $g(\Gamma(R)^{(2)}) \leq 1$, *then* $n = 3$ *and* $R_i \cong \mathbb{Z}_2$ *for every* $i$.

Now let us see some concrete examples for which $g(\Gamma_I(R)) = 1$.

**Example 5.3.7.** ([422, Example 4.11]) Let $R$ be a finite commutative ring and $I$ be a proper nonzero ideal of $R$. Then the following hold:

(a) Suppose that $R/I \cong \mathbb{Z}_2 \times \mathbb{Z}_2 \times \mathbb{Z}_2$ and $|I| = 2$. Then $g(\Gamma_I(R)) = 1$;

(b) Suppose that $R/I \cong \mathbb{Z}_{16}$ and $|I| = 2$. Then $g(\Gamma_I(R)) = 1$;

(c) Suppose that $\Gamma(R/I) = P_3$, $R/I$ is local, and $|I| = 3$. Then $g(\Gamma_I(R)) = 1$;

(d) Suppose that $R/I \cong \mathbb{Z}_9$ or $R/I \cong \mathbb{Z}_2[x]/(x^3)$ and $|I| = 3$. Then $g(\Gamma_I(R)) = 1$.

**Example 5.3.8.** ([422, Example 4.13]) Let $R$ be a finite commutative ring and $I$ be a proper nonzero ideal of $R$.

(a) If $R/I \cong \frac{\mathbb{Z}_4[x,y]}{(x^2,y^2,xy-2)}, \frac{\mathbb{Z}_2[x,y]}{(x^2,y^2)}$, or $\frac{\mathbb{Z}_4[x]}{(x^2)}$, then $g(\Gamma_I(R)) \geq 2$;

(b) If $R/I \cong \frac{\mathbb{Z}_2[x,y]}{(x^3,y^2-x^2,xy)}, \frac{\mathbb{Z}_4[x]}{(x^3,x^2-2x)}, \frac{\mathbb{Z}_4[x,y]}{(x^3,x^2-2,xy,y^2-2)}$, or $\frac{\mathbb{Z}_8[x]}{(x^2-4,2x)}$, then $g(\Gamma_I(R)) \geq 2$.

**Lemma 5.3.9.** ([422, Lemma 4.14]) *Let* $R$ *be a finite commutative ring and* $I$ *a proper nonzero ideal of* $R$. *If* $g(\Gamma(R/I)) \geq 2$, *then* $g(\Gamma_I(R)) \geq 2$.

**Lemma 5.3.10.** ([422, Lemma 4.15]) *Let* $R$ *be a finite commutative ring and* $I$ *a proper nonzero ideal of* $R$. *Suppose that the following hold:*

(a) $R/I$ *is local with unique maximal ideal* $\mathfrak{m}/I$;

(b) $|R/\mathfrak{m}| = 2$;

(c) $g(\Gamma(R/I)) = 1$.

*Then* $g(\Gamma_I(R)) \geq 2$.

**Theorem 5.3.11.** ([422, Theorem 4.16]) *Let* $R$ *be a finite commutative ring and* $I$ *be a proper nonzero ideal of* $R$. *Suppose that* $\omega(\Gamma(R/I)) \geq 4$ *or* $g(\Gamma(R/I)) \geq 1$. *Then* $g(\Gamma_I(R)) \geq 2$.

Thus, if $g(\Gamma_I(R)) \leq 1$, then $\omega(\Gamma(R/I)) \leq 3$. The next theorem characterizes all finite commutative rings $R$ with $\omega(\Gamma(R/I)) \leq 2$ and $g(\Gamma_I(R)) = 1$.

**Theorem 5.3.12.** ([422, Theorem 4.17]) *Let $R$ be a finite commutative ring and $I$ a proper nonzero ideal of $R$ which is not a prime ideal. Suppose that $\omega(\Gamma(R/I)) \leq 2$. Then $g(\Gamma_I(R)) = 1$ if and only if one of the following holds:*

(a) $R/I \cong \mathbb{Z}_3 \times \mathbb{Z}_3$ *and* $|I| \leq 2$;

(b) $R/I \cong \mathbb{Z}_2 \times \mathbb{Z}_2$ *and* $|I| \leq 4$;

(c) $R/I \cong \mathbb{Z}_2 \times \mathbb{Z}_3$ *and* $|I| \leq 3$;

(d) $R/I \cong \mathbb{Z}_2 \times \mathbb{Z}_4$ *and* $|I| \leq 2$;

(e) $R/I \cong \mathbb{Z}_2 \times \mathbb{Z}_2[x]/(x^2)$ *and* $|I| \leq 2$;

(f) $R/I \cong \mathbb{Z}_2 \times \mathbb{F}_q$ *with* $q \geq 4$ *and* $|I| \leq 2$;

(g) $R/I \cong \mathbb{Z}_4$ *and* $|I| \leq 7$;

(h) $R/I \cong \mathbb{Z}_2[x]/(x^2)$ *and* $|I| \leq 7$;

(i) $R/I \cong \mathbb{Z}_9$ *and* $|I| \leq 3$;

(j) $R/I \cong \mathbb{Z}_3[x]/(x^2)$ *and* $|I| \leq 3$;

(k) $R/I \cong \mathbb{Z}_8$ *and* $|I| \leq 3$;

(l) $R/I \cong \mathbb{Z}_2[x]/(x^3)$ *and* $|I| \leq 3$;

(m) $R/I \cong \mathbb{Z}_4[x]/(x^3, x^2 - 2)$ *and* $|I| \leq 3$.

The next result characterizes all finite commutative rings $R$ with $\omega(\Gamma(R/I)) = 3$ and $g(\Gamma_I(R)) = 1$.

**Theorem 5.3.13.** ([422, Theorem 4.18]) *Let $R$ be a finite commutative ring and $I$ be a proper nonzero ideal of $R$. Suppose that $\omega(\Gamma(R/I)) = 3$. Then $g(\Gamma_I(R)) = 1$ if and only if $|I| = 2$ and $R/I$ is isomorphic to one of the following rings:*

$$\mathbb{Z}_2 \times \mathbb{Z}_2 \times \mathbb{Z}_2,\ \mathbb{Z}_{16},\ \frac{\mathbb{Z}_2[x,y]}{(x^2, xy, y^2)},\ \frac{\mathbb{Z}_4[x]}{(x^2, 2x)},\ \frac{\mathbb{F}_4[x]}{(x^2)},\ \frac{\mathbb{Z}_4[x]}{(x^2+x+1)}.$$

The nonorientable genus of the ideal-based zero-divisor graph of a commutative ring was investigated by Tamizh Chelvam and Nithya in [383]. Recall that $\tilde{S}_k$ denotes the nonoriented surface with $k$ crosscaps, where $k$ is a nonnegative integer. Then the *crosscap number* of a graph $G$, denoted $\tilde{g}(G)$, is the minimum integer $n$ such that the graph $G$ can be embedded in $\tilde{S}_n$. For more details on the nonorientable genus, see Section 4.6. First, the authors determined all finite quotient rings $R/I$ and ideals $I$ of $R$ for which the crosscap number of $\Gamma_I(R)$ is at most one.

The following few results provide effective criterion for discussing crosscap one in the finite case.

**Lemma 5.3.14.** ([383, Lemma 2.6]) *Let $R$ be a commutative ring and $I$ a nonzero ideal of $R$ such that $R/I$ is a finite local ring with maximal ideal $\mathfrak{m}/I$, $(\mathfrak{m}/I)^2 \neq (0)$, and $|R/\mathfrak{m}| \geq 3$. Then $\tilde{g}(\Gamma_I(R)) > 1$.*

**Lemma 5.3.15.** ([383, Lemma 2.7]) *Let $R \cong \mathbb{Z}_2 \times S$, where $S$ is a finite commutative local ring. Then the following hold:*

(a) *If $|\Gamma(S)| \leq 1$, then $\Gamma(R)$ is planar and contains no cycles;*

(b) *If $|\Gamma(S)| > 1$, then $K_3$ and $K_{2,2}$ are subgraphs of $\Gamma(R)$.*

**Lemma 5.3.16.** ([383, Lemma 2.8]) *Let $I$ be a nonzero ideal of a commutative ring $R$ and $R/I \cong R_1 \times \cdots \times R_n$, where $R_i$ is a finite local ring for every $i$ and $n \geq 3$. If $\tilde{g}(\Gamma_I(R)) \leq 1$, then $n = 3$ and $R_i \cong \mathbb{Z}_2$ for every $i$.*

**Lemma 5.3.17.** ([383, Lemma 2.9]) *Let $R$ be a finite commutative ring and $I$ a nonzero ideal of $R$. Suppose the following hold:*

(a) *$R/I$ is local with maximal ideal $\mathfrak{m}/I$;*

(b) *$|R/\mathfrak{m}| = 2$;*

(c) *$\tilde{g}(\Gamma(R/I)) = 1$.*

*Then $\tilde{g}(\Gamma_I(R)) > 1$.*

The following theorem plays on important role in classifying all the finite quotient rings $R/I$ and ideals $I$ of $R$ for which the crosscap number of $\Gamma_I(R)$ is at most one.

**Theorem 5.3.18.** ([383, Theorem 2.10]) *Let $R$ be a finite commutative ring and $I$ a nonzero ideal of $R$. Suppose that $\omega(\Gamma(R/I)) \geq 4$ or $\tilde{g}(\Gamma(R/I)) \geq 1$. Then $\tilde{g}(\Gamma_I(R)) > 1$.*

Thus, we have $\tilde{g}(\Gamma_I(R)) \leq 1$ implies that $\omega(\Gamma(R/I)) \leq 3$. The following theorem determines all finite quotient rings $R/I$ and ideals $I$ of $R$ with $\omega(\Gamma(R/I)) \leq 2$ for which the crosscap number of $\Gamma_I(R)$ is at most one.

**Theorem 5.3.19.** ([383, Theorem 2.11]) *Let $R$ be a finite commutative ring and $I$ a nonzero ideal of $R$. Suppose that $\omega(\Gamma(R/I)) \leq 2$. Then $\tilde{g}(\Gamma_I(R)) \leq 1$ if and only if one of the following holds:*

(a) *$R/I \cong \mathbb{Z}_2 \times \mathbb{Z}_2$ and $|I| \leq 3$;*

(b) *$R/I \cong \mathbb{Z}_2 \times \mathbb{Z}_4$ and $|I| = 2$;*

(c) *$R/I \cong \mathbb{Z}_2 \times \mathbb{Z}_2[x]/(x^2)$ and $|I| = 2$;*

(d) *$R/I \cong \mathbb{Z}_2 \times \mathbb{F}_q$, $q \geq 3$, and $|I| = 2$;*

(e) *$R/I \cong \mathbb{Z}_4$ and $|I| \leq 6$;*

(f) *$R/I \cong \mathbb{Z}_2[x]/(x^2)$ and $|I| \leq 6$;*

(g) *$R/I \cong \mathbb{Z}_9$ and $|I| \leq 3$;*

(h) *$R/I \cong \mathbb{Z}_3[x]/(x^2)$ and $|I| \leq 3$;*

(i) *$R/I \cong \mathbb{Z}_8$ and $|I| = 2$;*

(j) $R/I \cong \mathbb{Z}_2[x]/(x^3)$ *and* $|I| = 2$;

(k) $R/I \cong \mathbb{Z}_4[x]/(x^2 - 2, x^3)$ *and* $|I| = 2$.

The following example illustrates the embedding of $\Gamma_I(R)$ in $\tilde{S}_1$.

**Example 5.3.20.** (a) Consider the ring $R = \mathbb{Z}_2 \times \mathbb{Z}_2 \times \mathbb{Z}_2 \times \mathbb{Z}_2$ and ideal $I = \{0\} \times \{0\} \times \{0\} \times \mathbb{Z}_2$. Label the vertices of $\Gamma_I(R)$ as $u_1 = (1,0,0,0)$, $u_2 = (0,1,0,0)$, $v_1 = (1,0,0,1)$, $v_2 = (0,1,0,1)$, $v_3 = (0,0,1,1)$, $v_4 = (0,1,1,0)$, $v_5 = (0,1,1,1)$, $v_6 = (1,0,1,0)$, $v_7 = (1,0,1,1)$, $v_8 = (1,1,0,0)$, and $v_9 = (1,1,0,1)$. Then the embedding of $\Gamma_I(R)$ in $\tilde{S}_1$ is given in Figure 5.7. Note that in this case, $R/I \cong \mathbb{Z}_2 \times \mathbb{Z}_2 \times \mathbb{Z}_2$.

(b) Let $R = \mathbb{Z}_{16} \times \mathbb{Z}_2$ and $I = \{0\} \times \mathbb{Z}_2$. Label the vertices of $\Gamma_I(R)$ as $u_1 = (4,0)$, $u_2 = (8,0)$, $u_3 = (12,0)$, $v_1 = (4,1)$, $v_2 = (8,1)$, $v_3 = (12,1)$, $v_4 = (2,0)$, $v_5 = (6,0)$, $v_6 = (10,0)$, $v_7 = (14,0)$, $v_8 = (2,1)$, $v_9 = (6,1)$, $v_{10} = (10,1)$, and $v_{11} = (14,1)$. Then the embedding of $\Gamma_I(R)$ in $\tilde{S}_1$ is given in Figure 5.8. In this case, $R/I \cong \mathbb{Z}_{16}$.

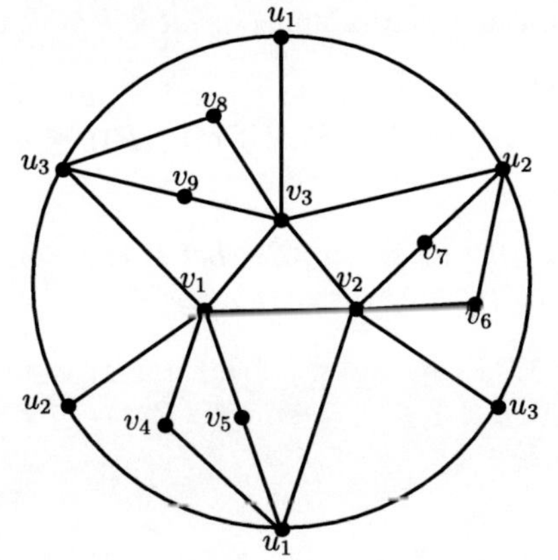


Figure 5.7: Embedding of $\Gamma_{\{0\}\times\{0\}\times\{0\}\times\mathbb{Z}_2}(\mathbb{Z}_2 \times \mathbb{Z}_2 \times \mathbb{Z}_2 \times \mathbb{Z}_2)$ in $\tilde{S}_1$

The following theorem determines all finite quotient rings $R/I$ with $\omega(\Gamma(R/I)) = 3$ and $\tilde{g}(\Gamma_I(R)) \leq 1$.

**Theorem 5.3.21.** ([383, Theorem 2.12]) *Let $R$ be a finite commutative ring and $I$ be a nonzero ideal of $R$. Suppose that $\omega(\Gamma(R/I)) = 3$. Then $\tilde{g}(\Gamma_I(R)) \leq 1$ if and only if $|I| = 2$ and $R/I$ is isomorphic to one of the following:*

$$\mathbb{Z}_2 \times \mathbb{Z}_2 \times \mathbb{Z}_2,\ \mathbb{Z}_{16},\ \frac{\mathbb{Z}_2[x,y]}{(x^2, xy, y^2)},\ \frac{\mathbb{Z}_4[x]}{(2x, x^2)},\ \frac{\mathbb{F}_4[x]}{(x^2)},\ \frac{\mathbb{Z}_4[x]}{(x^2+x+1)}.$$

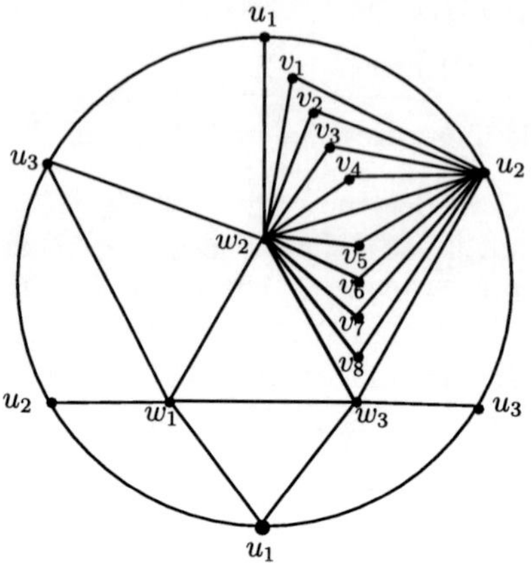


Figure 5.8: Embedding of $\Gamma_{\{0\}\times\mathbb{Z}_2}(\mathbb{Z}_{16}\times\mathbb{Z}_2)$ in $\tilde{S}_1$

We conclude the section with some sufficient conditions for a commutative ring (not necessarily finite) $R$ and a nonzero ideal $I$ of $R$ to have $\tilde{g}(\Gamma_I(R)) \leq 1$.

**Theorem 5.3.22.** ([383, Proposition 3.1]) *Let $P_1$ and $P_2$ be prime ideals of a commutative ring $R$ and $I = P_1 \cap P_2$. If $\tilde{g}(\Gamma_I(R)) \leq 1$, then $|P_1 \setminus I| \leq 3$ or $|P_2 \setminus I| \leq 3$.*

**Theorem 5.3.23.** ([383, Proposition 3.2]) *Let $I$ be a nonzero ideal of a commutative ring $R$. If $|Ass(R/I)| \leq 3$, then $\tilde{g}(\Gamma_I(R)) \leq 1$.*

**Theorem 5.3.24.** ([383, Proposition 3.3]) *Let $I$ be a nonzero ideal of a commutative ring $R$. If $\tilde{g}(\Gamma_I(R)) \leq 1$, then $\omega(\Gamma(R/I)) \leq 3$ and either $|I| \leq 3$ or $|V(\Gamma(R/I))| = 1$ with $|I| \leq 6$.*

## 5.4 Primal Ideal-based Zero-divisor Graphs

Atani and Darani [109] studied the properties of $\Gamma_I(R)$, where $I$ is a primal or weakly primal ideal of a commutative ring $R$. First, we recall the basic concepts of primal and weakly primal ideals of a commutative ring $R$. A proper ideal $P$ of $R$ is said to be *weakly prime* if $0 \neq ab \in P$ implies that $a \in P$ or $b \in P$ [61]. However, since (0) is always weakly prime (by definition), a weakly prime ideal need not be prime. An $a \in R$ is called *prime* (resp., *weakly prime*) to an ideal $I$ of $R$ if $ra \in I$ (resp., $0 \neq ra \in I$) (where $r \in R$) implies that $r \in I$. Denote by $S(I)$ (resp., $W(I)$) the set of elements of $R$ that are not prime (resp., are not weakly prime) to $I$. A proper ideal $I$ of $R$ is said to be *primal* if $S(I)$ forms an ideal (so (0) is not necessarily primal) [198]; this ideal is always a prime ideal, called the *adjoint ideal* $P$ of $I$. In this case, we also say that $I$ is a *P-primal ideal* of $R$. Note that if $r \in R$ and $a \in S(I)$, then clearly $ra \in S(I)$. So what we require for $I$ to be primal is that if $a$ and $b$ are not prime to $I$, then their difference is also not prime to $I$. For example, let $R = \mathbb{Z}$ and $I = 6\mathbb{Z}$. Then 2 and 3 are not prime to $I$, but $3 - 2$ is prime to $I$; so $I$ is not primal. A ring $R$ is said to be *primal* if the zero ideal is a primal ideal of $R$. Also, a proper ideal $I$ of $R$ is called *weakly primal* if the set $P = W(I) \cup \{0\}$ forms an ideal; this ideal is always a weakly prime ideal ([108, Proposition 4]). In this case, we also say that $I$ is a $P$-weakly primal ideal. If $R$ is not an integral domain, then (0) is a 0-weakly primal ideal of $R$ (by definition); so a weakly primal ideal need not be primal. For more details about primal and weakly primal ideals, see [108, 198]. The goal of Atani et al. [109] is to show that if $I$ is primal (resp., weakly primal), then $\Gamma_I(R)$ commutes with localization. That is, $S^{-1}(\Gamma_I(R)) = \Gamma_{S^{-1}I}(S^{-1}R)$, where $S$ is a multiplicatively closed subset of $R$ and $S^{-1}I = \{a/s \mid a \in I,\ s \in S\} \subseteq S^{-1}R$. (By abuse of notation, we will sometimes denote the vertices $V(G)$ of a graph $G$ simply by $G$.)

In this regard, Atani et al. [109] investigated the ideal-based zero-divisor graph with respect to primal ideals. The class of primal ideals is a large class. For example, all primary ideals and irreducible ideals are primal.

**Theorem 5.4.1.** ([109, Theorem 2.3]) *Let $I$ be an ideal of a commutative ring $R$. Then $I$ is a primal ideal of $R$ if and only if $\Gamma_I(R) \cup I$ is an (prime) ideal of $R$.*

Let $I, J$ be ideals of a ring $R$. It is natural to ask whether $\Gamma_I(R) = \Gamma_J(R)$ implies $I = J$. In the case of $P$-primal ideals, we have the following theorem.

**Theorem 5.4.2.** ([109, Theorem 2.5]) *Let $I$ and $J$ be $P$-primal ideals of a commutative ring $R$. Then $\Gamma_I(R) = \Gamma_J(R)$ if and only if $I = J$.*

**Theorem 5.4.3.** ([109, Theorem 2.6]) *Let $I$ be an ideal of a commutative ring $R$. Then $I$ is primary if and only if $\Gamma_I(R) = \sqrt{I} \setminus I$.*

**Theorem 5.4.4.** ([109, Theorem 2.7]) *Let $I$ be an ideal of a commutative ring $R$. Then*

(a) (0) *is a primal ideal of $R$ if and only $Z(R)$ is an (prime) ideal of $R$;*

(b) *$I$ is a primal ideal of $R$ if and only if $Z(R/I)$ is an ideal of $R/I$.*

By Theorem 5.1.27, we have $diam(\Gamma_I(R)) \leq 3$ for an ideal $I$ of a commutative ring $R$. But for a primal ideal we have the following results.

**Theorem 5.4.5.** ([109, Theorem 2.8]) *Let $R$ be a commutative Noetherian ring. If $I$ is a primal ideal of $R$, then $diam(\Gamma(R/I)) \leq 2$.*

**Theorem 5.4.6.** ([109, Corollary 2.13]) *Let $I$ be a primal radical ideal of a commutative Noetherian ring $R$. Then $diam(\Gamma_I(R)) \leq 2$.*

The following result gives the girth of $\Gamma_I(R)$ when $I$ is a quasi-primary ideal (i.e., $\sqrt{I}$ is a prime ideal).

**Theorem 5.4.7.** ([175, Theorem 2.5]) *Let $I$ be a quasi-primary ideal of a commutative ring $R$ with $\sqrt{I} \subsetneq S(I)$ and $|\sqrt{I} \setminus I| \geq 2$. Then $gr(\Gamma_I(R)) = 3$ or $\infty$.*

The following theorem due to Atani et al. [109] shows that the zero-divisor graph with respect to $P$-primal ideals commutes with localization.

**Theorem 5.4.8.** ([109, Theorem 2.15]) *Assume that $S$ is a multiplicatively closed subset of a commutative ring $R$, and let $I$ be a $P$-primal ideal of $R$ with $P \cap S = \emptyset$. Then $S^{-1}(\Gamma_I(R)) = \Gamma_{S^{-1}I}(S^{-1}R)$.*

**Theorem 5.4.9.** ([109, Theorems 2.16 and 2.18]) *Assume that $S$ is a multiplicatively closed subset of a commutative ring $R$ which consists of regular elements of $R$, and let $I$ be a $P$-primal ideal of $R$ with $P \cap S = \emptyset$. Then $diam(\Gamma_I(R)) = diam(\Gamma_{S^{-1}I}(S^{-1}R))$ and $gr(\Gamma_I(R)) = gr(\Gamma_{S^{-1}I}(S^{-1}R))$.*

Next, we study the diameter of $\Gamma_I(R)$ when $I$ is not a primal ideal. Let $I$ be an ideal of a ring $R$. An ideal $J$ of $R$ is called *prime to $I$* if $(I :_R J) = I$.

**Theorem 5.4.10.** ([109, Theorem 3.3]) *Let $I$ be a radical ideal of a commutative ring $R$, and suppose that $I$ is not a primal ideal of $R$ and $|Min(I)| \geq 3$. Then $diam(\Gamma_I(R)) = 3$.*

**Theorem 5.4.11.** ([109, Theorem 3.4]) *Assume that $I$ is a radical ideal of a commutative ring $R$ and $I$ is not a primal ideal of $R$. Then $diam(\Gamma_I(R)) \leq 2$ if and only if $|Min(I)| = 2$.*

**Theorem 5.4.12.** ([109, Theorem 3.7]) *Let $I$ be an ideal of a commutative ring $R$ which is not a radical ideal, and suppose that $I$ is not a primal ideal of $R$. Then $diam(\Gamma_I(R)) = 3$.*

In the next few results, we study the ideal-based zero-divisor graph with respect to weakly prime and weakly primal ideals.

**Theorem 5.4.13.** ([109, Theorem 4.1]) *Let $R$ be a finite commutative local ring. Then $\Gamma(R)$ is a complete graph if and only if every proper ideal of $R$ is weakly prime.*

For an ideal $I$ of a commutative ring $R$, let $Z_I(R) = \{r \in R \setminus I \mid ra = 0 \text{ for some } a \in R \setminus I\}$.

**Theorem 5.4.14.** ([109, Theorem 4.3]) *Let $I$ be a an ideal of a commutative ring $R$, and let $P$ be a weakly prime ideal of $R$ with $W(I) \subseteq P$ and $(P \setminus I) \cap Z_I(R) = \emptyset$. Then $\Gamma_I(R) = (P \setminus I) \cup Z_I(R)$ if and only if $I$ is a $P$-weakly primal ideal of $R$.*

Consider the ring $R = \mathbb{Z}_{24}$, $I = (8)$, and $P = (2)$. Then $I$ is a $P$-primal ideal of $R$. Hence, $\Gamma_I(R) = P \setminus I$. It is easy to check that $\Gamma_I(R) = Z_I(R)$. Note that $I$ is not a weakly prime ideal of $R$ because $0 \neq 2 \cdot 4 \in I$ with $2, 4 \notin I$. This example shows that if $\Gamma_I(R) = Z_I(R)$, then $I$ need not necessarily be weakly prime. But the converse holds:

**Theorem 5.4.15.** ([109, Theorem 4.4]) *Let $I$ be a weakly prime ideal of a commutative ring $R$. Then $\Gamma_I(R) = Z_I(R)$. In particular, $\Gamma_I(R)$ is a subgraph of $\Gamma(R)$.*

The next two results concern the ideal-based zero-divisor graph with respect to a $P$-weakly prime primal ideal.

**Theorem 5.4.16.** ([109, Theorem 4.6]) *Let $S$ be a multiplicatively closed subset of a commutative ring $R$, and let $I$ be a $P$-weakly primal ideal of $R$ with $P \cap S = \emptyset$. Then $S^{-1}(\Gamma_I(R)) = \Gamma_{S^{-1}I}(S^{-1}R)$.*

**Theorem 5.4.17.** ([109, Theorems 4.7 and 4.8]) *Let $S$ be a multiplicatively closed subset of a commutative ring $R$ which consists of regular elements of $R$, and let $I$ be a $P$-weakly primal ideal of $R$ with $P \cap S = \emptyset$. Then $diam(\Gamma_I(R)) = diam(\Gamma_{S^{-1}I}(S^{-1}R))$ and $gr(\Gamma_I(R)) = gr(\Gamma_{S^{-1}I}(S^{-1}R))$.*

## 5.5 Annihilating-ideal Graphs

The graph of zero-divisors for commutative rings has been generalized to the annihilating-ideal graph of commutative rings. The concept of the annihilating-ideal graph of a commutative ring was introduced by Behboodi and Rakeei in [139]. (However, this concept was also given in a 2002 unpublished manuscript by U. Vishne [415].) In ring theory, the structure of a ring is sometimes more closely tied to the behavior of its ideals than its elements. In fact, ideals often reveal structure that is obscured by only considering elements; so it is natural to define a graph whose vertices are ideals instead of elements. Therefore, the authors of [139] replaced the vertices of the zero-divisor graph by the nonzero annihilating-ideals. Hence annihilating-ideal graphs of rings provide an excellent setting for studying some aspects of algebraic properties of a commutative ring, especially, the ideal structure of a ring.

We denote the set of all ideals which are a subset of an ideal $J$ of $R$ by $\mathbb{I}(J)$. Therefore, the set of all ideals of $R$ is $\mathbb{I}(R)$. Note that $\mathbb{I}(R)$ is a commutative monoid with zero element under the binary operation of ideal multiplication. We call an ideal $I$ of $R$ an *annihilating-ideal* if there exists a nonzero ideal $J$ of $R$ such that $IJ = (0)$, and use the notation $\mathbb{A}(R)$ for the set of all annihilating-ideals of $R$. Note that $\mathbb{A}(R)^* = \mathbb{I}(R) \setminus \{0, R\}$ if $R$ is Artinian. Recall that an ideal $I$ of a ring $R$ is called an *annihilator ideal* if $I = Ann(x)$ for some $x \in R$. Let $soc(R)$ be the sum of all minimal ideals of $R$ (if there are no minimal ideals, this

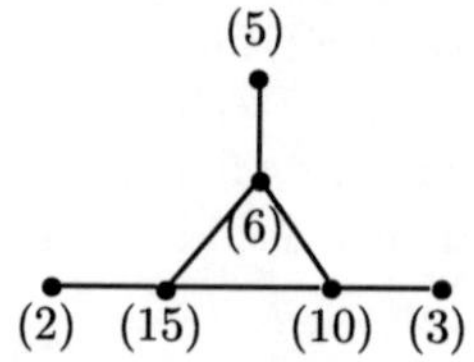


Figure 5.9: The annihilating-ideal graph of $\mathbb{Z}_{30}$

sum is defined to be zero). Also, if $X$ is an element or a subset of $R$, define the annihilator of $X$ in $R$ by $Ann(X) = \{r \in R \mid rX = (0)\}$.

The *annihilating-ideal graph* of $R$, denoted by $\mathbb{AG}(R)$, is the graph with vertices $\mathbb{A}(R)^* = \mathbb{A}(R) \setminus \{(0)\}$, where distinct vertices $I$ and $J$ are adjacent if and only if $IJ = (0)$. Thus $\mathbb{AG}(R)$ is the empty graph if and only if $R$ is an integral domain. Note that $\mathbb{AG}(R) = \Gamma(\mathbb{I}(R))$, i.e., $\mathbb{AG}(R)$ is the zero-divisor graph of the semigroup $\mathbb{I}(R)$ (see Section 6.2 for zero-divisor graphs of semigroups). In particular, $diam(\mathbb{AG}(R)) \in \{0, 1, 2, 3\}$ and $gr(\mathbb{AG}(R)) \in \{3, 4, \infty\}$ by Theorem 6.2.2. In this section, we discuss the interplay between the graph-theoretic properties of $\mathbb{AG}(R)$ and the ring-theoretic properties of $R$.

The annihilating-ideal graph of $\mathbb{Z}_{30}$ is shown in Fig 5.9.

First, we state some finiteness conditions for $\mathbb{AG}(R)$. For instance, the following result states that if $R$ is not an integral domain, then $\mathbb{AG}(R)$ has ACC (resp., DCC) on vertices if and only if $R$ is Noetherian (resp., Artinian). We say that the annihilating-ideal graph $\mathbb{AG}(R)$ has ACC (resp., DCC) on vertices if $R$ has ACC (resp., DCC) on $\mathbb{A}(R)^*$.

**Theorem 5.5.1.** ([139, Theorem 1.1]) *Let $R$ be a commutative ring that is not an integral domain. Then $\mathbb{AG}(R)$ has ACC (resp., DCC) on vertices if and only if $R$ is a Noetherian (resp., Artinian) ring.*

In view of above the theorem and Cohen's theorem, one may naturally ask: "If $Spec(R) \cap \mathbb{A}(R)^* \neq \emptyset$ and every prime ideal $P \in \mathbb{A}(R)^*$ is finitely generated, is $R$ Noetherian?". The answer is no! The following example gives a nonNoetherian ring $R$ for which every prime vertex of $\mathbb{AG}(R)$ is finitely generated.

**Example 5.5.2.** ([139, Example 1.2]) Let $K$ be a field, $R = K[\{X_i \mid i \in \mathbb{N}\}]/(X_1^2)$, and $P = (X_1)/(X_1^2)$. Then $P$ is a nonzero prime ideal of $R$ with $P^2 = (0)$. Thus $P \in \mathbb{AG}(R)$ and $P \subseteq Q$ for every $Q \in Spec(R)$. If $IJ = (0)$ in $R$ with $I \neq (0)$ and $J \neq (0)$, then $I$ and $J$ are both contained in $P$ since $(0)$ is $P$-primary. Thus $P = Z(R)$ is a principal ideal and $Spec(R) \cap \mathbb{A}(R)^* = \{P\}$, but $R$ is not Noetherian.

The following proposition shows that the commutative rings $R$ for which every nonzero proper ideal $I$ of $R$ is a vertex of $\mathbb{AG}(R)$ are abundant.

**Theorem 5.5.3.** ([139, Proposition 1.3]) *Let $R$ be a commutative Artinian ring. Then every nonzero proper ideal $I$ of $R$ is a vertex of $\mathbb{AG}(R)$.*

The following result shows that the set of all commutative rings for which the annihilating-ideal graphs are finite coincides with the set of all (Artinian) rings with only finitely many proper ideals. Moreover, for a commutative ring $R$, the graph $\mathbb{AG}(R)$ has $n$ vertices ($n \geq 1$) if and only if $R$ has only $n$ nonzero proper ideals.

**Theorem 5.5.4.** ([139, Theorem 1.4]) *Let $R$ be a commutative ring. Then the following statements are equivalent:*

(a) $\mathbb{AG}(R)$ *is a finite graph;*

(b) $R$ *has only finitely many ideals;*

(c) *Every vertex of* $\mathbb{AG}(R)$ *has finite degree.*

*Moreover, $\mathbb{AG}(R)$ has $n$ ($n \geq 1$) vertices if and only if $R$ has only $n$ nonzero proper ideals.*

The authors of [139] did not provide a nondomain ring $R$ such that $|\mathbb{A}(R)^*| < |\mathbb{I}(R)^*|$. The lack of such counterexamples, together with the fact that $|\mathbb{A}(R)^*| = |\mathbb{I}(R)^*|$ when $R$ is a commutative Artinian ring or $R$ is a decomposable commutative ring (see Theorem 5.5.6), motivates the following fundamental conjecture:

**Conjecture 5.5.5.** ([139]) Let $R$ be a commutative ring that is not an integral domain. Then the set of vertices of $\mathbb{AG}(R)$ and the set of nonzero proper ideals of $R$ have the same cardinality.

**Theorem 5.5.6.** ([139, Proposition 1.6]) *Let $R$ be a decomposable commutative ring. Then the set of vertices of $\mathbb{AG}(R)$ and the set of nonzero proper ideals of $R$ have the same cardinality.*

Note that every nontrivial ideal of a commutative Artinian ring $R$ is a vertex of $\mathbb{AG}(R)$. The following result shows that if every nonzero proper ideal of a commutative Noetherian ring $R$ is a vertex of $\mathbb{AG}(R)$, then $R$ is a semilocal ring.

**Theorem 5.5.7.** ([139, Proposition 1.7]) *Let $R$ be a commutative Noetherian ring. If all nontrivial ideals of $R$ are vertices of $\mathbb{AG}(R)$, then $R$ is a semilocal ring.*

Now a natural question is posed: "If $R$ is a commutative Noetherian ring and all nontrivial ideals of $R$ are vertices, then is $R$ an Artinian ring?". The answer is no. For example, let $R = F[[x^4, x^3y, xy^3, y^4]]/(x^4)$, where $x$ and $y$ are indeterminates. It is well known that all ideals of $R$ are vertices of $\mathbb{AG}(R)$, but $R$ is not Artinian. The following two theorems are proved by Aalipour, Akbari, Behboodi, Nikandish, Nikmehr, and Shaveisi [5].

**Theorem 5.5.8.** ([5, Theorem 18]) *Let $R$ be a commutative Noetherian ring such that all nontrivial ideals of $R$ are vertices of $\mathbb{AG}(R)$. Then $R$ is Cohen-Macaulay if and only if $R$ is Artinian.*

**Theorem 5.5.9.** ([5, Theorem 19]) *Let $R$ be a reduced commutative Noetherian ring. Then all nontrivial ideals of $R$ are vertices of $\mathbb{AG}(R)$ if and only if $R$ is a finite direct product of fields.*

The following result shows that for every commutative Noetherian ring $R$, if $R$ is not a domain, then at least one of the vertices of $\mathbb{AG}(R)$ is a prime ideal.

**Theorem 5.5.10.** ([139, Proposition 1.8]) *Let $R$ be a commutative Noetherian ring. Then either $\mathbb{A}(R)^* = \emptyset$ or $Spec(R) \cap \mathbb{A}(R)^* \neq \emptyset$.*

The following example shows that the Noetherian hypothesis is needed in Theorem 5.5.10.

**Example 5.5.11.** ([139, Example 1.9]) Let $K$ be a field and $R = K[\{X_i \mid i \in \mathbb{N}\}]/(\{X_i^2 \mid i \in \mathbb{N}\}) = K[\{x_i \mid i \in \mathbb{N}\}]$. Then $R$ is a zero-dimensional commutative local ring (but not Noetherian) with maximal ideal $\mathfrak{m} = (\{x_i \mid i \in \mathbb{N}\})$. Note that $Ann(\mathfrak{m}) \subseteq \bigcap x_i R = (0)$. Thus $Spec(R) \cap \mathbb{A}(R)^* = \emptyset$.

Theorem 5.5.10 is not true in general if we replace $Spec(R)$ with $Max(R)$. For instance, if $R = \mathbb{Z} \times \mathbb{Z}$, though $R$ has ACC on its ideals and $\mathbb{A}(R)^* \neq \emptyset$, then $Max(R) \cap \mathbb{A}^*(R) = \emptyset$. In the following result, we state a characterization of all nondomain commutative rings $R$ for which $Max(R) \cap \mathbb{A}(R) \neq \emptyset$.

**Theorem 5.5.12.** ([139, Proposition 1.12]) *Let $R$ be a nondomain commutative ring. Then $Max(R) \cap \mathbb{A}(R) \neq \emptyset$ if and only if $soc(R) \neq (0)$.*

The following result gives a characterization for commutative rings $R$ for which every nonzero proper principal ideal $I$ of $R$ is a vertex of $\mathbb{AG}(R)$.

**Theorem 5.5.13.** ([139, Proposition 1.13]) *Let $R$ be a commutative ring. Then every nonzero proper principal ideal of $R$ is a vertex of $\mathbb{AG}(R)$ if and only if every element in $R$ is a unit or a zero-divisor.*

In what follows, we give all commutative rings whose annihilating-ideal graphs have fewer than four vertices.

**Theorem 5.5.14.** ([139, Corollary 2.9]) *Let $R$ be a commutative ring. Then*

(a) *$\mathbb{AG}(R)$ is a graph with one vertex if and only if $R$ has only one nonzero ideal;*

(b) *$\mathbb{AG}(R)$ is a graph with two vertices if and only if $R = F_1 \oplus F_2$, where $F_1, F_2$ are fields or $R$ is an Artinian local ring with exactly two nonzero proper ideals $Z(R)$ and $Z(R)^2$;*

(c) $\mathbb{AG}(R)$ *is a graph with three vertices if and only if* $R$ *is an Artinian local ring with exactly three nonzero proper ideals* $Z(R)$, $Z(R)^2$, *and* $Z(R)^3$.

The following theorem due to Aalipour et al. [5] deals with annihilating-ideal graphs on four vertices.

**Theorem 5.5.15.** ([5, Theorem 12]) *Let* $R$ *be a commutative ring. If* $|V(\mathbb{AG}(R))| = 4$, *then* $\mathbb{AG}(R)$ *is* $P_4$, $K_4$, *or* $(K_1 + K_2) \vee K_1$, *where* $G_1 \vee G_2$ *denotes the join of* $G_1$ *and* $G_2$. *Moreover, the latter case occurs if and only if* $(R, \mathfrak{m})$ *is a local ring with* $\mathbb{I}(R) = \{\mathfrak{m}, \mathfrak{m}^2, \mathfrak{m}^3, \mathfrak{m}^4, R\}$.

A natural question one may ask is "what will happen if all vertices of $\mathbb{AG}(R)$ are prime (resp., maximal) ideals of $R$?". This is what we answer below in Theorem 5.5.16. In fact, the following two results characterize all commutative rings that satisfy these conditions.

**Theorem 5.5.16.** ([139, Theorem 2.10]) *Let* $R$ *be a commutative ring that is not an integral domain. If every vertex of* $\mathbb{AG}(R)$ *is a prime ideal of* $R$ *(i.e.,* $\mathbb{A}(R)^* \subseteq Spec(R)$*), then either* $R = F_1 \oplus F_2$ *for fields* $F_1$ *and* $F_2$ *or* $R$ *has only one nonzero proper ideal.*

The following is now immediate.

**Corollary 5.5.17.** ([139, Corollary 2.11]) *Let* $R$ *be a commutative ring that it is not an integral domain. Then the following statements are equivalent:*

(a) $\mathbb{A}(R)^* \subseteq Max(R)$, *i.e., every vertex of* $\mathbb{AG}(R)$ *is a maximal ideal of* $R$;

(b) $\mathbb{A}(R)^* = Max(R)$;

(c) $\mathbb{A}(R)^* = Spec(R)$;

(d) $\mathbb{A}(R)^* \subseteq Spec(R)$;

(e) *Either* $R = F_1 \oplus F_2$, *where* $F_1, F_2$ *are fields, or* $R$ *has only one nonzero proper ideal.*

Now, we state results about the diameter of annihilating-ideal graphs of commutative rings. Recall that for every commutative ring $R$, the zero-divisor graph $\Gamma(R)$ is a connected graph and $diam(\Gamma(R)) \leq 3$. Moreover, if $\Gamma(R)$ contains a cycle, then $gr(\Gamma(R)) \leq 4$. The following theorem says that the same results hold for the annihilating-ideal graph of a commutative ring.

**Theorem 5.5.18.** ([139, Theorem 2.1]) *Let* $R$ *be commutative ring. Then the annihilating-ideal graph* $\mathbb{AG}(R)$ *is connected and* $diam(\mathbb{AG}(R)) \leq 3$. *Moreover, if* $\mathbb{AG}(R)$ *contains a cycle, then* $gr(\mathbb{AG}(R)) \leq 4$.

The work on the diameter of an annihilating-ideal graph of a commutative ring was continued by Behboodi and Rakeei in [140]. The following result more or less summarizes the overall situation for the relationship between the diameter of $\mathbb{AG}(R)$ and $\Gamma(R)$.

**Theorem 5.5.19.** ([140, Proposition 1.1]) *Let $R$ be a commutative ring.*

(a) *If $diam(\Gamma(R)) = 0$, then $diam(\mathbb{AG}(R)) = 0$;*

(b) *If $diam(\Gamma(R)) = 1$, then $diam(\mathbb{AG}(R)) = 0$ or $1$;*

(c) *If $diam(\Gamma(R)) = 2$, then $diam(\mathbb{AG}(R)) = 1, 2$ or $3$;*

(d) *If $diam(\Gamma(R)) = 3$, then $diam(\mathbb{AG}(R)) = 3$;*

(e) *If $diam(\mathbb{AG}(R)) = 0$, then $diam(\Gamma(R)) = 0$ or $1$;*

(f) *If $diam(\mathbb{AG}(R)) = 1$, then $diam(\Gamma(R)) = 1$ or $2$;*

(g) *If $diam(\mathbb{AG}(R)) = 2$, then $diam(\Gamma(R)) = 2$;*

(h) *If $diam(\mathbb{AG}(R)) = 3$, then $diam(\Gamma(R)) = 2$ or $3$.*

The next two theorems characterize when $Z(R)$ is not an ideal of $R$ and either $diam(\mathbb{AG}(R)) = 2$ or $diam(\mathbb{AG}(R)) = 3$, respectively.

**Theorem 5.5.20.** ([140, Theorem 1.2]) *Let $R$ be a commutative ring such that $Z(R)$ is not an ideal of $R$. Then $diam(\mathbb{AG}(R)) = 2$ if and only if $R$ is reduced with exactly two minimal primes and at least three nonzero annihilating ideals.*

**Theorem 5.5.21.** ([140, Theorem 1.3]) *Let $R$ be a commutative ring such that $Z(R)$ is not an ideal. Then the following statements are equivalent:*

(a) *$diam(\mathbb{AG}(R)) = 3$;*

(b) *$diam(\Gamma(R)) = 3$;*

(c) *Either $R$ is not reduced or $R$ is a reduced ring with more than two minimal primes.*

Now we are in a position to give a complete characterization for the possible diameters of $\mathbb{AG}(R)$ when $Z(R)$ is not an ideal of $R$.

**Theorem 5.5.22.** ([140, Theorem 1.4]) *Let $R$ be a commutative ring such that $Z(R)$ is not an ideal of $R$. Then $1 \leq diam(\mathbb{AG}(R)) \leq 3$ and*

(a) *$diam(\mathbb{AG}(R)) = 1$ if and only if $R \cong F_1 \times F_2$, where $F_1$ and $F_2$ are fields;*

(b) *$diam(\mathbb{AG}(R)) = 2$ if and only if $R$ is reduced with exactly two minimal primes and at least three nonzero annihilating ideals;*

(c) *$diam(\mathbb{AG}(R)) = 3$ if and only if either $R$ is not a reduced ring or $R$ is a reduced ring with more than two minimal primes.*

The next theorem provides a sufficient condition for $\mathbb{AG}(R)$ to have diameter 3 when $R$ is a nonreduced commutative ring.

**Theorem 5.5.23.** ([140, Theorem 1.6]) *Let $R$ be a nonreduced commutative ring. If there is a pair of annihilating-ideals $I, J$ of $R$ such that $Ann(I + J) = (0)$, then $diam(\mathbb{AG}(R)) = 3$.*

It is easy to see that a reduced ring has the property that an ideal with a nonzero annihilator must be contained in at least one minimal prime.

**Lemma 5.5.24.** ([140, Lemma 1.8]) *Let $R$ be a commutative reduced ring with finitely many minimal primes. If $R$ has more than two minimal primes, then $diam(\mathbb{AG}(R)) = 3$.*

Now we are in position to state a complete characterization for the possible diameters of $\mathbb{AG}(R)$ when $R$ has only finitely many minimal primes (e.g., Noetherian rings).

**Theorem 5.5.25.** ([140, Theorem 1.9]) *Let $R$ be a commutative ring with only finitely many minimal primes. Then*

(a) *$diam(\mathbb{AG}(R)) = 0$ if and only if $Z(R)$ is the only nonzero proper ideal of $R$;*

(b) *$diam(\mathbb{AG}(R)) = 1$ if and only if $R = F_1 \oplus F_2$, where $F_1, F_2$ are fields, $Z(R)$ is an ideal of $R$ with $Z(R)^2 = (0)$, or $R$ is a local ring with exactly two nonzero proper ideals $Z(R)$ and $Z(R)^2$;*

(c) *$diam(\mathbb{AG}(R)) = 2$ if and only if either $R$ is reduced with exactly two minimal primes and at least three nonzero annihilating-ideals, or $R$ is not reduced, $Z(R)$ is an ideal whose square is not $(0)$, and for every pair of annihilating-ideals $I$ and $J$, $I + J$ is an annihilating-ideal;*

(d) *$diam(\mathbb{AG}(R)) = 3$ if and only if either $R$ is reduced with more than two minimal primes or $R$ is not reduced and there are annihilating-ideals $I \neq J$, such that $I + J$ is not an annihilating-ideal and either.*

Aliniaeifard, Behboodi, Mehdi-Nezhad, and Rahimi [50] studied the interplay between the diameter of annihilating-ideal graphs and zero-divisor graphs. We remark that if $R$ is a commutative ring, then the set of regular elements of $R$ forms a saturated multiplicatively closed subset of $R$. Hence the collection of zero-divisors of $R$ is the set-theoretic union of prime ideals. Thus $Z(R) = \cup_{i \in \Lambda} P_i$ with each $P_i$ a prime ideal.

**Theorem 5.5.26.** ([50, Lemma 2.3]) *Let $R$ be a commutative Noetherian ring with $\mathbb{AG}(R) \neq K_2$. Then $diam(\mathbb{AG}(R)) = 2$ if and only if $Z(R)$ is either the union of two primes with intersection $(0)$ or $Z(R)$ is a prime ideal such that $(Z(R))^2 \neq (0)$.*

The following theorem concerns the diameter of annihilating-ideal graphs of polynomials and power series over commutative rings.

**Theorem 5.5.27.** ([50, Theorem 2.4]) *Let $R$ be a Noetherian ring and $\mathbb{AG}(R) \neq K_2$. Then the following conditions are equivalent:*

(a) *$diam(\mathbb{AG}(R)) = 2$;*

(b) *diam*$(\mathbb{AG}(R[x])) = 2$;
(c) *diam*$(\mathbb{AG}(R[x_1, \ldots, x_n])) = 2$ *for every* $n \geq 1$;
(d) *diam*$(\mathbb{AG}(R[[x]])) = 2$;
(e) *diam*$(\Gamma(R)) = 2$;
(f) *diam*$(\Gamma(R[x])) = 2$;
(g) *diam*$(\Gamma(R[x_1, \ldots, x_n])) = 2$ *for every* $n \geq 1$;
(h) *diam*$(\Gamma(R[[x]])) = 2$;
(i) $Z(R)$ *is either the union of two primes with intersection* $(0)$, *or* $Z(R)$ *is prime and* $(Z(R))^2 \neq 0$.

Let $R$ be a commutative Noetherian ring with $\mathbb{AG}(R) \neq K_2$. From the above observations, $diam(\mathbb{AG}(R)) = diam(\mathbb{AG}(R[x])) = diam(\mathbb{AG}(R[x_1, \ldots, x_n])) = diam(\mathbb{AG}(R[[x]])) = diam(\Gamma(R)) = diam(\Gamma(R[x])) = diam(\Gamma(R[x_1, \ldots, x_n])) = diam(\Gamma(R[[x]])) = 2$.

If $R$ is a reduced commutative ring with more than two minimal prime ideals, then by Theorems 2.3.8 and 2.3.10, we have $gr(\Gamma(R)) = 3$. This fact is also true for the annihilating-ideal graph of a reduced ring $R$ with finitely many minimal primes $P_1, P_2, \ldots, P_n$ $(n \geq 3)$. The authors of [140] have not found any examples of a commutative reduced ring $R$ with infinitely many minimal primes for which $gr(\mathbb{AG}(R)) \neq 3$. The lack of such counterexamples, together with the above fact, motivates the following conjecture:

**Conjecture 5.5.28.** ([140]) *Let $R$ be a reduced commutative ring with more than two minimal primes. Then* $gr(\mathbb{AG}(R)) = 3$.

We conclude this section by studying the domination number of $\mathbb{AG}(R)$. In [298], Nikandish and Maimani studied the domination number of $\mathbb{AG}(R)$. The following theorem determines the domination number of the annihilating-ideal graphs of reduced rings with finitely many minimal primes.

**Theorem 5.5.29.** ([298, Theorem 2.4]) *Let $R$ be a reduced commutative ring with* $|Min(R)| < \infty$. *If* $\gamma(\mathbb{AG}(R)) > 1$, *then* $\gamma(\mathbb{AG}(R)) = \gamma_t(\mathbb{AG}(R)) = |Min(R)|$.

In the following theorem, we state the domination number of bipartite annihilating-ideal graphs.

**Theorem 5.5.30.** ([298, Theorem 2.7]) *Let $R$ be a commutative ring. If* $\mathbb{AG}(R)$ *is a bipartite graph, then* $\gamma(\mathbb{AG}(R)) \leq 2$.

The next theorem is on the domination number of the annihilating-ideal graphs of commutative Artinian rings.

**Theorem 5.5.31.** ([298, Theorem 2.8]) *Let $R$ be a commutative Artinian ring and* $R \not\cong F_1 \times F_2$, *where $F_1$ and $F_2$ are fields. Then* $\gamma(\mathbb{AG}(R)) = \gamma_t(\mathbb{AG}(R)) = |Min(R)|$.

The condition that $R$ be an Artinian ring in the previous theorem is necessary. For example, let $R = F[x, y, z]/(xy, xz, yz)$, where $F$ is a field and $x, y$, and $z$ are indeterminates. Then $\gamma(\mathbb{AG}(R)) = 2$, but $|Min(R)| > 2$.

The following theorem provides an upper bound for the domination number of the annihilating-ideal graph of a commutative Noetherian ring.

**Theorem 5.5.32.** ([298, Theorem 2.11]) *Let $R$ be a commutative Notherian ring. Then $\gamma(\mathbb{AG}(R)) \leq |Ass(R)| < \infty$.*

The inequality in the above theorem may be strict. For example, let $R = F[x, y]/(x^2, xy)$, where $F$ is a field and $x$ and $y$ are indeterminates. Then $Ass(R) = \{(x), (x, y)\}$, but $\gamma(\mathbb{AG}(R)) = 1$.

The following result provides the domination number of the annihilating-ideal graph of a direct product of rings.

**Theorem 5.5.33.** ([298, Theorem 2.14]) *Let $R$ be a commutative ring.*

(a) *If $R \cong F \times D$, where $F$ is a field and $D$ is an integral domain, then $\gamma(\mathbb{AG}(R)) = 1$;*

(b) *If $R \cong D_1 \times D_2$, where $D_1$ and $D_2$ are integral domains which are not fields, then $\gamma(\mathbb{AG}(R)) = 2$;*

(c) *If $R \cong R_1 \times D$, where $R_1$ is a commutative ring which is not integral domain and $D$ is an integral domain, then $\gamma(\mathbb{AG}(R)) = \gamma(\mathbb{AG}(R_1)) + 1$;*

(d) *If $R \cong R_1 \times R_2$, where $R_1$ and $R_2$ are commutative rings which are not integral domains, then $\gamma(\mathbb{AG}(R)) = \gamma(\mathbb{AG}(R_1)) + \gamma(\mathbb{AG}(R_2))$.*

## 5.6 Special Annihilating-ideal Graphs

Now, we consider some special classes of graphs that can be an annihilating-ideal graph of a commutative ring. In particular, we characterize the commutative rings $R$ whose graph $\mathbb{AG}(R)$ is complete, a path, a star, or bipartite. The first result characterizes all the commutative rings $R$ whose graph $\mathbb{AG}(R)$ has a vertex adjacent to every other vertex. This characterization is then applied to characterize the commutative rings $R$ whose graph $\mathbb{AG}(R)$ is complete or a star graph.

**Theorem 5.6.1.** ([139, Theorem 2.2]) *Let $R$ be a commutative ring. Then there is a vertex of $\mathbb{AG}(R)$ which is adjacent to every other vertex if and only if either $R = F \oplus D$, where $F$ is a field and $D$ is an integral domain, or $Z(R)$ is an annihilator ideal.*

Now, we state a characterization of all commutative rings with complete annihilating-ideal graphs.

**Theorem 5.6.2.** ([139, Theorem 2.7]) *Let $R$ be a commutative ring. Then $\mathbb{AG}(R)$ is a complete graph if and only if $R$ is one of the following three types of rings:*

(a) *$R \cong F_1 \times F_2$, where $F_1, F_2$ are fields;*

(b) *$Z(R)$ is an ideal of $R$ with $Z(R)^2 = (0)$;*

(c) *$R$ is a local ring with exactly two nonzero proper ideals $Z(R)$ and $Z(R)^2$.*

Let $R$ be a commutative ring. In Theorem 2.2.5, it is shown that the zero-divisor graph $\Gamma(R)$ is a complete graph if and only if either $R \cong \mathbb{Z}_2 \times \mathbb{Z}_2$ or $Z(R)$ is an ideal of $R$ and $Z(R)^2 = (0)$. Thus, by the above theorem, if $\Gamma(R)$ is a complete graph, then $\mathbb{AG}(R)$ is also a complete graph, but the converse is not true even if $Z(R)$ is an ideal of $R$. For example, let $R = \mathbb{Z}_{p^3}$, where $p$ is a prime number. Then $\mathbb{AG}(R)$ is a complete graph, but $Z(R)^2 \neq (0)$, and so $\Gamma(R)$ is not a complete graph.

Further, Aalipour et al. [5] studied the completeness of $\mathbb{AG}(R)$.

**Theorem 5.6.3.** ([5, Theorem 3]) *Let $R$ be a commutative ring. Then $\mathbb{AG}(R)$ is a complete graph if and only if either $\mathbb{AG}(R) = K_2$ or $Z(R)$ is an ideal of $R$ with $Z(R)^2 = (0)$. Moreover, in the first case, either $R \cong F_1 \times F_2$, where $F_1, F_2$ are fields, or $(R, \mathfrak{m})$ is a local ring with exactly two nontrivial ideals $\mathfrak{m}$ and $\mathfrak{m}^2$.*

**Theorem 5.6.4.** ([5, Theorem 7]) *Let $R$ be a commutative ring with $\mathbb{AG}(R) = K_n$ for some positive integer $n$. Then either $n \in \{1, 2\}$ or $(R, \mathfrak{m})$ is a finite local ring with $\mathfrak{m}^2 = (0)$. Moreover, in the latter case,*

$$n = \sum_{1 \leq k \leq d} \frac{(q^d - 1)(q^{d-1} - 1) \cdots (q^{d-k+1} - 1)}{(q - 1)(q^2 - 1) \cdots (q^k - 1)},$$

*where $q$ is the number of elements of the finite field $F = R/\mathfrak{m}$ and $d = dim_F(\mathfrak{m})$.*

**Theorem 5.6.5.** ([5, Theorem 8]) *Let $R$ be a commutative ring. If $\mathbb{AG}(R)$ is a regular graph of finite degree, then $\mathbb{AG}(R)$ is a complete graph.*

**Corollary 5.6.6.** ([5, Corollary 9]) *The annihilating-ideal graph of a commutative ring cannot be a cycle.*

The following theorem deals with commutative rings whose annihilating-ideal graph is a path.

**Theorem 5.6.7.** ([5, Theorem 11]) *Let $R$ be a commutative ring and $\mathbb{AG}(R) = P_n$. Then $n \leq 4$, and*

(a) *$\mathbb{AG}(R) \cong P_2$ if and only if $R \cong F_1 \times F_2$ or $(R, \mathfrak{m})$ is a local ring with $\mathbb{I}(R) = \{\mathfrak{m}, \mathfrak{m}^2, R\}$;*

(b) *$\mathbb{AG}(R) \cong P_3$ if and only if $(R, \mathfrak{m})$ is a local ring and $\mathbb{I}(R) = \{\mathfrak{m}, \mathfrak{m}^2, \mathfrak{m}^3, R\}$;*

(c) $\mathbb{AG}(R) \cong P_4$ *if and only if* $R \cong F \times S$, *where* $F$ *is a field and* $S$ *is a ring with exactly one nontrivial ideal.*

The following three results proved by Houyi Yu, Tongsuo Wu, and Weiping Gu [218] provide a list of all finite local principal ideal rings (PIRs) whose annihilating-ideal graph is a path on fewer than four vertices. In the following three results, if $f(x) = x^t + \sum_{i=0}^{t-1} a_i x^i \in \mathbb{Z}_p[x]$, we let $f_r(x)$ denote the polynomial $x^t + \sum_{i=0}^{t-1} a_i x^i \in \mathbb{Z}_{p^r}[x]$ with $r \in \{1, 2, 3, 4\}$.

**Theorem 5.6.8.** ([218, Lemma 3.1]) *Let* $(R, \mathfrak{m})$ *be a finite local PIR with residue field of characteristic* $p$. *Then* $\mathbb{AG}(R)$ *is an isolated vertex if and only if* $R$ *is isomorphic to one of the following homomorphic images of the polynomial ring* $\mathbb{Z}_{p^r}[x, y]$ $(r = 1, 2)$:

(a) $\dfrac{\mathbb{Z}_p[x, y]}{(g(x), y^2)}$,

(b) $\dfrac{\mathbb{Z}_{p^2}[x]}{(g_2(x) + pw(x))}$,

*where* $g(x)$ *is monic and irreducible in* $\mathbb{Z}_p[x]$ *and* $w(x)$ *is a polynomial of* $\mathbb{Z}_p[x]$ *such that* $deg(w(x)) < deg(g(x))$.

**Theorem 5.6.9.** ([218, Lemma 3.2]) *Let* $(R, \mathfrak{m})$ *be a finite local PIR with residue field of characteristic* $p$. *Then* $\mathbb{AG}(R) \cong K_2$ *if and only if* $R$ *is isomorphic to one of the following homomorphic images of the polynomial ring* $\mathbb{Z}_{p^r}[x, y]$ $(r = 1, 2, 3)$:

(a) $\dfrac{\mathbb{Z}_p[x, y]}{(g(x), y^3)}$,

(b) $\dfrac{\mathbb{Z}_{p^2}[x, y]}{(g_2(x) + pw_2(x), p - u_2(x)y^2, py)}$ *or*
$\dfrac{\mathbb{Z}_{p^2}[x, y]}{(g_2(x) - v_2(x)y + pw_2(x), p - u_2(x)y^2, py)}$,

(c) $\dfrac{\mathbb{Z}_{p^3}[x]}{(g_3(x) + pz_3(x) + p^2 w_3(x))}$,

*where* $g(x)$ *is monic and irreducible in* $\mathbb{Z}_p[x]$, $z(x), w(x), u(x)$, *and* $v(x)$ *are polynomials of* $\mathbb{Z}_p[x]$ *such that* $deg(t(x)) < deg(g(x))$ *for any* $t(x) \in \{z(x), w(x)\}$, $u(x) \not\equiv 0\ (mod(g(x)))$, *and* $v(x) \not\equiv 0\ (mod(g(x)))$.

**Theorem 5.6.10.** ([218, Lemma 3.3]) *Let* $(R, \mathfrak{m})$ *be a finite local PIR with residue field of characteristic* $p$. *Then* $\mathbb{AG}(R) \cong P_3$ *if and only if* $R$ *is isomorphic to one of the following homomorphic images of a polynomial ring* $\mathbb{Z}_{p^r}[x, y]$ $(r = 1, 2, 4)$:

(a) $\dfrac{\mathbb{Z}_p[x, y]}{(g(x), y^4)}$,

(b) $\dfrac{\mathbb{Z}_{p^2}[x,y]}{(g_2(x)+pw_2(x), y^3-pu_2(x), py)}$ *or*
$\dfrac{\mathbb{Z}_{p^2}[x,y]}{(g_2(x)-v_2(x)y^2+pw_2(x), y^3-pu_2(x), py)}$ *or*
$\dfrac{\mathbb{Z}_{p^2}[x,y]}{(g_2(x)-v_2(x)y+pw_2(x), y^3-pu_2(x), py)}$ *or*
$\dfrac{\mathbb{Z}_{p^2}[x,y]}{(g_2(x)-pv_2(x)y, y^2-pu_2(x)-pz_2(x)y)}$ *or*
$\dfrac{\mathbb{Z}_{p^2}[x,y]}{(g_2(x)-pw_2(x)-pt_2(x)y, y^2-pu_2(x)-pz_2(x)y)}$ *or*
$\dfrac{\mathbb{Z}_{p^2}[x,y]}{(g_2(x)-v_2(x)y-pt_2(x)y-pw_2(x), y^2-pu_2(x)-pz_2(x)y)}$;

(c) $\dfrac{\mathbb{Z}_{p^4}[x]}{(g_4(x)+pt_4(x)+p^2z_4(x)+p^3w_4(x))}$,

*where* $g(x)$ *is monic and irreducible in* $\mathbb{Z}_p[x]$, $t(x), w(x), z(x), u(x)$, *and* $v(x)$ *are polynomials of* $\mathbb{Z}_p[x]$ *such that* $deg(a(x)) < deg(g(x))$ *for any* $a(x) \in \{t(x), z(x), w(x)\}$, $u(x) \not\equiv 0 \ (mod(g(x)))$, *and* $v(x) \not\equiv 0 \ (mod(g(x)))$.

The next result is a characterization of the reduced commutative rings $R$ whose annihilating-ideal graph $\mathbb{AG}(R)$ is a star graph. Recall that a graph is called a *star graph* if it is a complete bipartite graph, one partition of which is an isolated vertex.

**Corollary 5.6.11.** ([139, Corollary 2.3]) *Let* $R$ *be a reduced commutative ring. Then the following statements are equivalent:*

(a) *There is a vertex of* $\mathbb{AG}(R)$ *which is adjacent to every other vertex;*

(b) $\mathbb{AG}(R)$ *is a star graph;*

(c) $R = F \oplus D$, *where* $F$ *is a field and* $D$ *is an integral domain.*

**Corollary 5.6.12.** ([139, Corollary 2.4]) *Let* $R$ *be a commutative Artinian ring. Then there is a vertex of* $\mathbb{AG}(R)$ *which is adjacent to every other vertex if and only if either* $R = F_1 \oplus F_2$, *where* $F_1, F_2$ *are fields, or* $R$ *is a local ring with nonzero maximal ideal.*

The following result gives a better characterization of the commutative Artinian rings with star annihilating-ideal graph.

**Theorem 5.6.13.** ([139, Theorem 2.6]) *Let* $R$ *be a commutative Artinian ring. Then* $\mathbb{AG}(R)$ *is a star graph if and only if either* $R = F_1 \oplus F_2$, *where* $F_1, F_2$ *are fields, or* $R$ *is a local ring with nonzero maximal ideal* $\mathfrak{m}$ *and one of the following cases holds:*

(a) $\mathfrak{m}^2 = (0)$, *and* $\mathfrak{m}$ *is the only nonzero proper ideal of* $R$*;*

(b) $\mathfrak{m}^3 = (0)$, $\mathfrak{m}^2$ *is the only minimal ideal of* $R$, *and for every distinct proper ideals* $I_1, I_2$ *of* $R$ *such that* $\mathfrak{m}^2 \neq I_i$ $(i = 1, 2)$, $I_1I_2 = \mathfrak{m}^2$;

(c) $\mathfrak{m}^4 = (0)$, $\mathfrak{m}^3 \neq (0)$, *and* $\mathbb{A}(R)^* = \{\mathfrak{m}, \mathfrak{m}^2, \mathfrak{m}^3\}$.

In view of Theorem 5.6.13, the structure of the corresponding nonlocal rings are quite clear, while the structure of the related Artinian local rings are yet to be investigated. In [218], Houyi Yu et al. improved the aforementioned result and clarified the structure of a commutative Artinian local ring $R$ whose annihilating-ideal graph is a star graph. For this purpose, recall that a ring $R$ is called a principal ideal ring (PIR) if each ideal of $R$ is a principal ideal. In particular, a commutative Artinian local ring $(R, \mathfrak{m})$ is a PIR if and only its maximal ideal $\mathfrak{m}$ is a principal ideal. The main results of the authors are the two theorems listed below.

**Theorem 5.6.14.** ([218, Theorem A]) *Let* $(R, \mathfrak{m})$ *be a commutative Artinian local ring. Then* $\mathbb{AG}(R)$ *is a star graph if and only if one of the following conditions holds:*

(a) $(R, \mathfrak{m})$ *is a PIR, where* $\mathfrak{m} \neq 0$ *and* $\mathfrak{m}$ *has nilpotency index less than or equal to 4. (This is equivalent to saying that there exists a* $\beta \in \mathfrak{m}$ *such that* $\mathfrak{m} = R\beta$, $\beta^s \neq 0$, *and* $\beta^{s+1} = 0$ *for some* $1 \leq s \leq 3$*);*

(b) *char*$(R) = 2$ *or char*$(R) = 4$, *and* $\mathfrak{m}$ *has a minimal generating set* $\{\beta_1, \beta_2\}$ *with* $\beta_1\beta_2 \neq 0$ *and* $\beta_1^2 = \beta_2^2 = 0$. *In this case,* $\mathfrak{m}^2 \neq 0$ *and* $\mathfrak{m}^3 = 0$.

In the following theorem, for a polynomial $g(x) = \sum_{i=0}^{t} a_i x^i \in \mathbb{Z}_4[x]$, let $\bar{g}(x) = \sum_{i=0}^{t} \varphi(a_i)x^i \in \mathbb{Z}_2[x]$, where $\varphi : \mathbb{Z}_4 \to \mathbb{Z}_2$ is the natural ring homomorphism.

**Theorem 5.6.15.** ([218, Theorem B]) *Let* $(R, \mathfrak{m})$ *be a finite commutative Artinian local ring which is not a PIR.*

(a) *If char*$(R) = 2$, *then* $\mathbb{AG}(R)$ *is a star graph if and only if* $R \cong F[x, y]/(x^2, y^2)$, *where* $F \cong R/\mathfrak{m}$;

(b) *If char*$(R) = 4$, *then* $\mathbb{AG}(R)$ *is a star graph if and only if there is a monic polynomial* $g(x)$, *polynomials* $u_i(x)$, $v_j(x)$ *in* $\{0\} \cup \mathbb{Z}_4[x] \setminus (2, g(x))$ $(1 \leq i \leq 3,\ 1 \leq j \leq 4)$, *where* $v_1(x) \neq 0$, $\bar{g}(x)$ *is irreducible in the polynomial ring* $\mathbb{Z}_2[x]$, *deg*$(u_i(x))$, *deg*$(v_j(x)) <$ *deg*$(g(x))$ *such that* $R \cong S_k/Q_k$ $(k = 1, 2)$, *where* $S_1 = \mathbb{Z}_4[x, y]$, $Q_1 = (g(x) - 2u_1(x) - u_2(x)y - 2u_3(x)y, y^2)$, $S_2 = \mathbb{Z}_4[x, y, z]$, $Q_2 = (g(x) - v_2(x)y - v_3(x)z - 2v_4(x), 2v_1(x) - yz, y^2, z^2)$.

The following corollary classifies all finite star graphs that can be realized as the annihilating-ideal graphs of finite commutative local rings.

**Corollary 5.6.16.** ([218, Corollary 3.4]) *A finite star graph* $G$ *is isomorphic to the annihilating-ideal graph of a finite local ring if and only if* $|V(G)| \in \{0, 1, 2, 3\} \cup \{2^n + 3 \mid n$ *a positive integer*$\}$.

Aalipour et al. [3] studied commutative rings with bipartite annihilating-ideal graphs. The following theorem is a characterization of nonreduced commutative rings with bipartite annihilating-ideal graphs.

**Theorem 5.6.17.** ([3, Theorem 20]) *Let $R$ be a nonreduced commutative ring and $\mathbb{AG}(R)$ be a triangle-free graph. Then $\mathbb{AG}(R)$ is bipartite and $R$ has at most two minimal ideals. Moreover, one of the following statements holds:*

(a) *$R$ has exactly two minimal ideals and $R \cong F \times S$, where $F$ is a field and $S$ is a commutative ring with exactly one nontrivial ideal;*

(b) *$R$ has exactly one minimal ideal, say $(x)$, $(x)^2 = (0)$, and $V_1 = N((x))$ and $V_1^c$ are the parts of $\mathbb{AG}(R)$. Furthermore, for $A = \{L \in V_1 \mid (x) \subseteq L\}$, the induced subgraph on $V_1^c \cup (V_1 \setminus A)$ is a complete bipartite graph and $deg(L) = 1$, for every $L \in A$.*

Let $R = D \oplus S$, where $D$ is an integral domain and $S$ is a commutative ring with exactly one nontrivial ideal. Then $\mathbb{AG}(R)$ is a bipartite graph satisfying part (b) of the previous theorem.

**Remark 5.6.18.** ([3, Remark 22]) For a reduced commutative ring $R$, if $\mathbb{AG}(R)$ is a triangle-free graph, then it is a complete bipartite graph. To see this, if $\mathbb{AG}(R)$ is a triangle-free graph, then $|Min(R)| \leq 2$. If $|Min(R)| = 1$, then $R$ is an integral domain. Thus we can assume that $Min(R) = \{P_1, P_2\}$. In this case, $\mathbb{AG}(R)$ is a complete bipartite graph with the parts $X = \{I \mid I \in A(R)^*,\ I \subseteq P_1\}$ and $Y = \{I \mid I \in A(R)^*,\ I \subseteq P_2\}$.

**Corollary 5.6.19.** ([3, Corollary 23]) *Let $R$ be a commutative ring. Then $\mathbb{AG}(R)$ is a tree if and only if $\mathbb{AG}(R)$ is isomorphic to either a star graph or path on 4 vertices.*

**Corollary 5.6.20.** ([3, Corollary 24,25]) *Let $R$ be a commutative ring. Then $\mathbb{AG}(R)$ is a complete bipartite graph if and only if $\mathbb{AG}(R)$ is a star graph or $R$ is a reduced ring and $|Min(R)| = 2$. Moreover, $\mathbb{AG}(R)$ is bipartite if and only if $\mathbb{AG}(R)$ is triangle-free.*

**Corollary 5.6.21.** ([3, Corollary 26]) *Let $R$ be a commutative ring.*

(a) *If $R$ is reduced, then $\mathbb{AG}(R)$ is a star graph if and only if $R \cong F \times D$, where $F$ is a field and $D$ is an integral domain;*

(b) *If $R$ is not reduced, then $\mathbb{AG}(R)$ is a star graph if and only if $\mathbb{AG}(R)$ is a bipartite graph and $Z(R) = Ann(x)$ for some $x \in R$.*

The next theorem provides some properties of rings whose annihilating-ideal graphs are bipartite.

**Theorem 5.6.22.** ([3, Theorem 27]) *Let $R$ be a commutative ring. If $\mathbb{AG}(R)$ is a bipartite graph, then one of the following holds:*

(a) $\mathbb{AG}(R)$ *is a star graph;*

(b) $\mathbb{AG}(R) = P_4$*;*

(c) $Nil(R) = Soc(R)$.

**Theorem 5.6.23.** ([3, Theorem 28]) *Let $R$ be a commutative ring with $\mathbb{AG}(R)$ a bipartite graph. If $|Min(R)| = 1$, then $\mathbb{AG}(R)$ is a star graph.*

Aliniaeifard et al. [50] characterized the commutative rings whose annihilating-ideal graphs are bipartite.

**Theorem 5.6.24.** ([50, Proposition 3.1]) *Let $R$ be a reduced commutative ring. Then the following statements are equivalent:*

(a) $Z(R)$ *is the union of two primes with intersection* $(0)$*;*

(b) $\mathbb{AG}(R)$ *is a complete bipartite graph.*

Using the bipartite nature of $\mathbb{AG}(R)$, the following four results characterize commutative rings $R$ with $gr(AG(R)) \geq 4$.

**Theorem 5.6.25.** ([50, Theorem 3.2]) *The following statements are equivalent for a reduced commutative ring $R$:*

(a) $gr(\mathbb{AG}(R)) = 4$*;*

(b) $\mathbb{AG}(R) = K_{|V_1|,|V_2|}$*, where* $|V_1|, |V_2| \geq 2$*;*

(c) $Z(R)$ *is the union of two primes $P_1$ and $P_2$ with intersection* $(0)$ *and* $|\mathbb{I}(P_1)|, |\mathbb{I}(P_2)| \geq 3$.

**Theorem 5.6.26.** ([50, Theorem 3.3]) *The following statements are equivalent for a commutative reduced ring $R$:*

(a) $\mathbb{AG}(R)$ *is nonempty with* $gr(\mathbb{AG}(R)) = \infty$*;*

(b) *There is a vertex which is adjacent to every other vertex of* $\mathbb{AG}(R)$*;*

(c) $R \cong F \oplus D$*, where $F$ is a field and $D$ is an integral domain;*

(d) $\mathbb{AG}(R)$ *is a star graph.*

**Theorem 5.6.27.** ([50, Theorem 3.4]) *The following statements are equivalent for a commutative ring $R$ that is not reduced:*

(a) $\mathbb{AG}(R)$ *is nonempty with* $gr(\mathbb{AG}(R)) = \infty$*;*

(b) *One of the following occurs:*

(i) *Either $R \cong F[x]/(x^2)$, where $F$ is a field, or $R \cong L$, where $L$ is a coefficient ring of characteristic $p^2$;*

(ii) *$R \cong R_1 \times R_2$ such that $R_1$ is a field and either $R_2 \cong F[x]/(x^2)$, where $F$ is a field or $R_2 \cong L$, where $L$ is a coefficient ring of characteristic $p^2$;*

(iii) *$Z(R)$ is an annihilating ideal and if $IJ = (0)$ and $I \neq J$, then $I = Ann(Z(R))$ or $J = Ann(Z(R))$;*

(c) *One of the following occurs:*

(i) $\mathbb{AG}(R) = K_1$;
(ii) $\mathbb{AG}(R) = P_4$;
(iii) $\mathbb{AG}(R) = K_{1,n}$ *for some* $n \geq 1$.

**Theorem 5.6.28.** ([50, Theorem 3.5]) *The following statements are equivalent for a commutative ring R that is not reduced:*

(a) $\mathbb{AG}(R)$ *is nonempty with* $gr(\mathbb{AG}(R)) = 4$;

(b) $R \cong R_1 \times R_2$, *where either* $R_1 \cong F[x]/(x^2)$, *where F is a field, or* $R_1 \cong L$, *where L is a coefficient ring of characteristic* $p^2$, *and* $R_2$ *is an integral domain which is not a field.*

Now, we see the cycle structure in the annihilating-ideal graph of a commutative ring. This was done by Aalipour et al. in [4]. The next theorem determines which trees can occur as the annihilating-graph of a commutative ring.

**Theorem 5.6.29.** ([4, Theorem 2], cf. Corollary 5.6.19) *Let R be a commutative ring. If* $\mathbb{AG}(R)$ *is a tree, then either* $\mathbb{AG}(R)$ *is a star graph or* $\mathbb{AG}(R) = P_4$. *Moreover,* $\mathbb{AG}(R) = P_4$ *if and only if* $R \cong F \times S$, *where F is a field and S is a ring with a unique nontrivial ideal; indeed,* $S \cong F[x]/(x^2)$, *where F is a field, or* $S \cong A/p^2A$, *where A is a discrete valuation ring of characteristic* 0 *with residue field of characteristic p for some prime p.*

**Theorem 5.6.30.** ([4, Theorem 5]) *Let R be a commutative ring. If* $|Min(R)| \geq 3$, *then* $\mathbb{AG}(R)$ *contains a cycle.*

**Theorem 5.6.31.** ([4, Theorem 8]) *Let R be a nonreduced commutative ring. If* $|Min(R)| = 2$, *then either* $\mathbb{AG}(R)$ *contains a cycle or* $\mathbb{AG}(R) = P_4$.

The annihilating-ideal graph of a nonreduced commutative ring $R$ with $|Min(R)| = 2$ need not be a complete bipartite graph or does not necessarily contain a triangle. To see this, let $R = D \oplus S$, where $D$ is any integral domain and $S$ is a ring with a unique nontrivial ideal. Then $\mathbb{AG}(R)$ is a bipartite graph which is not a complete bipartite graph.

**Theorem 5.6.32.** ([4, Theorem 10]) *Let R be a commutative ring such that* $|Min(R)| = 1$. *If* $\mathbb{AG}(R)$ *is a triangle-free graph, then* $\mathbb{AG}(R)$ *is a star graph.*

**Corollary 5.6.33.** ([4, Corollary 11]) *Let R be a commutative ring with* $|Min(R)| = 1$. *If* $\mathbb{AG}(R)$ *is a bipartite graph, then* $\mathbb{AG}(R)$ *is a star graph.*

In Section 3.6.1, we have seen necessary and sufficient conditions for the zero-divisor graph to be complemented. Visweswaran and Patel [416] provided necessary and sufficient conditions for $\mathbb{AG}(R)$ to be complemented under the assumption that $R$ is a zero-dimensional commutative semilocal ring with at least two nonzero annihilating ideals.

First of all, we state that if $R$ is a reduced commutative ring which is not an integral domain, then $\mathbb{AG}(R)$ is complemented.

**Theorem 5.6.34.** ([416, Proposition 2.2]) *Let $R$ be a commutative reduced ring which is not an integral domain. Then $\mathbb{AG}(R)$ is complemented. Moreover, $\mathbb{AG}(R)$ is uniquely complemented.*

Now, we present some necessary conditions for $\mathbb{AG}(R)$ to be complemented. We begin with the following lemma.

**Lemma 5.6.35.** ([416, Lemma 3.3]) *Let $R$ be a commutative ring. If $\mathbb{AG}(R)$ is complemented, then $Nil(R)^4 = (0)$.*

The following result provides some more necessary conditions on $R$ for $Nil(R)^3 \neq (0)$ and $\mathbb{AG}(R)$ to be complemented.

**Theorem 5.6.36.** ([416, Proposition 3.4]) *Let $R$ be a commutative ring with $Nil(R)^3 \neq (0)$. If $\mathbb{AG}(R)$ is complemented, then the following hold:*

(a) $Nil(R)^i \perp Nil(R)^3$ *for* $i = 1, 2$;

(b) $Nil(R)^3 = Rx$ *for every* $x \in Nil(R)^3 \setminus \{0\}$;

(c) $Nil(R)^2 = Ry$ *for every* $y \in Nil(R)^2 \setminus Nil(R)^3$;

(d) $Nil(R)^2 = Rz$ *for every* $z \in Nil(R) \setminus Nil(R)^2$;

(e) *If $I$ is an ideal of $R$ with $I \subseteq Nil(R)$, then $I \in \{(0), Nil(R), Nil(R)^2, Nil(R)^3\}$, and so $I \in \{(0), Rz, Rz^2, Rz^3\}$ for every $z \in Nil(R) \setminus Nil(R)^2$.*

We next consider commutative rings $R$ with $Nil(R)^3 = (0)$, but $Nil(R)^2 \neq (0)$, and determine some necessary conditions for $\mathbb{AG}(R)$ to be complemented.

**Theorem 5.6.37.** ([416, Proposition 3.5]) *Let $R$ be a commutative ring with $Nil(R)^3 = (0)$, but $Nil(R)^2 \neq (0)$. If $\mathbb{AG}(R)$ is complemented, then the following hold:*

(a) $Nil(R) \perp Nil(R)^2$;

(b) $Nil(R)^2 = Rx$ *for every* $x \in Nil(R)^2 \setminus \{0\}$;

(c) *If $ab \neq 0$ for some $a, b \in Nil(R)$, then $Ra \perp Nil(R)^2$ and $Rb \perp Nil(R)^2$;*

(d) *If $z^2 \neq 0$ for some $z \in Nil(R)$, then $Nil(R) = Ra$ for every $a \in Nil(R) \setminus Nil(R)^2$;*

(e) *If $z^2 = 0$ for every $z \in Nil(R)$, then $Nil(R)$ is not principal, but there exist $a, b \in Nil(R)$ such that $Nil(R) = Ra + Rb$.*

Next, we provide all zero-dimensional commutative local rings $R$ such that $\mathbb{AG}(R)$ is complemented. We begin with the following lemma.

**Lemma 5.6.38.** ([416, Lemma 4.2]) *Let $R$ be a commutative quasilocal ring with maximal ideal $\mathfrak{m}$. Suppose that $\mathfrak{m}^3 = (0)$, but $\mathfrak{m}^2 \neq (0)$. Then the following statements are equivalent:*

(a) $\mathbb{AG}(R)$ *is complemented;*

(b) *If* $z^2 \neq 0$ *for some* $z \in \mathfrak{m}$, *then* $\mathfrak{m}$ *is principal. If* $z^2 = 0$ *for every* $z \in \mathfrak{m}$, *then* $\mathfrak{m}$ *is not principal, but there are* $a, b \in \mathfrak{m}$ *such that* $\mathfrak{m} = Ra + Rb$*;*

(c) $I \perp \mathfrak{m}^2$ *for every nonzero proper ideal* $I$ *of* $R$ *with* $I \neq \mathfrak{m}^2$.

**Lemma 5.6.39.** ([416, Lemma 4.3]) *Let* $R$ *be a commutative quasilocal ring with maximal ideal* $\mathfrak{m}$ *such that* $\mathfrak{m}^3 = (0)$, *but* $\mathfrak{m}^2 \neq (0)$. *If* $\mathbb{AG}(R)$ *is complemented and* $\mathfrak{m}$ *is not principal, then* $char(R/\mathfrak{m}) = 2$, *and moreover,* $char(R) \in \{2, 4\}$.

The following theorem characterizes when $\mathbb{AG}(R)$ is complemented, where $R$ is any zero-dimensional commutative local ring with $\mathbb{AG}(R)$ admitting at least two vertices. Recall that a commutative local ring $(R, \mathfrak{m})$ is said to be a *special principal ideal ring* (SPIR) if $R$ is a principal ideal ring and $\mathfrak{m}$ is nilpotent.

**Theorem 5.6.40.** ([416, Theorem 4.8]) *Let* $R$ *be a zero-dimensional commutative quasilocal ring with maximal ideal* $\mathfrak{m}$. *Suppose that* $\mathbb{AG}(R)$ *has at least two vertices. Then* $\mathbb{AG}(R)$ *is complemented if and only if* (a) *and* (b) *hold, and moreover, either* (c) *or* (d) *holds, where* (a), (b), (c), *and* (d) *are given below:*

(a) $\mathfrak{m}^2 \neq (0)$*;*

(b) $\mathfrak{m}^4 = (0)$*;*

(c) $R$ *is a SPIR;*

(d) *If* $z^2 = 0$ *for every* $z \in \mathfrak{m}$, $\mathfrak{m}$ *is not principal, but there are* $a, b \in \mathfrak{m}$ *such that* $\mathfrak{m} = Ra + Rb$.

Let $R$ be a zero-dimensional commutative semilocal ring with more than one maximal ideal. The following theorem determines necessary and sufficient conditions for $\mathbb{AG}(R)$ to be complemented.

**Theorem 5.6.41.** ([416, Theorem 5.6]) *Let* $R$ *be a commutative semilocal ring which is not local, and let* $dim(R) = 0$. *Then the following statements are equivalent:*

(a) $\mathbb{AG}(R)$ *is complemented;*

(b) *Either* $R \cong F_1 \times \cdots \times F_n$, *where* $n \geq 2$ *and* $F_i$ *is a field for every* $i \in \{1, \ldots, n\}$, *or* $R \cong S \times F$, *where* $(S, \mathfrak{m})$ *is a SPIR with* $\mathfrak{m} \neq (0)$, *but* $\mathfrak{m}^2 = (0)$, *and* $F$ *is a field.*

The following corollary determines when $\mathbb{AG}(R)$ is complemented, where $R$ is a finite commutative semilocal ring which is not local.

**Corollary 5.6.42.** ([416, Theorem 5.7]) *Let* $R$ *be a a finite commutative semilocal ring which is not local. Then the following statements are equivalent:*

(a) $\mathbb{AG}(R)$ *is complemented;*

(b) *Either* $R \cong F_1 \times \cdots \times F_n$*, where* $n \geq 2$ *and* $F_i$ *is a field for every* $i \in \{1, \ldots, n\}$*, or* $R \cong S \times F$*, where* $(S, \mathfrak{m})$ *is a SPIR with* $\mathfrak{m} \neq (0)$*, but* $\mathfrak{m}^2 = (0)$*, and* $F$ *is a finite field.*

The following theorem determines when $\mathbb{AG}(R)$ is complemented, where $R$ is a nonreduced commutative ring. Recall that a prime ideal $P$ of $R$ is said to be a maximal $N$-prime of an ideal $I$ of $R$ if $P$ is maximal with respect to the property of being contained in $Z_R(R/I)$. It follows from [228, Theorem 1] that maximal $N$-primes of (0) always exist, and if $\{P_a\}_{\alpha\in\Lambda}$ is the set of all maximal $N$-primes of (0) in $R$, then $Z(R) = \cup_{\alpha\in\Lambda} P_\alpha$.

**Theorem 5.6.43.** ([416, Theorem 6.9]) *Let* $R$ *be a commutative ring which is not reduced. Let* $n \geq 2$ *and let* $\{P_1, \ldots, P_n\}$ *be the set of all maximal* $N$*-primes of* (0) *in* $R$*. Suppose that* $Nil(R) = \cap_{i=1}^n P_i$*. Then* $\mathbb{AG}(R)$ *is complemented if and only if either* $R$ *is isomorphic to* $F \times S$*, where* $F$ *is a field and* $(S, \mathfrak{m})$ *is a SPIR with* $\mathfrak{m} \neq (0)$*, but* $\mathfrak{m}^2 = (0)$*, or* $R$ *is isomorphic to* $S \times D$*, where* $(S, \mathfrak{m})$ *is a SPIR with* $\mathfrak{m} \neq (0)$*, but* $\mathfrak{m}^2 = (0)$*, and* $D$ *is an integral domain which is not a field.*

## 5.7 Coloring of Annihilating-ideal Graphs

Behboodi and Rakeei [140] studied the coloring of annihilating-ideal graphs of commutative rings, in particular, the interplay between the chromatic number $\chi(\mathbb{AG}(R))$ and the clique number $\omega(\mathbb{AG}(R))$. Recall from Section 3.1.1 that Beck showed that if $R$ is a commutative reduced ring or a principal ideal ring, then $\chi(\Gamma_0(R)) = \omega(\Gamma_0(R))$ and also proved that for $n = 2, 3$, or 4, $\chi(\Gamma_0(R)) = n \Leftrightarrow \omega(\Gamma_0(R)) = n$. Based on these positive results, Beck conjectured that $\chi(\Gamma_0(R)) = \omega(\Gamma_0(R))$ for every commutative ring $R$ when $\chi(\Gamma_0(R)) < \infty$. Then, Anderson and Naseer gave a counterexample to this conjecture. Now, we start this section with the following remark about the Anderson-Naseer counterexample in the context of $\mathbb{AG}(R)$.

**Example 5.7.1.** ([140, Proposition 2.1]) Let

$$R = \mathbb{Z}_4[x, y, z]/(x^2 - 2, y^2 - 2, z^2, 2x, 2y, 2z, xy, xz, yz - 2).$$

Then $\chi(\mathbb{AG}(R)) = \omega(\mathbb{AG}(R)) = 4$.

The next two results characterize the rings $R$ for which $\chi(\mathbb{AG}(R)) = 1$ or 2.

**Theorem 5.7.2.** ([140, Proposition 2.2]) *Let* $R$ *be a commutative ring. Then* $\chi(\mathbb{AG}(R)) = 1$ *if and only if* $R$ *has only one nonzero proper ideal.*

**Theorem 5.7.3.** ([140, Theorem 2.3]) *Let* $R$ *be commutative ring. Then the following statements are equivalent:*

(a) $\chi(\mathbb{AG}(R)) = 2$;

(b) $\mathbb{AG}(R)$ *is a bipartite graph with two nonempty parts;*

(c) $\mathbb{AG}(R)$ *is a complete bipartite graph with two nonempty parts;*

(d) *Either* $R$ *is a reduced ring with exactly two minimal primes or* $\mathbb{AG}(R)$ *is a star graph with more than one vertex.*

**Corollary 5.7.4.** ([140, Corollary 2.4]) *Let* $R$ *be a commutative Artinian ring. Then the following statements are equivalent:*

(a) $\chi(\mathbb{AG}(R)) = 2$;

(b) $\mathbb{AG}(R)$ *is a bipartite graph with two nonempty parts;*

(c) $\mathbb{AG}(R)$ *is a complete bipartite graph with two nonempty parts;*

(d) *Either* $R \cong F_1 \times F_2$ *for fields* $F_1$ *and* $F_2$, *or* $R$ *is a local ring such that* $\mathbb{AG}(R)$ *is a star graph with more than one vertex.*

**Corollary 5.7.5.** ([140, Corollary 2.5]) *Let* $R$ *be a commutative reduced ring. Then the following statements are equivalent:*

(a) $\chi(\mathbb{AG}(R)) = 2$;

(b) $\chi(\Gamma(R)) = 2$;

(c) $\mathbb{AG}(R)$ *is a bipartite graph with two nonempty parts;*

(d) $\mathbb{AG}(R)$ *is a complete bipartite graph with two nonempty parts;*

(e) $R$ *has exactly two minimal primes.*

Let $R = \mathbb{Z}_{p^3}$, where $p$ is prime. Then it is easy to check that $\chi(\mathbb{AG}(R)) = 2$, but, by Theorem 3.1.38, $\chi(\Gamma(R)) = p$. The following result shows that if $\chi(\Gamma(R))$ is finite, then $\chi(\mathbb{AG}(R))$ is also finite.

**Theorem 5.7.6.** ([140, Proposition 2.7]) *Let* $R$ *be a commutative ring. If* $\mathbb{AG}(R)$ *has an infinite clique, then* $\Gamma(R)$ *has an infinite clique.*

The next result characterizes all reduced commutative rings $R$ with $\chi(\mathbb{AG}(R))$ finite.

**Theorem 5.7.7.** ([140, Theorem 2.10]) *Let* $R$ *be a reduced commutative ring. Then the following statements are equivalent:*

(a) $\chi(\mathbb{AG}(R))$ *is finite;*

(b) $\omega(\mathbb{AG}(R))$ *is finite;*

(c) $\mathbb{AG}(R)$ *does not have an infinite clique;*

(d) $R$ *has finite number of minimal prime ideals.*

Now we can show that Beck's conjecture is true for annihilating-ideal graphs in the case of reduced rings.

**Corollary 5.7.8.** ([140, Corollary 2.11]) *Let $R$ be a reduced commutative ring. Then $\chi(\mathbb{AG}(R)) = \omega(\mathbb{AG}(R))$. Moreover, if $\chi(\mathbb{AG}(R))$ is finite, then $R$ has a finite number of minimal primes, and $\chi(\mathbb{AG}(R)) = \omega(\mathbb{AG}(R)) = |Min(R)|$, where $Min(R)$ is the set of minimal prime ideals in $R$.*

**Corollary 5.7.9.** ([140, Corollary 2.12]) *Let $R$ be a nonzero reduced commutative ring. Then $\chi(\mathbb{AG}(R))$ is finite if and only if $\chi(\Gamma(R))$ is finite. Furthermore, if they are finite, then $\chi(\mathbb{AG}(R)) = \chi(\Gamma(R))$.*

Next, we state a characterization of commutative Noetherian rings $R$ with $\omega(\mathbb{AG}(R)) < \infty$.

**Theorem 5.7.10.** ([140, Theorem 2.14]) *Let $R$ be a commutative Noetherian ring. Then $\omega(\mathbb{AG}(R))$ is finite if and only if for every ideal $I$ of $R$ with $I^2 = (0)$, $I$ has a finite number of $R$-submodules.*

The study of the clique number and chromatic number of annihilating-ideal graphs was continued by Aalipour et al. in [3, 5]. In fact, they provided a relationship between the clique number and chromatic number of the annihilating-ideal graphs.

**Theorem 5.7.11.** ([5, Theorem 13]) *Let $R$ be a commutative ring. Then $\omega(\mathbb{AG}(R)) = 2$ if and only if $\chi(\mathbb{AG}(R)) = 2$.*

Using the tool of intersection graphs, Devhare et al. [182] proved certain characterizations for annihilating-ideal graphs. In fact, they proved the following with regard to weakly perfectness of the annihilating-ideal graph and its complement of a reduced commutative ring.

**Theorem 5.7.12.** ([182, Corollary 4.2]) *Let $R$ be a reduced commutative ring such that $Z(R) \neq \{0\}$. If $\omega(C\mathbb{AG}(R)) < \infty$, then $\mathbb{AG}(R)$ and its complement $C\mathbb{AG}(R)$ are weakly perfect.*

Again using the tool of intersection graphs of ideals of a commutative ring, Devhare et al. [183] proved a characterization for the connectedness of the complement of the annihilating-ideal graph of reduced commutative Artinian rings.

**Theorem 5.7.13.** ([183, Corollary 3.7]) *The following statements hold for a reduced commutative ring $R$:*

(a) *If $R$ has a unique maximal ideal, then $IG(R)$ and $C\mathbb{AG}(R)$ are null;*

(b) *If $R$ has two maximal ideals, then $IG(R) = C\mathbb{AG}(R)) \cong K_1 \sqcup K_2$;*

(c) *$IG(R) = C\mathbb{AG}(R)$ is connected if and only if $R$ has at least three maximal ideals.*

Recall that a commutative Noetherian local ring $R$ is *Gorenstein* if it has finite injective dimension as an $R$-module, and a Gorenstein ring is a commutative Noetherian ring such that $R_P$ is Gorenstein for every prime ideal $P$ of $R$. Moreover, a commutative Artinian local ring $(R, \mathfrak{m})$ is Gorenstein if and only if $dim_{R/\mathfrak{m}} Ann(\mathfrak{m}) = 1$.

**Theorem 5.7.14.** ([5, Theorem 14]) *If $R$ is a commutative Artinian ring and $\omega(\mathbb{AG}(R)) = 2$, then $R$ is Gorenstein.*

**Theorem 5.7.15.** ([5, Theorem 15]) *Let $(R, \mathfrak{m})$ be a commutative local ring such that $\mathfrak{m}^k = (0)$ for some positive integer $k$. If $\omega(\mathbb{AG}(R))$ is finite, then either $R/\mathfrak{m}$ is a finite field or $\mathfrak{m}^i$ is a principal ideal of $R$ for every $i \geq \lceil \frac{k}{2} \rceil$.*

**Theorem 5.7.16.** ([5, Theorem 20]) *Let $R$ be a reduced commutative ring such that $\mathbb{AG}(R)$ has no infinite cliques. Then the set of ideals of the form $Ann(I)$, where $I$ is an ideal of $R$, is finite.*

Next, we concentrate on the vertex coloring of annihilating-ideal graphs of some rings of fractions. Recall that a *retract* of a graph $G$ is a subgraph $H$ of $G$ such that there is a homomorphism $\varphi : G \to H$ such that $\varphi(x) = x$ for every vertex $x$ of $H$. The homomorphism $\varphi$ is called a *retract (graph) homomorphism.*

**Theorem 5.7.17.** ([3, Theorem 1]) *Let $R$ be a commutative ring and $S$ a multiplicatively closed subset of $R$ containing no zero-divisors. Then $\omega(\mathbb{AG}(R_S)) \leq \omega(\mathbb{AG}(R))$. Moreover, $\mathbb{AG}(R_S)$ is a retract of $\mathbb{AG}(R)$ if $R$ is reduced. In particular, $\omega(\mathbb{AG}(R_S)) = \omega(\mathbb{AG}(R))$ when $R$ is reduced.*

**Corollary 5.7.18.** ([3, Corollary 2]) *Let $R$ be a reduced commutative ring. Then $\omega(\mathbb{AG}(T(R))) = \omega(\mathbb{AG}(R))$.*

**Corollary 5.7.19.** ([3, Corollary 3,4]) *Let $R$ be a commutative ring and $S$ a multiplicatively closed subset of $R$ containing no zero-divisors. Then $\chi(\mathbb{AG}(R_S)) \leq \chi(\mathbb{AG}(R))$. Moreover, if $R$ is reduced, then $\chi(\mathbb{AG}(R_S)) = \chi(\mathbb{AG}(R))$. In particular, if $R$ is reduced, then $\chi(\mathbb{AG}(T(R))) = \chi(\mathbb{AG}(R))$.*

Conjecture 5.5.28 states that for a reduced commutative ring $R$ with more than two minimal prime ideals, $gr(\mathbb{AG}(R)) = 3$. The following theorem settles this conjecture.

**Theorem 5.7.20.** ([3, Theorem 6]) *Let $R$ be a commutative ring and $P_1, \ldots, P_n$ be $n$ distinct prime ideals of $R$. Then there is a clique in $\mathbb{AG}(R)$ of size $n$. In particular, $gr(\mathbb{AG}(R)) = 3$.*

Thus, for a commutative ring $R$, we have $\omega(\mathbb{AG}(R)) \geq |Min(R)|$. The next theorem is a generalization of Corollary 5.7.8.

**Theorem 5.7.21.** ([3, Theorem 8]) *Let $R$ be a reduced commutative ring. Then $\chi(\mathbb{AG}(R)) = \omega(\mathbb{AG}(R)) = |Min(R)|$.*

**Theorem 5.7.22.** ([3, Theorem 11]) *Let $R$ be a commutative ring with $|Max(R)| < \infty$. If $\omega(\mathbb{AG}(R_\mathfrak{m}))$ is finite for every $\mathfrak{m} \in Max(R)$, then $\omega(\mathbb{AG}(R))$ is finite.*

**Theorem 5.7.23.** ([3, Theorem 12]) *Let $R$ be a commutative ring with $|Max(R)| < \infty$. If $\chi(\mathbb{AG}(R_\mathfrak{m}))$ is finite for every $\mathfrak{m} \in Max(R)$, then $\chi(\mathbb{AG}(R))$ is finite.*

In the next theorem, we state some formulas for the clique number and the chromatic number of the annihilating-ideal graph of a direct product of rings.

**Theorem 5.7.24.** ([3, Theorem 13]) *Let $S = S_1 \times \cdots \times S_k$, $T = T_1 \times \cdots \times T_\ell$, where $S_i$ is a commutative nondomain ring and $T_j$ is a reduced commutative ring for $i = 1, \ldots, k$ and $j = 1, \ldots, \ell$. If $R = S \times T$, then the following statements hold:*

(a) *If $T$ is not an integral domain, then $\omega(\mathbb{AG}(R)) = \omega(\mathbb{AG}(S)) + \omega(\mathbb{AG}(T))$;*

(b) *If every $T_i$ is an integral domain, then $\omega(\mathbb{AG}(R)) = \omega(\mathbb{AG}(S)) + \ell$;*

(c) $\sum_{i=1}^{k} \omega(\mathbb{AG}(S_i)) + max\left\{\ell, \sum_{i=1}^{\ell} \omega(\mathbb{AG}(T_i))\right\} \leq \omega(\mathbb{AG}(R))$;

(d) $\omega(\mathbb{AG}(S)) \leq \prod_{i=1}^{k}(\omega(\mathbb{AG}(S_i)) + 1) - 1$. *Moreover, the equality holds if and only if every $S_i$ has a maximum clique $C_i$ such that for every $I \in C_i$, $I^2 = (0)$.*

**Corollary 5.7.25.** ([3, Corollary 14]) *Let $R_1$ and $R_2$ be commutative rings and $R = R_1 \times R_2$. If $\mathbb{AG}(R)$ is a triangle-free graph, then exactly one of the following statements holds:*

(a) *Both $R_1$ and $R_2$ are integral domains;*

(b) *One $R_i$ is an integral domain and the other is a ring with a unique nontrivial ideal.*

*Moreover, $\mathbb{AG}(R)$ has no cycles if and only if either $R \cong F \times S$ or $R \cong F \times D$, where $F$ is a field, $S$ is a ring with a unique nontrivial ideal, and $D$ is an integral domain.*

Now, we state some similar results for the chromatic number of the annihilating-ideal graph of a direct product of commutative rings.

**Theorem 5.7.26.** ([3, Theorem 15]) *Let $S = S_1 \times \cdots \times S_k$, $T = T_1 \times \cdots \times T_\ell$, $S_i$ be a commutative nondomain, and $T_j$ be a reduced commutative ring for $i = 1, \ldots, k$ and $j = 1, \ldots, \ell$. If $R = S \times T$, then the following statements hold:*

(a) *If $T$ is not an integral domain, then $\chi(\mathbb{AG}(R)) = \chi(\mathbb{AG}(S)) + \chi(\mathbb{AG}(T))$;*

(b) *If every $T_i$ is an integral domain, then $\chi(\mathbb{AG}(R)) = \chi(\mathbb{AG}(S)) + \ell$;*

(c) $\sum_{i=1}^{k} \chi(\mathbb{AG}(S_i)) + max\left\{\ell, \sum_{i=1}^{\ell} \chi(\mathbb{AG}(T_i))\right\} \leq \chi(\mathbb{AG}(R))$;

(d) $\chi(\mathbb{AG}(S)) \leq \prod_{i=1}^{k}(\chi(\mathbb{AG}(S_i)) + 1) - 1$.

**Corollary 5.7.27.** ([3, Corollary 16]) *Let $R_1, \ldots, R_k$ be rings and $R = R_1 \times \cdots \times R_k$. Then*

(a) $\omega(\mathbb{AG}(R)) < \infty$ *if and only if* $\omega(\mathbb{AG}(R_i)) < \infty$ *for every* $i$, $1 \leq i \leq k$;

(b) $\chi(\mathbb{AG}(R)) < \infty$ *if and only if* $\chi(\mathbb{AG}(R_i)) < \infty$ *for every* $i$, $1 \leq i \leq k$.

In the sequel, the chromatic numbers of the two graphs $\mathbb{AG}(R)$ and $\Gamma(R)$ are compared. The following example shows that the chromatic numbers of the zero-divisor graph and the annihilating-ideal graph of a ring are not comparable in general.

**Example 5.7.28.** ([3, Example 17])

(a) *If* $R = F[x]/(x^2)$, *where* $F$ *is a finite field with at least three elements, then* $\omega(\mathbb{AG}(R)) = \chi(\mathbb{AG}(R)) = 1$, *but* $\chi(\Gamma(R)) \geq \omega(\Gamma(R)) \geq |F| - 1$;

(b) *If* $R = \mathbb{Z}_2 \times \mathbb{Z}_2$, *then* $\omega(\mathbb{AG}(R)) = \chi(\mathbb{AG}(R)) = \omega(\Gamma(R)) = \chi(\Gamma(R)) = 2$;

(c) *If* $R = \mathbb{Z}_2[x, y]/(x, y)^2$, *then* $\omega(\mathbb{AG}(R)) = \chi(\mathbb{AG}(R)) = 4$, *but* $\omega(\Gamma(R)) = \chi(\Gamma(R)) = 3$.

The following two theorems provide some relationships between the chromatic numbers of the annihilating-ideal graph and the zero-divisor graph.

**Theorem 5.7.29.** ([3, Theorem 18]) *Let* $R$ *be a commutative ring with* $\chi(\Gamma(R)) < \infty$. *Then* $\chi(\mathbb{AG}(R)) \leq 2^{\chi(\Gamma(R))} - 1$.

**Theorem 5.7.30.** ([3, Theorem 19]) *Let* $R$ *be a commutative ring with* $\chi(\Gamma(R)) < \infty$ *and every nonzero annihilating-ideal of* $R$ *is infinite. Then* $\chi(\mathbb{AG}(R)) \leq \chi(\Gamma(R))$.

## 5.8 Genus of Annihilating-ideal Graphs

In this section, we investigate the genus of the annihilating-ideal graph of a commutative ring. The following results are due to F. Aliniaeifard and Behboodi [47].

**Lemma 5.8.1.** ([47, Lemma 2.4]) *Let* $(R, \mathfrak{m})$ *be a commutative Artinian local ring with* $|R/\mathfrak{m}| = \infty, \mathfrak{m}^3 = (0)$, *and* $dim_{R/\mathfrak{m}} Ann(\mathfrak{m}) = 1$. *If* $g(\mathbb{AG}(R)) < \infty$, *then* $dim_{R/\mathfrak{m}} \mathfrak{m}/\mathfrak{m}^2 \leq 2$.

**Theorem 5.8.2.** ([47, Theorem 2.5]) *Let* $(R, \mathfrak{m})$ *be a commutative Artinian local ring. If* $g(\mathbb{AG}(R)) < \infty$, *then either* $R$ *has only finitely many ideals or* $R$ *is a Gorenstein ring with* $dim_{R/M} \mathfrak{m}/\mathfrak{m}^2 = 2$.

**Theorem 5.8.3.** ([47, Theorem 2.7]) *Let* $R$ *be a commutative Artinian ring. If* $g(\mathbb{AG}(R)) < \infty$, *then either* $R$ *has only finitely many ideals or* $(R, \mathfrak{m})$ *is a Gorenstein ring with* $dim_{R/\mathfrak{m}} \mathfrak{m}/\mathfrak{m}^2 = 2$.

**Lemma 5.8.4.** ([47, Lemma 2.8]) *Let* $(R, \mathfrak{m})$ *be a commutative Gorenstein ring,* $k$ *a positive integer, and* $q$ *a prime power. If* $\mathfrak{m}^2 \neq (0)$, $\mathfrak{m}^3 = (0)$, $|R/\mathfrak{m}| = q$, *and* $dim_{R/\mathfrak{m}}(\mathfrak{m}/\mathfrak{m}^2) = k > 6$, *then* $\mathbb{AG}(R)$ *contains a copy of* $K_{k-6,3}$.

In Theorem 4.5.1, it was shown that for every positive integer $k$, there are only finitely many finite commutative rings with $g(\Gamma(R)) = k$. But the result is not true if we replace $g(\Gamma(R))$ with $g(\mathbb{AG}(R))$. In fact, for every finite field $\mathbb{F}$, the ring $R = \mathbb{F} \times \mathbb{F} \times \mathbb{F} \times \mathbb{F}$ has a toroidal annihilating-ideal graph. It is proved that for all integers $q > 0$ and $k \geq 0$, there are only finitely many commutative Artinian rings $R$ such that $g(\mathbb{AG}(R)) = k$ and $|R/\mathfrak{m}| \leq q$ for every maximal ideal $\mathfrak{m}$ of $R$. We first state a particular case (when $q$ is a prime power) of the result.

**Theorem 5.8.5.** ([47, Theorem 2.10]) *For a prime power $q$ and an integer $k \geq 0$, there are only finitely many commutative Artinian local rings $(R, \mathfrak{m})$ satisfying the following conditions:*

(a) $g(\mathbb{AG}(R)) = k$;

(b) $|R/\mathfrak{m}| = q$.

**Corollary 5.8.6.** ([47, Corollary 2.11]) *For integers $q > 0$ and $k \geq 0$, there are only finitely many commutative Artinian rings $R$ satisfying the following conditions:*

(a) $g(\mathbb{AG}(R)) = k$;

(b) $|R/\mathfrak{m}| \leq q$ *for every maximal ideal* $\mathfrak{m}$ *of* $R$.

Next, we present some results concerning the genus of annihilating-ideal graphs of commutative Noetherian rings.

**Theorem 5.8.7.** ([47, Proposition 3.4]) *Let $R$ be a commutative Noetherian local ring. If $g(\mathbb{AG}(R)) < \infty$, then either $R$ is an integral domain or all nontrivial ideals of $R$ are vertices of $\mathbb{AG}(R)$.*

**Theorem 5.8.8.** ([47, Theorem 3.5]) *Let $R$ be a commutative Noetherian ring such that all nontrivial ideals of $R$ are vertices of $\mathbb{AG}(R)$. If $g(\mathbb{AG}(R)) < \infty$, then either $R$ is a Gorenstein ring or $R$ is an Artinian ring with only finitely many ideals.*

**Corollary 5.8.9.** ([47, Corollary 3.6]) *Let $R$ be a nondomain Noetherian ring. If $g(\mathbb{AG}(R)) < \infty$, then either $R$ is a Gorenstein ring or $R$ is an Artinian ring with only finitely many ideals.*

In [52], Aliniaeifard, Behboodi, and Li continued the investigation of the genus of the annihilating-ideal graph of a commutative Noetherian ring. In this regard, the first result shows that commutative Noetherian local rings whose annihilating-ideal graphs have positive genus are Artinian.

**Theorem 5.8.10.** ([52, Proposition 3.1]) *Let $R$ be a commutative local ring with $0 < g(\mathbb{AG}(R)) < \infty$. Then $R$ is Artinian if and only if $R$ is Noetherian.*

The next theorem shows that commutative Artinian rings whose annihilating-ideal graphs have positive genus have finitely many ideals.

**Theorem 5.8.11.** ([52, Theorem 3.2]) *Let $R$ be a commutative Artinian ring with $0 < g(\mathbb{AG}(R)) < \infty$. Then $R$ has only finitely many ideals.*

**Corollary 5.8.12.** ([52, Corollary 3.3]) *Let $R$ be a commutative Noetherian ring such that all nontrivial ideals of $R$ are vertices of $\mathbb{AG}(R)$. If $0 < g(\mathbb{AG}(R)) < \infty$, then $R$ is an Artinian ring with only finitely many ideals.*

The following result characterizes commutative Noetherian rings whose nonzero proper ideals have nonzero annihilators and annihilating-ideal graphs have finite genus (including genus zero).

**Theorem 5.8.13.** ([52, Theorem 3.4]) *Let $R$ be a commutative Noetherian ring such that all nontrivial ideals of $R$ are vertices of $\mathbb{AG}(R)$ and $g(\mathbb{AG}(R)) < \infty$. If $R$ is not a local ring, then $R$ is an Artinian ring with only finitely many ideals. Otherwise, either $(R, \mathfrak{m})$ is a Gorenstein ring such that $\mathfrak{m}^n \neq \mathfrak{m}^{n+1}$ (for every positive integer $n$) and $g(\mathbb{AG}(R)) = 0$, or $R$ is an local Artinian ring with maximal ideal $\mathfrak{m}$ such that $\mathfrak{m}^t \neq (0)$ and $\mathfrak{m}^{t+1} = (0)$. We have the following:*

(a) *If $t = 1$, then $R$ is either finite or a special principal ideal ring;*

(b) *If $t = 2$, then one of the following holds:*

   (i) *$R$ is finite;*

   (ii) *$R$ is a special principal ideal ring;*

   (iii) *$dim_{R/\mathfrak{m}}(\mathfrak{m}^2/\mathfrak{m}^3) = 1$, $dim_{R/\mathfrak{m}}(\mathfrak{m}/\mathfrak{m}^2) = 2$, $R$ has infinitely many ideals, and $g(\mathbb{AG}(R)) = 0$;*

(c) *If $t = 3$, then one of the following holds:*

   (i) *$R$ is finite;*

   (ii) *$R$ is a special principal ideal ring;*

   (iii) *$dim_{R/\mathfrak{m}}(\mathfrak{m}/\mathfrak{m}^2) = 2$, $dim_{R/\mathfrak{m}}(\mathfrak{m}^2/\mathfrak{m}^3) = dim_{R/\mathfrak{m}}(\mathfrak{m}^3/\mathfrak{m}^4) = 1$, $R$ has infinitely many ideals, and $g(\mathbb{AG}(R)) = 0$;*

(d) *If $t \geq 4$, then $R$ is either finite or a special principal ideal ring.*

One can refer to Visweswaran and Patel [417] for the study of the complement of the annihilating ideal graph of a commutative ring.

In [49], Aliniaeifard, Behboodi, Mehdi-Nezhad, and Rahimi have extended and studied the notion $\mathbb{AG}(R)$ to the more general setting of the annihilating-ideal graph of a commutative ring $R$ with respect to an ideal $I$ of $R$ by replacing (nonzero) ideals whose product is zero with ideals ($\nsubseteq I$) whose product lies in $I$, i.e., the annihilating-ideal graph of $R$ with respect to $I$, denoted by $\mathbb{AG}_I(R)$, is the graph whose vertices are the set $\{A \in N_I(R) \mid AB \subseteq I$ for some $B \in N_I(R)\}$ with distinct vertices $A$ and $B$ adjacent if and only if $AB \subseteq I$, where $N_I(R)$ denotes the set of all ideals of $R$ that are not contained in $I$. Thus, $\mathbb{AG}_I(R) = \mathbb{AG}(R)$ for $I = (0)$. Clearly, $\mathbb{AG}_I(R) = \emptyset$ if and only if $I$ is a nonzero prime ideal of $R$. They also discussed when $\mathbb{AG}_I(R)$ is bipartite and gave some results on the subgraphs and the parameters of $\mathbb{AG}_I(R)$.

# 5.9 Compressed Zero-divisor Graphs

Inspired by ideas from Mulay in [295], Spiroff and Wickham in [357] studied the graph of *equivalence classes of zero-divisors* of a commutative ring $R$, denoted by $\Gamma_E(R)$, which is constructed from classes of zero-divisors determined by annihilator ideals, rather than individual zero-divisors themselves. Anderson and LaGrange in [77] called this graph the *compressed zero-divisor graph.* Also see [80], [179], and [172].

Specifically, define an equivalence relation $\sim$ on $R$ by $r \sim s$ if and only if $Ann_R(r) = Ann_R(s)$, and let $[r]_R = \{s \in R \mid s \sim r\}$ and $R_E = R/\!\sim\, = \{[r]_R \mid r \in R\}$. It is easily checked that $[0]_R = \{0\}, [1]_R = R \setminus Z(R)$, and $[r]_R \subseteq Z(R)^*$ for every $r \in R \setminus ([0]_R \cup [1]_R)$. Moreover, $R_E$ becomes a commutative monoid under the multiplication $[r] \cdot [s]_R = [rs]_R$. Then $\Gamma_E(R)$ is the simple graph with vertices the elements of $R_E \setminus \{[0]_R, [1]_R\}$, and distinct vertices $[r]_R$ and $[s]_R$ are adjacent if and only if $[r]_R[s]_R = [0]_R$, if and only if $rs = 0$. Moreover, $\Gamma_E(R)$ is just the semigroup zero-divisor graph of $R_E$, i.e., $\Gamma_E(R) = \Gamma(R_E)$ (see Section 6.2 for zero-divisor graphs of semigroups). In particular, $diam(\Gamma_E(R)) \in \{0, 1, 2, 3\}$ and $gr(\Gamma_E(R)) \in \{3, 4, \infty\}$ for any commutative ring $R$ by Theorem 6.2.2.

## 5.9.1 Properties of Compressed Zero-divisor Graphs

We first collect some basic results about $\Gamma_E(R)$.

**Proposition 5.9.1.** ([172, Proposition 5.4]) *Let $R$ be a commutative Noetherian ring. If $\Gamma_E(R)$ is acyclic, then it is a star graph or a path of length* 3.

**Proposition 5.9.2.** ([172, Theorem 5.6]) *Let $R$ be a commutative Noetherian ring. If the girth of $\Gamma_E(R)$ is finite, then it is* 3.

**Proposition 5.9.3.** ([357, Proposition 1.4]) *The graph $\Gamma_E(R)$ is connected and $diam(\Gamma_E(R)) \leq 3$.*

While the above result holds for any commutative ring $R$ since $\Gamma(R_E)$ is a semigroup zero-divisor graph, the following results from [357] all include the riding hypothesis that $R$ is a commutative Noetherian ring and use associated primes.

**Proposition 5.9.4.** ([357, Proposition 1.5]) *Let $R$ be such that $\Gamma_E(R)$ has at least three vertices. Then $\Gamma_E(R)$ is not complete.*

**Proposition 5.9.5.** ([357, Proposition 1.7]) *Let $R$ be a ring such that $\Gamma_E(R)$ is complete $r$-partite. Then $r = 2$ and $\Gamma_E(R) = K_{n,1}$ for some $n \geq 1$.*

**Proposition 5.9.6.** ([357, Proposition 1.8]) *For any $R$, $\Gamma_E(R)$ is not a cycle graph.*

**Proposition 5.9.7.** ([357, Proposition 1.10]) *For any ring $R$, there are no finite regular graphs $\Gamma_E(R)$ with more than two vertices. (Note that a graph is called regular if the degrees of all vertices are equal.)*

**Proposition 5.9.8.** ([357, Proposition 2.4]) *Any ring $R$ such that $\Gamma_E(R)$ is a star graph with at least four vertices satisfies the following properties:*

(1) $Ass(R) = \{P\}$;

(2) $P^3 = 0$;

(3) $char(R) = 2, 4,$ *or* $8$.

**Corollary 5.9.9.** ([357, Corollary 2.5]) *If $R$ is a finite ring and $\Gamma_E(R)$ is a star graph with at least four vertices, then $R$ is a local ring.*

Recall that for a graph $G$, $N_G(a)$ denotes the set of all vertices in $G$ that are adjacent to $a$. If $|N_G(a)| \leq 1$, then the vertex $a$ is called an end.

**Proposition 5.9.10.** ([357, Proposition 3.2]) *If $R$ is a ring such that $|\Gamma_E(R)| > 3$, then no associated prime of $R$ is an end.*

Let $R$ be a commutative ring with $1 \neq 0$. The unique maximal ring of quotients of $R$ is denoted by $Q(R)$ (some authors call this ring the *complete ring of quotients of $R$*). Recall that $T(R)$ is called the total quotient ring of $R$ (i.e., let $S = R \setminus Z(R)$, then $T(R) = R_S$), and $\Gamma(T(R)) \cong \Gamma(R)$ by [81, Theorem 2.2].

**Theorem 5.9.11.** ([77, Theorem 3.2]) *Let $R$ be a commutative ring with $1 \neq 0$, and let $S$ be a subring of $Q(R)$ containing $R$. If $a, b \in R$, then $[a]_R = [b]_R$ if and only if $[a]_S = [b]_S$. In particular, the function $f : \Gamma_E(R) \to \Gamma_E(S)$ defined by $f([a]_R) = [a]_S$ is a well-defined isomorphism onto a subgraph of $\Gamma_E(S)$. Moreover, if $R \subseteq S \subseteq T(R)$, then $f$ is an isomorphism. Thus, if $S_1$ and $S_2$ are any two subrings of $T(R)$ containing $R$, then $\Gamma_E(S_1) \cong \Gamma_E(S_2)$.*

Let $G$ be a simple graph. Then as in [321], given $v, w \in V(G)$, define $v \equiv w$ if and only if $N_G(v) \setminus \{w\} = N_G(w) \setminus \{v\}$. Also, as in [80], define $v \approx w$ if and only if $N_G(v) = N_G(w)$. Note that if $v$ and $w$ are not adjacent, then $v \equiv w$ if and only if $v \approx w$. Given a commutative ring $R$ with $1 \neq 0$, let $r, s \in V(\Gamma(R))$. Then $r \equiv s$ in $\Gamma(R)$ if and only if $\text{Ann}(r) \setminus \{r, s\} = \text{Ann}(s) \setminus \{r, s\}$. In [321, p. 78], it is observed (without proof) that $\equiv$ is an equivalence relation on $V(G)$.

**Theorem 5.9.12.** ([80, Theorem 2.1], [321, p. 78]) *Let $G$ be a simple graph. Then the relation $\equiv$ on $V(G)$ is an equivalence relation.*

*Proof.* The relation $\equiv$ is obviously reflexive and symmetric. To prove it is transitive, let $u, v, w \in V(G)$ such that $u \equiv v$ and $v \equiv w$. Of course, the relation $u \equiv w$ holds trivially if $u$, $v$, and $w$ are not mutually distinct. Henceforth, assume that $u$, $v$, and $w$ are mutually distinct.

The relations $u \equiv v$ and $v \equiv w$ imply $N(u)\setminus\{v\} = N(v)\setminus\{u\}$ and $N(v)\setminus\{w\} = N(w)\setminus\{v\}$, respectively. Then $N(u)\setminus\{v,w\} = N(v)\setminus\{u,w\} = N(w)\setminus\{u,v\}$. Thus, to show that $u \equiv w$, it suffices to prove that $v \in N(u)$ if and only if $v \in N(w)$.

If $v \in N(u)$, then $u \in N(v)\setminus\{w\} = N(w)\setminus\{v\} \subseteq N(w)$. Thus, $w \in N(u)\setminus\{v\} = N(v)\setminus\{u\} \subseteq N(v)$. Therefore, $v \in N(w)$ as desired. A symmetric argument proves that $v \in N(u)$ whenever $v \in N(w)$. Hence, $u \equiv w$. □

For a simple graph $G$, let $G/\!\equiv$ be the graph whose vertices are the equivalence classes $\overline{v} = \{w \in V(G) \mid v \equiv w\}$ (for every $v \in V(G)$) such that distinct vertices $\overline{v}_1$ and $\overline{v}_2$ are adjacent in $G/\!\equiv$ if and only if some (and hence every) element of $\overline{v}_1$ is adjacent to some (and hence every) element of $\overline{v}_2$ in $G$. Let $\psi_G : G \to G/\!\equiv$ be the natural surjective mapping defined by $\psi_G(v) = \overline{v}$. Note that if $v_1$ and $v_2$ are adjacent vertices in $G$, then $\psi_G(v_1)$ and $\psi_G(v_2)$ are adjacent in $G/\!\equiv$ if and only if $\overline{v_1} \neq \overline{v_2}$. Of course, if $\psi_G(v_1)$ and $\psi_G(v_2)$ are adjacent in $G/\!\equiv$, then $v_1$ and $v_2$ are adjacent in $G$. Similarly, one can define $G/\!\approx$.

Observe that if $G$ is a complete graph, then $v \equiv w$ for every $v, w \in V(G)$.

**Lemma 5.9.13.** ([80, Lemma 2.2]) *Let $R$ be a commutative ring with $1 \neq 0$. Assume that $|V(\Gamma(R))| \geq 3$, and let $v, w \in V(\Gamma(R))$ such that $v \equiv w$. If $v \in N(w)$, then $v^2 = w^2 = 0$.*

*Proof.* If $v + w = 0$, then $v^2 = v(-w) = -(vw) = 0$ and $w^2 = (-v)w = -(vw) = 0$. Thus, assume that $v + w \neq 0$. Since $\Gamma(R)$ is connected and $|V(\Gamma(R))| \geq 3$, there exists an $x \in V(\Gamma(R))$ such that either $x \in N(v)\setminus\{w\}$ or $x \in N(w)\setminus\{v\}$. Either way, it follows that $x \in N(v)\setminus\{w\} = N(w)\setminus\{v\}$. Note that $w(x+v) = wx + wv = 0 + 0 = 0$.

If $w = x + v$, then $v^2 = v(w - x) = vw - vx = 0 - 0 = 0$, and the above comment shows that $w^2 = 0$. If $w \neq x + v$, then either $x + v = 0$ or $x + v \in N(w)\setminus\{v\} = N(v)\setminus\{w\}$. Therefore, either $v^2 = v(-x) = -(vx) = 0$ or $v^2 = 0 + v^2 = vx + v^2 = v(x+v) = 0$, respectively. It follows that either $v + w = 0$ or $v + w \in N(v)\setminus\{w\} = N(w)\setminus\{v\}$, which implies that $w^2 = 0 + w^2 = wv + w^2 = w(v+w) = 0$. □

By Lemma 5.9.13, $\approx$ and $\equiv$ are the same relation on $V(\Gamma(R))$ if $R$ is reduced and $R \ncong \mathbb{Z}_2 \times \mathbb{Z}_2$. Actually, it shows that they are the same relation even if $|\mathrm{Nil}(R)| = 2$ (cf. [321, Corollary 45]). The next lemma gives a partial converse to Lemma 5.9.13.

**Lemma 5.9.14.** ([80, Lemma 2.3]) *Let $R$ be a commutative ring with $1 \neq 0$. Assume that $|V(\Gamma(R))| \geq 3$, and let $v, w \in V(\Gamma(R))$ such that $v \equiv w$. If $v \neq w$ and $v \notin N(w)$ (i.e., if $v \approx w$), then $v^2 \neq 0$ and $w^2 \neq 0$.*

*Proof.* As in the proof of Lemma 5.9.13, there exists an $x \in N(v)\setminus\{w\} = N(w)\setminus\{v\}$. Since $v \notin N(w)$, it follows that $x \in N(v) = N(w)$. If $v^2 = 0$, then either $x+v=0$ or $x+v \in N(v)$. Note that $vw = 0+vw = xw+vw = (x+v)w$. Thus, if $x+v=0$, then $vw = (x+v)w = 0w = 0$, contradicting that $v \notin N(w)$. If $x+v \in N(v) = N(w)$, then $vw = (x+v)w = 0$, contradicting that $v \notin N(w)$. Therefore, $v^2 \neq 0$. A symmetric argument shows that $w^2 \neq 0$. □

Let $R$ be a commutative ring with $1 \neq 0$. Using Theorem 5.9.12, it can easily be shown that the relation $\Theta$ on $R$ defined by $r\Theta s$ if and only if $\mathrm{Ann}_R(r) \setminus \{r,s\} = \mathrm{Ann}_R(s) \setminus \{r,s\}$ is an equivalence relation. The next result shows that $\Theta$ and $\sim$ are the same relation for most rings. An equivalent result was proved by Redmond in [321, Corollary 43].

**Theorem 5.9.15.** ([80, Theorem 2.4], [321, Corollary 43]) *Let $R$ be a commutative ring with $1 \neq 0$. Assume that $R \ncong S$ for every $S \in \{\mathbb{Z}_2, \mathbb{Z}_4, \mathbb{Z}_2[x]/(x^2), \mathbb{Z}_2 \times \mathbb{Z}_2\}$. If $r, s \in R$, then $r \sim s$ if and only if $r\Theta s$, i.e., $Ann_R(r) = Ann_R(s)$ if and only if $Ann_R(r) \setminus \{r,s\} = Ann_R(s) \setminus \{r,s\}$.*

*Proof.* It is straightforward to check that the theorem holds if either $R$ is an integral domain with more than two elements or $R \cong S$ for some $S \in \{\mathbb{Z}_9, \mathbb{Z}_3[x]/(x^2)\}$. Henceforth, assume that $Z(R) \setminus \{0\} \neq \emptyset$ and $R \ncong S$ for every $S \in \{\mathbb{Z}_9, \mathbb{Z}_3[x]/(x^2)\}$.

The "only if" part of the theorem is clear. To prove the "if" statement, suppose that $r\Theta s$, i.e., $\mathrm{Ann}_R(r) \setminus \{r,s\} = \mathrm{Ann}_R(s) \setminus \{r,s\}$. The relation $r \sim s$ trivially holds if either $r \in R \setminus V(\Gamma(R))$ or $r = s$. So we may assume that $r \in V(\Gamma(R))$ and $r \neq s$. Then $s \in V(\Gamma(R))$ (because $|V(\Gamma(R))| \geq 3$) and $r \equiv s$ in $\Gamma(R)$.

First, suppose that $rs = 0$ (i.e., $r \in N(s)$). Then $r^2 = s^2 = 0$ by Lemma 5.9.13. Hence, $\{r,s\} \subseteq \mathrm{Ann}_R(r) \cap \mathrm{Ann}_R(s)$. Therefore, the equality $\mathrm{Ann}_R(r) \setminus \{r,s\} = \mathrm{Ann}_R(s) \setminus \{r,s\}$ implies that $\mathrm{Ann}_R(r) = \mathrm{Ann}_R(s)$, i.e., $r \sim s$.

Next, suppose that $rs \neq 0$ (i.e., $r \notin N(s)$). Then $r^2 \neq 0$ and $s^2 \neq 0$ by Lemma 5.9.14. Therefore, $\mathrm{Ann}_R(r) = \mathrm{Ann}_R(r) \setminus \{r,s\} = \mathrm{Ann}_R(s) \setminus \{r,s\} = \mathrm{Ann}_R(s)$, i.e., $r \sim s$. □

The next theorem shows that $\Gamma_E(R)$ is completely determined by, and may be computed from, $\Gamma(R)$ when $R \ncong \mathbb{Z}_2 \times \mathbb{Z}_2$. Recall that $\Gamma_E(R) = \Gamma(R_E)$.

**Theorem 5.9.16.** ([80, Theorem 2.5]) *Let $R$ be a commutative ring with $1 \neq 0$ that is not isomorphic to $\mathbb{Z}_2 \times \mathbb{Z}_2$. Then $\Gamma(R/{\sim}) \cong \Gamma(R)/{\equiv}$. In particular, $\Gamma(R/{\sim}) \cong \Gamma(R)/{\equiv}$ if $|V(\Gamma(R))| \neq 2$.*

*Proof.* It is straightforward to check that the theorem holds if $R \in \{\mathbb{Z}_4, \mathbb{Z}_2[X]/(X^2), \mathbb{Z}_9, \mathbb{Z}_3[X]/(X^2)\}$. Since $R \ncong \mathbb{Z}_2 \times \mathbb{Z}_2$, it can now be assumed that $|V(\Gamma(R))| \geq 3$ (e.g., see [325, 326]). For every $r \in V(\Gamma(R))$, set $\overline{r} = \{s \in R \mid r \equiv s\}$.

Let $r, s \in V(\Gamma(R))$. Of course, $r\Theta s$ in $R$ if and only if $r \equiv s$ in $\Gamma(R)$. By Theorem 5.9.15, $r\Theta s$ if and only if $r \sim s$. It follows that the mapping $\mu_R$ :

$\Gamma(R/\sim) \to \Gamma(R)/\equiv$ defined by $\mu_R([r]) = \overline{r}$ is well-defined and bijective (in fact, it is the identity mapping on $V(\Gamma(R/\sim)) = V(\Gamma(R)/\equiv)$). This mapping preserves and reflects adjacency relations since $[r]$ is adjacent to $[s]$ in $\Gamma(R/\sim)$ if and only if $[r] \neq [s]$ and $xy = 0$ for every $x \in [r]$ and $y \in [s]$, if and only if every element of $[r] = \overline{r}$ is adjacent to every element of $[s] = \overline{s}$ in $\Gamma(R)$, if and only if $\overline{r}$ is adjacent to $\overline{s}$ in $\Gamma(R)/\equiv$. Therefore, $\Gamma(R/\sim) \cong \Gamma(R)/\equiv$ (in fact, $\Gamma(R/\sim) = \Gamma(R)/\equiv$). The "in particular" statement is clear since $|V(\Gamma(\mathbb{Z}_2 \times \mathbb{Z}_2))| = 2$. □

**Corollary 5.9.17.** ([80, Corollary 2.6]) *Let $R$ and $S$ be commutative rings with $1 \neq 0$. If $\Gamma(R) \cong \Gamma(S)$ and $R, S \not\cong \mathbb{Z}_2 \times \mathbb{Z}_2$, then $\Gamma_E(R) \cong \Gamma_E(S)$. In particular, $\Gamma_E(R) \cong \Gamma_E(S)$ if $\Gamma(R) \cong \Gamma(S)$ and $|V(\Gamma(R))| \neq 2$.*

*Proof.* If $\Gamma(R) \cong \Gamma(S)$ and $R, S \not\cong \mathbb{Z}_2 \times \mathbb{Z}_2$, then $\Gamma_E(R) = \Gamma(R_E) = \Gamma(R/\sim) \cong (\Gamma(R)/\equiv) \cong (\Gamma(S)/\equiv) \cong \Gamma(S/\sim) = \Gamma(S_E) = \Gamma_E(S)$. The "in particular" statement is clear since $|V(\Gamma(\mathbb{Z}_2 \times \mathbb{Z}_2))| = 2$. □

### 5.9.2 Girth of Compressed Zero-divisor Graphs

The following result removes the Noetherian hypothesis in Proposition 5.9.2.

**Theorem 5.9.18.** ([80, Theorem 3.1]) *Let $R$ be a commutative ring with $1 \neq 0$. Then $\mathrm{gr}(\Gamma_E(R)) \in \{3, \infty\}$.*

*Proof.* It is clear that $\mathrm{gr}(\Gamma_E(R)) = \mathrm{gr}(\Gamma(R_E)) \in \{3, 4, \infty\}$, and $\mathrm{gr}(\Gamma_E(S)) \in \{3, \infty\}$ when $S$ is Noetherian by Proposition 5.9.2. Suppose, by way of contradiction, that $\mathrm{gr}(\Gamma_E(R)) = 4$. Then there is a 4-cycle $[a]_R - [b]_R - [c]_R - [d]_R - [a]_R$ in $\Gamma_E(R)$. Let $x \in R$ such that $x$ is an element of only one of $Ann_R(a)$ and $Ann_R(c)$ and $y \in R$ such that $y$ is an element of only one of $Ann_R(b)$ and $Ann_R(d)$. Now let $S$ be the subring of $R$ generated by $\{1, a, b, c, d, x, y\}$. Then $S$ is Noetherian and $\mathrm{gr}(\Gamma_E(S)) = 4$ since $\Gamma_E(S)$ contains the 4-cycle $[a]_S - [b]_S - [c]_S - [d]_S - [a]_S$, but no 3-cycle since any 3-cycle in $\Gamma_E(S)$ would also be a 3-cycle in $\Gamma_E(R)$. But this is a contradiction; so $\mathrm{gr}(\Gamma_E(R)) \neq 4$, and thus $\mathrm{gr}(\Gamma_E(R)) \in \{3, \infty\}$. □

The next theorem removes the Noetherian hypothesis in Theorem 5.9.1. We denote the graph obtained by introducing $m$ pendant vertices to one of the three vertices of the bipartition of order 3 in $K_{m,3}$ by $\overline{K}_{m,3}$.

**Theorem 5.9.19.** ([80, Theorem 3.2]) *Let $R$ be a commutative ring with $1 \neq 0$. Then $\mathrm{gr}(\Gamma_E(R)) = \infty$ if and only if $\Gamma_E(R)$ is either a star graph or a path on four vertices.*

*Proof.* Clearly, a star graph or a path on four vertices has infinite girth. Conversely, assume that $\mathrm{gr}(\Gamma_E(R)) = \infty$, but $\Gamma_E(R)$ is neither a star graph nor a path on four vertices. Since $\Gamma_E(R)$ is connected with $\mathrm{diam}(\Gamma_E(R)) \leq 3$, $\Gamma_E(R)$ contains a subgraph isomorphic to $\overline{K}_{1,3}$ with vertices $\{[a_1]_R, \ldots, [a_5]_R\}$ for $a_1, \ldots, a_5 \in R$, where $[a_4]_R$ and $[a_5]_R$ are the two ends adjacent to $[a_3]_R$. Since $[a_4]_R \neq [a_5]_R$, there is an $x \in R$ such that $x$ is in one, but not both, of $Ann_R(a_4)$ and $Ann_R(a_5)$. Let $S$ be the subring of $R$ generated by $\{1, a_1, \ldots, a_5, x\}$; then $S$ is Noetherian and $[a_1]_S, \ldots, [a_5]_S$ are distinct vertices in $\Gamma_E(S)$. Note that $\mathrm{gr}(\Gamma_E(S)) = \infty$ since any $n$-cycle in $\Gamma_E(S)$ would also be an $n$-cycle in $\Gamma_E(R)$. It is easily checked that $\Gamma_E(S)$ has a subgraph isomorphic to $\overline{K}_{1,3}$, which contradicts the fact that $\Gamma_E(S)$ must be either a star graph or a path on four vertices since $S$ is Noetherian with $\mathrm{gr}(\Gamma_E(S)) = \infty$. □

The next result shows that if $\Gamma_E(R)$ is finite, then $\Gamma_E(R) \cong \Gamma_E(S)$ for a commutative Noetherian ring $S$ with $1 \neq 0$ (in fact, a subring of $R$).

**Theorem 5.9.20.** ([80, Theorem 3.3]) *Let $R$ be a commutative ring with $1 \neq 0$ such that $\Gamma_E(R)$ is finite. Then there is a Noetherian subring $S$ of $R$ such that $\Gamma_E(S) \cong \Gamma_E(R)$.*

*Proof.* Let $|V(\Gamma_E(R))| = n < \infty$ and $V(\Gamma_E(R)) = \{[a_1]_R, \ldots, [a_n]_R\}$ for $a_1, \ldots, a_n \in R$. For all integers $i, j$ with $1 \leq i < j \leq n$, there is an $x_{ij} \in R$ such that $x_{ij}$ is in one, but not both, of $\mathrm{Ann}_R(a_i)$ and $\mathrm{Ann}_R(a_j)$. In fact, we may choose every $x_{ij} \in \{a_1, ..., a_n\}$ since $V(\Gamma_E(R)) = \{[a_1]_R, \ldots, [a_n]_R\}$. Let $S$ be the subring of $R$ generated by $\{1, a_1, \ldots, a_n\}$; then $S$ is Noetherian. Note that $S_E = \{[a_1]_S, \ldots, [a_n]_S\} \cup \{[0]_S, [1]_S\}$, $[a_i]_S = [a_j]_S$ if and only if $i = j$, and $[a_i]_S[a_j]_S = [0]_S$ if and only if $[a_i]_R[a_j]_R = [0]_R$. Thus $\Gamma_E(S) \cong \Gamma_E(R)$. □

The following theorem shows that when $\Gamma_E(R)$ is a finite star graph, then $\Gamma(R) \cong \Gamma_E(S)$ for a finite commutative local ring $S$ (cf. Proposition 5.9.8 and Corollary 5.9.9).

**Theorem 5.9.21.** ([80, Theorem 3.4]) *Let $R$ be a commutative ring with $1 \neq 0$ such that $\Gamma_E(R) \cong K_{1,\alpha}$ for some cardinal number $\alpha$.*

(1) *If $\alpha \geq 3$, then char$(R) \in \{2, 4, 8\}$ and $Z(R)$ is an associated prime ideal of $R$ with $Z(R)^3 = \{0\}$. If, in addition, $R = Z(R) \cup U(R)$ (e.g., $\dim(R) = 0$), then $R$ is quasilocal with maximal ideal $Z(R)$;*

(2) *There is a quasilocal commutative ring $S$ with maximal ideal $Z(S)$ and char$(S) \in \{2, 4, 8\}$ such that $\Gamma_E(S) \cong K_{1,\alpha}$;*

(3) *If $\alpha$ is finite, then there is a finite local commutative ring $S$ with char$(S) \in \{2, 4, 8\}$ such that $\Gamma_E(S) \cong K_{1,\alpha}$. If, in addition, $\alpha \geq 3$, then $S$ may be chosen to be a subring of $R$.*

*Proof.* (1) The first sentence follows from Proposition 5.9.8 (the proof of this result does not use the Noetherian hypothesis) and the fact that $Z(R)$ is an associated prime ideal of $R$ with $Z(R)^3 = \{0\}$. The "in addition" statement is clear since, in this case, $Z(R)$ is the unique maximal ideal of $R$.

(2) Let $T = R \setminus Z(R)$. Then $S = T(R) = R_T$ satisfies $S = Z(S) \cup U(S)$ and $\Gamma_E(S) \cong \Gamma_E(R)$ by Theorem 5.9.11. For $\alpha \geq 3$, $S$ is quasilocal with maximal ideal $Z(S)$ and $\mathrm{char}(S) \in \{2, 4, 8\}$ by part (1) above, and $\Gamma_E(S) \cong \Gamma_E(R) \cong K_{1,\alpha}$. For $\alpha < 3$, we have $\Gamma_E(\mathbb{Z}_4) \cong K_{1,0}, \Gamma_E(\mathbb{Z}_8) \cong K_{1,1}$, and $\Gamma_E(\mathbb{Z}_2[x, y]/(x^2, y^3, xy^2)) \cong K_{1,2}$.

(3) For $\alpha < 3$, use the three rings in the proof of part (2) above. So we may assume that $\alpha \geq 3$. Let $V(\Gamma_E(R)) = \{[a_1]_R, \dots, [a_{\alpha+1}]_R\}$ for $a_1, \dots, a_{\alpha+1} \in R$. Let $S$ be the subring of $R$ generated by $\{1, a_1, \dots, a_{\alpha+1}\}$; then $S$ is Noetherian. Moreover, $S$ is finite since $\mathrm{char}(S)$ is finite and $a_i a_j a_k = 0$ for all $1 \leq i, j, k \leq \alpha + 1$ by part (1) above, and $S$ is local by part (1) above. As in the proof of Theorem 5.9.20, we have $S_E = \{[a_1]_S, \dots, [a_{\alpha+1}]_S\} \cup \{[0]_S, [1]_S\}$, $[a_i]_S = [a_j]_S$ if and only if $i = j$, and $[a_i]_S[a_j]_S = [0]_S$ if and only if $[a_i]_R[a_j]_R = [0]_R$. Thus $\Gamma_E(S) \cong \Gamma_E(R)$. □

**Open Question** ([80, Question 3.6]). Let $R$ be a commutative ring with $1 \neq 0$ such that $\Gamma_E(R)$ is finite. Then is there a finite commutative ring $S$ with $1 \neq 0$ such that $\Gamma_E(R) \cong \Gamma_E(S)$?

## 5.10 Extended Zero-divisor Graphs

In [143], Bennis, Mikram, and Taraza gave another generalization of the zero-divisor graph. For a commutative ring $R$ with identity, the *extended zero-divisor graph of* $R$, denoted by $\overline{\Gamma}(R)$, is the simple graph with vertices $Z(R)^*$, and distinct vertices $x$ and $y$ are adjacent if and only if $x^m y^n = 0$ for positive integers $m$ and $n$ with $x^m \neq 0$ and $y^n \neq 0$. This graph has also been studied in [84], [144], and [135].

Clearly, $\Gamma(R)$ is a subgraph of $\overline{\Gamma}(R)$ and the two graphs coincide when $R$ is reduced. The next example shows that the graphs may be different when $R$ is not reduced.

**Example 5.10.1.** Let $R = \mathbb{Z}_8$. Then $R$ is not reduced, and it is easily checked that $\Gamma(R) = K_{1,2}$ and $\overline{\Gamma}(R) = K_3$. Moreover, $diam(\overline{\Gamma}(R)) = 1 < 2 = diam(\Gamma(R))$ and $gr(\overline{\Gamma}(R)) = 3 < \infty = gr(\Gamma(R))$.

The next theorem determines when $\overline{\Gamma}(R) = \Gamma(R)$ (cf. [84, Corollary 2.9(a)]).

**Theorem 5.10.2.** ([143, Theorem 2.1]) *Let $R$ be a commutative ring with $1 \neq 0$. Then $\overline{\Gamma}(R) = \Gamma(R)$ if and only if*

(1) *$x \in Nil(R)$ implies $x^2 = 0$, and*

(2) *$x \in R \setminus Nil(R)$ implies $Ann_R(x^2) = Ann_R(x)$.*

The following result holds since $\Gamma(R)$ is connected with $diam(\Gamma(R)) \leq 3$ (Theorem 2.4.1) and $\Gamma(R)$ is a subgraph of $\overline{\Gamma}(R)$ with the same vertices.

**Theorem 5.10.3.** ([143, Theorem 3.1]) *Let $R$ be a commutative ring with $1 \neq 0$. Then $\overline{\Gamma}(R)$ is connected and $diam(\overline{\Gamma}(R)) \leq 3$.*

The authors in [143] also determined when $\overline{\Gamma}(R)$ has a vertex adjacent to every other vertex (cf. Theorem 2.2.2) and when $\overline{\Gamma}(R)$ is a complete graph (cf. Theorem 2.2.5). Recall that $n_x$ is the degree of nilpotency of $x \in Nil(R)$.

**Theorem 5.10.4.** ([143, Theorem 3.2]) *Let $R$ be a commutative ring with $1 \neq 0$. Then there is a vertex $x$ of $\overline{\Gamma}(R)$ that is adjacent to every other vertex if and only if either $R \cong \mathbb{Z}_2 \times D$, where $D$ is an integral domain, or $Z(R) = \sqrt{Ann(x^{n_x - 1})}$.*

**Theorem 5.10.5.** ([143, Theorem 3.3]) *Let $R$ be a commutative ring with $1 \neq 0$. Then $\overline{\Gamma}(R)$ is a complete graph if and only if $R \cong \mathbb{Z}_2 \times \mathbb{Z}_2$, or $Z(R) = Nil(R)$ and $x^{n_x - 1}y^{n_y - 1} = 0$ for every $x, y \in Z(R)^*$.*

The next result shows that $gr(\overline{\Gamma}(R)) \in \{3, 4, \infty\}$ (since its proof showed that $\overline{\Gamma}(R)$ contains a 3-cycle or a 4-cycle and $gr(\Gamma(R)) \in \{3, 4, \infty\}$ by Theorem 2.3.3).

**Theorem 5.10.6.** ([143, Theorem 4.1]) *Let $R$ be a commutative ring with $1 \neq 0$. If $\overline{\Gamma}(R) \neq \Gamma(R)$, then $\overline{\Gamma}(R)$ contains a cycle.*

If $gr(\Gamma(R)) = 3$, then also $gr(\overline{\Gamma}(R)) = 3$; so the next two results, together with Theorems 2.3.8, 2.3.9, 2.3.10, and 2.3.11, essentially determine $gr(\overline{\Gamma}(R))$ (cf. the proof of [84, Theorem 2.15]).

**Theorem 5.10.7.** ([143, Theorem 4.5]) *Let $R$ be a commutative ring with $1 \neq 0$, $Nil(R) \neq 0$, and $gr(\Gamma(R)) = 4$. Then $\overline{\Gamma}(R) \neq \Gamma(R)$ implies that $gr(\overline{\Gamma}(R)) = 4$ and $\overline{\Gamma}(R)$ is a complete bipartite graph.*

**Theorem 5.10.8.** ([143, Theorem 4.6]) *Let $R$ be a commutative ring with $1 \neq 0$, $Nil(R) \neq 0$, and $gr(\Gamma(R)) = \infty$. Then exactly one of the following holds:*
(1) *$\overline{\Gamma}(R) = \Gamma(R)$ is a singleton or a star graph. In this case, $\overline{\Gamma}(R) = \infty$;*
(2) *$\Gamma(R) = \overline{K}_{1,3}$ (i.e., $R \cong \mathbb{Z}_2 \times \mathbb{Z}_4$ or $R \cong \mathbb{Z}_2 \times \mathbb{Z}_2[X]/(X^2)$). In this case, $gr(\overline{\Gamma}(R)) = 4$ and $\overline{\Gamma}(R) \neq \Gamma(R)$.*

Let $I$ be an ideal of a commutative ring $R$ with identity. In [84], Anderson and McClurkin introduced the *ideal-based extended zero-divisor graph of $R$ with respect to an ideal*, denoted by $\overline{\Gamma}_I(R)$, as the simple graph with vertices $\{x \in R \setminus I \mid xy \in I$ for some $y \in R \setminus I\}$, and distinct vertices $x$ and $y$ are adjacent if and only if $x^m y^n \in I$ for positive integers $m$ and $n$ with $x^m \notin I$ and $y^n \notin I$. Clearly, $\overline{\Gamma}_I(R) = \overline{\Gamma}(R)$ when $I = \{0\}$, and $\overline{\Gamma}_I(R) = \Gamma_I(R)$ when $I$ is a radical ideal of $R$.

The next two results generalize Theorems 5.10.2 and 5.10.3, respectively.

**Theorem 5.10.9.** ([84, Theorem 2.8]) *Let $R$ be a commutative ring with $1 \neq 0$ and $I$ an ideal of $R$. Then $\overline{\Gamma}_I(R) = \Gamma_I(R)$ if and only if*
(1) *$x \in \sqrt{I}$ implies $x^2 \in I$, and*
(2) *$x \in R \setminus \sqrt{I}$ implies $(I : x^2) = (I : x)$.*

**Theorem 5.10.10.** ([84, Theorem 2.11]) *Let $R$ be a commutative ring with $1 \neq 0$ and $I$ an ideal of $R$.*

(a) *$\overline{\Gamma}_I(R)$ is connected and $0 \leq diam(\overline{\Gamma}_I(R)) \leq diam(\Gamma_I(R)) \leq 3$;*

(b) *$gr(\overline{\Gamma}_I(R)) \in \{3, 4, \infty\}$, and moreover, $gr(\overline{\Gamma}_I(R)) \leq gr(\Gamma_I(R))$.*

# Chapter 6

# Zero-divisor Graph Generalizations

The concept of zero-divisor graph has been generalized to many other algebraic structures such as commutative semigroups, noncommutative rings, commutative semirings, modules over commutative rings, posets, and lattices. We first study the zero-divisor graph of noncommutative (i.e., not necessarily commutative) rings.

## 6.1 Noncommutative Rings

Let $R$ be a noncommutative ring and $Z(R), Z_L(R), Z_R(R), Z_T(R)$ denote the set of zero-divisors, left zero-divisors, right zero-divisors, and two-sided zero-divisors of $R$, respectively, i.e., $Z(R) = \{x \in R \mid xy = 0 \text{ or } yx = 0 \text{ for some } y \in R^*\}$, $Z_L(R) = \{x \in R \mid xa = 0 \text{ for some } a \in R^*\}$, $Z_R(R) = \{x \in R \mid bx = 0 \text{ for } b \in R^*\}$, and $Z_T(R) = \{x \in R \mid xa = bx = 0 \text{ for some } a, b \in R^*\}$. Note that when $R$ is commutative, all of these sets are equal. The concept of directed zero-divisor graphs of noncommutative rings was introduced and studied by Redmond [322]. Subsequently, several authors studied the properties of directed zero-divisor graphs of noncommutative rings [222, 396, 398]. Hereafter, in this section, by a ring $R$, we mean a noncommutative ring. For a ring $R$, the zero-divisor graph of $R$ is the directed graph $\overrightarrow{\Gamma}(R)$ with vertices $Z(R)^*$, where $x \to y$ is an edge between two distinct vertices $x$ and $y$ if and only if $xy = 0$. It is trivial that if $R$ is a commutative ring, then $y \to x$ is an edge whenever $x \to y$ is an edge. Therefore, if we consider the underlying undirected graph of $\overrightarrow{\Gamma}(R)$, this definition agrees with the usual definition of the zero-divisor of a commutative ring. Let us see some examples of directed zero-divisor graphs of noncommutative rings.



D. F. Anderson et al., *Graphs from Rings*, https://doi.org/10.1007/978-3-030-88410-9_6

**Example 6.1.1.** ([322])

Consider the ring $R = \left\{ \begin{pmatrix} a & b \\ 0 & 0 \end{pmatrix} \mid a \in \mathbb{Z}_4, b \in 2\mathbb{Z}_4 \right\}$.

Then $Z(R)^* = \{x_1, x_2, \ldots, x_7\}$, where $x_1 = \begin{pmatrix} 0 & 2 \\ 0 & 0 \end{pmatrix}$, $x_2 = \begin{pmatrix} 1 & 0 \\ 0 & 0 \end{pmatrix}$, $x_3 = \begin{pmatrix} 1 & 2 \\ 0 & 0 \end{pmatrix}$, $x_4 = \begin{pmatrix} 2 & 0 \\ 0 & 0 \end{pmatrix}$, $x_5 = \begin{pmatrix} 2 & 2 \\ 0 & 0 \end{pmatrix}$, $x_6 = \begin{pmatrix} 3 & 0 \\ 0 & 0 \end{pmatrix}$, $x_7 = \begin{pmatrix} 3 & 2 \\ 0 & 0 \end{pmatrix}$.

The corresponding directed zero-divisor graph $\overrightarrow{\Gamma}(R)$ is given below:

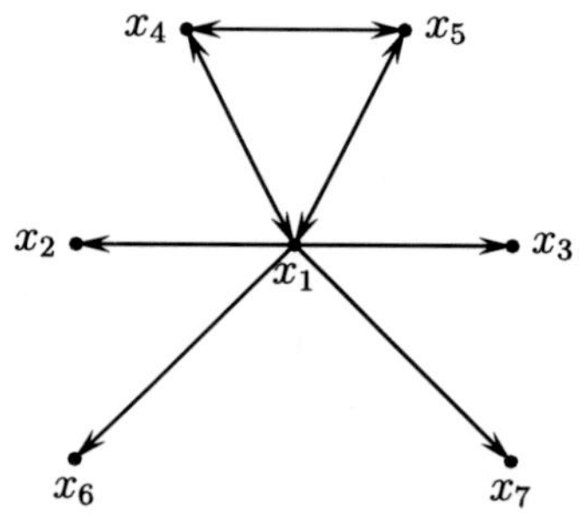


Figure 6.1: Directed zero-divisor graph

**Example 6.1.2.** ([322, Example 2.2]) Let $K$ be a field, $V = \oplus_{i=1}^{\infty} K$, and $R = Hom_K(V, V)$. Under point-wise addition and multiplication taken to be composition of functions, $R$ is an infinite noncommutative ring with identity. Let $\pi_1 : V \to V$ be defined by $(a_1, a_2, \ldots) \longmapsto (a_1, 0, 0, \ldots)$ and $f : V \to V$ be defined by $(a_1, a_2, \ldots) \longmapsto (0, a_1, a_2, \ldots)$. Then $\pi_1, f \in R$ and $\pi_1 f = 0$, while $f\pi_1 \neq 0$. Clearly $f$ is not a left zero-divisor in R (if $g \in R$ with $fg(a_1, a_2, \ldots) = (0, 0, \ldots)$ for every $(a_1, a_2, \ldots,) \in V$, then $g(a_1, a_2, \ldots) = (0, 0, \ldots)$ for every $(a_1, a_2, \ldots) \in V$). Hence, $\overrightarrow{\Gamma}(R)$ is nonempty and $\overrightarrow{\Gamma}(R)$ is not connected since there is no path leading from the vertex f to any other vertex of $\overrightarrow{\Gamma}(R)$.

Note that the structure of $\overrightarrow{\Gamma}(R)$ in Example 6.1.2 is still rather rich. By defining $\pi_j$ to be $(a_1, a_2, \ldots) \longmapsto (0, \ldots, 0, a_i, 0, \ldots)$, the vertices $\{\pi_j\}_{j=1}^{\infty}$ induce a complete subgraph of $\overrightarrow{\Gamma}(R)$.

The following theorem gives when $\overrightarrow{\Gamma}(R)$ is connected.

**Theorem 6.1.3.** ([322, Theorem 2.3]) *Let $R$ be a ring. Then $\overrightarrow{\Gamma}(R)$ is connected if and only if $Z_L(R) = Z_R(R)$. Further, if $\overrightarrow{\Gamma}(R)$ is connected, then diam($\overrightarrow{\Gamma}(R)$)* $\leq 3$.

**Corollary 6.1.4.** ([322, Corollary 2.4]) *Let $R$ be a left and right Artinian ring with a two-sided identity. Then $Z_L(R) = Z_R(R)$; so $\overrightarrow{\Gamma}(R)$ is connected.*

Note that when $R$ is a commutative ring, then $\overrightarrow{\Gamma}(R) = \Gamma(R)$. Hence we recover the following result on zero-divisor graphs as a corollary.

**Corollary 6.1.5.** *Let $R$ be a commutative ring with $Z(R)^* \neq \emptyset$. Then $\Gamma(R)$ is connected, diam$(\Gamma(R)) \leq 3$, and gr$(\Gamma(R)) \leq 4$ .*

Note that as in the case of commutative rings, $R$ is finite if and only if $\overrightarrow{\Gamma}(R)$ is finite [201, Theorem 1]. Further, when $R$ is finite, $\overrightarrow{\Gamma}(R)$ is connected and diam($\overrightarrow{\Gamma}(R)$)$\leq 3$. We may have $Z_R(R) = Z_L(R)$ for an infinite noncommutative ring $R$ as well. Let $\mathbb{H}$ be the ring of quaternions over the field of real numbers. Then $\mathbb{H}$ is an infinite noncommutative division ring. Let $R = \mathbb{H} \times \mathbb{H}$. The only nonzero zero-divisors of $R$ are of the form $(0, t)$ or $(t, 0)$ for $0 \neq t \in \mathbb{H}$. Note that all zero-divisors of $R$ are two-sided (i.e., $Z_L(R) = Z_R(R)$).

**Theorem 6.1.6.** ([322, Theorem 2.7]) *Let $R$ be a finite ring with no nonzero nilpotent elements and $\overrightarrow{\Gamma}(R) \neq \emptyset$. Then $\overrightarrow{\Gamma}(R)$ is not a tournament.*

Note that for a ring $R$, $M_n(R)$ denotes the ring of all $n \times n$ matrices over $R$. Unlike the zero-divisor graphs of commutative rings, directed zero-divisor graphs of noncommutative rings need not be connected. We proceed to see whether directed zero-divisor graphs of matrices over commutative rings are connected.

**Theorem 6.1.7.** ([222, Theorem 3.1]) *Let $R$ be a commutative ring. Then $\overrightarrow{\Gamma}(M_n(R))$ is connected and diam$(\overrightarrow{\Gamma}(M_n(R))) \leq 3$.*

Our next aim is to mention a relationship between the diameters of $\Gamma(R)$ and $\overrightarrow{\Gamma}(M_n(R))$.

**Proposition 6.1.8.** ([222, Proposition 3.1]) *Let $R$ be a commutative ring with $Z(R)^* \neq \emptyset$. Then diam$(\Gamma(R)) \leq$ diam$(\overrightarrow{\Gamma}(M_n(R)))$.*

Let $T(R)$ be the total quotient ring of a commutative ring $R$. Anderson et al. [81] proved that $\Gamma(R)$ and $\Gamma(T(R))$ were isomorphic as graphs, and as a consequence, it followed that these two graphs had the same diameter and girth. It is now observed that the same claim holds for $\overrightarrow{\Gamma}(M_n(R))$ and $\overrightarrow{\Gamma}(M_n(T(R)))$ as well.

**Theorem 6.1.9.** ([222, Theorem 3.2]) *Let $R$ be a commutative ring with total quotient ring. Then $\overrightarrow{\Gamma}(M_n(R)) \cong \overrightarrow{\Gamma}(M_n(T(R)))$.*

In view of the above, the following corollary holds.

**Corollary 6.1.10.** ([222, Corollary 3.1]) *Let $R$ be a commutative ring with total quotient ring. Then* $diam(\overrightarrow{\Gamma}(M_n(R))) = diam(\overrightarrow{\Gamma}(M_n(T(R))))$.

For a noncommutative ring $R$, the undirected zero-divisor graph $\Gamma(R)$ is the undirected graph with vertices $Z(R)^*$, and for distinct vertices $a$ and $b$, there is an edge connecting them if and only if $ab = 0$ or $ba = 0$. The following results concern the girth of the undirected zero-divisor graph $\Gamma(R)$ of a noncommutative ring $R$.

**Proposition 6.1.11.** ([222, Proposition 3.2]) *Let $R$ be a commutative ring. Then* $gr(\Gamma(M_n(R))) = 3$.

**Theorem 6.1.12.** ([322, Theorems 3.2 and 3.3]) *Let $R$ be a noncommutative ring with $Z(R)^* \neq \emptyset$. Then $diam(\Gamma(R)) \leq 3$, and $gr(\Gamma(R)) \leq 4$ if $\Gamma(R)$ contains a cycle.*

Next, we present a characterization of $diam(\overrightarrow{\Gamma}(M_n(R)))$ in terms of $\text{diam}(\Gamma(R))$.

**Lemma 6.1.13.** ([222, Lemma 4.1]) *Let $R$ be a commutative ring and $A \in Z(M_n(R))^*$. If all elements of $R$ are either invertible or zero-divisors, then there exists an invertible $P \in M_n(R)$ such that the first row of $PA$ (alternatively, the last column of $AP$) consists entirely of zero-divisors.*

**Lemma 6.1.14.** ([222, Lemma 4.2]) *Let $R$ be a commutative ring. If every finite set of zero-divisors of $R$ has a nonzero annihilator, then* $diam(\overrightarrow{\Gamma}(M_n(R))) = 2$.

**Corollary 6.1.15.** ([222, Corollary 4.1]) *Let $F$ is a field. Then* $diam(\overrightarrow{\Gamma}(M_n(F))) = 2$.

**Proposition 6.1.16.** ([222, Proposition 4.1]) *Let $R$ be an integral domain. Then* $diam(\overrightarrow{\Gamma}(M_n(R))) = 2$.

Now we state a characterization for the diameter of $(\overrightarrow{\Gamma}(M_n(R)))$ for rings with complete directed zero-divisor graphs.

**Theorem 6.1.17.** ([222, Theorem 4.1]) *Let $R$ be a commutative ring such that $R \not\cong \mathbb{Z}_2 \times \mathbb{Z}_2$, and let $diam(\Gamma(R)) = 1$. Then* $diam(\overrightarrow{\Gamma}(M_n(R))) = 2$.

**Theorem 6.1.18.** ([222, Theorem 4.2]) *Let $R$ be a commutative ring such that $T(R) = K_1 \times K_2$, where both $K_i$ are fields. Then* $diam(\overrightarrow{\Gamma}(M_n(R))) = 3$.

In the theorem below, when $Z(R)$ is an ideal in $R$, we determine the diameter of $\overrightarrow{\Gamma}(M_n(R)))$ for a commutative Noetherian ring $R$.

**Theorem 6.1.19.** ([222, Theorem 4.3]) *Let $R$ be a commutative Noetherian ring with $diam(\Gamma(R)) = 2$. If $Z(R)$ is a (prime) ideal of $R$, then $diam(\overrightarrow{\Gamma}(M_n(R))) = 2$.*

**Proposition 6.1.20.** ([222, Proposition 4.2]) *Let $R$ be a commutative McCoy ring with $diam(\Gamma(R)) = 2$. If $Z(R)$ is a (prime) ideal of $R$, then $diam(\overrightarrow{\Gamma}(M_n(R))) = 2$.*

### 6.1.1 Directed Zero-divisor Graphs of Finite Rings

Wu [434] obtained properties for the directed zero-divisor graph $\overrightarrow{\Gamma}(R)$ of finite noncommutative rings $R$, and they are presented below. In this connection, let us first present some properties of finite rings

**Lemma 6.1.21.** ([434, Lemma 2.1]) *Suppose that every right identity element of a finite ring $R$ is a two-sided identity. Then we have the following:*

(1) *Each left zero-divisor is also a right zero-divisor;*

(2) *If in addition $|R| \geq 5$, then for distinct $a, b \in R^*$ with $ab = 0$, there exists $c \in R^*$ such that $ca = 0$ and $c \neq a$.*

**Lemma 6.1.22.** ([434, Lemma 2.2]) *Suppose that $R$ is a ring with $|R| \leq 4$ and every proper one-sided identity element is a two-sided identity. Then for distinct $a, b \in R^*$ with $ab = 0$, we have $ba = 0$.*

**Proposition 6.1.23.** ([434, Proposition 2.3]) *Let $R$ is a finite ring with the property that every one-sided identity element is a two-sided identity. Then for any path $a \to b$ in $\overrightarrow{\Gamma}(R)$, there is a walk $c \to a \to b \to d$, where $c \neq a$ and $b \neq d$.*

The following characterizes all finite rings $R$ whose directed zero-divisor graph $\overrightarrow{\Gamma}(R)$ is connected.

**Theorem 6.1.24.** ([434, Theorem 2.4]) *Let $R$ be a finite ring. Then the following statements are equivalent:*

(1) *The directed zero-divisor graph $\overrightarrow{\Gamma}(R)$ is connected;*

(2) *Every one-sided identity element of $R$ is the two-sided identity of $R$;*

(3) *Either $R$ has a two-sided identity or $R$ has no proper one-sided identity;*

(4) *There exists no end-vertex (sink or source) in $\overrightarrow{\Gamma}(R)$.*

It may be noted that Theorem 6.1.24 actually holds for Artinian rings which are both right Artinian and left Artinian since Lemma 6.1.21 holds for Artinian rings. Obviously, every finite ring is an Artinian ring, but the converse is false. Now, let $R$ be a ring with a proper left identity element $e$. Let $I_e = ann_\ell(e) = \{a \in R \mid ae = 0\}$ and $R_e = \{a \in R \mid a = ae\}$. Then we have the following:

(1) $I_e$ is a two-sided ideal of $R$ with at least two elements;
(2) $R_e$ is a subring of $R$ and $e$ is the identity of $R_e$;
(3) $R = R_e \oplus I_e$ as left $R$-modules;
(4) There is a ring isomorphism $R_e \cong R/I_e$.

For the graph $\overrightarrow{\Gamma}(R)$, consider its edge set $E(\overrightarrow{\Gamma}(R))$. By definition of $\overrightarrow{\Gamma}(R)$, there are no loops. But if there is an element $a \in \overrightarrow{\Gamma}(R)$ such that $a^2 = 0$, we can naturally add a loop to $E(\overrightarrow{\Gamma}(R))$ whose vertex is $a$. If such a loop can appear in $\overrightarrow{\Gamma}(R)$, we shall temporarily add 2 to the number of the edges of $\overrightarrow{\Gamma}(R)$ in the next Proposition 6.1.25, just for convenience. For subsets $M, N$ of $Z(R)^*$, let $\Gamma(M, N)$ be the induced bipartite subgraph of $\overrightarrow{\Gamma}(R)$, and denote its edge set by $E(M, N)$.

**Proposition 6.1.25.** ([434, Proposition 2.5]) *For a ring $R$ with proper left identity element $e$, the graph $\overrightarrow{\Gamma}(R)$ has the following properties:*
(1) *For every $a \in I_e$, the out-degree of $a$ is $|R| + 1$;*
(2) *The number of vertices of $\overrightarrow{\Gamma}(R)$ is $|R| - 1 = |R_e||I_e| - 1$;*
(3) $|E(\overrightarrow{\Gamma}(R))| = |I_e|[|E(R_e^*, R_e^* \oplus I_e^*)| + |E(R_e^*, I_e^*)| + (2 - |I_e|)|E(R_e)| + |I_e| - 1]$.

The following proposition gives an upper bound for the distance between vertices in $\overrightarrow{\Gamma}(R)$ for a finite ring $R$.

**Proposition 6.1.26.** ([434, Proposition 2.6]) *Let $R$ be a finite ring with a proper one-sided identity. For distinct elements $x, y$ in $\overrightarrow{\Gamma}(R)$, $d(x, y) \leq 6$ if there is a directed path from $x$ to $y$.*

**Corollary 6.1.27.** ([434, Corollary 2.7]) *Let $R$ be a ring with a proper one-sided identity element $e$. Then $R_e$ is a subring of $R$, $e$ is a two-sided identity of $R_e$, and $diam(\overrightarrow{\Gamma}(R)) \leq 3 + diam(\overrightarrow{\Gamma}(R_e))$.*

Now we present an example for $\overrightarrow{\Gamma}(R)$.

**Example 6.1.28.** ([434, Example 2.8]) *For any field $F$, let $R$ be the $n \times n$ full matrix ring over $F (n > 1)$. Then the diameter of $\overrightarrow{\Gamma}(R)$ is 2.*

A few more examples are also given in [434]. The following proposition is true for both finite and infinite rings.

**Proposition 6.1.29.** ([434, Proposition 3.1]) *Let $R$ be any ring. Then*
(1) *If there is a source $b$ in $\overrightarrow{\Gamma}(R)$ with $b^2 = 0$, then $R = \{0, a, b, c\}$, where $a$ and $c$ are left identity elements, and $ba = 0 = bc$;*
(2) *If there is a sink $b$ in $\overrightarrow{\Gamma}(R)$ with $b^2 = 0$, then $R = \{0, a, b, c\}$, where $a$ and $c$ are right identity elements, and $ab = 0 = cb$.*

**Proposition 6.1.30.** ([434, Proposition 3.2]) *Let $R$ be a finite ring with proper left identity elements. If $|R| \geq 5$, then*

(1) *$\overrightarrow{\Gamma}(R)$ contains at least two sinks;*

(2) *For any source $r$ in $\overrightarrow{\Gamma}(R)$, $r^2 \neq 0$;*

(3) *$\overrightarrow{\Gamma}(R)$ contains no sources.*

**Corollary 6.1.31.** ([434, Corollary 3.3]) *Let $R$ be a finite ring with proper left identity elements. Assume that $R$ has at least five elements. For any left identity $e$ of $R$, if $ann_\ell(e) = \{0, b\}$ for some nonzero element $b$, then in $\overrightarrow{\Gamma}(R)$, the out-degree of $b$ is $|R| - 1$ and the in-degree of $b$ is positive.*

Similar to the above, we have the following:

**Proposition 6.1.32.** ([434, Proposition 3.4]) *Let $R$ be a finite ring with proper right identity elements. If $|R| \geq 5$, then*

(1) *$\overrightarrow{\Gamma}(R)$ contains at least two sources;*

(2) *For any source $r$ in $\overrightarrow{\Gamma}(R)$, $r^2 \neq 0$;*

(3) *$\overrightarrow{\Gamma}(R)$ contains no sinks.*

**Corollary 6.1.33.** ([434, Corollary 3.6]) *For any finite ring $R$ with at least five elements,*

(1) *$\overrightarrow{\Gamma}(R)$ cannot contain a sink and a source at the same time;*

(2) *$\overrightarrow{\Gamma}(R)$ is not a network for any finite ring.*

Further properties are given in [434].

## 6.2 Commutative Semigroups

The concept of zero-divisor graph of a commutative ring was generalized to the zero-divisor graph of a commutative semigroup by DeMeyer, McKenzie, and Schneider in [177]. Let $S$ be a (multiplicative) commutative semigroup with 0 (i.e., $0x = 0$ for every $x \in S$), and let $Z(S) = \{x \in S \mid xy = 0$ for some $0 \neq y \in S\}$ be the set of zero-divisors of $S$. As in [177], the *zero-divisor graph* of $S$ is the (simple) graph $\Gamma(S)$ with vertices $Z(S) \setminus \{0\}$, the set of nonzero zero-divisors of $S$, and distinct vertices $x$ and $y$ are adjacent if and only if $xy = 0$. Note that $\Gamma(S)$ is the empty graph if and only if $Z(S) \subseteq \{0\}$ ($Z(S) = \emptyset$ if $S = \{0\}$).

For more about the zero-divisor graph of a commutative semigroup and many more references, see the survey article by Anderson and Badawi [73]. In [155], Cannon, Neuerburg, and Redmond extended the concept of the zero-divisor graph to noncommutative semigroups. For more details on the zero-divisor graph of noncommutative semigroups, see [155]. In [252], LaGrange introduced the zero-divisor graph of a commutative groupoid.

Many of the zero-divisor graphs discussed in this book can be viewed as the zero-divisor graph of a suitable semigroup, and thus some of their properties are special cases of results given in this section. We start with some examples.

**Example 6.2.1.** Let $R$ be a commutative ring with $1 \neq 0$.

(1) The "usual" zero-divisor graph $\Gamma(R)$ defined in [83] (see Section 2.1) has vertices $Z(R) \setminus \{0\}$, and distinct vertices $x$ and $y$ are adjacent if and only if $xy = 0$. Thus, $\Gamma(R) = \Gamma(S)$, where $S = R$ or $S = Z(R)$ is considered as a multiplicative semigroup.

(2) Let $I$ be an ideal of $R$. As in [323] (see Section 5.1), the *ideal-based zero-divisor graph* of $R$ with respect to $I$ is the (simple) graph $\Gamma_I(R)$ with vertices $\{\, x \in R \setminus I \mid xy \in I$ for some $y \in R \setminus I \,\}$, and distinct vertices $x$ and $y$ are adjacent if and only if $xy \in I$. Thus, $\Gamma_I(R) = \Gamma(S)$, where $S = R/I$ is the Rees semigroup of (the multiplicative semigroup) $R$ with respect to $I$ (i.e., the ideal $I$ collapses to 0). In particular, $\Gamma_{\{0\}}(R) = \Gamma(R)$.

(3) Define an (congruence) equivalence relation $\sim$ on $R$ by $x \sim y \Leftrightarrow \mathrm{ann}_R(x) = \mathrm{ann}_R(y)$, and let $R_E = \{\, [x] \mid x \in R \,\}$ be the commutative monoid of (congruence) equivalence classes under the induced multiplication $[x][y] = [xy]$. Note that $[0] = \{0\}$ and $[1] = R \setminus Z(R)$; so $[x] \subseteq Z(R)^*$ for every $x \in R \setminus ([0] \cup [1])$. The *compressed zero-divisor graph* of $R$ is the (simple) graph $\Gamma_E(R)$ with vertices $R_E \setminus \{[0], [1]\}$, and distinct vertices $[x]$ and $[y]$ are adjacent if and only if $[x][y] = [0]$, if and only if $xy = 0$. Thus, $\Gamma_E(R) = \Gamma(R_E)$. This zero-divisor graph was first defined (using different notation) in [295] (see Section 5.9). The semigroup analog $\Gamma_E(S)$ has been studied in [179] and [193]. The monoid $R_E$ has also been studied in [78] and [79].

(4) Let $\sim$ be a multiplicative congruence relation on $R$ (i.e., $x \sim y \Rightarrow xz \sim yz$ for $x, y, z \in R$), and let $R/{\sim} = \{\, [x]_\sim \mid x \in R \,\}$ be the commutative monoid of congruence classes under the induced multiplication $[x]_\sim [y]_\sim = [xy]_\sim$. As in [82], the *congruence-based zero-divisor graph* of $R$ with respect to $\sim$ is the (simple) graph $\Gamma_\sim(R)$ with vertices $Z(R/{\sim}) \setminus \{[0]_\sim\}$, and distinct vertices $[x]_\sim$ and $[y]_\sim$ are adjacent if and only if $[xy]_\sim = [0]_\sim$, if and only if $xy \sim 0$. Thus, $\Gamma_\sim(R) = \Gamma(R/{\sim})$. The congruence-based zero-divisor graph includes the three above zero-divisor graphs as special cases. See Section 10.5 for more details about congruence-based zero-divisor graphs.

(5) Let $S$ be the semigroup of ideals of $R$ under the usual ideal multiplication. As in [139] (see Section 5.5), $\mathbb{AG}(R) = \Gamma(S)$ is called the *annihilating-ideal graph* of $R$ (this zero-divisor graph was first defined in [415]). Similarly, as in [181], define the *annihilating-ideal graph* of a commutative semigroup $S$ with 0 to be $\mathbb{AG}(S) = \Gamma(T)$, where $T$ is the semigroup of (semigroup) ideals of $S$ under the usual multiplication of (semigroup) ideals.

(6) Let $(S, \wedge)$ be a meet semilattice with least element 0. As in [301], the *zero-divisor graph* of $S$ is the (simple) graph $\Gamma(S)$ with vertices $Z(S) \setminus \{0\} = \{0 \neq x \in S \mid x \wedge y = 0 \text{ for some } 0 \neq y \in S\}$, and distinct vertices $x$ and $y$ are adjacent if and only if $x \wedge y = 0$. Recall that $S$ becomes a commutative (Boolean) semigroup $S'$ with 0 under the multiplication $xy = x \wedge y$; so $\Gamma(S) = \Gamma(S')$. Similar zero-divisor graphs have been defined for posets and lattices (see [195], [213], [251], [254], [268], and Theorem 6.2.14(1)).

However, not all "zero-divisor" graphs can be realized as $\Gamma(S)$ for a suitable commutative semigroup $S$ with 0. For example, as in [128] (see Section 10.2), the *annihilator graph* of a commutative ring $R$ with $1 \neq 0$ is the (simple) graph $AG(R)$ with vertices $Z(R) \setminus \{0\}$, and distinct vertices $x$ and $y$ are adjacent if and only if $ann_R(x) \cup ann_R(y) \neq ann_R(xy)$. Then $\Gamma(R)$ is a subgraph of $AG(R)$, and may be a proper subgraph (e.g., $\Gamma(\mathbb{Z}_8) = K_{1,2}$, while $AG(\mathbb{Z}_8) = K_3$). Thus, $AG(R)$ need not be a $\Gamma(S)$. Similarly, as in [22], one can define the *annihilator graph* $AG(S)$ of a commutative semigroup $S$ with 0 to be the (simple) graph with vertices $Z(S) \setminus \{0\}$, and distinct vertices $x$ and $y$ are adjacent if and only if $ann_S(x) \cup ann_S(y) \neq ann_S(xy)$. The zero-divisor graph $\Gamma(S)$ is a subgraph of $AG(S)$ when $S \neq Z(S)$ [22, Theorem 3.1]. However, for $S = Z(\mathbb{Z}_6)$, we have $S = Z(S)$, $\Gamma(S) = K_{1,2}$, and $AG(S) = CK_3$.

Similarly, the extended zero-divisor graph $\overline{\Gamma}(R)$ as in [143] (see Section 5.10) is not a zero-divisor semigroup graph. As in [84], we define the extended zero-divisor graph $\overline{\Gamma}(S)$ of a commutative semigroup $S$ with 0 to be the (simple) graph with vertices $Z(S) \setminus \{0\}$, and distinct vertices $x$ and $y$ are adjacent if and only if $x^m y^n = 0$ for positive integers $m$ and $n$ with $x^m \neq 0$ and $y^n \neq 0$. Moreover, $\Gamma(S)$ is a subgraph of $\overline{\Gamma}(S)$, and $\overline{\Gamma}(S)$ is a subgraph of $AG(S)$ if $S \neq Z(S)$. But as above, for $S = Z(\mathbb{Z}_6)$, we have $S = Z(S)$, $\overline{\Gamma}(S) = \Gamma(S) = K_{1,2}$, and $AG(S) = CK_3$.

As in [443], for a commutative semigroup $S$ with 0, let $\Gamma^*(S)$ be the (simple) graph with vertices $Z(S) \setminus \{0\}$, and distinct vertices $x$ and $y$ are adjacent if and only if $xSy = \{0\}$ (they used the notation $\overline{\Gamma}(S)$). Then $\Gamma(S)$ is a subgraph of $\Gamma^*(S)$, and may be a proper subgraph (e.g., if $S = \{0, 2, 4, 6\} \subseteq \mathbb{Z}_8$, then $\Gamma(S) = K_{1,2}$, while $\Gamma^*(S) = K_3$).

Next, we give some properties of the zero-divisor graph $\Gamma(S)$ of a commutative semigroup $S$ with 0 and are particularly interested in which graphs can be realized as $\Gamma(S)$ for some commutative semigroup $S$ with 0. We start with some basic properties of $\Gamma(S)$. Parts (1) - (3) of Theorem 7.4.23 were first proved for $\Gamma(R)$ (cf. [75], [83], [178], [295], and Chapter 2). Recall that the *core* of a graph $G$ is the union of the cycles in $G$.

**Theorem 6.2.2.** *Let $S$ be a commutative semigroup with* 0.

(1) ([177, Theorem 1.2]) $\Gamma(S)$ *is connected with* $diam(\Gamma(S)) \in \{0, 1, 2, 3\}$*;*

(2) ([177, Theorem 1.3]) *If* $\Gamma(S)$ *does not contain a cycle, then* $\Gamma(S)$ *is a connected subgraph of two star graphs whose centers are connected by a single edge;*

(3) ([177, Theorem 1.5]) *If* $\Gamma(S)$ *contains a cycle, then the core of* $\Gamma(S)$ *is a union of triangles and squares, and any vertex not in the core of* $\Gamma(S)$ *is an end In particular,* $gr(\Gamma(S)) \in \{3, 4, \infty\}$;

(4) ([176, Theorem 1(4)]) *For every pair* $x, y$ *of distinct nonadjacent vertices of* $\Gamma(S)$, *there is a vertex* $z$ *of* $\Gamma(S)$ *with* $N(x) \cup N(y) \subseteq N[z]$.

**Remark 6.2.3.** (1) In Theorem 6.2.2(4), it is easily shown that $N(x) \cup N(y) \subsetneq N[z]$ (for any such $z$), and either case $z \in N(x) \cup N(y)$ or $z \notin N(x) \cup N(y)$ may occur. Moreover, we can always choose $z = xy$, but there may be other choices for $z$.

(2) In [268], a (simple) connected graph which satisfies condition (4) of Theorem 6.2.2 is called a *compact graph.* In [268, Theorem 3.1], it was shown that a simple graph $G$ is the zero-divisor graph of a poset if and only if $G$ is a compact graph.

For small graphs, conditions (1), (3), and (4) of Theorem 6.2.2 actually characterize zero-divisor graphs.

**Theorem 6.2.4.** ([176, Theorem 2]) *Let* $G$ *be a (simple) graph with* $|V(G)| \leq 5$ *satisfying conditions* (1), (3), *and* (4) *of Theorem 6.2.2. Then* $G \cong \Gamma(S)$ *for some commutative semigroup* $S$ *with* 0.

**Example 6.2.5.** ([176, Example 2]). In view of Theorem 6.2.4, Figure 6.2 is a graph with six vertices which satisfies conditions (1), (3), and (4) of Theorem 6.2.2, but $G$ is not the zero-divisor graph of any commutative semigroup with 0. (Also, see [179, Figure 2, p. 3372].)

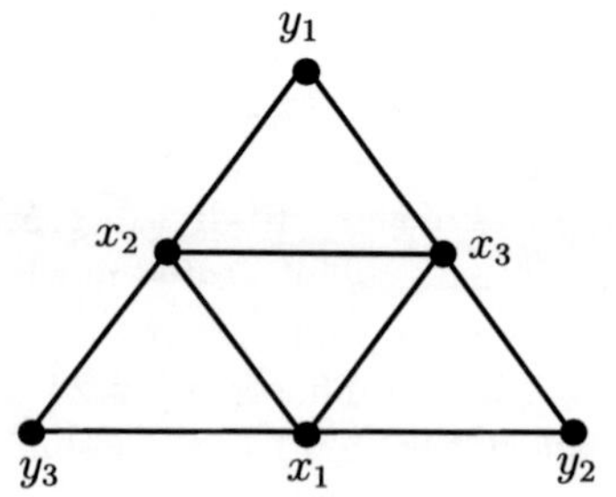


Figure 6.2: A graph with six vertices which satisfies conditions (1), (3), and (4) of Theorem 6.2.4, but $G$ is not the zero-divisor graph of any commutative semigroup with 0.

The next theorem gives several classes of graphs which can be realized as the zero-divisor graph of a commutative semigroup with 0. As to be expected, many more graphs can be realized as $\Gamma(S)$ for a commutative semigroup $S$ with 0 than as $\Gamma(R)$ for a commutative ring $R$ with $1 \neq 0$ (cf. [75], [83], [325], and [326]). For example, $K_n$ and $K_{1,n}$ (for an integer $n \geq 1$) can be realized as a $\Gamma(S)$ for every $n \geq 1$, but can be realized as a $\Gamma(R)$ if and only if $n + 1$ is a prime power (see Theorem 2.2.7, Example 2.2.8, and Theorem 2.2.10).

**Theorem 6.2.6.** ([176, Theorem 3]) *The following graphs are the zero-divisor graph of some commutative semigroup with* 0*:*

(1) *A complete graph or a complete graph together with an end;*

(2) *A complete bipartite graph or a complete bipartite graph together with an end;*

(3) *A refinement of a star graph;*

(4) *A graph which has at least one end and diameter* $\leq 2$*;*

(5) ([177, Theorem 1.3(2)]) *A graph which is the union of two star graphs whose centers are connected by a single edge.*

**Example 6.2.7.** ([176, Example 3]). By (3) and (5) of Theorem 6.2.6, the refinement of a star graph and the union of two star graphs whose centers are connected by an edge are each the zero-divisor graph of a commutative semigroup with 0. The graph in Figure 6.3 is also a refinement of the union of two star graphs with centers at vertex $a$ and vertex $b$. However, it is not the zero-divisor graph of any commutative semigroup with 0. The vertices $a$ and $f$ do not satisfy condition (4) of Theorem 6.2.2 since vertex $a$ is adjacent to $d$ and vertex $f$ is adjacent to $c$, but there is no vertex adjacent to both $c$ and $d$.

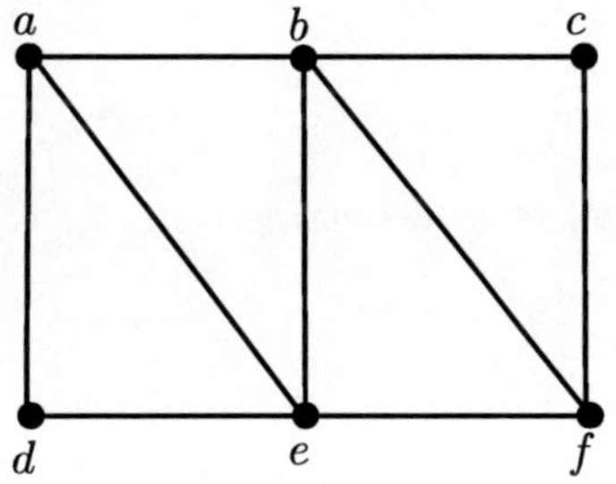


Figure 6.3: A graph which is a refinement of the union of two star graphs with centers at vertex a and vertex b. However, it is not the zero-divisor graph of any commutative semigroup with 0.

The following theorem gives necessary and sufficient conditions on the semigroup $S$ for $\Gamma(S)$ to be a refinement of a star graph (cf. [83, Theorem 2.5] for commutative rings).

**Theorem 6.2.8.** ([436, Theorem 1.1]) *Let $S$ be commutative semigroup with 0 and $Z(S) \neq \{0\}$. Then $\Gamma(S)$ is a refinement of a star graph if and only if either $Z(S)$ is an annihilator ideal (and hence a prime ideal) of $S$ or $Z(S) = A \cup B$, where $A \cong (\mathbb{Z}_2, \cdot)$, $A \cap B = \{0\}$, and $A$, $B$ are ideals of $S$.*

For a vertex $c$ of a graph $G$, let $G_c^*$ be the induced subgraph of $G$ with vertices $V(G_c^*) = V(G) \setminus \{ u \in V(G) \mid u = c$ or $u$ is an end vertex adjacent to $c \}$.

**Theorem 6.2.9.** ([436, Theorem 2.3]) *Let $S$ be a set with a commutative binary operation and a zero element 0 such that $S = \{0\} \cup \{c\} \cup T \cup S_1$ is the disjoint union of four nonempty subsets. Assume further that $Z(S) = S$, whose zero-divisor graph $\Gamma(S)$ is a refinement of a star graph with center $c$ such that $S_1 = V(G_c^*)$ and $G_c^*$ has at least two components. Then the following statements are equivalent:*

(1) *$S$ is a commutative zero-divisor semigroup (i.e., the binary operation is associative);*

(2) *$S_1^2 = \{0, c\}$, $T^2 \subseteq \{0, c\}$, $c^2 = 0$, and $ts_1 = c$ for every $t \in T$ and $s_1 \in S_1$;*

(3) *$S^2 = \{0, c\}$ and $S^3 = \{0\}$.*

Recall that a vertex $x$ of a graph $G$ has *degree* $m$, denoted by $\deg(x) = m$, if $|N(x)| = m$. For an integer $k \geq 1$, let $G_k$ be the induced subgraph of $G$ with vertices $V(G_k) = \{ x \in V(G) \mid \deg(x) \geq k \}$. For a commutative semigroup $S$ with 0 and integer $k \geq 1$, let $I_k = \{ x \in V(G) \mid \deg(x) \geq k \} \cup \{0\}$. Although the next two theorems from [176] were stated for nilpotent semigroups (i.e., $S^n = \{0\}$ for some positive integer $n$), their proofs show that they hold for nil semigroups (i.e., every element of $S$ is nilpotent).

**Theorem 6.2.10.** *Let $S$ be a commutative semigroup with 0.*

(1) ([176, Theorem 4]) *$I_k$ is a descending chain of ideals in $S$;*

(2) ([176, Corollary 1]) *The core of $\Gamma(S)$ together with $\{0\}$ is an ideal of $S$ whose zero-divisor graph is the core of $\Gamma(S)$;*

(3) ([176, Corollary 2]) *If $S$ is a nil semigroup, then $\Gamma(S)_k = \Gamma(I_k)$ for every integer $k \geq 1$;*

(4) ([176, Corollary 3]) *Let $G$ be a graph and assume that $G_k$ is not the zero-divisor graph of any commutative semigroup with 0 for some integer $k \geq 1$; Then $G$ is not the zero-divisor graph of any commutative nil semigroup;*

(5) ([176, Corollary 4]) *Let $G$ be a graph which is equal to its core, but is not the zero-divisor graph of any commutative semigroup with* 0*, and let $H$ be the graph obtained from $G$ by adding ends to $G$. Then $H$ is not the zero-divisor graph of any commutative semigroup with* 0.

Sharper results hold when $S$ is a nil semigroup. A well-known special case is for $\Gamma(R)$ when $Z(R) = \text{nil}(R)$ (e.g., when $R$ is a finite local ring).

**Theorem 6.2.11.** *Let $S$ be a commutative semigroup with* 0.

(1) ([176, Theorem 5]) *Assume that $S$ is a nil semigroup. Then*

   (a) *diam*$(\Gamma(S)) \in \{0, 1, 2\}$;

   (b) *Every edge in the core of $\Gamma(S)$ is the edge of a triangle in $\Gamma(S)$. In particular, gr*$(\Gamma(S)) \in \{3, \infty\}$;

(2) ([176, Corollary 5]) *If every element of $S$ has finite order and some edge in the core of $\Gamma(S)$ is the edge of a square, but not a triangle, then $S$ contains a nonzero idempotent element.*

In [179], the authors gave several criteria for a graph $G$ to be a zero-divisor graph in terms of the number of edges of $G$ and adding or removing edges from a given zero-divisor graph $\Gamma(S)$. In the next theorem, we are removing edges from $K_n$ (which has $n(n-1)/2$ edges).

**Theorem 6.2.12.** ([179, Theorem 2.5(1)]) *Let $G$ be a connected graph with $n$ vertices and $n(n-1)/2-p$ edges. Then $G$ is the zero-divisor graph of a commutative semigroup with* 0 *if $0 \leq p \leq \lceil n/2 \rceil + 1$ (i.e., if $G$ has at least $n(n-1)/2 - \lceil n/2 \rceil - 1$ edges).*

**Theorem 6.2.13.** ([179, Theorem 3.22]) *Let $G = \Gamma(S)$ be a zero-divisor graph with cycles for a commutative semigroup $S$ with* 0.

(1) *If $a$ is an end adjacent to $x$ in $G$, then adding another end adjacent to $x$ results in a zero-divisor graph;*

(2) *Removing an end from $G$ results in a zero-divisor graph.*

Not only is it of interest to know which graphs can be realized as $\Gamma(S)$ for some commutative semigroup $S$ with 0, but also, what are the choices for such semigroups $S$. The case for commutative semigroups $S$ with 0 and $\Gamma(S)$ is somewhat different than for commutative rings $R$ with $1 \neq 0$ and $\Gamma(R)$. Since $|R| \leq |Z(R)|^2$ when $Z(R) \neq \{0\}$, there are (up to isomorphism) only finitely many commutative rings with $1 \neq 0$ that have a given (nonempty) finite zero-divisor graph. However, for semigroups, one can always adjoin units; so if there is a commutative semigroup $S$ with 0 and $G \cong \Gamma(S)$, then for every cardinal number $n \geq |S|$, there is a commutative semigroup $S(n)$ with 0 (and $Z(S(n)) = Z(S)$) such that

$G \cong \Gamma(S(n))$ and $|S(n)| = n$. Thus, to determine which commutative semigroups with 0 realize a given graph $G$, we will restrict our attention to commutative zero-divisor semigroups (i.e., $S = Z(S)$).

While it is usually not true that $\Gamma(S) \cong \Gamma(T)$ implies that $S \cong T$ for commutative semigroups $S$ and $T$ with 0, we get better results when we restrict to certain classes of zero-divisor semigroups. We first consider the case when $S$ is reduced (i.e., $x^n = 0$ implies $x = 0$). The zero-divisor graph of reduced commutative semigroups with 0 has been studied in [77], [80], [193], [251], and [268]. The next theorem shows that this case reduces to Boolean semigroups (i.e., $x^2 = x$ for every element). Call a monoid $S$ with 0 a *zero-divisor monoid* if $S \setminus \{1\} = Z(S)$. Special cases of the next theorem have been proved in [268, Theorem 4.3] for (1) and [267, Theorem 4.2] for (2).

**Theorem 6.2.14.** (1) ([251, Corollary 1.2]) *The following statements are equivalent for a graph $G$ with at least two vertices:*

(a) *$G \cong \Gamma(S)$ for some reduced commutative semigroup $S$ with 0;*

(b) *$G \cong \Gamma(S)$ for some commutative Boolean semigroup $S$ with 0;*

(c) *$G \cong \Gamma(S)$ for some meet semilattice $S$;*

(2) ([77, Theorem 2.1]) *Let $S$ and $T$ be commutative Boolean zero-divisor monoids. Then $\Gamma(S) \cong \Gamma(T)$ if and only if $S \cong T$.*

We next give several classes of graphs for which one can actually determine all possible commutative zero-divisor semigroups with a given graph. However, we will just give the number (up to isomorphism) of such semigroups rather than list them all explicitly. For additional classes of graphs, see the discussion in [73].

In [435], the authors gave recursive formulas for the number (up to isomorphism) of commutative zero-divisor semigroups whose zero-divisor graphs are either $K_n$ or $K_n + 1$ (a $K_n$ with an end adjoined) and compute these numbers for $n \leq 10$. For example, there are (up to isomorphism) 139 commutative zero-divisor semigroups with zero-divisor graph $K_{10}$ and 7,101 with zero-divisor graph $K_{10} + 1$. We give the explicit formula for $K_n$; let $p(m, r)$ be the number of partitions $x_1 + \cdots + x_r = m$ of the integer $m$ with $x_1 \geq x_2 \geq \cdots \geq x_r \geq 1$.

**Theorem 6.2.15.** ([435, Theorem 2.2]) *For every integer $n \geq 1$, there are (up to isomorphism)*

$$1 + \sum_{r=1}^{n} \sum_{e=0}^{n-r} p(n - e, r)$$

*commutative zero-divisor semigroups whose zero-divisor graph is $K_n$.*

We next consider star graphs.

**Theorem 6.2.16.** ([410]) *Let $n \geq 1$ be an integer and $f(n)$ be the number (up to isomorphism) of commutative semigroups with $n$ elements. Then there are (up to isomorphism) $n + 2 + 2f(n-1) + 2f(n)$ commutative zero-divisor semigroups whose zero-divisor graph is $K_{1,n}$.*

**Theorem 6.2.17.** ([436, Theorem 2.13])

(1) *If $S$ is a nilpotent commutative semigroup with $\Gamma(S)$ a star graph, then $S^4 = \{0\}$;*

(2) *For every cardinal number $n \geq 2$, there is a unique (up to isomorphism) nilpotent commutative semigroup $S(n)$ such that $\Gamma(S(n)) = K_{1,n}$ and $S(n)^3 \neq \{0\}$.*

**Theorem 6.2.18.** *Let $n$ be an integer.*

(1) ([436, Theorem 3.6]) *For every $n \geq 2$, there are (up to isomorphism) $n + 2$ nilpotent commutative semigroups with 0 whose zero-divisor graph is the star graph $K_{1,n}$;*

(2) ([409, Theorem 2.1]) *There are (up to isomorphism) 12 commutative zero-divisor semigroups whose zero-divisor graph is the star graph $K_{1,2}$;*

(3) ([409, Theorem 2.2]) *There is (up to isomorphism) a unique commutative zero-divisor semigroup whose zero-divisor graph is the path graph $P_4$: $a - b - c - d$;*

(4) ([409, Theorem 2.5]) *There are (up to isomorphism) 35 commutative zero-divisor semigroups whose zero-divisor graph is the graph $K_{1,3}$;*

(5) ([409, Theorem 2.6]) *There are (up to isomorphism) 10 commutative zero-divisor semigroups whose zero-divisor graph is the graph $K_{2,2}$;*

(6) ([409, Theorem 2.7]) *There are (up to isomorphism) 31 commutative zero-divisor semigroups whose zero-divisor graph is the graph in Figure 3.*

By [436, p. 339], the number of commutative zero-divisor semigroups whose zero-divisor graph is $K_2$ (resp., $K_3, K_4$, and $K_3 + 1$) is 4 (resp., $7, 12$, and 22). Combining this with Theorem 6.2.18 gives all commutative zero-divisor semigroups whose zero-divisor graph has at most 4 vertices.

## 6.3 Commutative Semirings

Throughout this section, all semirings are assumed to be commutative semirings with nonzero identity. Also, $0x = 0$ for every $x$ in a semiring $R$. Recall that the *zero-divisor graph of a semiring* is the simple graph whose vertex set is the set of nonzero zero-divisors, and an edge is drawn between two distinct vertices if their product is zero. T In 2008, Atani [106] introduced and studied the zero-divisor graph with respect to ideals of a commutative semiring. The bulk of this section is devoted to stating results analogous to several well-known results for the ideal-based zero-divisor graph in the theory of rings (see Section 5.1). In fact, the main object of this section is to indicate the interplay of semiring-theoretic properties of $R$ with graph-theoretic properties of $\Gamma_I(R)$ for some ideal $I$ of $R$. It is noteworthy mentioning that many results about the ideal-based zero-divisor graph of a commutative ring can be easily extended to commutative semirings by a parallel argument.

Let $I$ be an ideal of a commutative semiring $R$ with nonzero identity. The zero-divisor graph with respect to $I$ of $R$, denoted by $\Gamma_I(R)$, is the simple graph with vertices $\{x \in R \setminus I \mid xy \in I \text{ for some } y \in R \setminus I\}$, where distinct vertices $x$ and $y$ are adjacent if and only if $xy \in I$. If $I = (0)$, then $\Gamma_I(R) = \Gamma(R)$; and $\Gamma_I(R) = \emptyset$ if and only if $I$ is a nonzero prime ideal.

An ideal $I$ of a semiring $R$ is called a *partitioning ideal* (= $Q$-ideal) if there is a subset $Q$ of $R$ such that $R = \cup\{q + I \mid q \in Q\}$, and if $q_1, q_2 \in Q$, then $(q_1 + I) \cap (q_2 + I) \neq \emptyset$ if and only if $q_1 + q_2 \in I$. Let $I$ be a $Q$-ideal of a semiring $R$, and let $R/I = \{q + I \mid q \in Q\}$. Then $R/I$ forms a semiring under the binary operations $\oplus$ and $\otimes$ defined as $(q_1 + I) \oplus (q_2 + I) = q_3 + I$, where $q_3 \in Q$ is the unique element such that $q_1 + q_2 + I \subseteq q_3 + I$, and $(q_1 + I) \otimes (q_2 + I) = q_4 + I$, where $q_4 \in Q$ is the unique element such that $q_1q_2 + I \subseteq q_4 + I$. This semiring $R/I$ is called the *quotient semiring* of $R$ by $I$. By definition of $Q$-ideal, there exists a unique $q_0 \in Q$ such that $0 + I \subseteq q_0 + I$. Then $q_0 + I$ is the zero element of $R/I$. Clearly, if $R$ is commutative, then so is $R/I$.

**Theorem 6.3.1.** ([106, Theorem 2.2]) *Let $I$ be a proper $Q$-ideal of a commutative semiring $R$. Then $\Gamma_I(R) = \emptyset$ if and only if $\Gamma(R/I) = \emptyset$.*

The next theorem investigates the relationship between $\Gamma_I(R)$ and $\Gamma(R/I)$ (compare the next result with Theorem 5.1.2).

**Theorem 6.3.2.** ([106, Theorem 2.4]) *Let $I$ be a proper $Q$-ideal of a commutative semiring $R$ and let $x = q_1 + a,\ y = q_2 + b \in R \setminus I$, where $q_1, q_2 \in Q$ and $a, b \in I$. Then*

(a) *If $q_1 + I$ is adjacent to $q_2 + I$ in $\Gamma(R/I)$, then $x$ is adjacent to $y$ in $\Gamma_I(R)$;*

(b) *If $x$ is adjacent to $y$ in $\Gamma_I(R)$ and $q_1 \neq q_2$, then $q_1 + I$ is adjacent to $q_2 + I$ in $\Gamma(R/I)$;*

(c) *If $x$ is adjacent to $y$ in $\Gamma_I(R)$ and $q_1 = q_2$, then $x^2, y^2 \in I$.*

**Theorem 6.3.3.** ([106, Theorems 2.6]) *Let $I$ be a proper $Q$-ideal of a commutative semiring $R$. Then $\Gamma_I(R)$ contains $|I|$ disjoint subgraphs isomorphic to $\Gamma(R/I)$.*

Next, we study the connectivity of $\Gamma_I(R)$.

**Theorem 6.3.4.** ([106, Theorem 2.7]) *Let $I$ be a nonzero proper $Q$-ideal of a commutative semiring $R$. Then $\Gamma_I(R)$ has no cut-points.*

**Theorem 6.3.5.** ([106, Theorem 2.8]) *Let $I$ be a proper $Q$-ideal of a commutative semiring $R$. Then*

(a) *If $\Gamma(R/I)$ is the graph on one vertex, then $\kappa(\Gamma_I(R)) = |I| - 1$;*

(b) *If $\Gamma(R/I)$ has at most two vertices, then $2 \leq \kappa(\Gamma_I(R)) \leq |I| \cdot \kappa(\Gamma(R/I))$.*

Now, we consider the girth of $\Gamma_I(R)$ when $R$ be a commutative semiring.

**Lemma 6.3.6.** ([106, Lemma 3.1]) *Let $I$ be a proper $Q$-ideal of a commutative semiring $R$. Then $gr(\Gamma_I(R)) \leq gr(\Gamma(R/I)$.*

**Theorem 6.3.7.** ([106, Proposition 3.2]) *Let $I$ be a proper $Q$-ideal of a commutative semiring $R$.*

(a) *If $|I| \geq 3$ and $\Gamma_I(R)$ contains a connected column, then $gr(\Gamma_I(R)) = 3$;*

(b) *Assume that $I \neq (0)$ and $\Gamma(R/I)$ has only one vertex. Then $gr(\Gamma_I(R)) = 3$ if $|I| \geq 3$, and $gr(\Gamma_I(R)) = \infty$ if $|I| = 2$;*

(c) *If $I$ has two elements, $\Gamma(R/I)$ has at least two vertices, and $\Gamma_I(R)$ has at least one connected column, then $gr(\Gamma_I(R)) = 3$.*

**Theorem 6.3.8.** ([106, Theorem 3.3]) *Let $I$ be a nonzero proper $Q$-ideal of a commutative semiring $R$ that is not a prime ideal. Then $gr(\Gamma_I(R)) = \infty$ if $\Gamma(R/I)$ has only one vertex and $|I| = 2$, $gr(\Gamma_I(R)) = 4$ if $gr(\Gamma(R/I)) > 3$ and $\Gamma_I(R)$ has no connected columns, and $gr(\Gamma_I(R)) = 3$ otherwise.*

Let $R$ be a commutative semiring. The Jacobson radical of $R$, denoted by $J(R)$, is the intersection of all the maximal $k$-ideals of $R$. Then $J(R)$ always exists and is a $k$-ideal.

**Lemma 6.3.9.** ([106, Lemma 3.4]) *Let $R$ be a commutative semiring and let $r \in R$.*

(a) *If $r$ is a nilpotent element of $R$, then it is not a semi-unit;*

(b) *If $r \in J(R)$, then $1 + ra$ is a semi-unit for every $a \in R$.*

**Lemma 6.3.10.** ([106, Lemma 3.5]) *Let $R$ be a commutative semiring.*

(a) *Let $I$ be an ideal in $R$ that is maximal among all annihilators of nonzero elements of $R$. Then $I$ is prime;*

(b) *Let $P_1, \ldots, P_n$ be prime $k$-ideals, and let $J$ be an ideal of $R$ contained in $\cup_{i=1}^n P_i$. Then $J \subseteq P_i$ for some $i$.*

**Lemma 6.3.11.** ([106, Lemma 3.6]) *Let $R$ be a commutative Noetherian semiring (in particular, $R$ could be a finite commutative semiring) and $I$ a nonzero ideal of $R$.*

(a) *Every maximal element of the family of $k$-ideals $A = \{(0 : x) \mid 0 \neq x \in I\}$ is an associated prime of $I$;*

(b) *The set of zero-divisors for $I$ is the union of all the associated primes of $I$.*

**Theorem 6.3.12.** ([106, Proposition 3.7]) *Let $R$ be an Artinian cancellative commutative semiring.*

(a) *Every element of $R$ is either a semi-unit or nilpotent;*

(b) *$R$ is a commutative local semiring.*

**Theorem 6.3.13.** ([106, Proposition 3.8]) *Let $R$ be an Artinian cancellative commutative semiring with a unique maximal $k$-ideal $P$. Then $P = (0 : x)$ for some $0 \neq x \in P$.*

**Theorem 6.3.14.** ([106, Theorems 3.9]) *Let $R$ be an Artinian cancellative commutative semiring (in particular, $R$ could be a finite cancellative commutative semiring) with a unique maximal $k$-ideal $P$. If $\Gamma(R)$ contains a cycle, then $gr(\Gamma(R)) = 3$.*

Next, we state certain properties of the ideal-based zero-divisor graph with respect to primary ideals of a commutative semiring.

**Theorem 6.3.15.** ([106, Proposition 4.1]) *Let $I$ be an ideal of a commutative semiring $R$. Then $I$ is a $P$-primary ideal of $R$ if and only if $\Gamma_I(R) = P \setminus I$.*

**Theorem 6.3.16.** ([106, Theorems 4.2]) *Let $I$ be an ideal of a commutative semiring $R$. Then $I$ is a primary ideal of $R$ if and only if $\Gamma_I(R) \cup I$ is an (prime) ideal of $R$.*

**Theorem 6.3.17.** ([106, Theorems 4.3]) *Let $I$ and $J$ be $P$-primary ideals of a commutative semiring $R$. Then $\Gamma_I(R) = \Gamma_J(R)$ if and only if $I = J$.*

**Theorem 6.3.18.** ([106, Theorems 4.5]) *Assume that $I$ is a $P$-primary $Q$-ideal of a commutative semiring $R$, and let $q_0$ be the unique element in $Q$ such that $q_0 + I$ is the zero in $R/I$. Then $\Gamma(R/I) \cup \{q_0 + I\}$ is a prime $k$-ideal of $R/I$.*

**Theorem 6.3.19.** ([106, Theorems 4.7]) *Let $I$ be a $P$-primary $Q$-ideal of a commutative Noetherian semiring $R$. Then $diam(\Gamma(R/I)) \leq 2$.*

Atani in [107] essentially studied zero-divisor graphs with respect to primal and nonprimal ideals of a commutative semiring. The main goal of his work was to generalize some of the results on the zero-divisor graph with respect to primal ideals of a commutative ring to a commutative semiring.

Let $R$ be a commutative semiring. An $a \in R$ is called prime to an ideal $I$ of $R$ if $ra \in I$ (where $r \in R$) implies that $r \in I$. Denote by $p(I)$ the set of elements of $R$ that are not prime to $I$. A proper ideal $I$ of $R$ is said to be *primal* if $p(I)$ forms an ideal (so 0 is not necessarily primal); this ideal is always a prime ideal, called the *adjoint ideal* $P$ of $I$. In this case, we say that $I$ is a $P$-primal ideal of $R$.

First, we study the ideal-based zero-divisor graph with respect to primal ideals of a commutative semiring. The class of primal ideals is a large class. For example, all primary ideals and irreducible ideals are primal.

**Lemma 6.3.20.** ([107, Lemma 2.1]) *Let $I$ be a proper ideal of a commutative semiring $R$. Then $I \subseteq p(I)$ and $V(\Gamma_I(R)) = p(I) \setminus I$. In particular, $V(\Gamma_I(R)) \cup I = p(I)$.*

**Lemma 6.3.21.** ([107, Lemma 2.2]) *Let $I$ and $P$ be ideals of a commutative semiring $R$ with $I \subseteq P$. Then $I$ is a $P$-primal ideal of $R$ if and only if $V(\Gamma_I(R)) = P \setminus I$.*

**Theorem 6.3.22.** ([107, Theorem 2.3]) *Let $I$ be an ideal of a commutative semiring $R$. Then $I$ is a primal ideal of $R$ if and only if $V(\Gamma_I(R)) \cup I$ is an (prime) ideal of $R$.*

**Theorem 6.3.23.** ([107, Theorem 2.4]) *Let $R$ be a commutative semiring. Then 0 is a primal ideal of $R$ if and only if $Z(R)$ is an (prime) ideal of $R$.*

**Theorem 6.3.24.** ([107, Theorem 2.5]) *Let $I$ and $J$ be $P$-primal ideals of a commutative semiring $R$. Then $\Gamma_I(R) = \Gamma_J(R)$ if and only if $I = J$.*

**Theorem 6.3.25.** ([107, Theorem 2.7]) *Assume that $I$ is a $P$-primal $Q$-ideal of a commutative semiring $R$, and let $q_0$ be the unique element in $Q$ such that $q_0 + I$ is the zero in $R/I$. Then $V(\Gamma(R/I)) \cup \{q_0 + I\}$ is a prime $k$-ideal of $R/I$.*

**Theorem 6.3.26.** ([107, Theorem 2.8]) *Let $I$ be a $P$-primal $Q$-ideal of a commutative Noetherian semiring $R$. Then $diam(\Gamma(R/I)) \leq 2$.*

**Theorem 6.3.27.** ([107, Theorem 2.12]) *If $I$ is a proper radical $Q$-ideal of a commutative semiring $R$, then $diam(\Gamma_I(R)) = diam(\Gamma(R/I))$.*

**Theorem 6.3.28.** ([107, Theorem 2.13]) *Let $I$ be a primal radical $Q$-ideal of a commutative Noetherian semiring $R$. Then $diam(\Gamma_I(R)) \leq 2$.*

In commutative ring theory, one of the essential questions is whether a zero-divisor graph with respect to an ideal commutes with localization, and in this case, what are the relationships between the diameters (resp., girths) of such graphs (see Section 5.1 for an affirmative answer to these questions). Atani [107] proved that the zero-divisor graph with respect to primal ideals commutes with the semiring of fractions of $R$ with respect to $S$, $diam(\Gamma_I(R)) = diam(\Gamma_{IR_S}(R_S))$, and $gr(\Gamma_I(R)) = gr(\Gamma_{IR_S}(R_S))$.

Let $R$ be a given commutative semiring, and let $S$ be the set of all multiplicatively cancelable elements of $R$ (so $1 \in S$). Clearly, the set $S$ is multiplicatively closed. Define a relation $\sim$ on $R \times S$ as follows: for $(a, s), (b, t) \in R \times S$, we write $(a, s) \sim (b, t)$ if and only if $ad = bc$. Then $\sim$ is an equivalence relation on $R \times S$. For $(a, s) \in R \times S$, denote the equivalence class of $\sim$ which contains $(a, s)$ by $a/s$, and denote the set of all equivalence classes of $\sim$ by $R_S$. Then $R_S$ can be given the structure of a commutative semiring under the operations $a/s + b/t = (ta + sb)/st$ and $(a/s)(b/t) = (ab)/st$ for every $a, b \in R$ and $s, t \in S$. This new semiring $R_S$ is called the *semiring of fractions* of $R$ with respect to $S$; its zero element is $0/1$, its multiplicative identity element is $1/1$, and every element of $S$ has a multiplicative inverse in $R_S$.

Hereafter, in this section we shall assume unless otherwise stated, that $S$ is the set of all multiplicatively cancelable elements of a semiring $R$

Now suppose that $I$ is an ideal of a commutative semiring $R$. The ideal generated by $I$ in $R_S$, that is, the set of all finite sums $s_1a_1 + \cdots + s_na_n$, where $a_i \in R_S$ and $s_i \in I$, is called the extension of $I$ to $R_S$, and it is denoted by $IR_S$. Again, if $J$ is an ideal of $R_S$, then by the contraction of $J$ in $R$, we mean $J \cap R = \{r \in R \mid r/1 \in J\}$, which is clearly an ideal of $R$.

Let $R$ be a commutative semiring, $X$ a nonempty subset of $R$, and $I$ an ideal of $R$. Set $XR_S = \{a/s \mid a \in X, s \in S\} \subseteq R_S$. We say that a zero-divisor graph with respect to $I$ commutes with the commutative semiring of fractions of $R$ with respect to $S$ if $\Gamma_I(R)R_S = \Gamma_{IR_S}(R_S)$.

**Theorem 6.3.29.** ([107, Proposition 2.15]) *Let $I$ be a $P$-primal ideal of a commutative semiring $R$ with $P \cap S = \emptyset$. Then $\Gamma_I(R)R_S = \Gamma_{IR_S}(R_S)$.*

**Theorem 6.3.30.** ([107, Theorem 2.16]) *Let $I$ be a $P$-primal ideal of a commutative semiring $R$ with $P \cap S = \emptyset$. Then $diam(\Gamma_I(R)) = diam(\Gamma_{IR_S}(R_S))$.*

**Theorem 6.3.31.** ([107, Theorem 2.17]) *Let $I$ be a $P$-primal ideal of a commutative semiring $R$ with $P \cap S = \emptyset$. Then $gr(\Gamma_I(R)) = gr(\Gamma_{IR_S}(R_S))$.*

Now we discuss the diameter of $\Gamma_I(R)$, where $I$ is not a primal ideal. Let $I$ be a $k$-ideal of a commutative semiring $R$. A prime $k$-ideal $P$ of $R$ is said to be a *minimal prime* $k$-ideal of $I$ if $J$ is a prime $k$-ideal in $R$ such that $I \subsetneq J \subseteq P$, then $J = P$. Denote by $Min_k(I)$ the set of minimal prime $k$-ideals of $R$ containing $I$. Note that $Min_k(0) = Min_k(R)$.

**Proposition 6.3.32.** ([107, Proposition 3.3]) *Let $I$ be a finitely generated $k$-ideal of a reduced commutative semiring $R$. Then $I$ is contained in a minimal prime $k$-ideal of $R$ if and only if $Ann(I) \neq 0$. In particular, $Z(R) = \cup_{P \in Min_k(0)} P$.*

**Proposition 6.3.33.** ([107, Proposition 3.4]) *Let $I$ be a $Q$-ideal of a commutative semiring $R$. Then $I$ is a radical ideal of $R$ if and only if $R/I$ is a reduced commutative semiring.*

**Theorem 6.3.34.** ([107, Theorem 3.6]) *Let $I$ be a radical $Q$-ideal of a commutative semiring $R$. Then $V(\Gamma_I(R)) \cup I = \cup_{P \in Min_k(I)} P$.*

**Theorem 6.3.35.** ([107, Proposition 3.7]) *Let $I$ be an ideal of a commutative semiring $R$. Then the following hold:*

(a) *If there are nonadjacent elements $a, b \in V(\Gamma_I(R))$ such that the ideal $J = (a, b)$ is prime to $I$, then $diam(\Gamma_I(R)) = 3$;*

(b) *$I$ is not primal if and only if there are elements $a, b \in V(\Gamma_I(R))$ such that the ideal $J = (a, b)$ is prime to $I$.*

To find the diameter of $\Gamma_I(R)$, we need the following theorem.

**Theorem 6.3.36.** ([107, Theorem 3.8])

(a) *Let $R$ be a reduced commutative semiring. If $|Min_k(0)| \geq 3$ and there are $0 \neq a, b \in Z(R)$ such that $(0 : (a, b)) = 0$, then $diam(\Gamma(R)) = 3$;*

(b) *Let $I$ be a $k$-ideal of a nonreduced commutative semiring $R$. If $Ann(I) \neq 0$ and $q \in R$ is nilpotent, then $(0 : qR + I) \neq 0$. In particular, if $a \in Z(R)$ and $q$ is nilpotent, then $a + q \in Z(R)$ and $(0 : (a, q)) \neq 0$.*

**Theorem 6.3.37.** ([107, Theorem 3.9]) *Let $I$ be a radical $Q$-ideal of a commutative semiring $R$, and suppose that $I$ is not a primal ideal of $R$ and $|Min_k(I)| \geq 3$. Then $diam(\Gamma_I(R)) = 3$.*

**Theorem 6.3.38.** ([107, Theorem 3.10]) *Let $I$ is a radical $Q$-ideal of a commutative semiring $R$ that is not a primal ideal. Then $diam(\Gamma_I(R)) \leq 2$ if and only if $|Min_k(I)| = 2$.*

**Theorem 6.3.39.** ([107, Proposition 3.11]) *Let $I$ be a $Q$-ideal of a commutative semiring $R$ which is not a radical ideal, and let $J$ be a $k$-ideal of $R$ which is not prime to $I$. If $x \in \sqrt{I}$, then the ideal $Rx + J$ is not prime to $I$.*

**Theorem 6.3.40.** ([107, Theorem 3.13]) *Let $I$ be a $Q$-ideal of a commutative semiring $R$ which is not a radical ideal, and suppose that $I$ is not a primal ideal of $R$. Then $diam(\Gamma_I(R)) = 3$.*

In 2015, Mehdi-Nezhad and Rahimi [288] extended the notion of the annihilating -ideal graph of a commutative ring with respect to an ideal to the commutative semiring case. We call an ideal $I$ of $R$ an annihilating ideal (or annihilator ideal) if there is a nonzero ideal $J$ of $R$ such that $IJ = (0)$. We use the notation $\mathbb{A}(R)$ for the set of all annihilating ideals of $R$. Notice that $R \notin \mathbb{A}(R)$ since $RI = I$ for any ideal $I$ of $R$. For the definition and study of the annihilating-ideal graph of a commutative ring and the annihilating-ideal graph of a commutative ring with respect to an ideal, refer back to Section 5.5.

Let $I$ be a proper ideal of a commutative semiring $R$ with nonzero identity. Let $N_I(R)$ be the set of all ideals of $R$ that are not contained in $I$. The annihilating-ideal graph of $R$ with respect to $I$, denoted by $\mathbb{AG}_I(R)$, is the graph whose vertex set is the set $\{A \in N_I(R) \mid AB \subseteq I \text{ for some } B \in N_I(R)\}$ with distinct vertices $A$ and $B$ adjacent if and only if $AB \subseteq I$. Clearly, $\mathbb{AG}_I(R) = \emptyset$ if and only if $I$ is a prime ideal of $R$.

Let $C_I(R)$ be the set of all ideals of $R$ that are contained in $I$. Note that $\mathbb{I}(R)$, the set of all ideals of $R$, is a semiring with zero $(= (0))$ and identity $(= R)$ under the multiplication and addition of ideals of $R$. It is clear that $C_I(R)$ is an ideal in the semiring $\mathbb{I}(R)$. Actually, $C_I(R)$ is a strongly $k$-ideal of $\mathbb{I}(R)$ since $A \subseteq A + B \subseteq I$ and $B \subseteq A + B \subseteq I$ for every $A, B \in C_I(R)$.

First of all, we discuss some basic properties of $\Gamma_I(R)$ and $\mathbb{AG}_I(R)$ and show how to construct ideals (resp., prime ideals) in $R$. In addition, we apply these results to the annihilating-ideal graph of $R$ with respect to $I$ as a special case of $\Gamma_I(R)$.

**Proposition 6.3.41.** ([288, Proposition 2.2]) *Let $A, B$, and $C$ be vertices of the graph $\mathbb{AG}_I(R)$, where $R$ is a commutative semiring (ring).*

(a) *Suppose $A \cap B$ is not contained in $I$. Then $(A \cap B)C$ is contained in $I$ whenever $AC$ or $BC$ is a subset of $I$;*

(b) *If $A \cap B$ is contained in $I$, then $A - B$ is an edge in $\mathbb{AG}_I(R)$ since $AB \subseteq A \cap B$;*

(c) *$(A + B)C$ is contained in $I$ whenever $AC$ and $BC$ are subsets of $I$.*

**Corollary 6.3.42.** ([288, Corollary 2.3]) *Let $I$ be an ideal of a commutative semiring (ring) $R$ such that the radical of $I$ is equal to $I$. If $\Gamma_I(R)$ contains two edges $a-c$ and $b-c$, then the ideals $(a, b)$ and $(c)$ are adjacent vertices in $\mathbb{AG}_I(R)$.*

**Lemma 6.3.43.** ([288, Lemma 2.5]) *Let $I$ be an ideal of a commutative semiring (ring) $R$ such that the radical of $I$ is equal to $I$. If $a$ (resp., $a - b$) is a vertex (resp., edge) in $\Gamma_I(R)$, then $(a)$ (resp., $(a) - (b)$) is a vertex (resp., edge) in $\mathbb{AG}_I(R)$. Conversely, If $(a)$ (resp., $(a) - (b)$) is a vertex (resp., edge) in $\mathbb{AG}_I(R)$, then $a$ (resp., $a - b$) is a vertex (resp., edge) in $\Gamma_I(R)$.*

The following result shows that for $I$ a radical ideal of $R$, $\Gamma_I(R)$ is isomorphic to a subgraph of $\mathbb{AG}_I(R)$, which naturally asserts some known results (properties) between them interchangeably.

**Proposition 6.3.44.** ([288, Proposition 2.6]) *Let $I$ be a proper radical ideal of a commutative semiring (ring) $R$.*

(a) $\mathbb{AG}_I(R)$ *contains an isomorphic copy of* $\Gamma_I(R)$*;*

(b) $\omega(\Gamma_I(R)) \leq \omega(\mathbb{AG}_I(R))$*;*

(c) $gr(\mathbb{AG}_I(R)) \leq gr(\Gamma_I(R))$*;*

(d) *If* $\mathbb{AG}_I(R)$ *is planar, then so is* $\Gamma_I(R)$*, or equivalently, the nonplanarity of* $\Gamma_I(R)$ *implies the nonplanarity of* $\mathbb{AG}_I(R)$*.*

**Proposition 6.3.45.** ([288, Proposition 2.7]) *Let* $I$ *be a nonzero proper ideal of a commutative ring* $R$*.*

(a) $\mathbb{AG}_I(R)$ *is the empty graph if and only if* $I$ *is a prime ideal of* $R$*;*

(b) *For any ideal* $I$ *of* $R$*,* $\mathbb{AG}_I(R)$ *contains a copy of* $\mathbb{AG}(R/I)$ *as a subgraph;*

(c) $\mathbb{AG}_I(R)$ *is the empty graph if and only if* $\mathbb{AG}(R/I)$ *is the empty graph.*

**Proposition 6.3.46.** ([288, Proposition 2.8]) *Let* $I$ *be a proper radical ideal of a commutative ring* $R$*. Then the following statements are equivalent:*

(a) $\mathbb{AG}_I(R)$ *is finite;*

(b) $\Gamma_I(R)$ *is finite;*

(c) $R$ *is finite or* $I$ *is prime;*

(d) *If* $R$ *be a finite ring, then* $|V(\mathbb{AG}_I(R))| \geq n|I|$*, where* $n$ *is the number of vertices of* $\Gamma(R/I)$*.*

**Proposition 6.3.47.** ([288, Proposition 2.9]) *Let* $R$ *and* $S$ *be finite commutative rings. Let* $I$ *and* $J$ *be ideals of* $R$ *and* $S$*, respectively, such that* $\sqrt{I} = I$ *and* $\sqrt{J} = J$*. If* $\Gamma_I(R) \cong \Gamma_J(S)$ *(nonempty graphs), then* $\mathbb{AG}(R/I) \cong \mathbb{AG}(R/J)$*.*

Note that in the above proposition, both graphs $\Gamma_I(R)$ and $\mathbb{AG}(R/I)$ are subgraphs of $\mathbb{AG}_I(R)$.

**Theorem 6.3.48.** ([288, Theorem 2.11]) *Let* $I$ *be an ideal of the commutative semiring (ring)* $R$ *such that* $a^2 \notin I$ *for every* $a \in R \setminus I$ *(in particular,* $R$ *could be reduced when* $I = (0)$*). Then there is a vertex which is adjacent to every other vertex of* $\Gamma_I(R)$ *if* $\Gamma_I(R)$ *is nonempty with* $gr(\Gamma_I(R)) = \infty$*.*

**Corollary 6.3.49.** ([288, Corollary 2.12]) *Let* $I$ *be an ideal of the commutative semiring (ring)* $R$ *such that* $A^2 \nsubseteq I$ *for every* $A \in \mathbb{I}(R) \setminus C_I$ *(in particular,* $R$ *could be reduced when* $I = (0)$*). Then there is a vertex which is adjacent to every other vertex of* $\mathbb{AG}_I(R)$ *if* $\mathbb{AG}_I(R)$ *is nonempty with* $gr(\mathbb{AG}_I(R)) = \infty$*.*

**Theorem 6.3.50.** ([288, Theorem 2.13], cf. Theorem 5.1.18) *Let* $I$ *be a nonzero* $k$*-ideal of a commutative semiring* $R$ *and* $a \in \Gamma_I(R)$ *be adjacent to every other vertex of* $\Gamma_I(R)$*. If there is a* $b \in I$ *such that* $a \neq a + b$*, then* $(I : a)$ *contains* $(I : x)$ *for every* $x \in R \setminus I$*. Moreover,* $(I : a)$ *is a prime ideal.*

Next, we present an example to show that the subtractive condition for $I$ in Theorem 6.3.50 is not a superficial assumption. Note that an ideal in a commutative ring $R$ is always subtractive and $a \neq a+b$ for every $a, b \in R$ with $b \neq 0$. Thus, the following example can only happen for a semiring which is not a ring.

**Example 6.3.51.** ([288, Example 2.18]) A commutative semiring $S$ is said to be idempotent if it is both additively and multiplicatively idempotent. Consider the idempotent semiring $S = \{0, 1, a\}$, where $1 + a = a + 1 = a$. Let $R = S \times S$ be the semiring of the direct product of the semirings. Clearly, $I = \{(0,0), (0,a)\}$ is an ideal of $R$, but not a $k$-ideal in $R$ since $(0,1) + (0,a) = (0,a) \in I$, but $(0,1) \notin I$. Now it can easily be seen that $\Gamma_I(R)$ is a star graph on 5 vertices with center $(0,1)$ and vertex set $V = \{(0,1), (1,0), (a,a), (a,0), (1,a)\}$.

Clearly, there is an element $(0,a) \in I$ such that $(0,1) \neq (0,1) + (0,a) = (0,a)$. Obviously, $V \cap I \neq (I : (0,1))$ since $(0,1)^2 = (0,1)$, and hence $(0,1) \notin (I : (0,1))$. Also, $(I : (0,1))$ does not contain $(I : (1,a))$ since $(0,1) \in (I : (1,a))$ and $(0,1) \notin (I : (0,1))$. Moreover, $(I : (0,1))$ is a prime ideal in $R$.

**Theorem 6.3.52.** ([288, Theorem 2.14]) *Let $I$ be a nonzero ideal of a commutative semiring (ring) $R$ and $A \in \mathbb{AG}_I(R)$.*

(a) *If $A$ is adjacent to every other vertex of $\mathbb{AG}_I(R)$ with $A \cap I \neq I$, then $(I : A)$ contains $(I : X)$ for every $X \in \mathbb{I}(R) \setminus C_I$. Moreover, $(I : A)$ is a prime ideal in $\mathbb{I}(R)$;*

(b) *If $A$ is adjacent to every other vertex of $\mathbb{AG}_I(R)$ with $A^2 \subseteq I$, then $(I : A)$ contains $(I : X)$ for each $X \in \mathbb{I}(R) \setminus C_I$. Moreover, $(I : A)$ is a prime ideal in $\mathbb{I}(R)$.*

**Theorem 6.3.53.** ([288, Theorem 2.15], cf. Theorem 5.1.19) *Let $I$ be an ideal of a commutative semiring (ring) $R$, and let $S$ be a maximal clique in $\Gamma_I(R)$ such that $x^2 \in I$ for every $x \in S$. Then $S \cup I$ is an ideal of $R$.*

**Corollary 6.3.54.** ([288, Corollary 2.16]) *Let $I$ be an ideal of a commutative semiring (ring) $R$, and let $S$ be a maximal clique in $\mathbb{AG}_I(R)$ such that $X^2 \subseteq I$ for every $X \in S$. Then $S \cup C_I$ is an ideal of $\mathbb{I}(R)$.*

Next, we give an example to show that for every $x$ in the maximal clique $S$, $x^2 \in I$ is not a superficial assumption in Theorem 6.3.53.

**Example 6.3.55.** ([288, Example 2.17]) Consider the idempotent semiring $S = \{0, 1, a\}$, where $1 + a = a + 1 = a$. Let $R = S \times S$ be the semiring of the direct product of the semirings. Clearly, $I = \{(0,0), (a,0), (0,a), (a,a)\}$ is an ideal of $R$, and $\Gamma_I(R)$ is a square $(1,0) - (0,1) - (1,a) - (a,1) - (1,0)$ in which every edge can be regarded as a maximal clique. Let $M = \{(1,0), (0,1)\}$. Clearly $(1,0)^2 = (1,0) \notin I$. Now it is clear that $M \cup I$ is not an ideal in $R$ since $(1,0) + (0,a) = (1,a) \notin M \cup I$.

Next, we show that $\Gamma_I(R)$ (resp., $\mathbb{AG}_I(R)$) is connected with $diam(\Gamma_I(R)) \leq 3$ (resp., $diam(\mathbb{AG}_I(R)) \leq 3$). Furthermore, if $\Gamma_I(R)$ (resp., $\mathbb{AG}_I(R)$) contains a cycle, then $gr(\Gamma_I(R)) \leq 4$ (resp., $gr(\mathbb{AG}_I(R)) \leq 4$).

**Theorem 6.3.56.** ([288, Theorem 3.4]) *Let $I$ be a proper ideal of a commutative semiring (ring) $R$. Then $\Gamma_I(R)$ is connected and $diam(\Gamma_I(R)) \leq 3$. Furthermore, $gr(\Gamma_I(R)) \leq 4$ if $\Gamma_I(R)$ contains a cycle.*

**Corollary 6.3.57.** ([288, Corollary 3.5]) *Let $I$ be an ideal of a commutative semiring (ring) $R$. Then $\mathbb{AG}_I(R)$ is connected with $diam(\mathbb{AG}_I(R)) \leq 3$. Furthermore, $gr(\mathbb{AG}_I(R)) \leq 4$ if $\mathbb{AG}_I(R)$ contains a cycle.*

We are now interested in the girth of $\Gamma_I(R)$ and $\mathbb{AG}_I(R)$ when $I$ is an intersection of prime ideals of the commutative semiring (ring) $R$.

**Theorem 6.3.58.** ([288, Proposition 5.1]) *Let $I$ be an ideal of a commutative ring $R$.*

(a) *If $\mathbb{AG}(R/I)$ contains a cycle, then $gr(\mathbb{AG}_I(R)) \leq 4$;*

(b) *If $\mathbb{AG}_I(R)$ has no cycles, then $gr(\mathbb{AG}(R/I)) = \infty$.*

**Theorem 6.3.59.** ([288, Proposition 5.5]) *Let $I$ be a proper ideal of a commutative semiring (ring) $R$. If $R$ contains at least three distinct prime ideals of the form $(I : x), (I : y)$, and $(I : z)$ for some $x, y, z \in R$, then $gr(\Gamma_I(R)) = 3$.*

**Corollary 6.3.60.** ([288, Corollary 5.6]) *Let $I$ be a proper ideal of a commutative semiring (ring) $R$. If $\mathbb{I}(R)$ contains at least three distinct prime ideals of the form $(I : A), (I : B)$, and $(I : C)$ for some (proper) ideals $A, B$, and $C$ of $R$, then $gr(\mathbb{AG}_I(R)) = 3$.*

**Theorem 6.3.61.** ([288, Theorem 5.7], cf. Theorem 5.1.46) *Let $I$ be a nonzero proper ideal of a ring $R$ such that $I = P_1 \cap P_2$ for some prime ideals $P_1$ and $P_2$ of $R$. Then $gr(\Gamma_I(R)) = 4$.*

**Theorem 6.3.62.** ([288, Proposition 5.8]) *Let $I$ be a nonzero proper ideal of a commutative semiring (ring) $R$, and let $P_1$ and $P_2$ be prime ideals of $R$ such that $I = P_1 \cap P_2$ (resp., $C_{P_1}$ and $C_{P_2}$ are prime ideals of the commutative semiring $\mathbb{I}(R)$ and $C_I = C_{P_1} \cap C_{P_2}$). If $|P_1 \setminus P_2| \geq 2$ and $|P_2 \setminus P_1| \geq 2$ (resp., $|C_{P_1} \setminus C_{P_2}| \geq 2$ and $|C_{P_2} \setminus C_{P_1}| \geq 2$), then $gr(\Gamma_I(R)) = 4$ (resp., $gr(\mathbb{AG}_I(R)) = 4$).*

**Theorem 6.3.63.** ([288, Theorem 5.9]) *Let $I$ be an ideal of a commutative semiring (ring) $R$.*

(a) *If $gr(\Gamma_I(R)) = 4$ and $a^2 \notin I$ for every $a \in R \setminus I$ (in particular, $R$ could be a reduced commutative semiring (ring) when $I = (0)$), then $\Gamma_I(R)$ is a complete bipartite graph;*

(b) *If $\Gamma_I(R)$ is a complete bipartite graph, then $gr(\Gamma_I(R)) = 4$ or $\infty$.*

**Corollary 6.3.64.** ([288, Corollary 5.10]) *Let $I$ be an ideal of a commutative semiring (ring) $R$.*

(a) *If $gr(\mathbb{AG}_I(R)) = 4$ and $A^2 \notin C_I$ for every $A \in \mathbb{I}(R) \setminus C_I$ (in particular, $R$ could be a reduced commutative semiring (ring) when $I = (0)$), then $\mathbb{AG}_I(R)$ is a complete bipartite graph;*

(b) *If $\mathbb{AG}_I(R)$ is a complete bipartite graph, then $gr(\mathbb{AG}_I(R)) = 4$ or $\infty$.*

Next, we state conditions under which a vertex (resp., an edge) of $\mathbb{AG}_I(R)$ is not a cut-point (resp., bridge).

**Theorem 6.3.65.** ([288, Proposition 3.7]) *Let $I$ be a nonzero proper ideal of the commutative semiring (ring) $R$ and $X$ a vertex in $\mathbb{AG}_I(R)$.*

(a) *If $C_X \setminus C_I$ has more than one element, then $X$ is not a cut-point in $\mathbb{AG}_I(R)$;*

(b) *If $X$ does not contain $I$, then $X$ is not a cut-point of $\mathbb{AG}_I(R)$;*

(c) *If $X$ is not a principal ideal of $R$, then $X$ is not a cut-point in $\mathbb{AG}_I(R)$;*

(d) *Let $X = (x)$ be a principal ideal containing $I$ and $\sqrt{I} = I$. If $(x) \neq (x)^2$, then $X$ is not a cut-point in $\mathbb{AG}_I(R)$.*

**Theorem 6.3.66.** ([288, Proposition 3.9]) *Let $I$ be a nonzero ideal of a commutative semiring (ring) $R$ and $A \neq B$ two distinct vertices of the graph $\mathbb{AG}_I(R)$ with $A - B$ an edge in $\mathbb{AG}_I(R)$. Then the following hold:*

(a) *Assume that $A \nsubseteq B$ and $B \nsubseteq A$ with $A \cap B \nsubseteq I$. Then $A - B$ is not a bridge in $\mathbb{AG}_I(R)$;*

(b) *Assume that $I \nsubseteq A$ and $I \nsubseteq B$. Then $A - B$ is not a bridge in $\mathbb{AG}_I(R)$;*

(c) *Assume that $A^2 \subseteq I$ and $B^2 \subseteq I$ with $A \nsubseteq B$ and $B \nsubseteq A$. Then $A - B$ is not a bridge in $\mathbb{AG}_I(R)$;*

(d) *Assume that neither $A$ nor $B$ is a principal ideal of $R$ with $A \nsubseteq B$ and $B \nsubseteq A$. Then $A - B$ is not a bridge in $\mathbb{AG}_I(R)$;*

(e) *Assume that $A = (x)$ and $B = (y)$ are principal ideals of $R$ containing $I$ with $A \nsubseteq B$ and $B \nsubseteq A$. Let $\sqrt{I} = I$, $(x)^2 \neq (x)$, and $(y)^2 \neq (y)$. Then $A - B$ is not a bridge in $\mathbb{AG}_I(R)$.*

The next result gives some conditions for $\mathbb{AG}_I(R)$ to have no bridges via a cardinal number criterion.

**Theorem 6.3.67.** ([288, Proposition 3.10]) *Let $I$ be a nonzero ideal of a commutative semiring (ring) $R$ and $A \neq B$ two distinct vertices of $\mathbb{AG}_I(R)$ with $A - B$ an edge in $\mathbb{AG}_I(R)$. Then the following hold:*

(a) *Assume that $C_{A \cap B} \setminus C_I$ has more than one element. Then $A - B$ is not a bridge in $\mathbb{AG}_I(R)$;*

(b) *Assume that* $A \notin C_B$ *and* $B \notin C_A$. *If* $|C_{A \cap B} \setminus C_I| = 1$, *then* $A - B$ *is not a bridge in* $\mathbb{AG}_I(R)$;

(c) *Let* $C_{A \cap B} \setminus C_I = \emptyset$. *Assume* $A^2 \subseteq I$ *and* $B^2 \subseteq I$. *Then* $A - B$ *is not a bridge in* $\mathbb{AG}_I(R)$;

(d) *Assume that* $A \in C_B$ *and* $I \notin C_B$ *or* $B \in C_A$ *and* $I \notin C_A$. *Then* $A - B$ *is not a bridge in* $\mathbb{AG}_I(R)$.

We now present a commutative semiring whose ideal-based zero-divisor graph has a cut-point and more than one bridge, which is in contrast to the ring case. Note that Theorem 5.2.20 says that $\Gamma_I(R)$ has no cut-points for every nonzero proper ideal $I$ of a ring $R$. But in contrast to the ring case, we give a semiring whose ideal-based zero-divisor graph has a cut-point.

**Example 6.3.68.** ([288, Example 3.11]) Let $S = \{0, 1, a\}$, where $1 + a = a + 1 = a$, and let $R = S \times S$ be the semiring of the direct product of the semirings. Consider the ideal $I = \{(0,0), (0,a)\}$. Then $\Gamma_I(R)$ is a star graph on 5 vertices with center $(0,1)$. Clearly, the center of $\Gamma_I(R)$ is a cut-point and every edge of $\Gamma_I(R)$ is a bridge, which is in contrast to Theorem 5.2.20 and Theorem 5.2.22.

In the following, we see that if $I = P_1 \cap P_2$ with $P_1$ and $P_2$ prime ideals of $R$, where $C_{P_1}$ and $C_{P_2}$ are prime ideals of $\mathbb{I}(R)$, then $\Gamma_I(R)$ is a complete bipartite graph.

**Theorem 6.3.69.** ([288, Theorem 4.1], cf. Theorem 5.2.8) *Let* $I$ *be a nonzero ideal of a commutative semiring (ring)* $R$. *Then*

(a) *If* $P_1$ *and* $P_2$ *are prime ideals of* $R$ *and* $I = P_1 \cap P_2 \neq (0)$, *then* $\Gamma_I(R)$ *is a complete bipartite graph;*

(b) *If* $I \neq 0$ *is an ideal of* $R$ *with* $\sqrt{I} = I$, *then* $\Gamma_I(R)$ *is a complete bipartite graph if and only if there are prime ideals* $P_1$ *and* $P_2$ *of* $R$ *such that* $I = P_1 \cap P_2$.

**Corollary 6.3.70.** ([288, Corollary 4.2]) *Let* $I$ *be a nonzero ideal of a commutative semiring (ring)* $R$. *Then*

(a) *If* $P_1$ *and* $P_2$ *are prime ideals of* $R$ *with* $I = P_1 \cap P_2 \neq 0$, *then* $\mathbb{AG}_I(R)$ *is a complete bipartite graph;*

(b) *Let* $I$ *be a nonzero ideal of* $R$ *with* $\sqrt{I} = I$. *Then* $\mathbb{AG}_I(R)$ *is a complete bipartite graph if and only if there are prime ideals* $\mathbb{P}_1$ *and* $\mathbb{P}_2$ *of* $\mathbb{I}(R)$ *such that* $C_I = \mathbb{P}_1 \cap \mathbb{P}_2$.

**Theorem 6.3.71.** ([288, Theorem 4.6], cf. Theorem 5.2.9) *Let* $I$ *be a nonzero proper* $k$*-ideal of a commutative semiring (ring)* $R$. *If* $\Gamma_I(R)$ *is a complete* $r$*-partite graph for* $r \geq 3$, *then at most one of the parts has more than one vertex.*

**Corollary 6.3.72.** ([288, Corollary 4.8]) *Let $I$ be a nonzero proper ideal of a commutative semiring (ring) $R$. If $\mathbb{AG}_I(R)$ is a complete $r$-partite graph for $r \geq 3$, then at most one of the parts has more than one vertex. If $V_i$ and $V_j$ $(i \neq j)$ are two parts such that $V_i = \{A\}$ and $V_j = \{B\}$, then either $A^2 \subseteq I$ or $B^2 \subseteq I$. Furthermore, if $B \nsubseteq A$, then $A^2 \subseteq I$.*

The following theorem presents a characterization of all regular graphs which can be realized as the zero-divisor graph of a decomposable commutative semiring. A semiring $R$ is said to be *indecomposable* if $R$ cannot be represented as $R_1 \times R_2$, where $R_1$ and $R_2$ are two semirings, otherwise, $R$ is said to be *decomposable*. Recall that a graph $G$ is *regular* if the degrees of all vertices of $G$ are the same.

**Theorem 6.3.73.** ([288, Theorem 4.9], cf. Theorem 2.2.13) *Let $R$ be a finite commutative semiring such that $\Gamma(R)$ is a regular graph. If $R$ is a decomposable semiring, then $\Gamma(R)$ is a complete bipartite graph.*

Next, we study the clique number of $\Gamma_I(R)$ and $\mathbb{AG}_I(R)$ when $I$ is an intersection of prime ideals of the semiring (ring) $R$.

**Theorem 6.3.74.** ([288, Theorem 5.11], cf. Theorem 5.2.18) *Let $I$ be an ideal of a commutative semiring (ring) $R$ such that $I = \bigcap_{1 \leq i \leq n} P_i$, and for every $1 \leq j \leq n$, $I \neq \bigcap_{1 \leq i \leq n, i \neq j} P_i$, where the $P_i$'s are prime ideals of $R$. Then $\omega(\Gamma_I(R)) = n$.*

**Corollary 6.3.75.** ([288, Corollary 5.12])

(a) *If $I = \bigcap_{1 \leq i \leq n} P_i$ and $J = \bigcap_{1 \leq j \leq m} Q_j$, where the $P_i$'s and $Q_j$'s are prime ideals of a commutative semiring (ring) $R$ such that $\Gamma_I(R) \cong \Gamma_J(R)$, then $m = n$;*

(b) *If $R$ is a commutative semiring (ring) with a finite (more than one) number of maximal ideals, then $\omega(\Gamma_{J(R)}(R)) = |Max(R)|$.*

**Corollary 6.3.76.** ([288, Corollary 5.13]) *Let $I$ be an ideal of a commutative semiring (ring) $R$ such that $I = \bigcap_{1 \leq i \leq n} P_i$ and for every $1 \leq j \leq n$, $I \neq \bigcap_{1 \leq i \leq n, i \neq j} P_i$, where the $P_i$'s are prime ideals of $R$. Then $\omega(\mathbb{AG}_I(R)) = n$.*

**Corollary 6.3.77.** ([288, Corollary 5.14]) *Let $R$ be a commutative semiring (ring). Let $I = \bigcap_{1 \leq i \leq n} P_i \neq (0)$ and $J = \bigcap_{1 \leq j \leq m} Q_j \neq (0)$ such that for every $1 \leq k \leq n$, $I \neq \bigcap_{1 \leq i \leq n, i \neq k} P_i$ and for every $1 \leq k \leq m$, $J \neq \bigcap_{1 \leq i \leq n, j \neq k} Q_j$, where the $P_i$'s and $Q_j$'s are prime ideals of $R$. If $\mathbb{AG}_I(R) \cong \mathbb{AG}_J(R)$, then $m = n$.*

A lower bound for $\omega(\Gamma_I(R))$ can be given using elements $a \in R \setminus I$ with $a^n \in I$ for some integer $n$. Similarly, a lower bound can be obtained for $\omega(\Gamma_I(R))$ using nilpotent elements of $R$ when $I = (0)$. Note that since a finite additively cancellative semigroup is a group, we can extend Theorem 3.1.46, which is for commutative rings, to an additively cancellative infinite commutative semiring as follows.

**Theorem 6.3.78.** ([288, Theorem 5.15], cf. Theorem 3.1.46) *Let $I$ be a $k$-ideal of an additively cancellative infinite commutative semiring $R$, and let $n$ be the least positive integer such that $x^n \in I$ for some $x \in R \setminus I$. Then*

(a) *If $n = 2k$, then $\omega(\Gamma_I(R)) \geq 2^k - 1$;*

(b) *If $n = 2k + 1$, then $\omega(\Gamma_I(R)) \geq 2^k$.*

**Corollary 6.3.79.** (cf. Corollary 3.1.47) *Let $R$ be an additively cancellative infinite commutative semiring. If there is no bound on the index of nilpotence of elements of $R$, then $\omega(\Gamma(R)) = \infty$.*

## 6.4 Modules

Barnard [133] first introduced the notion of multiplication modules in 1981, and then El-Bast and Smith [190] obtained various properties of multiplication modules in 1988. Multiplication modules are natural generalizations of commutative rings, and hence it is natural to generalize zero-divisor graphs of commutative rings to those of multiplication modules. The concept of the zero-divisor graph for a ring has been extended to module theory by many researchers.

Throughout this section, $R$ is a commutative ring with identity and $M$ is an (unitary) $R$-module. Recall that, for $x \in M$, the *residual* of $Rx$ by $M$, denoted by $[x : M]$, is the set of elements $r \in R$ such that $rM \subseteq Rx$. The *annihilator of an $R$-module $M$*, denoted by $Ann_R(M)$, is $[0 : M]$. Let $T(M) = \{m \in M \mid Ann(m) = 0\}$ be the set of *torsion elements* of $M$. It is clear that if $R$ is an integral domain, then $T(M)$ is a submodule of M which is called the *torsion submodule* of $M$. If $T(M) = 0$ then $M$ is said to be *torsionfree*, and $M$ is a *torsion module* if $T(M) = M$. An $R$-module $M$ is a *multiplication module* if for every $R$-submodule $K$ of $M$, there is an ideal $I$ of $R$ such that $K = IM$.

### 6.4.1 Torsion Graphs of Modules

Motivated by the work of Anderson et al. [81], Ghalandarzadeh et al. [202] introduced the concept of torsion-graph for modules as a natural generalization of zero-divisor graphs for rings. The *torsion graph* of $M$, denoted by $\Gamma(M)$, is a simple graph whose vertices are the nonzero torsion elements of $M$, and two distinct vertices $x$ and $y$ are adjacent if and only if $[x : M][y : M]M = 0$. Thus, $\Gamma(M)$ is the empty graph if and only if $M$ is a torsionfree $R$-module. The torsion graph investigates the interplay of module properties of $M$ in relation to the properties of $\Gamma(M)$. Also, the torsion graph helps us study the algebraic properties of modules using graph-theoretical tools. First, we give some examples of the torsion graph of modules.

**Example 6.4.1.** ([203, Example 2.1])

(a) Let $M_1$ be a torsionfree module, $M_2$ an $R$-module, and $M = M_1 \oplus M_2$. Then $T(M)^* = \{(0, m_2) \mid m_2 \in T(M_2)^*\}$ and $[(0, m_2) : M] = 0$. Hence $\Gamma(M)$ is a complete graph.

(b) The torsion graphs for several modules are given in Figure 6.4. Note that these examples show that nonisomorphic modules may have the same torsion graph. Therefore, all connected graphs with fewer than four vertices may be regarded as $\Gamma(M)$.

$\Gamma_{\mathbb{Z}_4}(\mathbb{Z}_4)$

$\Gamma_{\mathbb{Z}}(\mathbb{Z} \oplus \mathbb{Z}_3)$ or $\Gamma_{\mathbb{Z}_9}(\mathbb{Z}_9)$

$\Gamma_{\mathbb{Z}_6}(\mathbb{Z}_2 \oplus \mathbb{Z}_3)$ or $\Gamma_{\mathbb{Z}_8}(\mathbb{Z}_8)$

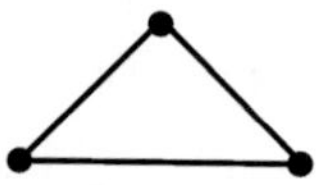

$\Gamma_{\mathbb{Z}}(\mathbb{Z} \oplus \mathbb{Z}_4)$ or $\Gamma_{\mathbb{Z}}(\mathbb{Z}_2 \oplus \mathbb{Z}_2)$

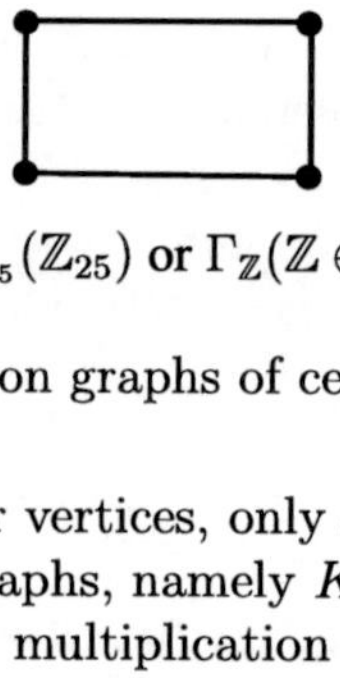

$\Gamma_{\mathbb{Z}_{25}}(\mathbb{Z}_{25})$ or $\Gamma_{\mathbb{Z}}(\mathbb{Z} \oplus \mathbb{Z}_5)$

Figure 6.4: The torsion graphs of certain modules

(c) Of the eleven graphs with four vertices, only six graphs are connected. Of these six graphs, only three graphs, namely $K_4$, $C_4$, $K_{1,3}$, may be realized as $\Gamma(M)$ when $M$ is a faithful multiplication $R$-module. One part is clear from Example 2.1.3 since every ring $R$ is a multiplication $R$-module. We next explain the proof that a graph $G$ with vertices $\{a, b, c, d\}$ and edges $a - b$, $b - c$, $c - d$ cannot be realized as $\Gamma(M)$. Suppose that there is a commutative ring $R$ and faithful multiplication $R$-module $M$ with $T(M)^* = \{a, b, c, d\}$ together with only the above torsion relations. Observe that $[a : M][b : M]M = 0 = [b : M][c : M]M$, and so $[b : M]a = 0 = [b : M]c$. Thus $[b : M][a + c : M]M = 0$. Hence $a + c \in T(M)^*$, and so $a + c$ must be either $a, b, c$ or $d$. If $a + c = a$ or $a + c = c$, then $a = 0$ or $c = 0$ and so we have a contradiction. Also, if $a + c = d$, then $[d : M][b : M]M = 0$, which is a contradiction. Therefore, $a + c = b$ is the only possibility. Similarly, $b + d = c$. Hence $b = a + c = a + b + d$; so $[a : M][d : M]M = 0$, which is a contradiction. The proofs for the other two nonrealizable connected graphs on four vertices are similar.

The next theorem shows that if $M$ is a multiplication $R$-module and $S = R \setminus Z(M)$, then $\Gamma(M) \cong \Gamma(S^{-1}M)$, which is a generalization of Theorem 3.3.6.

**Theorem 6.4.2.** ([202, Theorem 2.1]) *Let $M$ be a faithful multiplication $R$-module and $S = R \setminus Z(M)$. Then $\Gamma(M)$ and $\Gamma(S^{-1}M)$ are isomorphic.*

**Corollary 6.4.3.** ([202, Corollary 2.2]) *Let $M$ and $N$ be multiplication $R$-modules with $S^{-1}M \cong S^{-1}N$. Then $\Gamma(M) \cong \Gamma(N)$. In particular $\Gamma(M) \cong \Gamma(N)$ when $S^{-1}M = S^{-1}N$.*

We know that $\Gamma(M)$ may be infinite (i.e., the $R$-module $M$ may have infinitely many torsion elements). But an interesting case occurs when $\Gamma(M)$ is finite, because in the finite case, drawing the graph is possible. The next theorem shows that $\Gamma(M)$ is finite (except when $\Gamma(M)$ is empty) if and only if $M$ is finite.

**Theorem 6.4.4.** ([203, Theorem 2.4]) *Let $R$ be a commutative ring and $M$ a faithful multiplication $R$-module. Then $\Gamma(M)$ is finite if and only if either $M$ is finite or $M$ is a torsionfree $R$-module. If $1 \leq |\Gamma(M)| < \infty$, then $M$ is finite and not torsionfree.*

Note that the multiplication condition is not superfluous. For, let $M = \mathbb{Z} \oplus \mathbb{Z}_3$ as a $\mathbb{Z}$-module. Clearly $M$ is not a finite multiplication module, but $T(M)^* = \{(0,1),(0,2)\}$, and so $\Gamma(M)$ is finite.

Example 6.4.1 gives several $R$-modules with $diam(\Gamma(M)) = 0, 1$, or 2. In $R = \mathbb{Z}_2 \oplus \mathbb{Z}_4$, the path $(0,1) - (1,0) - (0,2) - (1,2)$ shows that $diam(\Gamma_R(R)) = 3$. Now we see that the torsion graphs of faithful $R$-modules are all connected with diameter $\leq 3$.

**Theorem 6.4.5.** ([203, Theorem 2.6]) *Let $R$ be a commutative ring and $M$ be a faithful $R$-module. Then $\Gamma(M)$ is connected and $diam(\Gamma(M)) \leq 3$. Moreover, if $\Gamma(M)$ contains a cycle, then $gr(\Gamma(M)) \leq 7$.*

Note that the faithful condition is not superfluous. For, let $M = \mathbb{Z}_2 \oplus \mathbb{Z}_3$ as a $\mathbb{Z}$-module; then $\Gamma(M)$ is not connected. The following theorem shows that $gr(\Gamma(M)) \leq 4$, which is a generalization of Theorem 2.3.2.

**Theorem 6.4.6.** ([203, Theorem 2.8]) *Let $R$ be a commutative ring and $M$ a multiplication $R$-module. If $\Gamma(M)$ contains a cycle, then $gr(\Gamma(M)) \leq 4$.*

Rad, Ghalandarzadeh, and Shirinkam [316] established some more results on the diameter of torsion graphs over a multiplication module. Moreover, they investigated the relationship between the diameter of $\Gamma(M)$ and $\Gamma(R)$. Recall that a commutative ring $R$ is called *Bézout* if every finitely generated ideal $I$ of $R$ is principal.

**Theorem 6.4.7.** ([316, Theorem 3.5]) *Let $R$ be a Bézout ring and $M$ be a multiplication $R$ module such that $|M| \neq 4$ and $M \neq T(M)$. Then $diam(\Gamma(M)) = diam(\Gamma(R))$.*

The results obtained in Theorems 6.4.8 and 6.4.9 give the diameter of $\Gamma(M)$ in terms of the minimal prime submodules of a multiplication module $M$ over a von Neumann regular ring.

**Theorem 6.4.8.** ([316, Theorem 3.7]) *Let $R$ be a von Neumann regular ring and $M$ a multiplication $R$-module. If $M$ has more than two minimal prime submodules and $T(M)$ is not a submodule of $M$, then $diam(\Gamma(M)) = 3$.*

**Theorem 6.4.9.** ([316, Theorem 3.8]) *Let $R$ be a von Neumann regular ring and $M$ a multiplication $R$-module. If $T(M)$ is not a submodule of $M$, then $diam(\Gamma(M)) \leq 2$ if and only if $M$ has exactly two minimal prime submodules.*

As an immediate consequence of Theorem 6.4.7 and Theorem 6.4.9, we obtain the following corollary.

**Corollary 6.4.10.** ([316, Corollary 3.9]) *Let $R$ be a von Neumann regular ring and $M$ a multiplication $R$-module. If $T(M)$ is not a submodule of $M$, then $M$ has exactly two minimal prime submodules if and only if $R$ has exactly two minimal prime ideals.*

Next, the authors of [203] investigated when $\Gamma(M)$ belongs to some standard class of graphs.

**Theorem 6.4.11.** ([203, Theorem 2.2]) *Let $R$ be a commutative ring and $M$ a multiplication $R$-module. Then $\Gamma(M)$ cannot be an $n$-gon for $n \geq 5$.*

Let $M_1$ be an $R_1$-module and $M_2$ be an $R_2$-module. Then $M = M_1 \times M_2$ is an $R = R_1 \times R_2$-module with coordinate-wise multiplication. It is clear that $\Gamma(M)$ is a complete bipartite graph with partition sets $V_1 = \{(m_1, 0) \mid m_1 \in M_1^*\}$ and $V_2 = \{(0, m_2) \mid m_2 \in M_2^*\}$.

We next determine when $\Gamma(M)$ is a complete graph.

**Theorem 6.4.12.** ([203, Theorem 2.9]) *Let $R$ be a ring and $M$ a multiplication $R$-module. Then there is a vertex of $\Gamma(M)$ which is adjacent to every other vertex if and only if either $M = M_1 \oplus M_2$ is a faithful $R$-module, where $M_1$ and $M_2$ are two submodules of $M$ such that $M_1$ has only two elements, $M_2$ is finitely generated with $T(M) = \{(x, 0), (0, m_2) \mid x \in M_1, m_2 \in M_2\}$, or $T(M) = IM$, where $I$ is an annihilator ideal of $R$ (and hence if $T(M) \neq M$, then $T(M)$ is a prime submodule).*

If $R$ is reduced, $M$ a faithful multiplication $R$-module, and $\Gamma(M)$ has a vertex adjacent to every other vertex, then $M$ must have the form $M = M_1 \oplus M_2$, where $M_1$ has only two elements and $M_2$ is finitely generated.

**Corollary 6.4.13.** ([203, Corollary 2.10]) *Let $R$ be a commutative ring and $M$ a finite multiplication $R$-module. Then there is a vertex of $\Gamma(M)$ that is adjacent to every other vertex if and only if either $M = M_1 \oplus M_2$ is a faithful $R$-module, where $M_1, M_2$ are two submodules of $M$ such that $M_1$ has only two elements and $M_2$ is simple, or $R$ is a local ring.*

By definition, $\Gamma(M)$ is complete if and only if $[x : M][y : M]M = 0$ for all distinct $x, y \in T(M)^*$. Except for the case when $M = M_1 \oplus M_2$, with $M_1, M_2$ having only two elements, the next theorem shows that we must also have $[x : M]^2 M = 0$ for every $x \in T(M)$ when $\Gamma(M)$ is complete. So, except for the mentioned case, $Nil(M)$ is detected by complete graphs, because $T(M) = Nil(M)$.

**Theorem 6.4.14.** ([203, Theorem 2.11]) *Let $R$ be a commutative ring and $M$ a multiplication $R$-module. Then $\Gamma(M)$ is complete if and only if either $M = M_1 \oplus M_2$ is faithful, with submodules $M_1, M_2$ having only two elements, or $[x : M][y : M]M = 0$ for every $x, y \in T(M)$.*

**Corollary 6.4.15.** ([203, Corollary 2.12]) *Let $M$ be a finite multiplication $R$-module. If $\Gamma(M)$ is complete, then either $M = M_1 \oplus M_2$, where $M_1, M_2$ are two submodules of $M$ such that $M_1, M_2$ has only two elements, or $R$ is a local ring.*

The work on when $\Gamma(M)$ is a complete graph was continued by Rad et al. in [316].

**Theorem 6.4.16.** ([316, Theorem 3.1]) *Let $R$ be a commutative ring and $M$ a multiplication $R$-module with $|M| \neq 4$. If $\Gamma(M)$ is a complete graph, then $\Gamma(R)$ is a complete graph. The converse is true if we further assume that $M$ is faithful.*

The multiplication condition in the above theorem is not superfluous. Let $R = \mathbb{Z}$ and $M = \mathbb{Z} \oplus \mathbb{Z}_6$. So $M$ is not a multiplication $R$-module. Also, $\Gamma(M)$ is a complete graph, but $V(\Gamma(R)) = \emptyset$.

**Corollary 6.4.17.** ([316, Corollary 3.3]) *Let $R$ be a commutative ring and $M$ a faithful multiplication $R$-module with $|M| \neq 4$. If $\Gamma(R)$ is a complete graph, then $R \simeq \mathbb{Z}_2 \times \mathbb{Z}_2$ or $Nil(M) = Nil(R)M = Z(R)M = T(M)$.*

**Corollary 6.4.18.** ([316, Corollary 3.4]) *Let $R$ be a commutative ring and $M$ a faithful multiplication $R$-module with $|M| \neq 4$. If $\Gamma(R)$ is a complete graph, then $|Min(M)| = 1$.*

Next, we look at the complemented nature of $\Gamma(M)$. Recall that a vertex $w$ is called a *complement* of a vertex $v$ if $v$ is adjacent to $w$ and no vertex is adjacent to both $v$ and $w$. In this case, we write $v \perp w$. In module-theoretic terms, for a multiplication $R$-module $M$, this is the same as saying $v \perp w$ in $\Gamma(M)$ if and only if $v, w \in T(M)^*$ and $Ann(w)M \cap Ann(v)M \subseteq \{0, v, w\}$. Moreover, a graph $G$ is *complemented* if every vertex has a complement, and it is *uniquely complemented* if it is complemented and any two complements of the vertices set are adjacent to the same vertices.

**Theorem 6.4.19.** ([202, Theorem 3.3]) *Let $R$ be a commutative ring and $M$ a reduced multiplication $R$-module. If $\Gamma(M)$ is complemented, then $S^{-1}M$ is von Neumann regular, where $S = R \setminus Z(M)$.*

The next example shows that $S^{-1}M$ is von Neumann regular, but $M$ is not von Neumann regular, in spite of $\Gamma(M) \cong \Gamma(S^{-1}M)$.

**Example 6.4.20.** ([202, Example 3.4])

(a) Let $M = \mathbb{Z}\times n\mathbb{Z}$ and $R = \mathbb{Z}\times\mathbb{Z}$. It is clear that $\Gamma(M)$ is a complete bipartite graph with partition sets $V_1 = \{(m_1, 0) \mid m_1 \in \mathbb{Z}^*\}$ and $V_2 = \{(0, m_2) \mid m_2 \in (n\mathbb{Z})^*\}$. Therefore $\Gamma(M)$ is complemented. Also, $M$ is a faithful multiplication R-module because $M = R(1, n)$. A simple check yields that $M$ is reduced; thus $S^{-1}M$ is von Neumann regular by Theorem 6.4.19. But $M$ is not von Neumann regular since $N = R(2, 2n)$ and $I = [N : M]$.

(b) Let $R = \mathbb{Z}_2 \times \mathbb{Z}$ and $M = R$ as an $R$-module. So $M$ is a faithful multiplication $R$-module. Clearly $M$ is reduced and $\Gamma(M)$ is an infinite star graph with center $(1, 0)$. Thus, $\Gamma(M)$ is complemented, and by Theorem 6.4.19, $S^{-1}M$ is von Neumann regular, but $M$ is not von Neumann regular.

**Corollary 6.4.21.** ([202, Corollary 3.5]) *Let $R$ be a commutative ring and $M$ a cyclic faithful reduced $R$-module. Then the following statements are equivalent:*

(a) *$S^{-1}M$ is von Neumann regular, where $S = R \setminus Z(M)$;*

(b) *$\Gamma(M)$ is uniquely complemented;*

(c) *$\Gamma(M)$ is complemented.*

**Corollary 6.4.22.** ([202, Corollary 3.6]) *Let $R$ be a commutative ring and $M$ a reduced multiplication $R$-module with $T(M) \neq M$. Then the following statements are equivalent:*

(a) *$S^{-1}M$ is von Neumann regular, where $S = R \setminus Z(M)$;*

(b) *$\Gamma(M)$ is uniquely complemented;*

(c) *$\Gamma(M)$ is complemented.*

Next, we look at some properties of $M$ when $M$ is a multiplication $R$-module with $Nil(M) \neq 0$. First we give the following key lemma.

**Lemma 6.4.23.** ([202, Lemma 3.7]) *Let $R$ be a commutative ring and $M$ a multiplication $R$-module with $Nil(M) \neq 0$. Then*

(a) *If $\Gamma(M)$ is complemented, then either $8 \leq |M| \leq 16$ or $|M| \geq 17$ and $Nil(M) = \{0, x\}$ for some $0 \neq x \in M$;*

(b) *If $\Gamma(M)$ is uniquely complemented and $|M| \geq 17$, then any complement of the nonzero element $x \in Nil(M)$ is an end vertex.*

Note that if $M$ is a faithful multiplication $R$-module such that $\Gamma(M)$ is complemented and $|Nil(M)| > 2$, then $8 \leq |M| \leq 16$ and $\Gamma(M)$ is a star graph with at most 5 edges. So it is uniquely complemented. Also, if $\Gamma(M)$ is not uniquely complemented, then $Nil(M) = \{0, x\}$, where $x$ is an element of $M$ such that $x[x : M] = 0$. Hence $x = \beta m$ for some $m \in M$ and $\beta \in [x : M]$.

**Theorem 6.4.24.** ([202, Proposition 3.9]) *Let $R_1$ and $R_2$ be commutative rings, $R = R_1 \times R_2$, and $M = M_1 \times M_2$ a multiplication $R$-module, where $M_1$ is reduced and $Nil(M_2) \neq 0$. If $\Gamma(M)$ is complemented, but not uniquely complemented, then $M_1$ is torsionfree, $Nil(M_2)$ is the unique maximal submodule of $M_2$ with $|Nil(M_2)| = 2$ and $|M_2| = 4$. Furthermore, if $D(M_2) \neq M_2$, then the converse is true, where $D(M) = \{m \in M \mid [m : M][m' : M]M = 0$ for some $0 \neq m' \in M\}$.*

Let $R = \mathbb{Z} \times \mathbb{Z}_2[x]$ and $M = \mathbb{Z} \times \mathbb{Z}_2[x]/(x^2)$ be an $R$-module. By Theorem 6.4.24, $\Gamma(M)$ is complemented, but not uniquely complemented.

**Theorem 6.4.25.** ([202, Theorem 3.11]) *Let $R$ be a commutative ring and $M$ be a multiplication $R$-module. If $\Gamma(M)$ is complemented, but not uniquely complemented, then $M = M_1 \oplus M_2$, where $M_1, M_2$ are submodules of $M$.*

Clearly star graphs are uniquely complemented. The next theorem shows that for a multiplication $R$-module $M$ with $Nil(M) \neq 0$, if $\Gamma(M)$ is uniquely complemented, then $\Gamma(M)$ is a star graph; it extends Theorem 3.6.17.

**Theorem 6.4.26.** ([202, Theorem 3.12]) *Let $R$ be a commutative ring and $M$ a multiplication $R$-module with $Nil(M) \neq 0$. If $\Gamma(M)$ is a uniquely complemented graph, then either $\Gamma(M)$ is a star graph with at most six edges or $\Gamma(M)$ is an infinite star graph with center $x$, where $Nil(M) = \{0, x\}$.*

**Corollary 6.4.27.** ([202, Corollary 3.13]) *Let $R$ be a commutative ring and $M$ a multiplication $R$-module. If $\Gamma(M)$ is uniquely complemented, then either $\Gamma(M)$ is a star graph or $S^{-1}M$ is von Neumann regular, where $S = R \setminus Z(M)$. Moreover, for a faithful cyclic $R$ module $M$, the converse is true.*

**Corollary 6.4.28.** ([202, Corollary 3.14]) *Let $R$ be a commutative ring and $M$ a multiplication $R$-module with $T(M) \neq M$. Then $\Gamma(M)$ is uniquely complemented if and only if either $\Gamma(M)$ is a star graph or $S^{-1}M$ is von Neumann regular, where $S = R \setminus Z(M)$.*

Ghalandarzadeh, Malakooti Rad, and Shirinkam [204] studied a special induced subgraph of $\Gamma(M)$. The symbol $\bar{\Gamma}(M)$ refers to the induced subgraph $\Gamma(M)$ with vertex set $\{m \in T(M)^* \mid Ann(m)M \neq 0\}$. If $M$ is a multiplication $R$-module, then it is shown that $M$ is finite if and only if $\bar{\Gamma}(M)$ is finite. The following are some of the fundamental results on $\bar{\Gamma}(M)$.

**Theorem 6.4.29.** ([204, Theorem 2.2]) *Let $R$ be a commutative ring and $M$ a multiplication $R$-module. Then $\bar{\Gamma}(M)$ is finite if and only if either $M$ is finite or $V(\bar{\Gamma}(M)) = \emptyset$.*

**Corollary 6.4.30.** ([204, Corollary 2.3]) *Let $R$ be a commutative ring and $M$ a multiplication $R$-module. Then $\Gamma(M)$ is finite if and only if either $M$ is finite or $M$ a torsionfree $R$-module.*

The multiplication condition is not superfluous. For, let $R = \mathbb{Z}$ and $M = \mathbb{Z} \oplus \mathbb{Z}_n$. Clearly $M$ is faithful and not finite, but $\bar{\Gamma}(M) = K_{n-1}$ is finite.

The following theorem shows that for every $R$-module $M$, $\bar{\Gamma}(M)$ is connected with diameter $\leq 3$.

**Theorem 6.4.31.** ([204, Theorem 2.5]) *Let $R$ be a commutative ring and $M$ an $R$-module. Then $\bar{\Gamma}(M)$ is connected with $diam(\bar{\Gamma}(M)) \leq 3$. Moreover, if $\bar{\Gamma}(M)$ contains a cycle, then $gr(\bar{\Gamma}(M)) \leq 7$.*

As an immediate consequence, we have the following result.

**Corollary 6.4.32.** ([204, Corollary 2.6]) *Let $R$ ba a commutative ring and $M$ a faithful $R$-module. Then $\Gamma(M)$ is connected with $diam(\Gamma(M)) \leq 3$.*

Note that the faithful condition is not superfluous. For, let $R = \mathbb{Z}$ and $M = \mathbb{Z}_2 \oplus \mathbb{Z}_3$. Then $\Gamma(M)$ is not connected.

For a multiplication $R$-module $M$ with $|M| \geq 5$, we prove that if $\bar{\Gamma}(M)$ is complete, then $Nil(M) = V(\bar{\Gamma}(M)) \cup \{0\}$.

**Theorem 6.4.33.** ([204, Theorem 2.9]) *Let $R$ be a commutative ring and $M$ a multiplication $R$-module. If $\bar{\Gamma}(M)$ is complete, then either $|M| = 4$ or $Nil(M) = V(\bar{\Gamma}(M)) \cup \{0\}$.*

**Example 6.4.34.** ([204, Example 2.10]) Let $R = \mathbb{Z}$ and $M = \mathbb{Z}_{p^2}$, where $p > 2$ is a prime number. It is clear that $\bar{\Gamma}(M) = K_{p-1}$ is a complete graph. So $Nil(M) = (p)$ by Theorem 6.4.33.

Note that $M$ can be viewed as a submodule of $S^{-1}M$. Next, we see that $\bar{\Gamma}(S^{-1}M)$ and $\bar{\Gamma}(M)$ are isomorphic by showing that there is a bijection map between equivalence classes of vertex sets $\bar{\Gamma}(S^{-1}M)$ and $\bar{\Gamma}(M)$ such that the corresponding equivalence classes have the same cardinality.

**Theorem 6.4.35.** ([204, Theorem 3.1]) *Let $R$ be a commutative ring, $M$ an $R$-module such that $Ann(x) = Ann([x : M]M)$ for every $x \in T(M)$, and $S = R \setminus Z(M)$. Then $\bar{\Gamma}(M) \cong \bar{\Gamma}(S^{-1}M)$.*

**Corollary 6.4.36.** ([204, Theorem 3.2]) *Let $R$ be a commutative ring, $M$ a multiplication $R$-module, and $S = R \setminus Z(M)$. Then $\Gamma(M) \cong \Gamma(S^{-1}M)$.*

**Corollary 6.4.37.** ([204, Corollary 3.3]) *Let $M$ and $N$ be multiplication $R$-modules with $S^{-1}M \cong S^{-1}N$. Then $\Gamma(M) \cong \Gamma(N)$. In particular, $\Gamma(M) \cong \Gamma(N)$ when $S^{-1}M = S^{-1}N$.*

Next, we determine the uniquely complemented nature of $\bar{\Gamma}(M)$.

**Lemma 6.4.38.** ([204, Lemma 4.3]) *Let $R$ be a commutative ring and $M$ a reduced multiplication $R$-module. Then $\bar{\Gamma}(M)$ is uniquely complemented if and only if $\bar{\Gamma}(M)$ is complemented.*

**Theorem 6.4.39.** ([204, Theorem 4.5]) *Let $R$ be a Bézout ring, $M$ a reduced multiplication $R$-module, and $S = R \setminus Z(M)$. If $\bar{\Gamma}(M)$ is complemented, then $S^{-1}M$ is von Neumann regular.*

**Corollary 6.4.40.** ([204, Corollary 4.7]) *Let $R$ be a von Neumann regular ring, $M$ a multiplication $R$-module, and $S = R \setminus Z(M)$. If $\bar{\Gamma}(M)$ is complemented, then $S^{-1}M$ is von Neumann regular.*

**Corollary 6.4.41.** ([204, Corollary 4.8]) *Let $R$ be a commutative ring, $M$ a reduced cyclic $R$-module, and $S = R \setminus Z(M)$. If $\bar{\Gamma}(M)$ is complemented, then $S^{-1}M$ is von Neumann regular.*

**Corollary 6.4.42.** ([204, Corollary 4.11]) *Let $R$ be a commutative ring and $M$ a cyclic reduced $R$-module. Then the following statements are equivalent:*

(a) *$S^{-1}M$ is von Neumann regular, where $S = R \setminus Z(M)$;*

(b) *$\bar{\Gamma}(M)$ is uniquely complemented;*

(c) *$\bar{\Gamma}(M)$ is complemented.*

**Corollary 6.4.43.** ([204, Corollary 4.12]) *Let $R = \mathbb{Z}$ and $M = \mathbb{Z}_{p_1 p_2 \cdots p_n}$, where $p_i$, $1 \leq i \leq n$, are distinct prime numbers. Then $\bar{\Gamma}(M)$ is uniquely complemented and $S^{-1}M$ is von Neumann regular.*

Note that we may have $S^{-1}M$ is von Neumann regular, but $M$ is not von Neumann regular in spite of $\Gamma(M) \cong \Gamma(S^{-1}M)$. For example, consider the ring $R = \mathbb{Z}_2 \times \mathbb{Z}$ and $M = R$ as an $R$-module. So $M$ is a faithful multiplication $R$-module. Clearly, $M$ is reduced and $\Gamma(M)$ is an infinite star graph with center $(1, 0)$. Thus $\Gamma(M)$ is complemented; $S^{-1}M$ is von Neumann regular by Corollary 6.4.41, but $M$ is not von Neumann regular.

**Theorem 6.4.44.** ([204, Proposition 4.15]) *Let $R$ be a commutative ring and $M$ a multiplication $R$-module with $Nil(M) \neq 0$.*

(a) *If $\bar{\Gamma}(M)$ is complemented, then either $|M| \leq 16$ or $|M| > 16$ and $Nil(M) = \{0, x\}$ for some $0 \neq x \in M$;*

(b) *If $\bar{\Gamma}(M)$ is uniquely complemented with $|M| > 16$, then any complement of a $0 \neq x \in Nil(M)$ is an end.*

Clearly, star graphs are uniquely complemented. The next theorem shows that for a multiplication $R$-module $M$ with $Nil(M) \neq 0$, if $\bar{\Gamma}(M)$ is uniquely complemented, then $\bar{\Gamma}(M)$ is a star graph.

**Theorem 6.4.45.** ([204, Theorem 4.18]) *Let $R$ be a Bézout ring and $M$ be a multiplication $R$-module with $Nil(M) \neq 0$. If $\bar{\Gamma}(M)$ is a uniquely complemented graph, then either $\bar{\Gamma}(M)$ is a star graph with at most six edges or $\bar{\Gamma}(M)$ is an infinite star graph with center $x$, where $Nil(M) = \{0, x\}$.*

**Corollary 6.4.46.** ([204, Corollary 4.19]) *Let $R$ be a commutative ring and $M$ a cyclic $R$-module with $Nil(M) \neq 0$. If $\bar{\Gamma}(M)$ is a uniquely complemented graph, then either $\bar{\Gamma}(M)$ is a star graph with at most six edges or $\bar{\Gamma}(M)$ is an infinite star graph with center $x$, where $Nil(M) = \{0, x\}$.*

**Corollary 6.4.47.** ([204, Corollary 4.20]) *Let $R$ be a Bézout ring and $M$ a multiplication $R$-module. If $\bar{\Gamma}(M)$ is uniquely complemented, then either $\bar{\Gamma}(M)$ is a star graph or $S^{-1}M$ is von Neumann regular. Moreover, for a cyclic $R$-module $M$, the converse is true.*

**Corollary 6.4.48.** ([204, Corollary 4.21]) *Let $R$ be a commutative ring and $M$ a cyclic $R$-module. Then $\bar{\Gamma}(M)$ is uniquely complemented if and only if either $\bar{\Gamma}(M)$ is a star graph or $S^{-1}M$ is von Neumann regular.*

In [317], Rad, Yassemi, Ghalandarzadeh, and Safari investigated the torsion-graph for modules as a natural generalization of the zero-divisor graph for rings. In their work, the torsion graph $\Gamma_R(M)$ of $M$ is a simple graph whose vertices are nonzero torsion elements of $M$, and two distinct elements $x$ and $y$ are adjacent if and only if $Ann(x) \cap Ann(y) \neq 0$. Thus $\Gamma_R(M)$ is the empty graph if and only if $M$ is a torsionfree $R$-module. Clearly, if $R$ is an integral domain or $Ann(M) \neq 0$, then $\Gamma_R(M)$ is complete.

In [317], it was further shown that if $\Gamma_R(M)$ is finite (when $\Gamma_R(M)$ is not empty), then $M$ is finite. Specifically, if $\Gamma_R(M)$ has an isolated vertex, then $M \cong M_1 \oplus M_2$, where $M_1, M_2$ are simple submodules of $M$. Also, they proved that $\Gamma_R(M)$ is connected with $diam(\Gamma_R(M)) \leq 3$ if one of the following hold:

(a) $\Gamma_R(R)$ is a complete graph;

(b) $R$ is a von Neumann regular ring, and $R \ncong Ann(x) \oplus Ann(y)$ for every $x, y \in T(M)^*$;

(c) $Nil(R) \neq 0$.

Finally, they proved that if $\Gamma_R(M)$ contains a cycle, then $gr(\Gamma_R(M)) = 3$.

Rad continued the study on the torsion graph $\Gamma_R(M)$ in [315] and considered the problem of when the torsion graph $\Gamma_R(M)$ is planar. In fact, it was shown that if the torsion graph of a multiplication $R$-module $M$ is planar, then $M$ is both Noetherian and Artinian.

### 6.4.2 Zero-divisor Graphs of Modules

There are several generalizations of zero-divisor graphs to modules $M$ over a commutative ring $R$, which for $M = R$, all coincide with the zero-divisor graph of $R$. The main objective is to establish connections between module-theoretic properties with properties of the associated graphs.

In [138], Behboodi defined three undirected simple graphs $\Gamma^*({}_RM) \subseteq \Gamma({}_RM) \subseteq \Gamma_*({}_RM)$, which for $M = R$, all coincide with the zero-divisor graph of $R$.

Let $R$ be a commutative ring with identity and $M$ be an $R$-module. According to [138], the graph $\Gamma(_RM)$ is a simple graph with vertex set $M$, and two distinct vertices $x, y \in M$ are adjacent if and only if $(xR :_R M)(yR :_R M)M = 0$, which is a direct generalization of the zero-divisor graph. The main objective of his work was to study the interplay of module-theoretic properties of $M$ with graph-theoretic properties of the these graphs.

In [258], Lee and Varmazyar investigated the concept of zero-divisor graphs of multiplication modules over commutative rings as a natural generalization of zero-divisor graphs of commutative rings. Let $R$ be a commutative ring with identity and $M$ be a nonzero unitary $R$-module. Let $Z(_RM) = \{x \in M \mid Rx * Ry = 0$ for some nonzero element $y$ of $M\}$. The authors of [258] associate a zero-divisor graph $\Gamma(_RM)$ to a multiplication module $M$ over a commutative ring $R$ with vertices being elements of $Z(_RM)^*$, and two distinct vertices $x, y$ are adjacent if and only if $Rx * Ry = 0$ in $M$. In [258], the authors showed that the diameter of $\Gamma(_RM)$ is bounded above for every multiplication module $M$. Moreover, when $M_1$ and $M_2$ are finitely generated multiplication $R$-modules satisfying certain condition on $R$, they calculate the diameter of $\Gamma(_R(M_1 \oplus M_2))$. Finally, they considered the relationship between the proper subgroups of the group $(\mathbb{Z}_n, +)$ and the zero-divisor graph $\Gamma(_{\mathbb{Z}}\mathbb{Z}_n)$.

In [331], Safaeeyan, Baziar, and E. Momtahan introduced a new generalization of the zero-divisor graph, which is more natural than the aforementioned generalizations. As any suitable generalization, this one reveals some properties which are untouched in the literature, even for the zero-divisor graph (see Theorem 6.4.52). In general, proofs of results are simpler than those proofs given for the counterpart results on the zero-divisor graph. Let $R$ be a commutative ring with identity and $M$ be a nonzero unitary right $R$-module. For $0 \neq x, y \in M$, we say that $x * y = y * x = 0$ provided that $x(yR : M) = 0$ or $y(xR : M) = 0$. For an $R$-module $M$, by $Z(M)$ we mean the set of all $x \in M$ such that $x * y = 0$ for some nonzero $y \in M$. The authors of [331] associated an undirected graph $\Gamma(M_R)$ to $M$ with vertices $Z(M)^*$ such that for distinct elements $x, y \in Z(M)^*$, the vertices $x$ and $y$ are adjacent provided that $x * y = 0$. It is easy to see that for each $x \in R$, $(xR : R) = xR$. Then for $0 \neq x, y \in R$, $xy = 0$ if and only if $x * y = 0$. Thus $\Gamma(M_R)$ is exactly a generalization of the classic zero-divisor graph.

Now, we are going to explain the relationship between the generalization of the zero-divisor graph $\Gamma(_RM)$ introduced in [138] and the graph $\Gamma(M_R)$. First of all, it is worth mentioning that $\Gamma(M_R)$ is a subgraph of $\Gamma(_RM)$, that is, if $x, y \in Z(M)^*$ are adjacent in $\Gamma(M_R)$, or equivalently either $x(yR : M) = 0$ or $y(xR : M) = 0$, then $(xR : M)(yR : M)M = 0$. However, the converse is not the case as we observe in the following example.

**Example 6.4.49.** ([331, Example 3.1]) Let $M = \mathbb{Z}_2 \oplus \mathbb{Z}_4$ as a $\mathbb{Z}$-module. Then the graph $\Gamma(_RM)$ is $K_7$. However, $\Gamma(M_R)$ is different from $K_7$.

However, if $M$ is a multiplication module, then $\Gamma({}_RM) = \Gamma(M_R)$. Let $(xR : M)(yR : M)M = 0$. As such, $(yR : M)M = yR$ and $(xR : M)M = xR$; so both $x(yR : M) = 0$ and $y(xR : M) = 0$.

Many well-known results for the zero-divisor graph of a commutative ring have been generalized to $\Gamma(M_R)$. The first result is a generalization of Theorem 2.3.1.

**Theorem 6.4.50.** ([331, Theorem 1.5]) *Let $R$ be a commutative ring and $M$ an $R$-module. Then $\Gamma(M_R)$ is connected with $diam(\Gamma(M_R)) \leq 3$.*

The following theorem is a generalization of Theorem 2.3.3.

**Theorem 6.4.51.** ([331, Theorem 1.6]) *Let $R$ be a commutative ring and $M$ an $R$-module. If $\Gamma(M_R)$ contains a cycle, then $gr(\Gamma(M_R)) \leq 4$.*

When does $\Gamma(M_R)$ contain a cycle? The next result gives a partial answer to this question. As we see, it happens when $\Gamma(M_R)$ contains a path of length 4.

**Theorem 6.4.52.** ([331, Proposition 1.7]) *Let $R$ be a commutative ring and $M$ an $R$-module. If $\Gamma(M_R)$ contains a path of length 4, then $\Gamma(M_R)$ contains a cycle.*

It is well known that $R$ is an integral domain if and only if the zero-divisor graph $\Gamma(R)$ is empty. The following is a natural generalization of this fact.

**Theorem 6.4.53.** ([331, Theorem 1.9]) *Let $R$ be a commutative ring and $M$ an $R$-module. Then the following are equivalent:*

(a) *$\Gamma(M_R)$ is the empty graph;*

(b) *$M$ is a uniform $R$-module, $Ann(M)$ is a radical ideal, and $Z(M)^* \neq M \setminus \{0\}$;*

(c) *$Ann(M)$ is a prime ideal and $Z(M)^* \neq M \setminus \{0\}$.*

Note that the zero-divisor graph of $\mathbb{Z}_3 \oplus \mathbb{Z}_3$ is a complete graph. It is not difficult to see that this holds for $M = \mathbb{Z}_p \oplus \mathbb{Z}_p$, where $p$ is any prime number, with $|Z(M)^*| = p^2 - 1$. According to the next result, this is also holds for every module which is a direct sum of two isomorphic simple submodules.

**Theorem 6.4.54.** ([331, Theorem 2.1]) *Let $R$ be a commutative ring, $S$ and $S'$ two isomorphic simple $R$-modules, and $M = S \oplus S'$. Then $Z(M)^* = M \setminus \{0\}$ and $\Gamma(M_R)$ is a complete graph.*

**Theorem 6.4.55.** ([331, Proposition 2.2]) *Let $R$ be a commutative ring. Then $R$ is a field if and only if $\Gamma(M_R)$ is a complete graph for every $R$-module $M$.*

The next result gives a partial answer to the question "when is $\Gamma(M_R)$ a complete bipartite graph?".

**Theorem 6.4.56.** ([331, Corollary 2.4]) *Let $R$ be a commutative ring and $M = M_1 \oplus M_2$, where $M_1$ and $M_2$ are nonisomorphic simple submodules of $M$. Then $\Gamma(M_R)$ is a complete bipartite graph.*

**Theorem 6.4.57.** ([331, Proposition 2.7]) *Let $R$ be a commutative ring and $M$ a reduced $R$-module with $Z(M)^* \neq M \setminus \{0\}$. If $\Gamma(M_R)$ is a bipartite graph, then the following hold:*

(a) *$\Gamma(M_R)$ is a complete bipartite graph;*

(b) *The uniform dimension of $M$ is 2.*

**Corollary 6.4.58.** ([331, Corollary 2.8]) *Let $R$ be a commutative ring and $M$ a reduced $R$-module with $Z(M)^* \neq M \setminus \{0\}$. Then $gr(\Gamma(M_R)) = \infty$ if and only if $\Gamma(M_R)$ is a star graph.*

In Theorem 2.3.8, it was proved that for a reduced commutative ring $R$, $gr(\Gamma(R)) = 4$ if and only if $\Gamma(R) = K_{m,n}$ with $m, n \geq 2$. The following result states the analog for $\Gamma(M_R)$.

**Theorem 6.4.59.** ([331, Proposition 2.10]) *Let $R$ be a commutative ring and $M$ an $R$-module. If $gr(\Gamma(M_R)) = 4$, then $\Gamma(M_R)$ is a bipartite graph with parts $V_1$ and $V_2$ such that $|V_1|, |V_2| \geq 2$. The converse is true if $M$ is a reduced module with $Z(M)^* \neq M \setminus \{0\}$.*

Motivated by the torsion graph, Nikmehr and Khojasteh [299] defined a new undirected graph of a module over a commutative ring with respect to proper ideals. Let $R$ be a commutative ring, $M$ an $R$-module, and $I$ a proper ideal of $R$. The graph structure of $M$ over $R$ with respect to $I$, denoted by $\Gamma_I({}_RM)$ or simply $\Gamma_I(M)$, is a simple graph with the vertex set $M \setminus \{0\}$, and two distinct vertices are adjacent if and only if $[x : M][y : M] \subseteq I$. Clearly, the zero-divisor graph of $R$ is a subgraph of $\Gamma_0(R)$. Also, if $Ann_R(M) \subseteq I$, then the torsion graph of $M$ is a subgraph of $\Gamma_I(M)$.

First, we determine the diameter and the girth of $\Gamma_I(M)$. We start with the following simple example.

**Example 6.4.60.** ([299, Example 2]) First, let $R = \mathbb{Z}$, $M = \mathbb{Z}_2 \times \mathbb{Z}_2$, and $e_1 = (1, 0)$, $e_2 = (0, 1)$. Then $[e_1 :_{\mathbb{Z}} \mathbb{Z}_2 \times \mathbb{Z}_2] = [e_2 :_{\mathbb{Z}} \mathbb{Z}_2 \times \mathbb{Z}_2] = 2\mathbb{Z}$ and $[e_1 + e_2 :_{\mathbb{Z}} \mathbb{Z}_2 \times \mathbb{Z}_2] = \mathbb{Z}$.

Next, let $R = M = \mathbb{Z}_4$. Then $[1 :_{\mathbb{Z}_4} \mathbb{Z}_4] = [3 :_{\mathbb{Z}_4} \mathbb{Z}_4] = \mathbb{Z}_4$ and $[2 :_{\mathbb{Z}_4} \mathbb{Z}_4] = 2\mathbb{Z}_4$.

The corresponding graphs $\Gamma_{4\mathbb{Z}}({}_{\mathbb{Z}}\mathbb{Z}_2 \times \mathbb{Z}_2)$, $\Gamma_{2\mathbb{Z}}({}_{\mathbb{Z}}\mathbb{Z}_2 \times \mathbb{Z}_2)$, and $\Gamma_{2\mathbb{Z}_4}({}_{\mathbb{Z}_4}\mathbb{Z}_4)$ are given in Figure 6.5.

Let $R$ be a commutative ring and $M$ an $R$-module. If $I$ and $J$ are proper ideals of $R$ with $I \subseteq J$, then $\Gamma_I(M)$ is a subgraph of $\Gamma_J(M)$.

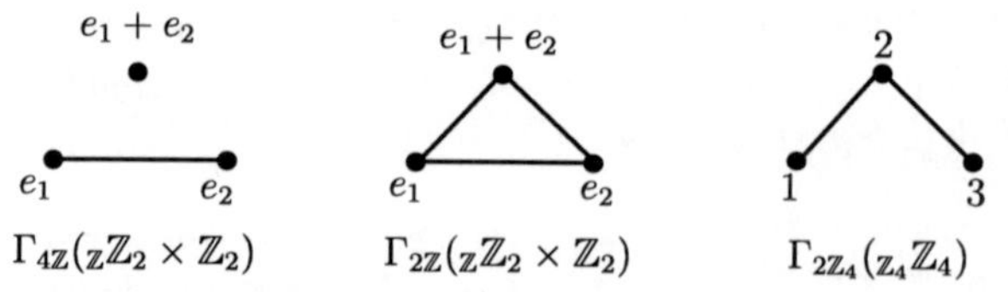


Figure 6.5: Examples of $\Gamma_I(M)$

**Theorem 6.4.61.** ([299, Theorem 1]) *Let $R$ be a commutative ring, $I$ a proper ideal of $R$, and $M$ an $R$-module. If $Ann_R(M) \subseteq I$ and $\Gamma_I(M)$ has no isolated vertices, then $\Gamma_I(M)$ is connected and $diam(\Gamma_I(M)) \leq 3$.*

Let $R$ be a commutative ring and $M$ an $R$-module. If $I$ is an ideal of $R$ with $IM = 0$, then $M$ can be regarded as an $R/I$-module. So we deduce the next theorem.

**Theorem 6.4.62.** ([299, Theorem 2]) *Let $R$ be a commutative ring and $M$ an $R$-module. If $I$ and $J$ are two proper ideals of $R$ and $I \subseteq Ann_R(M) \subseteq J$, then $\Gamma_{J/I}({}_{R/I}M)$ is a subgraph of $\Gamma_J({}_RM)$.*

Next, we state a relationship between $\Gamma_{Nil(R)}(M)$ and $\Gamma_{Nil(S^{-1}R)}({}_{S^{-1}R}S^{-1}M)$.

**Theorem 6.4.63.** ([299, Theorem 3]) *Let $R$ be a commutative ring, $M$ an $R$-module, and $S = R \setminus Z(R)$. If $Z(M) \subseteq Z(R)$ and $\Gamma_{Nil(S^{-1}R)}({}_{S^{-1}R}S^{-1}M)$ is connected, then $\Gamma_{Nil(R)}(M)$ is a connected graph and $diam(\Gamma_{Nil(R)}(M)) \leq diam(\Gamma_{Nil(S^{-1}R)}({}_{S^{-1}R}S^{-1}M))$. Moreover, if $\Gamma_{Nil(S^{-1}R)}({}_{S^{-1}R}S^{-1}M)$ has no isolated vertices, then $\Gamma_{Nil(R)}(M)$ has no isolated vertices.*

The following theorem gives the possible values for the girth of $\Gamma_I(M)$.

**Theorem 6.4.64.** ([299, Theorem 4]) *Let $R$ be a commutative ring, $I$ a proper ideal of $R$, and $M$ be an $R$-module. If $Ann_R(M) \subseteq I$, then $gr(\Gamma_I(M)) \in \{3, 4, \infty\}$.*

Next, we see the relationship between the chromatic number and the clique number of $\Gamma_{Nil(R)}(M)$.

**Theorem 6.4.65.** ([299, Theorem 5]) *Let $R$ be a commutative ring and $M$ an $R$-module. If $Spec(R)$ and $\omega(\Gamma_{Nil(R)}(M))$ are finite, then $\chi(\Gamma_{Nil(R)}(M)) \leq |Spec(R)| + \omega(\Gamma_{Nil(R)}(M))$.*

**Corollary 6.4.66.** ([299, Corollary 2]) *Let $R$ be a commutative ring and $M$ an $R$-module. If $I_1, \ldots, I_n$ are prime ideals of $R$, $I = \cap_{i=1}^{n} I_i$, and $\omega(\Gamma_I(M))$ is finite, then $\chi(\Gamma_I(M)) \leq n + \omega(\Gamma_I(M))$. In particular, if $Max(R)$ and $\omega(\Gamma_{J(R)}(M))$ are finite, then $\chi(\Gamma_{J(R)}(M)) \leq |Max(R)| + \omega(\Gamma_{J(R)}(M))$.*

**Corollary 6.4.67.** ([299, Corollary 3]) *Let $(R, \mathfrak{m})$ be a commutative local ring and $M$ be an $R$-module. If $\omega(\Gamma_{\mathfrak{m}}(M))$ is finite, then $\chi(\Gamma_{\mathfrak{m}}(M)) \leq 1 + \omega(\Gamma_{\mathfrak{m}}(M))$.*

An $R$-module $M$ is called *secondary* if $M \neq 0$ and for every $a \in R$, $aM = M$ or $a^t \in Ann_R(M)$ for some positive integer $t$. An $R$-module $M$ is called a *cyclic* $R$-module, if $M = Rx$, for some $x \in M$, and $x$ is called a *generator* of $M$.

In the following theorem, we study $\Gamma_I(M)$ when $M$ is a secondary $R$-module.

**Theorem 6.4.68.** ([299, Theorem 6]) *Let $R$ be a commutative ring, $M$ a secondary $R$-module, $P$ a prime ideal of $R$, and $A$ the set of all generators of $M$. If $Ann_R(M) \subseteq P$, then the following hold:*

(a) *If $A = \emptyset$, then $\Gamma_P(M) = K_{|M^*|}$;*

(b) *If $A \neq \emptyset$, then $C\Gamma_P(M) = K_{|A|} \cup CK_{|M^* \setminus A|}$.*

By the previous theorem, we have the next corollary.

**Corollary 6.4.69.** ([299, Corollary 4]) *Let $R$ be a commutative ring, $M$ a secondary $R$-module, and $P$ a prime ideal of $R$. If $Ann_R(M) \subseteq P$, then $diam(\Gamma_P(M)) \in \{1, 2, \infty\}$ and $gr(\Gamma_P(M)) \in \{3, \infty\}$.*

**Theorem 6.4.70.** ([299, Lemma 3]) *Let $R$ be a commutative ring, $I$ a proper ideal of $R$, and $M$ an $R$-module. If $[x : M] \subseteq I$ for some $0 \neq x \in M$, then the following hold:*

(a) *$gr(\Gamma_I(M)) = \infty$ if and only if $\Gamma_I(M)$ is a star graph;*

(b) *If $\Gamma_I(M)$ contains a cycle, then $gr(\Gamma_I(M)) = 3$.*

**Theorem 6.4.71.** ([299, Theorem 7]) *Let $R$ be a commutative ring, $M$ a secondary $R$-module, $P$ a prime ideal of $R$, and $A$ the set of all generators of $M$. If $Ann_R(M) \subseteq P$, then the following hold:*

(a) *If $A = \emptyset$, then $\omega(\Gamma_P(M)) = |M| - 1$, $\alpha(\Gamma_P(M)) = 1$, and $\chi(\Gamma_P(M)) = \omega(\Gamma_P(M))$, where $\alpha(G)$ denotes the independence number of $G$;*

(b) *If $A \neq \emptyset$, then $\omega(\Gamma_P(M)) = |M \setminus A|$, $\alpha(\Gamma_P(M)) = |A|$, and $\chi(\Gamma_P(M)) = \omega(\Gamma_P(M))$.*

Let $R$ be a commutative ring, $M$ a secondary $R$-module, $P$ a prime ideal of $R$, and $Ann_R(M) \subseteq P$. If $\omega(\Gamma_P(M))$ is finite, then by Theorem 6.4.71, $M \setminus A$ is finite, where $A$ is the set of all generators of $M$. This yields that $M$ has only finitely many submodules. Therefore, if $\omega(\Gamma_P(M))$ is finite, then $M$ has only finitely many submodules.

Let $R$ be a commutative ring, $M$ an $R$-module, and $P$ a prime ideal of $R$. Now, we consider the adjacency of a vertex in $\Gamma_P(M)$. If $\Gamma_P(M)$ has at least one edge, then $\Gamma_P(M)$ has a vertex, say $x$, such that $x$ is adjacent to every other vertex of $\Gamma_P(M)$. In fact, if $a$ and $b$ are adjacent in $\Gamma_P(M)$, then we have $[a : M][b : M] \subseteq P$. Thus $[a : M] \subseteq P$ or $[b : M] \subseteq P$ since $P$ is a prime ideal. If $[a : M] \subseteq P$, then $a$ is adjacent to every vertex of $\Gamma_P(M)$. Similarly, if $[b : M] \subseteq P$, then $b$ is adjacent to every other vertex of $\Gamma_P(M)$.

**Lemma 6.4.72.** ([299, Lemma 4]) *Let $R$ be a commutative ring and $I$ a proper ideal of $R$. If $M$ is a simple $R$-module and $Ann_R(M) \subseteq I$, then $\Gamma_I(M)$ is totally disconnected.*

**Lemma 6.4.73.** ([299, Lemma 5]) *Let $R$ be a commutative ring and $M$ an $R$-module. If $M = M_1 \oplus M_2$ and $Ann_R(M) \subseteq I$, then every element of $M_1^*$ is adjacent to every element of $M_2^*$.*

From the previous lemma, we have the following result.

**Theorem 6.4.74.** ([299, Theorem 8]) *Let $R$ be a commutative ring, $I$ a proper ideal of $R$, $M$ a semisimple $R$-module, and $Ann_R(M) \subseteq I$. Then $\Gamma_I(M)$ is a forest if and only if $\Gamma_I(M)$ is a union of isolated vertices or a star.*

In [136], the authors associated two different graphs to an $R$-module $M$ with respect to its first dual, $M^* = Hom(M, R)$. Though they are not necessarily generalizations of the zero-divisor graph, there are some deep interrelationships between these two graphs and the classical one.

### 6.4.3 Annihilator Ideal-based Graphs of Modules

Anderson, Ghalandarzadeh, Shirinkam, and Rad [76] investigated an annihilator ideal-based zero-divisor graph of an $R$-module. Let $R$ be a commutative ring with nonzero identity and $M$ be a unitary $R$-module. The annihilator ideal-based zero-divisor graph is a graph $\Gamma_I(R)$, where we replace the ideal $I$ with the ideal $Ann(M)$ for the $R$-module $M$. Here the annihilator ideal-based zero-divisor graph $\Gamma_{Ann(M)}(R)$ is a simple graph, whose vertices are the set $\{a \in R \setminus Ann(M) \mid abM = 0$ for some $b \in R \setminus Ann(M)\}$, and distinct vertices $a$ and $b$ are adjacent if and only if $abM = 0$. First of all, we investigate the relationship between the diameter of $\Gamma_{Ann(M)}(R)$ and the minimal prime ideals of $Ann(M)$ over a multiplication $R$-module $M$. It is easy to verify that if $M$ is reduced, then $I = Ann(M)$ is a radical ideal of $R$, and hence $R/I$ is a reduced ring. But the converse is not true. For example, let $R = \mathbb{Z}$ and $M = \mathbb{Z} \oplus \mathbb{Z}_4$. Then $Ann(M) = Ann(\mathbb{Z}) \cap Ann(\mathbb{Z}_4) = (0) \cap 4\mathbb{Z} = (0)$ is a radical ideal of $\mathbb{Z}$. However, $M$ is not reduced since $Ann((0, 1 + 4\mathbb{Z})) = 4\mathbb{Z}$ is not a radical ideal of $\mathbb{Z}$.

**Lemma 6.4.75.** ([76, Lemma 2.3]) *Let $R$ be a commutative ring, $M$ a reduced multiplication $R$-module, and $I$ an ideal of $R$. If $I \subseteq P$ for some $P \in Min(Ann(M))$, then $I \subseteq Z(M)$.*

**Lemma 6.4.76.** ([76, Proposition 2.4]) *Let $R$ be a commutative ring and $M$ a reduced $R$-module. Then $V(\Gamma_{Ann(M)}(R)) \cup Ann(M) = \cup_{P \in Min(Ann(M))} P$.*

**Theorem 6.4.77.** ([76, Theorem 2.5]) *Let $R$ be a commutative ring and $M$ a reduced multiplication $R$-module. If $R$ has more than two minimal prime ideals of $Ann(M)$ and $R\alpha + R\beta \not\subseteq Z(M)$ for some $\alpha, \beta \in V(\Gamma_{Ann(M)}(R))$, then $diam(\Gamma_{Ann(M)}(R)) = 3$.*

Note that the condition $|Min(Ann(M))| > 2$ is not superfluous. For example, let $R = \mathbb{Z} \times \mathbb{Z} = M$. One can easily check that $M$ is a reduced multiplication $R$-module and $Ann(M) = \{0\}$. Thus $\Gamma_{Ann(M)}(R) = \Gamma(R)$. Also, we have $R\alpha + R\beta \not\subseteq Z(M)$ for $\alpha = (1, 0)$, $\beta = (0, 1) \in V(\Gamma_{Ann(M)}(R))$ and $Min(Ann(M)) = \{0 \times \mathbb{Z}, \mathbb{Z} \times 0\}$. One can easily check that $\Gamma_{Ann(M)}(R)$ is a complete bipartite graph and $diam(\Gamma_{Ann(M)}(R)) \neq 3$. So the condition $|Min(Ann(M))| > 2$ is not superfluous.

**Theorem 6.4.78.** ([76, Theorem 2.7]) *Let $R$ be a commutative ring, $M$ a reduced multiplication $R$-module, and $R\alpha + R\beta \not\subseteq Z(M)$ for some $\alpha, \beta \in V(\Gamma_{Ann(M)}(R))$. Then $diam(\Gamma_{Ann(M)}(R)) \leq 2$ if and only if $R$ has exactly two minimal prime ideals of $Ann(M)$.*

**Theorem 6.4.79.** ([76, Theorem 2.8]) *Let $R$ be a commutative ring and $M$ a multiplication $R$-module with $Nil(M) \neq 0$. If there are $\alpha, \beta \in V(\Gamma_{Ann(M)}(R))$ such that $R\alpha + R\beta \not\subseteq Z(M)$, then $diam(\Gamma_{Ann(M)}(R)) = 3$.*

Next, we determine when $\Gamma_{Ann(M)}(R)$ is complete.

**Theorem 6.4.80.** ([76, Theorem 3.2]) *Let $R$ be a commutative ring and $M$ a reduced $R$-module. Then $\Gamma_{Ann(M)}(R)$ is a complete (nonempty) graph if and only if $R \cong \mathbb{Z}_2 \times \mathbb{Z}_2$ and $M \cong M_1 \times M_2$ for $M_1$ and $M_2$ nonzero $\mathbb{Z}_2$-modules.*

Note that the above theorem fails if we do not assume that $M$ is a reduced $R$-module. Let $R = \mathbb{Z}_4 \times \mathbb{Z}_n$ and $I = \{0\} \times \mathbb{Z}_n$. Then $R/I \cong \mathbb{Z}_4$; so $\Gamma(R/I) = K_1$. Thus $\Gamma_I(R) = K_n$. So for $M = R/I$, we have $Ann(M) = I$. Hence $\Gamma_{Ann(M)}(R) = K_n$. For $n = 1$, let $R = M = \mathbb{Z}_4$. Then $I = Ann(M) = 0$; so $\Gamma_{Ann(M)}(R) = \Gamma(\mathbb{Z}_4) = K_1$. So for every $n \geq 1$, there is a ring $R$ and an $R$-module $M$ such that $\Gamma_{Ann(M)}(R) = K_n$.

**Theorem 6.4.81.** ([76, Theorem 3.4]) *Let $R$ be a commutative ring and $M$ be a reduced $R$-module. If $\Gamma(R)$ is complete, then either $\Gamma_{Ann(M)}(R)$ is complete or the vertex sets of $\Gamma_{Ann(M)}(R)$ and $\Gamma(R)$ are disjoint.*

**Corollary 6.4.82.** ([76, Corollary 3.5]) *Let $R$ be a commutative ring and $M$ a reduced $R$-module. If $\Gamma(R)$ is complete, then either $\Gamma_{Ann(M)}(R)$ is complete or $Z(R)M = 0$.*

**Theorem 6.4.83.** ([76, Corollary 3.6]) *Let $R$ be a commutative ring and $M$ an $R$-module.*

(a) $diam(\Gamma_{Ann(M)}(R)) = 0$ *if and only if* $Nil(M) \neq 0$*,* $R$ *is isomorphic to either* $\mathbb{Z}_4$ *or* $\mathbb{Z}_2[x]/(x^2)$*, and* $Ann(M) = 0$*;*

(b) $diam(\Gamma_{Ann(M)}(R)) = 1$ *if and only if either* (i) $M$ *is reduced,* $R \cong \mathbb{Z}_2 \times \mathbb{Z}_2$*, and* $M \cong M_1 \times M_2$ *for* $M_1$ *and* $M_2$ *nonzero* $\mathbb{Z}_2$*-modules, or* (ii) $Nil(M) \neq 0$ *and* $xyM = 0$ *for every distinct pair of vertices* $x$ *and* $y$*.*

*Let* $M$ *be a multiplication* $R$*-module. Suppose that there are two distinct vertices* $\alpha$ *and* $\beta$ *such that* $R\alpha + R\beta \not\subseteq Z(M)$*;*

(c) $diam(\Gamma_{Ann(M)}(R)) = 2$ *if and only if either* (i) $R$ *has exactly two minimal prime ideals of* $Ann(M)$*,* $M$ *is reduced, and* $\Gamma_{Ann(M)}(R)$ *has at least three vertices, or* (ii) *for every distinct pair of vertices* $\alpha$ *and* $\beta$*, there is a vertex which is adjacent to both* $\alpha$ *and* $\beta$*;*

(d) $diam(\Gamma_{Ann(M)}(R)) = 3$ *if and only if either* (i) $R$ *has more than two minimal prime ideals of* $Ann(M)$ *and* $M$ *is reduced, or* (ii) $Nil(M) \neq 0$*.*

Ghalandarzadeh, Shirinkam, and Rad [205] investigated some more properties of a multiplication $R$-module $M$ associated with the annihilator ideal-based zero-divisor graph $\Gamma_{Ann(M)}(R)$. First of all, they were interested in the complemented nature of $\Gamma_{Ann(M)}(R)$.

**Lemma 6.4.84.** ([205, Lemma 2.1]) *Let $R$ be a von Neumann regular ring and $M$ be an $R$-module. Then $\Gamma_{Ann(M)}(R)$ is complemented.*

The converse of the above lemma is not true. Let $R = \mathbb{Z}$ and $M = \mathbb{Z}_{30}$. Then $Ann(M) = 30\mathbb{Z}$ and $\Gamma_{Ann(M)}(R)$ is a complemented graph. But $\mathbb{Z}$ is not a von Neumann regular ring.

**Lemma 6.4.85.** ([205, Proposition 2.3]) *Let $R$ be a commutative ring and $M$ a multiplication $R$-module. Then $M$ is a von Neumann regular module if and only if every cyclic submodule of $M$ is pure in $M$.*

**Theorem 6.4.86.** ([205, Theorem 2.4]) *Let $R$ be a von Neumann regular ring, $M$ a multiplication $R$-module, and $S = R \setminus Z(M)$. Then $S^{-1}M$ is a von Neumann regular module.*

Thus every multiplication $R$-module is a von Neumann regular module whenever $R$ is a von Neumann regular ring.

The converse of the above theorem is not true. Let $R = \mathbb{Z}$, $M = \mathbb{Z}_{30}$, and $S = R \setminus Z(M)$. Note that $M = \mathbb{Z}_{30}$ is a multiplication $R$-module. But $R = \mathbb{Z}$ is not a von Neumann regular ring. Also, $S^{-1}M$ is a von Neumann regular module.

We next consider the case when $M$ is a reduced $R$-module.

**Theorem 6.4.87.** ([205, Proposition 2.7]) *Let $R$ be a commutative ring and $M$ a reduced $R$-module. Then $\Gamma_{Ann(M)}(R)$ is uniquely complemented if and only if $\Gamma_{Ann(M)}(R)$ is complemented.*

**Theorem 6.4.88.** ([205, Proposition 2.8]) *Let $R$ be a commutative ring and $M$ a finitely generated von Neumann regular multiplication $R$-module. Then $\Gamma_{Ann(M)}(R)$ is a complemented graph.*

As an immediate consequence, we obtain the following result.

**Corollary 6.4.89.** ([205, Corollary 2.9]) *Let $R$ be a von Neumann regular ring and $M$ a multiplication $R$-module. Then*

(a) *$M$ is a von Neumann regular module;*

(b) *$\Gamma_{Ann(M)}(R)$ is complemented;*

(c) *$\Gamma_{Ann(M)}(R)$ is uniquely complemented.*

The following few theorems in this section are somewhat more delicate in their characterization of a multiplication $R$-module.

**Theorem 6.4.90.** ([205, Theorem 3.2]) *Let $R$ be a commutative ring with $|\bar{R}| > 4$ and $M$ a multiplication $R$-module, where $\bar{R} = R/Ann(M)$. If $\Gamma_{Ann(M)}(R)$ is a complete graph, then either $M = Z(\bar{R})M$ or $Nil(M) = Z(\bar{R})M$.*

**Lemma 6.4.91.** ([205, Lemma 3.3]) *Let $R$ be a commutative ring, $M$ a multiplication $R$-module, and $0 \neq x \in Nil(M)$. If $\Gamma_{Ann(M)}(R)$ is complemented, then there is an $\alpha \in (x : M)$ such that $\alpha^t M = 0$, $\alpha^{t-i} M \neq 0$, and $\alpha^{t-1} \perp \alpha^i$, where $t = 2, 3$ and $1 \leq i \leq t-2$.*

**Lemma 6.4.92.** ([205, Lemma 3.4]) *Let $R$ be a commutative ring, $M$ a multiplication $R$-module with $Nil(M) \neq 0$, and $\bar{R} = R/Ann(M)$. If $\Gamma_{Ann(M)}(R)$ is complemented and $t = 3$ in Lemma 6.4.91, then $|\bar{R}| = 8$ and $\Gamma_{Ann(M)}(R)$ is a star graph.*

**Lemma 6.4.93.** ([205, Lemma 3.5]) *Let $R$ be a commutative ring, $M$ a multiplication $R$-module with $Nil(M) \neq 0$, and $\bar{R} = R/Ann(M)$. If $\Gamma_{Ann(M)}(R)$ is complemented and $t = 2$ in Lemma 6.4.91, then either $|M| = |\bar{R}| = 9$ and $\Gamma_{Ann(M)}(R)$ is a star graph, or $|M| > 9$ and $Nil(M) = \{0, x\}$ for some $0 \neq x \in M$.*

**Theorem 6.4.94.** ([205, Theorem 3.6]) *Let $R$ be a commutative ring, $M$ a multiplication $R$-module with $Nil(M) \neq 0$, and $\bar{R} = R/Ann(M)$. If $\Gamma_{Ann(M)}(R)$ is complemented, then either $|M| = |\bar{R}| = 8$, $|M| = |\bar{R}| = 9$, and $\Gamma_{Ann(M)}(R)$ is a star graph, or $|M| > 9$ and $Nil(M) = \{0, x\}$ for some $0 \neq x \in M$.*

The following example shows that the converse of the above theorem is not true. For example, consider the ring $R = \mathbb{Z}$ and $M = \mathbb{Z}_{20}$. Then $M$ is a multiplication module, $Nil(M) = \{0, 10\}$, and $|M| > 9$. But $\Gamma_{Ann(M)}(R)$ is not a complemented graph.

**Theorem 6.4.95.** ([205, Theorem 3.8]) *Let $R$ be a commutative ring and $M$ a multiplication $R$-module. Suppose that $|M| > 9$ and $Nil(M) \neq 0$. If $\Gamma_{Ann(M)}(R)$ is uniquely complemented, then $\Gamma_{Ann(M)}(R)$ is an infinite star graph.*

Let $R = \mathbb{Z}$ and $M = \mathbb{Z}_{30}$. Then $M$ is a reduced multiplication module, and so $Nil(M) = 0$. Also, $|M| > 9$ and $\Gamma_{Ann(M)}(R)$ is a uniquely complemented graph. But $\Gamma_{Ann(M)}(R)$ is not an infinite star graph. Therefore, the condition $Nil(M) \neq 0$ is a necessary condition in the above theorem.

**Theorem 6.4.96.** ([205, Theorem 3.10]) *Let $R = R_1 \times R_2$ be a commutative ring with $|\bar{R}| > 9$, and $M = M_1 \times M_2$ be a multiplication $R$-module with $|M| > 9$, where $\bar{R} = R/Ann(M)$. Let $M_1$ be a reduced $R_1$-module, $Nil_{R_2}(M_2) \neq 0$, and $\Gamma_{Ann(M)}(R)$ be complemented. Then $Ann_{R_1}(M_1)$ is a prime ideal, $|M_2| = 4$, and $\bar{R}_2 \cong \mathbb{Z}_4$ or $\bar{R}_2 \cong \mathbb{Z}_2[x]/(x^2)$.*

The zero-divisor graph has also been extended to other algebraic structures such as posets, lattices, and near-rings.

Halaš and Jukl [213] introduced the zero-divisor graphs of posets and gave an affirmative answer to Beck's question. The study of the zero-divisor graphs of posets was then continued by Xue and Liu in [437]. In 2010, a different method of associating a zero-divisor graph to a poset $P$ was proposed by Lu and Wu in [268]. The graph defined by them is slightly different from the one defined in [213, 437], where the vertex set of the graph consists of all the elements of $P$. The vertex set of the graph defined in [268] consists of all nonzero zero-divisors of $P$. Lu and Wu [268] proved that the diameter of such a graph is either 1, 2, or 3, while its girth is either 3, 4, or $\infty$, and the authors of [53] characterized zero-divisor graphs of posets in terms of their diameter and girth. Furthermore, Afkhami et al. [15] completely investigated the planarity of zero-divisor graphs of posets. Then Alizadeh et al. [54] characterized complete partite zero-divisor graphs of posets via the ideals of the posets. Moreover, Joshi [225] introduced the zero-divisor graph of a poset with respect to an ideal $I$, proved a form of Beck's conjecture, and characterized all complete bipartite zero-divisor graphs.

Estaji and Khashyarmanesh [195] extended the concept of zero-divisor graph to an arbitrary finite bounded lattice. For more work on zero-divisor graph of lattices, see [302, 301, 425]. The zero-divisor graph was also extended to near-rings, see [155, 426].

## 6.5 Domination in Directed Zero-divisor Graphs

In this section, we review some results concerning domination parameters in the zero-divisor graph of a commutative ring. The study of dominating sets in zero-divisor graph of a commutative ring was initiated by Redmond [324]. The following is the main result in this connection.

**Theorem 6.5.1.** ([324, Theorem 5.1]) *Let $R$ be a commutative Artinian ring that is not an integral domain. If the radius of $\Gamma(R)$ is at most 1, then the domination number of $\Gamma(R)$ is 1. If the radius is 2, then the domination number is equal to the number of factors in the Artinian decomposition of $R$. (In particular, the domination number is finite and at least two.)*

One immediate consequence of the above theorem is that if the radius of $\Gamma(R)$ is 2, then the domination number is equal to the number of distinct maximal ideals of $R$. However, this need not be true if the radius of $\Gamma(R)$ is 1. For any field $F$, $\Gamma(\mathbb{Z}_2 \times F)$ is a star graph, which has radius 1, but $\mathbb{Z}_2 \times F$ has two distinct maximal ideals. The following corollary shows that a more precise relationship between the domination number and the number of maximal ideals occurs in the finite case.

**Corollary 6.5.2.** ([324, Corollary 5.2]) *Let $R$ be a finite commutative ring that is not a field. If $\Gamma(R)$ is not a star graph, then $R$ has $\gamma(\Gamma(R))$ distinct maximal ideals. If $\Gamma(R)$ is a star graph, then either $R$ has 2 distinct maximal ideals or $R$ is isomorphic to one of the following five local rings: $\mathbb{Z}_9$, $\mathbb{Z}_3[x]/(x^2)$, $\mathbb{Z}_8$, $\mathbb{Z}_2[x]/(x^3)$, or $\mathbb{Z}_4[x]/(2x, x^2-2)$. (In other words, if $\Gamma(R)$ is a star graph, then $R$ has $\gamma(\Gamma(R))$ distinct maximal ideals if $R$ is local and $\gamma(\Gamma(R))+1$ distinct maximal ideals if $R$ is reduced.)*

**Corollary 6.5.3.** ([324, Corollary 5.3]) *Let $R$ be a finite commutative ring that is not a field. If $R \ncong \mathbb{Z}_2 \times F$ for any finite field $F$, then the domination number of $\Gamma(R)$ equals the number of distinct maximal ideals of $R$. If $R \cong \mathbb{Z}_2 \times F$ for some finite field $F$, then the domination number is one less than the number of maximal ideals of $R$.*

It is interesting to note that these dominating sets are connected, showing that the connected domination number of $\Gamma(R)$ equals the domination number of $\Gamma(R)$. The dominating set in zero-divisor graphs has been implicitly studied by Jafari Rad, Jafari, and Mojdeh in [223]. They first determined the domination number for the zero-divisor graph of the product of two commutative rings.

**Proposition 6.5.4.** ([223, Proposition 2.2]) *Let $R$ be an integral domain. Then $\gamma(\Gamma(\mathbb{Z}_2 \times R)) = 1$.*

A *semi-total dominating set* in $\Gamma(R)$ is a subset $S \subseteq Z(R)$ such that $S$ is a dominating set for $\Gamma(R)$ and for any $x \in S$ there is a vertex $y \in S$ (not necessarily distinct) such that $xy = 0$. The *semi-total domination number* $\gamma_{st}(\Gamma(R))$ of $\Gamma(R)$ is the minimum cardinality of a semi-total dominating set in $\Gamma(R)$. (Note that for a commutative ring $R$, $\gamma(\Gamma(R)) \leq \gamma_{st}(\Gamma(R)) \leq 2\gamma(\Gamma(R))$.) For a commutative ring $R$, let

$$a(R) = \begin{cases} 1 & \text{if } Z(R) = 0; \\ \gamma_{st}(\Gamma(R)) & \text{otherwise.} \end{cases}$$

**Theorem 6.5.5.** ([223, Theorem 2.6]) *Let $R_1$, $R_2$ be commutative rings with $\mathbb{Z}_2 \notin \{R_1, R_2\}$. Then $\gamma(\Gamma(R_1 \times R_2)) = a(R_1) + a(R_2)$.*

**Corollary 6.5.6.** ([223, Corollary 2.7, 2.8]) *Let $R_1$, $R_2$ be commutative rings and $\mathbb{Z}_2 \notin \{R_1, R_2\}$. Then $\gamma(\Gamma(R_1 \times R_2)) = \gamma_t(\Gamma(R_1 \times R_2)) = \gamma_c(\Gamma(R_1 \times R_2)) = \gamma_{st}(\Gamma(R_1 \times R_2))$.*

**Corollary 6.5.7.** ([223, Corollary 2.11]) *Let $R_1, \ldots, R_n$ be local commutative Artinian rings and $R = R_1 \times \cdots \times R_n$, where $R \ncong F$ or $\mathbb{Z}_2 \times F$ for a field $F$. Then $\gamma(\Gamma(R)) = n$.*

The study of properties of dominating sets of zero-divisor graphs was continued by Mojdeh and Rahimi [293].

**Theorem 6.5.8.** ([293, Proposition 8]) *Let $R = R_1 \times \cdots \times R_n$, where $n \geq 2$ and $R_i$ is an integral domain for every $i = 1, \ldots, n$. Then*

(a) *$\gamma(\Gamma(R)) = n$ if $n \geq 3$;*

(b) *$\gamma(\Gamma(R)) = 2$ if $n = 2$ and $min\{|R_1|, |R_2|\} \geq 3$;*

(c) *$\gamma(\Gamma(R)) = 1$ if $n = 2$ and $min\{|R_1|, |R_2|\} = 2$.*

**Corollary 6.5.9.** ([293, Corollary 9]) *For every positive integer $k$, there is a commutative ring whose zero-divisor graph has domination number $k$.*

**Theorem 6.5.10.** ([293, Theorem 11]) *Let $R$ be a commutative Artinian ring (in particular, $R$ could be a finite commutative ring). Suppose that $R = R_1 \times \cdots \times R_n$, where $R_i$ is a local ring for every $i = 1, \ldots, n$. Then*

(a) *$\gamma(\Gamma(R)) = n$ if $n \geq 3$;*

(b) *$\gamma(\Gamma(R)) = 2$ if $n = 2$ and $min\{|R_1|, |R_2|\} \geq 3$, or $|R_1| = 2$ and the maximal ideal of $R_2$ is nonzero;*

(c) *$\gamma(\Gamma(R)) = 1$ if $n = 1$ and the maximal ideal of $R_2$ is nonzero;*

(d) *$\gamma(\Gamma(R)) = 1$ if $n = 2$, $|R_1| = \mathbb{Z}_2$, and $R_2$ is a field.*

In the next corollary, we state certain necessary and sufficient conditions for $\gamma(\Gamma(\mathbb{Z}_n)) = k$, where $n = p_1^{t_1} \cdots p_k^{t_k}$ for distinct primes $p_1, \ldots, p_k$.

**Corollary 6.5.11.** ([293, Corollary 13]) *For any fixed integer $k \geq 1$, let $n = p_1^{t_1} \cdots p_k^{t_k}$ for distinct primes $p_1, \ldots, p_k$ and positive integers $t_1, \ldots, t_k$. Then*

(a) $\gamma(\Gamma(\mathbb{Z}_n)) = k \geq 3$ *if and only if* $n = p_1^{t_1} \cdots p_k^{t_k}$ *for distinct primes* $p_1, \ldots, p_k$ *and positive integers* $t_1, \ldots, t_k$*;*

(b) $\gamma(\Gamma(\mathbb{Z}_n)) = 2$ *if and only if* $n = p_1^{t_1} p_2^{t_2}$*, where either* $t_1 \geq 2$ *or* $t_2 \geq 2$*; or* $t_1 = t_2 = 1$ *and* $p_1, p_2 \geq 3$*;*

(c) $\gamma(\Gamma(\mathbb{Z}_n)) = 1$ *if and only if* $n = p_1^{t_1}$*, where* $t_1 \geq 2$*; or* $n = p_1 p_2$*, where either* $p_1 = 2$*,* $p_2 \geq 3$ *or* $p_1 \geq 3$*,* $p_2 = 2$*.*

**Theorem 6.5.12.** ([293, Theorem 15]) *Let* $R$ *be a reduced finite commutative ring which is not a field. If* $\gamma(\Gamma(R)) \neq 1$*, then* $\gamma(\Gamma(R))$ *is equal to the number of minimal prime ideals of* $R$*. In addition, if* $R$ *has* $k \geq 3$ *minimal prime ideals, then* $\gamma(\Gamma(R)) = k$*.*

Further, Kiani, Maimani, and Nikandish [234] investigated the domination, total domination, and semi-total domination numbers of zero-divisor graphs of commutative Noetherian rings.

As mentioned in earlier sections of this chapter, the graph of zero-divisors for commutative rings has been generalized to the ideal-based zero-divisor graph and annihilating-ideal graph of commutative rings. Also, the zero-divisor graph has been generalized to commutative semirings and modules over commutative rings. In 2001, Redmond [323] introduced a generalization of the zero-divisor graph called the ideal-based zero-divisor graph for commutative rings. In this connection, Mojdeh and Rahimi [293] studied the domination number of the zero-divisor graph with respect to an ideal. Actually, they explored the relationship between the domination numbers of $\Gamma_I(R)$ and $\Gamma(R/I)$.

**Theorem 6.5.13.** ([293, Theorem 19]) *Let* $R$ *be a commutative ring. Let* $S$ *be a nonempty subset of* $R \setminus I$*. If* $S$ *is a dominating set of* $\Gamma_I(R)$*, then* $S + I = \{s + I \mid s \in S\}$ *is a dominating set of* $\Gamma(R \setminus I)$*.*

The converse of Theorem 6.5.13 is not necessarily true. For example, let $R = \mathbb{Z}_6 \times \mathbb{Z}_3$ and $I = \{0\} \times \mathbb{Z}_3$. Then it is clear that $S = \{(3,0) + I\}$ is a dominating set of $\Gamma(R/I)$, but $\{(3,1),(4,1)\}$ is a dominating set of $\Gamma_I(R)$ and $\Gamma_I(R)$ cannot be dominated by any set of one vertex. Readers may refer to [234] for some more relationships between the domination numbers of $\Gamma_I(R)$ and $\Gamma(R/I)$.

Recall that Redmond [322] also extended the zero-divisor graph concept to directed zero-divisor graphs of noncommutative rings.

Let us review some definitions and notation from domination parameters of directed graphs. Let $D = (V, A)$ be a digraph (directed graph) with vertex set $V$ and arc set $A$. The in-degree and out-degree of a vertex $v$ are respectively denoted by $id(v)$ and $od(v)$. For a subset $S$ of vertices of $D$, the out-neighborhood $N^+(S)$ of $S$ consists of all those vertices $w$ in $D \setminus S$ such that $(v, w)$ is an arc of $D$ for some $v \in S$. The in-neighborhood $N^-(S)$ consists of all those vertices $u \in D \setminus S$

such that $(u, v)$ is an arc of $D$ for some $v \in S$. For a digraph $D = (V, A)$, a subset $S$ of $V$ is called an *out-dominating set* of $D$ if for every $v \in V \setminus S$, there is a $u \in S$ such that $(u, v) \in A$. The out-dominating set of a digraph $D$ is commonly called a *dominating set* of $D$. A subset $S$ of $V$ is called an *in-dominating set* of $D$ if for every $v \in V \setminus S$, there exists a $u \in S$ such that $(v, u) \in A$. A subset $S$ of $V$ is called a *twin dominating set* if $S$ is both an out-dominating and an in-dominating set. A dominating set $S$ of $V$ is called *independent* if the subdigraph induced by $S$ has no arcs. A dominating set $S$ of $V$ is called a *total dominating set* if the induced subdigraph $\langle S \rangle$ has no isolated vertices. A dominating set $S$ of $V$ is called an *open dominating set* of $D$ if for every $v \in V$, there exists a $u \in S$ such that $(u, v) \in A$.

The *out-domination number* (resp., *upper out-domination number*) of a digraph $D$, denoted by $\gamma^+(D)$ (resp., $\Gamma^+(D)$), is the minimum(resp., maximum) cardinality of an out-dominating set of $D$. In a similar way, one can define the *in-domination number* $\gamma^-$, the *twin domination number* $\gamma^*$, the *independent domination number* $\gamma_i$, the *open domination number* $\gamma_o$, the *total domination number* $\gamma_t$, and the *weakly connected domination number* $\gamma_{wc}$. A out-dominating set $S$ in a digraph $D$ with cardinality $\gamma^+$ is called an $\gamma^+$*- set* of $D$. The *irredundance number*, $ir(D)$, and the *upper irredundance number*, $IR(D)$, are respectively, the minimum and maximum cardinalities of a maximal irredundant set. An irredundant set $S$ in a digraph $D$ with cardinality $ir(D)$ is called an *ir(D)-set* of $D$. The *out-domatic (resp., in-domatic) number* $d^+(D)$ (resp., $d^-(D)$) of a digraph $D$ is the maximum number of elements in a partition of $V(D)$ into out-dominating (resp., in-dominating) sets. The *total domatic number* $d_t(D)$ of a digraph $D$ is the maximum number of elements in a partition of $V(D)$ into total dominating sets. A digraph $D$ is *domatically full* if $d^+(D) = 1 + \delta(D)$. The *reinforcement number* $r(D)$ of a digraph $D$ is the minimum number of extra arcs whose addition to $D$ results in a graph $D'$ with $\gamma^+(D') < \gamma^+(D)$.

Tamizh Chelvam and Selvakumar [396] obtained the values for certain domination parameters for the directed zero-divisor graph $D$ on $M_2(\mathbb{Z}_p)$. Some of their results are given below.

**Theorem 6.5.14.** ([396, Theorem 2.1]) *Let $D$ be the directed zero-divisor graph of $M_2(\mathbb{Z}_p)$ for $p$ prime. Then $ir(D) = IR(D) = p + 1$.*

**Proposition 6.5.15.** ([396, Proposition 2.1]) *Let $D$ be the directed zero-divisor graph of $M_2(\mathbb{Z}_p)$ for $p$ prime. Then $\gamma_t(D) = \gamma_{wc}(D) = \gamma_o(D) = p + 1$.*

**Theorem 6.5.16.** ([396, Theorem 2.2]) *Let $D$ be the directed zero-divisor graph of $M_2(\mathbb{Z}_p)$ for $p$ prime. If $\Omega$ is a minimal dominating set of $D$, then $\Omega$ is independent if and only if $A^2 = 0$ for all $A \in \Omega$.*

**Theorem 6.5.17.** ([396, Theorem 2.3]) *Let $D$ be the directed zero-divisor graph of $M_2(\mathbb{Z}_p)$ for $p$ prime. Then $d^+(D) = d^-(D) = d_t(D) = p^2 - 1$.*

**Proposition 6.5.18.** ([396, Proposition 3.1]) *Let $D$ be the directed zero-divisor graph of $R = M_2(\mathbb{Z}_p)$ for $p$ prime. Then $\gamma^*(D) = \gamma_p(D) = \gamma_e(D) = p + 1$.*

**Corollary 6.5.19.** ([396, Lemma 3.1]) *Let $D$ be the directed zero-divisor graph of $R = M_2(\mathbb{Z}_p)$ for $p$ prime. The number of efficient dominating sets in $D$ is $p - 1$.*

For a digraph $D$, let $\eta(D) = n - max\{|N+[\Omega]| \mid \Omega \subseteq V(D), |\Omega| = \gamma^+(D) - 1\}$, where $n$ is the number of vertices in $D$. We call $\Omega \subseteq V(D)$ an *$\eta$-set* if $\eta(D) = n - |N + [\Omega]|$ in $D$.

For a commutative ring $R$, let $G = U(R)$ and $X$ be the set of all nonzero nonunits of $R$. The group action on $X$ by $G$ given by $(g, x) \longrightarrow gx$ (resp., $(g, x) \longrightarrow xg^{-1}$) from $G \times X$ to $X$ is called the left (resp., right) regular action. If $\varphi : G \times X \longrightarrow X$ is the left (resp., right) regular action, then for every $x \in X$, $O_l(x) = \{\varphi(g, x) = gx \mid \text{ for all } g \in G\}$ (resp., $O_r(x) = \{\varphi(g, x) = xg^{-1} \mid \text{ for all } g \in G\}$) is called the left (resp., right) orbit of $x$. For a prime number $p$, let $R = M_2(\mathbb{Z}_p)$. Then $O_l(M_i)$ is the left orbit of the matrix $M_i \in M_2(\mathbb{Z}_p)$.

**Theorem 6.5.20.** ([396, Theorem 3.1]) *Let $D$ be the directed zero-divisor graph of $R = M_2(\mathbb{Z}_p)$. If $Z(R)^* = \cup_{i=0}^{p} O_l(M_i)$, then $\Omega$ is a $\eta$-set of $D$ if and only if*

(a) *every element of $\Omega$ belongs to a different orbit of $Z(R)^*$;*

(b) *$S^2 \neq 0$ for every $S \in \Omega$;*

(c) *$\Omega$ is independent.*

In 2014, Tamizh Chelvam and Selvakumar [398] studied the domination parameters of the directed zero-divisor graph of $M_2(\mathbb{F})$, where $\mathbb{F}$ is a finite field.

**Theorem 6.5.21.** ([398, Theorem 2.2]) *Let $\mathbb{F}$ be a finite field with $|\mathbb{F}| = p^m$, where $p$ is a prime and $m \geq 1$, and $D = \overrightarrow{\Gamma}(M_2(\mathbb{F}))$. Then $ir(D) = IR(D) = p^m + 1$.*

**Proposition 6.5.22.** ([398, Proposition 2.8]) *Let $\mathbb{F}$ be a finite field with $|\mathbb{F}| = p^m$, where $p$ is a prime and $m \geq 1$, and $D = \overrightarrow{\Gamma}(M_2(\mathbb{F}))$. Then $\gamma_t(D) = \gamma_{wc}(D) = \gamma_o(D) = p^m + 1$.*

**Proposition 6.5.23.** ([398, Proposition 2.9]) *Let $\mathbb{F}$ be a finite field with $|\mathbb{F}| = p^m$, where $p$ is a prime and $m \geq 1$, and $D = \overrightarrow{\Gamma}(M_2(\mathbb{F}))$. If $\Omega$ is an open dominating set of $D$, then $\Omega$ contains no nilpotent elements.*

**Theorem 6.5.24.** ([398, Theorem 3.1]) *Let $\mathbb{F}$ be a finite field with $|\mathbb{F}| = p^m$, where $p$ is a prime and $m \geq 1$, and let $\Omega$ be a minimal dominating set of $\overrightarrow{\Gamma}(M_2(\mathbb{F}))$. Then $\Omega$ is independent if and only if $a^2 = 0$ for every $a \in \Omega$.*

**Theorem 6.5.25.** ([398, Theorem 3.2]) *Let $\mathbb{F}$ be a finite field with $|\mathbb{F}| = p^m$, where $p$ is a prime number and $m \geq 1$, and $D = \overrightarrow{\Gamma}(M_2(\mathbb{F}))$. Then $d^+(D) = d^-(D) = d_t(D) = p^{2m} - 1$, and hence $D$ is domatically full.*

**Theorem 6.5.26.** ([398, Theorem 3.3]) *Let $\mathbb{F}$ be a finite field with $|\mathbb{F}| = p^m$, where $p$ is a prime number and $m \geq 1$, and $D = \overrightarrow{\Gamma}(M_2(\mathbb{F}))$. Then $\gamma^*(D) = \gamma_p(D) = \gamma_e(D) = p^m + 1$.*

**Corollary 6.5.27.** ([398, Corollary 3.4]) *Let $\mathbb{F}$ be a finite field with $|\mathbb{F}| = p^m$, where $p$ is a prime number and $m \geq 1$. If $\Omega$ is a minimal dominating set of $\overrightarrow{\Gamma}(M_2(\mathbb{F}))$, then $\Omega$ is perfect (and so efficient) if and only if $a^2 = 0$ for every $a \in \Omega$.*

The domination parameters of zero-divisor graphs of matrix rings over a commutative ring have been discussed in a recent article by Heidar Jafari and Jafari Rad [216]. The remaining part of this section lists results from [216].

**Theorem 6.5.28.** ([216, Lemma 2.5]) *Let $R$ be a commutative ring $R$. Then $\gamma_o(\overrightarrow{\Gamma}(M_n(R))) = \gamma_i(\overrightarrow{\Gamma}(M_n(R)))$.*

**Theorem 6.5.29.** ([216, Lemma 2.6]) *Let $R$ be a commutative ring and $A$ an out-dominating set for $\overrightarrow{\Gamma}(M_n(R))$. Then there is an out-dominating set $B$ for $\overrightarrow{\Gamma}(M_n(R))$ such that $|B| \leq |A|$ and every element of $B$ has rank 1.*

**Theorem 6.5.30.** ([216, Corollary 2.8]) *Let $R = M_n(\mathbb{F})$, where $\mathbb{F}$ is a finite field. Then*

$$S = \{A = (a_{ij})_{n\times n} \mid a_{ij} = \delta_{1j}(\lambda_j), \text{ where } \lambda_1, \ldots, \lambda_n \in F \text{ and } \lambda_j = 1 \text{ for some } j\}$$

*is a $\gamma_o(\overrightarrow{\Gamma}(R))$-set.*

**Theorem 6.5.31.** ([216, Corollary 2.9]) *Let $\mathbb{F}$ be a finite field with $|\mathbb{F}| = q$. Then $\gamma_o(\overrightarrow{\Gamma}(M_n(\mathbb{F}))) = \gamma_i(\overrightarrow{\Gamma}(M_n(\mathbb{F}))) = \frac{q^n-1}{q-1}$ for every integer $n \geq 1$.*

**Theorem 6.5.32.** ([216, Theorem 3.6]) *Let $(R, \mathfrak{m})$ be a commutative local ring with $R/\mathfrak{m}$ finite. Then $\gamma_o(\overrightarrow{\Gamma}(M_n(R))) \leq \gamma_o(\overrightarrow{\Gamma}(M_n(R/\mathfrak{m})))$.*

**Theorem 6.5.33.** ([216, Theorem 3.7]) *Let $(R, \mathfrak{m})$ be a finite commutative local ring with $\mathfrak{m}$ cyclic as an $R$-module. Then $\gamma_o(\overrightarrow{\Gamma}(M_n(R))) = \gamma_o(\overrightarrow{\Gamma}(M_n(R/\mathfrak{m})))$.*

**Theorem 6.5.34.** ([216, Theorem 3.10]) *Let $R = R_1 \times \cdots \times R_t$, where every $R_i$ is a commutative local ring with principal maximal ideal. Then*

$$\gamma_o(\overrightarrow{\Gamma}(M_n(R))) = \gamma_o(\overrightarrow{\Gamma}(M_n(R_1))) + \cdots + \gamma_o(\overrightarrow{\Gamma}(M_n(R_t))).$$

# Chapter 7

# Total Graphs of Commutative Rings

One of the associated graphs to a ring is the total graph. In the case of zero-divisor graphs, multiplication of the ring is used for adjacency in the graph. In variation to this, a graph using the addition of the ring is constructed and is called the total graph of commutative rings. The next several chapters of this book are devoted to this notion of total graph.

As seen earlier, over the past several years, there has been considerable attention in the literature to associating graphs with commutative rings (and other algebraic structures) and studying the interplay between ring-theoretic and graph-theoretic properties. The total graph (as in [67]) has been investigated in [36], [34], [25], [32], [271], [277], [312], [100], and [348]; and several variants of the total graph have been studied in [8], [70], [72], [74], [111], [134], [365], [362], [363], [364], [366], [101], and [232]. The goal of next three chapters is to give many of the main results on the total graph of a commutative ring.

In [67], D. F. Anderson and A. Badawi introduced the concept of total graph of a commutative ring.

## 7.1 Definition of Total Graph

**Definition 7.1.1.** Let $R$ be a commutative ring with nonzero identity. The *total graph* of $R$, denoted by $T(\Gamma(R))$, is the simple graph with all elements of $R$ as vertices, and for two distinct $x, y \in R$, the vertices $x$ and $y$ are adjacent if and only if $x + y \in Z(R)$. Let $Reg(\Gamma(R))$ be the (induced) subgraph of $T(\Gamma(R))$ with vertices $Reg(R) = R \setminus Z(R)$, which is called the *regular graph of* $R$, and let $Z(\Gamma(R))$ be the (induced) subgraph of $T(\Gamma(R))$ with vertices $Z(R)$.

Note that if $A$ is a subring of a commutative ring $B$, then $T(\Gamma(A))$ need not be an induced subgraph of $T(\Gamma(B))$. Although $x, y \in A$ are adjacent in $T(\Gamma(B))$ if they are adjacent in $T(\Gamma(A))$ since $Z(A) \subseteq Z(B)$, they may be adjacent in



D. F. Anderson et al., *Graphs from Rings*, https://doi.org/10.1007/978-3-030-88410-9_7

$T(\Gamma(B))$, but not adjacent in $T(\Gamma(A))$. In fact, $T(\Gamma(A))$ is an induced subgraph of $T(\Gamma(B))$ if and only if $Z(B) \cap A = Z(A)$.

**Example 7.1.2.** (i) Let $R = \mathbb{Z}_6$. Then $Z(R) = \{0, 2, 3, 4\}$, and the total graph of $\mathbb{Z}_6$ is given in Figure 7.1.

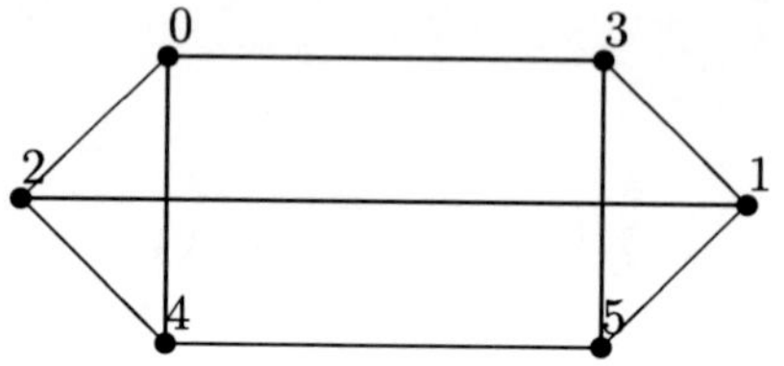


Figure 7.1: $T(\Gamma(\mathbb{Z}_6))$

(ii) Let $R = \mathbb{Z}_9$. Then $Z(R) = \{0, 3, 6\}$, and the total graph of $\mathbb{Z}_9$ is given in Figure 7.2.

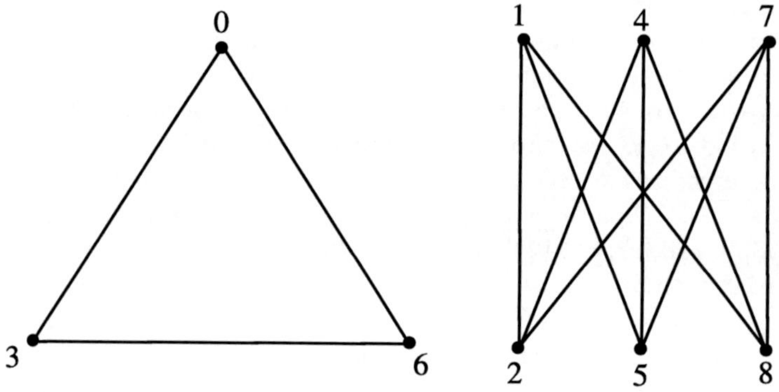


Figure 7.2: $T(\Gamma(\mathbb{Z}_9))$

(iii)Let $R = \mathbb{Z}_{15}$. Then $Z(R) = \{0, 3, 5, 6, 9, 10, 12\}$, and the total graph of $\mathbb{Z}_{15}$ is given in Figure 7.3.

## 7.2 Basic Properties of Total Graphs

The study of $T(\Gamma(R))$ breaks naturally into two cases depending on whether or not $Z(R)$ is an ideal of $R$.

First, we study the case when $Z(R)$ is an ideal of $R$ (i.e., when $Z(R)$ is closed under addition). In this case, $Z(R)$ is actually a prime ideal of $R$, and

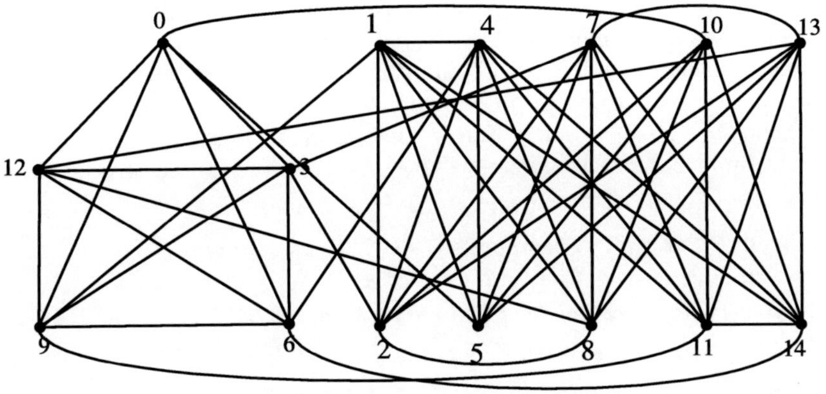


Figure 7.3: $T(\Gamma(\mathbb{Z}_{15}))$

hence $R/Z(R)$ is an integral domain. Moreover, if $R$ is a finite commutative ring and $Z(R)$ is an ideal of $R$, then $R$ is local with $Z(R) = Nil(R)$ its unique maximal ideal. The main goal of this section is a general structure theorem (Theorem 7.2.1) for $Reg(\Gamma(R))$ when $Z(R)$ is an ideal of $R$. But first, we record the trivial observation that if $Z(R)$ is an ideal of $R$, then $Z(\Gamma(R))$ is a complete subgraph of $T(\Gamma(R))$ and is disjoint from $Reg(\Gamma(R))$. Thus we will concentrate on the subgraph $Reg(\Gamma(R))$ throughout this section (recall that $Reg(\Gamma(R))$ is the (induced) subgraph of $T(\Gamma(R))$ with vertices $Reg(R)$).

**Theorem 7.2.1.** ([67, Theorem 2.2]) *Let $R$ be a commutative ring with $Z(R)$ an ideal of $R$, $|Z(R)| = \alpha$, and $|R/Z(R)| = \beta$.*

(1) *If $2 \in Z(R)$, then $Reg(\Gamma(R))$ is the union of $\beta - 1$ disjoint $K_\alpha$'s;*

(2) *If $2 \notin Z(R)$, then $Reg(\Gamma(R))$ is the union of $(\beta - 1)/2$ disjoint $K_{\alpha,\alpha}$'s.*

*Proof.* (1) Assume that $2 \in Z(R)$, and let $x \in Reg(R)$. Then each coset $x+Z(R)$ is a complete subgraph of $Reg(\Gamma(R))$ since $(x+z_1)+(x+z_2) = 2x+z_1+z_2 \in Z(R)$ for every $z_1, z_2 \in Z(R)$ since $2 \in Z(R)$ and $Z(R)$ is an ideal of $R$. Note that distinct cosets form disjoint subgraphs of $Reg(\Gamma(R))$ since if $y + z_1$ and $x + z_2$ are adjacent for some $y \in Reg(R)$ and $z_1, z_2 \in Z(R)$, then $x + y = (x + z_1) + (y + z_2) - (z_1 + z_2) \in Z(R)$, and hence $x - y = (x + y) - 2y \in Z(R)$ since $Z(R)$ is an ideal of $R$ and $2 \in Z(R)$. But then $x + Z(R) = y + Z(R)$. Thus $Reg(\Gamma(R))$ is the union of $\beta - 1$ disjoint (induced) subgraphs $x + Z(R)$, each of which is a $K_\alpha$, where $\alpha = |Z(R)| = |x + Z(R)|$.

(2) Next assume that $2 \notin Z(R)$, and let $x \in Reg(R)$. Then no two distinct elements in $x+Z(R)$ are adjacent since $(x+z_1)+(x+z_2) \in Z(R)$ for $z_1, z_2 \in Z(R)$ implies that $2x \in Z(R)$, and hence $2 \in Z(R)$, a contradiction. Also, the two

cosets $x + Z(R)$ and $-x + Z(R)$ are disjoint, and every element of $x + Z(R)$ is adjacent to every element of $-x + Z(R)$. Thus $(x + Z(R)) \cup (-x + Z(R))$ is a complete bipartite (induced) subgraph of $Reg(\Gamma(R))$. Furthermore, if $y + z_1$ is adjacent to $x + z_2$ for some $y \in Reg(R)$ and $z_1, z_2 \in Z(R)$, then $x + y \in Z(R)$, and hence $y + Z(R) = -x + Z(R)$. Thus $Reg(\Gamma(R))$ is the union of $(\beta - 1)/2$ disjoint (induced) subgraphs $(x + Z(R)) \cup (-x + Z(R))$, each of which is a $K_{\alpha,\alpha}$, where $\alpha = |Z(R)| = |x + Z(R)|$. □

**Theorem 7.2.2.** ([67, Theorem 2.4]) *Let $R$ be a commutative ring with $Z(R)$ an ideal of $R$. Then*

(1) *$Reg(\Gamma(R))$ is complete if and only if either $R/Z(R) \cong \mathbb{Z}_2$ or $R \cong \mathbb{Z}_3$;*

(2) *$Reg(\Gamma(R))$ is connected if and only if either $R/Z(R) \cong \mathbb{Z}_2$ or $R/Z(R) \cong \mathbb{Z}_3$;*

(3) *$Reg(\Gamma(R))$ is totally disconnected if and only if $R$ is an integral domain with $char(R) = 2$.*

*Proof.* Let $|Z(R)| = \alpha$ and $|R/Z(R)| = \beta$.

(1) By Theorem 7.2.1, $Reg(\Gamma(R))$ is complete if and only if $Reg(\Gamma(R))$ is a single $K_\alpha$ or $K_{1,1}$. If $2 \in Z(R)$, then $\beta - 1 = 1$. Thus $\beta = 2$, and hence $R/Z(R) \cong \mathbb{Z}_2$. If $2 \notin Z(R)$, then $\alpha = 1$ and $(\beta - 1)/2 = 1$. Thus $Z(R) = \{0\}$ and $\beta = 3$; so $R \cong R/Z(R) \cong \mathbb{Z}_3$.

(2) By Theorem 7.2.1, $Reg(\Gamma(R))$ is connected if and only if $Reg(\Gamma(R))$ is a single $K_\alpha$ or $K_{\alpha,\alpha}$. Thus either $\beta - 1 = 1$ if $2 \in Z(R)$ or $(\beta - 1)/2 = 1$ if $2 \notin Z(R)$; so $\beta = 2$ or $\beta = 3$, respectively. Thus $R/Z(R) \cong \mathbb{Z}_2$ or $R/Z(R) \cong \mathbb{Z}_3$, respectively.

(3) $Reg(\Gamma(R))$ is totally disconnected if and only if it is a disjoint union of $K_1$'s. So by Theorem 7.2.1, $R$ must be an integral domain with $2 \in Z(R)$, i.e., $char(R) = 2$. □

It is also easy to compute the diameter and girth of $Reg(\Gamma(R))$ using Theorem 7.2.1.

**Theorem 7.2.3.** ([67, Theorem 2.5]) *Let $R$ be a commutative ring with $Z(R)$ an ideal of $R$. Then*

(1) *$diam(Reg(\Gamma(R))) = 0, 1, 2,$ or $\infty$. In particular, $diam(Reg(\Gamma(R))) \leq 2$ if $Reg(\Gamma(R))$ is connected;*

(2) *$gr(Reg(\Gamma(R))) = 3, 4,$ or $\infty$. In particular, $gr(Reg(\Gamma(R))) \leq 4$ if $Reg(\Gamma(R))$ contains a cycle.*

*Proof.* (1) Suppose that $Reg(\Gamma(R))$ is connected. Then $Reg(\Gamma(R))$ is a singleton, a complete graph, or a complete bipartite graph by Theorem 7.2.1. Thus $diam(Reg(\Gamma(R))) \leq 2$.

(2) Suppose that $Reg(\Gamma(R))$ contains a cycle. Since $Reg(\Gamma(R))$ is a disjoint union of either complete or complete bipartite graphs by Theorem 7.2.1, it must contain either a 3-cycle or a 4-cycle. Thus $gr(Reg(\Gamma(R))) \leq 4$. □

The next theorem gives a more explicit description of the diameter and girth of $Reg(\Gamma(R))$.

**Theorem 7.2.4.** ([67, Theorem 2.6]) *Let $R$ be a commutative ring with $Z(R)$ an ideal of $R$.*

(1) (a) *$diam(Reg(\Gamma(R))) = 0$ if and only if $R \cong \mathbb{Z}_2$;*
(b) *$diam(Reg(\Gamma(R))) = 1$ if and only if either $R/Z(R) \cong \mathbb{Z}_2$ and $R \ncong \mathbb{Z}_2$ (i.e., $R/Z(R) \cong \mathbb{Z}_2$ and $|Z(R)| \geq 2$), or $R \cong \mathbb{Z}_3$;*
(c) *$diam(Reg(\Gamma(R))) = 2$ if and only if $R/Z(R) \cong \mathbb{Z}_3$ and $R \ncong \mathbb{Z}_3$ (i.e., $R/Z(R) \cong \mathbb{Z}_3$ and $|Z(R)| \geq 2$);*
(d) *Otherwise, $diam(Reg(\Gamma(R))) = \infty$;*

(2) (a) *$gr(Reg(\Gamma(R))) = 3$ if and only if $2 \in Z(R)$ and $|Z(R)| \geq 3$;*
(b) *$gr(Reg(\Gamma(R))) = 4$ if and only if $2 \notin Z(R)$ and $|Z(R)| \geq 2$;*
(c) *Otherwise, $gr(Reg(\Gamma(R))) = \infty$;*

(3) (a) *$gr(T(\Gamma(R))) = 3$ if and only if $|Z(R)| \geq 3$;*
(b) *$gr(T(\Gamma(R))) = 4$ if and only if $2 \notin Z(R)$ and $|Z(R)| = 2$;*
(c) *Otherwise, $gr(T(\Gamma(R))) = \infty$.*

*Proof.* These results all follow directly from the definitions and Theorem 7.2.1. □

The following examples illustrate the previous theorems.

**Example 7.2.5.** ([67, example 2.7])

(i) Let $m \geq 2$ be an integer. Then $Z(\mathbb{Z}_m)$ is an ideal of $\mathbb{Z}_m$ if and only if $m = p^n$ for some prime $p$ and integer $n \geq 1$. So suppose that $Z(\mathbb{Z}_m)$ is an ideal of $\mathbb{Z}_m$. Thus $Reg(\Gamma(\mathbb{Z}_m))$ is connected if and only if either $m = 2^n$ or $m = 3^n$ for some integer $n \geq 1$. Moreover, $Reg(\Gamma(\mathbb{Z}_m))$ is complete if and only if either $m = 2^n$ for some integer $n \geq 1$ or $m = 3$. An analogous result holds for any principal ideal domain.

(ii) Let $K$ be a field with $|K| = \alpha$, $n \geq 2$ an integer, and $R = K[x]/(x^n)$. Then R is local with maximal ideal $Z(R) = Nil(R) = (x)/(x^n)$, $R/Z(R) \cong K$, $|R| = \alpha^n$, $|Z(R)| = \alpha^{n-1}$, $|Reg(R)| = \alpha^{n-1}(\alpha - 1)$, and $|R/Z(R)| = \alpha$. If $char(K) = 2$, then $Reg(\Gamma(R))$ is the union of $\alpha - 1$ disjoint complete graphs $K_m$, where $m = \alpha^{n-1}$. If $char(K) \neq 2$, then $Reg(\Gamma(R))$ is the union of $\frac{\alpha-1}{2}$ disjoint complete bipartite graphs $K_{m,m}$ , where $m = \alpha^{n-1}$. Thus $Reg(\Gamma(R))$ is connected if and only if $K \cong \mathbb{Z}_2$ or $K \cong \mathbb{Z}_3$, and $Reg(\Gamma(R))$ is complete if and only if $K \cong \mathbb{Z}_2$.

For the special case when $K = \mathbb{F}_{p^k}$, we have that $Reg(\Gamma(R))$ is the union of $2^k - 1$ disjoint $K_m$'s, where $m = 2^{k(n-1)}$, when $p = 2$; and $Reg(\Gamma(R))$ is the union of $\frac{p^k-1}{2}$ disjoint $K_{m,m}$'s, where $m = p^{k(n-1)}$ when $p \neq 2$.

(iii) Let $m \geq 2$ be an integer and $R = \mathbb{Z}(+)\mathbb{Z}_m$. Then $Z(R)$ is an ideal of $R$ if and only if $m = p^n$ for some prime $p$ and integer $n \geq 1$. Moreover, $Z(R) = p\mathbb{Z}(+)\mathbb{Z}_{p^n}$ and $R/Z(R) \cong \mathbb{Z}_p$ when $m = p^n$ and $n \geq 1$. So in this case, $Reg(\Gamma(R))$ is connected if and only if $p = 2$ or $3$, and $Reg(\Gamma(R))$ is complete if and only if $p = 2$. For any $0 \neq a \in \mathbb{Z}_m$, the 3-cycle $(1,0) - (-1,0) - (1,a) - (1,0)$ shows that $gr(Reg(\Gamma(R))) = 3$ when $m = 2^n$; and the 4-cycle $(1,0) - (-1,0) - (1-p,0) - (p-1,0) - (1,0)$ shows that $gr(Reg(\Gamma(R))) = 4$ when $m = p^n$ and $p \neq 2$.

Specifically, let $R_1 = \mathbb{Z}(+)\mathbb{Z}_2$ and $R_2 = \mathbb{Z}(+)\mathbb{Z}_3$. Then $Reg(\Gamma(R_1))$ is complete with $diam(Reg(\Gamma(R_1))) = 1$ and $gr(Reg(\Gamma(R_1))) = 3$, and $Reg(\Gamma(R_2))$ is connected with $diam(Reg(\Gamma(R_2))) = 2$ and $gr(Reg(\Gamma(R_2))) = 4$.

Many of the earlier results of this section can also be easily proved directly without recourse to Theorem 7.2.1. Two such cases are given below.

**Theorem 7.2.6.** ([67, Theorem 2.8]) *Let $R$ be a commutative ring with $Z(R)$ an ideal of $R$.*

(1) *Let $G$ be an induced subgraph of $Reg(\Gamma(R))$, and let $x$ and $y$ be distinct vertices of $G$ that are connected by a path in $G$. Then there is a path in $G$ of length at most 2 between $x$ and $y$. In particular, if $Reg(\Gamma(R))$ is connected, then $diam(Reg(\Gamma(R))) \leq 2$;*

(2) *Let $x$ and $y$ be distinct regular elements of $R$ that are connected by a path. If $x + y \notin Z(R)$ (i.e., if $x$ and $y$ are not adjacent), then $x - (-x) - y$ and $x - (-y) - y$ are paths of length 2 between $x$ and $y$ in $Reg(\Gamma(R))$.*

*Proof.* (1) It suffices to show that if $x_1, x_2, x_3$, and $x_4$ are distinct vertices of $G$ and there is a path $x_1 - x_2 - x_3 - x_4$ from $x_1$ to $x_4$, then $x_1$ and $x_4$ are adjacent. Now $x_1 + x_2, x_2 + x_3, x_3 + x_4 \in Z(R)$ implies $x_1 + x_4 = (x_1 + x_2) - (x_2 + x_3) + (x_3 + x_4) \in Z(R)$ since $Z(R)$ is an ideal of $R$. Thus $x_1$ and $x_4$ are adjacent.

(2) Suppose that $x + y \notin Z(R)$. Then there is a $z \in Reg(R)$ such that $x - z - y$ is a path of length 2 by part (1) above (note that necessarily $z \in Reg(R)$ since $x, y \in Reg(R)$). Thus $x + z, z + y \in Z(R)$, and hence $x - y = (x + z) - (z + y) \in Z(R)$ since $Z(R)$ is an ideal of $R$. Also, $x \neq -x$ and $y \neq -x$ since $x + y \notin Z(R)$. Thus $x - (-x) - y$ and $x - (-y) - y$ are paths of length 2 between $x$ and $y$ in $Reg(\Gamma(R))$. □

It has been observed that $Z(\Gamma(R))$ is always connected and $T(\Gamma(R))$ is never connected when $Z(R)$ is an ideal of $R$. Several new criteria for when $Reg(\Gamma(R))$ is connected are given below.

**Theorem 7.2.7.** ([67, Theorem 2.9]) *Let $R$ be a commutative ring with $Z(R)$ an ideal of $R$. Then the following statements are equivalent:*

(1) *$Reg(\Gamma(R))$ is connected;*

(2) *Either* $x + y \in Z(R)$ *or* $x - y \in Z(R)$ *for every* $x, y \in Reg(R)$;

(3) *Either* $x + y \in Z(R)$ *or* $x + 2y \in Z(R)$ *for every* $x, y \in Reg(R)$. *In particular, either* $2x \in Z(R)$ *or* $3x \in Z(R)$ *(but not both) for every* $x \in Reg(R)$;

(4) *Either* $Z(R) \cong \mathbb{Z}_2$ *or* $R/Z(R) \cong \mathbb{Z}_3$.

*Proof.* (1) $\Rightarrow$ (2) Suppose that $Reg(\Gamma(R))$ is connected, and let $x, y \in Reg(R)$. If $x = y$, then $x - y \in Z(R)$. Hence assume that $x \neq y$. If $x + y \notin Z(R)$, then $x - (-y) - (y)$ is a path from $x$ to $y$ by Theorem 7.2.6(2), and thus $x - y \in Z(R)$.

(2) $\Rightarrow$ (3) Let $x, y \in Reg(R)$, and suppose that $x + y \notin Z(R)$. Since $(x + y) - y = x \notin Z(R)$, thus $x + 2y = (x + y) + y \in Z(R)$ by hypothesis. In particular, if $x \in Reg(R)$, then either $2x \in Z(R)$ or $3x \in Z(R)$. Both $2x$ and $3x$ cannot be in $Z(R)$ since then $x = 3x - 2x \in Z(R)$, a contradiction.

(3) $\Rightarrow$ (1) Let $x, y \in Reg(R)$ be distinct elements of $R$ such that $x+y \notin Z(R)$. Then $x+2y \in Z(R)$ by hypothesis. Since $Z(R)$ is an ideal of $R$ and $x+2y \in Z(R)$, we conclude that $2y \notin Z(R)$. Thus $3y \in Z(R)$ by hypothesis. Since $x+y \notin Z(R)$ and $3y \in Z(R)$, we conclude that $x \neq 2y$, and hence $x - 2y - y$ is a path from $x$ to $y$ in $Reg(\Gamma(R))$. Thus $Reg(\Gamma(R))$ is connected.

(2) $\Rightarrow$ (4) Let $x \in Reg(R)$. Then either $x - 1 \in Z(R)$ or $x + 1 \in Z(R)$ by hypothesis, and thus either $x + Z(R) = 1 + Z(R)$ or $x + Z(R) = -1 + Z(R)$. If $2 \in Z(R)$, then $R/Z(R) \cong \mathbb{Z}_2$; otherwise, $R/Z(R) \cong \mathbb{Z}_3$.

(4) $\Rightarrow$ (2) This is clear. □

Next, we concentrate on the remaining case when $Z(R)$ is not an ideal of $R$. Since $Z(R)$ is always closed under multiplication by elements of $R$, this just means that there are distinct $x, y \in Z(R)^*$ such that $x + y \in Reg(R)$. In this case, $Z(\Gamma(R))$ is always connected (but never complete), $Z(\Gamma(R))$ and $Reg(\Gamma(R))$ are never disjoint subgraphs of $T(\Gamma(R))$, and $|Z(R)| \geq 3$. First it is shown that $T(\Gamma(R))$ is connected when $Reg(\Gamma(R))$ is connected. However, an example is given to show that the converse fails.

**Theorem 7.2.8.** ([67, Theorem 3.1]) *Let* $R$ *be a commutative ring with* $Z(R)$ *not an ideal of* $R$.

(1) $Z(\Gamma(R))$ *is connected with* $diam(Z(\Gamma(R))) = 2$;

(2) *Some vertex of* $Z(\Gamma(R))$ *is adjacent to a vertex of* $Reg(\Gamma(R))$. *In particular, the subgraphs* $Z(\Gamma(R))$ *and* $Reg(\Gamma(R))$ *of* $T(\Gamma(R))$ *are not disjoint;*

(3) *If* $Reg(\Gamma(R))$ *is connected, then* $T(\Gamma(R))$ *is connected.*

*Proof.* (1) Every $x \in Z(R)^*$ is adjacent to 0. Thus $x - 0 - y$ is a path in $Z(\Gamma(R))$ of length two between any two distinct $x, y \in Z(R)^*$. Moreover, there are nonadjacent $x, y \in Z(R)^*$ since $Z(R)$ is not an ideal of $R$; so $diam(Z(\Gamma(R))) = 2$.

(2) Since $Z(R)$ is not an ideal of $R$, there are distinct $x, y \in Z(R)^*$ such that $x + y \in Reg(R)$. Then $-x \in Z(R)$ and $x + y \in Reg(R)$ are adjacent vertices in $T(\Gamma(R))$ since $-x + (x + y) = y \in Z(R)$. The "in particular" statement is clear.

(3) Suppose that $Reg(\Gamma(R))$ is connected. Since $Z(\Gamma(R))$ is also connected by part (1) above, it is sufficient to show that there is a path from $x$ to $y$ in $T(\Gamma(R))$ for every $x \in Z(R)$ and $y \in Reg(R)$. By part (2) above, there are adjacent vertices $z$ and $w$ in $Z(\Gamma(R))$ and $Reg(\Gamma(R))$, respectively. Since $Z(\Gamma(R))$ is connected, there is a path from $x$ to $z$ in $Z(\Gamma(R))$; and since $Reg(\Gamma(R))$ is connected, there is a path from $w$ to $y$ in $Reg(\Gamma(R))$. As $z$ and $w$ are adjacent in $T(\Gamma(R))$, there is a path from $x$ to $y$ in $T(\Gamma(R))$. Thus $T(\Gamma(R))$ is connected. □

**Example 7.2.9.** ([67, Example 3.2]) Let $R = \mathbb{Q}[x](+)(\mathbb{Q}(x)/\mathbb{Q}[x])$. Then one can easily show that $Z(R) = (\mathbb{Q}[x] \setminus \mathbb{Q}^*)(+)(\mathbb{Q}(x)/\mathbb{Q}[x])$ is not an ideal of $R$ and $Reg(R) = U(R) = \mathbb{Q}^*(+)(\mathbb{Q}(x)/\mathbb{Q}[x])$. Thus $T(\Gamma(R))$ is connected with $\mathrm{diam}(T(\Gamma(R))) = 2$ (by Theorem 7.2.10 below) since $R = ((X, 0), (X + 1, 0))$ with $(X, 0), (X + 1, 0) \in Z(R)$. However, $Reg(\Gamma(R))$ is not connected since there is no path from $(1, 0)$ to $(2, 0)$ in $Reg(\Gamma(R))$. It has already been observed that $Z(\Gamma(R))$ is connected with $diam(Z(\Gamma(R))) = 2$.

The next result determines when $T(\Gamma(R))$ is connected and computes $diam(T(\Gamma(R)))$. In particular, $T(\Gamma(R))$ is connected if and only if $diam(T(\Gamma(R))) < \infty$.

**Theorem 7.2.10.** ([67, Theorems 3.3 and 3.4]) *Let $R$ be a commutative ring with $Z(R)$ not an ideal of $R$. Then $T(\Gamma(R))$ is connected if and only if $1 = z_1 + \cdots + z_n$ for some $z_1, \ldots, z_n \in Z(R)$. Furthermore, suppose that $T(\Gamma(R))$ is connected and let $n$ be the least integer such that $1 = z_1 + \cdots + z_n$ for some $z_1, \ldots, z_n \in Z(R)$. Then $diam(T(\Gamma(R))) = n$. In particular, if $R$ is a finite commutative ring and $Z(R)$ is not an ideal of $R$, then $T(\Gamma(R))$ is connected and $diam(T(\Gamma(R))) = 2$.*

*Proof.* Suppose that $T(\Gamma(R))$ is connected. Then there is a path $0 - b_1 - b_2 - \cdots - b_n - 1$ from 0 to 1 in $T(\Gamma(R))$. Thus $b_1, b_1 + b_2, \ldots, b_{n-1} + b_n, b_n + 1 \in Z(R)$. Hence $1 \in (b_1, b_1 + b_2, \ldots, b_{n-1} + b_n, b_n + 1) \subseteq (Z(R))$. Conversely, suppose that $1 = z_1 + \cdots + z_n$ for some $z_1, \ldots, z_n \in Z(R)$. We first show that there is a path from 0 to $x$ in $T(\Gamma(R))$ for every $0 \neq x \in R$. By hypothesis, $x = a_1 + \cdots + a_n$ for some $a_1, \ldots, a_n \in Z(R)$. Let $b_0 = 0$ and $b_k = (-1)^{n+k}(a_1 + \cdots + a_k)$ for each integer $k$ with $1 \leq k \leq n$. Then $b_k + b_{k+1} = (-1)^{n+k+1}a_{k+1} \in Z(R)$ for each integer $k$ with $0 \leq k \leq n-1$, and thus $0 - b_1 - b_2 - \cdots - b_{n-1} - b_n = x$ is a path from 0 to $x$ in $T(\Gamma(R))$ of length at most $n$. Now let $0 \neq z, w \in R$. Then by the preceding argument, there are paths from $z$ to 0 and 0 to $w$ in $T(\Gamma(R))$. Hence there is a path from $z$ to $w$ in $T(\Gamma(R))$; so $T(\Gamma(R))$ is connected.

For the "furthermore" statement, we first show that any path from 0 to 1 in $T(\Gamma(R))$ has length $\geq n$. Suppose that $0 - b_1 - b_2 - \cdots - b_{m-1} - 1$ is a path from 0 to 1 in $T(\Gamma(R))$ of length $m$. Thus $b_1, b_1 + b_2, \ldots, b_{m-2} + b_{m-1}, b_{m-1} + 1 \in Z(R)$, and hence $1 \in (b_1, b_1 + b_2, \ldots, b_{m-2} + b_{m-1}, b_{m-1} + 1) \subseteq (Z(R))$. Thus $m \geq n$.

Now, let $x$ and $y$ be distinct elements in $R$. We show that there is a path from $x$ to $y$ in $T(\Gamma(R))$ with length $\leq n$. Let $1 = z_1 + \cdots + z_n$ for some $z_1, \ldots, z_n \in Z(R)$, and let $z = y + (-1)^{n+1}x$. Define $d_0 = x$ and $d_k = (-1)^{n+k}z(z_1 + \cdots + z_k) + (-1)^k x$ for every integer $k$ with $1 \leq k \leq n$. Then $d_k + d_{k+1} = (-1)^{n+k+1}zz_{k+1} \in Z(R)$ for every integer $k$ with $0 \leq k \leq n-1$ and $d_n = z + (-1)^n x = y$. Thus $x - d_1 - \cdots - d_{n-1} - y$ is a path from $x$ to $y$ in $T(\Gamma(R))$ with length at most $n$. In particular, we conclude that a shortest path between 0 and 1 in $T(\Gamma(R))$ has length $n$, and thus $diam(T(\Gamma(R))) = n$.

For the "in particular" statement, suppose that $R$ is finite and $Z(R)$ is not an ideal of $R$. Then $x + y \in Reg(\Gamma(R))$ for some $x, y \in Z(R)$. Since every regular element of a finite commutative ring is a unit, we conclude that $R = (x, y)$, and thus $diam(T(\Gamma(R))) = 2$. □

**Corollary 7.2.11.** ([67, Corollary 3.5]) *Let $R$ be a commutative ring with $Z(R)$ not an ideal of $R$, and suppose that $T(\Gamma(R))$ is connected.*

(1) *$diam(T(\Gamma(R))) = d(0, 1)$;*

(2) *If $diam(T(\Gamma(R))) = n$, then $diam(Reg(\Gamma(R))) \geq n - 2$.*

*Proof.* (1) This is clear from the proof of Theorem 7.2.10.

(2) Since $n = diam(T(\Gamma(R))) = d(0, 1)$ by part (1) above, let $0 - s_1 - \cdots - s_{n-1} - 1$ be a shortest path from 0 to 1 in $T(\Gamma(R))$. Clearly $s_1 \in Z(R)$. If $s_i \in Z(R)$ for some integer $i$ with $2 \leq i \leq n-1$, then we can construct the path $0 - s_i - \cdots - s_{n-1} - 1$ from 0 to 1 which has length less than $n$, a contradiction. Thus $s_i \in Reg(R)$ for each integer $i$ with $2 \leq i \leq n-1$. Hence $s_2 - \cdots - s_{n-1} - 1$ is a shortest path from $s_2$ to 1 in $Reg(R)$, and it has length $n - 2$. Thus $diam(Reg(\Gamma(R))) \geq n - 2$. □

Let $a, b \in Reg(R)$. Then $d_{T(\Gamma(R)}(a, b)$ is the length of a shortest path in $T(\Gamma(R))$ from $a$ to $b$, and $d_{Reg(\Gamma(R)}(a, b)$ is the length of a shortest path in $Reg(\Gamma(R))$ from $a$ to $b$.

In view of Corollary 7.2.11(2), let $R$ be Noetherian and $m$ be a natural number. In the following result, it is shown that if $a, b \in Reg(R)$ and $d_{T(\Gamma(R))}(a, b) = m$, then $d_{Reg(\Gamma(R))}(a, b) = m$.

**Theorem 7.2.12.** ([36, Theorem 1]) *Let $R$ be a commutative Noetherian ring and $m$ be a natural number. If $a, b \in Reg(R)$ and $d_{T(\Gamma(R))}(a, b) = m$, then $d_{Reg(\Gamma(R))}(a, b) = m$.*

*Proof.* First assume that $m = 2$. If there exists a regular element $c$ such that $c$ is adjacent to both $a$ and $b$, then we are done. Thus suppose that no regular element is adjacent to both $a$ and $b$. Since $R$ is a commutative Noetherian ring, then by [105, Proposition 4.7, p. 53], there are prime ideals $P_1, \ldots, P_n$ such that $Z(R) = \bigcup_{r=1}^{n} P_r$. Since $a$ and $b$ are not adjacent, we have $a + b \in Reg(R)$. Since $d_{T(\Gamma(R))}(a, b) = 2$, there exists $z \in Z(R)$ such that $a + z, b + z \in Z(R)$. If $a - b \in Z(R)$, then $a, -b, b$ is a path in $Reg(\Gamma(R))$. Therefore we can assume that $a - b \in Reg(R)$. Set $a_1 = a + z, a_2 = b + z, a_3 = z$, and $a_4 = a_3 + a_1a_2$.

If $a_4 \in Reg(R)$, then $a, a_4, b$ is a path in $Reg(\Gamma(R))$, a contradiction. Now, assume that $a_4 \in Z(R)$. It is easy to see that for every i, $i = 1, \ldots, n$, we have $|P_i \cap \{a_1, a_2, a_3, a_4\}| \leq 1$. Without loss of generality, assume that $a_i \in P_i$ for $i = 1, 2, 3, 4$. Clearly, $a + a_4 \in P_1$ and $b + a_4 \in P_2$ . Now, we inductively construct $a_i, i > 4$, such that $a + a_i \in P_1$ , $b + a_i \in P_2$ , $a_i \notin \bigcup_{r=1}^{i-1} P_r$, and $a_i \in P_i$. Suppose that we have constructed $a_1, \ldots, a_{i-1}$. Now, define $a_i = a_{i-1} + a_1 a_2 \cdots a_{i-2}$. Since $a + a_{i-1} \in P_1$ , $b + a_{i-1} \in P_2$ , $a_1 \in P_1$, and $a_2 \in P_2$ , we have $a + a_i \in P_1$ and $b + a_i \in P_2$ . Since $a_i$ is adjacent to $a$ and $b$, we have $a_i \in Z(R) = \bigcup_{r=1}^{n} P_r$. Thus there is some $j$ such that $a_i \in P_j$. We claim that $j \geq i$. By way of contradiction, suppose that $j < i$. If $j \leq i - 2$, then $a_{i-1} \in P_j$ , which is a contradiction. Thus $j = i-1$, $(a_1 \cdots a_{i-3})a_{i-2} \in P_{i-1}$, and $a_{i-2} + a_1 \cdots a_{i-3} \in P_{i-1}$. So $a_1 \cdots a_{i-3}, a_{i-2} \in P_{i-1}$. Repeating this argument, we have $a_1 a_2 a_3, a_4 \in P_{i-1}$. This implies that $a_4$ and $a_t$ are contained in $P_{i-1}$ for some $t$, $1 \leq t \leq 3$, a contradiction. Therefore $a_i \notin \bigcup_{r=1}^{i-1} P_r$, and the claim is proved. Without loss of generality, assume that $j = i$. Since the number of prime ideals which cover $Z(R)$ is $n$, we have $a_{n+1} \notin Z(R)$, a contradiction.

Next, assume that $m = 3$. As above, there are prime ideals $P_1, \ldots, P_n$ such that $Z(R) = \bigcup_{i=1}^{n} P_i$. Because $a$ and $b$ are not adjacent, $a + b \in Reg(R)$. Since $d_{T(\Gamma(R))}(a, b) = 3$, there are $z_1, z_2 \in R$ such that $a + z_1, z_1 + z_2, b + z_2 \in Z(R)$. If $z_1$ or $z_2$ is regular, then by the previous case, we have nothing to prove. Thus let $z_1, z_2 \in Z(R)$. If $a - b \in Z(R)$, then $a, -b, b$ is a path in $T(\Gamma(R))$, a contradiction. Therefore we can assume that $a - b \in Reg(R)$. If $a + z_1 + z_2 \in Reg(R)$, then $a, -(a + z_1 + z_2), z_2, b$ is a path in $T(\Gamma(R))$ and by the case $m = 2$, there is a path of length 3 between $a$ and $b$ in $Reg(\Gamma(R))$. Thus we can suppose that $a + z_1 + z_2 \notin Reg(R)$. Similarly, we can assume that $b + z_1 + z_2 \notin Reg(R)$. Set $a_1 = a + z_1 + z_2, a_2 = b + z_1 + z_2$, and $a_3 = z_1 + z_2$. Without loss of generality, assume that $a_1 \in P_1$; so $a_2, a_3 \notin P_1$. Also, $a_2$ and $a_3$ are not contained in the same prime ideal. Without loss of generality, assume that $a_2 \in P_2$ and $a_3 \in P_3$. Set $a_4 = a_3 + a_1 a_2$. Clearly, $a_4 \notin \bigcup_{i=1}^{3} P_i$, $a + a_4 \in P_1$, and $b + a_4 \in P_2$. Therefore $a, a_4, b$ is a path in $T(\Gamma(R))$, a contradiction. Thus $d_{Reg(\Gamma(R))}(a, b) = 3$, as desired.

Now, by induction on $r$, we show that if $a, b \in Reg(R)$ and $a, z_1, \ldots, z_r, b$ is a shortest path between $a$ and $b$ in $T(\Gamma(R))$, then there is a path $a, b_1, \ldots, b_n, b$ in $Reg(\Gamma(R))$, where $n = r$. If $r = 1$ or $r = 2$, then as before, one can replace the $z_i$'s with regular elements. Therefore assume that $r \geq 3$. If there is a $j$, $1 \leq j \leq r$, such that $z_j \in Reg(R)$, then using induction (two times), we obtain a path of length at most $r + 1$ between $a$ and $b$ in $Reg(\Gamma(R))$. Thus we may assume that all $z_i$ are zero-divisors. If $z_1 + z_3 \in Z(R)$, then we obtain the path $a, z_1, z_3, z_4, \ldots, z_r, b$, a contradiction. So $z_1 + z_3 \in Reg(R)$. Now, consider the path $a, z_1, -(z_1 + z_3), z_3, \ldots, z_r, b$, and the proof is complete. □

In light of Corollary 7.2.11(2) and Theorem 7.2.12, we have the following result.

**Corollary 7.2.13.** ([36, Corollary 1]) *Let $R$ be a commutative Noetherian ring with $diam(T(\Gamma(R))) = n$. Then $n-2 \leq diam(Reg(\Gamma(R))) \leq n$.*

Note that there are some commutative Noetherian rings $R$ for which the equality $diam(Reg(\Gamma(R))) = diam(T(\Gamma(R)))$ does not hold. For instance, $diam(Reg(\Gamma(\mathbb{Z}_6))) = 1$, but $diam(T(\Gamma(\mathbb{Z}_6))) = 2$.

**Corollary 7.2.14.** ([67, Corollary 3.6]) *Let $R$ be a commutative ring. If $R$ has a nontrivial idempotent, then $T(\Gamma(R))$ is connected with $diam(T(\Gamma(R))) = 2$.*

*Proof.* Let $e \in R\backslash\{0,1\}$ be idempotent. Then $R = (e, 1-e)$ with $e, 1-e \in Z(R)$; so the claim is clear by Theorem 7.2.10. □

**Corollary 7.2.15.** ([67, Corollary 3.7]) *Let $\{R_\alpha\}_{\alpha\in\Lambda}$ be a family of commutative rings with $|\Lambda| \geq 2$, and let $R = \prod_{\alpha\in\Lambda} R_\alpha$. Then $T(\Gamma(R))$ is connected with $diam(T(\Gamma(R))) = 2$.*

*Proof.* This follows directly from Corollary 7.2.14 since in this case, $R$ has a nontrivial idempotent. □

Recall that the *eccentricity* of a vertex $v$ in a graph $G$ is defined by $e(v) = max\{d(v,y) \mid y \in V(G)\}$, and the *radius* of $G$ is given by $r(G) = min\{e(x) \mid x \in V(G)\}$.

**Theorem 7.2.16.** ([312, Theorem 2.2]) *Let $R$ be a commutative ring with $Z(R)$ not an ideal of $R$, and let $n \geq 2$ be the smallest integer such that $R = (z_1, \ldots, z_n)$, for some $z_1, \ldots, z_n \in Z(R)$. Then $r(T(\Gamma(R)))= n$. In particular, if $R$ is finite, then $r(T(\Gamma(R)))= 2$.*

*Proof.* We know that diam$(T(\Gamma(R)))= n$; so we only have to prove that $r \leq n-1$ is not possible. Assume that $r \leq n-1$. So, there is an $x \in R$ such that $e(x) \leq n-1$, i.e.,

$$(\forall y \subset R) \quad d(x,y) \leq n-1.$$

In particular, $d(x, 1+x) \leq n-1$ and $d(x, 1-x) \leq n-1$. Let

$$x \text{---} s_1 \text{---} s_2 \text{---} \cdots \text{---} s_{k-2} \text{---} 1 + (-1)^{k-1}x$$

be a path of length $k-1 \leq n-1$ in $T(\Gamma(R))$. We get $k-1$ zero-divisors

$$x + s_1, s_1 + s_2, \ldots, s_{k-3} + s_{k-2}, s_{k-2} + 1 + (-1)^{k-1}x$$

from which it follows that

$$(x + s_1, s_1 + s_2, \ldots, s_{k-3} + s_{k-2}, s_{k-2} + 1 + (-1)^{k-1}x) \subseteq (Z(R)) = R.$$

Since

$$1 \in (x + s_1, s_1 + s_2, \ldots, s_{k-3} + s_{k-2}, s_{k-2} + 1 + (-1)^{k-1}x),$$

we must have

$$(x + s_1, s_1 + s_2, \ldots, s_{k-3} + s_{k-2}, s_{k-2} + 1 + (-1)^{k-1}x) = R.$$

It follows that $R$ may be generated by $k-1$ zero-divisors, which is a contradiction $(k-1 < n)$. □

**Corollary 7.2.17.** *Let $R$ be a commutative ring with $Z(R)$ not an ideal of $R$, $diam(T(\Gamma(R))) = m < \infty$, and $a \in R$. Then there is a $b \in R$ such that $d_{T(\Gamma(R))}(a, b) = m$.*

**Theorem 7.2.18.** ([67, Theorem 3.10]) *Let $R$ be a commutative ring with $Z(R)$ not an ideal of $R$. Then $T(\Gamma(T(R)))$ is connected with $diam(T(\Gamma(T(R)))) = 2$. In particular, if $R$ is finite and $Z(R)$ is not an ideal of $R$, then $T(\Gamma(R))$ is connected with $diam(T(\Gamma(R))) = 2$.*

*Proof.* Let $T = T(R)$. Since $Z(R)$ is not an ideal of $R$, there are $z_1, z_2 \in Z(R)$ such that $s = z_1 + z_2 \in Reg(R)$. Thus $z_1/s + z_2/s = 1$ in $T$; so $Z(T)$ is not an ideal of $T$. Hence $T = (z_1/s, z_2/s)T$, and thus $T(\Gamma(T))$ is connected with $diam(T(\Gamma(R))) = 2$ by Theorem 7.2.10. The "in particular" statement is clear (and has already been observed in Theorem 7.2.10) since $T(R) = R$ when $R$ is finite. □

The following result is related to the previous theorem.

**Theorem 7.2.19.** ([67, Theorem 3.11]) *Let $P_1$ and $P_2$ be prime ideals of a commutative ring $R$ with $xy = 0$ for some $x \in P_1 \setminus P_2$ and $y \in P_2 \setminus P_1$, and $S = R \setminus (P_1 \cup P_2)$. Then $T(\Gamma(R_S))$ is connected with $diam(T(\Gamma(R_S))) = 2$.*

*Proof.* Since $x \notin P_2$, $y \notin P_1$, and $s \notin P_1 \cup P_2$ for every $s \in S$, we have $sx \neq 0$ and $sy \neq 0$ for every $s \in S$. Thus $x/s$ and $y/s$ are nonzero zero-divisors in $R_S$ for all $s \in S$. Note that $t = x + y \in S$, and hence is a unit in $R_S$, since $t \notin P_1 \cup P_2$. Thus $R_S = (x/t, y/t)R_S$, and hence $T(\Gamma(R_S))$ is connected with $diam(T(\Gamma(R_S))) = 2$ by Theorem 7.2.10. □

The following is an example of a commutative ring $R$ such that neither $Reg(\Gamma(R))$ nor $T(\Gamma(R))$ is connected, but $T(\Gamma(R_S))$ is connected for some multiplicative subset $S$ of $R$ with $S \neq R \setminus Z(R)$.

**Example 7.2.20.** ([67, Example 3.12]) Let $R = \mathbb{Z}[X_1, X_2, X_3]/(X_1X_2X_3) = \mathbb{Z}[x_1, x_2, x_3]$, $P_1 = (x_1)$ and $P_2 = (x_2)$ be prime ideals of $R$, $x = x_1$, and $y = x_2x_3$. Then $xy = 0$ and $x \in P_1 \setminus P_2$ and $y \in P_2 \setminus P_1$. Let $S = R \setminus (P_1 \cup P_2) \supsetneq R \setminus Z(R)$. Then $T(\Gamma(R_S))$ is connected with $diam(T(\Gamma(R_S))) = 2$ by Theorem 7.2.19. By Theorem 7.2.10, $T(\Gamma(R))$ is not connected since $(Z(R)) \subsetneq R$ and $Z(R)$ is not an ideal of $R$, and $Reg(\Gamma(R))$ is not connected since there is no path from 1 to 2 in $Reg(\Gamma(R))$ (or use Theorem 7.2.8(3)).

If $Z(R)$ is not an ideal of $R$, then $diam(Z(\Gamma(R))) = 2$. Moreover, we have $2 \leq diam(T(\Gamma(R))) < \infty$ when $T(\Gamma(R))$ is connected. In the following example, for every integer $n \geq 2$, we construct a commutative ring $R_n$ such that $Z(R_n)$ is not an ideal of $R_n$ and $T(\Gamma(R_n))$ is connected with $diam(T(\Gamma(R_n))) = n$.

**Example 7.2.21.** ([67, Example 3.8]) For an integer $n \geq 2$, let $D = \mathbb{Z}[x_1, \ldots, x_{n-1}]$, $K$ be the quotient field of $D$, $P_0 = (x_1 + x_2 + \cdots + x_{n-1})$, $P_i = (x_i)$ for every integer $i$ with $1 \leq i \leq n-2$, and $P_{n-1} = (x_{n-1} + 1)$. Then $P_0, P_1, \ldots, P_{n-1}$ are distinct prime ideals of $D$. Let $F = P_0 \cup P_1 \cup \cdots \cup P_{n-1}$; then $S = D \setminus F$ is a multiplicative subset of $D$. Set $R_n = D(+)(K/D_S)$. Then $Z(R_n) = F(+)(K/D_S)$. Since $(1,0) = (-x_1 - x_2 - \cdots - x_{n-1}, 0) + (x_1, 0) + (x_2, 0) + (x_3, 0) + \cdots + (x_{n-1} + 1, 0)$ is the sum of $n$ zero-divisors of $R_n$, by construction we conclude that $n$ is the least integer $m \geq 2$ such that $R_n$ is generated by $m$ zero-divisors of $R_n$. Hence $T(\Gamma(R_n))$ is connected with $diam(T(\Gamma(R_n))) = n$ by Theorem 7.2.10.

Example 7.2.9 shows that we may have $diam(T(\Gamma(R)))$ finite whereas $diam(Reg(\Gamma(R)))$ is infinite (if $R$ is Notherian, see Corollary 7.2.13). Examples given below show that we may also have either $diam(T(\Gamma(R))) = diam(Reg(\Gamma(R)))$ or $diam(T(\Gamma(R))) > diam(Reg(\Gamma(R)))$ when $Z(R)$ is not an ideal of $R$.

**Example 7.2.22.** ([67, Example 3.9]) (a) Let $R = \mathbb{Z}_5 \times \mathbb{Z}_5$. Then $diam(T(\Gamma(R))) = 2$ by Theorem 7.2.10 (or Corollary 7.2.15), and it is easy to check that $diam(Reg(\Gamma(R))) = 2$. Thus $diam(T(\Gamma(R))) = diam(Reg(\Gamma(R)))$.

(b) Let $R = \mathbb{Z}_2 \times \mathbb{Z}_3$. By Theorem 7.2.10 (or Corollary 7.2.15), diam$(T(\Gamma(R))) = 2$, and it is easy to check that $diam(Reg(\Gamma(R))) = 1$. Thus $diam(T(\Gamma(R))) > diam(Reg(\Gamma(R)))$.

We next determine the girth of $Z(\Gamma(R))$, $Reg(\Gamma(R))$, and $T(\Gamma(R))$ when $Z(R)$ is not an ideal of $R$. Recall that $|Z(R)| \geq 3$ if $Z(R)$ is not an ideal of $R$. We first state a lemma.

**Lemma 7.2.23.** ([67, Lemma 3.13]) *Let $R$ be a commutative ring with $Z(R)$ not an ideal of $R$. Then $2Z(R) = \{0\}$ if and only if $char(R) = 2$.*

*Proof.* If $char(R) = 2$, then clearly $2Z(R) = \{0\}$.

Conversely, suppose that $2z = 0$ for every $z \in Z(R)$. Since $Z(R)$ is not an ideal of $R$, there are distinct $x, y \in Z(R)$ such that $z = x + y \in Reg(R)$. Then $2z = 2x + 2y = 0$; so $2 = 0$ since $z \in Reg(R)$, i.e., $char(R) = 2$. □

**Theorem 7.2.24.** ([67, Theorem 3.14]) *Let $R$ be a commutative ring with $Z(R)$ not an ideal of $R$.*

(1) *Either $gr(Z(\Gamma(R))) = 3$ or $gr(Z(\Gamma(R))) = \infty$. Moreover, if $gr(Z(\Gamma(R))) = \infty$, then $R \cong \mathbb{Z}_2 \times \mathbb{Z}_2$; so $Z(\Gamma(R))$ is a $K_{1,2}$ star graph with center 0;*

(2) *$gr(T(\Gamma(R))) = 3$ if and only if $gr(Z(\Gamma(R))) = 3$ (if and only if $R \ncong \mathbb{Z}_2 \times \mathbb{Z}_2$);*

(3) *$gr(T(\Gamma(R))) = 4$ if and only if $gr(Z(\Gamma(R))) = \infty$ (if and only if $R \cong \mathbb{Z}_2 \times \mathbb{Z}_2$);*

(4) *If $char(R) = 2$, then $gr(Reg(\Gamma(R))) = 3$ or $\infty$. In particular, $gr(Reg(\Gamma(R))) = 3$ if $char(R) = 2$ and $Reg(\Gamma(R))$ contains a cycle;*

(5) *$gr(Reg(\Gamma(R))) = 3, 4,$ or $\infty$. In particular, $gr(Reg(\Gamma(R))) \leq 4$ if $Reg(\Gamma(R))$ contains a cycle.*

*Proof.* (1) If $x + y \in Z(R)$ for distinct $x, y \in Z(R)^*$, then $0 - x - y - 0$ is a 3-cycle in $Z(\Gamma(R))$; so $gr(Z(\Gamma(R))) = 3$. Otherwise, $x + y \in Reg(R)$ for every distinct $x, y \in Z(R)^*$. So in this case, every $x \in Z(R)^*$ is adjacent to 0, and no two distinct $x, y \in Z(R)^*$ are adjacent. Thus $Z(\Gamma(R))$ is a star graph with center 0; so $gr(Z(\Gamma(R))) = \infty$.

Let $Z(R) = \cup_{\alpha \in \Lambda} P_\alpha$, where every $P_\alpha$ is a prime ideal of $R$ ([228, p.3]). Then $|\Lambda| \geq 2$ since $Z(R)$ is not an ideal of $R$. Assume that $gr(Z(\Gamma(R))) = \infty$. Then $x + y \in Reg(R)$ for all distinct $x, y \in Z(R)^*$, and thus every $|P_\alpha| = 2$. Hence the intersection of any two distinct $P_\alpha$'s is $\{0\}$, and thus $|\Lambda| = 2$. So let $Z(R) = P_1 \cup P_2$ for prime ideals $P_1, P_2$ of $R$ with $P_1 \cap P_2 = \{0\}$ and $|P_1| = |P_2| = 2$. Hence $|Z(R)| = 3$, and thus $R$ is also finite [200, Theorem 1]. So $P_1$ and $P_2$ are the only prime (maximal) ideals of $R$. By the Chinese Remainder Theorem, we have $R \cong R/P_1 \times R/P_2 \cong \mathbb{Z}_2 \times \mathbb{Z}_2$.

(2) We need only show that $gr(Z(\Gamma(R))) = 3$ when $gr(T(\Gamma(R))) = 3$. If $2z \neq 0$ for some $z \in Z(R)^*$, then $0 - z - (-z) - 0$ is a 3-cycle in $Z(\Gamma(R))$. Thus we may assume that $2z = 0$ for every $z \in Z(R)$, and hence $char(R) = 2$ by Lemma 7.2.23. Let $a - b - c - a$ be a 3-cycle in $T(\Gamma(R))$. Then $z = a + b, w = a + c, b + c \in Z(R)^*$. Moreover, $z + w = (a + b) + (a + c) = 2a + (b + c) = b + c \in Z(R)$. Thus $0 - z - w - 0$ is a 3-cycle in $Z(\Gamma(R))$; so $g(Z(\Gamma(R))) = 3$.

(3) Suppose that $gr(Z(\Gamma(R))) = \infty$. Then $R \cong \mathbb{Z}_2 \times \mathbb{Z}_2$ by part (1) above; so $gr(T(\Gamma(R))) = 4$. Conversely, suppose that $gr(T(\Gamma(R))) = 4$. Then $gr(Z(\Gamma(R))) = \infty$ by parts (1) and (2) above.

(4) Suppose that $char(R) = 2$ and $Reg(\Gamma(R))$ contains a cycle $C$. Then $C$ contains two distinct vertices $x, y \in Reg(R)$ with $x \neq 1, y \neq 1$, and $x + y \in Z(R)$. Suppose that $R$ contains a $0 \neq w \in Nil(R)$. If $w = wx = wy$, then $x + 1$ and $y + 1$ are nonzero zero-divisors of $R$, and thus $1 - x - y - 1$ is a 3-cycle in $Reg(\Gamma(R))$. If either $wx \neq w$ or $wy \neq w$, then either $1 - (w + 1) - (wx + 1) - 1$ or $1 - (w + 1) - (wy + 1) - 1$ is a 3-cycle in $Reg(\Gamma(R))$. If $R$ is reduced, then $x^2 + y^2 = (x + y)^2 \neq 0$. Hence $x^2 \neq y^2$, and thus $x^2 - xy - y^2 - x^2$ is a 3-cycle in $Reg(\Gamma(R))$. Hence $gr(Reg(\Gamma(R))) = 3$.

(5) By part (4) above, we may assume that $char(R) \neq 2$. Suppose that $Reg(\Gamma(R))$ contains a cycle $C$. Then $C$ contains two distinct vertices $x, y \in Reg(R)$ such that $y \neq -x$ and $x + y \in Z(R)$. Thus $x - y - (-y) - (-x) - x$ is a 4-cycle in $Reg(\Gamma(R))$; so $gr(Reg(\Gamma(R))) \leq 4$. □

The next example shows that the 3 possibilities for $gr(Reg(\Gamma(R)))$ when $Z(R)$ is not an ideal of $R$ from Theorem 7.2.24(5) may occur when $gr(Z(\Gamma(R))) = gr(T(\Gamma(R))) = 3$. However, if $\mathrm{gr}(Z(\Gamma(R))) = \infty$ and $Z(R)$ is not an ideal of $R$, then $R \cong \mathbb{Z}_2 \times \mathbb{Z}_2$ by Theorem 7.2.24(1), and thus $gr(Reg(\Gamma(R))) = \infty$ and $gr(T(\Gamma(R))) = 4$. In particular, $gr(Z(\Gamma(R))) = 3$ when $R$ is not reduced and $Z(R)$ is not an ideal of $R$.

**Example 7.2.25.** ([67, Example 3.15]) (a) Let $R = \mathbb{Z}_2 \times \mathbb{Z}_3$. Then it is easy to check that $gr(Z(\Gamma(R))) = gr(T(\Gamma(R))) = 3$ and $gr(Reg(\Gamma(R))) = \infty$.

(b) Let $R = \mathbb{Z}_3 \times \mathbb{Z}_4$. Then it is easy to check that $gr(Reg(\Gamma(R))) = gr(Z(\Gamma(R))) = gr(T(\Gamma(R))) = 3$.

(c) Let $R = \mathbb{Z}_3 \times \mathbb{F}_4$. Then it is easy to check that $Reg(\Gamma(R))$ is a $K_{3,3}$. Thus $Reg(\Gamma(R))$ (and hence $T(\Gamma(R))$) is connected with $gr(Reg(\Gamma(R))) = 4$. It is also easy to check that $gr(T(\Gamma(R))) = gr(Z(\Gamma(R))) = 3$.

**Theorem 7.2.26.** ([70, Theorem 4.7]) *Let $R$ be a commutative ring. Then $gr(T(\Gamma(R))) \in \{3, 4, \infty\}$. Moreover,*

(1) $gr(T(\Gamma(R))) = \infty$ *if and only if either $R$ is an integral domain or $R$ is isomorphic to $\mathbb{Z}_4$ or $\mathbb{Z}_2[x]/(x^2)$;*

(2) $gr(T(\Gamma(R))) = 4$ *if and only if $R$ is isomorphic to $\mathbb{Z}_2 \times \mathbb{Z}_2$, and*

(3) $gr(T(\Gamma(R))) = 3$ *otherwise.*

*Proof.* By Theorem 7.2.4(3) and Theorem 7.2.24, $gr(T(\Gamma(R))) = 3$ unless $R \cong \mathbb{Z}_2 \times \mathbb{Z}_2$ or $|Z(R)| \leq 2$. If $R \cong \mathbb{Z}_2 \times \mathbb{Z}_2$, then $gr(T(\Gamma(R))) = 4$. If $|Z(R)| \leq 2$, then $R$ is either an integral domain or isomorphic to $\mathbb{Z}_4$ or $\mathbb{Z}_2[x]/(x^2)$. In each of these cases, $gr(T(\Gamma(R))) = \infty$. The result now follows. □

## 7.3 Polynomial Rings and Idealization

Note that $Z(R) \subseteq Z(R[x]) \subseteq Z(R)[x]$ always holds. The second inclusion may be proper, for example, $2+3x \in Z(\mathbb{Z}_6)[x] \setminus Z(\mathbb{Z}_6[x])$. It is clear that the first inclusion may be proper as well. The well-known McCoy's Theorem gives a description of the set of zero-divisors in a polynomial ring: $f(x) \in Z(R[x])$ if and only if there is an $r \in R^*$ such that $rf(x) = 0$. Therefore, not only the coefficients have to be zero-divisors, but the ideal generated by these coefficients has a nonzero annihilator. A commutative ring $R$ is a *McCoy ring* if every finitely generated ideal $I \subseteq Z(R)$ has the property that $Ann(I) \neq 0$. It is well known that the polynomial ring $R[x]$ is always a McCoy ring.

The structure of the total graph $T(\Gamma(R[x]))$ of a polynomial ring depends on whether or not $Z(R[x])$ is an ideal of $R[x]$. According to [269, Theorem 3.3], $Z(R[x])$ is an ideal of $R[x]$ if and only if $R$ is a McCoy ring with $Z(R)$ an ideal of $R$. This allows us to characterize the structure of the total graph of polynomial rings. First, suppose that $Z(R[x])$ is an ideal of $R[x]$. It is evident that in this case the subgraph of zero-divisors $Z(\Gamma(R[x]))$ is complete. One has $Z(R[x]) = Z(R)[x]$; consequently

$$\frac{R[x]}{Z(R[x])} = \frac{R[x]}{Z(R)[x]} \cong (R/Z(R))\,[x].$$

The right-hand side is a polynomial ring which clearly cannot be isomorphic to $\mathbb{Z}_2$ or to $\mathbb{Z}_3$; so $Reg(\Gamma(R[x]))$ is not connected by Theorem 7.2.2. From the previous discussion, one can characterize the structure of the total graph of a polynomial ring in which the zero-divisors form an ideal.

**Theorem 7.3.1.** ([312, Theorem 3.1]) *Let $R$ be a McCoy ring with $Z(R)$ an ideal of $R$. Then the total graph $T(\Gamma(R[x]))$ is not connected, the induced subgraph $Z(\Gamma(R[x]))$ is complete, and $Reg(\Gamma(R[x]))$ is not connected.*

**Remark 7.3.2.** ([312, Remark 3.1]) Since $R[x]$ is always a McCoy ring and $R[x,y] = R[x][y]$, from the hypothesis that $R$ is a McCoy ring and $Z(R)$ is an ideal of $R$, it follows that $Reg(\Gamma(R[x,y]))$ is not connected.

Let us now concentrate on the case when $Z(R[x])$ is not an ideal of $R[x]$.

**Lemma 7.3.3.** ([312, Lemma 3.1]) *Let $R$ be a commutative ring with $Z(R)$ not an ideal of $R$. Then*

$$(Z(R[x])) = R[x] \quad \text{if and only if} \quad (Z(R)) = R.$$

*Proof.* Suppose that $(Z(R[x])) = R[x]$. Therefore, there are polynomials $f_1(x), \dots, f_n(x) \in Z(R[x])$ such that $f_1(x) + \cdots + f_n(x) = 1$. It follows that $z_1 + \cdots + z_n = 1$, where $z_1, \dots, z_n$ are constant coefficients of the previous polynomials. Since $Z(R[x]) \subseteq Z(R)[x]$, all coefficients of these polynomials are zero-divisors. Therefore, $z_1, \dots, z_n \in Z(R)$ as well. So, $R = (Z(R)) = (z_1, \dots, z_n)$. The other implication is trivial. □

Using Lemma 7.3.3 and Theorem 7.2.10, as well as Theorem 7.2.16, we have the following result.

**Theorem 7.3.4.** ([312, Theorem 3.2]) *Let $R$ be a commutative ring with $Z(R)$ not an ideal of $R$. Then $T(\Gamma(R[x]))$ is connected if and only if $T(\Gamma(R))$ is connected, i.e., there are $z_1, \dots, z_n \in Z(R)$ with $R = (Z(R)) = (z_1, \dots, z_n)$. If $n$ is the minimal number of such generators, then*

$$diam(T(\Gamma(R[x]))) = r(T(\Gamma(R[x]))) = n\,.$$

Although rings of formal power series share some properties with polynomial rings, they do differ when it comes to zero-divisors. For example, McCoy's Theorem does not hold for $R[[x]]$. In [196], Fields presented an example of formal power series, with an invertible coefficient, which is nonetheless a proper zero-divisor. This example shows that $Z(R[[x]]) \subseteq Z(R)[[x]]$ need not hold. It is clear that the reverse inclusion may not hold. The reason for this lies in the nilpotent elements. Namely, if the ring $R$ is *reduced*, it has been shown [207] that $f(x) \in Z(R[[x]])$ if and only if there is a $z \in Z(R)^*$ with $zf(x) = 0$. The obvious question one might ask is whether $Z(R[[x]])$ is an ideal of $R[[x]]$. Suppose that $Z(R)$ is an ideal of $R$. It is clear that $Z(R[[x]]) \subseteq Z(R)[[x]]$. The equality in this case holds if and only if the ring $R$ is a *countably McCoy* ring, i.e., if every countably generated ideal $I \subseteq Z(R)$ has a nonzero annihilator. The following lemma summarizes the above discussion.

**Lemma 7.3.5.** ([312, Lemma 3.2]) *Let $R$ be a reduced, countably McCoy ring with $Z(R)$ an ideal of $R$. Then $Z(R[[x]])$ is an ideal of $R[[x]]$ and $Z(R[[x]]) = Z(R)[[x]]$.*

Under these conditions,

$$\frac{R[[x]]}{Z(R[[x]])} = \frac{R[[x]]}{Z(R)[[x]]} \cong \left(\frac{R}{Z(R)}\right)[[x]].$$

The right-hand side is not isomorphic to $\mathbb{Z}_2$ or to $\mathbb{Z}_3$; so one has a result analogous to the one for polynomial rings.

**Theorem 7.3.6.** ([312, Theorem 3.3]) *Let $R$ be a reduced, countably McCoy ring with $Z(R)$ an ideal of $R$. Then $T(\Gamma(R[[x]]))$ is not connected, the induced subgraph $Z(\Gamma(R[[x]]))$ is complete, and $Reg(\Gamma(R[[x]]))$ is not connected.*

Using the same reasoning as in the polynomial case, one can prove that for a reduced commutative ring $R$ with $Z(R)$ an ideal of $R$

$$(Z(R[[x]])) = R[[x]] \quad \text{if and only if} \quad (Z(R)) = R\,.$$

**Theorem 7.3.7.** ([312, Theorem 3.4]) *Let $R$ be a reduced commutative ring with $Z(R)$ not an ideal of $R$. Then $T(\Gamma(R[[x]]))$ is connected if and only if $T(\Gamma(R))$ is connected, i.e., there are $z_1, \ldots, z_n \in Z(R)$such that $R = (Z(R)) = (z_1, \ldots, z_n)$. If $n \geq 2$ is the smallest number of such generators, then*

$$diam(T(\Gamma(R[[x]]))) = r(T(\Gamma(R[[x]]))) = n\,.$$

Recall that for an $R$-module $M$, the *idealization of $M$ over $R$* is the commutative ring formed from $R \times M$ by defining addition and multiplication as $(r, m) + (s, n) = (r + s, m + n)$ and $(r, m)(s, n) = (rs, rn + sm)$, respectively. A standard notation for this "idealized ring" is $R(+)M$; see [219] for basic properties of rings resulting from the idealization construction. The zero-divisor graph $\Gamma(R(+)M)$ has been studied in [85] and [120].

Let $M$ be an $R$-module. We state some results about the graphs of the idealization $R(+)M$. In the first result, we assume that $Z(R)(+)M = Z(R(+)M)$. Note that $Z(R)(+)M \subseteq Z(R(+)M)$ always holds, but the inclusion may be proper since $Z(\mathbb{Z}(+)\mathbb{Z}_2) = 2\mathbb{Z}(+)\mathbb{Z}_2$. However, equality holds if either $M$ is an ideal of $R$ or $R$ is an integral domain and $M$ is torsion free.

First, the following example ([312, Example 3.2]) shows that it is possible that the graph $T(\Gamma(R))$ is not connected while the graph $T(\Gamma(R(+)M))$ is connected.

**Example 7.3.8.** ([312, Example 3.2]) Let $R = \mathbb{Z}[X]/(X^2) = \mathbb{Z}[x] = \{a + bx \mid a, b \in \mathbb{Z}\}$, $M = \mathbb{Z}_{10}$. It is easy to check that $M$ is an $R$-module with the action defined by $(a + bx)m = am$. The set $Z(R) = \{ax \mid a \in \mathbb{Z}\}$ of zero-divisors of $R$ is a (principal) ideal of $R$; so $T(\Gamma(R))$ is not connected and $Z(\Gamma(R))$ is complete. We also have $R/Z(R) \cong \mathbb{Z}$. Therefore, $Reg(\Gamma(R))$ is not connected.

On the other hand, $Z(M) = P \cup Q$, where $P = 2\mathbb{Z} + (x)$ and $Q = 5\mathbb{Z} + (x)$ are prime ideals, is not an ideal of $R$. Note that $Z(R) \subseteq Z(M)$. It is easy to see that the set of zero-divisors $Z(R(+)M) = \{(z, m) \mid z \in P \cup Q, m \in \mathbb{Z}_{10}\}$ fails to be an ideal of $R(+)M$. For example, let $z_1 = (2, 0)$, $z_2 = (5, 0)$. Then $z_1, z_2 \in Z(R(+)M)$, but $z_1 + z_2 = (7, 0) \in Reg(R(+)M)$. Since $(3, 5)z_1 + (-1, 0)z_2 = (1, 0)$, we have $(Z(R)) = (z_1, z_2) = R(+)M$, and $T(\Gamma(R(+)M))$ is connected with diameter 2. The subgraph $Z(\Gamma(R(+)M))$ is also connected with diameter 2 as well. The subgraph $Reg(\Gamma(R(+)M))$ is actually complete, since for $(s_1, m_1), (s_2, m_2) \in Reg(\Gamma(R(+)M))$, it holds that $s_1 + s_2 \in P$, and therefore $(s_1 + s_2, m_1 + m_2) \in Z(\Gamma(R(+)M))$.

**Theorem 7.3.9.** ([312, Theorem 3.5]) *Let $R$ be a commutative ring and $M$ an $R$-module with $Z(M) \subseteq Z(R)$. Then the following conditions are equivalent:*

(1) *$Z(R(+)M)$ is an ideal of $R(+)M$;*

(2) *$Z(R)$ is an ideal of $R$.*

*Moreover, $Z(R)(+)M = Z(R(+)M)$.*

*Proof.* First assume that $Z(R)$ is an ideal of $R$. Since $Z(M) \subseteq Z(R)$, we have $Z(R) \cup Z(M) = Z(R)$, and therefore $Z(R(+)M) = Z(R)(+)M$. Note that the set on the right-hand side is an ideal.

Assume that $Z(R(+)M)$ is an ideal of $R(+)M$, and let $z_1, z_2 \in Z(R)$. Then $(z_1, 0), (z_2, 0) \in Z(R(+)M)$; so $(z_1 + z_2, 0) \in Z(R(+)M)$. Thus $z_1 + z_2 \in Z(R) \cup Z(M) = Z(R)$. Likewise, if $r \in R$ and $z \in Z(R)$, then $(r, 0) \in R(+)M$ and $(z, 0) \in Z(R(+)M)$. Consequently, $(r, 0)(z, 0) = (rz, 0) \in Z(R(+)M)$, and we have $rz \in Z(R) \cup Z(M) = Z(R)$. ☐

**Theorem 7.3.10.** ([312, Theorem 3.6]) *Let $R$ be a commutative ring with $Z(R)$ an ideal of $R$ and $M$ an $R$-module with $Z(M) \subseteq Z(R)$. Then $T(\Gamma(R(+)M))$ is disconnected, $Z(\Gamma(R(+)M))$ is complete, and $Reg(\Gamma(R(+)M))$ is connected if and only if $Reg(\Gamma(R))$ is connected.*

*Proof.* From Theorem 7.3.9, it follows that $T(\Gamma(R(+)M))$ is disconnected and $Z(\Gamma(R(+)M))$ is complete. Furthermore,

$$\frac{(R(+)M)}{(Z(R(+)M))} = \frac{(R(+)M)}{(Z(R)(+)M)} \cong \frac{R}{Z(R)}(+)0 \cong \frac{R}{Z(R)}.$$

Therefore, $Reg(\Gamma(R(+)M))$ is connected if and only if $Reg(\Gamma(R))$ is connected. □

**Theorem 7.3.11.** ([67, Theorem 3.16]) *Let $R$ be a commutative ring with $Z(R)$ not an ideal of $R$ and $M$ an $R$-module with $Z(R(+)M) = Z(R)(+)M$.*

(1) *$T(\Gamma(R(+)M))$ is connected if and only if $T(\Gamma(R))$ is connected;*

(2) *$diam(T(\Gamma(R(+)M))) = diam(T(\Gamma(R)))$.*

*Proof.* (1) Suppose that $T(\Gamma(R(+)M))$ is connected. Let $x, y \in R$ be distinct. Then $(x, 0), (y, 0) \in R(+)M$; so there is a path $(x, 0)-(s_1, t_1)-\cdots-(s_n, t_n)-(y, 0)$ from $(x, 0)$ to $(y, 0)$ in $T(\Gamma(R(+)M))$. Since $Z(R(+)M) = Z(R)(+)M$, we have $x - s_1 - \cdots - s_n - y$ is a path from $x$ to $y$ in $T(\Gamma(R))$. Thus $T(\Gamma(R))$ is connected and $diam(T(\Gamma(R(+)M))) \geq diam(T(\Gamma(R)))$.

Conversely, suppose that $T(\Gamma(R))$ is connected (so $diam(T(\Gamma(R))) \geq 2$ by Theorem 7.2.10). Let $(x, a), (y, b) \in R(+)M$ be distinct. Then there is a path $x - s_1 - \cdots - s_n - y$ from $x$ to $y$ in $T(\Gamma(R))$. Since $Z(R)(+)M \subseteq Z(R(+)M)$, we have that $(x, a) - (s_1, 0) - \cdots - (s_n, 0) - (y, b)$ is a path from $(x, a)$ to $(y, b)$ in $T(\Gamma(R(+)M))$ (if $x = y$, then use the path $(x, a) - (-x, 0) - (y, b)$). Thus $T(\Gamma(R(+)M))$ is connected and $diam(T(\Gamma(R(+)M))) \leq diam(T(\Gamma(R)))$. (Observe that the hypothesis that $Z(R(+)M) = Z(R)(+)M$ is not needed in this direction.)

(2) This follows directly from the proof of part (1) above. □

In view of the (proof of the) above theorem, we have the following corollary.

**Corollary 7.3.12.** ([67, Corollary 3.17]) *Let $R$ be a commutative ring with $Z(R)$ not an ideal of $R$ and $M$ an $R$-module. If $T(\Gamma(R))$ is connected, then $T(\Gamma(R(+)M))$ is connected with $diam(T(\Gamma(R(+)M))) \leq diam(T(\Gamma(R)))$.*

The following is an example of a commutative ring $R$ with $Z(R)$ not an ideal of $R$, both $T(\Gamma(R))$ and $T(\Gamma(R(+)M))$ are connected, but $diam(T(\Gamma(R))) < diam(T(R(+)M)))$. Thus the hypothesis that $Z(R(+)M) = Z(R)(+)M$ is needed in Theorem 7.3.11(2), and the inequality in Corollary 7.3.12 may be strict.

**Example 7.3.13.** ([67, Example 3.18]) Let $R = R_3$ be the ring constructed in Example 7.2.21, and let $M = T(R)/R$. Since $(x_2, 0), (x_2+1, 0) \in Z(R(+)M)$, we have $R(+)M = ((x_2, 0), (x_2+1, 0))$, and hence $diam(T(\Gamma(R(+)M))) = 2$ by Theorem 7.2.10. However, $diam(T(\Gamma(R))) = 3$ as in Example 7.2.21, and so $diam(T(\Gamma(R(+)M))) < diam(T(\Gamma(R)))$.

The girth of $T(\Gamma(R(+)M))$ and its subgraphs $Z(\Gamma(R(+)M))$ and $Reg(\Gamma(R(+)M))$ are investigated in the next result.

**Theorem 7.3.14.** ([67, Theorem 3.19]) *Let $R$ be a commutative ring and $M$ be a nonzero $R$-module.*

(1) $gr(Reg(\Gamma(R(+)M))) = 3, 4,$ *or* $\infty$. *In particular,* $gr(Reg(\Gamma(R(+)M))) \leq 4$ *if* $Reg(\Gamma(R(+)M))$ *contains a cycle;*

(2) *If* $|M| \geq 3$, *then* $gr(Z(\Gamma(R(+)M))) = gr(T(\Gamma(R(+)M))) = 3$ *and* $gr(Reg(\Gamma(R(+)M))) \leq 4$;

(3) *If* $|M| = |R| = 2$ *(i.e., if $R$ and $M$ are both isomorphic to $\mathbb{Z}_2$), then* $gr(Z(\Gamma(R(+)M))) = gr(Reg(\Gamma(R(+)M))) = g(T(\Gamma(R(+)M))) = \infty$;

(4) *If* $2 = |M| < |R|$ *(i.e., if $R \ncong \mathbb{Z}_2$ and $M \cong \mathbb{Z}_2$), then* $gr(Z(\Gamma(R(+)M))) = gr(T(\Gamma(R(+)M))) = 3$ *and* $gr(Reg(\Gamma(R(+)M))) = 3$ *or* $\infty$. *In particular,* $gr(Reg(\Gamma(R(+)M))) = 3$ *if* $Reg(\Gamma(R(+)M))$ *contains a cycle.*

*Proof.* (1) This follows directly from Theorems 7.2.3 and 7.2.24(4).

(2) Let $a$ and $b$ be distinct nonzero elements of $M$. Then the 3-cycle $(0,0) - (0,a) - (0,b) - (0,0)$ shows that $gr(Z(\Gamma(R(+)M))) = gr(T(\Gamma(R(+)M))) = 3$. If $char(R) = 2$, then $(1,0) - (1,a) - (1,b) - (1,0)$ is a 3-cycle; so $gr(Reg(\Gamma(R(+)M))) = 3$. If $char(R) \neq 2$, then $(1,0) - (-1,0) - (1,a) - (-1,a) - (1,0)$ is a 4-cycle; so $gr(Reg(\Gamma(R))) \leq 4$.

(3) This is easy to check.

(4) Let $M = \{0, m\}$ and $I = Ann_R(m)$. Then $R/I \cong M \cong \mathbb{Z}_2$. Note that $I$ is a maximal ideal of $R$ and $I(+)M \subsetneq Z(R(+)M)$. Let $0 \neq r \in I$. Then the 3-cycle $(0,0) - (r,0) - (0,m) - (0,0)$ shows that $gr(Z(\Gamma(R(+)M))) = gr(T(\Gamma(R(+)M))) = 3$. Suppose that $Reg(\Gamma(R(+)M))$ contains a cycle $C$. Then $C$ contains three distinct vertices $x = (r_1, a), y = (r_2, b), z = (r_3, c) \in Reg(R(+)M)$. Since $r_1, r_2, r_3 \notin I$, we have $r_1 + I = r_2 + I = r_3 + I = 1 + I$, and thus $r_1 + r_2, r_2 + r_3, r_3 + r_1 \in I \subsetneq Z(R)$. Hence $x + y, y + z, z + x \in Z(R)(+)M \subseteq Z(R(+)M)$; so $x - y - z - x$ is a 3-cycle in $Reg(\Gamma(R(+)M))$. Thus $gr(Reg(\Gamma(R(+)M))) = 3$. □

The following example shows that, unlike the case for the diameter in Theorem 7.3.11, we can have both $T(\Gamma(R))$ and $T(\Gamma(R(+)M))$ connected and $Z(R(+)M) = Z(R)(+)M$, but $gr(T(\Gamma(R))) \neq gr(T(\Gamma(R(+)M)))$ (the inequality $gr(T(\Gamma(R))) \geq gr(T(\Gamma(R(+)M)))$ always holds). We also give examples to illustrate the possible values for $gr(Reg(\Gamma(R(+)M)))$ in parts (2) and (4) of Theorem 7.3.14.

**Example 7.3.15.** ([67, Example 3.20]) (a) Let $R = M = \mathbb{Z}_2 \times \mathbb{Z}_2$. Then $Z(R)$ is not an ideal of $R$, $T(\Gamma(R))$ and $T(\Gamma(R(+)M))$ are both connected, and $Z(R(+)M) = Z(R)(+)M$. However, $gr(T(\Gamma(R))) \neq gr(T(\Gamma(R(+)M)))$ since $gr(T(\Gamma(R(+)M))) = 3$ by Theorem 7.3.14(2) and $gr(T(\Gamma(R))) = 4$.

(b) It is clear that $gr(Z(\Gamma(R(+)M))) \leq gr(Z(\Gamma(R)))$ and $gr(Reg(\Gamma(R(+)M))) \leq gr(Reg(\Gamma(R)))$. However, both inequalities may be strict, even if $Z(R(+)M) = Z(R)(+)M$. For example, if $R = M = \mathbb{Z}_3$; then $gr(Z(\Gamma(R))) = gr(Reg(\Gamma(R))) = \infty$, $gr(Z(\Gamma(R(+)M))) = 3$, and $gr(Reg(\Gamma(R(+)M))) = 4$.

(c) If $|M| \geq 3$, then $gr(Reg(\Gamma(R(+)M))) = 3$ or $4$ by Theorem 7.3.14(2). Both values are possible. For example, we have $gr(Reg(\Gamma(\mathbb{Z}_3(+)\mathbb{Z}_3))) = 4$ and $gr(Reg(\Gamma(\mathbb{Z}_2(+)\mathbb{F}_4))) = 3$.

(d) If $2 = |M| < |R|$, by Theorem 7.3.14(4), $gr(Reg(\Gamma(T(+)M))) = 3$ or $\infty$. Both values are possible. For example, we have $gr(Reg(\Gamma((\mathbb{Z}_2 \times \mathbb{Z}_2)(+)\mathbb{Z}_2))) = \infty$ and $gr(Reg(\Gamma(\mathbb{Z}_4(+)\mathbb{Z}_2))) = 3$.

Let $D$ be a PID, $m \in D$ a nonzero nonunit, and $R = D(+)(D/mD)$. Then $Z(R)$ is not an ideal of $R$ if and only if $m = q_1^{m_1} \cdots q_n^{m_n}$, where the $q_i$'s are distinct nonassociate primes of $D$, $n \geq 2$, and every $m_i \geq 1$.

**Theorem 7.3.16.** ([67, Theorem 3.21]) *Let $D$ be a PID (e.g., $\mathbb{Z}$) and $m = q_1^{m_1} \cdots q_n^{m_n}$, where the $q_i$'s are distinct nonassociate primes of $D$, $n \geq 2$, and every $m_i \geq 1$, and let $R = D(+)/(D/mD)$.*

(1) *$Z(R) = ((q_1) \cup \cdots \cup (q_n))(+)(D/mD)$ is not an ideal of $R$;*

(2) *$Reg(\Gamma(R))$ is connected with $diam(Reg(\Gamma(R))) = 2$ and $gr(Reg(\Gamma(R))) = 3$;*

(3) *$Z(\Gamma(R))$ is connected with $diam(Z(\Gamma(R))) = 2$ and $gr(Z(\Gamma(R))) = 3$;*

(4) *$T(\Gamma(R))$ is connected with $diam(T(\Gamma(R))) = 2$ and $gr(T(\Gamma(R))) = 3$.*

*Proof.* (1) This is clear.

(2) Let $x = (r_1, a), y = (r_2, b) \in Reg(R)$ with $x+y \notin Z(R)$. Then $r_1, r_2 \notin (q_i)$ for every integer $i$ with $1 \leq i \leq n$ by part (1) above. By the Chinese Remainder Theorem, there is a $z \in D$ such that $z + r_1 \in (q_1), z + r_2 \in (q_2), \ldots, z + r_2 \in (q_n)$. By construction, $z \notin (q_i)$ for every integer $i$ with $1 \leq i \leq n$, and hence $(z, a) \in Reg(R)$ by part (1) above. Thus $x - (z, a) - y$ is a path from $x$ to $y$ in $Reg(\Gamma(R))$ of length 2; so $diam(Reg(\Gamma(R))) = 2$. Now, let $d = (m + 1, 0), c = (-1, 0) \in R$. Clearly $d, c \in Reg(R)$ and $c + d \in Z(R)$ by part (1) above. Again, by the Chinese Remainder Theorem, there is a $w \in D$ with $w + (m + 1) \in (q_1), w - 1 \in (q_2), \ldots, w-1 \in (q_n)$. By construction, $w \notin (q_i)$ for every integer $i$ with $1 \leq i \leq n$. Thus $(w, 0) \in Reg(R)$ by part (1) above, and hence $d - (w, 0) - c - d$ is a 3-cycle in $Reg(\Gamma(R))$; so $gr(Reg(\Gamma(R))) = 3$.

(3) $Z(\Gamma(R))$ is connected with $diam(Z(\Gamma(R))) = 2$ by Theorem 7.2.8(1), and $gr(Z(\Gamma(R))) = 3$ by Theorem 7.3.14(2) since $|D/mD| \geq 3$.

(4) Since $(q_1, 0), (q_2, 0) \in Z(R)$ and $R = ((q_1, 0), (q_2, 0))$, we have that $T(\Gamma(R))$ is connected with $diam(T(\Gamma(R))) = 2$ by Theorem 7.2.10. By part (3) above, $gr(Z(\Gamma(R))) = 3$, and hence $gr(T(\Gamma(R))) = 3$. □

## 7.4 Properties of Total Graphs

In this section, we study the Euler, Hamiltonian, coloring, connectivity, and domination properties of the total graph of a commutative ring. We start with the Eulerian and Hamiltonian nature of the total graphs.

### 7.4.1 Eulerian and Hamiltonian Graphs

Recall that a graph $G$ is said to be *Eulerian* if and only if $G$ is connected and every vertex in $G$ is of even degree; and a graph $G$ is said to be *Hamiltonian* if $G$ has a cycle, say $C$, such that each vertex of $G$ appears exactly once in $C$ (except the initial vertex of $C$, since the initial vertex of a cycle is also the terminal vertex).

Let $G_1 = (V_1, E_1)$ and $G_2 = (V_2, E_2)$ be two graphs with disjoint vertex sets $V_i$ and edge sets $E_i$. The *Cartesian product* of $G_1$ and $G_2$, denoted by $G = G_1 \Box G_2$, has vertex set $V_1 \times V_2$, and $(x, y)$ is adjacent to $(x', y')$ if $x = x'$ and $y$ is adjacent $y'$ in $G_2$, or $y = y'$ and $x$ is adjacent to $x'$ in $G_1$.

**Remark 7.4.1.** Note that if $R$ is infinite and not an integral domain, then $Z(R)$ is infinite, see Theorem 2.2.1. Therefore, $R$ is infinite if and only if $deg_{T(\Gamma(R))}(v) = \infty$ for every $v \in V(T(\Gamma(R)))$.

The following result is an analogue of Remark 7.4.1.

**Theorem 7.4.2.** ([36, Theorem 2]) *Let $R$ be a commutative ring and $I_i$ be proper ideals of $R$ for $i = 1, \ldots, n$. If $R \setminus \bigcup_{i=1}^{n} I_i$ is finite, then $R$ is finite.*

**Corollary 7.4.3.** ([36, Corollary 2]) *Let $R$ be a commutative Noetherian ring. If $Reg(R)$ is finite, then $R$ is finite.*

**Remark 7.4.4.** If in Theorem 7.4.2, the number of ideals is infinite, then the assertion is not true. To see this, consider the direct product of infinitely many $\mathbb{Z}_2$. The set of zero-divisors of this ring is a union of prime ideals (it is well known that the set of zero-divisors of every commutative ring is a union of a family of prime ideals), but it has just one regular element.

**Lemma 7.4.5.** ([36, Lemma 1]) *Let $R$ be a commutative Noetherian ring with $|Reg(R)| \leqslant 2$. Then $R$ is isomorphic to one of the following rings:*

$$\mathbb{Z}_3,\ \mathbb{Z}_4,\ \mathbb{Z}_2[x]/(x^2),\ \mathbb{Z}_2{}^r,\ \mathbb{Z}_2{}^r \times \mathbb{Z}_3,\ \mathbb{Z}_2{}^r \times \mathbb{Z}_4,\ \mathbb{Z}_2{}^r \times \mathbb{Z}_2[x]/(x^2),$$

*where $r$ is a positive integer.*

Observe that if $R_1$ and $R_2$ are two commutative rings with $R_1 \cong R_2$, then $T(\Gamma(R_1)) \cong T(\Gamma(R_2))$. However, the converse is not true. Let $R_1 = \mathbb{Z}_4$ and $R_2 = \mathbb{Z}_2[x]/(x^2)$. Then $T(\Gamma(\mathbb{Z}_4)) \cong T(\Gamma(\mathbb{Z}_2[x]/(x^2))) = 2K_2$, but $\mathbb{Z}_4 \ncong \mathbb{Z}_2[x]/(x^2)$.

**Lemma 7.4.6.** ([277, Lemma 1.1]) *Let $R$ be a finite commutative ring and let $x$ a vertex of $T(\Gamma(R))$. Then either $deg(x) = |Z(R)|$ or $deg(x) = |Z(R)| - 1$. In particular, if $2 \in Z(R)$, then $T(\Gamma(R))$ is a $(|Z(R)| - 1)$-regular graph.*

*Proof.* If $x$ is adjacent to $y$, then $x + y = a \in Z(R)$, and hence $y = a - x$ for some $a \in Z(R)$. We have two cases:
*Case 1.* Suppose $2x \in Z(R)$. Then $x$ is adjacent to $a-x$ for every $a \in Z(R)\setminus\{2x\}$. Thus the degree of $x$ is $|Z(R)| - 1$. In particular, if $2 \in Z(R)$, then $T(\Gamma(R))$ is a $(|Z(R)| - 1)$-regular graph.
*Case 2.* Suppose $2x \notin Z(R)$. Then $x$ is adjacent to $a - x$ for every $a \in Z(R)$. Thus the degree of $x$ is $|Z(R)|$. □

The following result is a consequence of Lemma 7.4.6.

**Theorem 7.4.7.** ([100, Lemma 5.1(i)], [348, Lemma 3.1]) *Let $R$ be a finite commutative ring with $Z(R)$ not an ideal of $R$. Then $T(\Gamma(R))$ is Eulerian if and only if $2 \in Z(R)$ and $|Z(R)|$ is odd.*

**Corollary 7.4.8.** *$T(\Gamma(\mathbb{Z}_n))$ is not Eulerian for $n > 2$.*

*Proof.* Let $R = \mathbb{Z}_n$. If $Z(R)$ is an ideal, then $T(\Gamma(R))$ is not connected, and hence it is not Eulerian. Thus assume that $Z(R)$ is not an ideal of $R$ and $T(\Gamma(R))$ is Eulerian. Then $2 \in Z(R)$ and $|Z(R)|$ is odd by Theorem 7.4.7. Since $2 \in Z(R)$, $n$ must be an even integer. Since $|Z(R)| = n - \varphi(n)$, where $\varphi(n) = |\{a \in \mathbb{Z}_n \mid gcd(a, n) = 1\}|$, and $\varphi(n)$ is even, we have $|Z(R)|$ is even, a contradiction. Thus $T(\Gamma(R))$ is not Eulerian. □

**Lemma 7.4.9.** *Let $R$ be a finite commutative ring with $2 \in Z(R)$ and $|Z(R)|$ odd. Then $char(R) = 2$ (i.e., $1 + 1 = 0$ in $R$).*

*Proof.* Since $2 \in Z(R)$, $2w = 0$ for some $0 \neq w \in Z(R)$. Hence $|w| = 2$ in the group $(R, +)$. Thus $2 \mid |R|$. Hence $|R|$ is even. Since $|Z(R)|$ is odd, we have $|U(R)| = |R| - |Z(R)|$ is odd. Since the group $(U(R), \cdot)$ has an odd number of elements, there exists a $v \in U(R)$ such that $v = v^{-1}$. Hence $2v = 0$, and thus $2 = 1 + 1 = 0$ in $R$. □

**Theorem 7.4.10.** ([348, Theorem 3.3]) *Let $R$ be a finite commutative ring. Then $T(\Gamma(R))$ is Eulerian if and only if $R$ is isomorphic to a direct sum of two or more finite fields of even orders, i.e., $R \cong \bigoplus_{i=1}^{k} \mathbb{F}_{2^{t_i}}$ for some $k \geq 2$.*

*Proof.* If $R \cong \bigoplus_{i=1}^{k} \mathbb{F}_{2^{t_i}}$ for some $k \geq 2$, then $T(\Gamma(R))$ is Eulerian by Theorem 7.4.7 and Lemma 7.4.9.

Conversely, suppose that $T(\Gamma(R))$ is Eulerian. Then $Z(R)$ is not an ideal of $R$ since $T(\Gamma(R))$ is connected. Hence $R \cong R_1 \oplus \cdots \oplus R_k$, where $k \geq 2$ and every $R_i$ is a local ring. Meanwhile, by Lemma 7.4.9, we have $1+1=0$, which means that $1_{R_i}+1_{R_i}=0$ for every $i=1,\ldots,k$. Hence, by Lemma 7.4.9, it can be inferred that the $|R_i|$'s are all even. If $|Z(R_i)| \geq 2$ for some $i=1,\ldots,k$, then $2||Z(R_i)|$, and hence $|U(R_i)| = |R_i| - |Z(R_i)|$ is even. Therefore, $|U(R)| = |U(R_1)|\cdots|U(R_k)|$ is also even, i.e., $|Z(R)|$ is even, which is a contradiction because $T(\Gamma(R))$ is an Eulerian graph and by Theorem 7.4.7, $|Z(R)|$ must be odd. Thus, we must have $|Z(R_i)| = 1$ for every $i=1,\ldots,k$. Therefore, the $R_i$'s are all finite fields of even orders, i.e., $R \cong \bigoplus_{i=1}^{k} \mathbb{F}_{2^{t_i}}$. □

The study of the Hamiltonian nature of the total graph of a finite commutative ring was initiated by Akbari et al. in [36].

**Lemma 7.4.11.** ([36, Lemma 2]) *Let $(R_1, \mathfrak{m}_1)$ and $(R_2, \mathfrak{m}_2)$ be two finite commutative local rings with char$(R_1/\mathfrak{m}_1) \neq 2$ and char$(R_2/\mathfrak{m}_2) \neq 2$. Then $T(\Gamma(R_1 \times R_2))$ and Reg$(\Gamma(R_1 \times R_2))$ are Hamiltonian graphs.*

**Lemma 7.4.12.** ([36, Lemma 3]) *Let $(R_1, \mathfrak{m}_1)$ and $(R_2, \mathfrak{m}_2)$ be two finite commutative local rings with char$(R_1/\mathfrak{m}_1) = 2$ and char$(R_2/\mathfrak{m}_2) \neq 2$. Then the following hold:*

(i) *If $R_1 \times R_2$ is not isomorphic to $\mathbb{Z}_6$, then Reg$(\Gamma(R_1 \times R_2))$ is a Hamiltonian graph;*

(ii) *$T(\Gamma(R_1 \times R_2))$ is a Hamiltonian graph.*

**Lemma 7.4.13.** ([36, Lemma 4]) *Let $(R_1, \mathfrak{m}_1)$ and $(R_2, \mathfrak{m}_2)$ be two finite commutative local rings with char$(R_1/\mathfrak{m}_1) = 2$ and char$(R_2/\mathfrak{m}_2) = 2$. Then the following hold:*

(i) *If $R_1 \times R_2$ is not isomorphic to any of the following rings, $\mathbb{Z}_2 \times \mathbb{Z}_2$, $\mathbb{Z}_2 \times \mathbb{Z}_4$, $\mathbb{Z}_2 \times \mathbb{Z}_2[x]/(x^2)$, then Reg$(\Gamma(R_1 \times R_2))$ is a Hamiltonian graph;*

(ii) *$T(\Gamma(R_1 \times R_2))$ is a Hamiltonian graph.*

**Lemma 7.4.14.** ([36, Lemma 5]) *Let $R_1$ and $R_2$ be two finite commutative rings. Then the following hold:*

(i) *If $T(\Gamma(R_1))$ is a Hamiltonian graph, then $T(\Gamma(R_1 \times R_2))$ is a Hamiltonian graph;*

(ii) *If Reg$(\Gamma(R_1))$ is a Hamiltonian graph, then Reg$(\Gamma(R_1 \times R_2))$ is a Hamiltonian graph.*

**Theorem 7.4.15.** ([36, Theorem 3]) *Let $R$ be a finite commutative ring with $Z(R)$ not an ideal. Then the following hold:*

(i) $T(\Gamma(R))$ *is a Hamiltonian graph;*

(ii) $Reg(\Gamma(R))$ *is a Hamiltonian graph if and only if* $R$ *is not isomorphic to any of the rings:* $\mathbb{Z}_2^{n+1}, \mathbb{Z}_2^n \times \mathbb{Z}_3, \mathbb{Z}_2^n \times \mathbb{Z}_4, \mathbb{Z}_2^n \times \mathbb{Z}_2[x](x^2)$*, where* $n$ *is a positive integer.*

*Proof.* (i) Let $R \cong R_1 \times R_2 \times \cdots \times R_n$, where $R_i$ is a local ring ($1 \leqslant i \leqslant n$). Since $Z(R)$ is not an ideal, then $n \geqslant 2$. If $n = 2$, then the result holds by Lemmas 7.4.11, 7.4.12, and 7.4.13. If $n > 2$, then set $R' = R_3 \times \cdots \times R_n$. Thus $T(\Gamma((R_1 \times R_2) \times R'))$ is a Hamiltonian graph by Lemma 7.4.14.

(ii) We have $Reg(R) \cong Reg(R_1) \times Reg(R_2) \times \cdots \times Reg(R_n)$, where $R_i$ is a local ring for $i = 1, \ldots, n$. Since $Z(R)$ is not an ideal, then $n \geqslant 2$. If $n = 2$, then the result holds by Lemmas 7.4.11, 7.4.12, and 7.4.13. If $n > 2$, then set $R' = R_3 \times \cdots \times R_n$. Then $Reg(\Gamma((R_1 \times R_2) \times R'))$ is a Hamiltonian graph by Lemma 7.4.14.

Conversely, if $R$ isomorphic to one of the rings given in the statement of theorem, then $|Reg(R)| \leq 2$ and so $Reg(\Gamma(R))$ is not a Hamiltonian graph. □

**Remark 7.4.16.** ([36, Remark 2]) If $R$ is a finite commutative local ring, then $Z(R)$ is an ideal and $T(\Gamma(R))$ is not connected. Thus $T(\Gamma(R))$ is not a Hamiltonian graph.

In the following theorem, Tamizh Chelvam and Asir [100] proved Theorem 7.4.15 with a weaker assumption that $diam(T(\Gamma(R))) = 2$ (cf. Theorem 7.2.10).

**Theorem 7.4.17.** ([100, Theorem 5.2]) *Let* $R$ *be a commutative ring with* $diam(T(\Gamma(R))) = 2$ *(and hence* $T(\Gamma(R))$ *is connected). Then* $T(\Gamma(R))$ *is Hamiltonian.*

*Proof.* By the assumption that $diam(T(\Gamma(R))) = 2$, $T(\Gamma(R))$ is connected, and so by Theorem 7.2.10, $R = (z_1, z_2, \ldots, z_n)$ for some $z_1, z_2, \ldots, z_n \in Z(R)$. We claim that $R$ is generated by two elements $z_1, z_2 \in Z(R)$. If not, $n \geq 3$ and by Theorem 7.2.10, $diam(T(\Gamma(R))) \geq 3$, a contradiction. Let $P_i$ be the ideal generated by $z_i$ for $i = 1, 2$. Since $z_1, z_2$ generates $R$, we have $R = P_1 + P_2$.

Let $H$ be the spanning subgraph of $T(\Gamma(R))$ in which two distinct vertices $x, y \in R$ are adjacent if $x + y \in P_1$. By noting that $P_1$ is an ideal contained in $Z(R)$, we obtain in the same way as in Theorem 7.2.1 that $H$ is a subgraph of $T(\Gamma(R))$. Further

(i) If $2 \in P_1$, then $H$ is the union of disjoint complete graphs.

(ii) If $2 \notin P_1$, then $H$ is the union of one complete graph together with disjoint complete bipartite graphs.

**Claim 1.** In $T(\Gamma(R))$, if a vertex in a coset $a + P_1$ is adjacent to some vertex in another coset $b + P_1$, then every vertex in $a + P_1$ is adjacent to at least one vertex in $b + P_1$.

Let $x \in a + P_1$ and $y \in b + P_1$ such that $x$ and $y$ are adjacent in $T(\Gamma(R))$. Then $x + y = z \in Z(R)$, and so $z \in a + b + P_1$. Let $x_1 \in a + P_1$. Take $y_1 \in R$ with $x_1 + y_1 = z$. Thus $x_1 \in a + P_1$ is adjacent to $y_1 \in b + P_1$.

**Claim 2.** In $T(\Gamma(R))$, every vertex in any coset of $P_1$ in $R$ is adjacent to at least one vertex in another coset of $P_1$ in $R$.

Since $R = (z_1, z_2)$, $z_1 \in P_1$, and $z_2 \in P_2$, there are $x \in P_1$ and $y \in P_2$ with $x + y = 1$. Define a map $f : P_2 \to R/P_1$ by $f(s) = s + P_1$ for $s \in P_2$. Clearly $f$ is a ring homomorphism, $ker(f) = P_1 \cap P_2$, and $f(y) = 1 + P_1$. Also, for every nonzero $r + P_1 \in R/P_1$, $f(yr) = r + P_1$, and so $f$ is surjective. Hence $P_2/P_1 \cap P_2 \cong R/P_1$. From this, every coset of $P_1$ in $R$ contains at least one element of $P_2$. Since $\langle P_2 \rangle \subseteq T(\Gamma(R))$ is complete and by Claim 1, every vertex in any coset $a + P_1$ is adjacent to at least one vertex in another coset $b + P_1$.

Let $R/P_1 = \{P_1, x_1 + P_1, x_2 + P_1, \ldots\}$ and $z_i \in (x_i + P_1) \cap P_2$ for $i \geq 1$. Start with the vertex $0 \in P_1$, traverse all the vertices of $P_1$, and pass to the next component of $H$, as this is possible by Claim 2. Continuing in this way, we get a Hamiltonian path which ends at $z_\ell$ in the last coset of $P_1$ in $R$ ($\ell$ may be infinite). By taking the edge joining 0 and $z_\ell$ at the end of the path constructed so for, we get a Hamiltonian cycle in $T(\Gamma(R))$. □

Recall that every Artinian ring $R \cong R_1 \times \cdots \times R_m$, where every $R_i$ is a local ring and $Reg(R) = Reg(R_1) \times \cdots \times Reg(R_m)$. From this, if $R$ is a nonlocal Artinian ring, then $((0, \ldots, 0), (0, -1, \ldots, -1), (1, \ldots, 1)) = P_3 \subseteq T(\Gamma(R))$. Note that if $Z(R)$ is not an ideal of $R$ and $T(\Gamma(R))$ is connected, then by Corollary 7.2.11, $diam(T(\Gamma(R))) = d(0, 1)$. Ultimately, if $R$ is an Artinian ring with $Z(R)$ not an ideal of $R$, then $diam(T(\Gamma(R))) = d(0, 1) = 2$. Hence we have the following result.

**Corollary 7.4.18.** ([100, Corollary 5.3]) *Let $R$ is a commutative Artinian ring with $Z(R)$ not an ideal of $R$. Then $T(\Gamma(R))$ is Hamiltonian.*

Let $R$ be a finite commutative ring such that $Z(R)$ is not an ideal of $R$. Then $T(\Gamma(R))$ is connected and $\text{diam}(T(\Gamma(R))) = 2$ by Theorem 7.2.10. Hence Theorem 7.4.15(i) is just a consequence of Theorem 7.4.17 or Corollary 7.4.18.

**Corollary 7.4.19.** (see Theorem 7.4.15(i)) *If $R$ is a finite commutative ring such that $Z(R)$ is not an ideal of $R$, then $T(\Gamma(R))$ is Hamiltonian.*

In the remainder of this section, we discuss results related to the degree of vertices in total graphs. Let $V(G)$ be the set of all vertices of a graph $G$, and let $v \in V(G)$. Then $deg(v)$ is the number of edges of $G$, where $v$ is a vertex in each one of them (i.e., the number of edges of $G$ incident with $v$), $\delta(G) = min\{deg(v) \mid v \in V(G)\}$, $\Delta(G) = max\{deg(v) \mid v \in V(G)\}$, $N(v) = \{u \in V(G) \mid u \text{ is adjacent to } v\}$, and $N[v] = N(v) \cup \{v\}$. Then $v$ is an *isolated* vertex of $G$ if every edge of $G$ is not incident with $v$, and $G$ is an *$m$-regular* graph if $deg(x) = m$ for every vertex $x$ of $G$.

**Lemma 7.4.20.** ([100, Lemma 2.8]) *Let $R$ be a finite commutative ring. Then*

(i) *$T(\Gamma(R))$ has no vertex of degree $|R|-1$;*

(ii) *If $R$ is not an integral domain, then $T(\Gamma(R))$ has no isolated vertices.*

*Proof.* (i) Since $1 \notin Z(R)$, $|Z(R)| \le |R|-1$. If $2 \in Z(R)$, then by Lemma 7.4.6 $deg(v) = |Z(R)|-1 \le |R|-2$ for every $v \in T(\Gamma(R))$. If $2 \notin Z(R)$, then by the fact that $1 \neq 0$, we have $|Z(R)| \le |R|-2$, and so by Lemma 7.4.6, $deg(v) \le |Z(R)| \le |R|-2$ for every $v \in T(\Gamma(R))$. Hence in either case, $T(\Gamma(R))$ has no vertex of degree $|R|-1$.

(ii) This follows directly from Lemma 7.4.6 since in this case $|Z(R)| \ge 2$. □

**Lemma 7.4.21.** ([100, Lemma 2.9]) *Let $R$ be a finite commutative ring. Then $T(\Gamma(R))$ contains an isolated vertex $a \in R$ if and only if $R$ is a field and $2a = 0$ (i.e., $a = -a$.)*

*Proof.* Let $R$ be a field and $a = -a$ for some $a \in R$. Since $Z(R) = \{0\}$ and $T(\Gamma(R))$ has no loops, $a$ is an isolated vertex in $T(\Gamma(R))$.

Conversely, assume that $a \in R$ is an isolated vertex in $T(\Gamma(R))$ and $R$ is not a field. Then by Lemma 7.4.20(ii), $T(\Gamma(R))$ has no isolated vertices, a contradiction. Suppose $a \neq -a$. Then $a + (-a) = 0$ and $Z(R) = \{0\}$; so $deg(a) = 1$, a contradiction. □

**Corollary 7.4.22.** ([100, Corollary 2.10]) *Let $R$ be a finite commutative ring.*

(i) *$T(\Gamma(R))$ is totally disconnected if and only if $R \simeq \mathbb{F}_{2^k}$ for some $k \in \mathbb{N}$;*

(ii) *The subgraph induced by $R^*$ in $T(\Gamma(R))$ is 1-regular if and only if $R$ is a field and $a \neq -a$ for all $a \in R^*$.*

**Remark 7.4.23.** ([100, Remark 2.11]) In view of Lemma 7.4.20 and Lemma 7.4.21, if $R$ is a finite commutative ring which is not an integral domain, then $1 \le deg(v) \le |R|-2$ for every $v \in V(T(\Gamma(R)))$. Also this bound is sharp. For if $R = \mathbb{Z}_4$, then $deg(v) = 1$ for every $v \in R$, whereas if $R = \mathbb{Z}_2 \times \mathbb{Z}_2$, then $deg(v) = |R|-2$ for every $v \in R$.

In view of the above remark, we characterize rings for which the degree of vertices attains either the lower bound or the upper bound.

**Proposition 7.4.24.** ([100, Proposition 2.12]) *Let $R$ be a finite commutative ring that is not a field.*

(i) *$deg(v) = 1$ for some $v \in V(T(\Gamma(R)))$ if and only if $R \cong \mathbb{Z}_4$ or $\mathbb{Z}_2[x]/(x^2)$;*

(ii) *$deg(v) = |R|-2$ for some $v \in V(T(\Gamma(R)))$ if and only if $R \cong \mathbb{Z}_2^m$, where $m$ is a positive integer.*

*Proof.* (i) If $deg(v) = 1$ for some $v \in V(T(\Gamma(R)))$, then $|Z(R)| \leq 2$ by Lemma 7.4.6, and so either $R \cong \mathbb{Z}_4$ or $\mathbb{Z}_2[x]/(x^2)$. The other part is trivial.

(ii) If $deg(v) = |R| - 2$ for some $v \in V(T(\Gamma(R)))$, then $|Z(R)| = |R| - 1$ or $|R| - 2$ by Lemma 7.4.6, and so $Reg(R) \leq 2$. By Lemma 7.4.5, if $Reg(R) \leq 2$, then $R$ is isomorphic to one of the following rings:

$$\mathbb{Z}_3,\ \mathbb{Z}_4,\ \mathbb{Z}_2[x]/(x^2),\ \mathbb{Z}_2^r,\ \mathbb{Z}_2^r \times \mathbb{Z}_3,\ \mathbb{Z}_2^r \times \mathbb{Z}_4,\ \mathbb{Z}_2^r \times \mathbb{Z}_2[x]/(x^2),$$

where $r$ is a positive integer. By the general assumption of the statement, $R$ cannot be $\mathbb{Z}_3$. Note that for the rings $\mathbb{Z}_4$ and $\mathbb{Z}_2[x]/(x^2)$, $deg(v) = 1$ for every $v \in V(T(\Gamma(R)))$. If $R = \mathbb{Z}_2^r$, then $|Reg(R)| = 1$ and $2 \in Z(R)$. From this, $deg(v) = |R| - 2$ for every $v \in R$. For the rings $\mathbb{Z}_2^r \times \mathbb{Z}_3, \mathbb{Z}_2^r \times \mathbb{Z}_4, \mathbb{Z}_2^r \times \mathbb{Z}_2[x]/(x^2)$, $|Z(R)| = |R| - 2$ and $2 \in Z(R)$. These together imply that $deg(v) = |R| - 3$ for all $v \in R$. All these observations show that if $deg(v) = |R| - 2$, then $R \cong \mathbb{Z}_2^r$. □

**Corollary 7.4.25.** ([100, Corollary 2.13]) *Let $R$ be a finite commutative ring that is not a field and $R \ncong \mathbb{Z}_4$, $\mathbb{Z}_2[x]/(x^2)$, $\mathbb{Z}_2^r$ for some $r \in \mathbb{N}$. Then $2 \leq deg(v) \leq |R| - 3$ for every $v \in V(T(\Gamma(R)))$.*

If $R = \mathbb{Z}_9$, then $deg(v) = 2$ for every $v \in Z(R)$. If $R = \mathbb{Z}_6$, then $deg(v) = 3 = |R| - 3$ for every $v \in R$. These examples show that the bounds given in Corollary 7.4.25 are sharp.

**Lemma 7.4.26.** ([100, Lemma 2.15]) *Let $R$ be a finite commutative ring and let $G = T(\Gamma(R))$. If $a$ and $b$ are associates in $R$, then $deg_G(a) = deg_G(b)$.*

*Proof.* Since $a$ and $b$ are associates in $R$, there is a unit $u \in R$ such that $a = ub$. If $b \in Z(R)$, then $a = ub \in Z(R)$. If $b \notin Z(R)$, then $a \notin Z(R)$. By Lemma 7.4.6, in either case, $deg(a) = deg(b)$. □

Recall that the eccentricity of a vertex $v$ in $G$ is defined by

$$e(v) = max\{d(v, y) \mid y \in V(G)\}$$

and the radius of $G$ is given by

$$r(G) = min\{e(x) \mid x \in V(G)\}.$$

A graph $G$ is said to be *self-centered* if $e(x) = r(G)$ for every $x \in V(G)$.

**Proposition 7.4.27.** ([100, Proposition 2.16]) *Let $R$ be a finite commutative ring. Then $T(\Gamma(R))$ is self-centered if and only if $Z(R)$ is not an ideal of $R$.*

*Proof.* Suppose $Z(R)$ is not an ideal of $R$. By Theorem 7.2.10, $diam(T(\Gamma(R))) = 2$. Now by Lemma 7.4.20, $T(\Gamma(R))$ has no vertex of degree $|R| - 1$, and so $e(v) = 2$ for every $v \in R$, i.e., $T(\Gamma(R))$ is self-centered. Conversely, suppose that $Z(R)$ is an ideal of $R$. Then by Theorem 7.2.1, $T(\Gamma(R))$ is disconnected, which is a contradiction to $T(\Gamma(R))$ is self-centered. □

### 7.4.2 Colorings

Now we deal with the coloring concepts in the case of the total graph of commutative rings. Recall that a *clique* in a graph $G$ is a subset of pairwise adjacent vertices and the supremum of the size of cliques in $G$, denoted by $\omega(G)$, is called the clique number of $G$. By $\chi(G)$, we denote the *chromatic number* of $G$, i.e., the minimum number of colors which can be assigned to the vertices of $G$ in such a way that every two adjacent vertices have different colors. A coloring of the vertices such that any two adjacent vertices have different colors is called a *proper vertex coloring*.

Suppose that $R$ is an infinite commutative ring which is not a domain. Then for every $0 \neq x \in Z(R)$, $R/Ann(x) \cong Rx$ as $R$-modules. Since both $Rx$ and $Ann(x)$ form a clique for $Z(\Gamma(R))$, we deduce that $Z(\Gamma(R))$ has an infinite clique and so $T(\Gamma(R))$ does. Hence, to determine the clique number and the chromatic number of $T(\Gamma(R))$, we assume that $R$ is a finite ring.

First of all, we list the characterization of rings $R$ whose regular graphs $Reg(\Gamma(R))$ are complete (cf. Theorem 7.2.2(1)).

**Theorem 7.4.28.** ([31, Theorem 1]) *Let $R = \prod_{i\in I} R_i$, where $R_i$ is a commutative ring and $Z(R_i)$ is an ideal of $R_i$ for every $i \in I$. Then the following hold:*

(i) *If $2 \in Z(R)$, then $Reg(\Gamma(R))$ is a complete graph if and only if $R_i/Z(R_i) \cong \mathbb{Z}_2$ for some $i \in I$;*

(ii) *If $2 \notin Z(R)$, then $Reg(\Gamma(R))$ is a complete graph if and only if $R_i \cong \mathbb{Z}_3$ for every $i \in I$.*

**Theorem 7.4.29.** ([31, Theorem 2]) *Let $R$ be a reduced commutative Noetherian ring and $Z(R)$ be an ideal of $R$. If $Reg(\Gamma(R))$ is a complete graph, then $R \cong \mathbb{Z}_2$ or $R \cong \mathbb{Z}_3$.*

**Theorem 7.4.30.** ([31, Corollary 1]) *Let $R$ be a commutative semilocal ring with $2 \notin Z(R)$. If $Reg(\Gamma(R))$ is a complete graph, then $R \cong \mathbb{Z}_3^r$, where $r$ is the number of maximal ideals of $R$.*

Now, we list results concerning the chromatic number and the clique number of the regular graph of a commutative ring.

**Theorem 7.4.31.** ([31, Theorem 6]) *Let $R$ be a commutative ring with $2 \notin Z(R)$. If $Z(R) = \cup_{i=1}^{n} P_i$, where $P_1, \ldots, P_n$ are prime ideals of $R$ and $Z(R) \neq \cup_{i\neq j} P_i$ for $j = 1, \ldots, n$, then $\omega(Reg(\Gamma(R))) = \chi(Reg(\Gamma(R))) = 2^n$.*

**Theorem 7.4.32.** ([31, Theorem 7]) *Let $R$ be a commutative Noetherian ring with $2 \notin Z(R)$. If $Reg(R) = U(R)$, then $R$ is a semilocal ring and $\omega(Reg(\Gamma(R)))$ $= \chi(Reg(\Gamma(R))) = 2^n$, where $n$ is the number of maximal ideals of $R$. In particular, if $R$ is a commutative Artinian ring and $2 \notin Z(R)$, then $\omega(Reg(\Gamma(R))) = \chi(Reg(\Gamma(R))) = 2^n$, where $n$ is the number of maximal ideals of $R$.*

**Theorem 7.4.33.** ([31, Theorem 8]) *Let $R$ be a commutative Noetherian ring and $M$ a finitely generated $R$-module.*

*If $2 \notin Z(R) \cup Z(M)$, then $\omega(Reg(\Gamma(R(+)M))) = \chi(Reg(\Gamma(R(+)M))) \leq 2^{n+m}$, where $n$ is the number of prime ideals associated with 0 and $m$ is the number of associated prime ideals of $M$.*

Similarly to the previous results, Aalipoura and Akbari [1] obtained the clique number and chromatic number of $Reg(\Gamma(R))$ for a finite commutative ring $R$ with $2 \in Z(R)$.

**Theorem 7.4.34.** ([1, Theorem 20]) *Let $R$ be a finite commutative ring and $\mathfrak{m}$ a maximal ideal of $R$ of maximum size. If $R/\mathfrak{m}$ has characteristic 2, then*

$$\chi(Reg(\Gamma(R))) = \omega(Reg(\Gamma(R))) = \frac{|Reg(R)|}{|R/\mathfrak{m}| - 1}.$$

Now, we study the chromatic number of the regular graph of a commutative ring $R$ with $R/Nil(R)$ finite.

**Theorem 7.4.35.** ([31, Theorem 9]) *Let $R$ be a commutative ring with $2 \notin Z(R)$. If $R/Nil(R)$ is finite, then $\chi(Reg(\Gamma(R))) \leq |R/Nil(R)| - 1$.*

**Theorem 7.4.36.** ([31, Theorem 11]) *Let $R$ be a commutative Noetherian ring. If $R/Nil(R)$ is finite, then $R$ is finite.*

Next, we reduce the problem of coloring of the total graph of a finite commutative ring to the problem of coloring of the total graph of a finite reduced commutative ring.

**Theorem 7.4.37.** ([1, Lemma 8]) *If*

$$\chi(T(\Gamma(R))) = \omega(T(\Gamma(R))) = \begin{cases} 4 & \text{if } R \cong \mathbb{Z}_3 \times \mathbb{Z}_3; \\ \max\{|\mathfrak{m}| \mid \mathfrak{m} \in Max(R)\} & \text{otherwise} \end{cases}$$

*holds for every finite reduced ring, then it holds for every finite ring.*

**Theorem 7.4.38.** ([1, Lemma 12]) *Let $n \geq 2$ be a positive integer and $F_1, \dots, F_n$ be finite fields with $|F_1| \leq \cdots \leq |F_n|$. Suppose that one of the following conditions holds:*

(i) *The field $F_1$ has even characteristic;*

(ii) *Every $F_i$ has odd characteristic and $F_1 \times F_2 \ncong \mathbb{Z}_3 \times \mathbb{Z}_3$. Then*

$$\chi(T(\Gamma(F_1 \times \cdots \times F_n)) = |F_2| \cdots |F_n|.$$

**Theorem 7.4.39.** ([1, Theorem 13]) *Let $R$ be a finite commutative ring such that one of the following conditions holds:*

(i) *The residue field of $R$ of minimum size has even characteristic;*

(ii) *Every residue field of $R$ has odd characteristic and $R/J(R)$ has no summand isomorphic to $\mathbb{Z}_3 \times \mathbb{Z}_3$.*

*Then the following equalities hold:*

$\chi(T(\Gamma(R))) = \omega(T(\Gamma(R))) = \chi(Z(\Gamma(R))) = \omega(Z(\Gamma(R))) = \max\{|\mathfrak{m}| \mid \mathfrak{m} \in Max(R)\}$.

After proving some positive results, Aalipoura and Akbari [1] conjectured the following statement.

**Conjecture:** The following equalities hold for every finite commutative ring $R$:

$$\chi(T(\Gamma(R))) = \omega(T(\Gamma(R))) = \begin{cases} 4 & \text{if } R \cong \mathbb{Z}_3 \times \mathbb{Z}_3; \\ \max\{|\mathfrak{m}| \mid \mathfrak{m} \in Max(R)\} & \text{otherwise.} \end{cases}$$

A set of vertices in a graph $G$ is called *independent* if no two of its vertices are adjacent. The *independence number* of $G$, denoted by $\alpha(G)$, is the largest size of a set of independent vertices in $G$. Note that the independence number $\alpha(G) = \omega(CG)$, where $CG$ is the complement of $G$.

The following result gives the independence number of the total graph of a local ring.

**Theorem 7.4.40.** ([412, Lemma 4.1]) *Let $R$ be a finite local ring with maximal ideal $\mathfrak{m}$, $|\mathfrak{m}| = \lambda$, and $|R/\mathfrak{m}| = \mu$. If $2 \in Z(R)$, then $\alpha(T(\Gamma(R))) = \mu$. If $2 \notin Z(R)$, then $\alpha(T(\Gamma(R))) = \lambda\left(\frac{\mu-1}{2}\right) + 1$.*

Let $R$ be a finite commutative ring which is not local. Since the independence number of the total graph is the same as the clique number of the complement of the total graph, we note that "$x$ is not adjacent to $y$ in the total graph" is equivalent to "$x+y$ is a unit element of $R$". The clique number of the complement of total graphs was studied by Maimani et al. in [274].

Let $R$ be a commutative ring and $U(R)$ be the group of units of $R$. The *unit graph* of $R$, denoted by $G(R)$, is the graph obtained by letting all the elements of $R$ be the vertices and defining distinct vertices $x$ and $y$ to be adjacent if and only if $x + y \in U(R)$.

**Theorem 7.4.41.** ([274, Theorem 2.2]) *If $R$ is a commutative ring, then the unit graph $G(R)$ is weakly perfect.*

An explicit formula for the clique number of $G(R)$, i.e., independence number of $T(\Gamma(R))$, is given below (see [274, Remark 3.8]):

For a finite commutative ring $R$, we may write $R \cong R_1 \times \cdots \times R_n$, where every $R_i$ is a local ring with maximal ideal $\mathfrak{m}_i$. If $2 \in Z(R)$, by reordering, we may assume that $R_1, \dots, R_\ell$ all have characteristic equal to 2, $|R_1|/|\mathfrak{m}_1| \leq \cdots \leq |R_\ell|/|\mathfrak{m}_\ell|$, and $R_{\ell+1}, \dots, R_n$ all have characteristic not equal to 2. Then

$$\omega(G(R)) = \alpha(T(\Gamma(R))) = \begin{cases} |R_1|/|\mathfrak{m}_1| & \text{if } 2 \in Z(R); \\ \frac{1}{2^n}\prod_{i=1}^{n}(|R_i| - |\mathfrak{m}_i|) + n & \text{otherwise.} \end{cases}$$

### 7.4.3 Connectivity

Now we look into the connectivity of the total graph of a commutative ring. A *vertex-cut* of $G$ is a subset $S \subseteq V(G)$ such that $G - S$ is disconnected. An *edge-cut* of $G$ is a subset $T \subseteq E(G)$, such that the graph $G - T$ is disconnected. The *vertex-connectivity* of $G$ is defined by $\kappa(G) = min\{n \geq 0 \mid$ there is a vertex-cut $S \subseteq V(G)$ such that $|S| = n\}$ if $G$ has a finite vertex-cut, and $\kappa(G) = \infty$ otherwise. If $\kappa(G) = k$, then $G$ is called a *$k$-connected* graph. Note that a graph is $k$-connected if any two of its vertices can be joined by $k$ independent paths. Similarly, the *edge-connectivity* of $G$ is defined by: $\lambda(G) = min\{n \geq 0 \mid$ there exists a edge-cut $T \subseteq E(G)$ such that $|T| = n\}$ if $G$ has a finite edge-cut, and $\lambda(G) = \infty$ otherwise. Clearly, for any graph G, $\kappa(G) \leq \lambda(G) \leq \delta(G)$, where $\delta(G)$ is the minimum degree of $G$.

The following theorem due to Menger is used to obtain a lower bound for $\kappa(T(\Gamma(R_1 \times R_2)))$, where $R_1$ and $R_2$ are finite commutative rings.

**Theorem 7.4.42.** ([427, Corollary 4.2.19]) *Let $G$ be a graph. Then $\kappa(G) \geqslant k$ if and only if there are at least $k$ internally vertex disjoint $(x, y)$-paths for every $x, y \in V(G)$.*

**Theorem 7.4.43.** ([36, Theorem 4]) *Let $R = R_1 \times R_2$ be a finite commutative ring. Then $\kappa(T(\Gamma(R))) \geqslant |R_1| + |R_2| - 4$.*

*Proof.* Let $(a, b)$ and $(a', b')$ be two distinct elements of $R$. If $a \neq a'$, $b' \neq \pm b, \lambda \notin \{b, -b, b', -b'\}$, then consider the paths $(a, b), (-a, \lambda), (-a', -\lambda), (a', b')$ for $\lambda \in R_2$. If $\eta \in R_1$, $(a, b) \neq (\eta, -b)$ and $(a', b') \neq (-\eta, -b')$, then consider the paths $(a, b), (\eta, -b), (-\eta, -b'), (a', b')$. If $(a, b) \neq (\eta, -b)$ and $(a', b') = (-\eta, -b')$, then consider the paths $(a, b), (\eta, -b), (a', b')$. If $(a, b) = (\eta, -b)$ and $(a', b') \neq (-\eta, -b')$, then consider the paths $(a, b), (-\eta, -b'), (a', b')$. If $(a, b) = (\eta, -b)$ and $(a', b') = (-\eta, -b')$ for some $\eta$, then consider the paths $(a, b), (a', b')$ and $(a, b), (\eta, -b), (a', b')$ for $\eta \neq a$. So there are at least $|R_1| + |R_2| - 4$ disjoint paths joining $(a, b)$ and $(a', b')$.

Let $a \neq a'$, $b' \neq b$, and $b' = -b$. Note that the paths $(a, b), (-a, \lambda), (-a', -\lambda), (a', -b)$ for $\lambda \in R_2 \setminus \{\pm b\}$ and the paths $(a, b), (\eta, -b), (-\eta, b), (a', -b)$ for $\eta \in R_1 \setminus \{-a, a'\}$ are $|R_1| + |R_2| - 4$ disjoint paths.

Let $a \neq a'$ and $b = b'$. Consider the paths $(a, b), (-a, \lambda), (-a', -\lambda), (a', b)$, where $\lambda \in R_2 \setminus \{b, -b\}$ and $(a, b), (\eta, -b), (a', b)$ for $\eta \in R_1 \setminus \{a, a'\}$.

If $a = a'$, then $b \neq b'$ since $(a, b)$ and $(a', b')$ are distinct, and the proof is the same as the case $a \neq a'$ and $b = b'$. □

The following result gives an upper bound for $\kappa(T(\Gamma(R)))$.

**Theorem 7.4.44.** ([319, Theorem 3.10]) *Let $R$ be a finite commutative ring. Then $\kappa(T(\Gamma(R))) \le |Z(R)| - |Nil(R)|$.*

The next result gives the exact value of the edge-connectivity of the total graph for a finite ring $R$ when $Z(R)$ is not an ideal of $R$.

**Theorem 7.4.45.** ([319, Theorem 3.11]) *Let $R$ be a finite commutative ring with $Z(R)$ not an ideal of $R$. Then $\lambda(T(\Gamma(R))) = |Z(R)| - 1$.*

### 7.4.4 Domination

Now we focus our attention on domination properties of total graphs of commutative rings. Let $G$ be a simple undirected graph with vertices set $V$. A subset $S$ of $V$ is called a *dominating set* of $G$ if for every $a \in V \setminus S$, there is a $b \in S$ such that $a$ and $b$ are adjacent in $G$. The domination number $\gamma(G)$ is the minimum size of a dominating set of $G$. The first volume of the two-volume book by Haynes, Hedetniemi, and Slater [214, 215] provides a comprehensive introduction to domination in graphs.

Shekarriz et al. [348] and Tamizh Chelvam et al. [363] have independently proved a result that gives the domination number of the total graph of a commutative Artinian ring. Note that Shekarriz et al. proved the result for finite rings, whereas Tamizh Chelvam et al. proved it for the more general class of commutative Artinian rings.

**Theorem 7.4.46.** ([348, Theorem 4.1], [363, Theorem 2.5]) *Let $R$ be a commutative Artinian ring. Then $\gamma(T(\Gamma(R))) = min\{|R/\mathfrak{m}| \mid \mathfrak{m}$ is a maximal ideal of $R\}$, except when $R$ is an integral domain, where $\gamma(T(\Gamma(R))) = \frac{|R|-k}{2} + k$, where $k = |\{a \in R \mid a = -a\}|$.*

*Proof.* If $R$ is an integral domain, then clearly $\gamma(T(\Gamma(R))) = \frac{|R|-k}{2} + k$, where $k = |\{a \in R \mid a = -a\}|$. Now assume that $R$ is not an integral domain. Let $n = min\{|R/\mathfrak{m}| \mid \mathfrak{m}$ is a maximal ideal of $R\}$, and let $I$ be a maximal ideal of $R$ with $|R/I| = n$ and $|I| = \lambda$. Let $H$ be the spanning subgraph of $T(\Gamma(R))$ in which two distinct vertices $x, y \in R$ are adjacent if $x + y \in I$. One can obtain, in the same way as in Theorem 7.2.1, $H$ is a spanning subgraph of $G$ and further

$$H = \begin{cases} K_\lambda \cup \underbrace{K_\lambda \cup K_\lambda \cup \cdots \cup K_\lambda}_{(n-1)\ copies} & \text{if } 2 \in I; \\ K_\lambda \cup \underbrace{K_{\lambda,\lambda} \cup K_{\lambda,\lambda} \cup \cdots \cup K_{\lambda,\lambda}}_{(\frac{n-1}{2})\ copies} & \text{if } 2 \notin I. \end{cases}$$

Note that each connected component of $H$ corresponds to a coset of $I$ in $R$. Take one element in each of the cosets of $I$ in $R$; they form a minimum dominating set of

$H$, and so $\gamma(H) = n$. Since $H$ is a spanning subgraph of $T(\Gamma(R))$, $\gamma(T(\Gamma(R))) \leq n$.

Since $R$ is a commutative Artinian ring, let $R \cong R_1 \times \cdots \times R_m$, where every $R_i$ is an Artinian local ring with maximal ideal $\mathfrak{m}_i$. Note that $\mathfrak{m}_i$ is the unique maximal ideal of $R_i$, and so $\gamma(T(\Gamma(R_i))) = |R_i/\mathfrak{m}_i|$ for every $i$. Let $I_i = R_1 \times \cdots \times R_{i-1} \times \mathfrak{m}_i \times R_{i+1} \times \cdots \times R_m$ and $n_i = |R/I_i|$. It is easy to see that the $I_i$'s $(1 \leq i \leq m)$ are the only maximal ideals of $R$, and so $|I| = \underset{1\leq i\leq m}{max}\{|I_i|\}$. From this, $n = \underset{1\leq i\leq m}{min}\gamma(T(\Gamma(R_i)))$. Suppose that $S = \{(x_{11}, x_{12}, \ldots, x_{1m}), (x_{21}, x_{22}, \ldots, x_{2m}), \ldots, (x_{(n-1)1}, x_{(n-1)2}, \ldots, x_{(n-1)m})\}$ is a dominating set of $T(\Gamma(R))$. Let $S_k = \{x_{1k}, x_{2k}, \ldots, x_{(n-1)k}\}$ for $1 \leq k \leq m$. Since $\gamma(T(\Gamma(R_j))) = n_j > n-1$, $S_j$ is not a dominating set of $T(\Gamma(R_j))$, and so there is a $y_j \in R_j$ such that $y_j$ is not dominated by $S_j$ for every $j = 1, 2, \ldots, m$. Therefore $(y_1, y_2, \ldots, y_m) \in R$ is not dominated by $S$, a contradiction to $S$ being a dominating set of $T(\Gamma(R))$. Hence $\gamma(T(\Gamma(R))) = n$. □

**Corollary 7.4.47.** ([362, Theorem 2.3]) (i) *If $n$ is a composite integer, then $\gamma(T(\Gamma(\mathbb{Z}_n))) = p$, where $p$ is the smallest prime divisor of $n$.*

(ii) *For every $n, k \in \mathbb{N}$, $\gamma(T(\Gamma(\mathbb{Z}_n[x]/(x^k)))) = \gamma(T(\Gamma(\mathbb{Z}_n[x,y]/(x^k, xy, y^k))))$ $= p$, where $p$ is the smallest prime divisor of $n$.*

The next result provides the connected domination number $\gamma_c$ and total domination number $\gamma_t$ of a total graph.

**Theorem 7.4.48.** ([363, Theorem 3.3]) *Let $R$ be a commutative Artinian ring, $Z(R)$ not an ideal of $R$, and $R = (Z(R))$. Then*

$$\gamma_c(T(\Gamma(R))) = \gamma_t(T(\Gamma(R))) = min\{|R/\mathfrak{m}| \mid \mathfrak{m} \textit{ is a maximal ideal of } R\}.$$

*Proof.* Let $n = min\{|R/\mathfrak{m}| \mid \mathfrak{m}$ is a maximal ideal of $R\}$. If $Z(R)$ is not an ideal of $R$ and $R = (Z(R))$, then $T(\Gamma(R))$ is connected by Theorem 7.2.10. Let $I$ be a maximal ideal of $R$ and $x_1 \in I$. Since $T(\Gamma(R))$ is connected, there is a vertex $x_2 \in a_1 + I$ for some $a_1 \in R \setminus I$ with $x_2$ adjacent to $x_1$. Again, by connectedness of $T(\Gamma(R))$, there is a coset $a_2 + I$ for some $a_2 \notin I$ as well as $a_2 \notin a_1 + I$ such that at least one element of $a_2 + I$ is adjacent to either a vertex in $I$ or in $a_1 + I$, say $I$.

If there is an $a \in a_i + I$ which is adjacent to some $b \in a_j + I$ with $a \notin a_j + I$, then every vertex in $a_i + I$ is adjacent to at least one vertex in $a_j + I$. For, if $a + b = c$ for some $c \in Z(R)$, then $c \in a_i + a_j + I$. Let $d_1 \in a_i + I$. Take $d_2 \in R$ such that $d_1 + d_2 = c$. From this, $d_2 \in a_j + I$ and $d_1$ is adjacent to $d_2$. Therefore, every vertex in $a_i + I$ is adjacent to at least one vertex in $a_j + I$.

Thus $x_1$ is adjacent to some vertex $x_3 \in a_2 + I$. Similarly, we can select coset representatives $x_i$, for $4 \leq i \leq \mu$, in distinct cosets of $I$ in $R$ other than $I$, $a_1 + I$, and $a_2 + I$ such that $(x_1, x_2, \ldots, x_\mu) \subseteq T(\Gamma(R))$ is connected. Then

$\{x_1, x_2, \ldots, x_\mu\}$ is a $\gamma_c$-set of $T(\Gamma(R))$, and so $\gamma_c(T(\Gamma(R))) = n$. Since $\gamma(G) \leq \gamma_t(G) \leq \gamma_c(G)$ for any graph $G$, we have $\gamma_t(T(\Gamma(R))) = n$. □

Let $H = \{d \mid d$ is a dominating set of $T(\Gamma(R))\}$. The *intersection graph of dominating sets*, denoted by $IT(R)$, is a simple undirected graph with vertex set $H$, and two distinct vertices $a$ and $b$ in $H$ are adjacent if an only if $a \cap b \neq \emptyset$ (see [366] and [101]). To study the intersection graph of dominating sets in total graphs, one needs to characterize all $\gamma$-sets of $T(\Gamma(R))$.

**Theorem 7.4.49.** ([366, Theorem 3.1]) *Let $R$ be a commutative Artinian ring which is not an integral domain and $\mu = min\{|R/\mathfrak{m}| \mid \mathfrak{m}$ is a maximal ideal of $R\}$ is finite. Assume that $I_i$'s $(i = 1, 2, \ldots, t)$ are the maximal ideals of $R$ such that $|R/I_i| = \mu$ and $R/I_i = \{a_{ij} + I_i \mid 1 \leq j \leq \mu\}$ for every $i$, $1 \leq i \leq t$. Then $\{x_1, x_2, \ldots, x_\mu\}$ is a $\gamma$-set of $T(\Gamma(R))$ if and only if $x_j \in a_{ij} + I_i$ for $j = 1, \ldots, \mu$ and some fixed $i$.*

*Proof.* Let $S = \{x_1, ..., x_\mu\} \subseteq V(T(\Gamma(R)))$, where $x_j \in a_{ij} + I_i$ for $j = 1, \ldots, \mu$ and some fixed $i$. For $i = 1, 2, \ldots, t$, let $H_i$ be the spanning subgraph of $T(\Gamma(R))$ in which two distinct vertices $x, y \in R$ are adjacent if $x + y \in I_i$. By the proof of Theorem 7.4.46, a set with exactly one element from each of the cosets of $I_i$ in $R$ is a dominating set of $T(\Gamma(R))$, and so $S$ is a $\gamma$-set of $T(\Gamma(R))$.

Conversely, assume that $S = \{x_1, ..., x_\mu\}$ is a $\gamma$-set of $T(\Gamma(R))$. Let $R \cong R_1 \times \cdots \times R_n$, where every $R_\ell$ is a local ring. Let $Z(R_\ell) = m_\ell$ and $|R_\ell/m_\ell| = \mu_\ell$ for $1 \leq \ell \leq n$. Since every maximal ideal of $R$ has the form $R_1 \times \cdots \times R_{s-1} \times m_s \times R_{s+1} \times \cdots \times R_n$ for some $s$, we have $t \leq n$. Assume that $I_i = R_1 \times \cdots \times R_{i-1} \times m_i \times R_{i+1} \times \cdots \times R_n$ for $1 < i \leq t$. For $r = 1, 2, \ldots, \mu$, let $x_r = (b_{r1}, b_{r2}, \ldots, b_{rn})$, where $b_{r\ell} \in R_\ell$ and $S_k = \{b_{1k}, b_{2k}, \ldots, b_{\mu k}\}$ for $1 \leq k \leq n$.

Suppose for every $i = 1, 2, \ldots, t$, there are $p$ and $q$ such that $p \neq q$ and $x_p, x_q \in a_{ij} + I_i$ for some $j$ with $1 \leq j \leq \mu$. Then there are $b_{pe}$ and $b_{qf}$ such that $b_{pe}, b_{qf} \in a_{ij} + Z(R_i)$, and so $S_i$ is not a dominating set of $T(\Gamma(R_i))$. Thus for every $i = 1, 2, \ldots, t$, there is a $c_i \in R_i$ such that $c_i$ is not dominated by $S_i$. Since $\mu_\ell > \mu$ for every $\ell \geq t+1$, $S_\ell$ is not a dominating set of $T(\Gamma(R_\ell))$, and so there are $c_\ell \in R_\ell$ such that $c_\ell$ is not dominated by $S_\ell$. Since $Reg(R) \cong Reg(R_1) \times \cdots \times Reg(R_n)$, $(c_1, c_2, \ldots, c_n) \in V(T(\Gamma(R)))$ is not dominated by $S$, a contradiction. □

### 7.4.5 Intersection Graphs of Dominating Sets

In this section, we study the intersection graph of gamma sets in the total graph of a commutative ring. Let $A$ be a set and $S$ a collection of nonempty subsets of $A$. The intersection graph of $S$ is the graph whose vertices are the elements of $S$, and two vertices are adjacent if the subsets have a nonempty intersection [287].

Since, by Theorem 7.4.49, we characterized all $\gamma$-sets of the total graph for a commutative Artinian ring, we can define the intersection graph of gamma sets in the total graph corresponding to a commutative Artinian ring. The intersection graph of gamma sets in the total graph of a commutative ring $R$, denoted by $IT(R)$, is the undirected graph with vertex set the collection of all $\gamma$-sets of $T(\Gamma(R))$, and two distinct vertices $u$ and $v$ are adjacent if and only if $u \cap v \neq \emptyset$. Now, we investigate the interplay between the graph-theoretic properties of $IT(R)$ and the ring-theoretic properties of $R$. Readers may refer to [157, 158] for the intersection graph of ideals of rings.

Note that there are infinitely many commutative rings $R$ with no (proper) ideal $I$ such that $|R/I|$ is finite. For example, let $R = \mathbb{Z} \times \mathbb{Z}[x]/(x^2)$. Then $R$ is infinite, and $|R/I|$ is infinite for every (proper) ideal $I$ of $R$. Because of this, throughout the intersection graph section, we assume that $R$ is a commutative Artinian ring with at least one ideal $I$ such that $|R/I|$ is finite.

**Example 7.4.50.** Figure 7.4 gives the intersection graphs of gamma sets in $T(\Gamma(R))$ for several commutative rings $R$. Note that in Figure 7.4, an element $(x, y)$ of $\mathbb{Z}_2 \times \mathbb{Z}_2$ is denoted as $xy$.

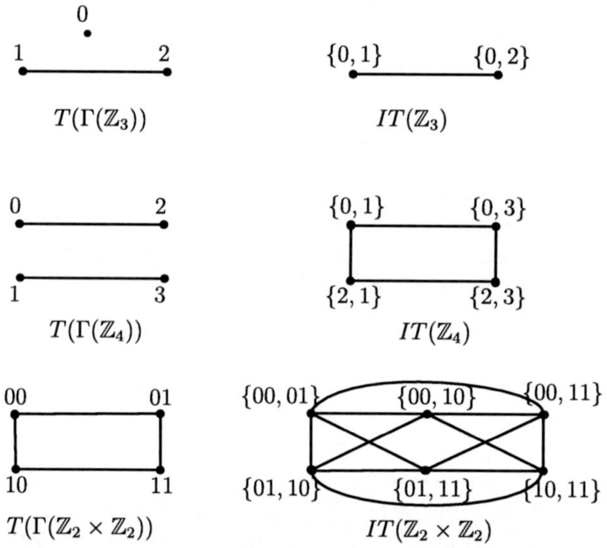


Figure 7.4: Some total graphs and intersection graphs

Also, note that although $\mathbb{Z}_4$ is not isomorphic to $\mathbb{Z}_2[x]/(x^2)$, the corresponding intersection graphs of gamma sets are the same.

In what follows, information about the diameter and girth of $IT(R)$ are given.

**Theorem 7.4.51.** ([366, Theorem 3.1]) *Let $R$ be a commutative Artinian ring with $|R| \geq 4$ and $min\{|R/\mathfrak{m}| \mid \mathfrak{m}$ is a maximal ideal of $R\}$ is finite.*

(i) *$IT(R)$ is connected with $diam(IT(R)) \leq 2$;*

(ii) $gr(IT(R)) \in \{3,4\}$. *In particular,* $gr(IT(R)) = 4$ *if and only if either* $R \cong \mathbb{Z}_4$ *or* $R \cong \mathbb{Z}_2[x]/(x^2)$.

*Proof.* Let $\mu = min\{|R/\mathfrak{m}| \mid \mathfrak{m}$ is a maximal ideal of $R\}$, and let $I_i (1 \le i \le t)$ be ideals of $R$ with $|I_i| = \lambda$ and $|R/I_i| = \mu$.

(i) If $R$ is an integral domain, then $IT(R)$ is a complete graph, and so $diam(IT(R)) = 1$. On the other hand, let $u = \{x_1, \ldots, x_\mu\}$ and $v = \{y_1, \ldots, y_\mu\}$ be two vertices in $IT(R)$. Let $u$ be a $\gamma$-set of $T(\Gamma(R))$ with respect to $I_i$ for some $1 \le i \le t$. If $x_j = y_k$ for some $1 \le j, k \le \mu$, then $u$ and $v$ are adjacent. If $x_j \ne y_k$ for every $1 \le j, k \le \mu$, let $a \in R$ such that $y_1 \in a + I_i$. Then by Theorem 7.4.49, there is an $x_j \in u$ such that $x_j \in a + I_i$ for some $j$, $1 \le j \le \mu$. Now $w = \{x_1, \ldots, x_{j-1}, y_1, x_{j+1}, \ldots, y_\mu\} \in V(IT(R))$, and so $u - w - v$ is a path in $IT(R)$. Hence $d(u, v) = 2$.

(ii) If $R$ is an integral domain, then $gr(IT(R)) = 3$. Now, assume that $R$ is not an integral domain. Let $u = \{x_1, \ldots, x_\mu\} \in V(IT(R))$ and $u$ be a $\gamma$-set of $T(\Gamma(R))$ with respect to $I_i$ for some $1 \le i \le t$. Let $y_j \in x_j + I_i$ and $y_j \ne x_j$ for $j = 1, \ldots, \mu$. If $\mu \ge 3$, then $v = \{y_1, x_2, \ldots, x_\mu\}$, $w = \{x_1, \ldots, x_{\mu-1}, y_\mu\}$ are two distinct vertices in $IT(R)$ and $u - v - w$ is a cycle in $IT(R)$. If $\lambda \ge 3$, then take $z_j \in x_j + I_i$ (different from $x_j$ and $y_j$) for $j = 1, \ldots, \mu$. From this, $v = \{y_1, x_2, \ldots, x_\mu\}, w = \{z_1, x_2, \ldots, y_\mu\} \in V(IT(R))$ and $u - v - w$ is a cycle in $IT(R)$. Hence $gr(IT(R)) = 3$ if $\mu \ge 3$ or $\lambda \ge 3$. Now consider the case $\lambda = 2$ and $\mu = 2$. Thus $R \cong \mathbb{Z}_4$, $\mathbb{Z}_2[x]/(x^2)$, or $\mathbb{Z}_2 \times \mathbb{Z}_2$. Hence $gr(IT(R)) = 4$ if and only if $R \cong \mathbb{Z}_4$ or $\mathbb{Z}_2[x]/(x^2)$. □

**Theorem 7.4.52.** ([366, Theorem 3.2]) *Let* $R$ *be a commutative Artinian ring with* $|R| \ge 4$ *and* $min\{|R/\mathfrak{m}| \mid \mathfrak{m}$ *is a maximal ideal of* $R\}$ *is finite. Then*

(i) $IT(R)$ *is a regular graph;*

(ii) $IT(R)$ *is a complete graph if and only if* $R$ *is an integral domain;*

(iii) $IT(R)$ *is a bipartite graph if and only if either* $R \cong \mathbb{Z}_4$ *or* $R \cong \mathbb{Z}_2[x]/(x^2)$;

(iv) $IT(R)$ *is a cycle if and only if either* $R \cong \mathbb{Z}_4$ *or* $R \cong \mathbb{Z}_2[x]/(x^2)$.

Next, we discuss the genus of the intersection graph of gamma sets in the total graph of a commutative ring. First, we need the following lemma, which gives the cardinality of the vertex set of $IT(R)$ and the degree of every vertex in $IT(R)$.

**Lemma 7.4.53.** ([366, Lemma 3.2]) *Let* $R$ *be a finite commutative ring and* $I_i$ *be ideals in* $R$ *such that* $|R/I_i| = min\{|R/\mathfrak{m}| \mid \mathfrak{m}$ *is a maximal ideal of* $R\}$ $1 \le i \le t$. *Let* $|I_i| = \lambda$, $|R/I_i| = \mu$, *and* $k = |\{x \in R \mid 2x = 0\}|$. *Then*

(i)

$$|V(IT(R))| = \begin{cases} 2^{\frac{|R|-k}{2}} & if\ R\ is\ an\ integral\ domain; \\ \lambda^\mu & if\ t = 1; \\ 2\lambda^\mu - \mu!(\frac{\lambda}{\mu})^\mu & if\ t = 2; \end{cases}$$

*Moreover, if* $t \geq 3$*, then* $|V(IT(R))| \geq 2\lambda^{\mu} - \mu!(\frac{\lambda}{\mu})^{\mu} > \lambda^{\mu}$;

(ii) *If* $t = 1$*, then* $deg(v) = \sum_{i=1}^{\mu}[\lambda^{\mu-i}(\lambda-1)^{i-1}] - 1$ *for every* $v \in V(IT(R))$. *Further, if* $\mu = 2$*, then* $deg(v) = |R| - 2$ *for all* $v$;

(iii) *If* $t \geq 2$*, then* $deg(v) \geq \sum_{i=1}^{\mu}[\lambda^{\mu-i}(\lambda-1)^{i-1}] - 1 + (\frac{\lambda}{\mu})\lambda^{\mu-2}$ *for every* $v \in V(IT(R))$.

**Theorem 7.4.54.** ([366, Theorem 5.4]) *Let* $R$ *be a finite commutative ring. Then*

(i) $IT(R)$ *is planar if and only if* $R$ *is isomorphic to either* $\mathbb{Z}_3$, $\mathbb{Z}_4$, $\mathbb{Z}_5$, $\mathbb{Z}_2[x]/(x^2)$, $\mathbb{Z}_2 \times \mathbb{Z}_2$*, or* $\mathbb{F}_{2^n}$ *for some positive integer* $n$;

(ii) $IT(R)$ *is toroidal if and only if* $R \cong \mathbb{Z}_6$;

(iii) $g(IT(R)) = 2$ *if and only if* $R \cong \mathbb{Z}_7$.

*Proof.* Let $I$ be an ideal of $R$ with

$$|R/I| = \min\{|R/\mathfrak{m}| \mid \mathfrak{m} \text{ is a maximal ideal of } R\} = \mu,$$

and let $|I| = \lambda$.

(i) We have two cases:

**Case 1**. Suppose that $R$ is an integral domain (field).

If $|R| = 2^n$ for some $n \in \mathbb{N}$, then $R \cong \mathbb{F}_{2^n}$, and so $k = |\{x \in R \mid 2x = 0\}| = |R|$. By Lemma 7.4.53(i), $IT(R) = K_1$, and so $IT(R)$ is planar.

If $|R| = p^n$ for some prime $p > 3$ and $n \in \mathbb{N}$, then $k = |\{x \in R \mid 2x = 0\}| = 1$. By Lemma 7.4.53(i), $IT(R) = K_{2^{\frac{|R|-1}{2}}}$. By Kuratowskis's Theorem, $IT(R)$ is planar if and only if $|R| \leq 5$, i.e., $IT(R)$ is planar if and only if $R$ is either $\mathbb{Z}_3$ or $\mathbb{Z}_5$. Hence, in this case, $IT(R)$ is planar if and only if $R$ is isomorphic to either $\mathbb{Z}_3$, $\mathbb{Z}_5$, or $\mathbb{F}_{2^n}$ for some $n \in \mathbb{N}$.

**Case 2**. Suppose that $R$ is not an integral domain.

Assume that $IT(R)$ is planar. Note that $\delta(IT(R)) \leq 5$ by Lemma 7.5.5. If $|R| \geq 6$, then $\lambda\mu \geq 6$, and so $\lambda^{\mu-1} \geq \lambda(\mu-1) \geq 6-\lambda$. By Lemma 7.4.53(i), $deg(v) \geq \lambda^{\mu-1} + (\lambda-1)\lambda^{\mu-2} + (\lambda-1)^{\mu-1} - 1 \geq 6-\lambda+\lambda-1+\lambda-1-1 = \lambda+3$. If $\lambda \geq 3$, then $deg(v) \geq 6$ for every $v \in V(IT(R))$, which is a contradiction to $\delta(IT(R)) \leq 5$. Therefore $\lambda = 2$, and so $|R| = 4$, a contradiction to the assumption that $|R| \geq 6$. Hence $|R| \leq 5$. Thus $R \cong \mathbb{Z}_4$, $\mathbb{Z}_2[x]/(x^2)$, or $\mathbb{Z}_2 \times \mathbb{Z}_2$.

Conversely, $IT(\mathbb{Z}_4)$ and $IT(\mathbb{Z}_2[x]/(x^2))$ are planar. Also, $IT(\mathbb{Z}_2 \times \mathbb{Z}_2)$ is planar as shown in the Figure 7.5. Hence in this case, $IT(R)$ is planar if and only if $R$ is either $\mathbb{Z}_4$, $\mathbb{Z}_2[x]/(x^2)$, or $\mathbb{Z}_2 \times \mathbb{Z}_2$.

(ii) Assume that $IT(R)$ is toroidal. If $R$ is an integral domain, then by (i), $|R| \geq 7$ and $|R| \neq 2^n$ for $n \in \mathbb{N}$. But $K_{2^3} \subseteq IT(R)$ and $g(K_8) = 2$. Therefore $g(IT(R)) \geq 2$, a contradiction. Hence $R$ is not an integral domain. Since the genus of $IT(R)$ is one, $\delta(IT(R)) \leq 6$ by Lemma 7.5.5. If $|R| \geq 7$, then

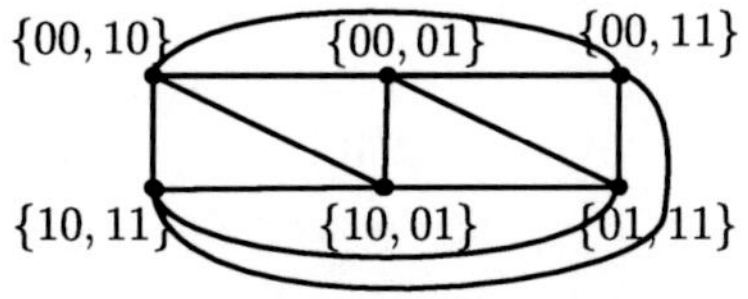


Figure 7.5: Planar embedding of $IT(\mathbb{Z}_2 \times \mathbb{Z}_2)$

$deg(v) \geq \lambda + 4$. If $\lambda \geq 3$, then $deg(v) \geq 7$ for every $v \in V(IT(R))$, which is contradiction to $\delta(IT(R)) \leq 6$. Therefore $\lambda = 2$, and so $|R| = 4$, a contradiction. Hence $|R| \leq 6$, and by $(i)$, $|R| = 6$. There is only one commutative ring of order 6 which is not an integral domain, namely $\mathbb{Z}_6$. The embedding of $IT(\mathbb{Z}_6)$ in a torus is given in Figure 7.6.

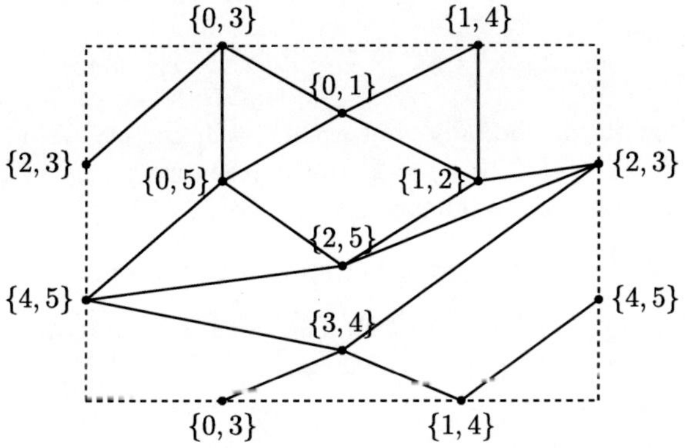


Figure 7.6: Embedding of $IT(\mathbb{Z}_6)$ in a Torus

(iii) Assume that $g(IT(R)) = 2$. If $R$ is an integral domain, then by (i), $|R| \geq 7$ and $|R| \neq 2^n$ for every $n \in \mathbb{N}$. If $|R| = 7$, then $IT(R) = K_{2^3}$ and so $g(IT(R)) = 2$. If $|R| \geq 9$, then $K_{2^4} \subseteq IT(R)$, and so $g(IT(R)) \geq 13$, a contradiction. Suppose that $R$ is not an integral domain. Note that $\delta(IT(R)) \leq 6 + \frac{12}{|V(IT(R))|}$ by Lemma 7.5.5. By (i) and (ii), $|R| \geq 7$. Thus $|V(IT(R))| \geq 7$, and so $\delta(IT(R)) \leq 7$. Since $|R| \geq 7$, $deg(v) \geq \lambda + 5$. If $\lambda \geq 3$, then $deg(v) \geq 8$ for every $v \in V(IT(R))$, which is a contradiction to $\delta(IT(R)) \leq 7$. Therefore $\lambda = 2$, and so $|R| = 4$, a contradiction. Hence $g(IT(R)) = 2$ if and only if $R \cong \mathbb{Z}_7$. □

The next result exhibits a relationship between the genus of the total graph $T(\Gamma(R))$ and the corresponding intersection graph $IT(R)$.

**Theorem 7.4.55.** ([366, Theorem 5.5]) *Let $R$ be a finite commutative ring. Then $g(T(\Gamma(R))) \leq g(IT(R))$.*

*Proof.* If $R$ is a field, then $T(\Gamma(R))$ is planar. Further, $IT(R) = K_\ell$ for some $\ell \in \mathbb{Z}^+$. This gives $g(T(\Gamma(R))) \leq g(IT(R))$. Now assume that $R$ is not a field. Let $I$ be an ideal of $R$ with $|R/I| = min\{|R/\mathfrak{m}| \mid \mathfrak{m}$ is a maximal ideal of $R\} = \mu$ and $|I| = \lambda$. Then $g(T(\Gamma(R))) \leq \left\lceil \frac{(|R|-3)(|R|-4)}{12} \right\rceil$ and $g(IT(R)) \geq \lambda \left\lceil \frac{(\lambda^{\mu-1}-3)(\lambda^{\mu-1}-4)}{12} \right\rceil$. To conclude the proof, it is enough to show that $\lambda^\mu - 3\lambda \geq |R| - 3$ and $\lambda^\mu - 4\lambda \geq |R| - 4$, i.e., $\lambda^{\mu-1} - 3 \geq \mu - \frac{3}{\lambda}$ and $\lambda^{\mu-1} - 4 \geq \mu - \frac{4}{\lambda}$. If $\lambda = 2$, then $R \cong \mathbb{Z}_4$ or $\mathbb{Z}_2[x]/(x^2)$, and so $g(T(\Gamma(R))) = g(IT(R)) = 0$. Hence $\lambda \geq 3$. Now it remains to show that $\lambda^{\mu-1} - 2 \geq \mu$ and $\lambda^{\mu-1} - 3 \geq \mu$. If $\lambda \geq 5$ or $\mu \geq 3$, then $\lambda^{\mu-1} - 3 \geq \mu$; so consider the case $\lambda = 3, 4$ and $\mu = 2$. Thus $|R| = 6$ or 8.

If $R$ is local, then $|R| = 8$. There are 5 local rings of order 8 (except $\mathbb{F}_8$). They are

$$\frac{\mathbb{Z}_2[x]}{(x^3)}, \frac{\mathbb{Z}_2[x,y]}{(x,y)^2}, \frac{\mathbb{Z}_4[x]}{(2x,x^2)}, \frac{\mathbb{Z}_4[x]}{(2x,x^2-2)}, \mathbb{Z}_8.$$

If $R$ is not local, then $R$ is one of the following rings:

$$\mathbb{Z}_6,\ \mathbb{Z}_2 \times \mathbb{F}_4,\ \mathbb{Z}_2 \times \mathbb{Z}_4,\ \mathbb{Z}_2 \times \mathbb{Z}_2[x]/(x^2).$$

Of these 9 rings, $\mathbb{Z}_6$ is the only ring with $\lambda = 3$; the others have $\lambda = 4$. Note that $0 = g(T(\Gamma(\mathbb{Z}_6))) \leq g(IT(\mathbb{Z}_6)) = 1$. In the remaining cases, $T(\Gamma(R))$ is either planar or toroidal, whereas by Theorem 7.4.54, $g(IT(R)) \geq 3$. □

The next theorem concerns the Eulerian and Hamiltonian nature of $IT(R)$.

**Theorem 7.4.56.** ([101, Theorem 2.1]) *Let $R$ be a commutative Artinian ring with $|R| \geq 4$, and assume that $I$ is the unique ideal of $R$ such that $|R/I| = min\{|R/\mathfrak{m}| \mid \mathfrak{m}$ is a maximal ideal of $R\}$. Then $IT(R)$ is Eulerian if and only if $R$ is not a field.*

**Theorem 7.4.57.** ([101, Theorem 2.2]) *Let $R$ be a commutative Artinian ring with $|R| \geq 4$, and assume that $I$ is an ideal of $R$ with $|R/I| = min\{|R/\mathfrak{m}| \mid \mathfrak{m}$ is a maximal ideal of $R\}$. Then $IT(R)$ is a Hamiltonian graph.*

We recall that a graph $G$ with $n \geq 3$ vertices is called *pancyclic* if $G$ contains cycles of all lengths from 3 to $n$. Also, $G$ is called *vertex-pancyclic* if every vertex of $G$ belongs to every cycle of length $l$ for $3 \leq l \leq n$.

**Theorem 7.4.58.** ([101, Theorem 2.3, Corollary 2.1]) *Let $R$ be a commutative Artinian ring with $|R| \geq 4$, and assume that $I$ is an ideal of $R$ with $|R/I| = min\{|R/\mathfrak{m}| \mid \mathfrak{m}$ is a maximal ideal of $R\}$.*

(i) *$IT(R)$ is pancyclic if and only if either $R \cong \mathbb{Z}_4$ or $R \cong \mathbb{Z}_2[x]/(x^2)$;*

(ii) *$IT(R)$ is vertex-pancyclic if and only if neither $R \cong \mathbb{Z}_4$ nor $R \cong \mathbb{Z}_2[x]/(x^2)$ (i.e., $IT(R)$ is not pancyclic).*

We recall that a *perfect graph* is a graph in which the chromatic number of every induced subgraph equals the size of the largest clique of that subgraph.

**Theorem 7.4.59.** ([101, Theorems 4.1, 4.2]) *Let $R$ be a finite commutative ring. Then*

(i) $\chi(IT(R)) = \omega(IT(R))$;

(ii) *$IT(R)$ is perfect if and only if either $R$ is an integral domain (field) or $R$ has a unique ideal $I$ with $|R/I| = min\{|R/\mathfrak{m}| \mid \mathfrak{m}$ is a maximal ideal of $R\} = 2$ or $R \cong \mathbb{Z}_2 \times \mathbb{Z}_2$.*

*Proof.* (i) If $R$ is an integral domain, then $IT(R)$ is complete, and so $\chi(IT(R)) = \omega(IT(R))$. Hence assume that $R$ is not an integral domain. Let $I_1, \ldots, I_t$ be ideals of $R$ such that $|R/I_i| = min\{|R/\mathfrak{m}| \mid \mathfrak{m}$ is a maximal ideal of $R\} = \mu$ and $|I_i| = \lambda$ for $i = 1, \ldots, t$. Let $v = \{x_1, \ldots, x_\mu\} \in V(IT(R))$. By Theorem 7.4.49, there is an integer $\ell$ with $1 \leq \ell \leq t$ and $x_q \notin x_p + I_\ell$ for every $1 \leq p \neq q \leq \mu$. Let $I_\ell = \{a_1, a_2, \ldots, a_\lambda\}$. Clearly $X = \{\{x_1 + a_1, \ldots, x_\mu + a_1\}, \{x_1 + a_2, \ldots, x_\mu + a_2\}, \ldots, \{x_1 + a_\mu, \ldots, x_\mu + a_\lambda\}\}$ is an independent set in $IT(R)$ containing $v$; so we can assign a single color to all the vertices of $X$. Since $v \in V(IT(R))$ is arbitrary and $|X| = \lambda$, $\chi(IT(R)) \leq \frac{|V(IT(R))|}{\lambda}$. Note that $\chi(IT(R)) \geq \omega(IT(R))$ is always true. Hence $\chi(IT(R)) = \omega(IT(R)) = \frac{|V(IT(R))|}{\lambda}$.

(ii) If $R$ is an integral domain, then $IT(R)$ is complete and so is perfect. So assume that $R$ is not an integral domain. Let $I$ be an ideal in $R$ such that $|R/I| = min\{|R/\mathfrak{m}| \mid \mathfrak{m}$ is a maximal ideal of $R\} = \mu$ and $|I| = \lambda$.

If $\lambda = 2$, then $R$ is isomorphic to either $\mathbb{Z}_4$, $\mathbb{Z}_2[x]/(x^2)$, or $\mathbb{Z}_2 \times \mathbb{Z}_2$, and so $IT(R)$ is perfect.

Assume that $\lambda \geq 3$. Let $I = \{a_1 = 0, a_2, \ldots, a_\lambda\}$ and $R/I = \{I, x_1 + I, \ldots, x_{\mu-1} + I\}$. If $\mu \geq 3$, then the subgraph induced by the vertices $\{\{a_1, x_1, \ldots, x_{\mu-1}\}, \{a_2, x_1, \ldots, x_{\mu-1}\}, \{a_2, x_1 + a_3, \ldots, x_{\mu-1} + a_3\}, \{a_3, x_1 + a_2, x_2 + a_3, \ldots, x_{\mu-1} + a_3\}, \{a_1, x_1 + a_2, \ldots, x_{\mu-1} + a_2\}\}$ is $C_5$ and is a subgraph of both $IT(R)$ and $CIT(R)$. Thus, by the strong perfect graph theorem, $IT(R)$ is not a perfect graph. Let $\mu = 2$ and $x_1 + I = \{b_1, \ldots, b_\lambda\}$. Now, assume that $I$ is the only ideal of $R$ with $|R/I| = min\{|R/\mathfrak{m}| \mid \mathfrak{m}$ is a maximal ideal of $R\} = 2$. Note that every vertex of $IT(R)$ has the form $\{a_i, b_j\}$, and if two vertices are adjacent, then they have exactly one element in common. From this, no induced subgraph of $IT(R)$ is an odd cycle, and so $IT(R)$ is perfect. If $R$ has at least two ideals $I$ and $J$ with $|R/I| = |R/J| = 2$, then by Theorem 7.4.49, there are $i$ and $j$ ($1 \leq i, j \leq \lambda$) such that $\{a_i, a_j\} \subseteq I$ and $\{a_i, a_j\}$ is a $\gamma$-set of $T(\Gamma(R))$ with respect to $J$, i.e., $\{a_i, a_j\} \in V(IT(R))$. Let $a_k \in I$ with $k \neq i$ and $k \neq j$. Now the subgraph induced by $\{\{a_i, b_1\}, \{a_k, b_1\}, \{a_k, b_2\}, \{a_j, b_2\}, \{a_i, a_j\}\}$ is $C_5$ and is a subgraph of both $IT(R)$ and $CIT(R)$, and so $IT(R)$ is not a perfect graph, i.e., when $\lambda \geq 3$, $IT(R)$ is perfect if and only if $R$ has a unique ideal $I$ with $|R/I| = min\{|R/\mathfrak{m}| \mid \mathfrak{m}$ is a maximal ideal of $R\} = 2$. □

In what follows, we give the vertex connectivity of $IT(R)$. The intersection graph of gamma sets in $T(\Gamma(R))$ represents very reliable networks, which means that the vertex connectivity of $IT(R)$ equals the degree of regularity of $IT(R)$.

**Theorem 7.4.60.** ([101, Theorem 4.4]) *Let $R$ be a finite commutative ring. Then the vertex connectivity $k(IT(R)) = \delta(IT(R))$.*

## 7.5 Genus of Total Graphs

In this section, we discuss the genus of the total graph of a commutative ring. Recall some basic concepts of genus of a graph from Section 4.1. For nonnegative integers $g$ and $k$, let $S_g$ denote the sphere with $g$ handles and $\tilde{S}_k$ the sphere with $k$ crosscaps attached to it. It is well known that every connected compact surface is homeomorphic to $S_g$ or $\tilde{S}_k$ for some nonnegative integers $g$ and $k$ (cf. [284, Theorem 5.1]). The genus of a graph $G$, denoted by $g(G)$, is the minimum integer $n$ such that the graph can embedded in $S_n$. Similarly, the *crosscap number* (nonorientable genus) $\tilde{g}(G)$ is the minimum $k$ such that $G$ can be embedded in $\tilde{S}_k$. A graph $G$ is called *planar* if $g(G) = 0$ (i.e., the graph's vertices can be placed on a plane such that no edges cross), and *toroidal* if $g(G) = 1$ (i.e., the graph's vertices can be placed on a torus such that no edges cross). Moreover, a nonorientable genus 1 graph is called a *projective graph*. Note that if $H$ is a subgraph of a graph $G$, then $g(H) \leq g(G)$ and $\tilde{g}(H) \leq \tilde{g}(G)$.

Let $G_1 = (V_1, E_1)$ and $G_2 = (V_2, E_2)$ be two graphs with disjoint vertices set $V_i$ and edges set $E_i$. The *union* of $G_1$ and $G_2$ is the graph $G_1 \cup G_2$ whose vertex set is $V_1 \cup V_2$ and whose edge set is $E_1 \cup E_2$. The Cartesian product of $G_1$ and $G_2$ is denoted by $G = G_1 \square G_2$ with vertices set $V_1 \times V_2$ and $(x, y)$ is adjacent to $(x', y')$ if $x = x'$ and $y$ is adjacent $y'$ in $G_2$ or $y = y'$ and $x$ is adjacent to $x'$ in $G_1$. For any graph G, the disjoint union of $k$ copies of G is denoted by $kG$.

In the following results, we give some well-known formulas (see Section 5.1).

**Theorem 7.5.1.** ([429, Theorem 6.37,6.38],[428, Theorem 2]) *The following statements hold:*

(i) *For $n \geq 3$, we have $g(K_n) = \left\lceil \frac{(n-3)(n-4)}{12} \right\rceil$;*

(ii) *For $m, n \geq 2$, we have $g(K_{m,n}) = \left\lceil \frac{(m-2)(n-2)}{4} \right\rceil$;*

(iii) *Let $G_1$ and $G_2$ be two graphs, and for each $i$, let $p_i$ be the number of vertices of $G_i$. Then $max\{p_1 g(G_2) + g(G_1), p_2 g(G_1) + g(G_2)\} \leq g(G_1 \times G_2)$.*

**Corollary 7.5.2.** ([429]) *If $G$ is a graph with $n$ vertices, then $g(G) \leq \lceil \frac{(n-3)(n-4)}{12} \rceil$.*

**Lemma 7.5.3.** ([330, Theorem 1.1]) *For all integers $n \not\equiv 5$ or $9$ $(mod\ 12)$ with $n \geq 2$, $g(K_n \times K_2) = \left\lceil \frac{(n-2)(n-3)}{6} \right\rceil$. If $n = 5$, then $g(K_n \times K_2) = 2$.*

**Lemma 7.5.4.** (Euler's Formula) *If $G$ is a finite connected graph with $n$ vertices, $m$ edges, and genus $g$, then $n - m + f = 2 - 2g$, where $f$ is the number of faces created when $G$ is minimally embedded on a surface of genus $g$.*

**Lemma 7.5.5.** ([430, Proposition 2.1]) *If $G$ is a graph with $n$ vertices and genus $g$, then*

$$\delta(G) \leq 6 + \frac{12g - 12}{n}.$$

*Moreover, equality holds if and only if the embedding is a triangulation of the surface.*

**Lemma 7.5.6.** ([89, Lemma 2.1]) *If $G$ is a graph with $n$ vertices, $m$ edges, girth $gr$, and genus $g$, then*

$$\frac{m(gr-2)}{2gr} - \frac{n}{2} + 1 \leq g.$$

**Lemma 7.5.7.** ([292, p 114]) *If $G = G_1 \cup G_2$, where $G_1, G_2$ are edge-disjoint and $G_1 \cap G_2$ contains $k$ vertices, then*

$$g(G) \leq g(G_1) + g(G_2) + k - 1.$$

According to Theorem 7.5.1, $g(K_n) = 0$ for $1 \leq n \leq 4$, $g(K_n) = 1$ for $5 \leq n \leq 7$, and for other values of $n$, $g(K_n) \geq 2$.

**Lemma 7.5.8.** ([277, Lemma 1.3]) *The total graph of $\mathbb{F}_2 \times \mathbb{F}_q$ is isomorphic to $K_2 \times K_q$. Furthermore, for any positive integer $m$ and $q > 2$,*

$$g(T(\Gamma(\mathbb{F}_{2^m} \times \mathbb{F}_q))) \geq 2^m \left\lceil \frac{(q-3)(q-4)}{12} \right\rceil.$$

*Proof.* We induct on $m$. If $m = 1$, let $V_0 = \{(0, y) \mid y \in \mathbb{F}_q\}$ and $V_1 = \{(1, y) \mid y \in \mathbb{F}_q\}$. Then the subgraphs $G_i$ of the total graph of $\mathbb{F}_2 \times \mathbb{F}_q$ induced by each of the $V_i$ is $K_q$. Now for every $y \in \mathbb{F}_q$, there is an edge between $(0, y) \in G_0$ and $(1, -y) \in G_1$. Furthermore, these are the only other edges in the total graph. Identifying $(1, -y)$ with $(1, y)$, we can rcplace $G_1$ with an isomorphic copy $G_1'$; under this isomorphism the edge between $(0, y) \in G_0$ and $(1, -y) \in G_1$ is the edge between $(0, y) \in G_0$ and $(1, y) \in G_1'$. Thus the total graph of $\mathbb{F}_2 \times \mathbb{F}_q$ has vertex set $\{(x, y) \mid x \in \mathbb{F}_2 \text{ and } y \in \mathbb{F}_q\}$, with an edge between $(x, y)$ and $(x', y')$ if $x = x'$ and $y \neq y'$, or $y = y'$ and $x \neq x'$. That is, it is the graph $K_2 \times K_q$. Parts (a) and (c) of Theorem 7.5.1 now yield

$$g(T(\Gamma(\mathbb{F}_2 \times \mathbb{F}_q))) = g(K_2 \times K_q) \geq 2 \left\lceil \frac{(q-3)(q-4)}{12} \right\rceil.$$

If $m > 1$, we can partition $\mathbb{F}_{2^m}$ into two sets, $S_1$ and $S_2$, each of cardinality $2^{m-1}$; let $f$ be a bijection from $S_1$ to $S_2$. Since each element of a field of characteristic 2 is its own inverse, the subgraph of $T(\Gamma(\mathbb{F}_{2^m} \times \mathbb{F}_q))$ induced by $S_i \times \mathbb{F}_q$ is isomorphic to $T(\Gamma(\mathbb{F}_{2^{m-1}} \times \mathbb{F}_q))$. For every $y \in \mathbb{F}_q$ and $s \in S_1$, the element $(s, y)$ is adjacent to $(f(s), -y)$. We thus have a copy of $K_2 \times T(\Gamma(\mathbb{F}_{2^{m-1}} \times \mathbb{F}_q))$ as a subgraph of

$T(\Gamma(\mathbb{F}_{2^m} \times \mathbb{F}_q))$. Part (c) of Theorem 7.5.1 and the induction hypothesis now yield

$$\begin{aligned} g(T(\Gamma(\mathbb{F}_{2^m} \times \mathbb{F}_q))) &\geq g(K_2 \times T(\Gamma(\mathbb{F}_{2^{m-1}} \times \mathbb{F}_q))) \\ &\geq 2g(T(\Gamma(\mathbb{F}_{2^{m-1}} \times \mathbb{F}_q))) \\ &\geq 2^m \left\lceil \frac{(q-3)(q-4)}{12} \right\rceil . \end{aligned}$$

□

A *subdivision* of a graph is a graph obtained from it by replacing edges with pairwise internally disjoint paths. A remarkably simple characterization of planar graphs was given by Kuratowski in 1930. Kuratowski's Theorem says that a graph is planar if and only if it contains no subdivision of $K_5$ or $K_{3,3}$. In addition, every planar graph has a vertex $v$ such that $deg(v) \leq 5$.

**Theorem 7.5.9.** ([277, Theorem 1.4]) *For every positive integer $m$, there are only finitely many finite commutative rings $R$ whose total graph has genus $m$.*

*Proof.* Let $R$ be a finite commutative ring. If $R$ is local, then $Z(R)$ is the maximal ideal of $R$ and $|R| \leq |Z(R)|^2$. If $R$ is not local, then $R = R_1 \times R_2 \times \cdots \times R_n$, where each of the $R_i$'s is a local ring and $n \geq 2$. Suppose that $|R_1| \leq |R_2| \leq \cdots \leq |R_n|$ and set $R_1^* = 0 \times R_2 \times \cdots \times R_n$. Since $|R| = |R_1||R_1^*|$, we conclude that $|R| \leq |R_1^*|^2$. Let $S$ denote either $Z(R)$ if $R$ is local or $R_1^*$ if $R$ is not local. Then every pair of elements of $S$ are adjacent in $T(\Gamma(R))$, and hence we have a complete graph $K_{|S|}$ in the structure of $T(\Gamma(R))$. This implies that $g(K_{|S|}) \leq m$. Therefore $\lceil((|S|-3)(|S|-4))/12\rceil \leq m$, and so $|S| \leq \left(7 + \sqrt{49 + 48(m-1)}\right)/2$, and hence $|R| \leq \left(\left(7 + \sqrt{49 + 48(m-1)}\right)/2\right)^2$. □

**Theorem 7.5.10.** ([277, Theorem 1.5]) *Let $R$ be a finite commutative ring with $T(\Gamma(R))$ planar. Then the following hold:*

(i) *If $R$ is a local ring, then $R$ is a field or $R$ is isomorphic to the one of the 9 following rings:*

$$\mathbb{Z}_4, \frac{\mathbb{Z}_2[x]}{(x^2)}, \frac{\mathbb{Z}_2[x]}{(x^3)}, \frac{\mathbb{Z}_2[x,y]}{(x,y)^2}, \frac{\mathbb{Z}_4[x]}{(2x, x^2)}, \frac{\mathbb{Z}_4[x]}{(2x, x^2-2)}, \mathbb{Z}_8, \frac{\mathbb{F}_4[x]}{(x^2)}, \frac{\mathbb{Z}_4[x]}{(x^2+x+1)};$$

(ii) *If $R$ is not a local ring, then $R$ is isomorphic to $\mathbb{Z}_2 \times \mathbb{Z}_2$ or $\mathbb{Z}_6$.*

*Proof.* Any planar graph has a vertex $v$ with $\deg(v) \leq 5$. So if the total graph of $R$ is planar, then $\delta(\ T(\Gamma(R))) \leq 5$. By Lemma 7.4.6, $\delta(\ T(\Gamma(R))) = |Z(R)|$ or $|Z(R)| - 1$, and hence $|Z(R)| \leq 6$.

(i) Assume that $R$ is a local ring, and let $n = |Z(R)|$ and $m = |R/Z(R)|$. If $2 \in Z(R)$, then $T(\Gamma(R)) \cong mK_n$ by Theorem 7.2.1. Hence $|Z(R)| \leq 4$.

Also, $|R| = 2^k$ since $2 \in Z(R)$. Thus $|R| = 16, 8, 4$, or 2. According to Corbas and Williams [167], there are two nonisomorphic rings of order 16 with maximal ideals of order 4, namely $\mathbb{F}_4[x]/(x^2)$ and $\mathbb{Z}_4[x]/(x^2+x+1)$; so for these rings, we have $T(\Gamma(R)) \cong 4K_4$. Since $K_4$ is planar, we conclude that the total graphs of these rings are planar. In [167], it is also shown that there are 5 local rings of order 8 (except $\mathbb{F}_8$). In all of these rings, we have $|Z(R)| = 4$, and hence $T(\Gamma(R)) \cong 2K_4$. Also, there are two nonisomorphic local rings of order 4; these are $\mathbb{Z}_4$ and $\mathbb{Z}_2[x]/(x^2)$. For both, we have $T(\Gamma(R)) \cong 2K_2$, and thus they are planar. Note that if $|Z(R)| = 1$, then $R$ is a field, and hence the total graph is planar. If $2 \notin Z(R)$, then $T(\Gamma(R)) \cong K_n \cup \left(\frac{m-1}{2}\right) K_{n,n}$ by Theorem 7.2.1. This implies that $n \leq 2$, and thus $R$ either has order 4 or is a field.

(ii) Suppose that $R$ is not a local ring. Since $R$ is finite, there are finite local rings $R_i$ such that $R = R_1 \times \cdots \times R_t$, where $t \geq 2$. Since $|Z(R)| \leq 6$, we have the following rings:

$$\mathbb{Z}_2 \times \mathbb{Z}_2, \mathbb{Z}_6, \mathbb{Z}_2 \times \mathbb{F}_4, \mathbb{Z}_2 \times \mathbb{Z}_4, \mathbb{Z}_2 \times \mathbb{Z}_2[x]/(x^2), \mathbb{Z}_2 \times \mathbb{Z}_5, \mathbb{Z}_3 \times \mathbb{Z}_3, \mathbb{Z}_3 \times \mathbb{F}_4.$$

The total graph of $\mathbb{Z}_2 \times \mathbb{Z}_2$ is isomorphic to the cycle $C_4$, and this graph is planar. By Lemma 7.5.8, the total graph of $\mathbb{Z}_6 \cong \mathbb{Z}_2 \times \mathbb{Z}_3$ is isomorphic to $K_2 \times K_3$, which is also planar.

Let $R$ be a commutative ring with $|R| = n$. The subgraph of the total graph of $\mathbb{Z}_2 \times R$ induced by $\{0\} \times R$ is a copy of $K_n$. The edge $(1,0) - (0,0)$, together with the paths $(1,0) - (1,-r) - (0,r)$ for every $r \in R$, yield a subdivision of $K_{n+1}$ in the total graph of $\mathbb{Z}_2 \times R$. Thus the total graphs of $\mathbb{Z}_2 \times \mathbb{F}_4$, $\mathbb{Z}_2 \times \mathbb{Z}_4$, $\mathbb{Z}_2 \times \mathbb{Z}_2[X]/(X^2)$, and $\mathbb{Z}_2 \times \mathbb{Z}_5$ are not planar. Also, the total graph of $\mathbb{Z}_3 \times R$ contains a subgraph which is isomorphic to $K_{3,n}$ (consider the induced subgraph $\langle S \rangle$, where $S = \{(1,r) \mid r \in R\} \cup \{(2,r) \mid r \in R\}$). Thus the total graphs of $\mathbb{Z}_3 \times \mathbb{Z}_3$ and $\mathbb{Z}_3 \times \mathbb{F}_4$ are not planar. □

**Theorem 7.5.11.** ([277, Theorem 1.6]) *Let $R$ be a finite commutative ring with $T(\Gamma(R))$ toroidal. Then the following statements hold:*

(i) *If $R$ is a local ring, then $R$ is isomorphic to $\mathbb{Z}_9$ or $\mathbb{Z}_3[x]/(x^2)$;*

(ii) *If $R$ is not a local ring, then $R$ is isomorphic to one of the following rings:*

$$\mathbb{Z}_2 \times \mathbb{F}_4, \mathbb{Z}_3 \times \mathbb{Z}_3, \mathbb{Z}_2 \times \mathbb{Z}_4, \mathbb{Z}_2 \times \mathbb{Z}_2[x]/(x^2), \mathbb{Z}_2 \times \mathbb{Z}_2 \times \mathbb{Z}_2$$

*Proof.* For any graph $G$ with $\nu$ vertices and genus $g$, we have $\delta(G) \leq 6 + (12g - 12)/\nu$. If $g(G) = 1$, then $\delta(G) \leq 6$, and equality holds if and only if $G$ is a triangulation of the torus and 6-regular (see [430, Proposition 2.1]). If $R$ is a finite ring with toroidal total graph, then $\delta(T(\Gamma(R))) \leq 6$, and by Lemma 7.4.6 $\delta(T(\Gamma(R))) = |Z(R)|$ or $|Z(R)| - 1$. Thus $|Z(R)| \leq 7$.

(i) Let $R$ be a local ring. If $2 \in Z(R)$, then $T(\Gamma(R))$ is a disjoint union of copies of the complete graph $K_n$, where $|Z(R)| = n$. Hence $5 \leq n \leq 7$. But in this case, $|Z(R)|$ is a power of 2, and thus there are no such local rings. Now suppose that $2 \notin Z(R)$. Then $T(\Gamma(R)) \cong K_n \cup \binom{m-1}{2} K_{n,n}$, where $n = |Z(R)|$ and $m = |R/Z(R)|$. Thus $3 \leq n \leq 4$, and since $2 \notin Z(R)$, we must have $n = 3$. There are two local rings, $\mathbb{Z}_9$ and $\mathbb{Z}_3[x]/(x^2)$, such that the cardinality of the set of the zero-divisors is 3; both of these rings have total graph $K_3 \cup K_{3,3}$ which is toroidal.

(ii) Assume that $R$ is not a local ring. Since $|Z(R)| \leq 7$, we have the following rings for $R$ by Theorem 1.5(b):

$$\mathbb{Z}_2\times\mathbb{F}_4, \mathbb{Z}_2\times\mathbb{Z}_4, \mathbb{Z}_2\times\mathbb{Z}_2[x]/(x^2), \mathbb{Z}_2\times\mathbb{Z}_5, \mathbb{Z}_3\times\mathbb{Z}_3, \mathbb{Z}_3\times\mathbb{F}_4, \mathbb{Z}_3\times \mathbb{Z}_5, \mathbb{F}_4\times\mathbb{F}_4, \mathbb{Z}_2\times\mathbb{Z}_2\times\mathbb{Z}_2.$$

By Theorem 7.5.10, $g(T(\Gamma(\mathbb{Z}_2 \times \mathbb{F}_4))), g(T(\Gamma(\mathbb{Z}_2 \times \mathbb{Z}_4))), g(T(\Gamma(\mathbb{Z}_3 \times \mathbb{Z}_3)))$ are all at least 1. The embeddings in Figure 7.7 parts (a), (b), and (c) show explicitly that $g(T(\Gamma(\mathbb{Z}_2 \times \mathbb{F}_4))) = g(T(\Gamma(\mathbb{Z}_2 \times \mathbb{Z}_4))) = g(T(\Gamma(\mathbb{Z}_3 \times \mathbb{Z}_3))) = 1$. Since $T(\Gamma(\mathbb{Z}_2 \times \mathbb{Z}_2[x]/(x^2))) \cong T(\Gamma(\mathbb{Z}_2 \times \mathbb{Z}_4))$, then $g(T(\Gamma(\mathbb{Z}_2 \times \mathbb{Z}_2[x]/(x^2)))) = 1$.

If we partition the elements of $\mathbb{Z}_2 \times \mathbb{Z}_2 \times \mathbb{Z}_2$ as $V_1 = \{(0,0,0),(1,1,1)\}$, $V_2 = \{(1,0,0),(0,1,1)\}$, $V_3 = \{(0,1,0),(1,0,1)\}$, and $V_4 = \{(0,0,1),(1,1,0)\}$, it is clear that $T(\Gamma(\mathbb{Z}_2 \times \mathbb{Z}_2 \times \mathbb{Z}_2)) = K_{2,2,2,2}$. Hence $g(T(\Gamma(\mathbb{Z}_2 \times \mathbb{Z}_2 \times \mathbb{Z}_2))) = 1$ by [173, Corollary 4].

By Lemma 7.5.8, $g(T(\Gamma(\mathbb{Z}_2 \times \mathbb{Z}_5))) \geq 2$, and thus is not toroidal. By (the proof of) Lemma 7.5.8, $g(T(\Gamma(\mathbb{F}_4\times\mathbb{F}_4))) \geq 2g(T(\Gamma(\mathbb{Z}_2\times\mathbb{F}_4)))$, and so by Theorem 7.5.10 $T(\Gamma(\mathbb{F}_4 \times \mathbb{F}_4))$ is not toroidal.

Finally, Figure 7.7 part (d) shows a subgraph of $T(\Gamma(\mathbb{Z}_3 \times \mathbb{F}_4))$ that is a subdivision of $K_{5,4}$, and thus by Theorem 7.5.1 $g(T(\Gamma(\mathbb{Z}_3 \times \mathbb{F}_4))) \geq 2$. Further, $T(\Gamma(\mathbb{Z}_3 \times \mathbb{Z}_5))$ is a super graph of $T(\Gamma(\mathbb{Z}_3 \times \mathbb{F}_4))$, and so $g(T(\Gamma(\mathbb{Z}_3 \times \mathbb{Z}_5))) \geq g(T(\Gamma(\mathbb{Z}_3 \times \mathbb{F}_4))) \geq 2$ □

The next theorem gives a lower bound for the genus of the total graph of a finite commutative ring.

**Theorem 7.5.12.** ([365, Theorem 3.2]) *Let $R$ be a finite commutative ring, $I$ an ideal contained in $Z(R)$, $|I| = \lambda \geq 3$, and $|R/I| = \mu$. Then the following are true:*

(i) *If $2 \in I$, then $g(T(\Gamma(R))) \geq \mu \left\lceil \frac{(\lambda-3)(\lambda-4)}{12} \right\rceil$;*

(ii) *If $2 \notin I$, then $g(T(\Gamma(R))) \geq \left\lceil \frac{(\lambda-3)(\lambda-4)}{12} \right\rceil + \binom{\mu-1}{2} \left\lceil \frac{(\lambda-2)^2}{4} \right\rceil$.*

*Proof.* Let $H$ be the spanning subgraph of $T(\Gamma(R))$, where two distinct vertices $x, y \in R$ are adjacent if $x + y \in I$. By noting that $I$ is an ideal contained in $Z(R)$, we obtain in the same way as in Theorem 7.2.1 that $H$ is a subgraph of

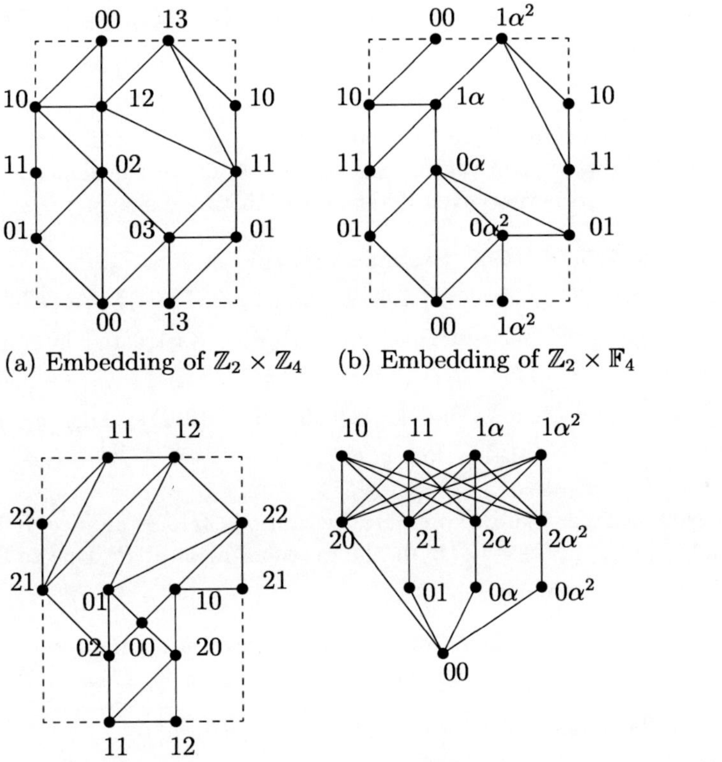


(a) Embedding of $\mathbb{Z}_2 \times \mathbb{Z}_4$ (b) Embedding of $\mathbb{Z}_2 \times \mathbb{F}_4$

(c) Embedding of $\mathbb{Z}_3 \times \mathbb{Z}_3$ (d) A subgraph of $T(\Gamma(\mathbb{Z}_3 \times \mathbb{F}_4))$ that is a subdivision of $K_{5,4}$

Figure 7.7: Embeddings in the torus and a subgraph of $T(\Gamma(\mathbb{Z}_3 \times \mathbb{F}_4))$

$G$, and further

$$H = \begin{cases} K_\lambda \cup \underbrace{K_\lambda \cup \cdots \cup K_\lambda}_{(\mu-1)\ copies} & \text{if } 2 \in I \\ K_\lambda \cup \underbrace{K_{\lambda,\lambda} \cup K_{\lambda,\lambda} \cup \cdots \cup K_{\lambda,\lambda}}_{(\frac{\mu-1}{2})\ copies} & \text{if } 2 \notin I. \end{cases}$$

Now, by Theorem 7.5.1, we have

$$g(H) = \begin{cases} \mu \left\lceil \frac{(\lambda-3)(\lambda-4)}{12} \right\rceil & \text{if } 2 \in I \\ \left\lceil \frac{(\lambda-3)(\lambda-4)}{12} \right\rceil + (\frac{\mu-1}{2}) \left\lceil \frac{(\lambda-2)^2}{4} \right\rceil & \text{if } 2 \notin I. \end{cases}$$

Since $H$ is a spanning subgraph of $T(\Gamma(R))$, $g(H) \leq g(T(\Gamma(R)))$, and so the proof is complete. □

Further, note that if we take $I$ as an ideal with $|R/I| = min\{|R/\mathfrak{m}| \mid \mathfrak{m}$ is a maximal ideal of $R\}$, then the bound in Theorem 7.5.12 is better. The following example shows that in various cases, the bounds for the genus of the total graph of a commutative ring given in Theorem 7.5.12 are sharp.

**Example 7.5.13.** ([365, Example 3.3]) (*a*) Let $R = \mathbb{Z}_6$ and $I = \{0, 2, 4\}$. Then $g(T(\Gamma(R))) = 0$ by Theorem 7.5.10, and so $g(T(\Gamma(R))) = \mu \left\lceil \frac{(\lambda-3)(\lambda-4)}{12} \right\rceil$.

(*b*) Let $R = \mathbb{Z}_{10}$ and $I = \{0, 2, 4, 6, 8\}$. As treated later in the proof of Theorem 7.5.21, $g(T(\Gamma(R))) = 2$, and so $g(T(\Gamma(R))) = \mu \left\lceil \frac{(\lambda-3)(\lambda-4)}{12} \right\rceil$.

(*c*) Let $R = \mathbb{Z}_3 \times \mathbb{Z}_3$ and $I = \{(0, 0), (0, 1), (0, 2)\}$. Then $g(T(\Gamma(R))) = 1$ by Theorem 7.5.11, and therefore $g(T(\Gamma(R))) = \left\lceil \frac{(\lambda-3)(\lambda-4)}{12} \right\rceil + (\frac{\mu-1}{2}) \left\lceil \frac{(\lambda-2)^2}{4} \right\rceil$.

Note that for a finite commutative ring $R$, $Z(R)$ is an ideal if and only if $R$ is local, and in this case, $Z(R)$ is the maximal ideal of $R$. Due to this, we restate Theorem 7.2.1 as follows:

**Corollary 7.5.14.** ([365, Corollary 3.4]) *Let $R$ be a finite commutative ring with $Z(R)$ an ideal of $R$, $|Z(R)| = \alpha$, and $|R/Z(R)| = \beta$. Then the following hold:*

(i) *If $2 \in Z(R)$, then $g(T(\Gamma(R))) = \beta \left\lceil \frac{(\alpha-3)(\alpha-4)}{12} \right\rceil$;*

(ii) *If $2 \notin Z(R)$, then $g(T(\Gamma(R))) = \left\lceil \frac{(\alpha-3)(\alpha-4)}{12} \right\rceil + (\frac{\beta-1}{2}) \left\lceil \frac{(\alpha-2)^2}{4} \right\rceil$.*

By Theorem 7.5.12, one can also obtain Lemma 7.5.8.

**Remark 7.5.15.** ([365, Remark 3.6]) Let $R = \mathbb{F}_{2^4} \times \mathbb{F}_9$. Then $g(T(\Gamma(R))) \geq 209$ by Theorem 7.5.12, but $g(T(\Gamma(R))) \geq 48$ by Lemma 7.5.8. Therefore the bound for the genus given by Theorem 7.5.12 is tighter than the bound given in Lemma 7.5.8.

Now we obtain some better lower bounds for the genus of the total graph of a commutative ring $R$ for some special classes of rings.

**Theorem 7.5.16.** ([365, Theorem 3.7]) *Let $R$ be a finite commutative ring, $I$ an ideal with maximal cardinality among the proper ideals in $R$, $|R/I| = \mu$, and $2 \leq \lambda \not\equiv 5, 9$ (mod 12). If $Z(R)$ is not an ideal of $R$ and $2 \in I$, then*

$$g(T(\Gamma(R))) \geq \begin{cases} \frac{\mu}{2} \left\lceil \frac{(\lambda-2)(\lambda-3)}{6} \right\rceil & \text{if } \mu \text{ is even;} \\ (\frac{\mu-1}{2}) \left\lceil \frac{(\lambda-2)(\lambda-3)}{6} \right\rceil + \left\lceil \frac{(\lambda-3)(\lambda-4)}{12} \right\rceil & \text{if } \mu \text{ is odd.} \end{cases}$$

*Proof.* Since $Z(R)$ is not an ideal of $R$, there is a $z \in Z(R) \setminus I$. Note that $x_i + I$ and $z + x_i + I$ are distinct for every $x_i \in R$. We can pair the cosets of $I$ as $\{x_i + I, z + x_i + I\}$, and every coset appears in exactly one pair. (In particular, $|R/I|$ is even.) Every pair induces a copy of $K_2 \times K_\lambda$ in the total graph, and the copies have no vertices in common. The bound then follows from Lemma 7.5.3. □

Recall that every finite commutative ring is a direct product of local rings. Using this, we obtain another lower bound for the genus of the total graph of a finite commutative ring. Moreover, this lower bound on the genus of a direct product depends on the order in which one "decomposes" the total graph. Further, using a relationship between the total graph of a product of rings and the Cartesian product of total graphs of the components, we obtain the following theorem.

**Theorem 7.5.17.** ([365, Theorem 3.8]) *Let $R$ be a finite commutative ring and $R \cong R_1 \times R_2 \times \cdots \times R_m$, where every $R_i$ is a finite local ring. Let $G_i = T(\Gamma(R_i))$, $k_i = g(G_i)$, $\ell_1 = max\{|R_1|k_2 + k_1, |R_2|k_1 + k_2\}$, and $\ell_j = max\{(\sum_{i=1}^{j} |R_i|)k_{j+1} + \ell_{j-1}, |R_j|\ell_{j-1} + k_{j+1}\}$ for $j = 2, \ldots, m-1$. Then $\ell_{m-1} \leq g(T(\Gamma(R)))$.*

*Proof.* First we prove that the Cartesian (graph) product of the total graphs is a subgraph of the total graph of a Cartesian (ring) product. We proceed by induction. For the base case, suppose that $R \cong R_1 \times R_2$. Let $G_i = T(\Gamma(R_i))$, $G = T(\Gamma(R_1 \times R_2))$, and $H = G_1 \Box G_2$. For $x_i, y_i \in R_i$, if $(x_1, x_2)(y_1, y_2) \in E(H)$, then either $x_1y_1 \in E(G_1)$ and $x_2 = y_2$, or $x_2y_2 \in E(G_2)$ and $x_1 = y_2$. Thus $x_1 + y_1 \in Z(R_1)$ or $x_2 + y_2 \in Z(R_2)$. Since $Reg(R_1 \times R_2) = Reg(R_1) \times Reg(R_2)$, in both cases $(x_1, x_2) + (y_1, y_2) \in Z(R_1 \times R_2)$, and so $(x_1, x_2)(y_1, y_2) \in E(G)$. From this, $H$ is a spanning subgraph of $G$. By induction, we get that $T(\Gamma(R_1)) \times \cdots \times T(\Gamma(R_m))$ is a spanning subgraph of $T(\Gamma(R_1 \times \cdots \times R_m))$.

Let $G_i = T(\Gamma(R_i))$, $|Z(R_i)| = \lambda_i$, and $|R_i/Z(R_i)| = \mu_i$ for $i = 1, \ldots, m$. As per the notations of the statement and by Corollary 7.5.14, for $i = 1, \ldots, m$, we have

$$k_i = \begin{cases} \mu_i \left\lceil \frac{(\lambda_i-3)(\lambda_i-4)}{12} \right\rceil & \text{if } 2 \in Z(R_i); \\ \left\lceil \frac{(\lambda_i-3)(\lambda_i-4)}{12} \right\rceil + \left(\frac{\mu_i-1}{2}\right) \left\lceil \frac{(\lambda_i-2)^2}{4} \right\rceil & \text{if } 2 \notin Z(R_i). \end{cases}$$

Since $T(\Gamma(R_1)) \Box \cdots \Box T(\Gamma(R_m))$ is a subgraph of $T(\Gamma(R_1 \times \cdots \times R_m))$, we have $g(T(\Gamma(R_1)) \Box \cdots \Box T(\Gamma(R_m))) \leq g(T(\Gamma(R_1 \times \cdots \times R_m)))$. From this and by the repeated application of Theorem 7.5.1(c), we have $\ell_{m-1} \leq g(T(\Gamma(R_1)) \Box \cdots \Box T(\Gamma(R_m))) \leq g(T(\Gamma(R)))$. □

The following example illustrates the use of Theorem 7.5.17.

**Example 7.5.18.** ([365, Example 3.9]) (*a*) Let $R = \mathbb{Z}_9 \times \mathbb{Z}_3[x]/(x^2)$. By Theorem 7.5.11, $k_1 = g(T(\Gamma(\mathbb{Z}_9))) = 1$ and $k_2 = g(T(\Gamma(\mathbb{Z}_3[x]/(x^2)))) = 1$. Thus $g(T(\Gamma(R))) \geq 10$ by Theorem 7.5.17.

(*b*) Let $R = \mathbb{Z}_9 \times \mathbb{Z}_9 \times \mathbb{Z}_9$. Then $k_i = g(T(\Gamma(\mathbb{Z}_9))) = 1$ for $i = 1, 2, 3$. Then $\ell_1 = 10$ and $\ell_2 = \max\{28, 91\} = 91$; also $g(T(\Gamma(R))) \geq 91$ by Theorem 7.5.17.

The following theorem gives an upper bound for the genus of the total graph of a commutative ring.

**Theorem 7.5.19.** ([365, Theorem 3.10]) *Let $R$ be a finite commutative ring and $Z(R) = \cup_{i=1}^m P_i$, where the $P_i$'s are ideals of $R$. Let $|P_i| = \alpha_i$ and $|R_i/P_i| = \beta_i$ for $i = 1, \ldots, m$. Suppose that $2 \in P_t$ for every $1 \leq t \leq j$ and $2 \notin P_t$ for every $j+1 \leq t \leq m$, where $j = 0$ if $2 \notin Z(R)$. Then*

$$g(T(\Gamma(R))) \leq (m-1)(|R|-1) + \sum_{t=1}^{j} \beta_t \left\lceil \frac{(\alpha_t-3)(\alpha_t-4)}{12} \right\rceil + \sum_{t=j+1}^{m} \left\{ \left(\frac{\beta_t - 1}{2}\right) \left\lceil \frac{(\alpha_t-2)^2}{4} \right\rceil + \left\lceil \frac{(\alpha_t-3)(\alpha_t-4)}{12} \right\rceil \right\}.$$

*(If $\alpha_i = 2$ for some $i$, then we take $\left\lceil \frac{(\alpha_i-3)(\alpha_i-4)}{12} \right\rceil = 0$.)*

*Proof.* Let $H_i$ be the spanning subgraph of $T(\Gamma(R))$, where two distinct vertices $x, y \in R$ are adjacent if $x + y \in P_i$ for $i = 1, \ldots, m$. Then, as in proof of Theorem 7.5.12,

$$H_i = \begin{cases} K_{\alpha_i} \cup \underbrace{K_{\alpha_i} \cup \cdots \cup K_{\alpha_i}}_{\beta_i - 1 \text{ copies}} & \text{if } 2 \in P_i; \\ K_{\alpha_i} \cup \underbrace{K_{\alpha_i,\alpha_i} \cup \cdots \cup K_{\alpha_i,\alpha_i}}_{\frac{\beta_i - 1}{2} \text{ copies}} & \text{if } 2 \notin P_i. \end{cases}$$

Since $Z(R) = \cup_{i=1}^m P_i$, we have $T(\Gamma(R)) = \cup_{i=1}^m H_i$. Deleting duplicate edges if necessary, we can write $T(\Gamma(R)) = \cup_{i=1}^m H_i'$, where $H_i' \subseteq H_i$. The bound then follows from Lemma 7.5.7. □

The following example shows that in some cases the upper bound in Theorem 7.5.19 is better than the one given in Corollary 7.5.2.

**Example 7.5.20.** ([365, Example 3.11]) (*a*) Let $R = \mathbb{Z}_{26}$. By Theorem 7.5.19, $g(T(\Gamma(R))) \leq 41$, whereas by Corollary 7.5.2, $g(T(\Gamma(R))) \leq 43$.

(*b*) Let $R = \mathbb{Z}_{55}$. By Theorem 7.5.19, $g(T(\Gamma(R))) \leq 117$, whereas by Corollary 7.5.2, $g(T(\Gamma(R))) \leq 221$.

The following theorem gives a characterization of all commutative rings whose total graph has genus two.

**Theorem 7.5.21.** ([365, Theorem 4.3]) *Let $R$ be a finite commutative ring. Then $g(T(\Gamma(R))) = 2$ if and only if $R$ is isomorphic to either $\mathbb{Z}_{10}$ or $\mathbb{Z}_3 \times \mathbb{F}_4$.*

*Proof.* Let $|R| = n$, $|Z(R)| = \alpha$, and $|R/Z(R)| = \beta$. Assume that $g(T(\Gamma(R))) = 2$.

If $R$ is local, then Corollary 7.5.14 applies with $\beta$ a prime power and $\alpha$ a power of $\beta$. However, in this case neither possibility for the equation $g(T(\Gamma(R))) = 2$ has a solution. Hence $R$ is not a local ring.

It is well known that $R = R_1 \times \cdots \times R_m$, where the $R_i$'s are finite local rings and $m \geq 2$. For further use in the proof, we list below all finite commutative rings $R$ with $|Z(R)| = \alpha \leq 7$.

$$\mathbb{Z}_2 \times \mathbb{Z}_2,\ \mathbb{Z}_6,\ \mathbb{Z}_2 \times \mathbb{F}_4,\ \mathbb{Z}_2 \times \mathbb{Z}_4,\ \mathbb{Z}_2 \times \mathbb{Z}_2[x]/(x^2), \mathbb{Z}_2 \times \mathbb{Z}_5, \mathbb{Z}_3 \times \mathbb{Z}_3,$$

$$\mathbb{Z}_3 \times \mathbb{F}_4,\ \mathbb{Z}_3 \times \mathbb{Z}_5,\ \mathbb{F}_4 \times \mathbb{F}_4,\ \mathbb{Z}_2 \times \mathbb{Z}_2 \times \mathbb{Z}_2. \tag{7.1}$$

Note that by Lemma 7.5.5, $\delta(G) \leq 6 + \frac{12g-12}{n}$. Since $g(T(\Gamma(R))) = 2$, $\delta(T(\Gamma(R))) \leq 6 + \frac{12}{n}$.

**Case 1**. Suppose $n \geq 13$; then $\delta(G) \leq 6$, and so by Lemma 7.4.6, $\alpha \leq 7$. Among the rings in list (7.1), there are two rings, viz., $\mathbb{Z}_3 \times \mathbb{Z}_5$ and $\mathbb{F}_4 \times \mathbb{F}_4$ satisfying the condition $n \geq 13$. Note that by Theorem 7.5.10 and Theorem 7.5.11, $g(T(\Gamma(\mathbb{Z}_3 \times \mathbb{Z}_5))) \geq 2$ and $g(T(\Gamma(\mathbb{F}_4 \times \mathbb{F}_4))) \geq 2$. Thus in this case, $R$ may be isomorphic to either $\mathbb{Z}_3 \times \mathbb{Z}_5$ or $\mathbb{F}_4 \times \mathbb{F}_4$.

**Case 2**. Suppose $7 \leq n \leq 12$; then $\delta(G) \leq 7$, and so by Lemma 7.4.6, $\alpha \leq 8$.

If $\alpha \leq 7$, then by list (7.1), we have the following candidates for $R$:

$$\mathbb{Z}_2 \times \mathbb{F}_4,\ \mathbb{Z}_2 \times \mathbb{Z}_4,\ \mathbb{Z}_2 \times \mathbb{Z}_2[x]/(x^2), \mathbb{Z}_2 \times \mathbb{Z}_5,\ \mathbb{Z}_3 \times \mathbb{Z}_3,\ \mathbb{Z}_3 \times \mathbb{F}_4,\ \mathbb{Z}_2 \times \mathbb{Z}_2 \times \mathbb{Z}_2.$$

Since $g(T(\Gamma(R))) = 2$, by Theorem 7.5.10 and Theorem 7.5.11, $R$ may be isomorphic to either $\mathbb{Z}_2 \times \mathbb{Z}_5$ or $\mathbb{Z}_3 \times \mathbb{F}_4$.

If $\alpha = 8$, then obviously $9 \leq n \leq 12$. Note that there are only two rings, viz., $R \cong \mathbb{Z}_3 \times \mathbb{Z}_2[x]/(x^2)$ and $\mathbb{Z}_3 \times \mathbb{Z}_4$ satisfying these conditions.

Hence if $g(T(\Gamma(R))) = 2$, then $R$ may be isomorphic to one of the 6 following rings:

$$\mathbb{Z}_2 \times \mathbb{Z}_5,\ \mathbb{Z}_3 \times \mathbb{F}_4,\ \mathbb{Z}_3 \times \mathbb{Z}_4,\ \mathbb{Z}_3 \times \mathbb{Z}_2[x]/(x^2),\ \mathbb{Z}_3 \times \mathbb{Z}_5,\ \mathbb{F}_4 \times \mathbb{F}_4.$$

Consider the ring $\mathbb{Z}_2 \times \mathbb{Z}_5 \cong \mathbb{Z}_{10}$. Then $Z(\mathbb{Z}_{10}) = P_1 \cup P_2$, where $P_1 = \{0, 2, 4, 6, 8\}$ and $P_2 = \{0, 5\}$. Thus by the proof of Theorem 7.5.16, $T(\Gamma(\mathbb{Z}_{10})) = K_5 \Box K_2$, and so by Lemma 7.5.3, $g(T(\Gamma(\mathbb{Z}_{10}))) = 2$.

Consider the ring $\mathbb{Z}_3 \times \mathbb{F}_4$. By Theorem 7.5.10 and Theorem 7.5.11, $T(\Gamma(\mathbb{Z}_3 \times \mathbb{F}_4))$ is neither planar nor toroidal, and so $g(T(\Gamma(\mathbb{Z}_3 \times \mathbb{F}_4))) > 1$. In Figure 7.8, we exhibit an embedding of $T(\Gamma(\mathbb{Z}_3 \times \mathbb{F}_4))$ in $S_2$. Therefore $g(T(\Gamma(\mathbb{Z}_3 \times \mathbb{F}_4))) = 2$. Note that in Figure 7.8, an element $(x, y)$ of $\mathbb{Z}_3 \times \mathbb{F}_4$ is denoted as $xy$.

Consider the ring $\mathbb{Z}_3 \times \mathbb{Z}_4 \cong \mathbb{Z}_{12}$. Here $Z(\mathbb{Z}_{12}) = \{0, 2, 3, 4, 6, 8, 9, 10\}$. Let $H_1$ be the spanning subgraph of $T(\Gamma(\mathbb{Z}_{12}))$ in which two distinct vertices $x, y \in \mathbb{Z}_{12}$ are adjacent if $x+y \in Z(\mathbb{Z}_{12}) - \{9\}$. Note that $H_1 = K_6 \Box K_2$, and by Lemma 7.5.3,

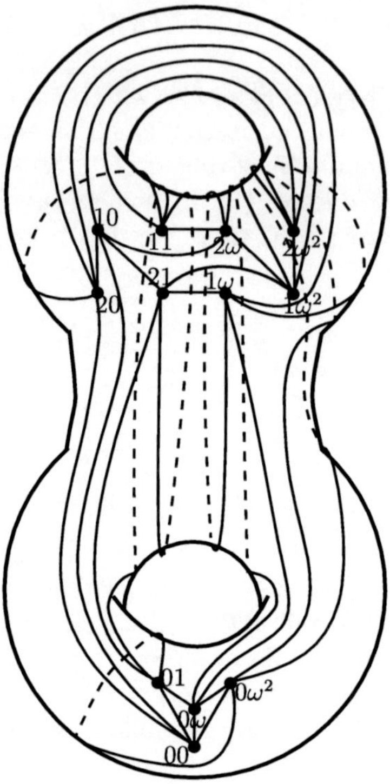


Figure 7.8: Embedding of $T_{\Gamma}(\mathbb{Z}_3 \times \mathbb{F}_4)$ in $S_2$

$g(H_1) = 2$. By Euler's Formula (Lemma 7.5.4), a cellular embedding of $H_1$ in $S_2$ has 22 faces. Let $f$ be the number of faces when $H_1$ is embedded in $S_2$ and $f_i$ be the number of $i$-gons in the embedding of $H_1$ in $S_2$. As noticed in the proof of [330, Theorem 1.1],

$$f_4 + 2f_5 + 3f_6 + \cdots = 2|E| - 3f = 6 \tag{7.2}$$

and $3f_3 \leq 2n^2 - 4n = 48$ implies $f_3 \leq 16$.

Since $f = f_3 + f_4 + \cdots = 22$, we get that

$$f_4 + f_5 + \cdots \geq 6. \tag{7.3}$$

By equations (7.2) and (7.3), $f_4 = 6$ and $f_5 = f_6 = \cdots = 0$. Consider the edges in $E\left(T(\Gamma(\mathbb{Z}_{12}))\right) - E(H_1)$. These edges are precisely the pairs of vertices $x, y \in \mathbb{Z}_{12}$ satisfying $x + y = 9$. Note that the vertex 0 is adjacent to 9. Suppose 0 and 9 are not in the same region of a cellular embedding of $H_1$ in $S_2$. Then the edge joining 0 and 9 must cross an edge in an extension to an embedding of $T(\Gamma(\mathbb{Z}_{12}))$ in $S_2$, and so $g(T(\Gamma(\mathbb{Z}_{12}))) \geq 3$. Suppose 0 and 9 are in the same region of a cellular embedding of $H_1$ in $S_2$. Since 0 (or 9) is adjacent to only one odd (or even) vertex in $H_1$, the region containing 0 and 9 is the rectangle formed by the vertices $\{0, 3, 9, 6\}$. Then the edge joining 0 and 9 can be drawn in the cellular embedding of $T(\Gamma(\mathbb{Z}_{12}))$ in $S_2$. If 3 and 6 are not vertices of any other face with 4 edges, then the edge joining 3 and 6 must cross a edge in embedding of $T(\Gamma(\mathbb{Z}_{12}))$ in $S_2$, and so $g(T(\Gamma(\mathbb{Z}_{12}))) \geq 3$. Suppose 3 and 6 are vertices of another rectangle face; then the rectangle must be formed by the same set of vertices $\{0, 3, 9, 6\}$. Note that $K_6 \cup K_6$ is a subgraph of $H_1$, and so the embedding of $K_6 \cup K_6$ is obtained by embedding each copy of $K_6$ in a torus (by Theorem 7.5.1). Since every vertex in one copy of $K_6$ in $H_1$ is adjacent to a vertex in the other copy of $K_6$, the cellular embedding of $H_1$ in $S_2$ has 6 distinct rectangle faces. Since $f_4 = 6$, the rectangle formed by $\{0, 3, 9, 6\}$ cannot occur more than once, a contradiction.

Let $R = \mathbb{Z}_3 \times \mathbb{Z}_2[x]/(x^2)$. Define a mapping $f : V(\mathbb{Z}_3 \times \mathbb{Z}_2[x]/(x^2)) \rightarrow V(\mathbb{Z}_{12})$ by $f((0,(x^2))) = 0$, $f((1,(x^2))) = 2$, $f((2,(x^2))) = 4$, $f((0,(x + x^2))) = 6$, $f((1, x+(x^2))) = 8$, $f((2, x+(x^2))) = 10$, $f((0, 1+(x^2))) = 1$, $f((1, 1+(x^2))) = 3$, $f((2, 1+(x^2))) = 5$, $f((0, 1+x+(x^2))) = 7$, $f((1, 1+x+(x^2))) = 9$, $f((2, 1+x+(x^2))) = 11$. Clearly $f$ is a graph-isomorphism, and so $T(\Gamma(\mathbb{Z}_3 \times \mathbb{Z}_2[x]/(x^2))) \cong T(\Gamma(\mathbb{Z}_{12}))$. Thus $g(T(\Gamma(\mathbb{Z}_3 \times \mathbb{Z}_2[x]/(x^2)))) = g(T(\Gamma(\mathbb{Z}_{12}))) \geq 3$.

If $R = \mathbb{Z}_3 \times \mathbb{Z}_5 \cong \mathbb{Z}_{15}$, then the proper ideal with maximal cardinality is $I = \{0, 3, 6, 9, 12\}$ and $2 \notin I$. Therefore $|I| = \lambda = 5$ and $|R/I| = \mu = 3$. Thus $g(T(\Gamma(R))) \geq 4$ by Theorem 7.5.12.

Let $R = \mathbb{F}_4 \times \mathbb{F}_4$. Take $\mathbb{F}_4 = \{0, 1, \omega, \omega^2\}$. Then $Z(R) = \{(0,0), (0,1), (0,\omega), (0,\omega^2), (1,0), (\omega,0), (\omega^2,0)\}$. Let $H_2$ be the spanning subgraph of $T(\Gamma(R))$, where two distinct vertices $x, y \in R$ are adjacent if $x + y \in \{(0,0), (0,1), (0,\omega), (0,\omega^2), (1,0)\}$. Actually, $H_2 = (K_4 \square K_2) \cup (K_4 \square K_2)$, and by Lemma 7.5.3, $g(H_2) = 2$. By Euler's Formula, a cellular embedding of $H_2$ in $S_2$ has 14 faces and the number of edges in $E(T(\Gamma(R))) - E(H_2)$ is 16. Suppose $T(\Gamma(R))$ has a cellular embedding in $S_2$ and consider the induced embedding of the subgraph $H_2 = 2(K_4 \square K_2)$. Since $H_2$ is the disjoint union of graphs, the embedding of $H_2$ is obtained by embedding each copy of $K_4 \square K_2$ in a torus (by Lemma 7.5.3), and then attaching a cylinder to each torus, interior to some face in the respective embeddings. Let $L = K_4 \square K_2$. Any embedding of $L$ in a torus has 8 vertices, 16 edges, and 8 faces. Since $\sum f_t = 8$ and $\sum t f_t = 16$, we have $\sum t f_{t+1} = 8$. If an embedding of $L$ has a face with 8 edges, then it can have no other faces, a contradiction. So every face in an embedding of $L$ in a torus has at most 7 vertices. In particular, for every face of every embedding of $L$, there exists a vertex in $L$ which is not on the boundary of the face. Considering the degree of each vertex, every vertex in one copy of $L$ in $H_2$ is adjacent, in $T(\Gamma(R))$, to some vertex in the other copy of $L$. This requires an edge crossing, and so we have a contradiction. □

Next, we are interested in the classification of the projective total graphs of commutative rings. Recall that a nonorientable genus 1 graph is called a *projective graph*. Glover, Huneke, and Wang [209] have constructed a list of 103 graphs which are irreducible for the projective plane. Afterward, Archdeacon [90] showed that their list is complete. Hence a graph embeds in the projective plane if and only if it contains no subgraph homeomorphic to one of the graphs in the list of the 103 graphs in [209].

The following results about crosscap numbers and minimum degree of graphs are very useful in the classification of projective total graphs.

**Lemma 7.5.22.** ([233, Lemma 2.2]) *Let $G$ be a connected graph with $n \geq 3$ vertices and crosscap number $k$. Then*

$$\delta(G) \leq 6 + \frac{6k-12}{n}.$$

The following theorem provides a characterization of the commutative local rings $R$ with $T(\Gamma(R))$ projective.

**Theorem 7.5.23.** ([233, Theorem 2.3]) *Let $R$ be a finite commutative ring such that $T(\Gamma(R))$ is a projective graph. Then the following hold:*

(a) *If $R$ is a local ring, then $R$ is isomorphic to $\mathbb{Z}_9$ or $\mathbb{Z}_3[x]/(x^2)$;*

(b) *If $R$ is not a local ring, then $R$ is isomorphic to $\mathbb{Z}_3 \times \mathbb{Z}_3$.*

Theorem 7.5.23, in conjunction with Theorem 7.5.11, shows that if $R$ is a commutative ring whose total graph is projective, then $T(\Gamma(R))$ is also toroidal. In view of this, Khashyarmanesh et al. [233] conjectured the following statement as in the case of zero-divisor graph.

**Conjecture 7.5.24.** *The inequality $\tilde{g}(T(\Gamma(R))) \geq g(T(\Gamma(R)))$ holds for every commutative ring $R$.*

In continuation to the conjecture, Asir and Mano [95] enumerated all isomorphism classes of commutative rings whose total graph has crosscap two. The relevant result with proof is given below.

**Theorem 7.5.25.** *Let $R$ be a finite commutative ring. Then $\tilde{g}(T(\Gamma(R))) = 2$ if and only if $R$ is isomorphic to one of the following 4 rings:*

$$\mathbb{Z}_2 \times \mathbb{F}_4, \mathbb{Z}_2 \times \mathbb{Z}_4, \mathbb{Z}_2 \times \mathbb{Z}_2[x]/(x^2), \mathbb{Z}_2 \times \mathbb{Z}_5.$$

*Proof.* Assume that $\tilde{g}(T(\Gamma(R))) = 2$. Then $\delta(T(\Gamma(R))) \leq 6$, and so $\delta(T(\Gamma(R))) = |Z(R)|$ or $|Z(R)| - 1$, and hence $|Z(R)| \leq 7$.

(a) Assume that $R$ is a local ring. Let $\alpha = |Z(R)|$ and $\beta = |R/Z(R)|$. If $2 \in Z(R)$, then $T(\Gamma(R)) = \beta K_\alpha$. Hence, $\tilde{g}(T(\Gamma(R))) = 2$ if and only if $\alpha$ is either 5 or 6 and $\beta = 2$. But no such local ring exists. If $2 \notin Z(R)$, then $T(\Gamma(R)) = K_\alpha \cup (\frac{\beta-1}{2})K_{\alpha,\alpha}$. So $\tilde{g}(T(\Gamma(R))) = 2$ implies that $\alpha = 4$ and $\beta = 3$, which is not possible.

(b) Assume that $R$ is not a local ring. Since $|Z(R)| \leq 7$, we have the following candidates for $R$:

$$\mathbb{Z}_2 \times \mathbb{Z}_2, \mathbb{Z}_2 \times \mathbb{Z}_3, \mathbb{Z}_2 \times \mathbb{F}_4, \mathbb{Z}_2 \times \mathbb{Z}_4, \mathbb{Z}_2 \times \mathbb{Z}_2[x]/(x^2),$$

$$\mathbb{Z}_2 \times \mathbb{Z}_5, \mathbb{Z}_3 \times \mathbb{Z}_3, \mathbb{Z}_3 \times \mathbb{F}_4, \mathbb{Z}_3 \times \mathbb{Z}_5, \mathbb{F}_4 \times \mathbb{F}_4, \mathbb{Z}_2 \times \mathbb{Z}_2 \times \mathbb{Z}_2.$$

If $R \cong \mathbb{Z}_2 \times \mathbb{Z}_2$ or $\mathbb{Z}_2 \times \mathbb{Z}_3$, then by Proposition 7.5.10, $T(\Gamma(R))$ is planar. Also Proposition 7.5.23 gives that $T(\Gamma(\mathbb{Z}_3 \times \mathbb{Z}_3))$ is projective. The embeddings in Figures 7.9 and 7.10 show explicitly that $\tilde{g}(T(\Gamma(\mathbb{Z}_2 \times \mathbb{F}_4))) = \tilde{g}(T(\Gamma(\mathbb{Z}_2 \times \mathbb{Z}_4))) = \tilde{g}(T(\Gamma(\mathbb{Z}_2 \times \mathbb{Z}_5))) = 2$. Further, $T(\Gamma(\mathbb{Z}_2 \times \mathbb{Z}_4)) \cong T(\Gamma(\mathbb{Z}_2 \times \mathbb{Z}_2[x]/(x^2)))$, and so $\tilde{g}(T(\Gamma(\mathbb{Z}_2 \times \mathbb{Z}_2/(x^2)))) = 2$.

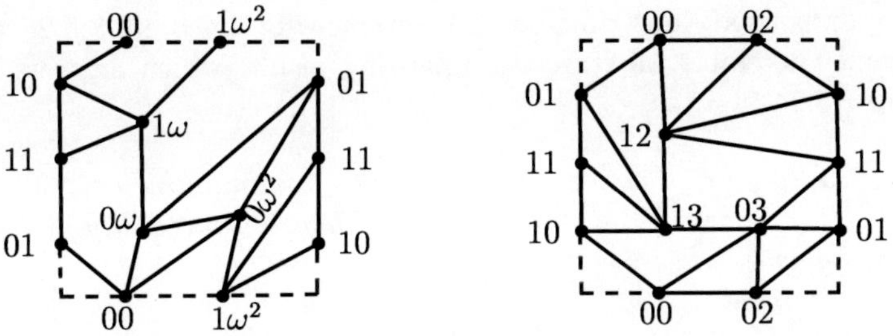


(a) Embedding of $T(\Gamma(\mathbb{Z}_2 \times \mathbb{F}_4))$ in $N_2$ (b) Embedding of $T(\Gamma(\mathbb{Z}_2 \times \mathbb{Z}_4))$ in $N_2$

Figure 7.9:

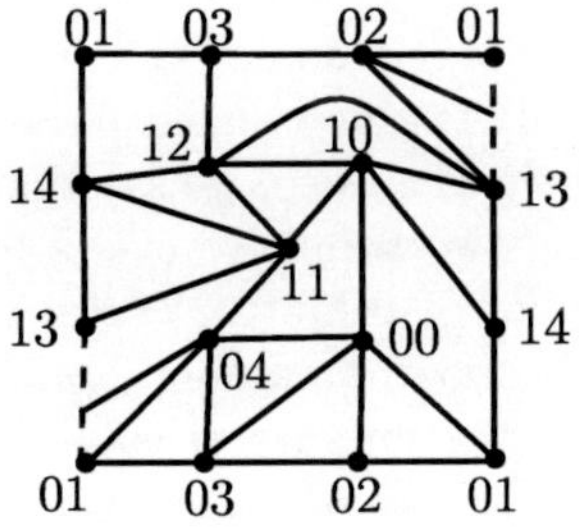


Figure 7.10: Embedding of $T(\Gamma(\mathbb{Z}_2 \times \mathbb{Z}_5))$ in $N_2$

If $R \cong \mathbb{Z}_3 \times \mathbb{F}_4$, then every vertex in the set $V_1 = \{(0,0),(1,0),(1,1),(1,\omega),(1,\omega^2)\}$ is adjacent to each vertex in the set $V_2 = \{(2,0),([(2,1),(0,1)]),([(2,\omega),$

$(0,\omega)]), ([(2,\omega^2),(0,\omega^2)])\}$, and so $K_{5,4}$ is a subgraph of the contraction of $T(\Gamma(R))$. Thus $\tilde{g}(T(\Gamma(\mathbb{Z}_3\times\mathbb{F}_4)))\geq 3$. Recall that, for $xy\in E(G)$, we denote the contracted edge by the vertex $[x,y]$.

If $R\cong\mathbb{Z}_3\times\mathbb{Z}_5$, then the set $V_1=\{1\}\times\mathbb{Z}_5$ is adjacent to every vertex in the set $V_2=\{2\}\times\mathbb{Z}_5$, and so $K_{5,5}$ is a subgraph of $T(\Gamma(R))$. Thus $\tilde{g}(T(\Gamma(\mathbb{Z}_3\times\mathbb{Z}_5)))\geq 5$.

Let $R\cong\mathbb{F}_4\times\mathbb{F}_4$. Let $H$ be the spanning subgraph of $T(\Gamma(R))$ in which two distinct vertices $x,y\in R$ are adjacent if $x+y\in\{(0,0),(0,1),(0,\omega),(0,\omega^2),(1,0)\}\subseteq Z(R)$. It is easy to verify that $H=(K_4\Box K_2)\cup(K_4\Box K_2)$. Now by the fact that $\tilde{g}(K_4\Box K_2)=2$ (see Figure 7.9(a)), we have $\tilde{g}(H)=4$, and so $\tilde{g}(T(\Gamma(R)))\geq 4$.

Let $R\cong\mathbb{Z}_2\times\mathbb{Z}_2\times\mathbb{Z}_2$. We partition the elements of $R$ by the four sets $V_1=\{(0,0,0),(1,1,1)\}, V_2=\{(1,0,0),(0,1,1)\}, V_3=\{(0,1,0),(1,0,1)\}$, and $V_4=\{(0,0,1),(1,1,0)\}$. It is easy to verify that every element of $V_i$ is adjacent to all the elements of $V_j$ for $i\neq j$ and $T(\Gamma(R))\cong K_{2,2,2,2}$. Thus $\tilde{g}(T(\Gamma(R)))=3$. □

By Theorem 7.5.10 and Theorem 7.5.11, we get that the orientable genus of the total graphs of all four rings in the statement of Theorem 7.5.25 are either one or two. Therefore, we state the following result, which answers a case of Conjecture 7.5.24.

**Theorem 7.5.26.** ([95, Theorem 4.5] *Let $R$ be a commutative ring with crosscap number of $T(\Gamma(R))$ less than or equal to 2. Then the inequality $\tilde{g}(T(\Gamma(R)))\geq g(T(\Gamma(R)))$ holds.*

Now we provide conditions so that total graphs are ring graphs or outer planar. Recall that an undirected graph is an *outerplanar graph* if it can be drawn in the plane without crossings in such a way that all of the vertices belong to the unbounded face of the drawing. There is a characterization of outerplanar graphs that says a graph is outerplanar if and only if it does not contain a subdivision of the complete graph $K_4$ or the complete bipartite graph $K_{2,3}$. Clearly, every outerplanar graph is planar.

Let $G$ be a graph with $n$ vertices and $q$ edges, and let $C$ be a cycle of $G$. A chord is any edge of $G$ joining two nonadjacent vertices in $C$. We say that $C$ is a *primitive cycle* if it has no chords. Also, we say that a graph $G$ has the *primitive cycle property* (PCP) if any two primitive cycles intersect in at most one edge. Let $frank(G)$ be the number of primitive cycles of $G$. The number $frank(G)$ is called the *free rank* of $G$ and the number $rank(G)=q-n+r$ is called the *cycle rank* of $G$, where $r$ is the number of connected components of $G$. These two numbers satisfy $rank(G)\leq frank(G)$ [208, Proposition 2.2]. In [208], the authors studied and classified the family of graphs where equality occurs. This family is precisely the family of *ring graphs*. The precise definition of a ring graph can be found in [208, Section 2].

Roughly speaking, ring graphs can be obtained starting with a cycle and subsequently attaching paths of length at least two that meet graphs already constructed in two adjacent vertices. Also, they showed that for the graph $G$, the following conditions are equivalent:

(i) $G$ is a ring graph,

(ii) $rank(G) = frank(G)$,

(iii) $G$ satisfies PCP and $G$ does not contain a subdivision of $K_4$ as a subgraph.

Clearly ring graphs are planar. In [208], the authors stated that the blocks of the graph $G$ are important for calculating the numbers $frank(G)$ and $rank(G)$. In fact, they proved the following lemma.

**Lemma 7.5.27.** ([208, Lemma 2.4] *Let $G$ be a graph and let $G_1, \ldots, G_r$ be its blocks. Then $rank(G) = frank(G)$ if and only if $rank(G_i) = frank(G_i)$ for every $1 \leq i \leq r$.*

In [208], the authors also showed that every outerplanar graph is a ring graph. In [16], Afkhami, Barati, and Khashyarmanesh provided conditions for a total graph to be a ring graph or outerplanar.

The following two results determine all finite commutative rings $R$ such that their total graphs $T(\Gamma(R))$ are ring graphs and also outerplanar.

**Theorem 7.5.28.** ([16, Theorem 4.1]) *Let $R$ be a finite commutative ring. Then $T(\Gamma(R))$ is a ring graph if and only if one of the following statements holds:*

(i) *If $R$ is a local ring, then $R$ is either a field or isomorphic to $\mathbb{Z}_4$ or $\mathbb{Z}_2[x]/(x^2)$;*

(i) *If $R$ is not local, then $R$ is isomorphic to $\mathbb{Z}_2 \times \mathbb{Z}_2$.*

*Proof.* First suppose that $T(\Gamma(R))$ is a ring graph. Therefore it is planar. Now we have the following cases:

**Case 1.** $R$ is local. Then, by Theorem 7.5.10, $R$ is a field or isomorphic to one of the following rings:

$$\mathbb{Z}_4, \frac{\mathbb{Z}_2[x]}{(x^2)}, \frac{\mathbb{Z}_2[x]}{(x^3)}, \frac{\mathbb{Z}_2[x,y]}{(x,y)^2}, \frac{\mathbb{Z}_4[x]}{(2x,x^2)},$$

$$\frac{\mathbb{Z}_4[x]}{(2x,x^2-2)}, \mathbb{Z}_8, \frac{\mathbb{F}_4[x]}{(x^2)}, \frac{\mathbb{Z}_4[x]}{(x^2+x+1)}.$$

Since $R$ is finite, every nonunit is a zero-divisor. Moreover, $Z(R)$ is the maximal ideal of $R$. Now, if $2 \notin Z(R)$, then since $T(\Gamma(R))$ is planar by Theorem 7.5.10, we have $|Z(R)| \leqslant 2$. Since $|R| \leq |Z(R)|^2$, $R$ is either a ring of order 4 or is a field; so $T(\Gamma(R))$ is a ring graph.

Suppose that $2 \in Z(R)$. If $|Z(R)| = 1$, then $R$ is a field. Hence $T(\Gamma(R))$ is totally disconnected; so it is a ring graph. Otherwise $|Z(R)| > 1$. Since $T(\Gamma(R))$ is planar, we have that $|Z(R)| \leqslant 4$. Also, since $2 \in Z(R)$, $|R| = 2^k$. Thus

$|R| = 2, 4, 8$, or 16. According to [167], there are two nonisomorphic commutative local rings of order 16 with maximal ideals of order 4, namely $\mathbb{F}_4[x]/(x^2)$ and $\mathbb{Z}_4[x]/(x^2+x+1)$. In these rings, we have $T(\Gamma(R)) \cong 4K_4$, and every block of $T(\Gamma(R))$ is $K_4$. Since $rank(K_4) \neq frank(K_4)$, we have that the numbers rank and frank of these graphs are not equal, which means that they are not ring graphs.

In [167], it is also shown that the commutative local rings of order 8 with $|Z(R)| = 4$ are isomorphic to one of the following rings

$$\frac{\mathbb{Z}_2[x]}{(x^3)}, \frac{\mathbb{Z}_2[x,y]}{(x,y)^2}, \frac{\mathbb{Z}_4[x]}{(2x,x^2)}, \frac{\mathbb{Z}_4[x]}{(2x,x^2-2)}, \mathbb{Z}_8.$$

In these rings, we have $T(\Gamma(R)) \cong 2K_4$, and so their ranks and free ranks are not equal. Hence they are not ring graphs.

In the remaining two rings $\mathbb{Z}_4$ and $\mathbb{Z}_2[x]/(x^2)$, we have $T(\Gamma(R)) \cong 2K_2$, and therefore they are ring graphs.

**Case 2.** $R$ is not local. Since $T(\Gamma(R))$ is planar and $R$ is finite, $R$ is isomorphic to $\mathbb{Z}_2 \times \mathbb{Z}_2$ or $\mathbb{Z}_6$ by Theorem 7.5.10. If $R \cong \mathbb{Z}_6$, then $T(\Gamma(R))$ contains a subdivision of $K_4$; so it is not a ring graph. If $R \cong \mathbb{Z}_2 \times \mathbb{Z}_2$, then $T(\Gamma(R)) \cong C_4$ which is a ring graph.

Conversely, it is easy to see that if one of the parts (i) or (ii) holds, then $T(\Gamma(R))$ is a ring graph. □

**Theorem 7.5.29.** ([16, Theorem 4.2]) *Let $R$ be a finite commutative ring. Then $T(\Gamma(R))$ is outerplanar if and only if $R$ is a field or isomorphic to $\mathbb{Z}_4$, $\mathbb{Z}_2[x]/(x^2)$, or $\mathbb{Z}_2 \times \mathbb{Z}_2$.*

*Proof.* Assume that $T(\Gamma(R))$ is outerplanar. Since outerplanar graphs are ring graphs, $R$ is isomorphic to $\mathbb{Z}_4$, $\mathbb{Z}_2[x]/(x^2)$, $\mathbb{Z}_2 \times \mathbb{Z}_2$, or $R$ is a field by Theorem 7.5.28. For $\mathbb{Z}_4$ and $\mathbb{Z}_2[x]/(x^2)$, we have $T(\Gamma(R)) \cong 2K_2$, and therefore they are outerplanar. Also, $T(\Gamma(\mathbb{Z}_2 \times \mathbb{Z}_2)) \cong C_4$ which is outerplanar. If $R$ is a field, then $T(\Gamma(R))$ is the union of a totally disconnected graph and a perfect matching, which is outerplanar.

The converse statement follows easily. □

# Chapter 8

# Graphs from Total Graphs

In this chapter, we study certain graphs obtained from total graphs of commutative rings. More specifically, we concentrate on the total graph without the zero element, the complement of the total graph, and its generalizations.

## 8.1 Total Graphs without the Zero Element

Let $Z(\Gamma(R))$ be the (induced) subgraph of $T(\Gamma(R))$ with $Z(R)$ as its set of vertices. Then $Z(\Gamma(R))$ is connected with $diam(Z(\Gamma(R))) \leq 2$ since $x - 0 - y$ is a path between any two vertices $x$ and $y$ in $Z(\Gamma(R))$. With this in mind, Anderson and Badawi [70] introduced the total graph without the zero element of a commutative ring. In this regard, we discuss the (induced) subgraphs $Z_0(\Gamma(R))$ of $Z(\Gamma(R))$ and $T_0(\Gamma(R))$ of $T(\Gamma(R))$ obtained by deleting 0 as a vertex. Specifically, $Z_0(\Gamma(R))$ (resp., $T_0(\Gamma(R)))$ has vertices $Z(R)^* = Z(R)\backslash\{0\}$ (resp., $R^* = R\backslash\{0\}$), and two distinct vertices $x$ and $y$ are adjacent if and only if $x + y \in Z(R)$. Note that $Z_0(\Gamma(R))$ is a finite nonempty graph if and only if $R$ is a finite ring that is not a field (cf. [83, Theorem 2.2]).

Let $R$ be a commutative ring, $Z(R)$ its set of zero-divisors, $Reg(R) = R\backslash Z(R)$ its set of regular elements, $Idem(R)$ its set of idempotent elements, $Nil(R)$ its ideal of nilpotent elements, $U(R)$ its group of units, and total quotient ring $T(R) = R_{Reg(R)}$. For any $A \subseteq R$, let $A^* = A \setminus \{0\}$. Let $Min(R)$ be the set of minimal prime ideals of $R$.

### 8.1.1 Basic Properties

In this section, we study the two graphs $Z_0(\Gamma(R))$ and $T_0(\Gamma(R))$. First we study the graph $Z_0(\Gamma(R))$; actually, it is shown that $Z_0(\Gamma(R))$ is connected unless $R$ is a reduced ring with exactly two minimal prime ideals. Moreover, if $Z_0(\Gamma(R))$ is connected, then $diam(Z_0(\Gamma(R))) \leq 2$. Further, $gr(Z_0(\Gamma(R))) \in \{3, \infty\}$. The case for $Z(\Gamma(R))$ is much simpler since every nonzero vertex in $Z(\Gamma(R))$ is adjacent to



D. F. Anderson et al., *Graphs from Rings*, https://doi.org/10.1007/978-3-030-88410-9_8

0. If $Z(R)$ is an ideal of $R$, then it is clear that $Z(\Gamma(R))$ is a complete graph; and if $Z(R)$ is not an ideal of $R$, then $Z(\Gamma(R))$ is connected with $diam(Z(\Gamma(R))) = 2$ by Theorem 7.2.8(1).

We begin with a lemma containing several results which we will use throughout this section.

**Lemma 8.1.1.** ([70, Lemma 2.1]) *Let $R$ be a commutative ring. Then the following are true:*

(1) *$Z(R)$ is a union of prime ideals of $R$;*

(2) *$P \subseteq Z(R)$ for every $P \in Min(R)$;*

(3) *$Z(R) = \cup\{ P \mid P \in Min(R) \}$ if $R$ is reduced;*

(4) *Let $x \in Z(R)$ and $y \in Nil(R)$. Then $x + y \in Z(R)$;*

(5) *If $P_1, P_2, P_3$ are distinct minimal prime ideals of $R$, then $P_1 \cap P_2 \cap P_3 \subsetneq P_1 \cap P_2$;*

*Proof.* For (1), see [228, Theorem 2]. Parts (2) and (3) may be found in [219, Theorem 2.1] and [219, Corollary 2.4], respectively.

(4) By (1) above, $x \in P \subseteq Z(R)$ for some $P \in Spec(R)$. Since $y \in Nil(R) \subseteq P$, it follows that $x + y \in P \subseteq Z(R)$.

(5) If $P_1 \cap P_2 = P_1 \cap P_2 \cap P_3$, then $P_1P_2 \subseteq P_1 \cap P_2 \subseteq P_3$. Thus either $P_1 \subseteq P_3$ or $P_2 \subseteq P_3$, a contradiction. □

Now we discuss the case when $R$ is not reduced.

**Theorem 8.1.2.** ([70, Theorem 2.2]) *Let $R$ be a nonreduced commutative ring. Then $Z_0(\Gamma(R))$ is connected with $diam(Z_0(\Gamma(R))) \in \{0, 1, 2\}$.*

*Proof.* Assume that $R$ is not reduced, and let $x, y \in Z(R)^*$ be distinct vertices of $Z_0(\Gamma(R))$. If either $x \in Nil(R)$ or $y \in Nil(R)$, then $x + y \in Z(R)$ by Lemma 8.1.1(4); so $x - y$ is an edge in $Z_0(\Gamma(R))$. Thus we may assume that $x \notin Nil(R)$, $y \notin Nil(R)$, and $x + y \notin Z(R)$. Let $0 \neq w \in Nil(R)$. Then $x - w - y$ is a path in $Z_0(\Gamma(R))$ by Lemma 8.1.1(4), and hence $diam(Z_0(\Gamma(R))) \leq 2$. □

Note that $Z_0(\Gamma(R))$ is a complete graph if and only if $Z(R)$ is an ideal of $R$, and in this case, $diam(Z_0(\Gamma(R))) \leq 1$. Also, $Z(R)$ is a union of prime ideals of $R$ by Lemma 8.1.1(1); so $Z(R)$ is an ideal of $R$ if and only if it is a prime ideal of $R$. Thus a nonreduced ring $R$ has $diam(Z_0(\Gamma(R))) = 0$ if and only if $|Z(R)^*| = 1$, if and only if $R \cong \mathbb{Z}_4$ or $R \cong \mathbb{Z}_2[x]/(x^2)$. Examples of nonreduced rings $R$ with either $diam(Z_0(\Gamma(R))) = 1$ or $diam(Z_0(\Gamma(R))) = 2$ are given in Example 8.1.8 (also see Theorem 8.1.7).

We next consider the case when $R$ is reduced. In this case, $R$ is an integral domain if and only if $|Min(R)| = 1$. If $R$ is an integral domain, then $Z_0(\Gamma(R))$ is the empty graph; so we assume that $|Min(R)| \geq 2$.

**Theorem 8.1.3.** ([70, Theorem 2.3]) *Let $R$ be a reduced commutative ring with $|Min(R)| = 2$. Then $Z_0(\Gamma(R))$ is not connected.*

*Proof.* Suppose that $R$ is reduced and $|Min(R)| = 2$. Let $P$ and $Q$ be the minimal prime ideals of $R$. Then $Nil(R) = P \cap Q = \{0\}$, and $Z(R) = P \cup Q$ by Lemma 8.1.1(3) since $R$ is reduced. Let $0 \neq x \in P$ and $0 \neq y \in Q$. Then $x + y \notin Z(R)$; so there can be no path in $Z_0(\Gamma(R))$ from any $a \in P^*$ to any $b \in Q^*$. Thus, $Z_0(\Gamma(R))$ is not connected. □

Note that the $P^*$ and $Q^*$ in the proof of Theorem 8.1.3 are the connected components of $Z_0(\Gamma(R))$, and each component is a complete subgraph of $Z_0(\Gamma(R))$. However, in this case, $Z(R)$ is not an ideal of $R$; so $Z(\Gamma(R))$ is connected with $diam(Z(\Gamma(R))) = 2$ when $R$ is reduced and $|Min(R)| = 2$.

**Theorem 8.1.4.** ([70, Theorem 2.4]) *Let $R$ be a reduced commutative ring that is not an integral domain. Then $Z_0(\Gamma(R))$ is connected if and only if $|Min(R)| \geq 3$. Moreover, if $Z_0(\Gamma(R))$ is connected, then $diam(Z_0(\Gamma(R))) \in \{1, 2\}$.*

*Proof.* Suppose that $Z_0(\Gamma(R))$ is connected and $R$ is reduced, but not an integral domain. Then $|Min(R)| \geq 3$ by Theorem 8.1.3.

Conversely, suppose that $R$ is reduced and $|Min(R)| \geq 3$. Let $x, y \in Z(R)^*$ with $x + y \notin Z(R)$ (thus $x \neq y$). Then there are minimal prime ideals $P_1$ and $P_2$ of $R$ with $x \in P_1$ and $y \in P_2$ by Lemma 8.1.1(3), and $P_1 \neq P_2$ since $x+y \notin Z(R)$. Since $|Min(R)| \geq 3$, there is a $Q \in Min(R)\backslash\{P_1, P_2\}$; so $P_1 \cap P_2 \neq \{0\}$ by Lemma 8.1.1(5). Pick $0 \neq z \in P_1 \cap P_2$. Then $x - z - y$ is a path in $Z_0(\Gamma(R))$ from $x$ to $y$. Thus $Z_0(\Gamma(R))$ is connected with $diam(Z_0(\Gamma(R))) \leq 2$, and $diam(Z_0(\Gamma(R))) \neq 0$ since $|Z(R)^*| \geq 2$. Hence $1 \leq diam(Z_0(\Gamma(R))) \leq 2$. □

**Corollary 8.1.5.** ([70, Corollary 2.5]) *Let $R$ be a reduced commutative ring with $3 \leq |Min(R)| < \infty$. Then $diam(Z_0(\Gamma(R))) = 2$. In particular, $diam(Z_0(\Gamma(R))) = 2$ when $R$ is a reduced Noetherian ring with $|Min(R)| \geq 3$.*

*Proof.* By Theorem 8.1.4, $1 \leq diam(Z_0(\Gamma(R))) \leq 2$. Also, $diam(Z_0(\Gamma(R))) \leq 1$ if and only if $Z(R)$ is a prime ideal of $R$. If $R$ is reduced with $Min(R)$ finite, then $Z(R)$ is a prime ideal of $R$ if and only if $Min(R) = \{Z(R)\}$ by Lemma 8.1.1(3) and the Prime Avoidance Lemma [228, Theorem 81]. But $|Min(R)| \geq 3$; so $diam(Z_0(\Gamma(R))) = 2$. The "in particular" statement is clear since $Min(R)$ is finite when $R$ is Noetherian [228, Theorem 88]. □

**Corollary 8.1.6.** ([70, Corollary 2.6]) *Let $R$ be a commutative ring. Then the following statements are equivalent:*

(1) *$Z_0(\Gamma(R))$ is not connected;*

(2) *$T(R)$ is a von Neumann regular ring with exactly two maximal ideals;*

(3) *$T(R)$ is isomorphic to $K_1 \times K_2$ for fields $K_1$ and $K_2$.*

*In particular, if $R$ is a finite ring, then $Z_0(\Gamma(R))$ is connected unless $R \cong K_1 \times K_2$ for finite fields $K_1$ and $K_2$.*

*Proof.* The proof follows directly from Theorems 8.1.3 and 8.1.4. The "in particular" statement is clear. □

Let $R$ be a reduced commutative ring with $|Min(R)| \geq 3$. By Corollary 8.1.5, $diam(Z_0(\Gamma(R))) = 2$ if $Min(R)$ is finite. Note that $diam(Z_0(\Gamma(R))) = 1$ if and only if $Z(R)$ is an (prime) ideal of $R$; so if $R$ is reduced with $|Min(R)| \geq 3$ and $diam(Z_0(\Gamma(R))) = 1$, then both $Min(R)$ and $Z(R)$ must be infinite. An example of a reduced local commutative ring $R$ with nonzero maximal ideal $Z(R)$ is given in [64, Example 3.13] (cf. [269, Example 5.1]). For this ring $R$, both $Min(R)$ and $Z(R)$ are infinite, and $Z_0(\Gamma(R))$ is connected with $diam(Z_0(\Gamma(R))) = 1$.

The next theorem summarizes results about $diam(Z_0(\Gamma(R)))$ when $R$ is a finite commutative ring. Note that $Max(R) = Min(R)$ when $R$ is a finite commutative ring.

**Theorem 8.1.7.** ([70, Theorem 2.8]) *Let $R$ be a finite commutative ring. Then $diam(Z_0(\Gamma(R))) \in \{0, 1, 2, \infty\}$. Moreover,*

(1) *$Z_0(\Gamma(R))$ is the empty graph if and only if $R$ is a field;*

(2) *$diam(Z_0(\Gamma(R))) = 0$ if and only if $R$ is isomorphic to $\mathbb{Z}_4$ or $\mathbb{Z}_2[x]/(x^2)$;*

(3) *$diam(Z_0(\Gamma(R))) = 1$ if and only if $R$ is a local ring with maximal ideal $\mathfrak{m}$ and $|\mathfrak{m}| \geq 3$;*

(4) *$diam(Z_0(\Gamma(R))) = 2$ if and only if either $|Max(R)| \geq 3$ or $R$ is not reduced with $|Max(R)| = 2$, and ;*

(5) *$diam(Z_0(\Gamma(R))) = \infty$ if and only if $R$ is reduced with $|Max(R)| = 2$.*

We next illustrate the above results by computing $diam(Z_0(\Gamma(R)))$ for $R = \mathbb{Z}_n$ and $R = \mathbb{Z}_{n_1} \times \cdots \times \mathbb{Z}_{n_k}$. The details are left to the reader; they follow directly from Theorem 8.1.7.

**Example 8.1.8.** (a) To compute $diam(Z_0(\Gamma(\mathbb{Z}_n)))$, let $R = \mathbb{Z}_n$ with $n \geq 2$ and $n$ not prime (note that $Z_0(\Gamma(\mathbb{Z}_n))$ is the empty graph if $n$ is prime). Then $diam(Z_0(\Gamma(\mathbb{Z}_4))) = 0$; $diam(Z_0(\Gamma(\mathbb{Z}_{p^m}))) = 1$ if either $p = 2$ and $m \geq 3$, or $p \geq 3$ is prime and $m \geq 2$; $diam(Z_0(\Gamma(\mathbb{Z}_{pq}))) = \infty$ for distinct primes $p$ and $q$; and $diam(Z_0(\Gamma(R))) = 2$ otherwise.

(b) To compute $diam(Z_0(\Gamma(\mathbb{Z}_{n_1} \times \cdots \times \mathbb{Z}_{n_k})))$, let $R = \mathbb{Z}_{n_1} \times \cdots \times \mathbb{Z}_{n_k}$ with $2 \leq n_1 \leq \cdots \leq n_k$ and $k \geq 2$. Then $diam(Z_0(\Gamma(\mathbb{Z}_p \times \mathbb{Z}_q))) = \infty$ for primes $p \leq q$; otherwise $diam(Z_0(\Gamma(R))) = 2$.

Next, we determine $gr(Z_0(\Gamma(R)))$. If $Z(R)$ is an ideal of $R$, then it is clear that $gr(Z_0(\Gamma(R))) = \infty$ if $|Z(R)| \leq 3$ and $gr(Z_0(\Gamma(R))) = 3$ if $|Z(R)| \geq 4$. Just as for the diameter, our answer depends on the number of minimal prime ideals of $R$. If $Z(R)$ is an ideal of $R$, then $gr(Z(\Gamma(R))) = \infty$ if $|Z(R)| \leq 2$ and $gr(Z(\Gamma(R))) = 3$ if $|Z(R)| \geq 3$. If $Z(R)$ is not an ideal of $R$, then $gr(Z(\Gamma(\mathbb{Z}_2 \times \mathbb{Z}_2))) = \infty$ and $gr(Z(\Gamma(R))) = 3$ if $R \not\cong \mathbb{Z}_2 \times \mathbb{Z}_2$ by Theorem 7.2.24(1) (also, see Theorem 8.1.11(1)).

We first handle the case when $R$ is not reduced.

**Theorem 8.1.9.** ([70, Theorem 3.1]) *Let $R$ be a nonreduced commutative ring. Then $gr(Z_0(\Gamma(R))) = \infty$ if and only if $R$ has a unique nonzero minimal prime ideal $P$ with $P = Nil(R) = Z(R)$ and $|P| \leq 3$ (i.e., $gr(Z_0(\Gamma(R))) = \infty$ if and only if $Nil(R) = Z(R)$ and $|Nil(R)| \leq 3$). Otherwise, $gr(Z_0(\Gamma(R))) = 3$. Moreover, $gr(Z_0(\Gamma(R))) = \infty$ if $|Z(R)| \leq 3$ and $gr(Z_0(\Gamma(R))) = 3$ if $|Z(R)| \geq 4$.*

*Proof.* Suppose that $|Min(R)| \geq 2$. Let $P$ and $Q$ be distinct minimal prime ideals of $R$. Then $\{0\} \subsetneq P \cap Q \subsetneq P$; so $|P \cap Q| \geq 2$, and thus $|P| \geq 4$. Let $x, y, z \in P^*$ be distinct. Then $x - y - z - x$ is a triangle in $Z_0(\Gamma(R))$; so $gr(Z_0(\Gamma(R))) = 3$. Now suppose that $Min(R) = \{P\}$, and thus $Nil(R) = P$. If $Nil(R) \subsetneq Z(R)$, then there is a prime ideal $Q$ of $R$ with $\{0\} \neq Nil(R) = P \subsetneq Q \subseteq Z(R)$ by Lemma 8.1.1(1). As above, $|Q| \geq 4$; so again $gr(Z_0(\Gamma(R))) = 3$. If $Nil(R) = Z(R)$, then $gr(Z_0(\Gamma(R))) = 3$ if $|Nil(R)| \geq 4$ and $gr(Z_0(\Gamma(R))) = \infty$ if $|Nil(R)| \leq 3$. The "moreover" statement follows directly from the above arguments. □

We next consider the case when $R$ is reduced.

**Theorem 8.1.10.** ([70, Theorem 3.2]) *Let $R$ be a reduced commutative ring that is not an integral domain. Then $gr(Z_0(\Gamma(R))) = \infty$ if and only if $Min(R) = \{P, Q\}$ with $\max\{|P|, |Q|\} \leq 3$. Otherwise, $gr(Z_0(\Gamma(R))) = 3$. In particular, $gr(Z_0(\Gamma(R))) = 3$ when $|Min(R)| \geq 3$.*

*Proof.* Suppose that $P_1, P_2, P_3$ are distinct minimal prime ideals of $R$. Then $\{0\} \subseteq P_1 \cap P_2 \cap P_3 \subsetneq P_1 \cap P_2 \subsetneq P_1$ by Lemma 8.1.1(5); so $|P_1 \cap P_2| \geq 2$, and thus $|P_1| \geq 4$. Let $x, y, z \in P_1^*$ be distinct. Then $x - y - z - x$ is a triangle in $Z_0(\Gamma(R))$; so $gr(Z_0(\Gamma(R))) = 3$ if $|Min(R)| \geq 3$. Thus we may assume that $|Min(R)| = 2$; say $Min(R) = \{P, Q\}$. As in the proof of Theorem 8.1.3, $P \cap Q = \{0\}$ and $Z(R) = P \cup Q$, and hence no $x \in P^*$ and $y \in Q^*$ are adjacent in $Z_0(\Gamma(R))$. Thus $gr(Z_0(\Gamma(R))) = 3$ if and only if either $|P| \geq 4$ or $|Q| \geq 4$. Otherwise, $gr(Z_0(\Gamma(R))) = \infty$. The " in particular" statement is clear. □

Using earlier mentioned results from Theorems 8.1.9 and 8.1.10, we can give explicit calculations for $gr(Z_0(\Gamma(R)))$.

**Theorem 8.1.11.** ([70, Theorem 3.3]) *Let $R$ be a commutative ring. Then $gr(Z_0(\Gamma(R))) \in \{3, \infty\}$. Furthermore, $Z_0(\Gamma(R))$ is the empty graph if and only if $R$ is an integral domain. For $R$ not an integral domain, $gr(Z_0(\Gamma(R))) = \infty$ if and only if $R$ is isomorphic to $\mathbb{Z}_4$, $\mathbb{Z}_2[x]/(x^2)$, $\mathbb{Z}_2 \times \mathbb{Z}_2$, $\mathbb{Z}_6$, $\mathbb{Z}_9$, $\mathbb{Z}_3 \times \mathbb{Z}_3$, or $\mathbb{Z}_3[x]/(x^2)$. Otherwise, $gr(Z_0(\Gamma(R))) = 3$.*

*Proof.* First, suppose that $R$ is not reduced. By Theorem 8.1.9, $gr(Z_0(\Gamma(R))) = \infty$ if and only if $\{0\} \neq Nil(R) = Z(R)$ and $|Z(R)| \leq 3$, and $gr(Z_0(\Gamma(R))) = 3$ otherwise. So in this case, $gr(Z_0(\Gamma(R))) = \infty$ if and only if $R$ is isomorphic to $\mathbb{Z}_4$, $\mathbb{Z}_2[x]/(x^2)$, $\mathbb{Z}_9$, or $\mathbb{Z}_3[x]/(x^2)$.

Next, suppose that $R$ is reduced and not an integral domain. Then by Theorem 8.1.10, $gr(Z_0(\Gamma(R))) = \infty$ if and only if $Min(R) = \{P, Q\}$ with $\max\{|P|, |Q|\} \leq 3$, and $gr(Z_0(\Gamma(R))) = 3$ otherwise. In the first subcase, we have $Z(R) = P \cup Q$ and $P \cap Q = \{0\}$ with $\max\{|P|, |Q|\} \leq 3$. In this case, $R$ is a reduced finite ring with two maximal ideals, each with two or three elements. Thus $gr(Z_0(\Gamma(R))) = \infty$ if and only if $R$ is isomorphic to $\mathbb{Z}_2 \times \mathbb{Z}_2$, $\mathbb{Z}_2 \times \mathbb{Z}_3 \cong \mathbb{Z}_6$, or $\mathbb{Z}_3 \times \mathbb{Z}_3$. The result now follows directly from the above two cases. □

We end this section with the analog of Example 8.1.8 for $gr(Z_0(\Gamma(R)))$ when $R = \mathbb{Z}_n$ or $R = \mathbb{Z}_{n_1} \times \cdots \times \mathbb{Z}_{n_k}$. The details are left to the reader; they follow directly from Theorem 8.1.11(2).

**Example 8.1.12.** ([70, Example 3.4]) (a) To compute $gr(Z_0(\Gamma(\mathbb{Z}_n)))$, let $R = \mathbb{Z}_n$ with $n \geq 2$ and $n$ not prime (note that $Z_0(\Gamma(\mathbb{Z}_n))$ is the empty graph if $n$ is prime). Then $gr(Z_0(\Gamma(R))) = \infty$ if either $n = 4$, $n = 6$, or $n = 9$. Otherwise, $gr(Z_0(\Gamma(R))) = 3$.

(b) To compute $gr(Z_0(\Gamma(\mathbb{Z}_{n_1} \times \cdots \times \mathbb{Z}_{n_k})))$, let $R = \mathbb{Z}_{n_1} \times \cdots \times \mathbb{Z}_{n_k}$ with $2 \leq n_1 \leq \cdots \leq n_k$ and $k \geq 2$. Then $gr(Z_0(\Gamma(R))) = \infty$ if either $n_1 = n_2 = 2$, $n_1 = 2$ and $n_2 = 3$, or $n_1 = n_2 = 3$. Otherwise, $gr(Z_0(\Gamma(R))) = 3$.

In the remainder of this section, we study the graph $T_0(\Gamma(R))$. It is shown that $diam(T_0(\Gamma(R))) = diam(T(\Gamma(R)))$ if and only if $|R| \geq 4$. (Note that $|R| \leq 3$ if and only if $R$ is isomorphic to $\mathbb{Z}_2$ or $\mathbb{Z}_3$.) We then explicitly compute $gr(T_0(\Gamma(R)))$. For $x, y \in R^*$, let $d_T(x, y)$ (resp., $d_{T_0}(x, y)$) denote the distance from $x$ to $y$ in $T(\Gamma(R))$ (resp., $T_0(\Gamma(R))$). We first show that these two distances are always equal.

**Lemma 8.1.13.** ([70, Lemma 4.1]) *Let $R$ be a commutative ring and $x, y \in R^*$. Then $x, y$ are connected by a path in $T_0(\Gamma(R))$ if and only if $x, y$ are connected by a path in $T(\Gamma(R))$. Moreover, $d_{T_0}(x, y) = d_T(x, y)$ and $diam(T_0(\Gamma(R))) \leq diam(T(\Gamma(R)))$.*

*Proof.* If $x, y$ are connected by a path in $T_0(\Gamma(R))$, then clearly $x, y$ are connected by a path in $T(\Gamma(R))$. Conversely, assume that $x - a_1 - \cdots - a_n - y$ is a shortest path from $x$ to $y$ in $T(\Gamma(R))$, and assume that $a_i = 0$ for some $i$ with $1 \leq i \leq n$. Then $a_{i-1}, a_{i+1} \in Z(R)^*$ and $a_{i-1} + a_{i+1} \in Reg(R)$ (let $a_0 = x$ and $a_{n+1} = y$). Let $z_i = -(a_{i-1} + a_{i+1})$. Then $x - a_1 - \cdots - a_{i-1} - z_i - a_{i+1} - \cdots - a_n - y$ is a shortest path from $x$ to $y$ in $T_0(\Gamma(R))$, and hence $x, y$ are connected by a path in $T_0(\Gamma(R))$. The "moreover" statement is clear. □

Recall that $T(\Gamma(R))$ is not connected if $Z(R)$ is an ideal of $R$. If $Z(R)$ is not an ideal of $R$, then $T(\Gamma(R))$ is connected if and only if $(Z(R)) = R$ (i.e., $1 = z_1 + \cdots + z_n$ for some $z_1, \ldots, z_n \in Z(R)$) by Theorem 7.2.10. Moreover, in this case, $diam(T(\Gamma(R))) = n$, where $n \geq 2$ is the least positive integer such that $R = (z_1, \ldots, z_n)$ for some $z_1, \ldots, z_n \in Z(R)$ by Theorem 7.2.10. Also, $diam(T(\Gamma(R))) = d_T(0, 1)$ by Corollary 7.2.11. Thus $T(\Gamma(R))$ is connected if and only if $diam(T(\Gamma(R))) < \infty$.

**Theorem 8.1.14.** ([70, Theorem 4.2]) *Let $R$ be a commutative ring. Then*

(1) *If $|R| \leq 3$, then $T_0(\Gamma(R))$ is connected, but $T(\Gamma(R))$ is not connected;*

(2) *If $|R| \geq 4$, then $T_0(\Gamma(R))$ is connected if and only if $T(\Gamma(R))$ is connected.*

*Proof.* (1) If $|R| \leq 3$, then $R \cong \mathbb{Z}_2$ or $R \cong \mathbb{Z}_3$. It is easily verified that (1) holds for these two rings.

(2) If $T(\Gamma(R))$ is connected, then $T_0(\Gamma(R))$ is also connected by Lemma 8.1.13. Conversely, assume that $T_0(\Gamma(R))$ is connected and $|R| \geq 4$. Then $R$ is not an integral domain; so there is an $x \in Z(R)^*$. Let $y \in R^*$. Then there is a path from $x$ to $y$ in $T_0(\Gamma(R))$. But $x$ is adjacent to 0 in $T(\Gamma(R))$; so there is a path from 0 to $y$ in $T(\Gamma(R))$. Thus $T(\Gamma(R))$ is also connected. □

**Corollary 8.1.15.** ([70, Corollary 4.3]) *Let $R$ be a commutative ring. Then $T_0(\Gamma(R))$ is connected if and only if either $(Z(R)) = R$ or $R$ is isomorphic to $\mathbb{Z}_2$ or $\mathbb{Z}_3$. Moreover, $T_0(\Gamma(R))$ is connected if and only if $diam(T_0(\Gamma(R))) < \infty$.*

*Proof.* This follows directly from Theorem 8.1.14 and the discussion preceding Theorem 8.1.14. □

In general, there is no relationship between the diameters of the graphs $Z_0(\Gamma(R))$ and $T_0(\Gamma(R))$. By Examples 8.1.8 and 8.1.18, we have

$diam(Z_0(\Gamma(\mathbb{Z}_8))) = 1 < \infty = diam(T_0(\Gamma(\mathbb{Z}_8)))$,

$diam(T_0(\Gamma(\mathbb{Z}_6))) = 2 < \infty = diam(Z_0(\Gamma(\mathbb{Z}_6)))$, and

$diam(Z_0(\Gamma(\mathbb{Z}_{12}))) = 2 = diam(T_0(\Gamma(\mathbb{Z}_{12})))$.

Our next goal is to show that $diam(T_0(\Gamma(R))) = diam(T(\Gamma(R)))$ when $|R| \geq 4$. In this regard, first we need the following lemma.

**Lemma 8.1.16.** ([70, Lemma 4.4]) *Let $R$ be a commutative ring with $diam(T(\Gamma(R))) = n < \infty$, and let $s \in R^*$ and $u \in U(R)$ be distinct.*

(1) *If $s \in Z(R)^*$, then $d_{T_0}(u, s) = d_T(u, s) \in \{n-1, n\}$;*

(2) *If $n$ is an even integer, then $d_{T_0}(u-s, s) = m = d_T(u-s, s)$ for some even integer $m \leq n$;*

(3) *If $n$ is an odd integer and $u \neq -s$, then $d_{T_0}(u+s, s) = m = d_T(u+s, s)$ for some odd integer $m \leq n$;*

(4) *If $n$ is an even integer, then $d_{T_0}(u-s, s) = n = d_T(u-s, s)$ for every $s \in Z(R)^*$;*

(5) *If $n$ is an odd integer, then $d_{T_0}(u+s,s) = n = d_T(u+s,s)$ for every $s \in Z(R)^*$.*

*Proof.* Observe that $n \geq 2$ by Theorem 7.2.10.

(1) Let $s - a_1 - \cdots - a_{m-1} - u$ be a shortest path from $s$ to $u$ in $T_0(\Gamma(R))$ of length $m$. Then $m = d_{T_0}(x,y) = d_T(x,y) \leq n$ by Lemma 8.1.13. Since $u \in (s, s+a_1, a_1+a_2, \ldots, a_{m-1}+u)$, we have $R = (s, s+a_1, a_1+a_2, \ldots, a_{m-1}+u)$. Since $R$ is generated by $m+1$ elements of $Z(R)$ and $diam(T(\Gamma(R))) = n$, we have $n \leq m+1$ by by Theorem 7.2.10. Thus $m \leq n \leq m+1$; so either $m = n-1$ or $m = n$.

(2) Let $n$ be an even integer. If $u-s = s$, then $d_{T_0}(u-s,s) = 0$. Thus we may assume that $u - s \neq s$, and hence $d_{T_0}(u-s,s) \geq 2$ since $(u-s)+s = u \notin Z(R)$. Let $m \geq 2$, and let $s - a_1 - \cdots - a_{m-1} - (u-s)$ be a shortest path from $s$ to $u-s$ in $T_0(\Gamma(R))$ of length $m$. Thus $m \leq n$. Suppose that $m$ is an odd integer. Since $u = (s+a_1) - (a_1+a_2) + \cdots - (a_{m-2}+a_{m-1}) + (a_{m-1}+(u-s))$, we have $R = (s+a_1, a_1+a_2, a_2+a_3, \ldots, a_{m-1}+(u-s))$ is generated by $m$ elements of $Z(R)$. Hence $n \leq m$ by Theorem 7.2.10; so $m = n$, which is a contradiction since $n$ is an even integer. Thus $d_{T_0}(u-s,s) = m = d_T(u-s,s)$ for some even integer $m \leq n$.

(3) Let $n$ be an odd integer and $s \neq -u$; so $u \neq u+s \in R^*$. If $u+2s \in Z(R)$, then $d_{T_0}(u+s,s) = 1$. Thus we may assume that $u+2s \notin Z(R)$, and hence $d_{T_0}(u+s,s) \geq 2$. Let $m \geq 2$, and let $s - a_1 - \cdots - a_{m-1} - (u+s)$ be a shortest path from $s$ to $u+s$ in $T_0(\Gamma(R))$ of length $m$. Thus $m \leq n$. Suppose that $m$ is an even integer. Since $-u = (s+a_1)-(a_1+a_2)+\cdots+(a_{m-2}+a_{m-1})-(a_{m-1}+(u+s))$, we have $R = (s+a_1, a_1+a_2, a_2+a_3, \ldots, a_{m-1}+(u+s))$ is generated by $m$ elements of $Z(R)$. Hence $n \leq m$ by Theorem 7.2.10; so $m = n$, which is a contradiction since $n$ is an odd integer. Thus $d_{T_0}(u+s,s) = m = d_T(u+s,s)$ for some odd integer $m \leq n$.

(4) Let $n$ be an even integer and $s \in Z(R)^*$. Then $u-s, s \in R^*$ are distinct and $(u-s)+s = u \notin Z(R)$; so $m = d_{T_0}(u-s,s)$ is an even positive integer by part (2) above. Let $s - a_1 - \cdots - a_{m-1} - (u-s)$ be a shortest path from $s$ to $u-s$ in $T_0(\Gamma(R))$ of length $m$. If $m = n$, then we are done; so assume that $m < n$. Since $u = 2s - (s+a_1) + (a_1+a_2) - \cdots - (a_{m-2}+a_{m-1}) + (a_{m-1}+(u-s))$, we have $R = (s, s+a_1, a_1+a_2, a_2+a_3, \ldots, a_{m-1}+(u-s))$ is generated by $m+1$ elements of $Z(R)$. Hence $n \leq m+1$ by Theorem 7.2.10. Thus $n = m+1$, which is a contradiction since $n$ is an even integer and $m+1$ is an odd integer. Thus $d_{T_0}(u-s,s) = n = d_T(u-s,s)$.

(5) Let $n$ be an odd integer and $s \in Z(R)^*$. Thus $u+s, s \in R^*$ are distinct and $2s+u \notin Z(R)$ (for if $2s+u \in Z(R)$, then $R = (s, 2s+u)$, and hence $diam(T(\Gamma(R))) = 2$ by Theorem 7.2.10); so $m = d_{T_0}(u+s,s) \geq 3$ is an odd integer by part (3) above. Let $s - a_1 - \cdots - a_{m-1} - (u+s)$ be a shortest path from $s$ to $u+s$ in $T_0(\Gamma(R))$ of length $m$. If $m = n$, then we are done; so assume that $m < n$. Since $-u = 2s-(s+a_1)+(a_1+a_2)-\cdots+(a_{m-2}+a_{m-1})-(a_{m-1}+(u+s))$,

we have $R = (s, s + a_1, a_1 + a_2, a_2 + a_3, \ldots, a_{m-1} + (u - s))$ is generated by $m+1$ elements of $Z(R)$. Hence $n \leq m + 1$ by Theorem 7.2.10. Thus $n = m + 1$, which is a contradiction since $n$ is an odd integer and $m + 1$ is an even integer. Hence $d_{T_0}(u + s, s) = n = d_T(u + s, s)$. □

**Theorem 8.1.17.** ([70, Theorem 4.5]) *Let $R$ be a commutative ring.*

(1) $diam(T_0(\Gamma(\mathbb{Z}_2))) = 0 < \infty = diam(T(\Gamma(\mathbb{Z}_2)))$;

(2) $diam(T_0(\Gamma(\mathbb{Z}_3))) = 1 < \infty = diam(T(\Gamma(\mathbb{Z}_3)))$;

(3) *If $|R| \geq 4$, then* $diam(T_0(\Gamma(R))) = diam(T(\Gamma(R)))$.

*Proof.* Parts (1) and (2) are easily verified; so we may assume that $|R| \geq 4$. Then $T(\Gamma(R))$ is connected if and only if $T_0(\Gamma(R))$ is connected by Theorem 8.1.14, and $diam(T_0(\Gamma(R))) \leq diam(T(\Gamma(R)))$ by Lemma 8.1.13. Thus $diam(T(\Gamma(R))) = \infty$ if and only if $diam(T_0(\Gamma(R))) = \infty$ by Corollary 8.1.15 and the remarks before Theorem 8.1.14. Hence we may assume that $diam(T(\Gamma(R))) = n < \infty$, and thus $R$ is not an integral domain. Let $z \in Z(R)^*$. If $n$ is an odd integer, then $d_T(1 + z, z) = n = d_{T_0}(1 + z, z)$ by Lemma 8.1.16(5), and hence $diam(T(\Gamma(R))) = diam(T_0(\Gamma(R))) = n$ by Lemma 8.1.13. If $n$ is an even integer, then $d_T(1 - z, z) = d_{T_0}(1 - z, z) = n$ by Lemma 8.1.16(4), and thus $diam(T(\Gamma(R))) = diam(T_0(\Gamma((R))) = n$ by Lemma 8.1.13.
Hence $diam(T_0(\Gamma(R))) = diam(T(\Gamma(R)))$ for all rings $R$ with $|R| \geq 4$. □

The next example follows directly from Theorem 8.1.17 and the discussion preceding Theorem 8.1.14.

**Example 8.1.18.** (a) $diam(T_0(\Gamma(\mathbb{Z}_2))) = 0, diam(T_0(\Gamma(\mathbb{Z}_3))) = 1$, and $diam(T_0(\Gamma(\mathbb{Z}_p))) = \infty$ when $p \geq 5$ is prime. Let $n = p_1^{m_1} \cdots p_k^{m_k}$ for distinct primes $p_i$ and $m_i \geq 1$. If $k = 1$ and $m_1 \geq 2$, then $diam(T_0(\Gamma(R))) = \infty$. If $k \geq 2$, then $diam(T_0(\Gamma(R))) = 2$.

(b) Let $R = \mathbb{Z}_{n_1} \times \cdots \times \mathbb{Z}_{n_k}$ with $2 \leq n_1 \leq \cdots \leq n_k$ and $k \geq 2$. Then $diam(T_0(\Gamma(R))) = 2$.

The girth of $T_0(\Gamma(R))$ is also easily determined. Recall from Theorem 7.2.4 that if $Z(R)$ is an ideal of $R$, then $gr(T(\Gamma(R))) = 3$ if and only if $|Z(R)| \geq 3$, $gr(T(\Gamma(R))) = 4$ if and only if $2 \notin Z(R)$ and $|Z(R)| = 2$, and $gr(T(\Gamma(R))) = \infty$ otherwise. (Note that if $|Z(R)| = 2$, then $R$ is isomorphic to $\mathbb{Z}_4$ or $\mathbb{Z}_2[x]/(x^2)$, and $2 \in Z(R)$ in either case. So, "the $gr(T(\Gamma(R))) = 4$" cannot actually happen when $Z(R)$ is an ideal of $R$.) If $Z(R)$ is not an ideal of $R$, then $gr(T(\Gamma(R))) = 4$ if and only if $R \cong \mathbb{Z}_2 \times \mathbb{Z}_2$, and $gr(T(\Gamma(R))) = 3$ otherwise by Theorem 7.2.24. Thus $gr(T(\Gamma(R))) \in \{3, 4, \infty\}$. Note that $gr(T(\Gamma(R))) \leq gr(T_0(\Gamma(R)))$ since $T_0(\Gamma(R))$ is a (induced) subgraph of $T(\Gamma(R))$. We next give explicit calculations for $gr(T(\Gamma(R)))$ and $gr(T_0(\Gamma(R)))$. These calculations show that $gr(T_0(\Gamma(R))) = gr(T(\Gamma(R)))$ unless $R$ is isomorphic to $\mathbb{Z}_2 \times \mathbb{Z}_2$, $\mathbb{Z}_9$, or $\mathbb{Z}_3[x]/(x^2)$.

**Theorem 8.1.19.** ([70, Theorem 4.8]) *Let $R$ be a commutative ring. Then $gr(T_0(\Gamma(R))) \in \{3, 4, \infty\}$. Moreover,*

(1) *$gr(T_0(\Gamma(R))) = \infty$ if and only if either $R$ is an integral domain or $R$ is isomorphic to $\mathbb{Z}_4$, $\mathbb{Z}_2[x]/(x^2)$, or $\mathbb{Z}_2 \times \mathbb{Z}_2$;*

(2) *$gr(T_0(\Gamma(R))) = 4$ if and only if $R$ is isomorphic to $\mathbb{Z}_9$ or $\mathbb{Z}_3[x]/(x^2)$; and*

(3) *$gr(T_0(\Gamma(R))) = 3$ otherwise.*

*Proof.* Note that $gr(T_0(\Gamma(R))) \leq gr(Z_0(\Gamma(R)))$ since $Z_0(\Gamma(R))$ is a (induced) subgraph of $T_0(\Gamma(R))$. Thus Theorem 8.1.19 follows directly from Theorem 8.1.11(2) since one can easily verify that the rings $\mathbb{Z}_4$, $\mathbb{Z}_2[x]/(x^2)$, $\mathbb{Z}_2 \times \mathbb{Z}_2$, $\mathbb{Z}_6$, $\mathbb{Z}_9$, $\mathbb{Z}_3 \times \mathbb{Z}_3$, and $\mathbb{Z}_3[x]/(x^2)$ have $gr(T_0(\Gamma(R)))$ equal to $\infty$, $\infty$, $\infty$, 3, 4, 3, and 4, respectively. □

We close this section with the analog of Example 8.1.8 for $gr(T_0(\Gamma(R)))$. It follows directly from Theorem 8.1.19.

**Example 8.1.20.** (a) To compute $gr(T_0(\Gamma(\mathbb{Z}_n)))$, let $R = \mathbb{Z}_n$ with $n \geq 2$. Then $gr(T_0(\Gamma(\mathbb{Z}_n))) = \infty$ if $n$ is prime, $gr(T_0(\Gamma(\mathbb{Z}_4))) = \infty, gr(T_0(\Gamma(\mathbb{Z}_9))) = 4$, and $gr(T_0(\Gamma(R))) = 3$ otherwise.

(b) To compute $gr(T_0(\Gamma(\mathbb{Z}_{n_1} \times \cdots \times \mathbb{Z}_{n_k})))$, let $R = \mathbb{Z}_{n_1} \times \cdots \times \mathbb{Z}_{n_k}$ with $2 \leq n_1 \leq \cdots \leq n_k$ and $k \geq 2$. Then $gr(T_0(\Gamma(\mathbb{Z}_2 \times \mathbb{Z}_2))) = \infty$, and $gr(T_0(\Gamma(R))) = 3$ otherwise.

### 8.1.2 Zero-divisor Paths and Regular Paths

Let $R$ be a commutative ring and $x, y \in R^*$ be distinct. We say that $x - a_1 - \cdots - a_n - y$ is a *zero-divisor path* from $x$ to $y$ if $a_1, \ldots, a_n \in Z(R)^*$ and $a_i + a_{i+1} \in Z(R)$ for every $0 \leq i \leq n$ (let $a_0 = x$ and $a_{n+1} = y$). We define $d_Z(x, y)$ to be the length of a shortest zero-divisor path from $x$ to $y$ ($d_Z(x, x) = 0$ and $d_Z(x, y) = \infty$ if there is no such path) and $diam_Z(R) = sup\{d_Z(x, y) \mid x, y \in R^*\}$. Thus $d_T(x, y) = d_{T_0}(x, y) \leq d_Z(x, y)$, for every $x, y \in R^*$. In particular, if $x, y \in R^*$ are distinct and $x + y \in Z(R)$, then $x - y$ is a zero-divisor path from $x$ to $y$ with $d_Z(x, y) = 1$. For any commutative ring $R$, we have $max\{diam(Z_0(\Gamma(R))), diam(T_0(\Gamma(R)))\} \leq diam_Z(R)$. However, if $R$ is a reduced local ring with $|Min(R)| \geq 3$, then $diam(Z_0(\Gamma(R))) \leq 2$ by Theorem 8.1.4, but $diam_Z(R) = \infty$ since there is no zero-divisor path from 1 to any $x \in Z(R)^*$ (cf. Theorem 8.1.21(1)). Also, $diam(T_0(\Gamma(\mathbb{Z}_{1225}))) = 2 < 3 = diam_Z(\mathbb{Z}_{1225})$ by Examples 8.1.18 and 8.1.25. Note that $diam_Z(\mathbb{Z}_2) = 0, diam_Z(\mathbb{Z}_3) = 1$, and $diam_Z(R) = \infty$ for any other integral domain $R$.

We first determine when there is a zero-divisor path between every two distinct elements of $R^*$.

**Theorem 8.1.21.** ([70, Theorem 5.1]) *Let $R$ be a commutative ring that is not an integral domain. Then there is a zero-divisor path from $x$ to $y$ for every $x, y \in R^*$ if and only if one of the following two statements hold:*

(1) *$R$ is reduced, $|Min(R)| \geq 3$, and $R = (z_1, z_2)$ for some $z_1, z_2 \in Z(R)^*$;*

(2) *$R$ is not reduced and $R = (z_1, z_2)$ for some $z_1, z_2 \in Z(R)^*$.*

*Moreover, if there is a zero-divisor path from $x$ to $y$ for every $x, y \in R^*$, then $R$ is not local and $diam_Z(R) \in \{2, 3\}$.*

*Proof.* Suppose that there is a zero-divisor path from $x$ to $y$ for every $x, y \in R^*$. First, assume that $R$ is reduced and not an integral domain. Since $Z_0(\Gamma(R))$ is connected if and only if $|Min(R)| \geq 3$ by Theorem 8.1.4, we have $|Min(R)| \geq 3$. Let $y \in Z(R)^*$. Then there is a zero-divisor path $1 - a_1 - \cdots - a_n - y$ from 1 to $y$ for some $a_1, \ldots, a_n \in Z(R)^*$. Thus $z = 1 + a_1 \in Z(R)^*$, and hence $R = (a_1, z)$. If $R$ is not reduced, then a similar argument, as in the reduced case, shows that $R = (z_1, z_2)$ for some $z_1, z_2 \in Z(R)^*$.

Conversely, assume that (1) holds. Thus $1 = w + z$ for some $w, z \in Z(R)^*$. Let $x, y \in R^*$ be distinct. Then $x = xw + xz$ and $y = yw + yz$. We consider two cases.

Case one: assume that $x, y \in Z(R)^*$. Then we are done by Theorem 8.1.4.

Case two: assume that $x \notin Z(R)$. Hence $xw, xz \in Z(R)^*$. Suppose that $x+y \notin Z(R)$. Then assume that either $xw = yw$ or $y = \pm yw$. Then $x-(-xw)-y$ is the desired zero-divisor path of length two from $x$ to $y$. Next, assume that $xw \neq yw$, $yw \neq 0$, and $y \neq \pm yw$. Then $x - (-xw) - (-yw) - y$ is the desired zero-divisor path of length three from $x$ to $y$. Finally, assume that $yw = 0$. Since $y \neq 0$ and $y = yw + yz$, we have $yz = y \neq 0$. Thus $x - (-xz) - y$ is the desired zero-divisor path of length two from $x$ to $y$. Now assume that (2) holds. Since $Z_0(\Gamma(R))$ is connected by Theorem 8.1.2, an argument similar to that in case two of the reduced case completes the proof.

Assume that there is a zero-divisor path from $x$ to $y$ for every $x, y \in R^*$ and that $R$ is not an integral domain. Then $R$ cannot be local since $R = (z_1, z_2)$ for some $z_1, z_2 \in Z(R)^*$ by (1) and (2) above. Clearly $diam_Z(R) \neq 0$. Let $z \in Z(R)^*$. Then $z, 1 - z \in R^*$ are distinct and $z + (1 - z) = 1 \notin Z(R)$; so $diam_Z(R) \geq 2$. The "moreover" statement now follows from the above proof. □

**Corollary 8.1.22.** ([70, Corollary 5.2]) *Let $R$ be a commutative ring. Then $diam_Z(R) \in \{0, 1, 2, 3, \infty\}$. Moreover, $diam_Z(R) \in \{2, 3, \infty\}$ except for $diam_Z(\mathbb{Z}_2) = 0$ and $diam_Z(\mathbb{Z}_3) = 1$.*

**Corollary 8.1.23.** ([70, Corollary 5.3]) *Let $R$ be a commutative ring such that $Z(R)$ is not an ideal of $R$. Then there is a zero-divisor path from $x$ to $y$ for every $x, y \in T(R)^*$ if and only if either $R$ is reduced with $|Min(R)| \geq 3$ or $R$ is not reduced.*

*Proof.* Since $Z(R)$ is not an ideal of $R$, there are $z_1, z_2 \in Z(R)^*$ such that $z_1 + z_2 \in Reg(R)$. Thus $T(R) = (z_1, z_2)$; so the corollary follows directly from Theorem 8.1.21. □

**Theorem 8.1.24.** ([70, Theorem 5.4]) (1) *Let* $R = R_1 \times R_2$ *for commutative local rings* $R_1$, $R_2$ *with maximal ideals* $\mathfrak{m}_1, \mathfrak{m}_2$, *respectively. If there are* $a_1 \in U(R_1)$ *and* $a_2 \in U(R_2)$ *with* $(2a_1, 2a_2) \in U(R)$ *and* $(3a_1, 3a_2) \notin Z(R)$, *then* $diam_Z(R) \in \{3, \infty\}$. *Moreover,* $diam_Z(R) = 3$ *if either* $R_1$ *or* $R_2$ *is not reduced;*

(2) *Let* $R = R_1 \times \cdots \times R_n$ *for commutative rings* $R_1, \ldots, R_n$ *with* $n \geq 3$. *Then* $diam_Z(R) = 2$.

*Proof.* (1) Let $a = (a_1, a_2), b = (2a_1, 2a_2) \in U(R)$. Then $a \neq b$ and $\mathrm{d}_Z(a, b) \neq 1$ since $a + b = (3a_1, 3a_2) \notin Z(R)$. Assume that there is an $f = (m_1, m_2) \in R^*$ such that $a - f - b$ is a zero-divisor path from $a$ to $b$. Thus $f \in Z(R)^*$; so either $m_1 \in \mathfrak{m}_1$ or $m_2 \in \mathfrak{m}_2$. If $m_1 \in \mathfrak{m}_1$, then $m_1 + a_1, m_1 + 2a_1 \in U(R_1)$. Hence $m_2 + a_2, m_2 + 2a_2 \in \mathfrak{m}_2$ since $a + f, b + f \in Z(R)$. But then $a_2 = (m_2 + 2a_2) - (m_2 + a_2) \in M_2$, a contradiction. In a similar manner, $m_2 \in \mathfrak{m}_2$ also leads to a contradiction; so no such $f$ exists. Thus $d_Z(a, b) \geq 3$; so $diam_Z(R) \in \{3, \infty\}$. The "moreover" statement now follows from Theorem 8.1.21.

(2) We have $diam_Z(R) \in \{2, 3\}$ by Theorem 8.1.21 since $|Min(R)| \geq n \geq 3$. Let $x = (x_1, \ldots, x_n), y = (y_1, \ldots, y_n) \in R^*$ with $x + y \notin Z(R)$. We may assume that $x_1 \neq 0$. Let $z = (-x_1, -y_2, 1, \ldots, 1, 0) \in Z(R)^*$. Then $x + z, y + z \in Z(R)$; so $x - z - y$ is the desired zero-divisor path from $x$ to $y$ of length 2. Hence $diam_Z(R) = 2$. □

The following example shows that all possible values for $diam_Z(R)$ given in Corollary 8.1.22 and Theorem 8.1.24 may be realized. The details are left to the reader.

**Example 8.1.25.** (a) We find $diam_Z(\mathbb{Z}_n)$. We have already observed that $diam_Z(\mathbb{Z}_2) = 0, diam_Z(\mathbb{Z}_3) = 1$, and $diam_Z(\mathbb{Z}_p) = \infty$ when $p \geq 5$ is prime. Let $R = \mathbb{Z}_n$ with $n \geq 2$ and $n$ not prime. Let $n = p_1^{m_1} \cdots p_k^{m_k}$ for distinct primes $p_i$ and $m_i \geq 1$. If either $k = 1$, or $k = 2$ and $m_1 = m_2 = 1$, then $diam_Z(R) = \infty$. If $k = 2, p_1, p_2 \geq 5$, and $m_1 + m_2 \geq 3$, then $diam_Z(R) = 3$. Otherwise, $diam_Z(R) = 2$.

(b) To find $diam_Z(\mathbb{Z}_{n_1} \times \cdots \times \mathbb{Z}_{n_k})$, let $R = \mathbb{Z}_{n_1} \times \cdots \times \mathbb{Z}_{n_k}$ with $2 \leq n_1 \leq \cdots \leq n_k$ and $k \geq 2$. If $k = 2$ and $n_1, n_2$ are prime, then $diam_Z(R) = \infty$. If $k = 2$ and $n_1 = p_1^{m_1}, n_2 = p_2^{m_2}$ for primes $p_1, p_2 \geq 5$ and $m_1 + m_2 \geq 3$, then $diam_Z(R) = 3$. Otherwise, $diam_Z(R) = 2$.

Let $x, y \in R^*$ be distinct. We say that $x - a_1 - \cdots - a_n - y$ is a *regular path* from $x$ to $y$ if $a_1, \ldots, a_n \in Reg(R)$ and $a_i + a_{i+1} \in Z(R)$ for every $0 \leq i \leq n$ (let $a_0 = x$ and $a_{n+1} = y$). We define $d_{reg}(x, y)$ to be the length of a shortest regular path from $x$ to $y$ ($d_{reg}(x, x) = 0$ and $d_{reg}(x, y) = \infty$ if there is no such path), and $diam_{reg}(R) = sup\{d_{reg}(x, y) \mid x, y \in R^*\}$. Thus $d_T(x, y) = d_{T_0}(x, y) \leq d_{reg}(x, y)$ for every $x, y \in R^*$. In particular, if $x, y \in R^*$ are distinct and $x + y \in Z(R)$, then $x - y$ is a regular path from $x$ to $y$ with $d_{reg}(x, y) = 1$. For any commutative ring $R$, we have $max\{\, diam(T_0(\Gamma(R))), diam(Reg(\Gamma(R)))\,\} \leq diam_{reg}(R)$. Note that

$diam(T_0(\Gamma(\mathbb{Z}_{60}))) = 2 < \infty = diam_{reg}(\mathbb{Z}_{60})$ and $diam(Reg(\Gamma(\mathbb{Z}_6))) = 1 < 2 = diam_{reg}(\mathbb{Z}_6)$. However, if $R$ is an integral domain, then $T_0(\Gamma(R)) = Reg(\Gamma(R))$; so all three diameters are equal. Moreover, $diam_{reg}(\mathbb{Z}_2) = 0, diam_{reg}(\mathbb{Z}_3) = 1$, and $diam_{reg}(R) = \infty$ for any other integral domain $R$. Hence $diam_Z(R) = diam_{reg}(R)$ for any integral domain $R$.

**Theorem 8.1.26.** ([70, Theorem 5.6]) *Let $R$ be a commutative ring with $diam(T_0(\Gamma(R))) = n < \infty$. Then*

(1) *Let $u \in U(R)$, $s \in R^*$, and $P$ be a shortest path from $s$ to $u$ of length $n-1$ in $T_0(\Gamma(R))$. Then $P$ is a regular path from $s$ to $u$;*

(2) *Let $u \in U(R)$, $s \in R^*$, and $P : s - a_1 - \cdots - a_n = u$ be a shortest path from $s$ to $u$ of length $n$ in $T_0(\Gamma(R))$. Then either $P$ is a regular path from $s$ to $u$, or $a_1 \in Z(R)^*$ and $a_1 - \cdots - a_n = u$ is a regular path from $a_1$ to $u$ of length $n - 1 = d_{T_0}(a_1, u)$.*

*Proof.* (1) If $n = 2$, then $P$ is a regular path from $s$ to $u$ by definition. Thus we may assume that $n > 2$. Since $d_{T_0}(z, u)$ is either $n-1$ or $n$ for every $z \in Z(R)^*$ by Lemma 8.1.16(1) and $d_{T_0}(s, u) = n-1$, we conclude that $P$ must be a regular path.

(2) Suppose that $P$ is not a regular path; so $a_i \in Z(R)^*$ for some $1 \leq i \leq n-1$. Since $d_{T_0}(z, u)$ is either $n-1$ or $n$ for every $z \in Z(R)^*$ by Lemma 8.1.16(1) and $d_{T_0}(s, u) = n$, we must have $a_1 \in Z(R)^*$ and $a_i \in Reg(R)$ for every $2 \leq i \leq n-1$. Thus $a_1 - \cdots - a_n - u$ is a regular path of length $n - 1 = d_{T_0}(a_1, u)$. □

We next determine when there is a regular path between every two distinct elements of $R^*$.

**Theorem 8.1.27.** ([70, Theorem 5.7]) *Let $R$ be a commutative ring.*

(1) *If $s \in Reg(R)$ and $w \in Nil(R)^*$, then there is no regular path from $s$ to $w$;*

(2) *If $R$ is reduced and local, then there is no regular path from any unit to any nonzero nonunit in $R$.*

*In particular, if there is a regular path from $x$ to $y$ for every $x, y \in R^*$, then either $R$ is reduced and not local or $R$ is isomorphic to $\mathbb{Z}_2$ or $\mathbb{Z}_3$.*

*Proof.* (1) Let $s \in Reg(R)$ and $w \in Nil(R)^*$. Since $a + w \in Reg(R)$ for every $a \in Reg(R)$ by Lemma 8.1.1(4), there is no regular path from $s$ to $w$.

(2) Let $\mathfrak{m}$ be the maximal ideal of $R$, $x \in U(R)$, and $0 \neq y \in \mathfrak{m}$. Suppose that there is a regular path $x - x_1 - \cdots - x_n - y$. Then $x + x_1 = z_1 \in Z(R) \subseteq \mathfrak{m}$; so $x_1 = -x + z_1 \in U(R)$. In a similar manner, each $x_i \in U(R)$. But then $x_n + y \in U(R)$, a contradiction.

The "in particular" statement is clear by parts (1) and (2) above and the remarks preceding Theorem 8.1.26. □

**Theorem 8.1.28.** ([70, Theorem 5.8]) *Let $R$ be a commutative ring. Then there is a regular path from $x$ to $y$ for every $x, y \in R^*$ if and only if $R$ is reduced, $Reg(\Gamma(R))$ is connected, and for every $z \in Z(R)^*$ there is a $w \in Z(R)^*$ such that $d_Z(z, w) > 1$ (possibly with $d_Z(z, w) = \infty$).*

*Proof.* Suppose that there is a regular path from $x$ to $y$ for every $x, y \in R^*$. Then $R$ is reduced by Theorem 8.1.27, and it is clear that $Reg(\Gamma(R))$ is connected. Let $z \in Z(R)^*$, and let $z - a_1 - \cdots - 1$ be a regular path from $z$ to 1. Then $a_1 \in Reg(R)$ and $w = -(z + a_1) \in Z(R)^*$. Thus $z \neq w$ and $z + w \notin Z(R)$; so $d_Z(z, w) > 1$.

Conversely, suppose that $R$ is reduced, $Reg(\Gamma(R))$ is connected, and for every $z \in Z(R)^*$ there is a $w \in Z(R)^*$ such that $d_Z(z, w) > 1$ (possibly with $d_Z(z, w) = \infty$). Let $x, y \in R^*$. If $x, y \in Reg(R)$, then there is nothing to prove. First, assume that $x \in Z(R)^*$ and $y \in Reg(R)$. Since $x \in Z(R)^*$, there is a $w \in Z(R)^*$ such that $d_Z(x, w) > 1$. Then $x + w \notin Z(R)$; so $x + u = -w \in Z(R)$ for some $u \in Reg(R)$. Since $Reg(\Gamma(R))$ is connected, let $u - u_1 - \cdots - y$ be a regular path from $u$ to $y$. Then $x - u - u_1 - \cdots - y$ is a regular path from $x$ to $y$. Next, assume that $x, y \in Z(R)^*$. Then again as above, there are $u, v \in Reg(R)$ such that $x + u \in Z(R)$ and $y + v \in Z(R)$. If $u = v$, then $x - u - y$ is a regular path from $x$ to $y$. So assume that $u \neq v$. Since $Reg(\Gamma(R))$ is connected, let $u - \cdots - v$ be a regular path from $u$ to $v$. Then $x - u - \cdots - v - y$ is a regular path from $x$ to $y$. □

In view of Theorem 8.1.3 and Theorem 8.1.28, we have the following result.

**Corollary 8.1.29.** ([70, Corollary 5.9]) *Let $R$ be a reduced commutative ring with $|Min(R)| = 2$. Then there is a regular path from $x$ to $y$ for every $x, y \in R^*$ if and only if $Reg(\Gamma(R))$ is connected.*

Recall from [191] that a commutative ring $R$ is a *p.p. ring* if every principal ideal of $R$ is projective. For example, a commutative von Neumann regular ring is a p.p. ring, and $\mathbb{Z} \times \mathbb{Z}$ is a p.p. ring that is not von Neumann regular. It was shown in [286, Proposition 15] that a commutative ring $R$ is a p.p. ring if and only if every element of $R$ is the product of an idempotent element and a regular element of $R$ (thus a commutative p.p. ring that is not an integral domain has nontrivial idempotents). We show that a commutative p.p. ring $R$ that is not an integral domain has $diam_{reg}(R) = 2$.

**Lemma 8.1.30.** ([70, Lemma 5.10]) *Let $R$ be commutative ring, $u, v \in Reg(R)$, and $e \in Idem(R)$. Then $eu + (1 - e)v \in Reg(R)$.*

*Proof.* Let $eu + (1 - e)v = w \in R$, and suppose that $cw = 0$ for some $c \in R$. Then $ew = e[eu + (1 - e)v] = eu$ and $(1 - e)w = (1 - e)[eu + (1 - e)v] = (1 - e)v$. Thus $ceu = cew = 0$ and $c(1 - e)v = c(1 - e)w = 0$, and hence $ce = c(1 - e) = 0$ since $u, v \in Reg(R)$. Thus $c = ce + c(1 - e) = 0$; so $eu + (1 - e)v = w \in Reg(R)$. □

**Theorem 8.1.31.** ([70, Theorem 5.11]) *Let $R$ be a commutative p.p. ring that is not an integral domain. Then there is a regular path from $x$ to $y$ for every $x, y \in R^*$. Moreover, $diam_{reg}(R) = diam(T_0(\Gamma(R))) = diam(T(\Gamma(R))) = 2$.*

*Proof.* Let $x, y \in R^*$ be distinct, and suppose that $x + y \notin Z(R)$. We consider three cases.

**Case 1:** Assume that $x, y \in Z(R)^*$. Since $x + y \notin Z(R)$, necessarily $x + y \in Reg(R)$, and thus $x - (-(x + y)) - y$ is the desired regular path of length two from $x$ to $y$.

**Case 2:** Assume that $x, y \in Reg(R)$. Since $R$ is a p.p. ring and not an integral domain, there is an $e \in Idem(R) \setminus \{0, 1\}$. Hence $w = -[(1 - e)x + ey] \in Reg(R)$ by Lemma 8.1.30. Since $e(1 - e) = 0$ and $e \notin \{0, 1\}$, we have $x + w = ex - ey = e(x - y) \in Z(R)$ and $y + w = (e - 1)x - (e - 1)y = (e - 1)(x - y) \in Z(R)$. Thus $x - w - y$ is the desired regular path of length 2 from $x$ to $y$.

**Case 3:** Assume that $x \in Reg(R)$ and $y \in Z(R)^*$. Hence $y = fu$ for some $f \in Idem(R) \setminus \{0, 1\}$ and $u \in Reg(R)$. Then $h = -[(1 - f)x + fu] \in Reg(R)$ by Lemma 8.1.30. Since $f(1 - f) = 0$ and $f \notin \{0, 1\}$, we have $x + h = fx - fu = f(x - u) \in Z(R)$ and $y + h = (f - 1)x \in Z(R)$. Thus $x - h - y$ is the desired regular path of length two from $x$ to $y$; so $diam_{reg}(R) \leq 2$.

For the "moreover" statement, we first note that $R$ has a nontrivial idempotent. Now by Corollary 7.2.14, $T(\Gamma(R))$ is connected with $diam(T(\Gamma(R))) = 2$. Thus $2 = diam(T(\Gamma(R))) = diam(T_0(\Gamma(R))) \leq diam_{reg}(R) \leq 2$ by Theorem 7.2.26 since $|R| \geq 4$; so we have the desired equality. □

**Corollary 8.1.32.** ([70, Corollary 5.12]) *Let $R$ be a commutative von Neumann regular ring that is not a field. Then there is a regular path from $x$ to $y$ for every $x, y \in R^*$. Moreover, $diam_{reg}(R) = 2$.*

**Corollary 8.1.33.** ([70, Corollary 5.13]) *Let $R$ be a commutative ring. If there is an $e \in Idem(R) \setminus \{0, 1\}$, then $Reg(\Gamma(R))$ is connected with $diam(Reg(\Gamma(R))) \in \{0, 1, 2\}$.*

*Proof.* Let $u, v \in Reg(R)$ be distinct, $u + v \notin Z(R)$, and $e \in Idem(R) \setminus \{0, 1\}$. Then $w = -eu + (1 - e)v \in Reg(R)$ by Lemma 8.1.30; so $u - w - v$ is the desired path form $u$ to $v$ in $Reg(\Gamma(R))$ of length two. Thus $Reg(\Gamma(R))$ is connected and $diam(Reg(\Gamma(R))) \leq 2$. □

One can easily verify that $diam(Reg(\Gamma(\mathbb{Z}_2 \times \mathbb{Z}_2))) = 0, diam(Reg(\Gamma(\mathbb{Z}_3 \times \mathbb{Z}_3))) = 1$, and $diam(Reg(\Gamma(\mathbb{Z}_5 \times \mathbb{Z}_5))) = 2$. Thus all possible values for $diam(Reg(\Gamma(R)))$ in Corollary 8.1.33 may be realized.

We next determine $diam_{reg}(R)$ for $R = \mathbb{Z}_n$ and $R = \mathbb{Z}_{n_1} \times \cdots \times \mathbb{Z}_{n_k}$. The details are left to the reader; they follow directly from Theorem 8.1.27 and Corollary 8.1.32.

**Example 8.1.34.** ([70, example 5.14]) (a) We find $diam_{reg}(\mathbb{Z}_n)$. We have already observed that $diam_{reg}(\mathbb{Z}_2) = 0, diam_{reg}(\mathbb{Z}_3) = 1$, and $diam_{reg}(\mathbb{Z}_p) = \infty$ when $p \geq 5$ is prime. Let $R = \mathbb{Z}_n$ with $n \geq 2$ and $n$ not prime. Then $diam_{reg}(R) = 2$ if $n$ is the product of (at least 2) distinct primes. Otherwise, $diam_{reg}(R) = \infty$.

(b) To find $diam_{reg}(\mathbb{Z}_{n_1} \times \cdots \times \mathbb{Z}_{n_k})$), let $R = \mathbb{Z}_{n_1} \times \cdots \times \mathbb{Z}_{n_k}$ with $2 \leq n_1 \leq \cdots \leq n_k$ and $k \geq 2$. Then $diam_{reg}(R) = 2$ if every $n_i$ is prime. Otherwise, $diam_{reg}(R) = \infty$.

The rings in Theorem 8.1.31 and Corollary 8.1.32 are reduced and not local. We next give an example of a reduced nonlocal ring $R$ that is not an integral domain such that there is no regular path from $x$ to $y$ for some $x, y \in R^*$.

**Example 8.1.35.** ([70, Example 5.15]) Let $I = 2x\mathbb{Z}[x]$ be an ideal of $\mathbb{Z}[x]$, and let $R = \mathbb{Z}[x]/I$. Then $R$ is reduced, not local, and $Z(R) = x\mathbb{Z}[x]/I \cup 2\mathbb{Z}[x]/I$. Note that $R \neq (Z(R))$; so $T_0(\Gamma(R))$ is not connected by Corollary 8.1.15. Thus there is no regular path from $a$ to $b$ for some $a, b \in R^*$. It may easily be shown that there is no regular path from $a = 1 + I$ to $b = x + I$.

We have $diam(T_0(\Gamma(R))) \leq min\{diam_Z(R), diam_{reg}(R)\}$ for any commutative ring $R$. Examples 8.1.18, 8.1.25, and 8.1.34 show that all three diameters may be different. For $n = 5^2 \cdot 7^2 = 1225$, we have $diam(T_0(\Gamma(\mathbb{Z}_n))) = 2 < 3 = diam_Z(\mathbb{Z}_n) < \infty = diam_{reg}(\mathbb{Z}_n)$. For $n = 2^2 \cdot 3 \cdot 5 = 60$, we have $diam(T_0(\Gamma(\mathbb{Z}_n))) = diam_Z(\mathbb{Z}_n) = 2 < \infty = diam_{reg}(\mathbb{Z}_n)$. Also, $diam(T_0(\Gamma(\mathbb{Z}_{35})))$ $= diam_{reg}(\mathbb{Z}_{35}) = 2 < \infty = diam_Z(\mathbb{Z}_{35})$.

We could also define $gr_Z(R)$ and $gr_{reg}(R)$ by only using cycles in $Z(R)^*$ and $Reg(R)$, respectively. However, this gives nothing new since $gr_Z(R) = gr(Z_0(\Gamma(R)))$ and $gr_{reg}(R) = gr(Reg(\Gamma(R)))$. We have already determined $gr(Z_0(\Gamma(R)))$ in Theorem 8.1.11, and $gr(Reg(\Gamma(R)))$ has been studied in Theorems 7.2.4 and 7.2.24. We end this section by giving $gr(Reg(\Gamma(R)))$ for $R = \mathbb{Z}_n$ and $R = \mathbb{Z}_{n_1} \times \cdots \times \mathbb{Z}_{n_k}$; details are left to the reader.

**Example 8.1.36.** ([70, Example 5.16]) Let $R = \mathbb{Z}_n$ with $n \geq 2$. Then $gr(Reg(\Gamma(R))) = \infty$ if $n = 4, n = 6$, or $n$ is prime; $gr(Reg(\Gamma(R))) = 4$ if $n = p^m$ with $p \geq 3$ prime and $m \geq 2$; and $gr(Reg(\Gamma(R))) = 3$ otherwise.

(b) Let $R = \mathbb{Z}_{n_1} \times \cdots \times \mathbb{Z}_{n_k}$ with $2 \leq n_1 \leq \cdots \leq n_k$ and $k \geq 2$. Then $gr(Reg(\Gamma(R))) = \infty$ if $n_{k-1} = 2$ and $n_k = 2, 3, 4$, or $6$. Otherwise, $gr(Reg(\Gamma(R))) = 3$.

## 8.2 The Regular Graph

Let $R$ be a commutative ring. Recall that the *regular graph of* $R$, denoted by $Reg(\Gamma(R))$, is the graph whose vertex set is $Reg(R)$, and two distinct vertices $x$ and $y$ are adjacent if and only if $x + y \in Z(R)$.

We start this section with the complete graph nature of the regular graph of a commutative ring.

**Theorem 8.2.1.** ([31, Theorem 1]) *Let $R = \prod_{i\in I} R_i$, where $R_i$ is a commutative ring and $Z(R_i)$ is an ideal of $R_i$ for every $i \in I$. Then the following hold:*

(1) *If $2 \in Z(R)$, then $Reg(\Gamma(R))$ is a complete graph if and only if $R_i/Z(R_i) \cong \mathbb{Z}_2$ for some $i \in I$;*

(2) *If $2 \notin Z(R)$, then $Reg(\Gamma(R))$ is a complete graph if and only if $R_i \cong \mathbb{Z}_3$ for every $i \in I$.*

Since every commutative Artinian ring is a direct product of finitely many Artinian local rings, the assertions of Theorem 8.2.1 hold for every Artinian ring.

**Theorem 8.2.2.** *Let $R$ be a commutative ring. Then*

(1) ([31, Theorem 2])*Let $R$ be a reduced Noetherian ring and $Z(R)$ an ideal of $R$. If $Reg(\Gamma(R))$ is a complete graph, then $R \cong \mathbb{Z}_2$ or $R \cong \mathbb{Z}_3$;*

(2) ([31, Theorem 3]) *If $2 \notin Z(R)$ and $Reg(\Gamma(R))$ is a complete graph, then $J(R) = 0$, and so $R$ is reduced;*

(3) ([31, Corollary 1]) *Let $R$ be a semilocal ring and $2 \notin Z(R)$. If $Reg(\Gamma(R))$ is a complete graph, then $R \cong \mathbb{Z}_3^r$, where $r$ is the number of maximal ideals of $R$.*

Next, we consider commutative rings whose regular graphs are forests. Recall that a *tree* is an undirected graph in which any two vertices are connected by exactly one simple path. In other words, any connected graph without simple cycles is a tree. A *forest* is a disjoint union of trees.

**Theorem 8.2.3.** ([31, Theorem 12]) *Let $R$ be a commutative ring. If $Reg(\Gamma(R))$ is a forest, then $Reg(\Gamma(R))$ is a disjoint union of copies of $K_1$ or a disjoint union of some copies of $K_2$. In particular, if $Reg(\Gamma(R))$ is a tree, then it has at most two vertices.*

**Corollary 8.2.4.** ([31, Corollary 8]) *Let $R$ be a commutative Noetherian ring. If $Reg(\Gamma(R))$ is a tree, then $R$ is isomorphic to one of the rings:*

$$\mathbb{Z}_3,\ \mathbb{Z}_4,\ \mathbb{Z}_2[x]/(x^2),\ \mathbb{Z}_2^r,\ \mathbb{Z}_2^r \times \mathbb{Z}_3,\ \mathbb{Z}_2^r \times \mathbb{Z}_4,\ \mathbb{Z}_2^r \times \mathbb{Z}_2[x]/(x^2)$$

*for some positive integer $r$.*

Let $R$ be a noncommutative ring. Then $Z_L(R) = \{x \in R \mid xa = 0$ for some $a \in R^*\}$ is the set of left zero-divisors, $Z_R(R) = \{x \in R \mid bx = 0,$ for some $b \in R^*\}$ is the set of right zero-divisors, and $Z(R) = Z_L(R) \cup Z_R(R)$ is the set of all zero-divisors of $R$. Now one can define $Reg(\Gamma(R))$ in the same way as for the commutative case. The vertices are all elements of the ring $Reg(R)$ and two elements $x, y \in Reg(R)$ are adjacent if and only if $x + y \in Z(R)$, see [312]. It is easy to show that in the case when $Z(R)$ is an ideal of $R$, one has the same properties as in the case of commutative rings.

The following four theorems extend the results proved for the regular graph of a commutative ring to the regular graph of an arbitrary ring.

**Theorem 8.2.5.** ([32, Theorem 2.1]) *Let $R$ be a ring (not necessarily commutative). Then $gr(Reg(\Gamma(R))) \in \{3, 4, \infty\}$.*

**Theorem 8.2.6.** ([32, Theorem 2.7]) *Let $R$ be a finite ring (not necessarily commutative). Then $Reg(\Gamma(R))$ is regular (i.e., all vertices have the same degree).*

The next theorem gives a characterization of rings whose regular graphs are complete.

**Theorem 8.2.7.** ([32, Theorems 3.4, 3.7, 3.8 and Corollary 3.13]) *Let $R$ be ring (not necessarily commutative). Then*

(1) *If $R$ is a left Artinian ring and $Reg(\Gamma(R))$ contains a vertex adjacent to every other vertex, then $Reg(\Gamma(R))$ is complete;*
(2) *If $2 \notin Z(R)$ and $Reg(\Gamma(R))$ is a complete graph, then $J(R) = 0$;*
(3) *If $R$ is a left Artinian ring and $2 \notin Z(R)$, then $Reg(\Gamma(R))$ is a complete graph if and only if $R \cong \mathbb{Z}_3^r$ for some positive integer $r$;*
(4) *If $R$ is a reduced left Noetherian ring and $2 \notin Z(R)$ such that $Reg(\Gamma(R))$ is a complete graph, then $R \cong \mathbb{Z}_3^r$ for some positive integer $r$.*

**Theorem 8.2.8.** ([32, Theorem 2.3]) *Let $R$ be a left Artinian ring and $Reg(\Gamma(R))$ be a tree. Then $R$ is isomorphic to one of the following rings:*

$$\mathbb{Z}_3,\ \mathbb{Z}_4,\ \mathbb{Z}_2[x]/(x^2),\ \mathbb{Z}_2^r,\ \mathbb{Z}_3 \times \mathbb{Z}_2^r,\ \mathbb{Z}_4 \times \mathbb{Z}_2^r,\ \mathbb{Z}_2[x]/(x^2) \times \mathbb{Z}_2^r,\ T_2(\mathbb{Z}_2),\ T_2(\mathbb{Z}_2) \times \mathbb{Z}_2^r,$$

*where $T_2(\mathbb{Z}_2)$ denotes the ring of $2 \times 2$ upper triangular matrices over $\mathbb{Z}_2$ and $r$ is a positive integer.*

Now we state a result concerning the chromatic number and the clique number of regular graphs.

**Theorem 8.2.9.** ([31, Theorem 6, Corollaries 5, 6, 7]) *Let $R$ be a commutative ring. Then*

(1) *If $2 \notin Z(R)$ and $Z(R) = \bigcup_{i=1}^{n} P_i$, where $P_1, \ldots, P_n$ are prime ideals of $R$ and $Z(R) \neq \bigcup_{i \neq j} P_i$ for $j = 1, \ldots, n$, then $\omega(Reg(\Gamma(R))) = \chi(Reg(\Gamma(R))) = 2^n$;*
(2) *If $R$ is Noetherian and $2 \notin Z(R)$, then $\omega(Reg(\Gamma(R))) = \chi(Reg(\Gamma(R))) = 2^n$, where $n$ is the minimum number of prime ideals whose union is $Z(R)$;*

(3) *If $R$ is a reduced Noetherian ring and $2 \notin Z(R)$, then* $\omega(Reg(\Gamma(R))) = \chi(Reg(\Gamma(R))) = 2^n$*, where $n$ is the number of minimal prime ideals of $R$;*

(4) *Let $R$ be a Noetherian ring. If $2 \notin Z(R)$, then* $\omega(Reg(\Gamma(R/Nil(R))) = \chi(Reg(\Gamma(R/Nil(R))) = 2^n$*, where $n$ is the number of minimal prime ideals of $R$.*

**Theorem 8.2.10.** ([31, Theorem 7]) *Let $R$ be a commutative Noetherian ring and $2 \notin Z(R)$. If $Reg(R) = U(R)$, then $R$ is a semilocal ring and $\omega(Reg(\Gamma(R))) = \chi(Reg(\Gamma(R))) = 2^n$, where $n$ is the number of maximal ideals of $R$. In particular, if $R$ is a commutative Artinian ring and $2 \notin Z(R)$, then $\omega(Reg(\Gamma(R))) = \chi(Reg(\Gamma(R))) = 2^n$, where $n$ is the number of maximal ideals of $R$.*

**Theorem 8.2.11.** ([31, Theorem 11]) *Let $R$ be a commutative Noetherian ring. If $R/Nil(R)$ is finite, then $R$ is finite.*

Let $R$ be a ring. Then $M_n(R), GL_n(R)$, and $T_n(R)$ denote the set of $n \times n$ matrices over $R$, the set of $n \times n$ invertible matrices over $R$, and the set of $n \times n$ upper triangular matrices over $R$, respectively. If $R$ is a commutative ring, then it is known that $A \in Z(M_n(R))$ if and only if $det(A) \in Z(R)$, and so $Z_L(M_n(R)) = Z_R(M_n(R)) = Z(M_n(R))$. Of course, this set is not an ideal of $M_n(R)$. If $F$ is a field with $char(F) \neq 2$, then $Reg(M_n(F)) = GL_n(F)$. The chromatic number and the clique number of the regular graph of $M_n(R)$ have been studied by various authors, see [34, 25, 414, 33]. We start with the diameter of $T(\Gamma(M_n(R)))$.

**Theorem 8.2.12.** ([312, Theorem 3.7]) *Let $R$ be a commutative ring. Then $T(\Gamma(M_n(R)))$ is connected and $diam(T(\Gamma(M_n(R)))) = 2$.*

The following theorem shows that for every field $F$ with $char(F) \neq 2$, $\omega(Reg(\Gamma(M_n(F))))$ is finite.

**Theorem 8.2.13.** ([34, Theorem 1]) *Let $F$ be a field with $char(F) \neq 2$ and $n$ be a positive integer. Then $\omega(Reg(\Gamma(M_n(F)))) < \infty$, and moreover,*

$$\omega(Reg(\Gamma(M_n(F)))) \leq \sum_{k=0}^{n} \frac{(n!)^2}{k![(n-k)!]^2}.$$

Let $F$ be a field, $char(F) \neq 2$, and $S \subseteq GL_n(F)$, where $n$ is a positive integer. By Theorem 8.2.13, if for all distinct elements $x, y \in S$, $x + y$ is singular, then $S$ is finite. The authors of [34] conjectured that this result is true if one replaces a field by a division ring.

Now we investigate the clique number of the regular graph of certain rings.

**Theorem 8.2.14.** ([34, Theorem 2]) *Let $F$ be a field with $char(F) \neq 2$. Then $\omega(Reg(\Gamma(M_2(F)))) = 5$.*

**Theorem 8.2.15.** ([34, Theorem 3]) *Let $D$ be a division ring with $char(D) \neq 2$. Then $diag(\pm 1, \ldots, \pm 1)$ (the set of all diagonal matrices with diagonal entries in $\{-1, 1\}$) forms a maximal clique for $Reg(\Gamma(M_n(D)))$.*

The next result investigates the clique number of cyclic subgroups of $GL_n(F)$, where $F$ is a field.

**Theorem 8.2.16.** ([25, Theorems 7 and 8, Corollary 4]) *Let $n$ be a positive integer.*

(1) *Let $F$ be a field containing a primitive $2^n$th root of unity. Then*

$$\max_{A \in GL_n(F)} \omega(Reg(\Gamma(A))) = 2^n;$$

(2) *Let $F$ be an algebraically closed field with $char(F) \neq 2$. Then*

$$\max_{A \in GL_n(F)} \omega(Reg(\Gamma(A))) = 2^n;$$

(3)

$$\max_{A \in GL_n(\mathbb{R})} \omega(Reg(\Gamma(A))) = \begin{cases} 2^{\frac{n}{2}} & n \text{ is even;} \\ 2^{\frac{n+1}{2}} & \text{otherwise.} \end{cases}$$

Akbari et al. [25] established that certain subgraphs of $Reg(\Gamma(M_n(F)))$ have finite chromatic number.

**Theorem 8.2.17.** ([25, Theorem 1]) *Let $F$ be a field with $char(F) \neq 2$ and $n$ a positive integer. Then $\chi(Reg(\Gamma(T_n(F)))) = \omega(Reg(\Gamma(T_n(F)))) = 2^n$.*

**Theorem 8.2.18.** ([25, Theorem 3]) *Let $F$ be a field and $n$ a positive integer. If $G$ is a solvable subgroup of $GL_n(F)$, then $\chi(Reg(\Gamma(G))) < \infty$.*

**Theorem 8.2.19.** ([25, Theorem 4]) *Let $X$ be a compact subspace of $GL_n(\mathbb{R})$ or $GL_n(\mathbb{C})$ in the Euclidean topology. Then $\chi(Reg(\Gamma(X))) < \infty$.*

Since $GL_n(\mathbb{R})$ and $GL_n(\mathbb{C})$ are a countable union of its compact subspaces, using a similar argument, we see that the chromatic numbers of $GL_n(\mathbb{R})$ and $GL_n(\mathbb{C})$ are at most countable.

**Theorem 8.2.20.** ([25, Corollary 1]) *For an integer $n > 0$, let $Reg(\Gamma'(M_n(\mathbb{Z})))$ be the subgraph of $Reg(\Gamma(M_n(\mathbb{Z})))$ induced on $GL_n(\mathbb{Z})$. Then $\chi(Reg(\Gamma'(M_n(\mathbb{Z}))))$ is finite.*

**Theorem 8.2.21.** ([25, Theorem 5]) *If $n$ is a positive integer and every finite subgraph of a graph has chromatic number at most $n$, then so does the full graph.*

**Theorem 8.2.22.** ([25, Corollary 2]) *Let $F$ be an infinite field. Suppose that $E$ is a pure transcendental extension of $F$, i.e., there are elements $\{x_i\}_{i\in I}$ of $F$ which are algebraically independent over $F$ and generate $E$ over $F$ as a field. Then $\chi(Reg(\Gamma(M_n(F)))) = \chi(Reg(\Gamma(M_n(E))))$ for every positive integer $n$, provided that $\chi(Reg(\Gamma(M_n(F)))) < \infty$. In particular,* $\chi(Reg(\Gamma(M_n(F)))) = \chi(Reg(\Gamma(M_n(F(x)))))$ *if* $\chi(Reg(\Gamma(M_n(F)))) < \infty$.

**Theorem 8.2.23.** ([25, Corollary 3]) *Let $p_1, p_2, \ldots$ be an infinite sequence of distinct prime numbers. Then*

$$\chi(Reg(\Gamma(M_n(\mathbb{Q})))) \leq \sup_{i\in\mathbb{N}} \chi(Reg(\Gamma(M_n(\mathbb{F}_{p_i}))))$$

*for every positive integer $n$. Here $\mathbb{F}_p$ is the finite field with $p$ elements.*

Tomon [414] proved that $\chi(Reg(\Gamma(M_n(\mathbb{F}_q)))) \geq (q/4)^{\lfloor n/2 \rfloor}$ holds for any odd prime power $q$.

The first result establishes a lower bound in the case $n = 2$.

**Theorem 8.2.24.** ([414, Theorem 2.1]) *Let $p$ be an odd prime and $q = p^m$, where $m$ is a positive integer. Then*

$$\chi(Reg(\Gamma(M_2(\mathbb{F}_q)))) > q/4.$$

**Theorem 8.2.25.** ([414, Corollary 2.3]) *Let $p \geq 3$ be a prime, $q = p^m$. For any positive integer $n$, we have*

$$\chi(Reg(\Gamma(M_n(\mathbb{F}_q)))) \geq (q/4)^{\lfloor n/2 \rfloor}.$$

Let $\mathbb{F}_p^{alg}$ denote the algebraic closure of the field $\mathbb{F}_p$. The result that $Reg(\Gamma(M_n(\mathbb{F}_p^{alg}))) = \infty$ is immediate from the previous result as $\mathbb{F}_p^{alg}$ contains a subfield isomorphic to $\mathbb{F}_{p^m}$ for every positive integer $m$.

**Theorem 8.2.26.** ([414, Theorem 2.4]) *Let $n \geq 2$ be an integer and $p \geq 3$ a prime. Then $\chi(Reg(\Gamma(M_n(\mathbb{F}_p^{alg})))) = \infty$.*

Hence, $\chi(Reg(\Gamma(M_n(F))))$ is not necessarily finite. The question whether $\chi(Reg(\Gamma(M_n(F))))$ is finite for $F = \mathbb{Q}$, $\mathbb{R}$, or $\mathbb{C}$ is still open. We finish this section by giving an upper bound for $\chi(Reg(\Gamma(M_2(\mathbb{F}_q))))$. As the maximal degree of $Reg(\Gamma(M_2(\mathbb{F}_q)))$ is at most $q^3 + q^2 - q$, we get the trivial bound that $\chi(Reg(\Gamma(M_2(\mathbb{F}_q)))) \leq q^3 + q^2 - q + 1$. Our next result shows that this can be improved.

**Theorem 8.2.27.** ([414, Theorem 2.5]) *Let $p \geq 3$ be an odd prime and $q = p^m$ for $m$ a positive integer. Then $\chi(Reg(\Gamma(M_2(\mathbb{F}_q)))) \leq 4q(q+1)$.*

## 8.3 Complements of Total Graphs

Let $CT(\Gamma(R))$ denote the *complement of the total graph* of a commutative ring $R$, i.e., $CT(\Gamma(R))$ is the simple undirected graph with $R$ as its vertex set, and two distinct vertices $x$ and $y$ in $CT(\Gamma(R))$ are adjacent if and only if $x+y \in Reg(R)$.

While we talk about the complement of the total graph, a similar graph has already been studied, namely, the unit graph of a ring (see [91]). The unit graph of a ring $R$, denoted by $G(R)$, is the graph obtained by letting all the elements of $R$ be vertices and defining distinct vertices $x$ and $y$ to be adjacent if and only if $x + y \in U(R)$. Note that if the ring $R$ is finite, then $CT(\Gamma(R))$ is nothing but the unit graph. In the case of an infinite ring, $G(R)$ is a subgraph of $CT(\Gamma(R))$, not necessarily a proper subgraph. For example, consider the ring $R = \mathbb{Q}[x](+)(\mathbb{Q}(x)/\mathbb{Q}[x])$. Then $R$ is infinite with $U(R) = Reg(R)$, and so $CT(\Gamma(R)) = G(R)$.

Using the language of the complement of a total graph, we have the following result. Note that a similar result was proved in the context of unit graphs (see [91, Proposition 2.4]).

**Lemma 8.3.1.** ([100, Corollary 2.5]) *Let $R$ be a commutative ring with $Reg(R)$ finite.*

(i) *If $2 \in Z(R)$, then $CT(\Gamma(R))$ is regular, and in this case, $deg(v) = |Reg(R)|$ for every $v \in V(CT(\Gamma(R)))$;*

(ii) *If $2 \notin Z(R)$, then $\Delta(CT(\Gamma(R))) = \delta(CT(\Gamma(R))) + 1$. Also, $deg(v) = \Delta(CT(\Gamma(R))) = |Reg(R)|$ if $v \in Z(R)$, and $deg(v) = \delta(CT(\Gamma(R)))$ if $v \notin Z(R)$.*

Recall that a *path graph* is a particularly simple example of a tree, namely a tree with two or more vertices that is not branched at all, that is, contains only vertices of degree 2 and 1. In particular, it has two terminal vertices (vertices that have degree 1), while all others (if any) have degree 2.

**Theorem 8.3.2.** ([100, Theorem 3.3]) *Let $R$ be a commutative ring.*

(i) *$CT(\Gamma(R))$ is a path if and only if $R \cong \mathbb{Z}_2$;*

(ii) *$CT(\Gamma(R))$ is complete if and only if $R$ is an integral domain and $char(R) = 2$;*

(iii) *$CT(\Gamma(R))$ is a star graph if and only if $R \cong \mathbb{Z}_2$ or $\mathbb{Z}_3$;*

(iv) *$CT(\Gamma(R))$ is a cycle if and only if either $R \cong \mathbb{Z}_4$, $\mathbb{Z}_2[x]/(x^2)$, or $\mathbb{Z}_6$;*

(v) *$CT(\Gamma(R))$ is a complete bipartite graph if and only if either $Z(R)$ is an ideal of $R$ with $R/Z(R) \cong \mathbb{Z}_2$ or $R \cong \mathbb{Z}_3$.*

*Proof.* (i) Suppose that $CT(\Gamma(R))$ is a path. Then $CT(\Gamma(R))$ contains a vertex of degree less than or equal to two, and so $|Reg(R)| \leq 2$. From this, $R$ is finite, and so by Proposition 7.4.24(*ii*), $R \cong \mathbb{Z}_2^m$ for some integer $m \geq 1$. Since $2 \in Z(\mathbb{Z}_2^m)$, $T(\Gamma(\mathbb{Z}_2^m))$ is regular, and so $CT(\Gamma(R)) = P_2$. This implies that $R \cong \mathbb{Z}_2$.

(ii) Assume that $CT(\Gamma(R))$ is complete. If $x \neq -x$ for some $x \in R$, then $x$ is not adjacent to $-x$ in $CT(\Gamma(R))$, a contradiction, i.e., $char(R) = 2$. If $R$ is not an integral domain, let $0 \neq z \in Z(R)$. From this, $z$ is not adjacent to 0, a contradiction. The proof of the other part is trivial.

(iii) Suppose that $CT(\Gamma(R))$ is a star graph. Then $CT(\Gamma(R))$ contains a vertex of degree one, and so $Reg(R)$ is finite. Then the degrees of the vertices in $CT(\Gamma(R))$ differ by at most 1. From this, $CT(\Gamma(R))$ is a star graph if and only if $CT(\Gamma(R))$ is either $K_{1,1}$ or $K_{1,2}$. Note that $CT(\Gamma(R)) = K_{1,1}$ if and only if $R \cong \mathbb{Z}_2$. If $CT(\Gamma(R)) = K_{1,2}$, then $Z(R) = \{0\}$, $|Reg(R)| = 2$, and $char(R) \neq 2$. From this, $R \cong \mathbb{Z}_3$.

(iv) The proof of the "if" part is trivial. Assume that $CT(\Gamma(R))$ is a cycle. Then $CT(\Gamma(R))$ contains a vertex of degree two, and so by Lemma 8.3.1, $2 \in Z(R)$ and $|Reg(R)| = 2$. From this, $R \cong \mathbb{Z}_4$, $\mathbb{Z}_2[x]/(x^2)$, $\mathbb{Z}_3 \times \mathbb{Z}_2^r$, $\mathbb{Z}_4 \times \mathbb{Z}_2^r$, *or* $\mathbb{Z}_2[x]/(x^2) \times \mathbb{Z}_2^r$, where $r$ is a positive integer. If $R = \mathbb{Z}_4$ or $\mathbb{Z}_2[x]/(x^2)$, then $CT(\Gamma(R)) = C_4$. Let $R = \mathbb{Z}_3 \times \mathbb{Z}_2^r$. Then the subgraph induced by the vertices $(0,0,\ldots), (1,1,\ldots), (1,0,0,\ldots), (0,1,1,\ldots), (2,0,0,\ldots)$, $(2,1,1,\ldots)$ is $C_6$ in $CT(\Gamma(R))$, and so $CT(\Gamma(R))$ is a cycle only if $R = \mathbb{Z}_2 \times \mathbb{Z}_3$. Similarly, if $R = \mathbb{Z}_4 \times \mathbb{Z}_2^r$ or $\mathbb{Z}_2[x]/(x^2) \times \mathbb{Z}_2^r$, then $C_4$ is a subgraph of $CT(\Gamma(R))$. Note that no ring isomorphic to $\mathbb{Z}_2^r \times \mathbb{Z}_4$ or $\mathbb{Z}_2^r \times \mathbb{Z}_2[x]/(x^2)$ can have 4 elements, and so $CT(\Gamma(R))$ is not a cycle.

(v) Assume that $CT(\Gamma(R))$ is a complete bipartite graph. Since 0 is adjacent to all other vertices in $Reg(R)$, $(Z(R), Reg(R))$ is the unique partition of the vertex set of $CT(\Gamma(R))$ such that $CT(\Gamma(R))$ is complete bipartite. Thus $Z(R)$ is closed under addition, and so $Z(R)$ is an ideal of $R$. If $2 \in Z(R)$, then by the structure of $T(\Gamma(R))$, $CT(\Gamma(R))$ is a complete bipartite graph if and only if $|R/Z(R)| = 2$, and so $R/Z(R) \cong \mathbb{Z}_2$. If $2 \notin Z(R)$, then $CT(\Gamma(R))$ is complete bipartite if and only if $|Z(R)| = 1$ and $|R/Z(R)| \leq 3$. Thus $R \cong \mathbb{Z}_3$. The other part can be easily verified. □

Theorem 8.3.2 generalized some of the results proved for the unit graph. For instance, Ashrafi et al. [91] characterized the rings $R$ for which $G(R)$ is a cycle. In particular, when $R$ is a commutative ring, it was proved that $G(R)$ is a cycle if and only if $R \cong \mathbb{Z}_4$ or $\mathbb{Z}_6$ [91, Theorem 3.2]. In this regard, note that $\mathbb{Z}_4 \ncong \mathbb{Z}_2[x]/(x^2)$ and $CT(\Gamma(\mathbb{Z}_2[x]/(x^2))) = G(\mathbb{Z}_2[x]/(x^2)) = C_4$. From this fact, and by Theorem 8.3.2(iv), the statement for the unit graph ([91, Theorem 3.2]) should be modified.

**Corollary 8.3.3.** ([100, Corollary 3.4]) *Let $R$ be a finite commutative ring. Then the unit graph $G(R)$ is a cycle graph if and only if $R$ is isomorphic to $\mathbb{Z}_4$, $\mathbb{Z}_2[x]/(x^2)$, or $\mathbb{Z}_6$.*

Note that [91, Theorem 3.5] characterized all commutative rings $R$ with a maximal ideal $\mathfrak{m}$ and $|R/\mathfrak{m}| = 2$ for which the unit graph $G(R)$ is a complete bipartite graph. However, in view of Theorem 8.3.2($v$), the following corollary generalizes the result for unit graphs.

**Corollary 8.3.4.** *Let $R$ be a finite commutative ring. Then the unit graph $G(R)$ is complete bipartite if and only if either $Z(R)$ is an ideal of $R$ with $R/Z(R) \cong \mathbb{Z}_2$ or $R \cong \mathbb{Z}_3$.*

The next result characterizes the graph $CT(\Gamma(R))$ when $|Reg(R)| = 2$.

**Theorem 8.3.5.** ([100, Theorem 4.1]) *Let $R$ be a commutative ring, $|Reg(R)| = 2$, and $|Z(R)| = \alpha$ (may be infinite). Then*

$$CT(\Gamma(R)) = \begin{cases} C_{|R|} & \textit{if } 2 \in Z(R) \textit{ and } \langle Z(R)^* \rangle \textit{ is connected;} \\ \bigcup_{i=1}^{r} C_{k_i} & \textit{if } 2 \in Z(R) \textit{ and } \langle Z(R)^* \rangle \textit{ is disconnected;} \\ K_{1,2} \cup \bigcup_{j=1}^{s} C_{\ell_j} & \textit{otherwise;} \end{cases}$$

*where $3 \leq k_i \leq \alpha - 2$ for $1 \leq i \leq r$ and $3 \leq \ell_j \leq \alpha - 1$ for $1 \leq j \leq s$.*

*Proof.* Let $Reg(R) = \{x, y\}$. Note that $deg(u) = 2$ for all $u \in Z(R)^*$ and $0$ is adjacent to both $x$ and $y$.

If $2 \in Z(R)$, then $deg(x) = deg(y) = |Reg(R)| = 2$. If $x$ and $y$ are adjacent in $CT(\Gamma(R))$, then either $x + y = x$ or $x + y = y$. This implies that either $x = 0$ or $y = 0$, which is a contradiction to $0 \notin Reg(R)$. Thus $x$ is adjacent to $u \in Z(R)^*$ and $y$ is adjacent to $v \in Z(R)^*$. Assume that $\langle Z(R)^* \rangle \subseteq CT(\Gamma(R))$ is connected. Since the degree of every vertex is two, we have $CT(\Gamma(R)) = C_{|R|}$. On the other hand, if $\langle Z(R)^* \rangle \subseteq CT(\Gamma(R))$ is disconnected, then $CT(\Gamma(R)) = C_{k_1} \cup C_{k_2} \cup \cdots \cup C_{k_r}$.

If $2 \notin Z(R)$, then $deg(x) = deg(y) = 1$, and so $\langle \{x, 0, y\} \rangle = K_{1,2} \subseteq CT(\Gamma(R))$. Suppose that $R$ contains more than 3 elements. Then the cycles will be part of $CT(\Gamma(R))$, and one can obtain them as described above. □

**Corollary 8.3.6.** ([100, Theorem 4.2]) *Let $R$ be a commutative ring with $|Reg(R)| = 2$. Then*

$$CT(\Gamma(R)) = \begin{cases} C_{|R|} & \textit{if } R \cong \mathbb{Z}_4,\ \mathbb{Z}_2[x]/(x^2),\ \mathbb{Z}_2 \times \mathbb{Z}_3; \\ \frac{|R|}{k} C_k & \textit{if } R \cong \mathbb{Z}_2^r \times \mathbb{Z}_3,\ \mathbb{Z}_2^s \times \mathbb{Z}_4,\ \mathbb{Z}_2^s \times \mathbb{Z}_2[x]/(x^2); \\ K_{1,2} & \textit{if } R \cong \mathbb{Z}_3; \end{cases}$$

*where $r(> 1)$, $k$, and $s$ are positive integers. Moreover, $k = 6$ if $R = \mathbb{Z}_2^r \times \mathbb{Z}_3$, and $k = 4$ if $R = \mathbb{Z}_2^s \times \mathbb{Z}_4$ or $\mathbb{Z}_2^s \times \mathbb{Z}_2[x]/(x^2)$.*

If $R$ is a finite commutative ring, then the unit graph $G(R)$ is connected if and only if $R$ does not have $\mathbb{Z}_2 \times \mathbb{Z}_2$ as a quotient (see [91, Theorem 5.7(d)]). Hence if $R$ is a finite commutative ring, then $CT(\Gamma(R))$ is connected if and only if $R$ does not have $\mathbb{Z}_2 \times \mathbb{Z}_2$ as a quotient. We are next interested in the connectedness of $CT(\Gamma(R))$ for infinite commutative rings.

**Theorem 8.3.7.** ([100, Theorem 4.6]) *Let $R$ be a commutative ring which is not an integral domain and $R \ncong \mathbb{Z}_2^r$, where $r \geq 1$ is an integer. If $\langle Z(R)^* \rangle \subseteq CT(\Gamma(R))$ is connected and $R$ satisfies any of the following conditions, then $CT(\Gamma(R))$ is connected:*

(i) *$R$ has a nonzero nilpotent element;*

(ii) *$Reg(R)$ contains more elements than the number of elements in $Z(R)^*$;*

(iii) *There is a $u \in Z(R)^*$ with $u = 1 + 1 + \cdots$ ;*

(iv) *There are two distinct elements $x, y \in Reg(R)$ such that $x + y \in Z(R)^*$;*

(v) *$R$ is a Noetherian ring.*

*Proof.* Note that in $CT(\Gamma(R))$, 0 is adjacent to every vertex in $Reg(R)$, and so $\langle Reg(R) \cup \{0\} \rangle$ is a connected subgraph of $CT(\Gamma(R))$. By the assumption, $\langle Z(R)^* \rangle$ is a connected subgraph of $CT(\Gamma(R))$. Therefore, to conclude the proof, it is enough to show there is an edge between the two induced subgraphs $\langle Reg(R) \cup \{0\} \rangle$ and $\langle Z(R)^* \rangle$.

(i) Let $0 \neq x \in Nil(R)$. Since $Nil(R) + Reg(R) \subseteq Reg(R)$, there is an edge joining $x$ with any element in $Reg(R)$.

(ii) Let $z \in Z(R)^*$. By the assumption, $z + Reg(R)$ contains at least one element from $Reg(R)$, say $y$. Then $z$ is adjacent to $-y \in Reg(R)$.

(iii) Let $\ell = 1 + 1 + \cdots$ be the least sum such that $\ell \in Z(R)^*$(such an element exists by the assumption). Then $\ell - 1 \in Reg(R)$, and so $\ell \in Z(R)^*$ is adjacent to $-(\ell - 1) \in Reg(R)$.

(iv) Clearly $x + y \in Z(R)^*$ is adjacent to $-x \in Reg(R)$.

(v) **Case 1.** If $|Reg(R)| = 2$, then by Theorem 8.3.5, $CT(\Gamma(R))$ is connected whenever $\langle Z(R)^* \rangle$ is connected.

**Case 2.** Suppose that $|Reg(R)| > 2$. By cases (iii) and (iv), it is enough to consider the case $2 \notin Z(R)^*$ and $x + y \notin Z(R)^*$ for every $x, y \in Reg(R)$. Note that the first implies that $2 \in Reg(R) \cup \{0\}$, and the latter implies that in the total graph $T(\Gamma(R))$, every regular element $x$ is adjacent to at most one regular element, namely $-x$. Also note that 0 is not adjacent to any of the regular elements in $T(\Gamma(R))$.

If $2 \in Reg(R)$, then $deg_{T(\Gamma(R))}(a) = |Z(R)|$ for every $a \in Reg(R)$. From this, $Reg(R) + Z(R)^* \subseteq Z(R)^*$, and so $d_{T(\Gamma(R))}(x, y) = 2$ for every $x, y \in Reg(R)$ with $x \neq -y$. Hence $d_{T(\Gamma(R))}(x, y) = 2$ and $d_{Reg_\Gamma(R)}(x, y) = \infty$, a contradiction to Theorem 7.2.12.

If $2 \notin Reg(R)$, then $2 = 0$. Therefore $deg_{T(\Gamma(R))}(a) = |Z(R)| - 1$ for every $a \in Reg(R)$. Let $x, y \in Reg(R)$. Since $R$ is not an integral domain, $Z(R)^* \neq \emptyset$ and $x$ and $y$ are adjacent to all elements in $Z(R)^*$. Thus $d_{T(\Gamma(R))}(x, y) = 2$ and $d_{Reg_\Gamma(R)}(x, y) = \infty$ for every $x, y \in Reg(R)$ with $x \neq -y$, a contradiction to Theorem 7.2.12. □

There do exist rings for which $CT(\Gamma(R))$ is connected even when $\langle Z(R)^* \rangle \subseteq CT(\Gamma(R))$ is disconnected and $R$ satisfies one or more conditions stated in Theorem 8.3.7. For example, when $R = \mathbb{Z}_{12}$, $\overline{2} = \overline{1} + \overline{1} \in Z(R)^*$ and $x + y \in Z(R)^*$ for every $x, y \in Reg(R)$. Here $R$ satisfies conditions (i), (iii), (iv), and (v). Note that $CT(\Gamma(R))$ is connected, whereas $\overline{6}$ is an isolated vertex in $\langle Z(R)^* \rangle$, and so $\langle Z(R)^* \rangle$ is disconnected.

By Corollary 7.2.11, $diam(T(\Gamma(R))) \leq diam(Reg(\Gamma(R))) + 2$. A similar result holds for $CT(\Gamma(R))$.

**Lemma 8.3.8.** ([100, Corollary 4.9]) *If $CT(\Gamma(R))$ and $\langle Z(R)^* \rangle \subseteq CT(\Gamma(R))$ are connected, then $2 \leq diam(CT(\Gamma(R))) \leq diam(\langle Z(R)^* \rangle) + 3$.*

*Proof.* Since $CT(\Gamma(R))$ is not complete, $diam(CT(\Gamma(R))) \geq 2$. Take $z \in Z(R)^*$ and $x \in Reg(R)$ such that $z$ and $x$ are not adjacent. Since $CT(\Gamma(R))$ is connected, there are vertices $a \in Reg(R)$ and $b \in Z(R)^*$ with $a$ and $b$ adjacent. Note that $d(z, a) \leq diam(\langle Z(R)^* \rangle)$ and $0$ is adjacent to $b$ and $x$. From this $diam(CT(\Gamma(R))) \leq diam(\langle Z(R)^* \rangle) + 3$. □

The next result characterizes all commutative rings $R$ according to the value of $gr(CT(\Gamma(R)))$.

**Theorem 8.3.9.** ([100, Theorem 4.10]) *Let $R$ be a commutative ring. Then $gr(CT(\Gamma(R))) \in \{3, 4, 6, \infty\}$. Further,*

(i) *$gr(CT(\Gamma(R))) = \infty$ if and only if $R \cong \mathbb{Z}_3$ or $\mathbb{Z}_2^r$, where $r$ is a positive integer;*

(ii) *$gr(CT(\Gamma(R))) = 6$ if and only if $R \cong \mathbb{Z}_2^r \times \mathbb{Z}_3$, where $r$ is a positive integer;*

(iii) *$gr(CT(\Gamma(R))) = 4$ if and only if $Reg(R) + Reg(R) \subseteq Z(R)$ and for any three elements $z_1, z_2, z_3 \in Z(R)$, $z_i + z_j \in Z(R)$ for some $i \neq j$, $1 \leq i, j \leq 3$. In particular, if $Z(R)$ is an ideal of $R$, then $gr(CT(\Gamma(R))) = 4$ if and only if $R/Z(R) \cong \mathbb{Z}_2$ and $2 \in Z(R)$;*

(iv) *$gr(CT(\Gamma(R))) = 3$, otherwise.*

*Proof.* Recall that if $|Reg(R)| = 1$, then $gr(CT(\Gamma(R))) = \infty$. When $R = \mathbb{Z}_3$, then $|Reg(R)| = 2$ and $gr(CT(\Gamma(\mathbb{Z}_3))) = \infty$. By Corollary 8.3.6, if $|Reg(R)| = 2$ and $R \ncong \mathbb{Z}_3$, then $gr(CT(\Gamma(R))) = 4$ or $6$. Hence assume that $Reg(R)$ contains at least three elements.

**Case 1.** If $R$ is an integral domain, then $0$ is adjacent to all the vertices of $R^*$. By the assumption that $|R^*| = |Reg(R)| \geq 3$, there are $v_1, v_2 (\neq -v_1) \in Reg(R)$

such that $v_1 + v_2 \in Reg(R)$. In this case, $\langle\{0, v_1, v_2\}\rangle = C_3 \subseteq CT(\Gamma(R))$, and so $gr(CT(\Gamma(R))) = 3$.

**Case 2.** Assume that $R$ is not an integral domain.

**Sub case 2.1.** $Z(R)$ is an ideal of $R$, $|Z(R)| = \alpha$, and $|R/Z(R)| = \beta$ (we allow $\alpha$ and $\beta$ to be infinite cardinals). Then

$$T(\Gamma(R)) = \begin{cases} K_\alpha \cup \underbrace{K_\alpha \cup K_\alpha \cup \cdots \cup K_\alpha}_{(\beta-1)\ copies} & if\ 2 \in Z(R); \\ K_\alpha \cup \underbrace{K_{\alpha,\alpha} \cup K_{\alpha,\alpha} \cup \cdots \cup K_{\alpha,\alpha}}_{(\frac{\beta-1}{2})\ copies} & if\ 2 \notin Z(R). \end{cases}$$

Suppose $2 \in Z(R)$. If $\beta \geq 3$, then consider three $K_\alpha$'s and take a vertex from each of them. Note that these three vertices form a cycle of length 3 in $CT(\Gamma(R))$. On the other hand, we have $\beta = 2$. Since $\alpha \geq 2$, by taking two distinct vertices in each $K_\alpha$, we get a cycle of length 4 in $CT(\Gamma(R))$.

Suppose $2 \notin Z(R)$. Since $\alpha \geq 2$ and $\beta \geq 3$, two distinct vertices in any one of the components $K_{\alpha,\alpha}$ together with a vertex in $Z(R)$ form a cycle of length 3 in $CT(\Gamma(R))$.

**Case 2.2.** $Z(R)$ is not an ideal of $R$. Let $x \in Reg(R)$. If $x + y \in Reg(R)$ for some $y \in Reg(R)$, then $\langle\{0, x, y\}\rangle = C_3 \subseteq CT(\Gamma(R))$. If $x + y \in Z(R)^*$ for every $-x \neq y \in Reg(R)$, then $\langle\{0, -x, x + y, -y\}\rangle = C_4 \subseteq CT(\Gamma(R))$.

Hence, if $Reg(R)$ contains at least three elements, then $gr(CT(\Gamma(R))) \leq 4$.

By taking various rings corresponding to various girth values, one can match the rings and the girth values as stated in the statement. □

Let $C(R)$ be a simple undirected graph with vertex set $R$, and for distinct $x, y \in R$, the vertices $x$ and $y$ are adjacent if and only if $x - y \in Z(R)$ [348]. It is natural for one to ask when $T(\Gamma(R))$ is isomorphic to $C(R)$. We have the following result.

**Theorem 8.3.10.** ([348, Theorem 5.2]) *Let $R$ be a finite commutative ring. Then the two graphs $T(\Gamma(R))$ and $C(R)$ are isomorphic if and only if at least one of the following conditions is true:*

(1) *$R \cong R_1 \times \cdots \times R_k$, $k \geq 1$, and every $R_i$ is a local ring of even order;*

(2) *$R \cong R_1 \times \cdots \times R_k$, $k \geq 2$, every $R_i$ is a local ring, and $min\{|R_i/\mathfrak{m}_i| \mid \mathfrak{m}_i$ is the maximal ideal of $R_i\} = 2$.*

## 8.4 Generalized Complements of Total Graphs

Let $R$ be a commutative ring, $U(R)$ the group of units of $R$, $G$ a multiplicative subgroup of $U(R)$, and $S$ a nonempty subset of $G$ such that $S^{-1} = \{s^{-1} \mid s \in S\} \subseteq S$.

When we compare the unit and unitary Cayley graphs, it seems that they have the same behavior, however in general, they are not isomorphic. This provided a motivation for Khashyarmanesh et al. [232] to introduce a generalization of these graphs. In [232], Khashyarmanesh et al. defined a graph $\Gamma(R, G, S)$ which generalizes the complement of the total graph of a finite ring.

**Definition 8.4.1.** ([232]) Let $R$ be a commutative ring, $G$ a multiplicative subgroup of $U(R)$, and $S$ a nonempty subset of $G$ such that $S^{-1} = \{s^{-1} \mid s \in S\} \subseteq S$. The *generalized unit and unitary Cayley graph* $\Gamma(R, G, S)$ is the simple graph with vertex set $R$, and two distinct vertices $x$ and $y$ are adjacent if and only if there is an $s \in S$ such that $x + sy \in G$. If we omit the word "distinct" from the definition of $\Gamma(R, G, S)$, then the corresponding $\Gamma(R, G, S)$ is denoted by $\overline{\Gamma}(R, G, S)$. For simplicity of notation, $\Gamma(R, U(R), S)$ is denoted by $\Gamma(R, S)$.

The graph $\Gamma(R, G, S)$ provides an "umbrella concept" which covers the unit graphs and the unitary Cayley graphs (see Section 10.1.1). Recall that Anderson et al. [67] introduced and studied the total graph $T(\Gamma(R))$ of a commutative ring $R$. Actually, the *total graph* of $R$ is the undirected graph whose vertices are the elements in $R$, and two distinct vertices $x$ and $y$ are adjacent if and only if $x + y \in Z(R)$. The complement of the total graph is denoted by $CT(\Gamma(R))$. Note that two distinct vertices $x$ and $y$ in $CT(\Gamma(R))$ are adjacent if and only if $x + y \in Reg(R)$.

In general, the co-maximal graph $\Gamma(R)$, unitary Cayley graph $Cay(R, U(R))$, unit graph $G(R)$, total graph $T(\Gamma(R))$, and the graph $\Gamma(R, G, S)$ are not isomorphic. But there are relationships between these graphs. For instance, if $S = \{1\}$, then $\Gamma(R, S)$ is same as the unit graph $G(R)$; and if $S = \{-1\}$, then $\Gamma(R, S)$ is the unitary Cayley graph $Cay(R, U(R))$ of $R$. Note that $\Gamma(R, G, S)$ is a subgraph of the co-maximal graph.

Further, if $\Gamma(R, S) \cong \Gamma(R, S')$ for $S, S' \subseteq U(R)$, then denote $\Gamma(R, U(R), S)$ by $\Gamma(R)$. If $R$ is finite and $S = \{1\}$, then $\Gamma(R, S)$ is the same as the complement of the total graph, $CT(\Gamma(R))$. If we omit the word distinct, we obtain the graph $\overline{\Gamma}(R, G, S)$, and this graph may have loops.

We begin with the following proposition.

**Proposition 8.4.2.** ([232, Proposition 2.1]) *The following statements hold.*

(a) *If $G = \{1\}$, then*

$$\Gamma(R, G, S) \cong \begin{cases} \frac{|R|}{2} K_2 & \text{if } 2 \notin U(R), \\ \frac{|R|-1}{2} K_2 \cup K_1 & \text{otherwise}. \end{cases}$$

(b) *If $G = \{1, a\}$, for some $1 \neq a \in R$ and $S = \{s\}$, then*

(i) *whenever $1 + s \notin U(R)$, the graph $\Gamma(R, G, S)$ is 2-regular; and,*

(ii) *whenever* $1+s \in U(R)$, *all vertices in* $\Gamma(R,G,S)$, *except two vertices, have degree* 2. *In particular, if* $R$ *is finite, then* $\Gamma(R,G,S) \cong P_n \cup H$, *where* $P_n$ *is a path of length* $n \geqslant 2$ *and* $H$ *is an empty graph or a* 2*-regular graph.*

*Proof.* (a) This follows from the fact that for every vertex $x$ of $\Gamma(R,\{1\},\{1\})$, $N(x) = \{1-x \mid 1-x \neq x\}$.

(b) The first claim follows from the fact that for every vertex $x$ of $\Gamma(R,\{1,a\}, \{s\})$, $N(x) = \{a-sx, 1-sx\}$. For the second one, note that there is a unique element $z \in R$ with $(1+s)z = 1$. Hence $N(z) = \{a-sz\}$ and $N(az) = \{1-saz\}$. Also, for a vertex $x$ with $x \neq z$ and $x \neq az$, $N(x) = \{a-sx, 1-sx\}$. The result now follows immediately from the fact that the two elements $a-sz$ and $1-saz$ do not belong to $\{z, az\}$. □

The following example shows that in the last part of Proposition 8.4.2, the graph $H$ may be nonempty.

**Example 8.4.3.** ([232, Example 2.2]) Let $R = \mathbb{Z}_{15}$, $G = \{1,4\}$, and $S = \{1\}$. Then it is not hard to see that $\Gamma(R,G,S) \cong P_4 \cup C_{10}$.

In view of the Proposition 8.4.2(a), we have the following result.

**Corollary 8.4.4.** ([232, Corollary 2.3]) *Let* $R$ *be a finite commutative ring. Then* $2 \in U(R)$ *if and only if* $|R|$ *is odd.*

Let $\Gamma_1$ and $\Gamma_2$ be two graphs without multiple edges. Recall that the *tensor product* $\Gamma = \Gamma_1 \otimes \Gamma_2$ is the graph with vertex set $V(\Gamma) = V(\Gamma_1) \times V(\Gamma_2)$, and two distinct vertices $(u_1,u_2)$ and $(v_1,v_2)$ of $\Gamma$ are adjacent if and only if $\{u_1,v_1\} \in E(\Gamma_1)$ and $\{u_2,v_2\} \in E(\Gamma_2)$.

**Remarks 8.4.5.** ([232, Remark 2.4])

(a) For every vertex $x$ of $\Gamma(R,G,S)$, we have $|G|-1 \leqslant \deg(x) \leqslant |G||S|$. Furthermore, $\deg(x) \geqslant |G|$ for every vertex $x$ of $\overline{\Gamma}(R,G,S)$.

(b) In the light of Proposition 8.4.2(a), the above inequalities imply that the graph $\Gamma(R,G,S)$ has an isolated vertex if and only if $G = \{1\}$ and $2 \in U(R)$.

(c) Suppose that $R_1$ and $R_2$ are commutative rings, and for each $i$ with $i = 1,2$, $G_i$ is a subgroup of $U(R_i)$. Also, assume that $S_i$ is a nonempty subset of $G_i$ with $S_i^{-1} \subseteq S_i$.

  (i) Then $\Gamma(R_1 \times R_2, G_1 \times G_2, S_1 \times S_2) \cong \overline{\Gamma}(R_1,G_1,S_1) \otimes \overline{\Gamma}(R_2,G_2,S_2)$.

  (ii) Furthermore, if $R_1 = R_2$, $G_1 \subseteq G_2$, and $S_1 \subseteq S_2$, then $\Gamma(R_1,G_1,S_1)$ is a subgraph of $\Gamma(R_2,G_2,S_2)$.

The next example shows that the bound obtained in Remark 8.4.5(a) is sharp.

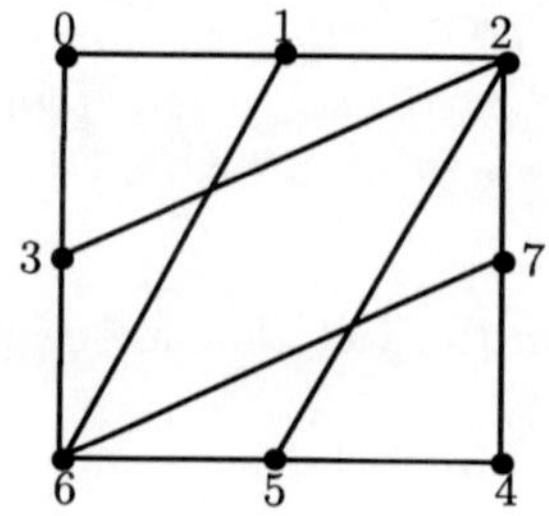


Figure 8.1: $\Gamma(\mathbb{Z}_8, \{1,3\}, \{1,3\})$

**Example 8.4.6.** ([232, Example 2.5]) Suppose that $R = \mathbb{Z}_8$ and $G = S = \{1,3\}$. Then $\deg(2) = \deg(6) = 4$ (see Figure 8.1).

**Theorem 8.4.7.** ([232, Theorem 2.6]) *Let $R$ be a finite commutative ring. Then the following statements hold:*

(a) *If $(R, \mathfrak{m})$ is a local ring of even order, then*

$$\Gamma(R, U(R), \{1\}) \cong \Gamma(R, U(R), \{-1\});$$

(b) *If $R$ has odd order, then $\Gamma(R, U(R), \{1\}) \ncong \Gamma(R, U(R), \{-1\})$.*

**Theorem 8.4.8.** ([232, Theorem 2.7]) *The graph $\Gamma(R, G, S)$ is a complete graph if and only if the following statements hold:*

(a) *$R$ is a field;*

(b) *$G = U(R)$; and,*

(c) *$|S| \geqslant 2$ or $S = \{-1\}$.*

*Proof.* First, assume that $\Gamma(R, G, S)$ is complete. Since every nonzero element $x$ of $R$ is adjacent to 0, there exists $s \in S$ such that $x + s0 = x \in G$, and so $G = R \setminus \{0\}$. This means that $R$ is a field with $G = U(R)$. Now, assume to the contrary that $S = \{s\}$ with $s \neq -1$. Then every $1 \neq x \in R$ is adjacent to 1, and hence $1 + sx$ is unit. On the other hand, since $s$ is unit, $s \notin J(R)$. Thus $1 + s$ is not unit, and so $s = -1$, which is the required contradiction.

Conversely, assume that there are two distinct nonadjacent elements $x$ and $y$ in $\Gamma(R, G, S)$. Without loss of generality, we may assume that $y \neq 0$. Thus $x + sy = 0$ for every $s \in S$. Hence $s = -y^{-1}x$, and so $|S| = 1$. Therefore $S = \{-1\}$, and thus $x = -sy = y$, which is impossible. □

**Proposition 8.4.9.** ([232, Proposition 2.8]) *The graph $\Gamma(R, G, S)$ is a refinement of a star graph with center 0 if and only if $R$ is a field with $G = U(R)$.*

**Corollary 8.4.10.** ([232, Corollary 2.9]) *$\Gamma(R,G,S)$ is a star graph with center 0 if and only if $R \cong \mathbb{Z}_2$ or $R \cong \mathbb{Z}_3$ with $S = \{1\}$.*

**Proposition 8.4.11.** ([232, Proposition 2.10]) *Let $R$ be a finite commutative ring and $\Gamma(R,G,S)$ a refinement of a star graph with nonzero center. If $S$ is singleton, then $\Gamma(R,G,S)$ is complete.*

**Theorem 8.4.12.** ([232, Theorem 2.11]) *Let $R$ be commutative ring and $-1 \in G$. Then $\Gamma(R,G,S)$ is connected if and only if every element of $R$ can be written as a sum of elements of $G$.*

*Proof.* ($\Rightarrow$) For every $0 \neq x \in R$, there is a walk

$$0 = x_0 - x_1 - x_2 - \cdots - x_n = x$$

between 0 and $x$. Hence, for every $i$ with $0 \leqslant i \leqslant n-1$, there are $s_i \in S$ and $g_i \in G$ such that $x_{i+1} = g_i - s_i x_i$. Now an elementary inductive argument on $i$ yields that for all $i$ with $0 \leqslant i \leqslant n-1$, $x_{i+1}$ can be written as a sum of elements of $G$.

($\Leftarrow$) Let $0 \neq x \in R$ and $s \in S$. Then, for each $1 \leqslant i \leqslant n$, there exists a $g_i \in G$ such that $sx = g_1 + g_2 + g_3 + \cdots + g_n$. So, we have the following walk between $x$ and 0.

$$x - (-sx + g_1) - (x - s^{-1}g_1 - s^{-1}g_2) - (-sx + g_1 + g_2 + g_3) -$$
$$(x - s^{-1}g_1 - s^{-1}g_2 - s^{-1}g_3 - s^{-1}g_4) - \cdots - 0.$$

Hence $\Gamma(R,G,S)$ is connected. □

Next, we study properties of $\Gamma(R,G,S)$ when $G = U(R)$. For simplicity of notation, we denote $\Gamma(R,U(R),S)$ by $\Gamma(R,S)$. Also, if there are no restrictions on $S$, then we denote $\Gamma(R,S)$ by $\Gamma(R)$.

**Remark 8.4.13.** ([232, Remark 3.1]) Suppose that $\{x_i + J(R)\}_{i \in I}$ is a complete set of coset representatives of $J(R)$. Note that if $x \in U(R)$ and $j \in J(R)$, then $x + j \in U(R)$. Hence, whenever $x_i$ and $x_j$ are adjacent vertices in $\Gamma(R,S)$, every element of $x_i + J(R)$ is adjacent to every element of $x_j + J(R)$. Moreover, if there are $s \in S$ with $(1+s)x_i \in U(R)$ for some $i$, then $x_i + J(R)$ is a clique in $\Gamma(R,S)$. Otherwise, $x_i + J(R)$ is a coclique in $\Gamma(R,S)$. For instance, if $x_i \notin U(R)$, then $x_i + J(R)$ is a coclique in $\Gamma(R,S)$.

**Proposition 8.4.14.** ([232, Proposition 3.2]) *Let $\mathfrak{m}$ be a maximal ideal of a commutative ring $R$ with $|R/\mathfrak{m}| = 2$. Then $\Gamma(R,S)$ is bipartite. Furthermore, if $R$ is a local ring, then $\Gamma(R,S)$ is a complete bipartite graph.*

**Lemma 8.4.15.** ([232, Lemma 3.3]) *Let $(R,\mathfrak{m})$ be a commutative local ring with maximal ideal $\mathfrak{m}$ such that $|R/\mathfrak{m}| > 2$. Then $N_{\overline{\Gamma(R,S)}}(x) \cap N_{\overline{\Gamma(R,S)}}(y) \neq \emptyset$ for every $x, y \in R$.*

**Lemma 8.4.16.** ([232, Lemma 3.4]) *Let $R$ be a commutative Artinian ring. Then the following statements hold:*

(a) *If $R/J(R)$ does not have a summand isomorphic to $\mathbb{Z}_2$, then*

$$diam(\Gamma(R,S)) \leq 2;$$

(b) *If $R/J(R)$ has exactly one summand isomorphic to $\mathbb{Z}_2$, then*

$$diam(\Gamma(R,S)) \leq 3.$$

*Furthermore, if $R$ is not local, then $diam(\Gamma(R,S)) = 3$;*

(c) *If $R/J(R)$ has $\mathbb{Z}_2 \times \mathbb{Z}_2$ as a summand, then $\Gamma(R,S)$ is disconnected. Furthermore, $diam(\Gamma(R,S)) = \infty$;*

(d) $diam(\Gamma(R,S)) \in \{1,2,3,\infty\}$.

Now we have a result that classifies $\Gamma(R,S)$ according to its diameter.

**Theorem 8.4.17.** ([232, Theorem 3.5]) *Let $R$ be a commutative Artinian ring.*

(i) *$diam(\Gamma(R,S)) = 1$ if and only if $R$ is a field with $|S| \geq 2$ or $S = \{-1\}$;*

(ii) *$diam(\Gamma(R,S)) = 2$ if and only if one of the following conditions hold:*
   (a) *$R$ is a field with $S = \{s\}$, where $s \neq -1$;*
   (b) *$R$ is not a field and $R/J(R)$ cannot have $\mathbb{Z}_2$ as a summand;*
   (c) *$(R, \mathfrak{m})$ is a local ring with $|R/M| = 2$ and $R \ncong \mathbb{Z}_2$;*

(iii) *$diam(\Gamma(R,S)) = 3$ if and only if $R/J(R)$ has exactly one summand isomorphic to $\mathbb{Z}_2$ and $R$ is not local;*

(iv) *$diam(\Gamma(R,S)) = \infty$ if and only if $R/J(R)$ has $\mathbb{Z}_2 \times \mathbb{Z}_2$ as a summand.*

The next result characterizes the commutative Artinian rings $R$ with $\Gamma(R,S)$ planar.

**Theorem 8.4.18.** ([232, Theorem 3.7]) *Let $R$ be a commutative Artinian ring. Then $\Gamma(R,S)$ is planar if and only if one of the following conditions hold:*

(i) *$R \cong \mathbb{Z}_2^\ell \times T$, where $\ell \geq 0$ and $T$ is isomorphic to one of the following rings: $\mathbb{Z}_2, \mathbb{Z}_3, \mathbb{Z}_4$, or $\mathbb{Z}_2[x]/(x^2)$;*

(ii) *$R \cong \mathbb{F}_4$;*

(iii) *$R \cong \mathbb{Z}_2^\ell \times \mathbb{F}_4$, where $\ell > 0$ with $S = \{1\}$;*

(iv) *$R \cong \mathbb{Z}_5$ with $S = \{1\}$;*

(v) *$R \cong \mathbb{Z}_3 \times \mathbb{Z}_3$ with $S = \{(1,1)\}$, $S = \{(1,-1)\}$, or $S = \{(-1,1)\}$.*

The following theorem characterizes all commutative Artinian rings with $\Gamma(R,S)$ toroidal (i.e, genus one).

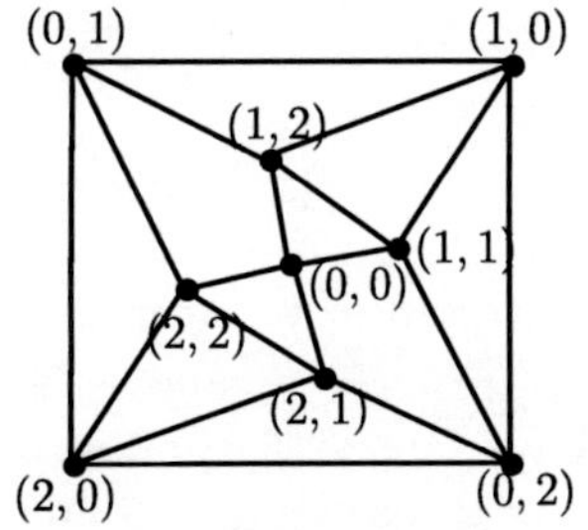


Figure 8.2: $\Gamma(\mathbb{Z}_3 \times \mathbb{Z}_3, \{(1,-1)\})$

**Theorem 8.4.19.** ([102, Theorem 4.2]) *Let $R$ be a commutative Artinian ring. Then $\Gamma(R,S)$ is toroidal if and only if one of the following conditions hold:*

(i) $R \cong \mathbb{Z}_5$ *with* $S \neq \{1\}$;

(ii) $R \cong \mathbb{Z}_7$;

(iii) $R \cong \mathbb{Z}_8, \mathbb{Z}_2[x]/(x^3), \mathbb{Z}_2[x,y]/(x^2,xy,y^2), \mathbb{Z}_4[x]/(x^2,2x)$, *or* $\mathbb{Z}_4[x]/(x^2-2,2x)$;

(iv) $R \cong \mathbb{Z}_2 \times \mathbb{F}_4$ *with* $S \neq \{(1,1)\}$;

(v) $R \cong \mathbb{Z}_9$ *with* $S = \{-1\}$, $S = \{2,5\}$, *or* $S = \{-1,2,5\}$;

(vi) $R \cong \mathbb{Z}_3[x]/(x^2)$ *with* $S = \{-1+(x^2)\}$, $S = \{-1+x+(x^2), -1-x+(x^2)\}$, *or* $S = \{-1+(x^2), -1+x+(x^2), -1-x+(x^2)\}$;

(vii) $R \cong \mathbb{Z}_3 \times \mathbb{Z}_3$ *with* $S = \{(-1,-1)\}$, $S = \{(1,1),(1,-1)\}$, $S = \{(1,1),(-1,1)\}$, $S = \{(-1,-1),(1,-1)\}$, *or* $S = \{(-1,-1),(-1,1)\}$;

(viii) $R \cong \mathbb{Z}_{10}$ *with* $S = \{1\}$ *or* $S = \{-1\}$;

(ix) $R \cong \mathbb{Z}_{12}$ *with* $S = \{1\}$, $S = \{-1\}$, $S = \{5\}$, $S = \{7\}$, $S = \{1,7\}$, *or* $S = \{5,-1\}$;

(x) $R \cong \mathbb{Z}_3 \times \mathbb{Z}_2[x](x^2)$ *with* $S = \{(1,1+(x^2))\}$, $S = \{(-1,1+x+(x^2))\}$, $S = \{(1,1+x+(x^2))\}$, $S = \{(-1,1+(x^2))\}$, $S = \{(1,1+(x^2)),(1,1+x+(x^2))\}$, *or* $S = \{(-1,1+(x^2)),(-1,1+x+(x^2))\}$.

*Proof.* Assume that the graph $\Gamma(R,S)$ is toroidal. Then $\delta(\Gamma(R,S)) \leq 6$ and $\delta(\Gamma(R,S)) \geq |U(R)| - 1$ by Lemma 7.5.5. Thus $|U(R)| \leq 7$. If $|J(R)| \geq 5$, then $|U(R)| \geq 5$. Since $J(R) + U(R) \subseteq U(R)$, $K_{5,5}$ is a subgraph of $\Gamma(R,S)$. This implies that $g(\Gamma(R,S)) \geq g(K_{5,5,}) = 3$, a contradiction. Thus $|J(R)| \leq 4$, i.e.,

$$|J(R)| \leq 4 \text{ and } |U(R)| \leq 7 \tag{8.1}$$

**Case 1.** Assume that $R$ is a field. Then $|R| \leq 8$ by equation (8.1). Also, by Theorem 8.4.18, $\Gamma(R,S)$ is planar whenever $|R| \leq 4$. Hence $|R| = 5, 7$, or 8. Since $g(\Gamma(\mathbb{Z}_5,\{1\})) = 0$ and $g(K_5) = g(K_7) = 1$, $g(\Gamma(\mathbb{Z}_5,S)) = 1$ for $S \neq \{1\}$, and $g(\Gamma(\mathbb{Z}_7)) = 1$. When $R \cong \mathbb{F}_8$, $char(R) = 2$, and so $\Gamma(\mathbb{F}_8) = K_8$, which has genus 2.

**Case 2.** Suppose $R$ is a local ring with maximal ideal $\mathfrak{m} \neq \{0\}$. Then $|\mathfrak{m}| \leq 4$ and $|R \setminus \mathfrak{m}| \leq 7$ by equation (8.1). Since $|R| \geq 4$ and $|R|$ is a prime power, but not prime, we have $|R|$ is either 8 or 9. According to Corbas et al. [167], there are 5 local rings $\mathbb{Z}_8$, $\frac{\mathbb{Z}_2[x]}{(x^3)}$, $\frac{\mathbb{Z}_2[x,y]}{(x^2,xy,y^2)}$, $\frac{\mathbb{Z}_4[x]}{(x^2,2x)}$, and $\frac{\mathbb{Z}_4[x]}{(x^2-2,2x)}$ of order 8 (except for $\mathbb{F}_8$). For all of these rings, we have $R/\mathfrak{m} \cong \mathbb{Z}_2$, and so $g(\Gamma(R,S)) = g(K_{4,4}) = 1$.

Again by [167], there are 2 local rings $\mathbb{Z}_9$ and $\mathbb{Z}_3[x]/(x^2)$ of order 9. First consider $\mathbb{Z}_9$. Clearly every element in $Z(\mathbb{Z}_9)$ is adjacent to all the elements of $U(\mathbb{Z}_9)$ in $\Gamma(\mathbb{Z}_9, \{1\})$, and so $K_{3,6}$ is a subgraph of $\Gamma(\mathbb{Z}_9, \{1\})$. Recall that the genus of $K_{3,6}$ is one, and hence one can fix an embedding of $K_{3,6}$ on the surface of a torus. By Euler's Formula, there are 9 faces in the embedding of $K_{3,6}$, say $\{F_1, \ldots, F_9\}$. Let $s_{F_i}$ be the length of the face $F_i$. Note that $\sum_{i=1}^{9} s_{F_i} = 36$ and $s_{F_i} \geq 4$ for every $i$. Thus $s_{F_i} = 4$ for every $i$. Further, the subgraph $H$ of $\Gamma(\mathbb{Z}_9, \{1\})$ induced by the vertices in $\{1 + Z(\mathbb{Z}_9)\}$ is $K_3$, $V(H) \subseteq V(K_{3,6})$, and $E(H) \cap E(K_{3,6}) = \emptyset$. Since $K_3$ cannot be embedded in a torus along with an embedding with only rectangles as faces, one cannot have an embedding of $H$ and $K_{3,6}$ together in a torus. This implies that $g(\Gamma(\mathbb{Z}_9, \{1\})) \geq 2$. It is easy to check that $\Gamma(\mathbb{Z}_9, \{-1\}) = K_{3,3,3}$; so $g(\Gamma(\mathbb{Z}_9, \{-1\})) = 1$. If $S = \{2, 5\}$ or $S = \{-1, 2, 5\}$, then $\Gamma(R, S) = \Gamma(R, \{-1\})$, and so $g(\Gamma(R, S)) = 1$. If $S = \{4, 7\}$ or $S = \{1, 4, 7\}$, then $\Gamma(R, S) = \Gamma(R, \{1\})$, and so $g(\Gamma(R, S)) \geq 2$. If $S$ is other than $\{1\}, \{-1\}, \{2, 5\}, \{-1, 2, 5\}, \{4, 7\}$, or $\{1, 4, 7\}$, then $\Gamma(R, S) = CK_3 + K_6 = K_9 - K_3$. According to Youngs [442], $g(K_9 - K_3) = 3$, and hence $g(\Gamma(R, S)) = 3$. Hence $g(\Gamma(\mathbb{Z}_9), S) = 1$ if and only if $S = \{-1\}$, $S = \{2, 5\}$, or $S = \{-1, 2, 5\}$. Thus we have $g(\Gamma(\mathbb{Z}_3[x]/(x^2), S)) = 1$ if and only if $S = \{-1 + (x^2)\}$, $S = \{-1 + x + (x^2), -1 - x + (x^2)\}$, or $S = \{-1 + (x^2), -1 + x + (x^2), -1 - x + (x^2)\}$.

**Case 3.** Let $R$ be a nonlocal ring. Then $R = R_1 \times \cdots \times R_m$, where the $R_i$'s are local rings and $m \geq 2$. Since $|J(R)| \leq 4$ and $|U(R)| \leq 7$, we have the following candidates for $R$:

(*a*) $\mathbb{Z}_2^{\ell} \times T$, where $\ell > 0$ and $T$ is isomorphic to one of the following rings:

$$\mathbb{Z}_2,\ \mathbb{Z}_3,\ \mathbb{F}_4,\ \mathbb{Z}_4,\ \frac{\mathbb{Z}_2[x]}{(x^2)},\ \mathbb{Z}_5,\ \mathbb{Z}_7,\ \mathbb{F}_8,\ \mathbb{Z}_8,\ \frac{\mathbb{Z}_2[x]}{(x^3)},\ \frac{\mathbb{Z}_2[x,y]}{(x^2,xy,y^2)},$$

$$\frac{\mathbb{Z}_4[x]}{(x^2,2x)},\ \frac{\mathbb{Z}_4[x]}{(x^2-2,2x)},\ \mathbb{Z}_9,\ \frac{\mathbb{Z}_3[x]}{(x^2)}.$$

(*b*) $\mathbb{Z}_2^{\ell} \times \mathbb{Z}_3 \times T$, where $\ell \geq 0$ and $T$ is isomorphic to one of the following rings:

$$\mathbb{Z}_3,\ \mathbb{F}_4,\ \mathbb{Z}_4,\ \mathbb{Z}_2[x]/(x^2).$$

(*c*) $\mathbb{Z}_2^{\ell} \times \mathbb{Z}_4 \times T$ or $\mathbb{Z}_2^{\ell} \times \mathbb{Z}_2[x](x^2) \times T$, where $\ell \geq 0$ and $T$ is isomorphic to one of the following rings:

$$\mathbb{F}_4,\ \mathbb{Z}_4,\ \mathbb{Z}_2[x]/(x^2).$$

For any ring $T$ and $\ell \geq 1$,

$$\Gamma(\mathbb{Z}_2^\ell \times T, \{1\} \times \cdots \times \{1\} \times S) = 2^{\ell-1}\Gamma(\mathbb{Z}_2 \times T, S). \tag{8.2}$$

**Case 3.1.** Suppose $R$ is isomorphic to one of the rings in (a). To characterize rings for which $g(\Gamma(R)) = 1$, by equation (8.2), it is sufficient to consider the rings with $\ell = 1$. By Theorem 8.4.18(i), the rings with $T$ isomorphic to $\mathbb{Z}_2$, $\mathbb{Z}_3$, $\mathbb{Z}_4$, or $\mathbb{Z}_2[x]/(x^2)$ are ruled out.

Let $R \cong \mathbb{Z}_2 \times \mathbb{F}_4$. By Theorem 8.4.18(iii), $S \neq \{(1,1)\}$. Since $\langle \{0\} \times \mathbb{F}_4 \rangle$ and $\langle \{1\} \times \mathbb{F}_4 \rangle)$ are totally disconnected in $\Gamma(R)$, $\Gamma(R)$ is a subgraph of $K_{4,4}$, and so $g(\Gamma(R)) = 1$. Thus $g(\Gamma(R,S)) = 1$ if and only if $R \cong \mathbb{Z}_2 \times \mathbb{F}_4$ with $S \neq \{(1,1)\}$.

Let $R \cong \mathbb{Z}_2 \times \mathbb{Z}_5 \cong \mathbb{Z}_{10}$. By Figure 8.3, $g(\Gamma(\mathbb{Z}_{10}, \{-1\})) = 1$. Note that $\Gamma(\mathbb{Z}_{10}, \{1\}) \cong \Gamma(\mathbb{Z}_{10}, \{-1\})$, and so $g(\Gamma(\mathbb{Z}_{10}, \{1\})) = 1$. If $|S| \geq 2$, then every even element in $\mathbb{Z}_{10}$ is adjacent to all odd elements except 5 in $\Gamma(\mathbb{Z}_{10}, S)$. Therefore $K_{5,4} \subseteq \Gamma(\mathbb{Z}_{10}, S)$, and hence $g(\Gamma(\mathbb{Z}_{10}, S)) \geq 2$. Thus $g(\Gamma(R,S)) = 1$ if and only if $R \cong \mathbb{Z}_{10}$ with $S = \{1\}$ or $S = \{-1\}$.

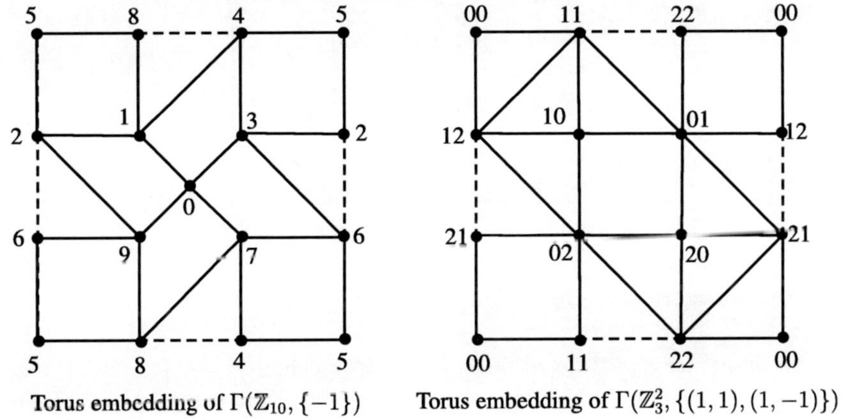


Figure 8.3: Torus embedding of $\Gamma(\mathbb{Z}_{10}, \{-1\})$ and $\Gamma(\mathbb{Z}_3^2, \{(1,1),(1,-1)\})$

If $R \cong \mathbb{Z}_2 \times \mathbb{F}_n$, where $n = 7$ or 8, then $K_{n,n} - nK_2 \subseteq \Gamma(R)$, and so $g(\Gamma(R)) \geq 3$.

Consider rings of the form $R \cong \mathbb{Z}_2 \times T$, where $T$ is isomorphic to one of $\mathbb{Z}_8$, $\frac{\mathbb{Z}_2[x]}{(x^3)}$, $\frac{\mathbb{Z}_2[x,y]}{(x^2,xy,y^2)}$, $\frac{\mathbb{Z}_4[x]}{(x^2,2x)}$, or $\frac{\mathbb{Z}_4[x]}{(x^2-2,2x)}$. Note that $\overline{\Gamma}(T) = K_{4,4}$ and $\overline{\Gamma}(\mathbb{Z}_2) = K_2$. Therefore $\Gamma(R) = 2K_{4,4}$, and so $g(\Gamma(R)) = 2$.

Let $R \cong \mathbb{Z}_2 \times \mathbb{Z}_9$. Since every vertex in $\{0\} \times Z(\mathbb{Z}_9)$ (or $\{1\} \times Z(\mathbb{Z}_9)$) is adjacent to all vertices of $\{1\} \times U(\mathbb{Z}_9)$ (or $\{0\} \times U(\mathbb{Z}_9)$) in $\Gamma(R)$, $2K_{3,6} \subseteq \Gamma(R)$, and so $g(\Gamma(R)) \geq 2$. If $R \cong \mathbb{Z}_2 \times \mathbb{Z}_3[x]/(x^2)$, then $g(\Gamma(R)) \geq 2$.

**Case 3.2.** Suppose that $R$ is isomorphic to one of the rings in (b).

First, let $R \cong \mathbb{Z}_2^\ell \times \mathbb{Z}_3 \times \mathbb{Z}_3$. Assume that $\ell = 0$. Then by Theorem 8.4.18(v), $S \neq \{(1,1)\}$, $S \neq \{(1,-1)\}$, and $S \neq \{(-1,1)\}$. Note that $\Gamma(R, \{(-1,-1)\})$

is a subgraph of $K_{3,3,3}$ and $g(K_{3,3,3}) = 1$. Thus, the genus of $\Gamma(R, \{(-1,-1)\})$ is 1. If $S = \{(1,1),(1,-1)\}$, then by Figure 8.3, $g(\Gamma(R,S)) = 1$. Note that $\Gamma(R, \{(1,1),(1,-1)\}) \cong \Gamma(R, \{(1,1),(-1,1)\})$, and so $g(\Gamma(R, \{(1,1),(-1,1)\})) = 1$. If $S = \{(1,1),(-1,-1)\}$, or $S = \{(1,-1),(-1,1)\}$, then every vertex in $\{(0,0),(0,1),(0,-1),(1,0),(-1,0)\}$ is adjacent to all vertices in $\{(1,1),(1,-1),(-1,1),(-1,-1)\}$, and so $g(\Gamma(R,S)) \geq g(K_{5,4}) = 2$. If $S = \{(-1,-1),(1,-1)\}$, then $g(\Gamma(R,S)) = 1$ by Figure 8.4. Since $\Gamma(R, \{(-1,-1),(1,-1)\}) \cong \Gamma(R, \{(-1,-1),(-1,1)\})$, $g(\Gamma(R, \{(-1,-1),(-1,1)\})) \geq 2$. If $|S| \geq 3$, then either $\{(1,1),(-1,-1)\}$ or $\{(1,-1),(-1,1)\}$ is a subset of $S$, and so $g(\Gamma(R,S)) \geq 2$. Hence $g(\mathbb{Z}_3 \times \mathbb{Z}_3, S) = 1$ if and only if $S = \{(-1,-1)\}$, $S = \{(1,1),(1,-1)\}$, or $S = \{(1,1),(-1,1)\}$.

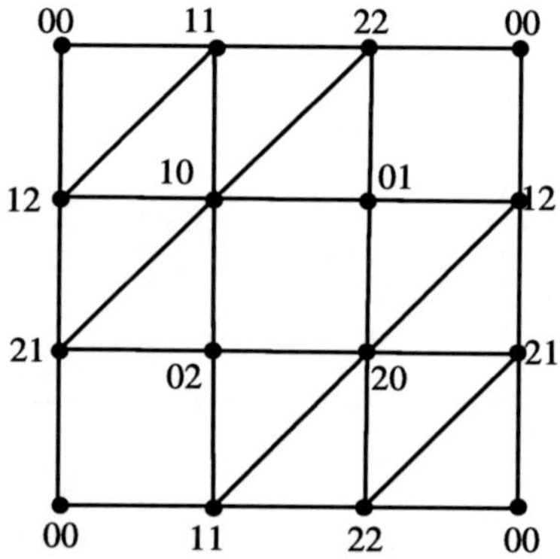


Figure 8.4: Torus Embedding of $\Gamma(\mathbb{Z}_3^2, \{(-1,-1),(1,-1)\})$

Secondly, let $\ell = 1$, i.e, $R \cong \mathbb{Z}_2 \times \mathbb{Z}_3 \times \mathbb{Z}_3$. When $S = \{(1,1,1)\}$, the graph in Figure 8.5 is a minor subgraph of $\Gamma(R,S)$. Note that the graph in Figure 8.5 is not toroidal because the vertices (0,1,1), (0,1,-1), (0,-1,1), (1,0,0), (1,1,0) would have to be embedded in one of the faces of $K_{3,3}$. Since the face containing the specified vertices together with the boundary would contain another subdivision of $K_{3,3}$, the cellular embedding leads to a contradiction to Kuratowki's Theorem. If the specified vertices are embedded in a torus embedding of $K_{3,3}$, we observe that they are attached to two consecutive vertices of the decagon; so that there exists a noncontractible curve which does not separate the vertices in $\{(0,1,1),\ (0,1,-1),\ (0,-1,1),\ (0,1,0),\ (1,0,1),\ (0,-1,0),(1,1,-1),\ (1,0,0),\ (1,1,0)\}$, again a contradiction to Kuratowski's Theorem. Hence $g(\Gamma(R,S)) \geq 2$. If $S = \{(1,1,-1)\}$, $S = \{(1,-1,1)\}$, or $S = \{(1,-1,-1)\}$, then it is easy to check that a minor subgraph of $\Gamma(R,S)$ is isomorphic to the graph in Figure 8.5. Thus $g(\Gamma(R)) \geq 2$. Hence by equation (8.2), $g(\Gamma(\mathbb{Z}_2^\ell \times \mathbb{Z}_3 \times \mathbb{Z}_3)) \geq 2$ when $\ell \geq 1$.

Let $R$ be a ring of the form $\mathbb{Z}_2^\ell \times \mathbb{Z}_3 \times \mathbb{F}_4$. When $\ell = 0$, $R \cong \mathbb{Z}_3 \times \mathbb{F}_4$. If

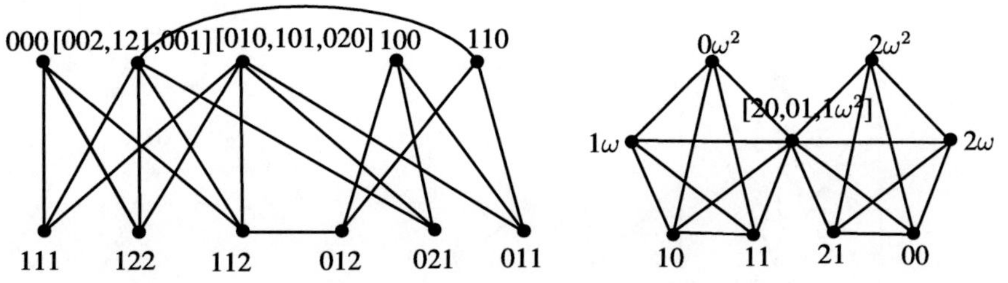


**Minor subgraph of $\Gamma(\mathbb{Z}_2 \times \mathbb{Z}_3^2, \{(1,1,1)\})$** **Minor subgraph of $\Gamma(\mathbb{Z}_3 \times \mathbb{F}_4, \{(1,1)\})$**

Figure 8.5: Minor subgraph of $\Gamma(\mathbb{Z}_2 \times \mathbb{Z}_3^2, \{(1,1,1)\})$ and $\Gamma(\mathbb{Z}_3 \times \mathbb{F}_4, \{(1,1)\})$

$|S| = 1$, then $S = \{(1,1)\}$ or $S = \{(-1,-1)\}$. If $S = \{(1,1)\}$, then by Figure 8.5, $2K_4 + K_1$ is a minor subgraph of $\Gamma(R,S)$. Note that $g(2K_4 + K_1) > 1$. From this, $g(\Gamma(R,S)) \geq 2$. If $S = \{(-1,-1)\}$, then the graph in Figure 8.6 is a minor of $\Gamma(R,S)$. By the argument given for Figure 8.5 above, one gets that $g(\Gamma(R,S)) \geq 2$. If $|S| \geq 2$, then $\Gamma(R,S)$ contains $K_3$, and so the girth of $\Gamma(R,S)$ is 3. Note that $deg((1,1)) \geq 7$ and $deg(x) \geq 6$ for every $x \in R \setminus \{(1,1)\}$ so that $|E(\Gamma(R,S))| \geq 3|V(\Gamma(R,S))| + 1$, a contradiction to $|E(\Gamma(R,S))| \leq 3|V(\Gamma(R,S))|$(Lemma 7.5.6). When $\ell \geq 1$, then $\delta(\Gamma(R,S)) \geq 6$, and so $|E(\Gamma(R,S))| \geq 3|V(\Gamma(R,S))|$. But by Lemma 7.5.6 and $\Gamma(R,S)$ is bipartite, we have $|E(\Gamma(R,S))| \leq 2|V(\Gamma(R,S))|$, a contradiction.

Consider the rings $\mathbb{Z}_2^\ell \times \mathbb{Z}_3 \times T$, where $T$ is isomorphic to either $\mathbb{Z}_4$ or $\mathbb{Z}_2[x]/(x^2)$. First, consider $R \cong \mathbb{Z}_2^\ell \times \mathbb{Z}_3 \times \mathbb{Z}_4$. If $\ell = 0$, then $R \cong \mathbb{Z}_{12}$. By Figure 8.6, $g(\Gamma(R, \{1\})) = 1$ and $g(\Gamma(R,S)) = 1$ whenever $S = \{-1\}$, $S = \{5\}$, or $S = \{7\}$. Note that $\Gamma(R, \{1,7\}) = \Gamma(R, \{1\})$ and $\Gamma(R, \{-1,5\}) = \Gamma(R, \{-1\})$. Thus $g(\Gamma(R, \{1,7\})) = g(\Gamma(R, \{-1,5\})) = 1$. If $|S| \geq 2$, then $deg(1) \geq 5$. Since the girth of $\Gamma(R,S)$ is 4, we have $g(\Gamma(R,S)) \geq 2$ by Lemma 7.5.6. Assume that $\ell \geq 1$. Since $\Gamma(\mathbb{Z}_3 \times \mathbb{Z}_4)$ is bipartite, $\Gamma(\mathbb{Z}_2^\ell \times \mathbb{Z}_3 \times \mathbb{Z}_4) = 2^\ell \Gamma(\mathbb{Z}_3 \times \mathbb{Z}_4)$, and hence $g(\Gamma(R)) \geq 2$. In conclusion, $g(\Gamma(\mathbb{Z}_2^\ell \times \mathbb{Z}_3 \times \mathbb{Z}_4, S)) = 1$ if and only if $\ell = 0$ and $S = \{(1,1)\}$, $S = \{(-1,-1)\}$, $S = \{(1,-1)\}$, $S = \{(-1,1)\}$, $S = \{(1,1),(1,-1)\}$, or $S = \{(-1,1),(-1,-1)\}$. Since $\Gamma(\mathbb{Z}_4) = \Gamma(\mathbb{Z}_2[x]/(x^2))$, $g(\Gamma(\mathbb{Z}_2^\ell \times \mathbb{Z}_3 \times \mathbb{Z}_2[x]/(x^2), S)) = 1$ if and only if $\ell = 0$ with $|S| = 1$ or $S = \{(1, 1 + (x^2)), (1, 1 + x + (x^2))\}$, or $S = \{(-1, 1 + (x^2)), (-1, 1 + x + (x^2))\}$.
**Case 3.3.** Let $R \cong \mathbb{Z}_2^\ell \times \mathbb{Z}_4 \times \mathbb{F}_4$ or $R \cong \mathbb{Z}_2^\ell \times \mathbb{Z}_2[x]/(x^2) \times \mathbb{F}_4$ with $\ell \geq 0$. Then $\Gamma(R)$ is bipartite and $deg(x) \geq 6$ for every $x \in R$. Therefore, $g(\Gamma(R)) \geq 2$ by Lemma 7.5.6.

Suppose $R \cong \mathbb{Z}_2^\ell \times R_1 \times R_2$, where $\ell \geq 0$ and $R_1$ and $R_2$ are isomorphic to either $\mathbb{Z}_4$ or $\mathbb{Z}_2[x]/(x^2)$. Note that $\overline{\Gamma}(\mathbb{Z}_4) = \overline{\Gamma}(\mathbb{Z}_2[x]/(x^2)) = K_{2,2}$. Then $\Gamma(R_1 \times R_2) = 2K_{4,4}$. Hence by equation (8.2), $g(\Gamma(R)) \geq 2$. □

Note that the $\Gamma(R, \{1\})$ is a subgraph of the complement of the total graph

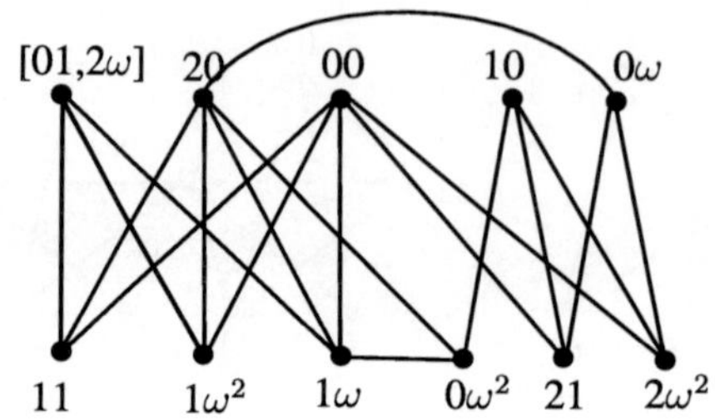


Minor subgraph of $\Gamma(\mathbb{Z}_3 \times \mathbb{F}_4, \{(-1,1)\})$

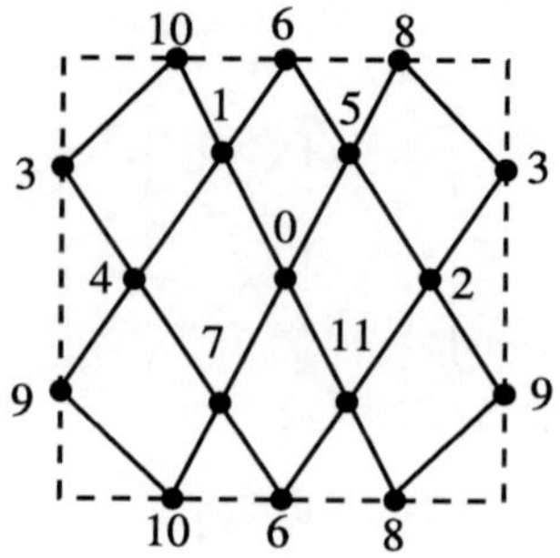


Torus embedding of $\Gamma(\mathbb{Z}_{12}, \{(1)\})$

Figure 8.6: Minor of $\Gamma(\mathbb{Z}_3 \times \mathbb{F}_4, \{(-1,1)\})$ and embedding of $\Gamma(\mathbb{Z}_{12}, \{(1)\})$

$CT(\Gamma(R))$. Using this fact, we have the following corollary, which characterizes the commutative rings with $CT(\Gamma(R))$ toroidal.

**Corollary 8.4.20.** ([102, Corollary 4.4]) *Let $R$ be a commutative Artinian ring. Then $CT(\Gamma(R))$ is toroidal if and only if $R$ is isomorphic to one of the rings:* $\mathbb{Z}_7$, $\mathbb{Z}_8$, $\frac{\mathbb{Z}_2[x]}{(x^3)}$, $\frac{\mathbb{Z}_2[x,y]}{(x^2,xy,y^2)}$, $\frac{\mathbb{Z}_4[x]}{(x^2,2x)}$, $\frac{\mathbb{Z}_4[x]}{(x^2-2,2x)}$, $\mathbb{Z}_{10}$, $\mathbb{Z}_{12}$, *or* $\mathbb{Z}_3 \times \frac{\mathbb{Z}_2[x]}{(x^2)}$.

The following theorem characterizes all commutative Artinian rings $R$ for which $\Gamma(R,S)$ has genus two.

**Theorem 8.4.21.** ([104, Theorem 3.5]) *Let $R$ be a commutative Artinian ring. Then $g(\Gamma(R,S)) = 2$ if and only if one of the following hold:*

(i) $R \cong \mathbb{F}_8$;

(ii) $R \cong \mathbb{Z}_9$ *with* $S = \{1\}$, $S = \{4,7\}$, *or* $S = \{1,4,7\}$;

(iii) $R \cong \mathbb{Z}_3[x]/(x^2)$ *with* $S = \{1+(x^2)\}$, $S = \{1+x+(x^2), 1-x+(x^2)\}$, *or* $S = \{1+(x^2), 1+x+(x^2), 1-x+(x^2)\}$;

(iv) $R \cong \mathbb{Z}_3 \times \mathbb{Z}_3$ *with* $S = \{(1,1),(-1,-1)\}$, $S = \{(1,-1),(-1,1)\}$, *or* $|S| \geq 3$;

(v) $R \cong \mathbb{Z}_{10}$ *with* $S \neq \{1\}$ *and* $S \neq \{-1\}$;

(vi) $R \cong \mathbb{Z}_3 \times \mathbb{F}_4$ *with* $S = \{(1,1)\}$ *or* $S = \{(-1,-1)\}$;

(vii) $R \cong \mathbb{Z}_2 \times \mathbb{Z}_2 \times \mathbb{F}_4$ *with* $S \neq \{(1,1,1)\}$;

(viii) $R \cong \mathbb{Z}_2 \times \mathbb{Z}_8, \mathbb{Z}_2 \times \frac{\mathbb{Z}_2[x]}{(x^3)}, \mathbb{Z}_2 \times \frac{\mathbb{Z}_2[x,y]}{(x^2,xy,y^2)}, \mathbb{Z}_2 \times \frac{\mathbb{Z}_4[x]}{(x^2,2x)}$, *or* $\mathbb{Z}_2 \times \frac{\mathbb{Z}_4[x]}{(x^2-2,2x)}$;

(ix) $R \cong \mathbb{Z}_4 \times \mathbb{Z}_4$, $\mathbb{Z}_2[x]/(x^2) \times \mathbb{Z}_2[x]/(x^2)$, *or* $\mathbb{Z}_4 \times \mathbb{Z}_2[x]/(x^2)$;

(x) $R \cong \mathbb{Z}_2 \times \mathbb{Z}_3 \times \mathbb{Z}_3$ *with* $S = \{(1,1,1)\}$, $S = \{(1,1,-1\}$, $S = \{(1,-1,1)\}$, *or* $S = \{(1,-1,-1)\}$;

(xi) $R \cong \mathbb{Z}_2 \times \mathbb{Z}_{10}$ *with* $S = \{(1,1)\}$ *or* $S = \{(1,-1)\}$;

(xii) $R \cong \mathbb{Z}_2 \times \mathbb{Z}_{12}$ *with* $S = \{(1,1)\}$, $S = \{(1,-1)\}$, $S = \{(1,5)\}$, $S = \{(1,7)\}$, $S = \{(1,1),(1,7)\}$, *or* $S = \{(1,5),(1,-1)\}$;

(xiii) $R \cong \mathbb{Z}_2 \times \mathbb{Z}_3 \times \mathbb{Z}_2[x]/(x^2)$ *with* $S = \{(1,1,1+(x^2))\}$, $S = \{(1,-1,1+x+(x^2))\}$, $S = \{(1,1,1+x+(x^2))\}$, $S = \{(1,-1,1+(x^2))\}$, $S = \{(1,1,1+(x^2)),(1,1,1+x+(x^2))\}$, *or* $S = \{(1,-1,1+(x^2)),(1,-1,1+x+(x^2))\}$.

*Proof.* Assume that $g(\Gamma(R,S)) = 2$. Then, by Lemma 7.5.5 $|U(R)| - 1 \leq \delta(T(\Gamma(R))) \leq 6 + \frac{12}{|R|}$. Thus $|U(R)| \leq 8$ when $8 \leq |R| \leq 12$ and $|U(R)| \leq 7$ when $|R| \geq 13$. Note that $g(K_n) \leq 1$ for all $n \leq 7$ so that $g(\Gamma(R,S))$ is at most one for all rings $R$ with $|R| \leq 7$. Thus $|R| \geq 8$. Suppose that $|J(R)| \geq 5$. Then $|U(R)| \geq 5$. Since $J(R)+U(R) \subseteq U(R)$, $K_{5,5}$ is a subgraph of $\Gamma(R,S)$ which has genus 3, a contradiction. Thus $|J(R)| \leq 4$. Hence

$$|R| \geq 8, |J(R)| \leq 4, \text{ and } |U(R)| \leq \begin{cases} 8 & \text{if } 8 \leq |R| \leq 12; \\ 7 & \text{if } |R| \geq 13. \end{cases} \tag{8.3}$$

**Case 1.** Suppose that $R$ is a local ring with maximal ideal $\mathfrak{m}$. Then by equation (8.3), $|R| = 8$ or 9.

If $R$ is a field, then $R$ is either $\mathbb{F}_8$ or $\mathbb{F}_9$. Note that $g(\Gamma(\mathbb{F}_8)) = g(K_8) = 2$ and $K_9 - 4K_2$ is a subgraph of $\Gamma(\mathbb{F}_9)$ which has genus 3, and so $g(\Gamma(\mathbb{F}_9)) \geq 3$.

Suppose that $R$ is not a field. By Theorem 8.4.19, $g(\Gamma(R,S)) = 1$ for all local rings $R$ of order 8. Thus $|R| = 9$. Then by Theorem 8.4.19, either $R \cong \mathbb{Z}_9$ with $S \neq \{-1\}$, $S \neq \{2,5\}$, and $S \neq \{-1,2,5\}$, or $R \cong \mathbb{Z}_3[x]/(x^2)$ with $S \neq \{-1+(x^2)\}$, $S \neq \{-1+x+(x^2), -1-x+(x^2)\}$, and $S \neq \{-1+(x^2), -1+x+(x^2), -1-x+(x^2)\}$. Note that by Figure 8.7, $g(\Gamma(\mathbb{Z}_9,\{1\})) = 2$. If $S = \{4,7\}$ or $S = \{1,4,7\}$, then $\Gamma(\mathbb{Z}_9,S) \cong \Gamma(\mathbb{Z}_9,\{1\})$, and so $g(\Gamma(\mathbb{Z}_9,S)) = 2$. If $S$ is other than $\{1\}, \{4,7\}$, or $\{1,4,7\}$, then $\Gamma(\mathbb{Z}_9,S) = K_9 - K_3$. By Youngs [442], $g(K_9 - K_3) = 3$, and so $g(\Gamma(\mathbb{Z}_9,S)) = 3$. Thus $g(\Gamma(\mathbb{Z}_9,S)) = 2$ if and only if $S = \{1\}$, $S = \{4,7\}$, or $S = \{1,4,7\}$. Thus we have $g(\Gamma(\mathbb{Z}_3[x]/(x^2),S)) = 2$ if and only if $S = \{1+(x^2)\}$, $S = \{1+x+(x^2), 1-x+(x^2)\}$, or $S = \{1+(x^2), 1+x+(x^2), 1-x+(x^2)\}$.

**Case 2.** Suppose that $R$ is not a local ring. Then by equation (8.3), Theorem 8.4.18, and Theorem 8.4.19, we have the following candidates for $R$:

(*a*) $\mathbb{Z}_2^\ell \times T$, where $\ell \geq 1$ and $T$ is isomorphic to one of the following rings:

$$\mathbb{F}_4,\ \mathbb{Z}_5,\ \mathbb{Z}_7,\ \mathbb{F}_8,\ \mathbb{Z}_8,\ \frac{\mathbb{Z}_2[x]}{(x^3)},\ \frac{\mathbb{Z}_2[x,y]}{(x^2,xy,y^2)},\ \frac{\mathbb{Z}_4[x]}{(x^2,2x)},\ \frac{\mathbb{Z}_4[x]}{(x^2-2,2x)},\ \mathbb{Z}_9,\ \frac{\mathbb{Z}_3[x]}{(x^2)}.$$

(*b*) $\mathbb{Z}_2^\ell \times \mathbb{Z}_3 \times T$, where $\ell \geq 0$ and $T$ is isomorphic to one of the following rings:

$$\mathbb{Z}_3,\ \mathbb{F}_4,\ \mathbb{Z}_4,\ \mathbb{Z}_2[x]/(x^2).$$

(*c*) $\mathbb{Z}_2^\ell \times \mathbb{Z}_4 \times T$ or $\mathbb{Z}_2^\ell \times \mathbb{Z}_2[x]/(x^2) \times T$, where $\ell \geq 0$ and $T$ is isomorphic to one of the following rings:

$$\mathbb{F}_4,\ \mathbb{Z}_4, \mathbb{Z}_2[x]/(x^2).$$

Note that for any ring $T$ and $\ell \geq 1$, we have

$$\Gamma(\mathbb{Z}_2^\ell \times T, \{1\} \times \cdots \times \{1\} \times S) = 2^{\ell-1}\Gamma(\mathbb{Z}_2 \times T, S). \tag{8.4}$$

**Case 2.1.** Assume that $R$ is isomorphic to one of the rings in (a).

First consider the rings $R \cong \mathbb{Z}_2^\ell \times \mathbb{F}_4$ with $\ell \geq 1$. If $\ell = 1$, then $g(\Gamma(R)) \leq 1$, and hence $\ell \geq 2$. Note that $g(\Gamma(\mathbb{Z}_2 \times \mathbb{F}_4, S)) = 1$ for every $S \neq \{(1,1)\}$. Applying this in equation (8.4), we have $g(\Gamma(R,S)) = 2$ if and only if $R \cong \mathbb{Z}_2^2 \times \mathbb{F}_4$ with $S \neq \{(1,1,1)\}$.

Suppose that $R \cong \mathbb{Z}_2^\ell \times \mathbb{Z}_5$. Let $\ell = 1$. If $S$ is either $\{1\}$ or $\{-1\}$, then $g(\Gamma(R,S)) = 1$ by Theorem 8.4.19. Therefore, assume that $S \neq \{1\}$ and $S \neq \{-1\}$. Now Figure 8.7 shows that $g(\Gamma(R,S)) = 2$. Suppose that $\ell \geq 2$. Then equation (8.4) and Theorem 8.4.19 give that $\ell = 2$ and $S = \{(1,1,1)\}$ or $S = \{(-1,-1,-1)\}$. Hence $g(\Gamma(R,S)) = 2$ if and only if $R \cong \mathbb{Z}_{10}$ with $S \neq \{1\}$ and $S \neq \{-1\}$, or $R \cong \mathbb{Z}_2^2 \times \mathbb{Z}_5$ with $S = \{(1,1,1)\}$ or $S = \{(-1,-1,-1)\}$.

If $R \cong \mathbb{Z}_2^\ell \times \mathbb{F}_n$, where $n = 7$ or 8, then $K_{n,n} - nK_2 \subseteq \Gamma(R)$, and so $g(\Gamma(R)) \geq 3$.

Consider rings of the form $\mathbb{Z}_2^\ell \times T$, where $T$ is isomorphic to one of the rings $\mathbb{Z}_8$, $\frac{\mathbb{Z}_2[x]}{(x^3)}$, $\frac{\mathbb{Z}_2[x,y]}{(x^2,xy,y^2)}$, $\frac{\mathbb{Z}_4[x]}{(x^2,2x)}$, or $\frac{\mathbb{Z}_4[x]}{(x^2-2,2x)}$. Note that $\overline{\Gamma}(T) = K_{4,4}$, and $\overline{\Gamma}(\mathbb{Z}_2) = K_2$. Now $g(\Gamma(\mathbb{Z}_2^\ell \times T)) = 2^\ell g(\Gamma(T))$ for every $\ell \geq 1$. Since $g(\Gamma(T)) = 1$, $g(\Gamma(\mathbb{Z}_2^\ell \times T)) = 2$ if and only if $\ell = 1$.

Consider rings of the form $\mathbb{Z}_2^\ell \times \mathbb{Z}_9$. Assume that $\ell = 1$, i.e., $R \cong \mathbb{Z}_2 \times \mathbb{Z}_9$. Let $\mathfrak{m} = Z(\mathbb{Z}_9)$. Since every vertex in $\{0\} \times \mathfrak{m}$, $\{0\} \times (1+\mathfrak{m})$, and $\{0\} \times (2+\mathfrak{m})$ is adjacent to all vertices in $\{1\} \times (1+\mathfrak{m})$, $\{1\} \times \mathfrak{m}$, and $\{1\} \times (2+\mathfrak{m})$, respectively in $\Gamma(R,\{1\})$, we have $3K_{3,3} \subseteq \Gamma(R,\{1\})$, and so $g(\Gamma(R,\{1\})) \geq 3$. Similarly, every vertex in $\{0\} \times \mathfrak{m}$, $\{0\} \times (1+\mathfrak{m})$, and $\{0\} \times (2+\mathfrak{m})$ is adjacent to all vertices in $\{1\} \times (1+\mathfrak{m})$, $\{1\} \times (2+\mathfrak{m})$, and $\{1\} \times \mathfrak{m}$, respectively in $\Gamma(R,\{-1\})$, so that $g(\Gamma(R\{-1\})) \geq 3$. If $S \neq \{1\}$ and $S \neq \{-1\}$, then either $\Gamma(R,\{1\})$ or $\Gamma(R,\{-1\})$ is a subgraph of $\Gamma(R,S)$, and so $g(\Gamma(R,S)) \geq 3$. Hence $g(\Gamma(\mathbb{Z}_2 \times \mathbb{Z}_9, S)) \geq 3$ for every $S$. Also, by equation (8.4), $g(\Gamma(\mathbb{Z}_2^\ell \times \mathbb{Z}_9)) \geq 3$ for every $\ell \geq 2$. Further, $g(\Gamma(\mathbb{Z}_2^\ell \times \frac{\mathbb{Z}_3[x]}{(x^2)})) \geq 3$ for every integer $\ell \geq 1$.

**Case 2.2.** Suppose that $R$ is isomorphic to one of the rings in $(b)$.

Consider rings of the form $\mathbb{Z}_2^\ell \times \mathbb{Z}_3 \times \mathbb{Z}_3$ with $\ell \geq 0$. Note that from Figure 8.8, $g(\Gamma(\mathbb{Z}_3 \times \mathbb{Z}_3, U(\mathbb{Z}_3 \times \mathbb{Z}_3))) = 2$, and so by Theorem 8.4.19, $g(\Gamma(\mathbb{Z}_3 \times \mathbb{Z}_3, S)) = 2$ if and only if $S = \{(1,1),(-1,-1)\}$, $S = \{(1,-1),(-1,1)\}$, or $|S| \geq 3$. Let $R = \mathbb{Z}_2 \times \mathbb{Z}_3 \times \mathbb{Z}_3$. By Figure 8.8, we have $g(\Gamma(R,\{(1,1,1)\})) = 2$. Note that $\Gamma(R,\{(1,1,1)\}) \cong \Gamma(R,\{(1,1,-1)\})$ by the isomorphism defined by $f((a,b,c)) = (a,b,-c)$ for $(a,b,c) \in R$. Similarly, one can easily verify that $\Gamma(R,S) \cong \Gamma(R,\{(1,1,1)\})$ for every $S$ with $|S| = 1$. Suppose $|S| \geq 2$. Then $gr(\Gamma(R)) = 4$. If $S_1 = \{(1,1,1),(1,1,-1)\}$, then $|E(\Gamma(R,S_1))| = 48$, and so $|E(\Gamma(R,S_1))| > 2|V(\Gamma(R,S_1))| + 4$, a contradiction to Lemma 7.5.6. Thus $g(\Gamma(R,S_1)) \geq 3$. If $S_2 = \{(1,1,1),(1,-1,1)\}$, then $\Gamma(R,S_1) \cong \Gamma(R,S_2)$, and so $g(\Gamma(R,S_2)) \geq 3$. If $|S| \geq 2$, then either $\Gamma(R,S_1)$ or $\Gamma(R,S_2)$ is a subgraph of

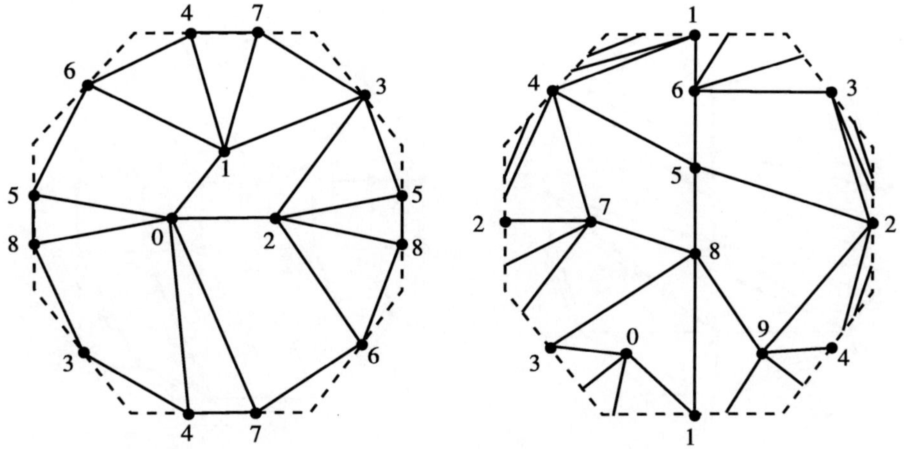


**Embedding of $\Gamma(\mathbb{Z}_9, \{(1)\})$ in $S_2$** **Embedding of $\Gamma(\mathbb{Z}_{10}, U(\mathbb{Z}_{10}))$ in $S_2$**

Figure 8.7: Embedding of $\Gamma(\mathbb{Z}_9, \{(1)\})$ and $\Gamma(\mathbb{Z}_{10}, U(\mathbb{Z}_{10}))$ in $S_2$

$\Gamma(R, S)$ so that $g(\Gamma(R, S)) \geq 3$. Thus $g(\Gamma(\mathbb{Z}_2 \times \mathbb{Z}_3 \times \mathbb{Z}_3, S)) = 2$ if and only if $|S| = 1$. Moreover, $g(\Gamma(\mathbb{Z}_2^\ell \times \mathbb{Z}_3 \times \mathbb{Z}_3, S)) \geq 3$ whenever $\ell \geq 2$ by equation (8.4).

Consider the rings $R \cong \mathbb{Z}_2^\ell \times \mathbb{Z}_3 \times \mathbb{F}_4$ with $\ell \geq 0$, and let $\mathbb{F}_4 = \{0, 1, \omega, \omega^2\}$. Suppose that $\ell = 0$. If $|S| = 1$, then $S = \{(1, 1)\}$ or $S = \{(-1, -1)\}$, so that $g(\Gamma(R, S)) = 2$ by Figure 8.9. If $|S| = 2$, then $S$ is $\{(1, \omega), (1, \omega^2)\}$, $\{(1, 1), (-1, 1)\}$, or $\{(-1, \omega), (-1, \omega^2)\}$. If $S = \{(1, \omega), (1, \omega^2)\}$, then every vertex in $\{0\} \times \mathbb{F}_4$ is adjacent to all vertices in $\{1, -1\} \times (\mathbb{F}_4 - \{0\})$, and so $K_{4,6}$ is a subgraph of $\Gamma(R, S)$. Let us take this $K_{4,6}$ as $H$. Fix an embedding of $H$ in $S_2$. Then there are 12 faces in the embedding of $H$, say $\{F_1, \ldots, F_{12}\}$. Let $s_{F_i}$ be the length of face $F_i$. Note that $\sum_{i=1}^{12} s_{F_i} = 48$ and $s_{F_i} \geq 4$ for every $i$. Thus $s_{F_i} = 4$ for every $i$. Also, $\langle \{1\} \times \mathbb{F}_4 \rangle = K_4$ in $E(\Gamma(R, S)) - E(H)$ and $V(K_4) \subseteq V(H)$. Since one cannot have an embedding of such $K_4$ and $H$ together in $S_2$, $g(\Gamma(R_1, S)) \geq 3$. If $S = \{(1, 1), (-1, 1)\}$ or $S = \{(-1, \omega), (-1, \omega^2)\}$, then $|E(\Gamma(R_1, S))| = 45$, and so $|E(\Gamma(R_1, S))| > 3|V(\Gamma(R_1, S))| + 6$, a contradiction to Lemma 7.5.6. Thus $g(\Gamma(R_1, S)) \geq 3$. If $\ell \geq 1$, then $\delta(\Gamma(R, S)) \geq 6$, and so $|E(\Gamma(R, S))| \geq 3|V(\Gamma(R, S))| > 2|V(\Gamma(R, S))| + 4$, a contradiction to Lemma 7.5.6. Hence $g(\Gamma(\mathbb{Z}_2^\ell \times \mathbb{Z}_3 \times \mathbb{F}_4, S)) = 2$ if and only if $\ell = 0$ and $|S| = 1$.

Consider the rings $\mathbb{Z}_2^\ell \times \mathbb{Z}_3 \times T$, where $\ell \geq 0$ and $T$ is isomorphic to either $\mathbb{Z}_4$ or $\mathbb{Z}_2[x]/(x^2)$. Let $R \cong \mathbb{Z}_3 \times \mathbb{Z}_4 \cong \mathbb{Z}_{12}$. Then by Theorem 8.4.19, $|S| \geq 2$, $S \neq \{1, 7\}$, and $S \neq \{5, -1\}$. Then $|E(\Gamma(R, S))| = 32$, and so $|E(\Gamma(R, S))| > 2|V(\Gamma(R, S))| + 4$, a contradiction. Since $\overline{\Gamma}(\mathbb{Z}_3 \times \mathbb{Z}_4)$ is bipartite, $g(\Gamma(\mathbb{Z}_2^\ell \times \mathbb{Z}_3 \times \mathbb{Z}_4, S)) = 2^\ell g(\Gamma(\mathbb{Z}_3 \times \mathbb{Z}_4, S))$. Hence by Theorem 8.4.19, $g(\Gamma(\mathbb{Z}_2^\ell \times \mathbb{Z}_3 \times \mathbb{Z}_4, S)) = 2$ if and only if $\ell = 1$ and $|S| = 1$, $S = \{(1, 1, 1), (1, 1, -1)\}$, or $S = \{(1, -1, 1), (1, -1, -1)\}$. Also, $g(\Gamma(\mathbb{Z}_2^\ell \times \mathbb{Z}_3 \times \mathbb{Z}_2[x]/(x^2), S)) = 1$ if

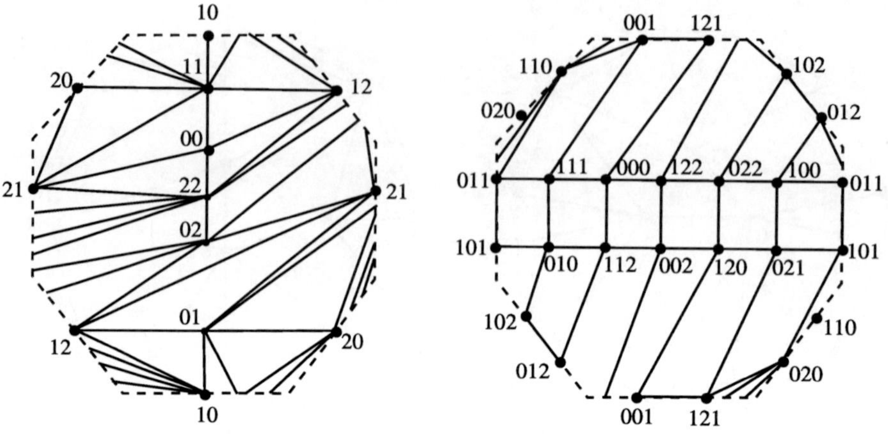


Embedding of $\Gamma(\mathbb{Z}_3^2, U(\mathbb{Z}_3^2))$ in $S_2$ Embedding of $\Gamma(\mathbb{Z}_2 \times \mathbb{Z}_3^2, \{(1,1,1)\})$ in $S_2$

Figure 8.8: Embedding of $\Gamma(\mathbb{Z}_3^2, U(\mathbb{Z}_3^2))$ and $\Gamma(\mathbb{Z}_2 \times \mathbb{Z}_3^2, \{(1,1,1)\})$ in $S_2$

and only if $\ell = 1$ with $|S| = 1$, $S = \{(1,1,1+(x^2)), (1,1,1+x+(x^2))\}$, or $S = \{(1,-1,1+(x^2)), (1,-1,1+x+(x^2))\}$.

**Case 2.3.** Let $R \cong \mathbb{Z}_2^\ell \times \mathbb{Z}_4 \times \mathbb{F}_4$ or $\mathbb{Z}_2^\ell \times \mathbb{Z}_2[x]/(x^2) \times \mathbb{F}_4$ with $\ell \geq 0$. Then $\delta(\Gamma(R)) \geq 6$, and hence $|E(\Gamma(R,S))| \geq 3|V(\Gamma(R,S))| > 2|V(\Gamma(R,S))| + 4$, a contradiction.

Suppose that $R \cong \mathbb{Z}_2^\ell \times R_1 \times R_2$, where $\ell \geq 0$ and $R_1$, $R_2$ are isomorphic to either $\mathbb{Z}_4$ or $\mathbb{Z}_2[x]/(x^2)$. Note that $\overline{\Gamma}(\mathbb{Z}_4) = \overline{\Gamma}(\mathbb{Z}_2[x]/(x^2)) = K_{2,2}$, so that $\Gamma(R_1 \times R_2) = 2K_{4,4}$. Hence $g(\Gamma(R)) = 2$ if and only if $\ell = 0$ by equation (8.4). □

**Corollary 8.4.22.** ([104, Corollary 3.6]) *Let $R$ be a commutative Artinian ring. Then the complement of the total graph, $CT(\Gamma(R))$, has genus two if and only if $R$ is isomorphic to one of the following rings:* $\mathbb{F}_8$, $\mathbb{Z}_9$, $\frac{\mathbb{Z}_3[x]}{(x^2)}$, $\mathbb{Z}_2 \times \mathbb{Z}_{10}$, $\mathbb{Z}_2 \times \mathbb{Z}_8$, $\mathbb{Z}_2 \times \frac{\mathbb{Z}_2[x]}{(x^3)}$, $\mathbb{Z}_2 \times \frac{\mathbb{Z}_2[x,y]}{(x^2,xy,y^2)}$, $\mathbb{Z}_2 \times \frac{\mathbb{Z}_4[x]}{(x^2,2x)}$, $\mathbb{Z}_2 \times \frac{\mathbb{Z}_4[x]}{(x^2-2,2x)}$, $\mathbb{Z}_2 \times \mathbb{Z}_3 \times \mathbb{Z}_3$, $\mathbb{Z}_2 \times \mathbb{Z}_{12}$, $\mathbb{Z}_2 \times \mathbb{Z}_3 \times \frac{\mathbb{Z}_2[x]}{(x^2)}$, $\mathbb{Z}_3 \times \mathbb{F}_4$, $\mathbb{Z}_4 \times \mathbb{Z}_4$, $\frac{\mathbb{Z}_2[x]}{(x^2)} \times \frac{\mathbb{Z}_2[x]}{(x^2)}$, *or* $\mathbb{Z}_4 \times \frac{\mathbb{Z}_2[x]}{(x^2)}$.

Finally, in this section, we study the ring graph and outerplanar graph nature of $\Gamma(R,G,S)$. In particular, we study when the unit graph or unitary graph is a ring graph or an outerplanar graph, respectively.

First, we want to characterize all rings $R$ such that the unit graph $G(R)$ is a ring graph (see Definition 10.1.7 for the definition of the unit graph). Since a ring graph is planar, it is sufficient to focus on the planar unit graphs. By [91, Theorem 5.14], we have that the unit graph $G(R)$ is planar if and only if $R$ is isomorphic to one of the following rings:

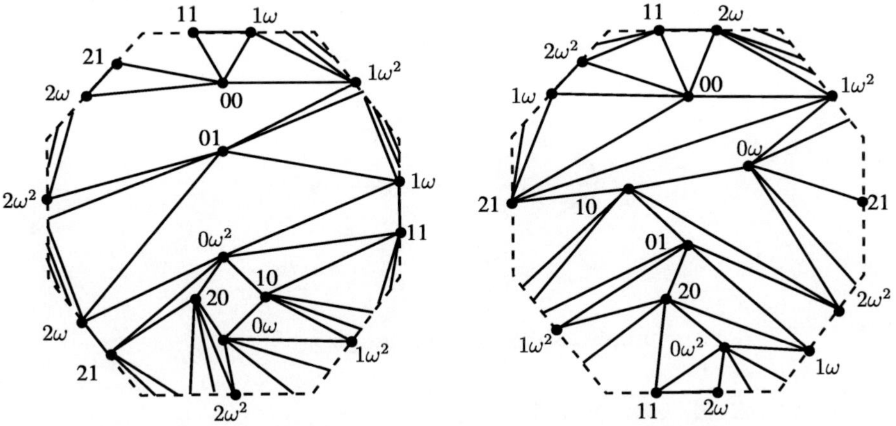


**Embedding of $\Gamma(\mathbb{Z}_3 \times \mathbb{F}_4, \{(1,1)\})$ in $S_2$** **Embedding of $\Gamma(\mathbb{Z}_3 \times \mathbb{F}_4, \{(2,1)\})$ in $S_2$**

Figure 8.9: Embedding of $\Gamma(\mathbb{Z}_3 \times \mathbb{F}_4, \{(2,1)\})$ and $\Gamma(\mathbb{Z}_3 \times \mathbb{F}_4, \{(1,1)\})$ in $S_2$

(i) $R \cong \mathbb{Z}_5$,

(ii) $R \cong \mathbb{Z}_3 \times \mathbb{Z}_3$,

(iii) $R \cong \underbrace{\mathbb{Z}_2 \times \cdots \times \mathbb{Z}_2}_{\ell} \times S$, where $\ell \geq 0$ and $S \cong \mathbb{Z}_2$, $S \cong \mathbb{Z}_3$, $S \cong \mathbb{Z}_4$, $S \cong \mathbb{F}_4$, or $S \cong \left\{ \begin{bmatrix} a & b \\ 0 & a \end{bmatrix} \mid a, b \in \mathbb{Z}_2 \right\}$.

**Theorem 8.4.23.** ([16, Theorem 2.1]) *The unit graph $G(R)$ is a ring graph if and only if $R$ is isomorphic to one of the following rings:*

(i) $R \cong \mathbb{Z}_5$,

(ii) $R \cong \underbrace{\mathbb{Z}_2 \times \cdots \times \mathbb{Z}_2}_{\ell} \times S$, *where* $\ell \geq 0$ *and* $S \cong \mathbb{Z}_2$, $S \cong \mathbb{Z}_3$, $S \cong \mathbb{Z}_4$, $S \cong \mathbb{F}_4$, *or* $S \cong \left\{ \begin{bmatrix} a & b \\ 0 & a \end{bmatrix} \mid a, b \in \mathbb{Z}_2 \right\}$.

*Proof.* First, assume that $G(R)$ is a ring graph. Since every ring graph is planar, we have that $G(R)$ is planar. Thus we have the following cases:

**Case 1.** $R \cong \mathbb{Z}_5$.
The unit graph $G(R)$ is pictured in Figure 8.10. One can easily see that $rank(G(R)) = frank(G(R)) = 4$; so $G(R)$ is a ring graph.

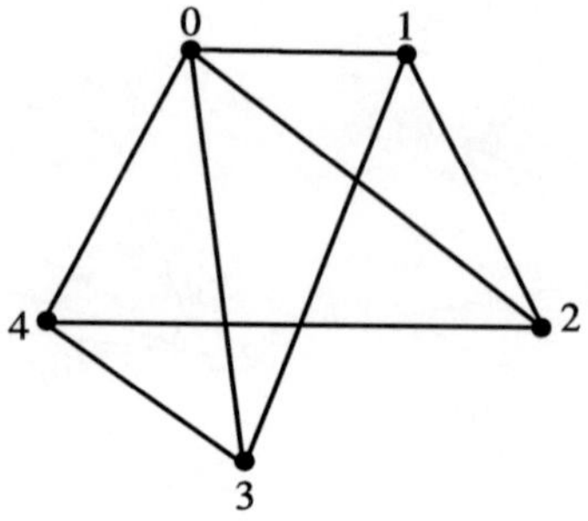


Figure 8.10: The unit graph of $\mathbb{Z}_5$

**Case 2.** $R \cong \mathbb{Z}_3 \times \mathbb{Z}_3$.
In this case, by considering the two cycles (1,0)-(1,2)-(0,0)-(1,1)-(1,0) and (1,0)-(1,2)-(0,2)-(2,2)-(2,0)-(2,1)-(0,1)-(1,1)-(1,0), we see that $G(R)$ does not satisfy PCP; so it is not a ring graph.

**Case 3.** $R \cong \underbrace{\mathbb{Z}_2 \times \cdots \times \mathbb{Z}_2}_{\ell} \times \mathbb{Z}_2$, where $\ell \geq 0$.
If $\ell = 0$, then there is nothing to prove. So we may assume that $\ell > 0$. We have $U(R) = \{(\underbrace{1, 1, \ldots, 1}_{\ell}, 1)\}$. Since $2 \notin U(R)$, $G(R)$ is a 1-regular graph, and so it is a perfect matching. Thus $q = r$ and $n = 2q$. Therefore $rank(G(R)) = frank(G(R)) = 0$; so $G(R)$ is a ring graph.

**Case 4.** $R \cong \underbrace{\mathbb{Z}_2 \times \cdots \times \mathbb{Z}_2}_{\ell} \times \mathbb{Z}_3$, where $\ell \geq 0$.
The case $\ell = 0$ is clear. Now, suppose that $\ell > 0$. Since $U(R) = \{(\underbrace{1, 1, \ldots, 1}_{\ell}, 1), (\underbrace{1, 1, \ldots, 1}_{\ell}, 2)\}$, $G(R)$ is a 2-regular graph. So $G(R)$ is the union of primitive cycles. In fact, every connected component of $G(R)$ is a primitive cycle. Therefore, we have $frank(G(R)) = r$. Since $q = n$, we conclude that $rank(G(R)) = r$. So $G(R)$ is again a ring graph.

**Case 5.** $R \cong \underbrace{\mathbb{Z}_2 \times \cdots \times \mathbb{Z}_2}_{\ell} \times \mathbb{Z}_4$
We have $U(R) = \{(\underbrace{1, 1, \ldots, 1}_{\ell}, 1), (\underbrace{1, 1, \ldots, 1}_{\ell}, 3)\}$. Since $2 \notin U(R)$, $G(R)$ is a 2-regular graph. As in Case 4, it is easy to see that $G(R)$ is a ring graph.

**Case 6.** $R \cong \underbrace{\mathbb{Z}_2 \times \cdots \times \mathbb{Z}_2}_{\ell} \times \mathbb{F}_4$.
Let $\mathbb{F}_4 = \{0, f_1, f_2, f_3\}$. Since $U(R) = \{(\underbrace{1, 1, \ldots, 1}_{\ell}, f_1), (\underbrace{1, 1, \ldots, 1}_{\ell}, f_2), (\underbrace{1, 1, \ldots, 1}_{\ell}, f_3)\}$, $G(R)$ is a 3-regular graph. Since $char(\mathbb{F}_4) = 2$, the sum of

every pair of distinct nonzero elements in $\mathbb{F}_4$ is nonzero. Therefore, the vertex $(a_1, a_2, \ldots, a_\ell, f_i)$ is adjacent to $(1-a_1, 1-a_2, \ldots, 1-a_\ell, f_j)$ for every $a_k \in \mathbb{Z}_2$ and $f_i, f_j \in \mathbb{F}_4$, where $1 \leq k \leq \ell$ and $1 \leq i \neq j \leq 3$. So every component of $G(R)$ has the form similar to that shown in the following figure.

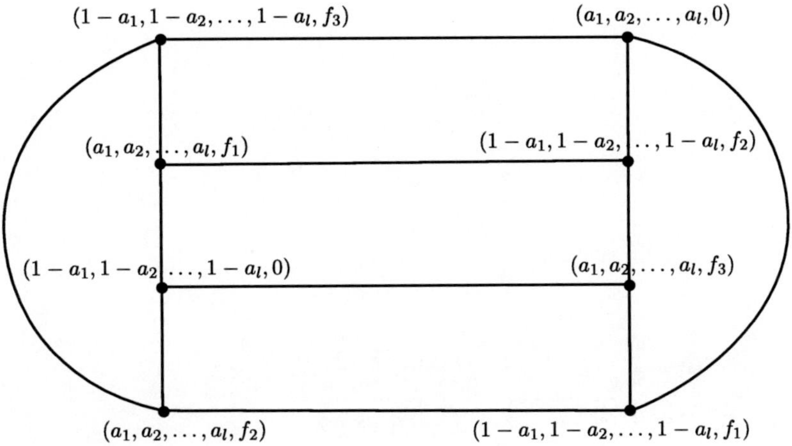


Figure 8.11:

Also, it is easy to see that every component of this graph is a block. Now, by Lemma 7.5.27, we must show that the rank and free rank of every component of $G(R)$ are equal. In view of Figure 8.11, we have $rank(G_i) - frank(G_i) = 5$ for every component $G_i$ of $G(R)$. Thus $G(R)$ is a ring graph.

**Case 7.** $R \cong \underbrace{\mathbb{Z}_2 \times \cdots \times \mathbb{Z}_2}_{\ell} \times \left\{ \begin{bmatrix} a & b \\ 0 & a \end{bmatrix} \mid a, b \in \mathbb{Z}_2 \right\}$.

In this case, we have

$$U(R) = \left\{ (\underbrace{1, 1, \ldots, 1}_{\ell}, \begin{bmatrix} 1 & 0 \\ 0 & 1 \end{bmatrix}), (\underbrace{1, 1, \ldots, 1}_{\ell}, \begin{bmatrix} 1 & 1 \\ 0 & 1 \end{bmatrix}) \right\}.$$

Since $2 \notin U(R)$, $G(R)$ is a 2-regular graph. Similarly to Case 3, $G(R)$ is a ring graph.

The converse statement follows easily. □

**Theorem 8.4.24.** ([16, Theorem 2.2]) *Let $R$ be a finite commutative ring. The unit graph $G(R)$ is outerplanar if and only if $R$ is one of the following rings:*

$$R \cong \underbrace{\mathbb{Z}_2 \times \cdots \times \mathbb{Z}_2}_{\ell} \times S, \text{ where } \ell \geq 0 \text{ and } S \cong \mathbb{Z}_2,\ S \cong \mathbb{Z}_3,\ S \cong \mathbb{Z}_4, \text{ or}$$

$$S \cong \left\{ \begin{bmatrix} a & b \\ 0 & a \end{bmatrix} \mid a, b \in \mathbb{Z}_2 \right\}.$$

*Proof.* Suppose that $G(R)$ is outerplanar. Since an outerplanar graph is a ring graph, we consider the unit graphs which are ring graphs. By Theorem 8.4.23, we have the following cases:

**Case 1.** $R \cong \mathbb{Z}_5$.
In view of Figure 8.12, $G(R)$ contains a subgraph isomorphic to $K_{2,3}$. Thus it is not outerplanar.

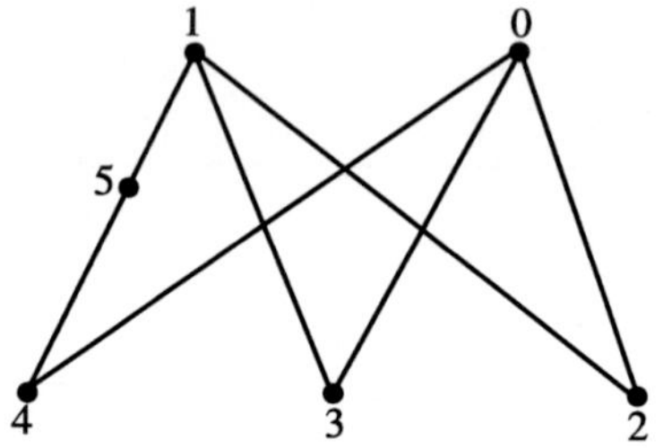


Figure 8.12:

**Case 2.** $R \cong \underbrace{\mathbb{Z}_2 \times \cdots \times \mathbb{Z}_2}_{\ell} \times \mathbb{Z}_2$, where $\ell \geq 0$.
The case $\ell = 0$ is clear. If $\ell > 0$, then $G(R)$ is a perfect matching; so clearly it is outerplanar.

**Case 3.** $R \cong \underbrace{\mathbb{Z}_2 \times \cdots \times \mathbb{Z}_2}_{\ell} \times S$, where $\ell \geq 0$ and $S \cong \mathbb{Z}_3$, $S \cong \mathbb{Z}_4$, or $S \cong \left\{ \begin{bmatrix} a & b \\ 0 & a \end{bmatrix} \mid a, b \in \mathbb{Z}_2 \right\}$. In these cases, $G(R)$ is a union of primitive cycles; so $G(R)$ is an outerplanar graph.

**Case 4.** $R \cong \underbrace{\mathbb{Z}_2 \times \cdots \times \mathbb{Z}_2}_{\ell} \times \mathbb{F}_4$.
In view of Figure 8.13, there is a subdivision of $K_{2,3}$ in $G(R)$, and so $G(R)$ is not outerplanar.

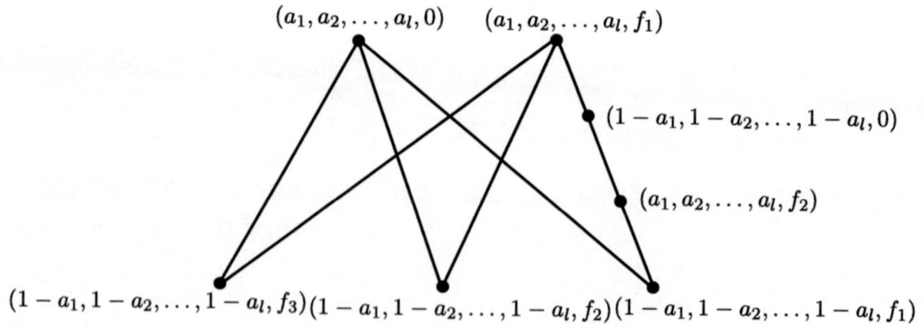


Figure 8.13:

Conversely, if

$$R \cong \underbrace{\mathbb{Z}_2 \times \cdots \times \mathbb{Z}_2}_{\ell} \times S, \text{ where } \ell \geq 0 \text{ and } S \cong \mathbb{Z}_2,\ S \cong \mathbb{Z}_3,\ S \cong \mathbb{Z}_4, \text{ or}$$

$$S \cong \left\{ \begin{bmatrix} a & b \\ 0 & a \end{bmatrix} \mid a, b \in \mathbb{Z}_2 \right\},$$

then one can easily check that $G(R)$ is outerplanar. □

The unitary graph $G_R = Cay(R, U(R))$ is defined to be the graph whose vertex set is $R$, with an edge between $x$ and $y$ if $x - y \in U(R)$. It is easy to see that $G_R$ is a $|U(R)|$-regular graph. Now, we provide a characterization of all finite rings whose $G_R$ are ring graphs and outerplanar. By [44, Theorem 8.2], $G_R$ is planar if and only if $R$ is one of the following rings:

$$\underbrace{\mathbb{Z}_2 \times \cdots \times \mathbb{Z}_2}_{\ell \geq 0},\ \underbrace{\mathbb{Z}_2 \times \cdots \times \mathbb{Z}_2}_{\ell \geq 0} \times \mathbb{Z}_3,\ \underbrace{\mathbb{Z}_2 \times \cdots \times \mathbb{Z}_2}_{\ell \geq 0} \times \mathbb{Z}_4, \text{ or } \underbrace{\mathbb{Z}_2 \times \cdots \times \mathbb{Z}_2}_{\ell \geq 0} \times \mathbb{F}_4.$$

**Theorem 8.4.25.** ([16, Theorem 3.1]) *Let $R$ be a finite commutative ring. Then $G_R$ is a ring graph if and only if $G_R$ is a planar graph.*

*Proof.* Since a ring graph is planar, it is sufficient to show that every planar unitary graph is a ring graph. We have the following cases:

**Case 1.** $R \cong \underbrace{\mathbb{Z}_2 \times \cdots \times \mathbb{Z}_2}_{\ell \geq 0}$.

The case $\ell = 0$ is clear. So we may assume that $\ell > 0$. Since $U(R) = \{(\underbrace{1, 1, \ldots, 1}_{\ell})\}$, $G_R$ is a perfect matching. Thus $frank(G_R) = rank(G_R) = 0$. Therefore $G_R$ is a ring graph.

**Case 2.** $R \cong \underbrace{\mathbb{Z}_2 \times \cdots \times \mathbb{Z}_2}_{\ell \geq 0} \times \mathbb{Z}_3$.

We have $U(R) = \{(\underbrace{1, 1, \ldots, 1}_{\ell}, 1), (\underbrace{1, 1, \ldots, 1}_{\ell}, 2)\}$. Hence $G_R$ is a 2-regular graph, and thus every connected component of $G_R$ is a primitive cycle. Hence $frank(G_R) = r$, and we have $q = n$. Therefore $frank(G_R) = rank(G_R) = r$; so $G_R$ is a ring graph.

**Case 3.** $R \cong \underbrace{\mathbb{Z}_2 \times \cdots \times \mathbb{Z}_2}_{\ell \geq 0} \times \mathbb{Z}_4$.

Similarly to Case 2, one can see that $G_R$ is a ring graph.

**Case 4.** $R \cong \underbrace{\mathbb{Z}_2 \times \cdots \times \mathbb{Z}_2}_{\ell \geq 0} \times \mathbb{F}_4$. Let $\mathbb{F}_4 = \{0, f_1, f_2, f_3\}$. Since $f_i - f_j \neq 0$ for every $1 \leq i \neq j \leq 3$, we have that $(a_1, a_2, \ldots, a_\ell, f_i)$ and $(1 - a_1, 1 - a_2, \ldots, 1 - a_\ell, f_j)$ are adjacent in $G_R$, where $a_k \in \mathbb{Z}_2$ and $1 \leq k \leq \ell$. So every connected component of the graph $G_R$ has a form similar to that in the following figure.

It is easy to check that every component of this graph is a block. In view of Figure 8.14, we have $rank(G_i) = frank(G_i) = 5$. So in this case, $G_R$ is a ring graph. □

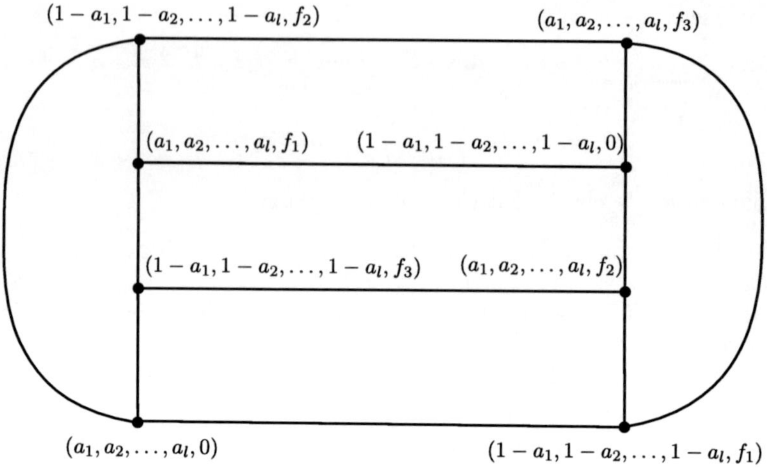


Figure 8.14:

**Theorem 8.4.26.** ([16, Theorem 3.2]) *Let $R$ be a finite commutative ring. The unitary graph $G_R$ is outerplanar if and only if $R$ is isomorphic to one of the following rings:*

$$\underbrace{\mathbb{Z}_2 \times \cdots \times \mathbb{Z}_2}_{\ell \geq 0}, \ \underbrace{\mathbb{Z}_2 \times \cdots \times \mathbb{Z}_2}_{\ell \geq 0} \times \mathbb{Z}_3, \ or \ \underbrace{\mathbb{Z}_2 \times \cdots \times \mathbb{Z}_2}_{\ell \geq 0} \times \mathbb{Z}_4.$$

*Proof.* Since every outerplanar graph is a ring graph, we need to check the outerplanarity of $G_R$, when $G_R$ is a ring graph. In conjunction with Theorem 8.4.25, we may assume that $R$ is isomorphic to one of the rings

$$\underbrace{\mathbb{Z}_2 \times \cdots \times \mathbb{Z}_2}_{\ell \geq 0}, \ \underbrace{\mathbb{Z}_2 \times \cdots \times \mathbb{Z}_2}_{\ell \geq 0} \times \mathbb{Z}_3, \ \underbrace{\mathbb{Z}_2 \times \cdots \times \mathbb{Z}_2}_{\ell \geq 0} \times \mathbb{Z}_4, \text{ or } \underbrace{\mathbb{Z}_2 \times \cdots \times \mathbb{Z}_2}_{\ell \geq 0} \times \mathbb{F}_4.$$

If $R$ is one of the rings $\underbrace{\mathbb{Z}_2 \times \cdots \times \mathbb{Z}_2}_{\ell \geq 0}$, $\underbrace{\mathbb{Z}_2 \times \cdots \times \mathbb{Z}_2}_{\ell \geq 0} \times \mathbb{Z}_3$, or $\underbrace{\mathbb{Z}_2 \times \cdots \times \mathbb{Z}_2}_{\ell \geq 0} \times \mathbb{Z}_4$, then $G_R$ is a 1-regular or a 2-regular graph. Thus $G_R$ is an outerplanar graph.

If $R \cong \underbrace{\mathbb{Z}_2 \times \cdots \times \mathbb{Z}_2}_{\ell \geq 0} \times \mathbb{F}_4$, then we can find a subdivision of $K_{2,3}$ (see Figure 8.13). So $G_R$ is not outerplanar.

The converse statement follows easily. □

## 8.5 Line Graphs of Total Graphs

In graph theory, one associates with a graph $G$ its line graph $L(G)$ so that the set of vertices of $L(G)$ is exactly the set of edges of $G$. Two vertices in $L(G)$

are adjacent if and only if the corresponding edges in $G$ have a common vertex. This allows one to investigate the properties of a graph $G$ which depend only on the edges of that graph as the properties of the graph $L(G)$ which depend only on its vertices. This is very useful for various problems in graph theory. For example, to a matching in $G$ there corresponds an independent set in $L(G)$. If $G$ is connected and if its line graph $L(G)$ is known, one may completely determine $G$ except in the case $L(G) = K_3$.

Knowing the structure of the total graph $T(\Gamma(R))$, it is natural to investigate the structure of its line graph $L(T(\Gamma(R)))$ and to look into relationships between them. In this section, we collect some results on the line graph of $T(\Gamma(R))$ as in [314]. If $x+y \in Z(R)$ for $x, y \in R$, then there is a vertex in the graph $L(T(\Gamma(R)))$, and we denote this vertex by $[x, y]$. From the definition of the graph $L(T(\Gamma(R)))$, it follows that the degree of every vertex of this graph depends on the number of zero-divisors, as well as on whether 2 is a zero-divisor in $R$.

**Theorem 8.5.1.** ([314, Theorem 2.7]) *Let $R$ be a finite commutative ring which is not local. Then $L(T(\Gamma(R)))$ is planar if and only if $R$ is isomorphic to $\mathbb{Z}_2 \times \mathbb{Z}_2$ or $\mathbb{Z}_2 \times \mathbb{Z}_3$.*

**Theorem 8.5.2.** ([314, Theorem 2.8]) *Let $R$ be a finite commutative local ring which is not a field. Then $L(T(\Gamma(R)))$ is planar if and only if $R$ is isomorphic to one of the following rings: $\mathbb{Z}_4$, $\frac{\mathbb{Z}_2[x]}{(x^2)}$, $\mathbb{Z}_8$, $\frac{\mathbb{Z}_2[x]}{(x^3)}$, $\frac{\mathbb{Z}_2[x,y]}{(x,y)^2}$, $\frac{\mathbb{Z}_4[x]}{(2x,x^2)}$, $\frac{\mathbb{Z}_4[x]}{(2x,x^2-2)}$, $\frac{\mathbb{F}_4[x]}{(x^2)}$, or $\frac{\mathbb{Z}_4[x]}{(x^2+x+1)}$.*

([314]): *The list of all finite commutative rings for which the line graph of the total graph is planar:*

$$\mathbb{Z}_2 \times \mathbb{Z}_2,\ \mathbb{Z}_2 \times \mathbb{Z}_3,\ \mathbb{Z}_4,\ \frac{\mathbb{Z}_2[x]}{(x^2)},\ \mathbb{Z}_8,\ \frac{\mathbb{Z}_2[x]}{(x^3)},\ \frac{\mathbb{Z}_2[x,y]}{(x,y)^2},\ \frac{\mathbb{Z}_4[x]}{(2x,x^2)},\ \frac{\mathbb{Z}_4[x]}{(2x,x^2-2)},\ \frac{\mathbb{F}_4[x]}{(x^2)},\ \frac{\mathbb{Z}_4[x]}{(x^2+x+1)}.$$

**Theorem 8.5.3.** ([314, Theorem 2.10]) *Let $R$ be a finite commutative ring which is not local. Then $L(T(\Gamma(R)))$ does not have genus 1.*

**Theorem 8.5.4.** ([314, Theorem 2.11]) *Let $R$ be a finite commutative local ring which is not a field. Then $L(T(\Gamma(R)))$ is toroidal if and only if $R$ is isomorphic to $\mathbb{Z}_9$ or $\mathbb{Z}_3[x]/(x^2)$.*

([314]): *The list of finite commutative rings such that the line graph of the total graph has genus 1 is $\mathbb{Z}_9$, $\mathbb{Z}_3[x]/(x^2)$.*

**Theorem 8.5.5.** ([314, Theorem 2.11]) *Let $n \geq 0$ be an integer. Then there are only finitely many commutative rings $R$ such that $g(L(T(\Gamma(R)))) = n$.*

It is interesting to note that for $|Z(R)| \geq 6$, there are no rings whose line graphs have genus 2 or 3.

# Chapter 9

# Generalized Total Graphs

In this chapter, we discuss some generalizations of total graphs of commutative rings. Actually, total graphs of commutative rings are generalized through ideals, multiplicatively closed sets, and multiplicative-prime sets. Further, we discuss the concept of total graphs corresponding to modules and commutative semirings.

## 9.1 Commutative Ring Generalizations

In this section, we present the notion of generalized total graphs of commutative rings. There are three major types of generalizations available; first we present the definitions of these three generalizations.

Let $I$ be a proper ideal of a commutative ring $R$. An $a \in R$ is *prime to $I$* if $ra \in I$ (where $r \in R$) implies that $r \in I$. Let $S(I)$ be the set of elements of $R$ that are not prime to $I$. Clearly, $S(I)$ is not empty since $I$ is a proper ideal of $R$.

A multiplicatively closed subset $S$ of $R$ is called *saturated* if $xy \in S$ for $x, y \in R$ implies $x \in S$ and $y \in S$.

A subset $H$ of $R$ is a *multiplicative-prime* subset of $R$ if the following two conditions hold: (i) $ab \in H$ for every $a \in H$ and $b \in R$; (ii) if $ab \in H$ for $a, b \in R$, then either $a \in H$ or $b \in H$.

**Definition 9.1.1.** (1) ([8], see Section 9.1.1) Let $R$ be a commutative ring, $I$ a proper ideal of $R$, and $S(I)$ the set of elements of $R$ that are not prime to $I$. The *total graph of $R$ with respect to $I$*, denoted by $T(\Gamma_I(R))$, is the simple graph with all elements of $R$ as vertices, and two distinct vertices $x$ and $y$ are adjacent if and only if $x + y \in S(I)$. If $I = \{0\}$, then $T(\Gamma_I(R))$ is just the total graph of $R$.

(2) ([134], see Section 9.1.2) Let $R$ be a commutative ring and $S$ a multiplicatively closed subset of $R$. Define a simple graph, denoted by $GT_S(R)$, with all elements of $R$ as vertices, and two distinct vertices $x$ and $y$ are adjacent if and only if $x + y \in S$. If $S = Z(R)$, then $GT_S(R) = T(\Gamma(R))$.



D. F. Anderson et al., *Graphs from Rings*, https://doi.org/10.1007/978-3-030-88410-9_9

(3) ([72], see Section 9.1.3) Let $R$ be a commutative ring and $H$ a nonempty proper subset of $R$ such that $R \setminus H$ is a saturated multiplicatively closed subset of $R$. The generalized total graph of $R$ is the simple graph $GT_H(R)$ with all elements of $R$ as vertices, and two distinct vertices $x$ and $y$ are adjacent if and only if $x + y \in H$. If $H = Z(R)$, then $GT_H(R)$ is just the total graph of $R$.

### 9.1.1 Ideal-based Generalized Total Graphs

We start this section with the concept of the total graph of a commutative ring with respect to a proper ideal as given by Abbasi and Habibi in [8]. Consider $T(\Gamma_I(R))$, the *total graph of a commutative ring $R$ with respect to proper ideal $I$*. Let $S(\Gamma_I(R))$ be the induced subgraph of $T(\Gamma_I(R))$ with vertices $S(I)$, and let $\overline{S}(\Gamma_I(R))$ be the induced subgraph $T(\Gamma_I(R))$ with vertices $R \setminus S(I)$.

The set $S(I)$ is not necessarily an ideal of $R$ since it is not always closed under addition. But $S(I)$ is a union of prime ideals of $R$ containing $I$; so whenever $xy \in S(I)$ for $x, y \in R$, then $x \in S(I)$ or $y \in S(I)$. Thus, if $S(I)$ is an ideal of $R$, then it is actually a prime ideal of $R$. Hence the study of $T(\Gamma_I(R))$ breaks naturally into two cases depending on whether $S(I)$ is an ideal of $R$.

We now explore the relationship between $T(\Gamma_I(R))$ and $T(\Gamma(R/I))$. It is easy to check that $Z(R/I) = \{a + I \mid a \in S(I)\}$ and $Reg(R/I) = \{a + I \mid a \notin S(I)\}$. Thus $Z(R/I)$ is an ideal of $R/I$ if and only if $S(I)$ is an ideal of $R$.

**Example 9.1.2.** ([8, Example 2.2]) Let $R = \mathbb{Z}_8$, $S = \mathbb{Z}_4 \times \mathbb{Z}_2$, $I = \{0, 2, 4, 6\} \subseteq R$, and $J = \{0\} \times \mathbb{Z}_2 \subseteq S$. It is easy to check that $S(I) = I$ and $S(J) = \{(0, 0), (0, 1), (2, 0), (2, 1)\}$. Then $T(\Gamma_I(R))$ and $T(\Gamma_J(S))$ are the union of 2 disjoint $K_4$'s. Now, $T(\Gamma(R/I))$ is a graph with two vertices, but $T(\Gamma(S/J))$ is a graph with four vertices.

**Theorem 9.1.3.** ([8, Theorem 2.3]) *Let $R$ be a commutative ring with proper ideal $I$, and let $x, y \in R$. Then*

(1) *If $x + I$ and $y + I$ are (distinct) adjacent vertices in $T(\Gamma(R/I))$, then $x$ is adjacent to $y$ in $T(\Gamma_I(R))$;*

(2) *If $x$ and $y$ are (distinct) adjacent vertices in $T(\Gamma_I(R))$ and $x + I \neq y + I$, then $x + I$ is adjacent to $y + I$ in $T(\Gamma(R/I))$;*

(3) *If $x$ is adjacent to $y$ in $T(\Gamma_I(R))$ and $x + I = y + I$, then $2x, 2y \in S(I)$ and all distinct elements of $x + I$ are adjacent in $T(\Gamma_I(R))$.*

**Corollary 9.1.4.** ([8, Corollary 2.4]) *Let $R$ be a commutative ring with proper ideal $I$. Then $T(\Gamma_I(R))$ contains $|I|$ disjoint subgraphs isomorphic to $T(\Gamma(R/I))$.*

According to the above corollary, there is a strong relationship between $T(\Gamma(R/I))$ and $T(\Gamma_I(R))$.

**Lemma 9.1.5.** ([8, Lemma 2.8]) *Let $R$ be a commutative ring with proper ideal $I$. Then $gr(T(\Gamma_I(R))) \leq gr(T(\Gamma(R/I)))$. If $T(\Gamma(R/I))$ contains a cycle, then so does $T(\Gamma_I(R))$, and therefore $gr(T(\Gamma_I(R))) \leq gr(T(\Gamma(R/I))) \leq 4$.*

Now we investigate the relationship between $T(\Gamma_I(R))$ and $T(\Gamma(R/I))$ with the assumption that $S(I)$ is an ideal of $R$. If $S(I)$ is an ideal of $R$, then, by definition, $S(\Gamma_I(R))$ is a complete subgraph of $T(\Gamma_I(R))$ and is disjoint from $\overline{S}(\Gamma_I(R))$.

**Theorem 9.1.6.** ([8, Theorem 3.3]) *Let $R$ be a commutative ring with proper ideal $I$ such that $S(I)$ is an ideal of $R$.*

(1) *Let $\Gamma$ be an induced subgraph of $\overline{S}(\Gamma_I(R))$ with $x$ and $y$ distinct vertices of $\Gamma$ that are connected by a path in $\Gamma$. Then there is a path in $\Gamma$ of length 2 between $x$ and $y$. In particular, if $\overline{S}(\Gamma_I(R))$ is connected, then $diam(\overline{S}(\Gamma_I(R))) \leq 2$;*

(2) *Let $x$ and $y$ be distinct elements of $\overline{S}(\Gamma_I(R))$ that are connected by a path. If $x + y \notin S(I)$ (that is, if $x$ and $y$ are not adjacent), then $x - (-x) - y$ and $x - (-y) - y$ are paths of length 2 between $x$ and $y$ in $\overline{S}(\Gamma_I(R))$.*

**Theorem 9.1.7.** ([8, Theorem 3.4]) *Let $R$ be a commutative ring with proper ideal $I$ such that $S(I)$ is an ideal of $R$. Then the following statements are equivalent:*

(1) *$\overline{S}(\Gamma_I(R))$ is connected;*

(2) *Either $x + y \in S(I)$ or $x - y \in S(I)$ for every $x, y \in R \setminus S(I)$;*

(3) *Either $x+y \in S(I)$ or $x+2y \in S(I)$ (but not both) for every $x, y \in R\setminus S(I)$. In particular, either $2x \in S(I)$ or $3x \in S(I)$ for every $x \in R \setminus S(I)$.*

**Theorem 9.1.8.** ([8, Theorem 3.5]) *Let $R$ be a commutative ring with proper ideal $I$ such that $S(I)$ is an ideal of $R$, and let $|S(I)| = \alpha$ and $|R/S(I)| = \beta$ (we allow $\alpha$ and $\beta$ to be infinite, then we have $\beta - 1 = (\beta - 1)/2 = \beta$).*

(1) *If $2 \in S(I)$, then $\overline{S}(\Gamma_I(R))$ is the union of $\beta - 1$ disjoint $K_\alpha$'s;*

(2) *If $2 \notin S(I)$, then $\overline{S}(\Gamma_I(R))$ is the union of $(\beta - 1)/2$ disjoint $K_{\alpha,\alpha}$'s.*

Note that if $S(I) = \{0\}$, then $R$ is an integral domain, and $2 \in S(I)$ if and only if $char(R) = 2$.

**Theorem 9.1.9.** ([8, Theorem 3.9]) *Let $R$ be a commutative ring with proper ideal $I$ such that $S(I)$ is an ideal of $R$. Then*

(1) *$\overline{S}(\Gamma_I(R))$ is complete if and only if $R/S(I) \cong \mathbb{Z}_2$ or $R \cong \mathbb{Z}_3$;*

(2) *$\overline{S}(\Gamma_I(R))$ is connected if and only if $R/S(I) \cong \mathbb{Z}_2$ or $R/S(I) \cong \mathbb{Z}_3$;*

(3) *$\overline{S}(\Gamma_I(R))$ (and hence $T(\Gamma_I(R))$ and $S(\Gamma_I(R))$) is totally disconnected if and only if $I = \{0\}$ and $R$ is an integral domain with $char(R) = 2$.*

The next few results give further explicit descriptions of the diameter and girth of $\overline{S}(\Gamma_I(R))$.

**Theorem 9.1.10.** ([8, Proposition 3.11]) *Let $R$ be a commutative ring with proper ideal $I$ such that $S(I)$ is an ideal of $R$. Then $diam(\overline{S}(\Gamma_I(R))) \in \{0, 1, 2, \infty\}$ and $gr(\overline{S}(\Gamma_I(R))) \in \{3, 4, \infty\}$.*

**Theorem 9.1.11.** ([8, Theorem 3.12]) *Let $R$ be a commutative ring with proper ideal $I$ such that $S(I)$ is an ideal of $R$.*

(1) *$diam(\overline{S}(\Gamma_I(R))) = 0$ if and only if $R \cong \mathbb{Z}_2$;*

(2) *$diam(\overline{S}(\Gamma_I(R))) = 1$ if and only if either $R/S(I) \cong \mathbb{Z}_2$ and $|S(I)| \geq 2$ or $R \cong \mathbb{Z}_3$;*

(3) *$diam(\overline{S}(\Gamma_I(R))) = 2$ if and only if $R/S(I) \cong \mathbb{Z}_3$ and $|S(I)| \geq 2$;*

(4) *Otherwise, $diam(\overline{S}(\Gamma_I(R))) = \infty$.*

**Theorem 9.1.12.** ([8, Theorem 3.15]) *Let $R$ be a commutative ring with proper ideal $I$ such that $S(I)$ is an ideal of $R$. Then*

(1) (a) *$gr(\overline{S}(\Gamma_I(R))) = 3$ if and only if $2 \in S(I)$ and $|S(I)| \geq 3$;*

(b) *$gr(\overline{S}(\Gamma_I(R))) = 4$ if and only if $2 \notin S(I)$ and $|S(I)| \geq 2$;*

(c) *Otherwise,. $gr(\overline{S}(\Gamma_I(R))) = \infty$;*

(2) (a) *$gr(T(\Gamma_I(R))) = 3$ if and only if $|S(I)| \geq 3$;*

(b) *$gr(T(\Gamma_I(R))) = 4$ if and only if $2 \notin S(I)$ and $|S(I)| = 2$;*

(c) *Otherwise, $gr(T(\Gamma_I(R))) = \infty$.*

We next study the remaining case when $S(I)$ is not an ideal of $R$. Since $S(I)$ is always closed under multiplication by elements of $R$, there are distinct $x, y \in S(I)^*$ such that $x + y \in R \setminus S(I)$. Thus $|S(I)| \geq 3$, and in this case, $S(\Gamma_I(R))$ and $\overline{S}(\Gamma_I(R))$ are never disjoint subgraphs. Also, we determine when $T(\Gamma_I(R))$ is connected and compute $diam(T(\Gamma_I(R)))$.

**Theorem 9.1.13.** ([8, Theorem 4.1]) *Let $R$ be a commutative ring with proper ideal $I$ such that $S(I)$ is not an ideal of $R$.*

(1) *$S(\Gamma_I(R))$ is connected with $diam(S(\Gamma_I(R))) = 2$;*

(2) *Some vertex of $S(\Gamma_I(R))$ is adjacent to a vertex of $\bar{S}(\Gamma_I(S))$; In particular, the subgraphs $S(\Gamma_I(R))$ and $\overline{S}(\Gamma_I(S))$ are not disjoint;*

(3) *If $\overline{S}(\Gamma_I(S))$ is connected, then $T(\Gamma_I(S))$ is connected.*

**Lemma 9.1.14.** ([8, Lemma 4.2]) *Let $R$ be a commutative ring with proper ideal $I$ such that $S(I)$ is not an ideal of $R$. Then $T(\Gamma_I(R))$ is connected if and only if $R = (a_1, \ldots, a_k)$ for some $a_1, \ldots, a_k \in S(I)$. In particular, if $R/I$ is a finite ring and $I \subseteq J(R)$, then $T(\Gamma_I(R))$ is connected.*

**Theorem 9.1.15.** ([8, Theorem 4.3]) *Let $R$ be a commutative ring with proper ideal $I$ such that $S(I)$ is not an ideal of $R$ and $R = (S(I))$. Let $n \geq 2$ be the least positive integer such that $R = (x_1, \ldots, x_n)$ for some $x_1, \ldots, x_n \in S(I)$ (that is, $T(\Gamma_I(R))$ is connected). Then $diam(T(\Gamma_I(R))) = n$. In particular, if $R/I$ is a finite ring and $I \subseteq J(R)$, then $diam(T(\Gamma_I(R))) = 2$.*

Clearly, if $R = (a_1, \ldots, a_k)$ for $a_1, \ldots, a_k \in S(I)$, then $R/I = (a_1 + I, \ldots, a_k + I)$, and hence $diam(T(\Gamma(R/I))) \leq diam(T(\Gamma_I(R)))$. Note that if $k \geq 2$ is the least positive integer such that $R = (a_1, \ldots, a_k)$, then $diam(T(\Gamma(R/I))) \geq diam(T(\Gamma_I(R))) - 1$.

**Theorem 9.1.16.** ([8, Theorem 4.5]) *Let $R$ be a commutative ring with proper ideal $I$ such that $S(I)$ is not an ideal of $R$. If $T(\Gamma_I(R))$ is connected, then*

(1) *$diam(T(\Gamma_I(R))) = d(0, 1)$;*

(2) *If $diam(T(\Gamma_I(R))) = n$, then $diam(\overline{S}(\Gamma_I(R))) \geq n - 2$.*

**Corollary 9.1.17.** ([8, Corollary 4.6]) *Let $\{R_\alpha\}_{\alpha \in \Lambda}$ be a family of commutative rings with $|\Lambda| \geq 2$, and let $R = \prod_{\alpha \in \Lambda} R_\alpha$. Let $I = \prod_{\alpha \in \Lambda} I_\alpha$, where $I_\alpha$ is a proper ideal of $R_\alpha$ for every $\alpha \in \Lambda$. Then $T(\Gamma_I(R))$ is connected with $diam(T(\Gamma_I(R))) = 2$.*

**Theorem 9.1.18.** ([8, Theorem 4.8]) *Let $R$ be a commutative ring with proper ideal $I$ such that $S(I)$ is not an ideal of $R$ and $S = R \backslash S(I)$. Then $T(\Gamma_{S^{-1}I}(S^{-1}R))$ is connected with $diam(T(\Gamma_{S^{-1}I}(S^{-1}R))) = 2$. In particular, if $R/I$ is a finite ring and $I \subseteq J(R)$, then $diam(T(\Gamma_{S^{-1}I}(S^{-1}R))) = 2$.*

**Theorem 9.1.19.** ([8, Theorem 4.9]) *Let $R$ be a commutative ring, $I$ an ideal of $R$, and $P_1$ and $P_2$ prime ideals of $R$ containing $I$. Suppose that $xy \in I$ for some $x \in P_1 \backslash P_2$ and $y \in P_2 \backslash P_1$. Then $diam(T(\Gamma_{S^{-1}I}(R_S))) = 2$, where $S = R \backslash P_1 \cup P_2$.*

The following theorem gives $gr(S(\Gamma_I(R)))$, $gr(\overline{S}(\Gamma_I(R)))$, and $gr(T(\Gamma_I(R)))$ when $S(I)$ is not an ideal of $R$.

**Theorem 9.1.20.** ([8, Theorem 4.10]) *Let $R$ be a commutative ring with proper ideal $I$ such that $S(I)$ is not an ideal of $R$. Then*

(1) *If $I \neq \{0\}$, then $gr(S(\Gamma_I(R))) = 3$. Otherwise, $gr(S(\Gamma_I(R))) = 3$ or $\infty$. Moreover, if $gr(S(\Gamma_I(R))) = \infty$, then $R \cong \mathbb{Z}_2 \times \mathbb{Z}_2$; so $S(\Gamma_I(R))$ is a $K_{1,2}$ star graph with center 0;*

(2) *$gr(T(\Gamma_I(R))) = 3$ if and only if $gr(S(\Gamma_I(R))) = 3$;*

(3) *The (induced) subgraph of $S(\Gamma_I(R))$ with vertices $\sqrt{I}$ is complete; hence $gr(S(\Gamma_I(R))) = 3$ when $|\sqrt{I}| \geq 3$;*

(4) *If $gr(T(\Gamma_I(R))) = 4$, then $gr(S(\Gamma_I(R))) = \infty$;*

(5) *If $2 \in I$, then $gr(\overline{S}(\Gamma_I(R))) = 3$ or $\infty$;*

(6) *If $2 \notin I$, then $gr(\overline{S}(\Gamma_I(R))) = 3, 4,$ or $\infty$.*

### 9.1.2 Total Graphs of Multiplicatively Closed Sets

Next, we study the second generalization of the total graph of a commutative ring. Let $R$ be a commutative ring. A subset $S$ of $R$ which is closed under multiplication is called *multiplicatively closed*. In 2012, Barati et al. [134] introduced the graph $GT_S(R)$ to a commutative ring $R$ and a multiplicatively closed subset $S$ of $R$. The graph $GT_S(R)$ is a simple undirected graph with all elements of $R$ as vertices, and two distinct vertices $x$ and $y$ are adjacent if and only if $x + y \in S$. Since the subsets $Z(R)$ and $U(R)$ of $R$ are multiplicatively closed, $GT_S(R)$ is a natural generalization of the total graph and the unit graph of $R$.

**Example 9.1.21.** ([134, Example 1.1]) Let $R = \mathbb{Z}_6$. Then $Z(R) = \{0, 2, 3, 4\}$ and $U(R) = \{1, 5\}$. The graphs $GT_S(R)$ for $S = Z(R) = \{0, 2, 3, 4\}$, $S = \{1, 3, 5\}$, and $S = U(R) = \{1, 5\}$ are shown in Figure 9.1.

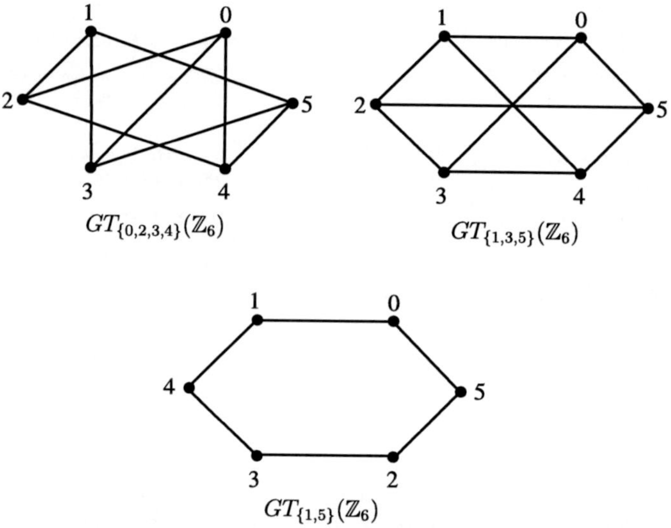


Figure 9.1: The generalized total graph of certain rings

First, we observe the relationship of the associated graphs $GT_S(R)$ with the total graph, unit graph, and some Cayley graphs. Recall that the *unit graph* of $R$ is the simple graph with all elements of $R$ as vertices, and two distinct vertices $x$ and $y$ are adjacent if and only if $x + y \in U(R)$. Let $H$ be a finite group with identity element $e$, $T$ a subset of $H$ such that $e \notin T$, and $T^{-1} = \{x^{-1} \mid x \in T\} \subseteq T$. Then the *Cayley graph* associated to $H$ and $T$, denoted by $Cay(H, T)$, is the simple graph with all elements of $H$ as vertices, and two distinct vertices $x$ and $y$ of $H$ are adjacent if and only if $xy^{-1} \in T$.

The following example shows that for a positive integer $n$, the Cayley graph $Cay(\mathbb{Z}_{2n}, \{1, -1\})$ is isomorphic to $GT_{\{1,-1\}}(\mathbb{Z}_{2n})$.

**Example 9.1.22.** ([134, Example 1.2]) Let $R = \mathbb{Z}_{2n}$ and $S = \{1, -1\} \subseteq R$. Then $S$ is a multiplicatively closed subset of $R$. Consider $(\mathbb{Z}_{2n}, +)$ as a group; so we can define $Cay(\mathbb{Z}_{2n}, \{1, -1\})$. We show that the two graphs $GT_{\{1,-1\}}(\mathbb{Z}_{2n})$ and $Cay(\mathbb{Z}_{2n}, \{1, -1\})$ are isomorphic. Let $f : GT_{\{1,-1\}}(\mathbb{Z}_{2n}) \to Cay(\mathbb{Z}_{2n}, \{1, -1\})$ be given by $f(x) = x$ if $x$ is even and $f(x) = -x$, otherwise. Clearly $f$ is a bijection. Now, we show that $f$ is a homomorphism. Suppose that $\{x, y\}$ is an edge in $GT_{\{1,-1\}}(\mathbb{Z}_{2n})$. So both $x$ and $y$ are neither even nor odd. Hence, without loss of generality, we may assume that $x$ is even and $y$ is odd and that the sum of $x$ and $y$ is 1 or $-1$. Therefore $x - (-y)$ is 1 or $-1$. This means that $f(x)$ and $f(y)$ are adjacent in $Cay(\mathbb{Z}_{2n}, \{1, -1\})$. Conversely, if $\{x, y\}$ is an edge in $Cay(\mathbb{Z}_{2n}, \{1, -1\})$, then $x - y$ is equal to 1 or $-1$. Thus, without loss of generality, we may assume that $x$ is odd and $y$ is even. Since $f(-x) = x$ and $f(y) = y$, we have that $-x - y$ is equal to 1 or $-1$, which implies that the corresponding vertices $x$ and $y$ are adjacent in $GT_{\{1,-1\}}(\mathbb{Z}_{2n})$.

Now we study some basic properties of $GT_S(R)$. We start with the degree of the vertices in $GT_S(R)$.

**Lemma 9.1.23.** ([134, Lemma 1.3]) *Let $R$ be a commutative ring and $S$ an arbitrary multiplicatively closed subset of $R$. Then, in the graph $GT_S(R)$,*

(1) *for every $x \in R$ with $x + x \notin S$, we have $deg(x) = |S|$, and*

(2) *for every $x \in R$ with $x + x \in S$, we have $deg(x) = |S| - 1$. In particular, if $2 \in S$, then $deg(x) = |S| - 1$ for all $x \in S$.*

Next, we characterize the graph $GT_S(R)$ with the assumption that $S$ is an ideal of $R$.

**Theorem 9.1.24.** ([134, Theorem 1.4]) *Let $R$ be a commutative ring and $S$ an ideal of $R$ with $|S| = \alpha$. Let $A = \{x + S \mid x \in R \setminus S \text{ and } 2x \in S\}$ and $B = \{x + S \mid x \in R \setminus S \text{ and } 2x \notin S\}$. Then $GT_S(R)$ is the disjoint union of $|A| + 1$ $K_\alpha$'s and $|B|/2$ $K_{\alpha,\alpha}$'s.*

**Example 9.1.25.** ([134, Example 1.5]) Let $R = \mathbb{Z}_{12}$ and $S = \{0, 6\} \subseteq R$. Then, using the notation in Theorem 9.1.24, $A = \{3 + S\}$ and $B = \{1 + S, 4 + S, 8 + S, 11 + S\}$. Note that $3 + S = 9 + S$. Therefore, the graph $GT_S(R)$ is the disjoint union of two $K_2$ components and two $K_{2,2}$ components.

As a consequence of Theorem 9.1.24, we have the following corollary which is an improved form of Theorem 7.2.1.

**Corollary 9.1.26.** ([134, Corollary 1.6]) *Let $S$ is an ideal of a commutative ring $R$ with $|S| = \alpha$ and $|R/S| = \beta$.*

(1) *If* $2 \in S$, *then* $GT_S(R)$ *is the union of* $\beta$ *disjoint* $K_\alpha$*'s;*

(2) *If* $2x \notin S$ *for every* $x \in R \backslash S$, *then* $GT_S(R)$ *is the union of* $K_\alpha$ *with* $(\beta-1)/2$ *disjoint* $K_{\alpha,\alpha}$*'s.*

Now, we find a necessary and sufficient condition for the connectedness of the graph $GT_S(R)$ when $S = -S$. Since $Z(R)$ and $U(R)$ satisfy this condition, the following theorem can be considered as an improved form of Theorem 7.2.10 and [273, Proposition 3.2].

**Theorem 9.1.27.** ([134, Theorem 1.7]) *Let* $R$ *be a commutative ring and* $S$ *a multiplicatively closed subset of* $R$ *such that* $S = -S$. *Then* $GT_S(R)$ *is connected if and only if* $(R, +)$ *is generated by* $S$.

Note that $S$ in Theorem 9.1.24 is not necessarily an ideal of $R$. For instance, if $R = \mathbb{Z}$ and $S = \{1, -1\}$, then $GT_S(R)$ is connected.

**Corollary 9.1.28.** ([134, Corollary 1.8]) *Let* $S$ *be a proper ideal of a commutative ring* $R$. *Then* $GT_S(R)$ *is disconnected.*

In the case that both subsets $S$ and $S^c = R \setminus S$ of $R$ are multiplicatively closed, we have the following description for the graphs associated to them. In this case, it is worthwhile to study the relationship between the associated graphs and the multiplicatively closed subsets $S$ and $S^c$.

**Theorem 9.1.29.** ([134, Theorem 1.9]) *Let* $R$ *be a commutative ring. If* $S$ *and* $S^c = R \setminus S$ *are two multiplicatively closed subsets of* $R$. *Then the complement of* $GT_S(R)$ *is isomorphic to* $GT_{S^c}(R)$.

If $R$ is finite, it is the disjoint union of $Z(R)$ and $U(R)$. Thus, as a consequence of Theorem 9.1.27, we observe the relationship between the unit graph and the total graph of a finite ring.

**Corollary 9.1.30.** ([134, Corollray 1.10]) *The complement of the unit graph of a finite commutative ring* $R$ *is isomorphic to its total graph.*

Let $R$ be a commutative ring and $M$ an $R$-module. As mentioned earlier, the idealization $R(+)M$ of $M$ over $R$ is a commutative ring. If $S$ is a multiplicatively closed subset of $R$, then it is easy to see that $\widehat{S} = S(+)M$ is a multiplicatively closed subset of $R(+)M$. The following theorem compares the diameter of $GT_{\widehat{S}}(R(+)M)$ with the diameter of $GT_S(R)$.

**Theorem 9.1.31.** ([134, Theorem 1.11]) *Let* $R$ *be a commutative ring and* $M$ *an* $R$*-module. Then* $GT_{\widehat{S}}(R(+)M)$ *is connected if and only if* $GT_S(R)$ *is connected. Moreover,* $diam(GT_{\widehat{S}}(R(+)M)) = diam(GT_S(R))$.

In general, $Z(R)(+)M \subseteq Z(R(+)M)$. Hence Theorem 9.1.31 is an improved from of Theorem 7.3.11, where the authors considered the case when $Z(R)(+)M = Z(R(+)M)$.

Next, we consider the case that $S$ is a saturated multiplicatively closed subset of $R$. A multiplicatively closed subset $S$ of $R$ is *saturated* if $xy \in S$ for $x, y \in R$ implies that $x \in S$ and $y \in S$. Note that the set of units $U(R)$ of $R$ is a saturated multiplicatively closed subset of $R$. On the other hand, if $S$ is a saturated multiplicatively closed subset of $R$, then $1 \cdot s \in S$ for $s \in S$; so $1 \in S$. Thus, for any $u \in U(R)$, $uu^{-1} = 1 \in S$, which implies that $u \in S$, and so $U(R) \subseteq S$. This means that $U(R)$ is the smallest saturated multiplicatively closed subset of $R$. Hereafter in this section, $S$ is a saturated multiplicatively closed subset of $R$, and so our results about the graph $GT_S(R)$ are natural generalizations of the corresponding results for the unit graph. The following results provide a characterization for the completeness of $GT_S(R)$.

**Lemma 9.1.32.** ([134, Proposition 2.1]) *The graph $GT_S(R)$ is complete if and only if $S = R$, or $char(R) = 2$ and $S = R \setminus \{0\}$.*

Note that if $0 \in S$, then $S = R$, and so $GT_S(R)$ is complete by Proposition 9.1.32. Hence in the rest of the consideration, we assume that $0 \notin S$. Now, in view of [228, Theorem 2], $R \setminus S = \bigcup P_i$ for some prime ideals $P_i$ of $R$. Put $I = \bigcap P_i$, and let $\tilde{S} = \{s + I \mid s \in S\}$.

**Lemma 9.1.33.** ([134, Lemma 2.3]) *Let $S$ be a saturated multiplicatively closed subset of a commutative ring $R$. Then the following statements hold in $GT_S(R)$:*

(1) *If $x \in R \setminus S$, then $deg(x) = |S|$;*

(2) *If $2 \notin S$, then $deg(x) = |S|$ for all $x \in R$.*

**Lemma 9.1.34.** ([134, Lemma 2.4]) *Let $R$ be a commutative ring, $S$ a saturated multiplicatively closed subset of $R$ with $0 \notin S$. Assume that $R \setminus S = \bigcup P_i$ for some prime ideals $P_i$ of $R$, $I = \bigcap P_i$, and $\tilde{S} = \{s + I \mid s \in S\}$. Let $x, y \in R$. Then the following statements are equivalent:*

(1) *$x$ is adjacent to $y$ in $GT_S(R)$;*

(2) *$x + I$ is adjacent to $y + I$ in $GT_{\tilde{S}}(R/I)$;*

(3) *Every element of $x + I$ is adjacent to every element of $y + I$ in $GT_S(R)$;*

(4) *There are $x + i \in x + I$ and $y + i' \in y + I$ which are adjacent in $GT_S(R)$.*

**Corollary 9.1.35.** ([134, Corollary 2.5]) *Let $R$ be a commutative ring, $S$ a saturated multiplicatively closed subset of $R$ with $0 \notin S$. Assume that $R \setminus S = \bigcup P_i$ for some prime ideals $P_i$ of $R$, $I = \bigcap P_i$, and $\tilde{S} = \{s + I \mid s \in S\}$. Let $x, y \in R$. Then $GT_S(R)$ is connected if and only if $GT_{\tilde{S}}(R/I)$ is connected.*

**Theorem 9.1.36.** ([134, Theorem 2.7]) *Let $R$ be a commutative ring, $S$ a saturated multiplicatively closed subset of $R$ with $0 \notin S$. Assume that $R \setminus S = \bigcup P_i$ for some prime ideals $P_i$ of $R$, $I = \bigcap P_i$, and $\tilde{S} = \{s + I \mid s \in S\}$. Let $x, y \in R$. Then the following statements hold:*

(1) *$gr(GT_S(R)) \leq gr(GT_{\tilde{S}}(R/I))$;*

(2) *$diam(GT_{\tilde{S}}(R/I)) \leq diam(GT_S(R))$;*

(3) *If $GT_{\tilde{S}}(R/I)$ is a complete graph, then $diam(GT_S(R)) \leq 2$;*

(4) *If $GT_{\tilde{S}}(R/I)$ is not a complete graph, then $diam(GT_{\tilde{S}}(R/I)) = diam(GT_S(R))$.*

The following example shows that we may have strict inequality in part (1) of Theorem 9.1.36. In our example, $GT_{\tilde{S}}(R/I)$ is a complete graph, but $GT_S(R)$ is not a complete graph by Theorem 9.1.36(3).

**Example 9.1.37.** ([134, Example 2.8]) Let $R = \mathbb{Z}_2[x]$ be the polynomial ring in an indeterminate $x$ with coefficients in $\mathbb{Z}_2$, and let $S = R \setminus (x)$. Hence $I = (x)$, and so $|R/I| = 2$. Let $V_1 = \{a_1x + \cdots + a_nx^n \mid n \in \mathbb{N} \text{ and } a_1, \ldots, a_n \in \mathbb{Z}_2\}$ and $V_2 = \{1 + b_1x + \cdots + b_mx^m \mid m \in \mathbb{N} \text{ and } b_1, \ldots, b_m \in \mathbb{Z}_2\}$. Clearly the sum of two elements of $V_1$ (or $V_2$) is in $(x)$. This implies that the vertices in the same part are not adjacent. Moreover, for every two vertices $f = a_1x + \cdots + a_nx^n \in V_1$ and $g = 1 + b_1x + \cdots + b_mx^m \in V_2$, we have $f + g \in S$. This means that $GT_S(R)$ is a complete bipartite graph, and so $diam(GT_S(R)) = 2$. On the other hand, $GT_{\tilde{S}}(R/I) = K_2$, and so $diam(GT_{\tilde{S}}(R/I)) = 1$. Hence the upper bound given in Theorem 9.1.36(3) is sharp.

Moreover, one can see that $gr(GT_{\tilde{S}}(R/I)) = \infty$ and $gr(GT_S(R)) = 4$. These facts show that the converse of Theorem 9.1.36(1) does not hold. Hence, in general, the equality in part (1) of Theorem 9.1.36 does not hold.

**Theorem 9.1.38.** ([134, Theorem 2.12]) *Let $R$ be a commutative ring, $S$ a saturated multiplicatively closed subset of $R$, and $M$ an $R$-module. Then we have the following statements:*

(1) *If $2 \notin S$, then $GT_{\widehat{S}}(R(+)M) = \bigoplus_{|M|^2} GT_S(R)$;*

(2) *If $2 \in S$, then $GT_{\widehat{S}}(R(+)M) = \bigoplus_{|M|^2} GT_S(R) \oplus (\bigoplus_{|S|} K_{|M|})$.*

**Proposition 9.1.39.** ([134, Proposition 2.13]) *Let $R$ be a commutative ring and $S$ a saturated multiplicatively closed subset of $R$ with $R \setminus S = \bigcup_{i=1}^{n} P_i$ such that $|R/P_i| = 2$ for some $i$. Then $GT_S(R)$ is a bipartite graph. Furthermore, $GT_S(R)$ is a complete bipartite graph if and only if $n = 1$.*

The next theorem gives the girth of $GT_S(R)$.

**Theorem 9.1.40.** ([134, Theorem 2.15]) *Let $R$ be a finite commutative ring and $S$ a saturated multiplicatively closed subset of $R$. Then $gr(GT_S(R)) \in \{3, 4, 6, \infty\}$.*

In the following, we provide some examples of saturated multiplicatively closed sets to show that each of the numbers 3, 4, 6, and $\infty$ in Theorem 9.1.40 can be realized as the girth of some graph.

**Example 9.1.41.** ([134, Example 2.16]) Let $R = \mathbb{Z}_6$. As in Example 9.1.21, $gr(GT_{Z(R)}(R)) = 3$, $gr(GT_{U(R)}(R)) = 6$, and $gr(GT_S(R)) = 4$, where $S = \{1, 3, 5\}$. For the saturated multiplicatively closed subset $S = \{-1, 1\}$ of $\mathbb{Z}$, one can easily show that $GT_S(R)$ is a path, and so $gr(GT_S(R)) = \infty$.

The next theorem characterizes the commutative rings $R$ with a saturated multiplicatively closed set $S$ such that $GT_S(R)$ has infinite girth, or equivalently, $GT_S(R)$ is a forest.

**Theorem 9.1.42.** ([134, Theorem 2.17]) *Let $R$ be a finite commutative ring and $S$ a saturated multiplicatively closed subset of $R$. Then $gr(GT_S(R)) = \infty$ if and only if one of the following statements holds:*

(1) $R = \mathbb{Z}_3$;

(2) $R = \mathbb{Z}_2 \times \mathbb{Z}_2 \times \cdots \times \mathbb{Z}_2$ *and* $|S| = 1$.

Using the proof of Theorem 9.1.42, one can obtain the following characterization for a commutative ring $R$ with saturated multiplicatively closed set $S$ such that $GT_S(R)$ is a forest. Recall that a matching in a bipartite graph $G$ is said to be complete if it contains all the vertices of $G$.

**Corollary 9.1.43.** ([134, Corollary 2.18]) *Let $R$ be a finite commutative ring such that $R \neq \mathbb{Z}_3$. Also, suppose that $S$ is a saturated multiplicatively closed subset of $R$. Then $GT_S(R)$ is forest if and only if it is a complete matching.*

In some special cases, we can find a better upper bound for the girth of the graph $GT_S(R)$. In the following theorem, we study the local case.

**Theorem 9.1.44.** ([134, Theorem 2.19]) *Let $(R, \mathfrak{m})$ be a commutative local ring and $S$ a saturated multiplicatively closed subset of $R$ such that $R \setminus S = \bigcup_{i \in A} P_i$ for prime ideals $P_i$ of $R$. Then*

(1) *If* $|A| \geq 2$, *then* $gr(GT_S(R)) = 3$;

(2) *If* $|A| = 1$, *then* $gr(GT_S(R)) \leq 4$.

**Lemma 9.1.45.** ([134, Lemma 2.20]) *Let $R$ be a finite commutative local ring and $S$ a saturated multiplicatively closed subset of $R$. Then $S = U(R)$.*

Let $R$ be a finite commutative ring. Then $R = R_1 \times R_2 \times \cdots \times R_k$ with every $R_i$ a finite local ring with maximal ideal $\mathfrak{m}_i$. Now, let $S$ be a saturated multiplicatively closed subset of $R$. It is not hard to see that $S = S_1 \times S_2 \times \cdots \times S_k$, where, for $1 \leq i \leq k$, $S_i = \{s_i \in R_i \mid (s_1, \ldots, s_{i-1}, s_i, s_{i+1}, \ldots, s_k) \in S\}$ is a saturated multiplicatively closed subset of $R_i$ or $S_i = R_i$.

**Theorem 9.1.46.** ([134, Proposition 2.21]) *Let* $R = R_1 \times R_2 \times \cdots \times R_k$*, where every* $(R_i, \mathfrak{m}_i)$ *is a finite commutative local ring with* $R_i/\mathfrak{m}_i \cong \mathbb{Z}_2$*, and let* $S = S_1 \times S_2 \times \cdots \times S_k$ *be a saturated multiplicatively closed subset of* $R$*. Then* $diam(GT_S(R)) \in \{1, 2, \infty\}$.

**Corollary 9.1.47.** ([134, Corollary 2.22]) *Let* $R = R_1 \times R_2 \times \cdots \times R_k$*, where every* $(R_i, \mathfrak{m}_i)$ *is a finite commutative local ring with* $R_i/\mathfrak{m}_i \cong \mathbb{Z}_2$*, and let* $S = S_1 \times S_2 \times \cdots \times S_k$ *be a saturated multiplicatively closed subset of* $R$*. Then* $GT_S(R)$ *is disconnected if and only if there are* $1 \leq i \neq j \leq n$ *such that* $S_i = U(R_i)$ *and* $S_j = U(R_j)$.

**Theorem 9.1.48.** ([134, Theorem 2.23]) *Let* $R$ *be a finite commutative ring and* $S$ *a saturated multiplicatively closed subset of* $R$*. Then* $diam(GT_S(R)) \in \{1, 2, 3, \infty\}$.

As an immediate consequence of the proof of Theorem 9.1.48, we have the following characterization for disconnected graphs.

**Corollary 9.1.49.** ([134, Corollary 2.24]) *Let* $R = R_1 \times \cdots \times R_n$ *be a finite commutative ring and* $S$ *a saturated multiplicatively closed subset of* $R$*. Then* $GT_S(R)$ *is disconnected if and only if there are* $i \neq j$*, where* $R_i$ *and* $R_j$ *have* $\mathbb{Z}_2$ *as a quotient and* $S_i = U(R_i)$ *and* $S_j = U(R_i)$.

### 9.1.3 Total Graphs of Multiplicative-prime Sets

Let $R$ be a commutative ring. A subset $H$ of $R$ is a *multiplicative-prime* subset of $R$ if the following two conditions hold: (i) $ab \in H$ for every $a \in H$ and $b \in R$; (ii) if $ab \in H$ for $a, b \in R$, then either $a \in H$ or $b \in H$. For example, $H$ is a multiplicative-prime subset of $R$ if $H$ is a prime ideal of $R$, $H$ is a union of prime ideals of $R$, $H = Z(R)$, or $H = R \setminus U(R)$. In fact, it is easily seen that $H$ is a multiplicative-prime subset of $R$ if and only if $R \setminus H$ is a saturated multiplicatively closed subset of $R$. Thus $H$ is a multiplicative-prime subset of $R$ if and only if $H$ is a union of prime ideals of $R$ [228, Theorem 2]. Note that if $H$ is a multiplicative-prime subset of $R$, then $Nil(R) \subseteq H \subseteq R \setminus U(R)$; and if $H$ is also an ideal of $R$, then $H$ is necessarily a prime ideal of $R$. In particular, if $R = Z(R) \cup U(R)$ (e.g., $R$ is finite), then $Nil(R) \subseteq H \subseteq Z(R)$.

Let $H$ be a multiplicative-prime subset of a commutative ring $R$. In [72], Anderson and Badawi defined the *generalized total graph* of $R$, denoted by $GT_H(R)$, to be the simple graph with all elements of $R$ as vertices, and distinct vertices $x$ and $y$ are adjacent if and only if $x + y \in H$. For $A \subseteq R$, let $GT_H(A)$ be the induced subgraph of $GT_H(R)$ with all elements of $A$ as the vertices. For example, $GT_H(R \setminus H)$ is the induced subgraph of $GT_H(R)$ with vertices $R \setminus H$. When $H = Z(R)$, $GT_H(R)$ is just the total graph. As to be expected, $GT_H(R)$ and $T(\Gamma(R))$ share many properties. However, the concept of generalized total graph, unlike the earlier concept of total graph, allows us to study graphs of integral domains.

The graphs $GT_H(R)$, where $R = \mathbb{Z}_6$, $H = Z(R) = \{0, 2, 3, 4\}$, and $H = (2) = \{0, 2, 4\}$ are given in Figure 9.2.

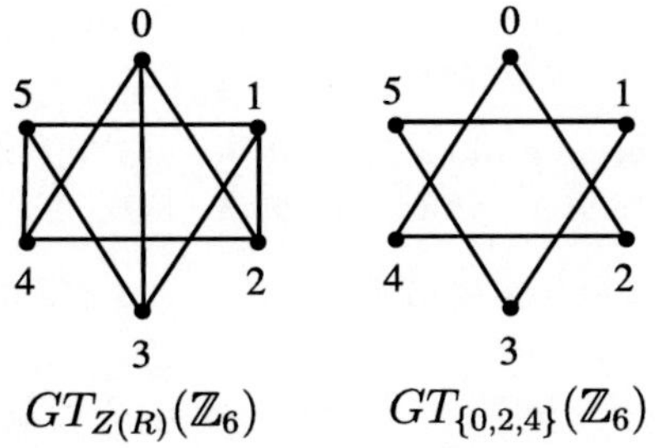


Figure 9.2: Generalized total graphs

Let $H$ be a multiplicative-prime subset of a commutative ring $R$. Since $H$ is a union of prime ideals of $R$, the study of $GT_H(R)$ breaks naturally into two cases depending on whether $H$ is an (prime) ideal of $R$. First, we handle the case when $H$ is an ideal of $R$.

The next theorem gives a complete description of $GT_H(R)$ when $H$ is an ideal of $R$. It also shows that nonisomorphic rings may have isomorphic graphs. We allow $\alpha$ and $\beta$ to be infinite cardinals; if $\beta$ is infinite, then $\beta - 1 = (\beta - 1)/2 = \beta$.

**Theorem 9.1.50.** ([72, Theorem 2.2]) *Let $H$ be a prime ideal of a commutative ring $R$, and let $|H| = \alpha$ and $|R/H| = \beta$.*

(1) *If $2 \in H$, then $GT_H(R \setminus H)$ is the union of $\beta - 1$ disjoint $K_\alpha$'s;*

(2) *If $2 \notin H$, then $GT_H(R \setminus H)$ is the union of $(\beta - 1)/2$ disjoint $K_{\alpha,\alpha}$'s.*

From the above theorem, we can easily deduce when $GT_H(R \setminus H)$ is complete or connected and can explicitly compute its diameter and girth. The first result of this kind determines when $GT_H(R \setminus H)$ is either complete or connected.

**Theorem 9.1.51.** ([72, Theorem 2.3]) *Let $H$ be a prime ideal of a commutative ring $R$.*

(1) *$GT_H(R \setminus H)$ is complete if and only if either $R/H \cong \mathbb{Z}_2$ or $R \cong \mathbb{Z}_3$;*

(2) *$GT_H(R \setminus H)$ is connected if and only if either $R/H \cong \mathbb{Z}_2$ or $R/H \cong \mathbb{Z}_3$;*

(3) *$GT_H(R \setminus H)$ (and hence $GT_H(H)$ and $GT_H(R)$) is totally disconnected if and only if $H = \{0\}$ (thus $R$ is an integral domain) and char$(R) = 2$.*

The next result computes both the diameter and girth of $GT_H(R \setminus H)$ when $H$ is a prime ideal of $R$.

**Theorem 9.1.52.** ([72, Theorem 2.4]) *Let $H$ be a prime ideal of a commutative ring $R$. Then the following conditions hold:*

(1) *$diam(GT_H(R \setminus H)) = 0, 1, 2,$ or $\infty$. In particular, $diam(GT_H(R \setminus H)) \leq 2$ if $GT_H(R \setminus H)$ is connected;*

(2) *$gr(GT_H(R \setminus H)) = 3, 4,$ or $\infty$. In particular, $gr(GT_H(R \setminus H)) \leq 4$ if $GT_H(R \setminus H)$ contains a cycle.*

The next theorem gives a more explicit description of the diameter and girth of $GT_H(R \setminus H)$ when $H$ is a prime ideal of $R$.

**Theorem 9.1.53.** ([72, Theorem 2.5]) *Let $H$ be a prime ideal of a commutative ring $R$.*

(1) (a) *$diam(GT_H(R \setminus H)) = 0$ if and only if $R \cong \mathbb{Z}_2$;*

(b) *$diam(GT_H(R \setminus H)) = 1$ if and only if either $R/H \cong \mathbb{Z}_2$ and $R \ncong \mathbb{Z}_2$ (i.e., $R/H \cong \mathbb{Z}_2$ and $|H| \geq 2$), or $R \cong \mathbb{Z}_3$;*

(c) *$diam(GT_H(R \setminus H)) = 2$ if and only if $R/H \cong \mathbb{Z}_3$ and $R \ncong \mathbb{Z}_3$ (i.e., $R/H \cong \mathbb{Z}_3$ and $|H| \geq 2$);*

(d) *Otherwise, $diam(GT_H(R \setminus H)) = \infty$;*

(2) (a) *$gr(GT_H(R \setminus H)) = 3$ if and only if $2 \in H$ and $|H| \geq 3$;*

(b) *$gr(GT_H(R \setminus H)) = 4$ if and only if $2 \notin H$ and $|H| \geq 2$;.*

(c) *Otherwise, $gr(GT_H(R \setminus H)) = \infty$;*

(3) (a) *$gr(GT_H(R)) = 3$ if and only if $|H| \geq 3$;*

(b) *$gr(GT_H(R)) = 4$ if and only if $2 \notin H$ and $|H| = 2$;*

(c) *Otherwise, $gr(GT_H(R)) = \infty$.*

The following examples illustrate the previous theorem.

**Example 9.1.54.** ([72, Example 2.6]) (a) Let $R = \mathbb{Z}$ and $H$ be a prime ideal of $R$. Then $GT_H(R \setminus H)$ is complete if and only if $H = 2\mathbb{Z}$, and $GT_H(R \setminus H)$ is connected if and only if either $H = 2\mathbb{Z}$ or $H = 3\mathbb{Z}$. Moreover, $diam(GT_H(R \setminus H)) = 1$ if and only if $H = 2\mathbb{Z}$, and $diam(GT_H(R \setminus H)) = 2$ if and only if $H = 3\mathbb{Z}$. Let $p \geq 5$ be a prime integer and $H = p\mathbb{Z}$. Then $GT_H(R \setminus H)$ is the the union of $(p-1)/2$ disjoint $K_{\alpha,\alpha}$'s; so $\mathrm{diam}(GT_H(R \setminus H)) = \infty$. Finally, $diam(GT_H(R \setminus H)) = \infty$ when $H = \{0\}$.

Also, $gr(GT_H(R \setminus H)) = \infty$ if $H = \{0\}$, $gr(GT_H(R \setminus H)) = 3$ if $H = 2\mathbb{Z}$, and $gr(GT_H(R \setminus H)) = 4$ otherwise. Moreover, $gr(GT_{\{0\}}(R)) = \infty$ and $gr(GT_H(R)) = 3$ for any nonzero prime ideal $H$ of $R$.

(b) Let $R = \mathbb{Z}_{pm} \times R_1 \times \cdots \times R_n$, where $m \geq 2$ is an integer, $p$ is a positive prime integer, and $R_1, \ldots, R_n$ are commutative rings. Then $H = p\mathbb{Z}_{pm} \times R_1 \times \cdots \times R_n$ is a prime ideal of $R$. The graph $GT_H(R \setminus H)$ is complete if and only if $p = 2$, and $GT_H(R \setminus H)$ is connected if and only if $p = 2$ or $p = 3$. Moreover,

$diam(GT_H(R \setminus H)) = 1$ if and only if $p = 2$, and $diam(GT_H(R \setminus H)) = 2$ if and only if $p = 3$. Assume that $p \geq 5$. Then $GT_H(R \setminus H)$ is the union of $(p-1)/2$ disjoint $K_{\alpha,\alpha}$'s, where $\alpha = m|R_1| \cdots |R_n|$; so $diam(GT_H(R \setminus H)) = \infty$.

Also, $gr(GT_H(R \setminus H)) = 3$ if $p = 2$ and $gr(GT_H(R \setminus H)) = 4$ otherwise. Moreover, $gr(GT_H(R)) = 3$ for any prime $p$.

We have already observed in Theorem 9.1.50 that $GT_H(H)$ is always connected and $GT_H(R)$ is never connected when $H$ is an ideal of $R$. The next theorem gives several new criteria for when $GT_H(R \setminus H)$ is connected.

**Theorem 9.1.55.** ([72, Theorem 2.8]) *Let $H$ be a prime ideal of a commutative ring $R$. Then the following statements are equivalent:*

(1) *$GT_H(R \setminus H)$ is connected;*

(2) *Either $x + y \in H$ or $x - y \in H$ for every $x, y \in R \setminus H$;*

(3) *Either $x + y \in H$ or $x + 2y \in H$ for every $x, y \in R \setminus H$. In particular, either $2x \in H$ or $3x \in H$ (but not both) for every $x \in R \setminus H$;*

(4) *Either $R/H \cong \mathbb{Z}_2$ or $R/H \cong \mathbb{Z}_3$.*

Now we consider the case when the multiplicative-prime subset $H$ is not an ideal of $R$. Since $H$ is always closed under multiplication by elements of $R$, this just means that $0 \in H$ and there are distinct $x, y \in H^*$ such that $x + y \in R \setminus H$. In this case, $GT_H(H)$ is always connected (but never complete), $GT_H(H)$ and $GT_H(R \setminus H)$ are never disjoint subgraphs of $GT_H(R)$, and $|H| \geq 3$.

**Theorem 9.1.56.** ([72, Theorem 3.1(3)]) *Let $R$ be a commutative ring and $H$ a multiplicative-prime subset of $R$ that is not an ideal of $R$. If $GT_H(R \setminus H)$ is connected, then $GT_H(R)$ is connected.*

Next, we determine when $GT_H(R)$ is connected and compute $diam(GT_H(R))$. In particular, $GT_H(R)$ is connected if and only if $diam(GT_H(R)) < \infty$.

**Theorem 9.1.57.** ([72, Theorem 3.2, Theorem 3.4]) *Let $R$ be a commutative ring and $H$ a multiplicative-prime subset of $R$ that is not an ideal of $R$. Then $GT_H(R)$ is connected if and only if $1 = z_1 + \cdots + z_n$ for some $z_1, \ldots, z_n \in H$. In particular, if $H$ is not an ideal of $R$ and either $dim(R) = 0$ (e.g., $R$ is finite) or $R$ is an integral domain with $dim(R) = 1$, then $GT_H(R)$ is connected. Furthermore, suppose that $G_H(R)$ is connected. Let $n \geq 2$ be the least positive integer such that $1 = z_1 + \cdots + z_n$ for some $z_1, \ldots, z_n \in H$. Then $diam(GT_H(R)) = n$. In particular, if $H$ is not an ideal of $R$ and either $dim(R) = 0$ (e.g., $R$ is finite) or $R$ is an integral domain with $dim(R) = 1$, then $diam(GT_H(R)) = 2$.*

**Theorem 9.1.58.** ([72, Corollary 3.5]) *Let $R$ be a commutative ring and $H$ a multiplicative-prime subset of $R$ that is not an ideal of $R$ such that $GT_H(R)$ is connected.*

(1) $diam(GT_H(R)) = d(0, 1)$;

(2) *If* $diam(GT_H(R)) = n$, *then* $diam(GT_H(R \setminus H)) \geq n - 2$.

The following is an example of a ring $R$ such that $GT_H(R)$ is connected, but $GT_H(R \setminus H)$ is not connected.

**Example 9.1.59.** ([72, Example 3.6]) (a) Let $R = \mathbb{Q}[X]$. Then $H = \mathbb{Q}[X] \setminus \mathbb{Q}^*$ is a multiplicative-prime subset of $R$ that is not an ideal of $R$. Thus $GT_H(H)$ is connected with $diam(GT_H(H)) = 2$. Moreover, $GT_H(R)$ is connected with $diam(GT_H(R)) = 2$ (by Theorem 9.1.56) since $R = (X, X+1)$ with $X$, $(X+1) \in H$. However, $GT_H(R \setminus H)$ is not connected since there is no path from 1 to 2 in $GT_H(R \setminus H)$. Thus the converse of Theorem 9.1.56 need not hold.

(b) Let $R = \mathbb{Z}$. Then $H = \mathbb{Z} \setminus U(\mathbb{Z})$ is a multiplicative-prime subset of $R$ that is not an ideal of $R$. Since $GT_H(R \setminus H)$ is clearly connected, $GT_H(R)$ is connected by Theorem 9.1.56.

**Theorem 9.1.60.** ([72, Theorem 3.9]) *Let $R$ be a commutative ring and $H$ a multiplicative-prime subset of $R$ that contains two co-maximal ideals of $R$. Then $GT_H(R)$ is connected with $diam(GT_H(R)) = 2$.*

The following is an example of a commutative ring $R$ with a multiplicative-prime subset $H$ such that neither $GT_H(R \setminus H)$ nor $GT_H(R)$ is connected, but $GT_{H_S}(R_S)$ is connected for some multiplicatively closed subset $S$ of $R$ with $S \neq R \setminus H$.

**Example 9.1.61.** ([72, Example 3.11]) Let $R = \mathbb{Z}[X]$, $H = X\mathbb{Z}[X] \cup 3\mathbb{Z}[X]$, and $S = \{1, (X+3), (X+3)^2, (X+3)^3, \ldots\} \subsetneq R \setminus H$. Then $H$ is a multiplicative-prime subset of $R$ that is not an ideal of $R$ and $(H) = (3, X) \subsetneq R$. Thus $GT_H(R)$ is not connected by Theorem 9.1.57, and hence $GT_H(R \setminus H)$ is not connected by Theorem 9.1.56. Since $X\mathbb{Z}[X]_S$, $3\mathbb{Z}[X]_S$ are co-maximal ideals of $R_S$ and $X\mathbb{Z}[X]_S$, $3\mathbb{Z}[X]_S \subseteq H_S$, the graph $GT_{H_S}(R_S)$ is connected with $diam(GT_{H_S}(R_S)) = 2$ by Theorem 9.1.58.

We next investigate the girth of $GT_H(H)$, $GT_H(R \setminus H)$, and $GT_H(R)$ when $H$ is not an ideal of $R$. Recall that $|H| \geq 3$ if $H$ is not an ideal of $R$.

**Theorem 9.1.62.** ([72, Theorem 3.14]) *Let $R$ be a commutative ring and $H$ a multiplicative-prime subset of $R$ that is not an ideal of $R$.*

(1) *Either* $gr(GT_H(H)) = 3$ *or* $gr(GT_H(H)) = \infty$. *Moreover, if* $gr(GT_H(H)) = \infty$, *then* $R \cong \mathbb{Z}_2 \times \mathbb{Z}_2$ *and* $H = Z(R)$; *so* $GT_H(H)$ *is a* $K_{1,2}$ *star graph with center* 0;

(2) $gr(GT_H(R)) = 3$ *if and only if* $gr(GT_H(H)) = 3$;

(3) $gr(GT_H(R)) = 4$ *if and only if* $gr(GT_H(H)) = \infty$ *(if and only if* $R \cong \mathbb{Z}_2 \times \mathbb{Z}_2$*)*;

(4) *If $char(R) = 2$, then $gr(GT_H(R \setminus H)) = 3$ or $\infty$. In particular, $gr(GT_H(R \setminus H)) = 3$ if $char(R) = 2$ and $GT_H(R \setminus H)$ contains a cycle;*

(5) *$gr(GT_H(R \setminus H)) = 3, 4$, or $\infty$. In particular, $gr(GT_H(R \setminus H)) \leq 4$ if $GT_H(R \setminus H)$ contains a cycle.*

## 9.2 Total Graphs of Modules

The total graph of a ring was generalized to a module in [111, 311, 9]. The study of the set of torsion elements of a module over a commutative ring can often be a frustrating one. Almost immediately one runs into the ugly issue of a profound lack of algebraic structure, highlighted by (typically) a lack of closure under addition. This unfortunate lack of algebraic structure is most disturbing in such an important subset of a module over a ring. In this section, we introduce and study the total torsion element graph of a module over a commutative ring with the tools and methods of graph theory.

### 9.2.1 Total Torsion Element Graphs

Let $M$ be a module over a commutative ring $R$. The total torsion element graph of $M$ over $R$ was introduced in [111, 311]. We use $T(M)$ to denote the set of torsion elements of $M$; that is, $T(M) = \{m \in M \mid rm = 0 \text{ for some } 0 \neq r \in R\}$. So, if $R$ is an integral domain, then $T(M)$ is a submodule of $M$. We will use $Tof(M) = M \setminus T(M)$ to denote the set of nontorsion elements of $M$. Recall that $M$ is said to be torsionfree if $T(M) = \{0\}$.

In a manner analogous to a commutative ring, the *total torsion element graph* of a module $M$ over a commutative ring $R$ is defined to be the simple graph $T(\Gamma(M))$ with the vertex set $M$, where two distinct vertices $m$ and $n$ are adjacent if and only if $m + n \in T(M)$. Let $Tof(\Gamma(M))$ be the (induced) subgraph of $T(\Gamma(M))$ with vertices $Tof(M)$, and let $Tor(\Gamma(M))$ be the (induced) subgraph of $T(\Gamma(M))$ with vertices $T(M)$. It is easy to check that $T(\Gamma(M)) = T(\Gamma(R))$ when $M = R$.

**Example 9.2.1.** ([311, Example 2.2]) Let $R = \mathbb{Z}_n \times \mathbb{Z}_m$ and $M = \mathbb{Z}_n$ an $R$-module, where the module structure is given by $(a, b) \cdot m = am$. Then $Ann(M) \neq \{(0, 0)\}$ since $(0, b) \in Ann(M)$ for every $b \in \mathbb{Z}_m$, and thus $T(\Gamma(M))$ is complete.

We start with the extreme cases $T(M) = M$ and $T(M) = \{0\}$.

**Theorem 9.2.2.** ([311, Theorem 2.1]) *Let $R$ be a commutative ring and $M$ an $R$-module. Then the total graph $T(\Gamma(M))$ is complete if and only if $T(M) = M$.*

Note that every finite abelian group $M$ is a torsion $\mathbb{Z}$-module. In particular, if $R = \mathbb{Z}$ and $M = \mathbb{Z}_n$, an $R$-module with the usual multiplication, then

$T(\Gamma(M)) \cong K_n$. Therefore, all finite complete graphs may be realized as total graphs of modules.

Let us look at the other extreme case, when the graph is totally disconnected. If $T(M) = \{0\}$, then the vertices $m_1$ and $m_2$ are adjacent if and only if $m_2 = -m_1$. If, in addition, $M \neq 0$, then $T(\Gamma(M))$ is a disconnected graph and its only edges are those that connect vertices $m_i$ and $-m_i$.

**Theorem 9.2.3.** ([311, Theorem 2.4]) *Let $R$ be a commutative ring and $M$ an $R$-module. Then $T(\Gamma(M))$ is totally disconnected if and only if $R$ has characteristic 2 and $M$ is torsionfree.*

The study of $T(\Gamma(M))$ breaks naturally into two cases depending on whether $T(M)$ is a submodule of $M$. First, we investigate some properties of the total torsion element graph of a module $M$ over a commutative ring $R$ such that $T(M)$ is a submodule of $M$. Our starting point is the following theorem.

**Theorem 9.2.4.** ([111, Theorem 2.1]) *Let $R$ be a commutative ring and $M$ an $R$-module such that $T(M)$ is a submodule of $M$.*

(1) *$Tor(\Gamma(M))$ is a complete (induced) subgraph of $T(\Gamma(M))$ and $Tor(\Gamma(M))$ is disjoint from $Tof(\Gamma(M))$;*

(2) *If $N$ is a submodule of $M$, then $T(\Gamma(N))$ is a (induced) subgraph of $T(\Gamma(M))$;*

(3) *If $(0 :_R M) \neq \{0\}$, then $T(\Gamma(M))$ is a complete graph.*

The next result gives a complete description of the structure of the total graph $T(\Gamma(M))$ when $T(M)$ is a submodule.

**Theorem 9.2.5.** ([311, Theorem 2.7], [111, Theorem 2.5]) *Let $R$ be a commutative ring and $M$ an $R$-module such that $T(M)$ is a proper submodule of $M$. Suppose $|T(M)| = \alpha$ and $|M/T(M)| = \beta$.*

(1) *If $2 \in Z(R)$, then $T(\Gamma(M))$ is a union of $\beta$ disjoint complete graphs $K_\alpha$;*

(2) *If $2 \notin Z(R)$, then $T(\Gamma(M))$ is a union of $\frac{\beta-1}{2}$ disjoint complete bipartite graphs $K_{\alpha,\alpha}$ and one complete graph $K_\alpha$.*

The question of under what conditions $T(M)$ is a submodule of $M$ and how is this related to the condition that $Z(R)$ is an ideal in $R$ naturally arises. If $Z(R) = \{0\}$, then $T(M)$ is obviously a submodule. The following theorem shows that a more general result holds.

**Theorem 9.2.6.** ([311, Theorem 2.9]) *Let $R$ be a commutative ring and $M$ an $R$-module. If $Z(R) = (z)$ is a principal ideal of $R$ with $z \in Nil(R)$, then $T(M)$ is a submodule of $M$.*

By the assumptions of the previous theorem, it follows that $Z(R) = Nil(R)$. Let $R$ be the polynomial ring $\mathbb{Z}[x]$. If we define a nonstandard multiplication on $R$ by $p(x) * q(x) = p(0)q(0)$, then $Z(R) = Nil(R) = (x)$. Thus the conditions of Theorem 9.2.6 are fulfilled. It also holds in the ring $\mathbb{Z}_{p^n}$ for any integer $n \geq 2$ and prime $p$. Here $Z(R) = Nil(R) = (p)$. When the zero-divisors form an ideal which is not principal, $T(M)$ need not be a submodule even if $Z(R) = Nil(R)$. Let us look at an example.

**Example 9.2.7.** ([311, Example 2.10]) Let $R$ be the commutative local ring $\mathbb{Z}_4[x]/(2x, x^2)$. Then $Z(R) = Nil(R) = (2, x)$, where $x$ is the corresponding class. Let $M = R/2R \oplus R/xR$, and take $m_1 = (1+2R, xR)$, $m_2 = (2R, 1+xR)$. Then $m_1, m_2 \in T(M)$ since $2m_1 = xm_2 = 0$, while $m_1+m_2 = (1+2R, 1+xR) \notin T(M)$.

The next theorem gives an explicit description of the diameter of $Tof(\Gamma(M))$.

**Theorem 9.2.8.** ([111, Theorem 2.7]) *Let $R$ be a commutative ring and $M$ an $R$-module such that $T(M)$ is a submodule of $M$.*

(1) *$diam(Tof(\Gamma(M))) = 0$ if and only if $T(M) = \{0\}$ and $|M| = 2$;*
(2) *$diam(Tof(\Gamma(M))) = 1$ if and only if either $T(M) \neq \{0\}$ and $|M/T(M)| = 2$ or $T(M) = \{0\}$ and $|M| = 3$;*
(3) *$diam(Tof(\Gamma(M))) = 2$ if and only if $T(M) \neq \{0\}$ and $|M/T(M)| = 3$;*
(4) *Otherwise, $diam(Tof(\Gamma(M))) = \infty$.*

The next result completely characterizes the girth of $T(\Gamma(M))$ and $Tof(\Gamma(M))$.

**Theorem 9.2.9.** ([111, Theorem 2.9]) *Let $R$ be a commutative ring and $M$ an $R$-module such that $T(M)$ is a submodule of $M$.*

(1) (a) *$gr(Tof(\Gamma(M))) = 3$ if and only if $2 \in Z(R)$ and $|T(M)| \geq 3$;*
    (b) *$gr(Tof(\Gamma(M))) = 4$ if and only if $2 \notin Z(R)$ and $|T(M)| \geq 2$;*
    (c) *Otherwise, $gr(Tof(\Gamma(M))) = \infty$;*
(2) (a) *$gr(T(\Gamma(M))) = 3$ if and only if $|T(M)| \geq 3$;*
    (b) *$gr(T(\Gamma(M))) = 4$ if and only if $2 \notin Z(R)$ and $|T(M)| = 2$;*
    (c) *Otherwise, $gr(T(\Gamma(M))) = \infty$.*

We now study the total torsion element graph of a module $M$ over a commutative ring $R$ such that $T(M)$ is not a submodule of $M$. Let $R = \mathbb{Z}_4$ and $M = \mathbb{Z}_{12}$. Then $M$ is an $R$-module with the usual operations, and $T(M) = \{0, 4, 6, 8\}$ is not a submodule of $M$. Clearly, $Tor(\Gamma(M))$ is connected with $diam(Tor(\Gamma(M))) = 2$. Moreover, since $6+10 = 4 \in T(M)$, the subgraphs $Tor(\Gamma(M))$ and $Tof(\Gamma(M))$ of $T(\Gamma(M))$ are not disjoint. The following few results deal with the connectedness and diameter of $T(\Gamma(M))$ when $T(M)$ is not a submodule of $M$.

**Theorem 9.2.10.** ([111, Theorem 3.1]) *Let $M$ be a module over a commutative ring $R$ such that $T(M)$ is not a submodule of $M$.*

(1) *$Tor(\Gamma(M))$ is connected with $diam(Tor(\Gamma(M))) = 2$;*

(2) *Some vertex of $Tor(\Gamma(M))$ is adjacent to a vertex of $Tof(\Gamma(M))$. In particular, the subgraphs $Tor(\Gamma(M))$ and $Tof(\Gamma(M))$ of $T(\Gamma(M))$ are not disjoint;*

(3) *If $Tof(\Gamma(M))$ is connected, then $T(\Gamma(M))$ is connected.*

**Theorem 9.2.11.** ([111, Theorem 3.2], [311, Theorem 3.3]) *Let $M$ be a module over a commutative ring $R$ such that $T(M)$ is not a submodule of $M$. Then $T(\Gamma(M))$ is connected if and only if $M = (T(M))$ (that is, $M = (a_1, \ldots, a_k)$ for some $a_1, \ldots, a_k \in T(M)$).*

**Theorem 9.2.12.** ([111, Theorem 3.3]) *Let $M$ be a module over a commutative ring $R$ such that $T(M)$ is not a submodule of $M$, and let $M = (T(M))$ (that is, $T(\Gamma(M))$ is connected). Let $n \geq 2$ be the least positive integer such that $M = (m_1, m_2, \ldots, m_n)$ for some $m_1, m_2, \ldots, m_n \in T(M)$. Then $diam(T(\Gamma(M))) \leq n$. In particular, if $M$ is a cyclic $R$-module, then $diam(T(\Gamma(M))) = n$.*

**Theorem 9.2.13.** ([111, Theorem 3.4]) *Let $M$ be a module over a commutative ring $R$ such that $T(M)$ is not a submodule of $M$, and let $M = (T(M))$ (that is, $T(\Gamma(M))$ is connected). Let $n \geq 2$ be the least positive integer such that $M = (m_1, m_2, \ldots, m_n)$ for some $m_1, m_2, \ldots, m_n \in T(M)$.*

(1) *If $M$ is a cyclic module with generator $m$, then $diam(T(\Gamma(M))) = d(0, m)$;*

(2) *If $diam(T(\Gamma(M))) = n$ and $M$ is a cyclic $R$-module with generator $m$, then $diam(Tof(\Gamma(M))) \leq n - 2$.*

**Theorem 9.2.14.** ([311, Theorem 3.4]) *Let $R$ be a commutative ring and $M$ an $R$-module. If $T(\Gamma(R))$ is connected, then $T(\Gamma(M))$ is connected.*

The next result deals with the girth of $T(\Gamma(M))$ and $Tor(\Gamma(M))$ when $T(M)$ is not a submodule of $M$.

**Theorem 9.2.15.** ([111, Theorem 3.5]) *Let $M$ be a module over a commutative ring $R$ such that $T(M)$ is not a submodule of $M$.*

(1) *Either $gr(Tor(\Gamma(M))) = 3$ or $gr(Tor(\Gamma(M))) = \infty$;*

(2) *$gr(T(\Gamma(M))) = 3$ if and only if $gr(Tor(\Gamma(M))) = 3$;*

(3) *If $gr(T(\Gamma(M))) = 4$, then $gr(Tor(\Gamma(M))) = \infty$;*

(4) *If $char(R) = 2$, $Tof(\Gamma(M))$ contains a cycle, and $0 \neq N(R) \subsetneq Nil(R)$, then $gr(Tof(\Gamma(M))) = 3$, where $N(R) = \{x \in R \mid x^2 = 0\}$;*

(5) *If $char(R) = 2$, $Nil(R) \neq 0$, and $Tof(\Gamma(M))$ contains a cycle, then $gr(Tof(\Gamma(M))) \leq 4$;*

(6) *If* $char(R) \neq 2$, *then* $gr(Tor(\Gamma(M))) = 3, 4$ *or* $\infty$.

The authors of [346] studied the domination number of $T(\Gamma(M))$. If $M$ is a module over a commutative ring $R$ such that $T(M) = M$, then the domination number $\gamma(T(\Gamma(M))) = 1$. First we find the domination number of $T(\Gamma(M))$ when $T(M)$ is a submodule of $M$.

**Theorem 9.2.16.** ([346, Theorem 3.5,3.6]) *Let $M$ be a module over a commutative ring $R$ such that $T(M)$ is the submodule and $|T(M)| = \alpha \neq 0$ and $|M/T(M)| = \beta$. Then $\gamma(T(\Gamma(M))) = \beta$. Moreover, if $T(M) = 0$, then $\gamma(T(\Gamma(M))) = \frac{\beta+1}{2}$.*

Now we investigate the domination number of the total graph of the module $\mathbb{Z}_n$ over the ring $\mathbb{Z}_m$. But generally, $\mathbb{Z}_n$ is not a module over $\mathbb{Z}_m$. We find a necessary condition for $\mathbb{Z}_n$ to be a $\mathbb{Z}_m$-module.

**Theorem 9.2.17.** ([346, Theorem 4.4]) *Let $m, n \in \mathbb{N}$. If $n = p_1^{\alpha_1} p_2^{\alpha_2} \dots p_\ell^{\alpha_\ell}$ is the prime factorization of $n$, then there exists a nonzero ring-homomorphism $f : \mathbb{Z}_m \longrightarrow \mathbb{Z}_n$ if and only if $p_i^{\alpha_i} | m$ for some $1 \leq i \leq \ell$.*

**Theorem 9.2.18.** ([346, Lemma 4.5]) *Let $m, n \in \mathbb{N}$ with $m < n$, $gcd(m, n) = d$, and $n/d = n_1$. Then $\mathbb{Z}_n$ is a nontrivial $\mathbb{Z}_m$-module if and only if $d | k(kn_1 - 1)$ for some $1 \leq k \leq d - 1$.*

**Corollary 9.2.19.** ([346, Corollary 4.6]) *Let $m, n \in \mathbb{N}$ with $m < n$ and* $\gcd(m, n) = 1$. *Then $\mathbb{Z}_n$ is a $\mathbb{Z}_m$-module only by the trivial zero multiplication.*

**Theorem 9.2.20.** ([346, Corollary 4.6,Theorem 4.8]) *Let $m, n \in \mathbb{N}$ with $gcd(m, n) = d > 1$ and $n/d = n_1$.*

(1) *If $d = 1$ and $m < n$, then $\mathbb{Z}_n$ is a $\mathbb{Z}_m$-module only by the trivial zero multiplication;*

(2) *If $d > 1$ and the prime factorization of $d$ is $d = p_1^{r_1} p_2^{r_2} \dots p_s^{r^s}$, then $\mathbb{Z}_n$ is an $\mathbb{Z}_m$-module if and only if $p_i \nmid n_1$ for some $1 \leq i \leq s$.*

**Theorem 9.2.21.** ([346, Theorem 4.7,Corollary 4.10]) *Let $m, n \in \mathbb{N}$.*

(1) *If $m < n$ and $gcd(m, n) = 1$, then $\mathbb{Z}_n$ is a module over $\mathbb{Z}_m$ and $\gamma(T(\Gamma(\mathbb{Z}_n))) = 1$;*

(2) *If $n < m$ and $n \mid m$, then $\mathbb{Z}_n$ is a module over $\mathbb{Z}_m$ and $\gamma(T(\Gamma(\mathbb{Z}_n))) = 1$.*

### 9.2.2 Total Graphs of Modules

In 2012, Abbasi and Habibi [9] introduced the concept of the total graph of a module with respect to a submodule. Let $R$ be a commutative ring, $M$ an $R$-module, and $N$ a proper submodule of $M$. The ideal $\{r \in R \mid rM \subseteq N\}$ will be denoted by $(N :_R M)$ or $(N : M)$. A proper submodule $N$ of an $R$-module $M$ is said to be a *prime submodule* if whenever $rm \in N$ for some $r \in R$ and $m \in M$, then either $m \in N$ or $r \in (N : M)$. Clearly, if $N$ is a prime submodule of $M$, then $P = (N : M)$ is a prime ideal of $R$. In this case, we say that $N$ is a $P$-prime submodule of $M$. Define $M(N) = \{m \in M \mid rm \in N \text{ for some } r \in R\backslash(N : M)\}$. Since $N$ is a proper submodule of $M$, $M(N) \supseteq N$, and hence $M(N)$ is not empty. Also, $M(N)$ is not necessarily a submodule of $M$ (not always closed under addition), but if $r \in R$ and $m \in M(N)$, then $rm \in M(N)$. It is clear that $M(N) = \{m \in M \mid Am \subseteq N \text{ for some ideal } A \not\subseteq (N : M)\}$. The *total graph of $M$ with respect to $N$*, denoted by $T(\Gamma_N(M))$, is the simple graph with vertices all elements of $M$, and two distinct vertices $x$ and $y$ are adjacent if and only if $x + y \in M(N)$. It is easy to check that $T(\Gamma(R)) = T(\Gamma_0(M))$ when $M = R$ and $T(\Gamma(M)) = T(\Gamma_0(M))$ when $M$ is faithful. Let $M(\Gamma_N(M))$ be the (induced) subgraph of $T(\Gamma_N(M))$ with vertices $M(N)$, and let $\overline{M}(\Gamma_N(M))$ be the (induced) subgraph $T(\Gamma_N(M))$ with vertices $M \setminus M(N)$.

Let $N$ be a proper submodule of $M$. Define $M(N) = \{m \in M \mid rm \in N \text{ for some } r \in R\backslash(N :_R M)\}$. It is easy to see that $M(N)$ is closed under scalar multiplication. However, $M(N)$ may not be an additive subgroup of $M$. Note that if $N$ is a prime submodule of $M$, then $M(N) = N$, and if $(N : M)$ is a prime ideal $R$, then $M(N)$ is a submodule of $M$.

First we investigate the primeness of $M(N)$ and the related ideal $S(N : M)$. We show that $U^{-1}(N :_R M) = (U^{-1}N : U^{-1}RU^{-1}M)$ when $U$ is a multiplicatively closed subset of $R$, $(N : M)$ is a prime ideal, and $M(N)$ is a proper submodule.

**Theorem 9.2.22.** ([9, Theorem 2.6]) *Let $R$ be a commutative ring, $M$ an $R$-module, and $N$ a proper submodule of $M$ such that $(N : M)$ is a prime ideal of $R$ and $M(N) \neq M$. Then $U^{-1}(N :_R M) = (U^{-1}N :_{U^{-1}R} U^{-1}M)$ for every multiplicatively closed subset $U$ of $R$ such that $U \cap (N : M) = \emptyset$.*

**Theorem 9.2.23.** ([9, Lemma 2.10]) *Let $R$ be a commutative ring, $M$ an $R$-module, and $N$ a proper submodule of $M$ such that $(N : M)$ is a prime ideal of $R$ and $M(N) \neq M$. Then $U^{-1}M(U^{-1}N) = U^{-1}(M(N))$ for every multiplicatively closed subset $U$ of $R$ such that $U \cap (N : M) = \emptyset$.*

As usual, the study of $T(\Gamma_N(M))$ breaks into two cases depending on whether $M(N)$ is a submodule of $M$. First, we investigate some properties of the total graph of an $R$-module $M$ with respect to a proper submodule $N$ under the assumption that $M(N)$ is a submodule of $M$. It follows directly from the definition that if $M(N) = M$, then $T(\Gamma_N(M))$ is complete. So, assume that $M(N) \neq M$. Since we assume that $M(N) \neq M$, $\overline{M}(\Gamma_N(M))$ is always nonempty.

**Theorem 9.2.24.** ([9, Theorem 3.2]) *Let $R$ be a commutative ring and $M$ an $R$-module such that $M(N)$ is a submodule of $M$.*

(1) *Let* $\Gamma$ *be an induced subgraph of* $\overline{M}(\Gamma_N(M))$, *and let* $m$ *and* $m'$ *be distinct vertices of* $\Gamma$ *which are connected by a path in* $\Gamma$. *Then there is a path in* $\Gamma$ *of length 2 between* $m$ *and* $m'$. *In particular, if* $\overline{M}(\Gamma_N(M))$ *is connected, then diam*$(\overline{M}(\Gamma_N(M))) \leq 2$;

(2) *Let* $m$ *and* $m'$ *be distinct elements of* $\overline{M}(\Gamma_N(M))$ *that are connected by a path. If* $m$ *and* $m'$ *are not adjacent, then* $m-(-m)-m'$ *and* $m-(-m')-m'$ *are paths of length 2 between* $m$ *and* $m'$ *in* $\overline{M}(\Gamma_N(M))$.

**Theorem 9.2.25.** ([9, Theorem 3.3]) *Let* $R$ *be a commutative ring and* $M$ *an* $R$*-module such that* $M(N)$ *is a submodule of* $M$. *Then the following statements are equivalent:*

(1) $\overline{M}(\Gamma_N(M))$ *is connected;*

(2) *Either* $m + m' \in M(N)$ *or* $m - m' \in M(N)$ *(but not both) for every* $m, m' \in M \setminus M(N)$;

(3) *Either* $m + m' \in M(N)$ *or* $m + 2m' \in M(N)$ *for every* $m, m' \in M \setminus M(N)$. *In particular, if (3) holds, then* $2m \in M(N)$ *or* $3m \in M(N)$ *for every* $m \in M \setminus M(N)$.

The next theorem gives the complete structure of $\overline{M}(\Gamma_N(M))$ when $M(N)$ is a proper submodule of $M$.

**Theorem 9.2.26.** ([9, Theorem 3.4]) *Let* $R$ *be a commutative ring,* $M$ *an* $R$*-module, and* $N$ *a proper submodule of* $M$ *such that* $M(N)$ *is a proper submodule of* $M$, *and let* $|M(N)| = \alpha$ *and* $|M/M(N)| = \beta$ *(we allow* $\alpha$ *and* $\beta$ *to be infinite).*

(1) *If* $2 \in S(N : M)$, *then* $\overline{M}(\Gamma_N(M))$ *is a disjoint union of* $\beta - 1$ *copies of* $K_\alpha$;

(2) *If* $2 \notin S(N : M)$, *then* $\overline{M}(\Gamma_N(M))$ *is a disjoint union of* $(\beta - 1)/2$ *copies of* $K_{\alpha,\alpha}$.

Next, we discuss the connectedness and diameter of $\overline{M}(\Gamma_N(M))$ when $M(N)$ is a proper submodule of $M$.

**Theorem 9.2.27.** ([9, Theorem 3.8]) *Let* $R$ *be a commutative ring,* $M$ *an* $R$*-module, and* $N$ *a proper submodule of* $M$ *such that* $M(N)$ *is a proper submodule of* $M$.

(1) $\overline{M}(\Gamma_N(M))$ *is complete if and only if* $|M/M(N)| = 2$ *or* $|M| = |M/M(N)| = 3$;

(2) $\overline{M}(\Gamma_N(M))$ *is connected if and only if* $|M/M(N)| = 2$ *or* $|M/M(N)| = 3$;

(3) $\overline{M}(\Gamma_N(M))$ *(and hence* $T(\Gamma_N(M))$ *and* $M(\Gamma_N(M))$*) are totally disconnected if and only if* $M(N) = \{0\}$ *and* $2 \in S(N : M)$.

**Theorem 9.2.28.** ([9, Theorem 3.9]) *Let $R$ be a commutative ring, $M$ an $R$-module, and $N$ a proper submodule of $M$ such that $M(N)$ is a submodule of $M$. Then $diam(\overline{M}(\Gamma_N(M))) = 0, 1, 2,$ or $\infty$. In particular, if $\overline{M}(\Gamma_N(M))$ is connected, then $diam(\overline{M}(\Gamma_N(M))) \leq 2$.*

**Theorem 9.2.29.** ([9, Theorem 3.10]) *Let $R$ be a commutative ring, $M$ an $R$-module, and $N$ a proper submodule of $M$ such that $M(N)$ is a submodule of $M$. Then the following conditions hold:*

(1) *$diam(\overline{M}(\Gamma_N(M))) = 0$ if and only if $M(N) = \{0\}$ and $|M| = 2$;*
(2) *$diam(\overline{M}(\Gamma_N(M))) = 1$ if and only if either $M(N) \neq \{0\}$ and $|M/M(N)| = 2$, or $M(N) = \{0\}$ and $|M| = 3$;*
(3) *$diam(\overline{M}(\Gamma_N(M))) = 2$ if and only if $M(N) \neq \{0\}$ and $|M/M(N)| = 3$;*
(4) *Otherwise, $diam(\overline{M}(\Gamma_N(M))) = \infty$.*

The next result completely characterizes the girth of $T(\Gamma_N(M))$ and $\overline{M}(\Gamma_N(M))$ when $M(N)$ is a proper submodule of $M$.

**Theorem 9.2.30.** ([9, Theorem 3.12]) *Let $R$ be a commutative ring, $M$ an $R$-module, and $N$ a proper submodule of $M$ such that $M(N)$ is a submodule of $M$.*

(1) (a) *$gr(\overline{M}(\Gamma_N(M))) = 3$ if and only if $2 \in S(N : M)$ and $|M(N)| \geq 3$;*
    (b) *$gr(\overline{M}(\Gamma_N(M))) = 4$ if and only if $2 \notin S(N : M)$ and $|M(N)| \geq 2$;*
    (c) *Otherwise, $gr(\overline{M}(\Gamma_N(M))) = \infty$;*
(2) (a) *$gr(T(\Gamma_N(M))) = 3$ if and only if $|M(N)| \geq 3$;*
    (b) *$gr(T(\Gamma_N(M))) = 4$ if and only if $2 \notin S(N : M)$ and $|M(N)| = 2$;*
    (c) *Otherwise, $gr(T(\Gamma_N(M))) = \infty$.*

Finally, we study $T(\Gamma_N(M))$ when $M(N)$ is not a submodule of $M$. In this case, $M(\Gamma_N(M))$ and $\overline{M}(\Gamma_N(M))$ are never disjoint. Also, we determine when $T(\Gamma_N(M))$ is connected and compute $diam(T(\Gamma_N(M)))$.

**Theorem 9.2.31.** ([9, Theorem 4.1]) *Let $R$ be a commutative ring, $M$ an $R$-module, and $N$ a proper submodule of $M$ such that $M(N)$ is not a submodule of $M$.*

(1) *$M(\Gamma_N(M))$ is connected with $diam(M(\Gamma_N(M))) = 2$;*
(2) *Some vertex of $M(\Gamma_N(M))$ is adjacent to a vertex of $\overline{M}(\Gamma_N(M))$. In particular, the subgraphs $M(\Gamma_N(M))$ and $\overline{M}(\Gamma_N(M))$ are not disjoint;*
(3) *If $\overline{M}(\Gamma_N(M))$ is connected, then $T(\Gamma_N(M))$ is connected.*

**Theorem 9.2.32.** ([9, Theorem 4.2]) *Let $R$ be a commutative ring, $M$ an $R$-module, and $N$ a proper submodule of $M$ such that $M(N)$ is not a submodule of $M$. Then $T(\Gamma_N(M))$ is connected if and only if $M = (M(N))$.*

**Theorem 9.2.33.** ([9, Theorem 4.3]) *Let $R$ be a commutative ring, $M$ an $R$-module, and $N$ a proper submodule of $M$ such that $M(N)$ is not a submodule of $M$, and let $n \geq 2$ be the least integer such that $M = (m_1, \ldots, m_n)$ for some $m_1, \ldots, m_n \in M(N)$ (that is, $T(\Gamma_N(M))$ is connected). If $n$ is an even integer, then $diam(T(\Gamma_N(M))) \leq n$. Otherwise, $diam(T(\Gamma_N(M))) \leq n+1$. In particular, if $M$ is a cyclic $R$-module, then $diam(T(\Gamma_N(M))) = n$ or $n+1$.*

**Theorem 9.2.34.** ([9, Theorem 4.4]) *Let $R$ be a commutative ring, $M$ an $R$-module, and $N$ a proper submodule of $M$ such that $M(N)$ is not a submodule of $M$, and let $n \geq 2$ be the least integer such that $M = (m_1, \ldots, m_n)$ for some $m_1, \ldots, m_n \in M(N)$.*

(1) *If $M$ is a cyclic module generated by $m$, then $diam(T(\Gamma_N(M))) = d(0, m)$ or $d(0, m) + 1$;*
(2) *If $diam(T(\Gamma_N(M))) = n$ or $n + 1$ and $M$ is a cyclic $R$-module, then $diam(M(\Gamma_N(M))) \geq n - 2$ or $n - 1$, respectively.*

The last theorem of this section investigates $gr(M(\Gamma_N(M)))$, $gr(\overline{M}(\Gamma_N(M)))$, and $gr(T(\Gamma_N(M)))$ when $M(N)$ is not a submodule of $M$.

**Theorem 9.2.35.** ([9, Theorem 4.5]) *Let $R$ be a commutative ring, $M$ an $R$-module, and $N$ a proper submodule of $M$ such that $M(N)$ is not a submodule of $M$.*

(1) *If $N \neq \{0\}$, then $gr(M(\Gamma_N(M))) = 3$. Otherwise, either $gr(M(\Gamma_N(M))) = 3$ or $gr(M(\Gamma_N(M))) = \infty$;*
(2) *$gr(T(\Gamma_N(M))) = 3$ if and only if $gr(M(\Gamma_N(M))) = 3$;*
(3) *The (induced) subgraph of $M(\Gamma_N(M))$ with vertices $N$ is complete. Hence $gr(M(\Gamma_N(M))) = 3$ when $|N| \geq 3$;*
(4) *If $gr(T(\Gamma_N(M))) = 4$, then $gr(M(\Gamma_N(M))) = \infty$;*
(5) *If $N \neq \{0\}$ and $2 \in (N : M)$, then $gr(\overline{M}(\Gamma_N(M))) = 3$ or $\infty$;*
(6) *If $2 \notin (N : M)$, then $gr(\overline{M}(\Gamma_N(M))) = 3, 4$ or $\infty$.*

## 9.3 Commutative Semirings

As a generalization of rings, semirings have proven to be useful tools in various disciplines. They have important applications in mathematics and theoretical computer science.

A commutative semiring $R$ is defined as an algebraic system $(R, +, \cdot)$ such that $(R, +)$ and $(R, \cdot)$ are commutative semigroups, connected by $a(b + c) = ab + ac$ for every $a, b, c \in R$, and there are $0, 1 \in R$ such that $r + 0 = r$, $r0 = 0r = 0$, and $r1 = 1r = r$ for every $r \in R$. Ideals of semirings play a central role in the structure theory and are useful for many purposes. However, they do not in general coincide with the ideals of rings and, for this reason, their use is somewhat limited in trying to obtain analogues of ring theorems for semirings. Indeed, many results in rings apparently have no analogues in semirings using only ideals. In order to overcome this deficiency, the authors defined a more restricted class of ideals in semirings, which are called "$k$-ideals" and "$Q$-ideals".

Throughout this section, $R$ is a commutative semiring with identity. In order to make this section easier to follow, we first recall various notions from semiring theory which will be used in the sequel. For further definitions related to semirings and ideals, see Section 1.1, Section 6.3, and [210].

- A semiring $R$ is said to be a *semidomain* if $a, b \in R$ with $ab = 0$ implies that $a = 0$ or $b = 0$.
- A semiring $R$ is called a *co-semidomain* if $a+b = 1$ for $a, b \in R$, then $a = 1$ or $b = 1$.
- A nonempty subset $I$ of a semiring $R$ is called a *co-ideal* if it is closed under multiplication and satisfies the condition $r + a \in I$ for every $a \in I$ and $r \in R$. A co-ideal $I$ of $R$ is called a *strong co-ideal* if $1 \in I$ (clearly, $0 \in I$ if and only if $I = R$).
- A proper co-ideal $P$ of a semiring $R$ is called *prime* if $x+y \in P$ for $x, y \in R$, then $x \in P$ or $y \in P$.
- A *subtractive ideal* (= $k$-ideal) $I$ is an ideal of a semiring $R$ such that if $x, x + y \in I$, then $y \in I$. Thus $\{0\}$ is a $k$-ideal of $R$, and every subtractive co-ideal is a strong co-ideal.
- An element $x$ of a semiring $R$ is called a *zero-sum element* in $R$ if $x + y = 0$ for some $y \in R$. Let $S(R)$ be the set of all zero-sum elements of $R$.
- A semiring $R$ is an *identity-semiring* (= I-semiring) if $1 + x = 1$ for every $x \in R$.
- A semiring $R$ is called an *R-cancellative semiring* if whenever $rs = rt$ for $s, t, r \in R$ with $r \neq 0$, then $s = t$.
- An ideal $I$ of a semiring $R$ is called a *partitioning ideal* (= $Q$-ideal) if there is a $Q \subseteq R$ such that
  (a) $R = \bigcup\{q + I \mid q \in Q\}$,
  (b) If $q_1, q_2 \in Q$, then $(q_1 + I) \cap (q_2 + I) \neq \emptyset$ if and only if $q_1 = q_2$.

### 9.3.1 Total Graphs of Semirings

Let $R$ be a commutative semiring and $Z(R)$ its set of zero-divisor elements. The *total graph of a commutative semiring*, denoted by $T(\Gamma(R))$, is the simple graph with vertex set $R$, and distinct vertices $x$ and $y$ are adjacent if and only if $x+y \in Z(R)$. Let $Reg(\Gamma(R))$ be the (induced) subgraph of $T(\Gamma(R))$ with vertices $Reg(R)$, and $Z(\Gamma(R))$ be the (induced) subgraph of $T(\Gamma(R))$ with vertices $Z(R)$.

First, we study the total graph $T(\Gamma(R))$ when $Z(R)$ is not an ideal of $R$.

**Theorem 9.3.1.** ([112, Theorem 3.2]) *Let $R$ be a commutative semiring.*

(1) *$T(\Gamma(R))$ is complete if and only if $Z(R) = R$;*

(2) *$T(\Gamma(R))$ is totally disconnected if and only if $Z(R) = S(R) = \{0\}$.*

**Theorem 9.3.2.** ([112, Theorem 3.4]) *Let $R$ be a semiring such that $Z(R)$ is not an ideal of $R$. Then $Z(\Gamma(R))$ is connected with $diam(Z(\Gamma(R))) = 2$.*

A semiring $R$ is called a *subtractive semiring* if every cyclic ideal of $R$ is a $k$-ideal.

**Theorem 9.3.3.** ([112, Theorem 3.9]) *Let $R$ be a subtractive semiring which is not a ring. If $|S(R)| \geq 3$, then $gr(Z(\Gamma(R))) = 3$.*

**Theorem 9.3.4.** ([112, Theorem 3.10]) *Let $R$ be a semiring such that $Z(R)$ is not an ideal of $R$. Then either $gr(Z(\Gamma(R))) = 3$ or $gr(Z(\Gamma(R))) = \infty$.*

**Theorem 9.3.5.** ([112, Theorem 3.12]) *Let $R$ be a semiring such that $Z(R)$ is not an ideal of $R$. Then $gr(Reg(\Gamma(R))) = 3$ or $\infty$.*

Let $R$ be a commutative semiring. The structure of the total graph $T(\Gamma(R))$ may be completely described when the zero-divisors form an ideal. The basic observation is that if $Z(R)$ is a $k$-ideal (resp., $Z(R)$ is not a $k$-ideal), then the subgraph $Z(\Gamma(R))$ is disjoint from $Reg(\Gamma(R))$ (resp., $Z(\Gamma(R))$ is not disjoint from $Reg(\Gamma(R))$). Thus we will concentrate on the subgraph $Reg(\Gamma(R))$.

**Theorem 9.3.6.** ([112, Theorem 4.6]) *Let $R$ be a commutative semiring such that $Z(R)$ is a $k$-ideal of $R$. If $r$ and $r'$ are distinct elements of $Reg(R)$ that are connected by a path with $r + r' \notin Z(R)$ (i.e., if $r$ and $r'$ are not adjacent), then there is a path in $Reg(\Gamma(M))$ of length at most 2 between $r$ and $r'$.*

**Theorem 9.3.7.** ([112, Theorem 4.7]) *Let $R$ be a commutative semiring.*

(1) *If $Z(R)$ is a $k$-ideal of $R$, then $Z(\Gamma(R))$ is disjoint from $Reg(\Gamma(R))$;*

(2) *If $Z(R)$ is not a $k$-ideal of $R$, then $Z(\Gamma(R))$ is not disjoint from $Reg(\Gamma(R))$.*

The next result provides a general structure theorem for $Reg(\Gamma(R))$ when $Z(R)$ is a $k$-ideal or $Z(R)$ is a $Q$-ideal.

**Theorem 9.3.8.** ([112, Theorem 4.8]) *Let $R$ be a commutative semiring and $|Z(R)| = \alpha$.*

(1) *If $Z(R)$ is a $k$-ideal of $R$ and $2 \in Z(R)$, then $Reg(\Gamma(R))$ is the union of disjoint complete subgraphs;*

(2) *If $Z(R)$ is a $k$-ideal of $R$ and $2 \notin Z(R)$, then $Reg(\Gamma(R))$ is the union of totally disconnected subgraphs and some connected subgraphs;*

(3) *If $Z(R)$ is a $Q$-ideal of $R$, $|Q \setminus Z(R)| = \beta$, and $2 \in Z(R)$, then $Reg(\Gamma(R))$ is the union of disjoint $K_\lambda$'s such that $\lambda \leq \alpha$;*

(4) *If $Z(R)$ is a $Q$-ideal of $R$, $|Q\backslash Z(R)| = \beta$, and $2 \notin Z(R)$, then $Reg(\Gamma(R))$ is the union of totally disconnected subgraphs and complete bipartite subgraphs.*

The following two results provide the diameter and girth of $Reg(\Gamma(R))$, respectively.

**Theorem 9.3.9.** ([112, Theorem 4.11]) *Let $R$ be a commutative semiring.*

(1) *If $Z(R)$ is a $k$-ideal of $R$, then $diam(Reg(\Gamma(R))) = 0$ if and only if $Z(R) = \{0\}$ and $|R| = 2$;*

(2) *Let $Z(R)$ be a $Q$-ideal of $R$.*

  (a) *$diam(Reg(\Gamma(R))) = 1$ if and only if $2 \in Z(R)$ and $|R/Z(R)| = 2$;*

  (b) *$diam(Reg(\Gamma(R))) = 2$ if and only if $|R/Z(R)| = 3$, $2 \notin Z(R)$, and $q + q' \in Z(R)$ for every $q, q' \in Q \setminus Z(R)$;*

  (c) *Otherwise, $diam(Reg(\Gamma(R))) = \infty$.*

**Theorem 9.3.10.** ([112, Theorem 4.15]) *Let $R$ be a commutative semiring such that $Z(R)$ is a $k$-ideal of $R$. Then the following statements hold:*

(1) *$gr(T(\Gamma(R))) = 3$ if and only if $|Z(R)| \geq 3$;*

(2) *$gr(T(\Gamma(R))) = 4$ if and only if $2 \notin Z(R)$, $|Z(R)| < 3$, and $r + r' \in Z(R)$ for some $r, r' \in Reg(R)$;*

(3) *Otherwise, $gr(T(\Gamma(R))) = \infty$.*

### 9.3.2 Identity Summand Graphs

An element $r$ of a commutative semiring $R$ with identity is said to be an *identity-summand* if there is a $1 \neq a \in R$ such that $r + a = 1$. In this section, we introduce and investigate the identity-summand graph of $R$ and the total graph of $R$ with respect to the identity-summand elements.

The *identity-summand graph* of $R$, denoted by $IS(R)$, is the simple graph whose vertices are the nonidentity identity-summands of $R$, and two distinct vertices are joined by an edge when the sum of the vertices is 1. Note that $IS(R) = \emptyset$ if and only if $R$ is a co-semidomain.

Let $R$ be a commutative semiring with identity. Let $S(R) = \{r \in R \mid r + a = 1 \text{ for some } 1 \neq a \in R\}$ be the set of all identity-summand elements of $R$ and $S^*(R) = S(R) \setminus \{1\}$. Recall that a semiring $R$ is said to be an *identity-semiring* (=$I$-semiring) if $1 + x = 1$ for every $x \in R$. One can easily show that $R$ is an $I$-semiring if and only if $\{1\}$ is a strong co-ideal of $R$.

**Example 9.3.11.** ([114, Example 2.2]) (a) Let $X = \{a, b, c\}$ and $R = (P(X), \cup, \cap)$, a semiring with $1_R = X$, where $P(X)$ is the set of all subsets of $X$. It is easy to see that $R$ is an $I$-semiring.

(b) Let $\mathbb{Z}^+$ be the set of all nonnegative integers and let $R = \mathbb{Z}^+ \times \mathbb{Z}^+$. Then $(R, +, \cdot)$ with the usual operations of addition and multiplication is a semiring which is not an $I$-semiring.

The following result investigates properties of $I$-semirings and their minimal prime co-ideals.

**Theorem 9.3.12.** ([114, Theorem 2.8]) *Let $R$ be an $I$-semiring.*

(1) *If $\{P_\alpha\}_{\alpha\in\Lambda}$ is the set of all prime co-ideals of $R$, then $\cap_{\alpha\in\Lambda} P_\alpha = \{1\}$;*

(2) *If $P_1, \ldots, P_n$ are the only distinct minimal prime co-ideals of $R$, then $\cap_{i=1}^n P_i = \{1\}$ and $\{1\} \neq \cap_{1\le i\le n, i\neq j} P_i$ for every $1 \le j \le n$.*

We now consider the following question: If $R$ is a semiring, is $IS(R)$ connected? The following is an example of a semiring $R$ where $IS(R)$ is not connected.

**Example 9.3.13.** ([114, Example 3.2]) Let $\mathbb{Z}^+$ be the set of all nonnegative integers and let $R = \mathbb{Z}^+ \times \mathbb{Z}^+ \times \mathbb{Z}^+$. Then $(R, +, \cdot)$ with the usual operations is a semiring and $S^*(R) = \{(1,0,0), (0,1,0), (0,0,1), (0,1,1), (1,0,1), (1,1,0)\}$. An inspection will show that $IS(R)$ is not connected.

It is important to be able to recognize when the identity-summand graph of a semirings is connected. The concept of "$R$ is an $I$-semiring" helps us to do this. The next theorem gives a more explicit description of the diameter of this graph.

**Theorem 9.3.14.** ([114, Theorem 3.3]) *Let $R$ be an $I$-semiring.*

(1) *$IS(R)$ is a connected graph with $diam(IS(R)) \le 3$;*

(2) *$IS(R)$ contains no loops;*

(3) *If $|S^*(R)| \ge 3$, then $IS(R)$ is not a complete graph;*

(4) *If $|S^*(R)| \ge 3$, then $diam(IS(R)) = 2$ or $3$.*

We give an example of a semiring $R$ for which $IS(R)$ is connected, but $R$ is not an $I$-semiring. Let $R = (\mathbb{Z}^+ \times \mathbb{Z}^+, +, \cdot)$. Then $S^*(R) = \{(1,0), (0,1)\}$ gives $IS(R)$ is connected, but $R$ is not an $I$-semiring.

The next result gives the girth of an identity-summand graph.

**Theorem 9.3.15.** ([114, Theorem 3.5]) *Let $R$ be an $I$-semiring. If $IS(R)$ contains a cycle, then $gr(IS(R)) \le 4$.*

Our next result characterizes the identity-summand graphs that are a cycle graph.

**Theorem 9.3.16.** ([114, Theorem 3.6]) *The only cycle identity-summand graph of a commutative $I$-semiring is $K_{2,2}$.*

Now we want to determine when the identity-summand graph $IS(R)$ is a complete bipartite graph. We start with the following definition.

Let $P$ be a prime co-ideal of an $I$-semiring $R$. We say that $P$ is an *associated prime co-ideal* of $R$ if there is an $x \in R$ with $P = (1 : x)$. The set of associated prime co-ideals of $R$ is denoted by *co-Ass*$(R)$.

The next result provides a finer relationship between vertices which are adjacent to all the other vertex and *co-Ass* of a semiring.

**Theorem 9.3.17.** ([114, Theorem 4.3]) *Let $R$ be an $I$-semiring and $a \in S^*(R)$ a vertex of $IS(R)$ which is adjacent to every other vertex in $IS(R)$. Then $(1 : a)$ is a maximal element in $\Delta = \{(1 : x) \mid x \in S^*(R)\}$ and $(1 : a) \in$ co-Ass$(R)$.*

**Theorem 9.3.18.** ([114, Theorem 4.4]) *Let $R$ be an $I$-semiring.*

(1) *If $IS(R)$ is a complete $r$-partite graph with $r \geq 3$, then at most one part has more than one vertex;*

(2) *If $IS(R)$ is a complete $r$-partite graph, then $IS(R)$ is a complete bipartite graph.*

**Theorem 9.3.19.** ([114, Theorem 4.5]) *Let $R$ be an $I$-semiring. Then $IS(R)$ is complete bipartite if and only if there are two distinct prime co-ideals $P_1$ and $P_2$ of $R$ such that $P_1 \cap P_2 = \{1\}$.*

**Theorem 9.3.20.** ([114, Theorem 4.6]) *Let $R$ be an $I$-semiring.*

(1) *If $P_1 = (1 : x_1)$ and $P_2 = (1 : x_2)$ are two distinct elements of co-Ass$(R)$, then $x_1 + x_2 = 1$;*

(2) *If co-Ass$(R) = \{P_1, P_2\}$, $|P_i| \geq 3$ $(i = 1, 2)$, and $P_1 \cap P_2 = \{1\}$, then $gr(IS(R)) = 4$;*

(3) *If $|$co-Ass$(R)| \geq 3$, then $gr(IS(R)) = 3$.*

**Theorem 9.3.21.** ([114, Theorem 4.8]) *Let $R$ be an $I$-semiring and $IS(R)$ a finite regular graph. Then $IS(R) = K_{n,n}$ for some $n \in \mathbb{N}$.*

**Theorem 9.3.22.** ([114, Theorem 4.9]) *Let $R_1$ and $R_2$ be $I$-co-semidomains, and let $R \subseteq R_1 \times R_2$ be an $I$-semiring with $|S^*(R)| \geq 3$. Then $IS(R)$ is a complete bipartite graph with diam$(IS(R)) = 2$.*

We next investigate the chromatic number and clique number of $IS(R)$.

**Theorem 9.3.23.** ([114, Theorem 5.2]) *Let $R$ be an $I$-semiring. Then the following are equivalent:*

(1) *The chromatic number $\chi(IS(R))$ is finite;*

(2) *The clique number $\omega(IS(R))$ is finite;*

(3) *The co-ideal $\{1\}$ is a finite intersection of prime co-ideals.*

**Theorem 9.3.24.** ([114, Theorem 5.4]) *Let $R$ be an $I$-semiring. Then $\omega(IS(R)) = |Min(R)|$, where $Min(R)$ is the set of minimal prime co-ideals of $R$.*

**Theorem 9.3.25.** ([114, Theorem 5.6]) *If $R$ is an $I$-semiring, then $\chi(IS(R)) = \omega(IS(R))$.*

The following theorem deals with the planarity of $IS(R)$.

**Theorem 9.3.26.** ([114, Theorem 6.1]) *Let $R$ be an $I$-semiring. If $|Min(R)| \geq 4$, then $IS(R)$ is not planar.*

**Remark 9.3.27.** ([114, Remark 6.3]) Let $R$ be an $I$-semiring.

(1) If $|Min(R)| = 1$, then $\{1\}$ is the only minimal prime co-ideal of $R$ by Theorem 9.3.12(2). Hence $R$ is a co-semidomain, and so $IS(R) = \emptyset$.

(2) If $|Min(R)| = 2$, then by Theorem 9.3.19, $IS(R) = K_{m,n}$ for some integers $m, n$, where $m = |P_1| - 1$ and $n = |P_2| - 1$. If $m, n \geq 3$, then $IS(R)$ is not planar.

(3) If $|Min(R)| \geq 4$, then $IS(R)$ is not planar by Theorem 9.3.26.

(4) Let $R = \mathbb{Z}^+ \times \mathbb{Z}^+ \times \mathbb{Z}^+$ be an $I$-semiring. Then $|Min(R)| = 3$ and $IS(R)$ is planar. However, there are $I$-semirings that have only three minimal prime co-ideals, and their identity-summand graphs are not planar. For example, let $X = \{a, b, c\}$ and $R = (P(X), \cup, \cap)$, a semiring with $1_R = X$. Then $S^*(R) = \{\{a\}, \{b\}, \{c\}, \{a, b\}, \{a, c\}, \{b, c\}\}$, and an inspection will show that $IS(R)$ is planar. So, it is not entirely clear for which semirings with $|Min(R)| = 3$, the graph $IS(R)$ is planar.

Let $R$ be a commutative semiring. As usual, an element $r \in R$ is an identity-summand of $R$ if there exists $1 \neq a \in R$ such that $r + a = 1$. Let $S(R)$ be to the set of elements of $R$ that are identity-summands and $S^*(R)$ the set of nonidentity identity-summands of $R$. Let $R$ be an $I$-semiring (i.e., $1 + r = 1$ for every $r \in R$). Studying $S(R)$ runs into the issue of a profound lack of algebraic structure, highlighted by a lack of closure under multiplication. This unfortunate lack of algebraic structure is the focus of our investigations in the rest of this section.

An $a \in R$ is called *co-regular* if $a$ is not an identity-summand element; so $Co\text{-}Reg(R) = R \setminus S(R)$.

Now we introduce the total graph of a semiring $R$ with respect to identity-summand elements. Let $R$ be an $I$-semiring. The *total graph of $R$ with respect to identity-summand elements*, denoted by $T(\Gamma(R))$, is the graph with all elements of $R$ as vertices, and distinct vertices $x$ and $y$ are adjacent if and only if $xy \in S(R)$. Also, $S(\Gamma(R))$ (resp., $Co\text{-}Reg(\Gamma(R))$, $S^*(\Gamma(R))$) denotes the subgraph of $T(\Gamma(R))$ with vertex set $S(R)$ (resp., $Co\text{-}Reg(R)$, $S^*(R)$).

In the following, we show that in general, for a semiring $R$, the total graph of identity-summand elements of $R$ and its subgraphs $S^*(\Gamma(R))$ and $Co\text{-}Reg(\Gamma(R))$ are not connected, but the subgraph $S(\Gamma(R))$ is always connected.

**Example 9.3.28.** ([115, Example 3.2]) (a) Let $R = (\mathbb{Z}^+ \times \mathbb{Z}^+, +, \cdot)$. Then $S^*(R) = \{(1,0),(0,1)\}$. It can be easily seen that $T(\Gamma(R))$, $S^*(\Gamma(R))$, and *Co-Reg*$(\Gamma(R))$ are not connected and $S(\Gamma(R))$ is connected.

(b) Let $R = (\mathbb{Z}^+ \times \mathbb{Z}^+ \times \mathbb{Z}^+, +, \cdot)$. Then $S^*(R) = \{(1,0,0),(0,1,0),(0,0,1),(1,1,0),(1,0,1),(0,1,1)\}$. It can be easily seen that $S^*(\Gamma(R))$ is connected and $R$ is not an $I$-semiring.

Among the subgraphs of $T(\Gamma(R))$ introduced above, $S(\Gamma(R))$ is connected in general. So we characterize the diameter and girth of $S(\Gamma(R))$ in the following theorems. After that, we use the "$I$-semiring" condition for $R$ to prove some important properties of $S(R)$, which improve our results about the total graph of $R$.

**Theorem 9.3.29.** ([115, Theorem 3.3,3.4]) *Let $R$ be a semiring. Then the following conditions hold:*

(1) *$S(\Gamma(R))$ is connected with $diam(S(\Gamma(R))) \in \{1,2\}$;*

(2) *$gr(S(\Gamma(R))) \in \{3,\infty\}$.*

**Lemma 9.3.30.** ([115, Lemma 3.6]) *Let $R$ be an $I$-semiring.*

(1) *For every $x, y \in R$, if $xy = 1$, then $x = 1$ and $y = 1$;*

(2) *If $\{1\} \neq S(R)$ is finite, then $S(R)$ is not a co-ideal of $R$;*

(3) *$S(R)$ is a union of prime co-ideals of $R$;*

(4) *Co-Reg$(R)$ is an ideal of $R$;*

(5) *$S(R)$ is a co-ideal of $R$ if and only if Co-Reg$(R)$ is a prime ideal of $R$.*

In the following theorem, we investigate the total graph of a commutative semiring based on identity-summand elements. It shows that the condition "$R$ is an $I$-semiring" is not enough to force $T(\Gamma(R))$ to be connected.

**Theorem 9.3.31.** ([115, Theorem 3.7]) *Let $R$ be an $I$-semiring. Then the following conditions hold:*

(1) *No element of Co-Reg$(R)$ is adjacent to any element of $S(R)$;*

(2) *Co-Reg$(\Gamma(R))$ is totally disconnected;*

(3) *$T(\Gamma(R))$ is always disconnected;*

(4) *$T(\Gamma(R))$ is the empty graph if and only if $R$ is a co-semidomain. Moreover, for every $I$-semiring $R$, Co-Reg$(\Gamma(R))$ is the empty graph.*

Note that if $R$ is an $I$-semiring, then $T(\Gamma(R))$ contains $|$*Co-Reg*$(R)|+1$ connected components. By the previous theorem, even if $R$ is an $I$-semiring, $T(\Gamma(R))$ is not connected. Hence, we investigate its connected subgraph $S(\Gamma(R))$.

**Theorem 9.3.32.** ([115, Theorem 3.9]) *Let $R$ be an $I$-semiring which is not a co-semidomain.*

(1) *$S(\Gamma(R))$ is complete if and only if $S(R)$ is a co-ideal of $R$;*
(2) *$diam(S(\Gamma(R))) = 1$ if and only if $S(R)$ is a co-ideal of $R$;*
(3) *$diam(S(\Gamma(R))) = 2$ if and only if $S(R)$ is not a co-ideal of $R$;*
(4) *$|S(R)| = 3$ if and only if $gr(S(\Gamma(R))) = \infty$;*
(5) *$|S(R)| \geq 4$ if and only if $gr(S(\Gamma(R))) = 3$.*

Let $R = (\mathbb{Z}^+, \gcd, \mathrm{lcm})$ with $\gcd(0,0) = 0$ and $\mathrm{lcm}(0,0) = 0$. It is clear that $S(R) = R \setminus \{0\}$ is a co-ideal of $R$ and $S(\Gamma(R))$ is a complete graph.

**Theorem 9.3.33.** ([115, Theorem 3.11]) *Let $R$ be an $I$-semiring.*

(1) *$\omega(S(\Gamma(R))) = \max\{|P_\alpha| \mid P_\alpha$ is a maximal element in the set of all co-ideals in which every element is an identity summand $\}$;*
(2) *$\omega(\Gamma(R)) \leq \omega(S(\Gamma(R)))$.*

In general, equality need not hold in Theorem 9.3.33. For example, let $R = (\{0, 1, 2, 3, 4, 5, 6, 10, 12, 15, 20, 30, 60\}, \gcd, \mathrm{lcm})$. We can easily see that $\omega(\Gamma(R)) = 3$ and $\omega(S(\Gamma(R))) = 5$.

**Theorem 9.3.34.** ([115, Theorem 3.13]) *Let $R$ be an $I$-semiring. Then $S(\Gamma(R))$ is not a cycle graph.*

Next, we refine our results on $diam(S^*(\Gamma(R)))$, $gr(S^*(\Gamma(R)))$, and the relationship between $S^*(\Gamma(R))$ and $S(\Gamma(R))$. At first, we prove one of the most important properties of $S(R)$ which will be used in the sequel. We prove this property of $S(R)$ for $I$-semirings, and we show that it need not be true when $R$ is not an $I$-semiring.

**Theorem 9.3.35.** ([115, Theorem 4.1]) *Let $R$ be an $I$-semiring. Then $S(R) = \cup P_\alpha$, where the $P_\alpha$'s are minimal prime co-ideals of $R$.*

In the following example, it is shown that the condition "$R$ is an $I$-semiring" in Theorem 9.3.35 cannot be omitted.

**Example 9.3.36.** ([115, Example 4.2]) Let $R = (\mathbb{Z}^+ \times \mathbb{Z}^+, +, \cdot)$. Let $I$ be a co-ideal of $R$ and $(a, b) \in I$. Then $(a + r, b + s) \in I$ for every $(r, s) \in R$. So every co-ideal of $R$ is infinite. It can easily be seen that $S(R) = \{(1, 1), (1, 0), (0, 1)\}$. So $S(R) \neq \cup P_\alpha$, where the $P_\alpha$'s are minimal prime co-ideals of $R$.

**Theorem 9.3.37.** ([115, Theorem 4.4]) *Let $R$ be an $I$-semiring which is not a co-semidomain. Then $S^*(\Gamma(R))$ is connected if and only if $|Min(R)| \neq 2$. Moreover, if $S^*(\Gamma(R))$ is connected, then $diam(S^*(\Gamma(R))) \in \{1, 2\}$.*

The next theorem shows the relationship between the prime co-ideals which are contained in $S(R)$ and the prime ideals which contain *Co-Reg*$(R)$. This theorem helps to characterize the relationship between $diam(\Gamma(R))$, $|Min(R)|$, and the connectivity of $S^*(\Gamma(R))$ in Theorems 9.3.39 and 9.3.41.

**Theorem 9.3.38.** ([115, Theorem 4.5]) *Let $R$ be an $I$-semiring and $S(R) = \cup_{i\in\Lambda} P_i$.*

(1) *$Q_i = R\backslash P_i$ is a prime $k$-ideal of $R$ for every $i \in \Lambda$;*

(2) *$Co\text{-}Reg(R) = \cap_{i\in\Lambda} Q_i$.*

**Theorem 9.3.39.** ([115, Proposition 4.6]) *Let $R$ be an $I$-semiring such that the diameter of the identity-summand graph $\Gamma(R)$ is 2.*

(1) *If there is an element in $S(R)$ that is contained in the unique minimal prime co-ideal of $R$, then $|Min(R)| = 2$;*

(2) *If $Min(R)$ is finite, then $|Min(R)| = 2$.*

The following example shows that Theorem 9.3.39 is not true when $diam(\Gamma(R)) = 3$.

**Example 9.3.40.** ([115, Example 4.7]) Let $R = (\{0, 1, 2, 3, 4, 5, 6, 10, 12, 15, 20, 30, 60\}, \gcd, \text{lcm})$ (take $\gcd(0, 0) = 0$ and $\text{lcm}(0, 0) = 0$). Then $S(R) = P_1 \cup P_2 \cup P_3$, where $P_1 = \{1, 2, 4, 5, 10, 20\}$, $P_2 = \{1, 2, 3, 4, 6, 12\}$, and $P_3 = \{1, 3, 5, 15\}$ are minimal prime co-ideals of $R$. Then $diam(\Gamma(R)) = 3$ and $P_3$ is the unique minimal prime co-ideal of $R$ which contains 15.

**Theorem 9.3.41.** ([115, Theorem 4.8]) *Let $R$ be an $I$-semiring which is not a co-semidomain. If $R$ has only finitely many minimal prime co-ideals, then*

(1) *$diam(\Gamma(R)) = 2$ if and only if $S^*(\Gamma(R))$ is not connected and $|S^*(R)| \geq 3$;*

(2) *$diam(\Gamma(R)) = 3$ if and only if $S^*(\Gamma(R))$ is connected.*

The next theorem shows that $gr(S^*(\Gamma(R))) \in \{3, \infty\}$ when $R$ is an $I$-semiring.

**Theorem 9.3.42.** ([115, Theorem 4.9]) *Let $R$ be an $I$-semiring. Then $gr(S^*(\Gamma(R))) \in \{3, \infty\}$.*

**Theorem 9.3.43.** ([115, Theorem 4.10]) *Let $R$ be an $I$-semiring which is not a co-semidomain and $S^*(\Gamma(R))$ be a complete graph. Then $Min(R)$ is infinite.*

**Theorem 9.3.44.** ([115, Theorem 4.11]) *Let $R$ be an $I$-semiring which is not a co-semidomain.*

(1) *If $|Min(R)| \neq 2$, then $diam(S^*(\Gamma(R))) = diam(S(\Gamma(R)))$;*

(2) *If $|S(R)| \neq 4, 5$, then $gr(S(\Gamma(R))) = gr(S^*(\Gamma(R)))$.*

The following example shows that if $|S(R)| = 4$ or 5, then we may have $gr(S(\Gamma(R))) \neq gr(S^*(\Gamma(R)))$.

**Example 9.3.45.** ([115, Example 4.12]) (a) Let $R = (\{0, 1, 2, 4, 5, 10, 20\}, \gcd, \text{lcm})$ with the assumption that $\gcd(0, 0) = 0$ and $lcm(0, 0) = 0$. Then $S(R) = \{1, 2, 4, 5\}$. We can easily see that $gr(S(\Gamma(R))) = 3$ and $gr(S^*(\Gamma(R))) = \infty$.

(b) Let $R = (\{0, 1, 2, 3, 4, 6, 9, 12, 18, 36\}, \gcd, lcm)$ with $\gcd(0, 0) = 0$ and $lcm(0, 0) = 0)$. Then $S(R) = \{1, 2, 3, 4, 9\}$, $gr(S(\Gamma(R))) = 3$, and $gr(S^*(\Gamma(R))) = \infty$.

The following theorem shows the relationship between the girth of the identity-summand graph $\Gamma(R)$ and $S^*(\Gamma(R))$.

**Theorem 9.3.46.** ([115, Theorem 4.13]) *Let $R$ be an $I$-semiring which is not a co-semidomain. Then $gr(S^*(\Gamma(R))) = gr(\Gamma(R))$ if and only if one of the following conditions hold:*

(1) $|Min(R)| \neq 2$;

(2) $Min(R) = \{P_1, P_2\}$ *with* $|P_1| + |P_2| \leq 5$.

# Chapter 10

# Other Graphs Associated with Rings

In the literature, one can find a number of different types of graphs attached to rings. In this chapter, we collect some of these graphs.

## 10.1 Graphs of Commutative Rings

In this section, we introduce a few graph structures that are constructed from commutative rings. Let us start with the well-known Cayley graphs.

### 10.1.1 Cayley Graphs

The definition of Cayley graph was introduced by Arthur Cayley in 1878 to explain the concept of abstract groups which are described by a set of generators. In the last 50 years, the theory of Cayley graphs has grown into a substantial branch of algebraic graph theory.

**Definition 10.1.1.** Let $G$ be a group with identity element $e$ and $\Omega$ a nonempty subset of $G$ such that $e \notin \Omega$ and $g^{-1} \in \Omega$ for every $g \in \Omega$. The *Cayley graph* $Cay(G, \Omega)$ is the simple graph with vertex set $G$, and two distinct vertices $g$ and $h$ are adjacent if and only if $g^{-1}h \in \Omega$. A *circulant graph* is the Cayley graph $Cay(\mathbb{Z}_n^+, \Omega)$ for some $\Omega \subseteq \mathbb{Z}_n \setminus \{0\}$ such that $\Omega = \{-x \mid x \in \Omega\}$.

Cayley graphs have found many useful applications in solving and understanding a variety of problems of scientific interest (see the survey article [231] and the monograph [229]).

In [230], the authors studied the Cayley graph of a semigroup.

**Definition 10.1.2.** ([230]) For a semigroup $H$ and a subset $S$ of $H$, the *Cayley graph* $Cay(H, S)$ of $H$ relative to $S$ is defined as the graph with vertex set $H$ and edge set $E(H, S)$ consisting of those ordered pairs $(x, y)$ such that $sx = y$ for



D. F. Anderson et al., *Graphs from Rings*, https://doi.org/10.1007/978-3-030-88410-9_10

some $s \in S$. In the undirected Cayley graph $\overline{Cay}(H, S)$, a vertex $x$ is adjacent to $y$ if and only if $(x, y)$ or $(y, x)$ is an element of $E(H, S)$.

Let $R$ be a commutative ring. Note that the subgraph of the undirected Cayley graph $\overline{Cay}(R^*, R^*)$ consisting of nonunit vertices is a subgraph of the complement of the co-maximal graph $\Gamma_c(R)$ (see Section 10.1.2). Moreover, the two subgraphs of the undirected Cayley graph $\overline{Cay}(H, S)$ and the co-maximal graph $\Gamma_c(R)$ consisting of unit elements of $R$ are isomorphic.

In 2009, Akhtar et al. [44] studied the unitary Cayley graph associated to an arbitrary finite commutative ring.

**Definition 10.1.3.** ([44]) Let $R$ be a commutative ring and $U(R)$ the set of units of $R$. The *unitary Cayley graph* of $R$, denoted by $Cay(R, U(R))$, is the graph whose vertex set is $R$, and two distinct vertices $x$ and $y$ are adjacent if and only if $x - y \in U(R)$.

This construction was introduced in [197], although it does not appear to have been considered in [237]. In [44], Akhtar et al. determined precisely its diameter, girth, eigenvalues, vertex and edge connectivity, and vertex and edge chromatic numbers. They also computed its automorphism group and settled a question of Klotz and Sander. For some other papers on unitary Cayley graphs, see [221, 235, 237, 318].

Let $R$ be a commutative ring and $R^+$ and $Z(R)^*$ be the additive group and the set of all nonzero zero-divisors of $R$, respectively. In [2], Aalipour et al. studied the Cayley graph $Cay(R^+, Z(R)^*)$ which is denoted by $\mathbb{CAY}(R)$. Also, $Reg(\mathbb{CAY}(R))$ denotes the induced subgraph on the regular elements of $R$. G. Aalipour et al. [2] showed that for every zero-dimensional nonlocal ring $R$, $\mathbb{CAY}(R)$ is a connected graph with diameter 2. Moreover, for a finite ring $R$, they obtained the vertex connectivity and edge connectivity of $\mathbb{CAY}(R)$. They showed that if $R$ is a Noetherian ring and $Reg(\mathbb{CAY}(R))$ has no infinite cliques, then $R$ is finite. Furthermore, for every finite ring $R$, the clique number and the chromatic number of $Reg(\mathbb{CAY}(R))$ are determined.

Grimaldi [211] defined a graph $G(\mathbb{Z}_n)$ based on the elements and units of $\mathbb{Z}_n$. The vertices of $G(\mathbb{Z}_n)$ are the elements of $\mathbb{Z}_n$, and distinct vertices $x$ and $y$ are adjacent if and only if $x + y$ is a unit of $\mathbb{Z}_n$. For a positive integer $m$, it follows that $G(\mathbb{Z}_{2m})$ is a $\varphi(2m)$-regular graph, where $\varphi$ is the Euler phi function. If $m \geq 2$, then $G(\mathbb{Z}_{2m})$ is the union of $\varphi(2m)/2$ Hamiltonian cycles. The odd case is not very easy, but the structure is clear and the results are similar to the even case.

In 2012, Afkhami et al. [20] associated a graph to a commutative ring which generalizes the Cayley graph.

**Definition 10.1.4.** ([20]) Let $R$ be a commutative ring. For a natural number $n$, the graph $\Gamma_R^n$ is the simple graph with $R^n \setminus \{0\}$ as the vertex set, and two distinct vertices $X$ and $Y$ are adjacent if and only if there is an $n \times n$ lower triangular matrix $A$ over $R$ with nonzero main diagonal entries such that $AX^T = Y^T$ or $AY^T = X^T$, where $B^T$ is the transpose of $B$.

When we consider the ring $R$ as a semigroup with respect to multiplication, then $\Gamma_R^1$ is the usual undirected Cayley graph $\overline{Cay}(H, S)$, where $H = S = R \backslash \{0\}$. The authors of [20] studied some basic properties of $\Gamma_R^n$ and investigated the genus of the generalized Cayley graph $\Gamma_R^n$.

Let $A = \{S_i \mid i \in I\}$ be an arbitrary family of sets. The intersection graph $G(A)$ is the graph whose vertices are $S_i$, $i \in I$, and in which the vertices $S_i$, $S_j$ are adjacent if and only if $S_i \neq S_j$ and $S_i \cap S_j \neq \emptyset$. It is more interesting to study the intersection graphs $G(A)$ when the elements of $A$ have an algebraic structure. These interdisciplinary studies allow us to obtain characterizations and representations of especial classes of algebraic structures in terms of graphs and vice versa.

### 10.1.2 Co-maximal and Unit Graphs

In the classes of graphs from rings, there are graphs constructed out of the additive operator "+" of $R$. In fact, Sharma and Bhatwadekar [347] initiated the study of graphs using the addition of a commutative ring $R$ by defining a graph, later named the "co-maximal graph" of $R$ (cf. [276]).

**Definition 10.1.5.** ([276, 347]) Let $R$ be a commutative ring. The *co-maximal graph* of $R$, denoted by $\Gamma_c(R)$, is the simple graph with vertex set $R$, and two distinct vertices $x$ and $y$ are adjacent if and only if $Rx + Ry = R$. Let $\Gamma_1(R)$ be the subgraph of $\Gamma_c(R)$ generated by the units of $R$, and $\Gamma_2(R)$ be the subgraph of $\Gamma_c(R)$ generated by nonunits.

To see an example of a co-maximal graph, consider the ring $R = \frac{\mathbb{Z}_2[x,y]}{(x^2,y^2,xy)} = \{0, 1, x, y, 1+x, 1+y, 1+x+y\}$. The graph $\Gamma_c(R)$ is given in Figure 10.1. Also, the graph $\Gamma_2(\mathbb{Z}_{12})$ is shown in Figure 10.2.

It is easy to see that the invertible elements in $R$ are adjacent to all elements and that elements from the Jacobson radical $J(R)$ are adjacent only to the invertible ones. Namely, it follows directly from the definition of the co-maximal graph that ring elements $a$ and $b$ are adjacent if and only if they are not contained in the same maximal ideal of $R$. In [276], Maimani et al. pointed out that the subgraph $\Gamma_2(R) \setminus J(R)$ induced on the subset $R \setminus (U(R) \cup J(R))$ is the key to the co-maximal graph $\Gamma(R)$. Sharma et al. [347] showed that $\chi(\Gamma_c(R)) < \infty$ if and only if $R$ is a finite ring. In this case, $\chi(\Gamma_c(R)) = \omega(\Gamma_c(R)) = |Max(R)| + |U(R)|$. The co-maximal graph of a commutative ring has been studied in [224, 290, 423].

H.-J. Wang [424] extend the above concept to noncommutative rings with identity as follows: He defined a graph on $R$ with vertices as the nonunits of $R$, and distinct vertices $a$ and $b$ are adjacent if and only if $Ra + Rb = R$. Wang [424]

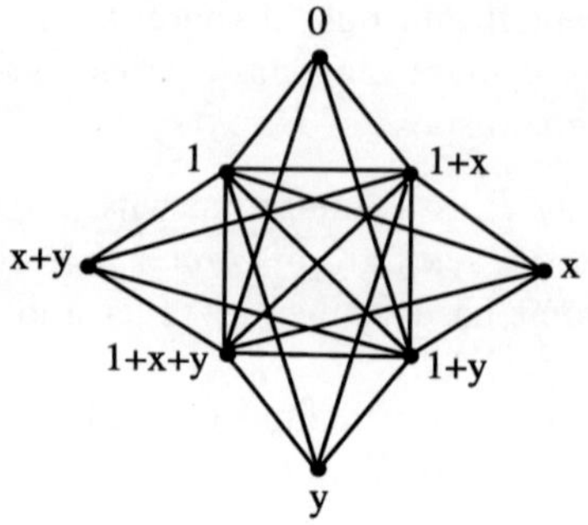


Figure 10.1: The co-maximal graph of $\frac{\mathbb{Z}_2[x,y]}{\langle x^2,y^2,xy\rangle}$

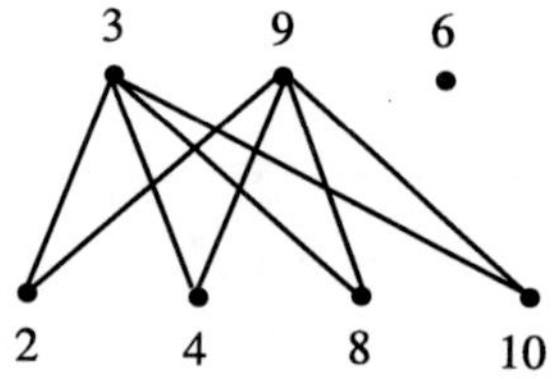


Figure 10.2: The graph $\Gamma_2(\mathbb{Z}_{12})$

proved that if $R$ is left Artinian, then $\Gamma_c(R)\backslash J(R)$ is connected, and if $\Gamma_c(R)\backslash J(R)$ is a forest, then $\Gamma_c(R) \setminus J(R)$ is a star graph.

In [440], a new kind of graph on a commutative ring $R$, namely the co-maximal ideal graph, was defined and studied.

**Definition 10.1.6.** ([424]) The *co-maximal ideal graph* of $R$, denoted by $\mathcal{C}(R)$, is the simple graph with vertices the proper ideals of $R$ which are not contained in the Jacobson radical of $R$, and two distinct vertices $I$ and $J$ are adjacent if and only if $I + J = R$.

The authors of [440] showed that this graph is a simple connected graph with diameter less than or equal to three, and both the clique number and the chromatic number of the graph are equal to the number of maximal ideals of $R$. For more details on the co-maximal ideal graph, see [289].

In 2010, Ashrafi et al. [91] introduced the unit graph of $R$, where $R$ is an arbitrary commutative ring and studied the properties of this graph.

**Definition 10.1.7.** ([91]) Let $R$ be a ring and $U(R)$ the set of units of $R$. The *unit graph* of $R$, denoted $G(R)$, is the graph with vertices all the elements of $R$, and distinct vertices $x$ and $y$ are adjacent if and only if $x + y \in U(R)$.

Figure 10.3 gives the unit graph of $\mathbb{Z}_5$.

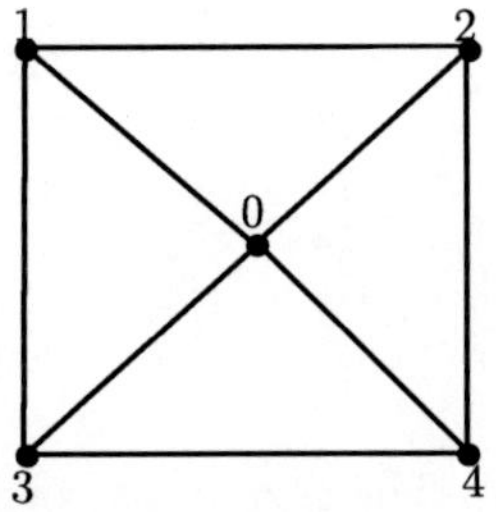


Figure 10.3: The unit graph of $\mathbb{Z}_5$

Let $R = S \times T$, where $S \cong \mathbb{Z}_2 \times \mathbb{Z}_2$ and $T$ is a ring. Then $G(R)$ is a union of two copies of vertex disjoint graphs $G(\mathbb{Z}_2 \times T)$, and therefore $G(R)$ is not connected. Note that the unit graph $G(R)$ is a complete graph if and only if $R$ is a division ring with $char(R) = 2$.

In [91], the authors investigated the basic properties of $G(R)$ and gave some results regarding connectedness, chromatic index, diameter, girth, and planarity for $G(R)$. Later, more properties of the unit graph of a ring and its applications were given in [273, 274, 275].

### 10.1.3 Cozero-divisor Graphs

In [18], Afkhami and Khashyarmanesh defined the cozero-divisor graph of a commutative ring.

**Definition 10.1.8.** ([18]) Let $R$ be a commutative ring. The *cozero-divisor graph* of $R$, denoted by $\Gamma'(R)$, is the graph with vertex set $W^*(R)$, which is the set of all nonzero nonunit elements of $R$, and two distinct vertices $x$ and $y$ are adjacent if and only if $x \notin Ry$ and $y \notin Rx$.

Some basic results on the structure of this graph were studied in [18]. If $R$ is a finite commutative ring, then $W^*(R) = Z(R)^*$. Moreover, if $R$ is a field, then $W^*(R) = \emptyset$, and hence $\Gamma'(R)$ is the empty graph. If $R = F_1 \times F_2$ is a finite commutative ring, where $F_1$ and $F_2$ are fields, then $\Gamma'(R)$ is a complete bipartite graph, and hence $\Gamma'(R) \cong K_{|F_1^*|,\, |F_2^*|}$.

Let $R = \mathbb{Z}_{p^k}$, where $p$ is prime and $k > 1$. Then $W^*(R) = \{\lambda p \mid \lambda \in \mathbb{Z}_{p^k}\}$ and $\Gamma'(R) \cong \overline{K_t}$, where $t = p^k - p^{k-1}(p-1) - 1$. Let $R = \mathbb{Z}_{12}$. Then $W^*(R) = \{2, 3, 4, 6, 8, 9, 10\}$, and $(2) = (10) = \{0, 2, 4, 6, 8, 10\}$, $(3) = (9) = \{0, 3, 6, 9\}$, and $(4) = (8) = \{0, 4, 8\}$. The cozero-divisor graph $\Gamma'(\mathbb{Z}_{12})$ is given in Figure 10.4.

A few main results for the cozero-divisor graph are listed below.

**Theorem 10.1.9.** ([18]) *Let $R$ be a commutative ring. Then $\Gamma'(R)$ is not complete if and only if there is an $a \in W^*(R)$ such that $|aR| > 2$.*

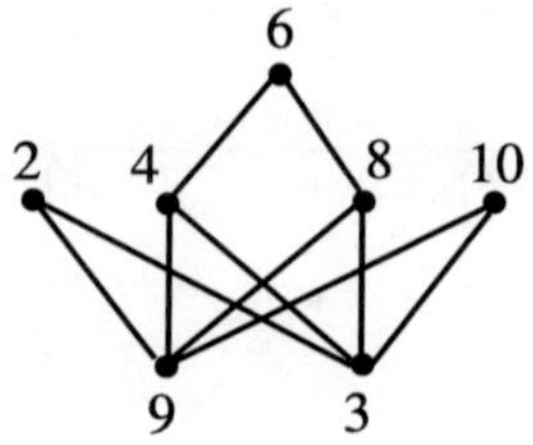


Figure 10.4: The cozero-divisor graph of $\mathbb{Z}_{12}$

**Theorem 10.1.10.** ([18]) *Let $R$ be a commutative ring. Then the following statements hold:*

(1) *$\Gamma'(R) \setminus J(R)$ is connected;*

(2) *If $R$ is a not a local ring, then $diam(\Gamma'(R) \setminus J(R)) \leq 2$.*

**Theorem 10.1.11.** ([18]) *Let $R$ be a nonlocal commutative ring such that for every $a \in J(R)$, there is an $\mathfrak{m} \in Max(R)$ and $b \in \mathfrak{m} \setminus J(R)$ with $a \notin bR$. Then $\Gamma'(R)$ is connected and $diam(\Gamma'(R)) \leq 3$.*

The next two results deal with the girth of co-zero divisor graphs.

**Theorem 10.1.12.** ([18]) *Let $R$ be a nonlocal commutative ring. Then $gr(\Gamma'(R) \setminus J(R)) \leq 5$ or $gr(\Gamma'(R) \setminus J(R)) = \infty$.*

**Theorem 10.1.13.** ([18]) *Let $R$ be a commutative ring with $|Max(R)| \geq 3$. Then $gr(\Gamma'(R)) = 3$.*

There are some relationships between the graphs $IS(R)$ and $\Gamma'(R)$. For instance, for commutative Artinian rings (in particular, for finite commutative rings) we always have $Z(R)^* = W^*(R)$. Also, if $\Gamma'(R)$ is complete, then $IS(R)$ is also complete; but not conversely. For example, $\Gamma(\mathbb{Z}_9)$ is complete, but $\Gamma'(\mathbb{Z}_9)$ is not complete. However, the converse is true for a reduced commutative ring.

In [19], Afkhami et al. studied some more properties of the cozero-divisor graph of a commutative ring. In particular, they characterized all commutative rings whose cozero-divisor graphs are double-star, unicyclic, a star, or a forest.

**Theorem 10.1.14.** ([19, Theorem 2.1]) *Let $R$ be a nonlocal finite commutative ring. If $\Gamma'(R)$ is a forest (contains no cycles), then $R \cong \mathbb{Z}_2 \times F$, where $F$ is a field.*

**Theorem 10.1.15.** ([19, Theorem 2.2]) *Let $R$ be a finite commutative ring.*

(i) *If $R$ is not local, then $\Gamma'(R)$ is a double-star graph if and only if $R \cong \mathbb{Z}_2 \times F$, where $F$ is a field;*

(ii) *If $R$ is local with principal maximal ideal $\mathfrak{m}$, then $\Gamma'(R)$ is a double-star graph if and only if $R$ is $\mathbb{Z}_4, \mathbb{Z}_2[x]/(x^2)$, or a field;*

(iii) *If $R$ is local with nonprincipal maximal ideal $\mathfrak{m}$ and $\Gamma'(R)$ is a double-star graph, then the minimal generating set of $\mathfrak{m}$ has two elements.*

Thus, if $R$ is a finite commutative ring and $\Gamma'(R)$ is a double-star graph, then either $R$ is local or $R \cong \mathbb{Z}_2 \times F$, where $F$ is a field.

**Theorem 10.1.16.** ([19, Theorem 2.3]) *Let $R$ be a nonlocal finite commutative ring. Then $\Gamma'(R)$ is a unicyclic graph if and only if $R$ is isomorphic to $\mathbb{Z}_2 \times \mathbb{Z}_4$, $\mathbb{Z}_2 \times \mathbb{Z}_2[x]/(x^2)$, or $\mathbb{Z}_3 \times \mathbb{Z}_3$.*

**Theorem 10.1.17.** ([19, Theorem 2.4]) *Let $R$ be a nonlocal finite commutative ring. Then $\Gamma'(R)$ is a union of cycle graphs if and only if $R \cong \mathbb{Z}_3 \times \mathbb{Z}_3$.*

**Theorem 10.1.18.** ([19, Theorem 2.5]) *Let $R$ is a commutative Noetherian ring. Then $\Gamma'(R)$ is totally disconnected if and only if $R$ is a local ring with principal maximal ideal.*

Further, Akbari et al. [24] continued the study of cozero-divisor graphs of a commutative ring. They were interested in the diameter of the cozero-divisor graph of the polynomial ring $R[x]$ and power series ring $R[[x]]$. They proved $diam(\Gamma'(R[x])) = 2$, and if $(R, \mathfrak{m})$ is a local ring with $\mathfrak{m} = 0$ (i.e., if $R$ is a field), then $\Gamma'(R[[x]])$ is totally disconnected, and otherwise, $diam(\Gamma'(R[[x]])) \leq 3$. Moreover, they proved that if $(R, \mathfrak{m})$ is a local ring with $\mathfrak{m} \neq 0$ and $\mathfrak{m}^2 = 0$, then $diam(\Gamma'(R[[x]])) = 2$, and if $R$ is a nonlocal ring, then $diam(\Gamma'(R[[x]])) \leq 3$.

### 10.1.4 Jacobson and Intersection Graphs

The Jacobson graph of a commutative ring was first introduced by Azimi et al. [124]. The definition of the Jacobson graph of a commutative ring is as follows:

**Definition 10.1.19.** ([124]) Let $R$ be a commutative ring. The *Jacobson graph* $\mathfrak{J}_R$ of $R$ is the graph with vertex set $R \setminus J(R)$ such that two distinct vertices $x$ and $y$ are adjacent if and only if $1 - xy \notin U(R)$. Note that if $R$ is a finite commutative ring, then $U(R) = R \setminus Z(R)$.

The graphs in Figure 10.5 are the Jacobson graphs of the rings indicated. Here, we abbreviate the ordered pair $(r, s)$ by $rs$.

Some of the properties of the Jacobson graph have been studied in detail in [124]. Several graph-theoretical properties as well as numerical invariants of these graphs were discovered. For instance, they proved that $\mathfrak{J}_R$ is complete if and only if $R$ is a local ring with associated residue field of order 2, and $\mathfrak{J}_R$ is a bipartite graph if and only if $R$ is a field or isomorphic to $\mathbb{Z}_4$, $\mathbb{Z}_2 \times \mathbb{Z}_2$, or $\mathbb{Z}_2[x]/(x^2)$. Moreover, the graph $\mathfrak{J}_R$ is a tree if and only if $R$ is isomorphic to

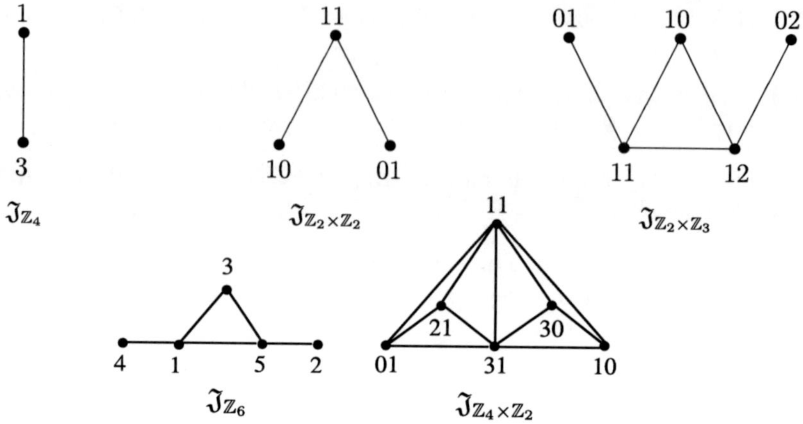


Figure 10.5: The Jacobson graphs of some specific rings

$\mathbb{Z}_2$, $\mathbb{Z}_4$, $\mathbb{Z}_2 \times \mathbb{Z}_2$, or $\mathbb{Z}_2[x]/(x^2)$. Also, they showed that the Jacobson graph of $R$ is connected whenever $R$ is a nonlocal ring, in which case, its diameter is at most 3; and they computed the girth, dominating number, independence number, and vertex chromatic number of Jacobson graphs. Moreover, they characterized all finite commutative rings with planar Jacobson graphs. The relevant result is stated below. Note that the ring $\mathbb{Z}_2 \times \mathbb{Z}_2[x]/(x^2)$ is missed in [124, Theorem 4.3].

**Theorem 10.1.20.** ([124, Theorem 4.3]) *Let $R$ be a finite commutative ring. Then $\mathfrak{J}_R$ is planar if and only if $R$ is a field or $R$ is isomorphic to one of the following rings:*

$$\mathbb{Z}_4,\ \mathbb{Z}_2\times\mathbb{Z}_2,\ \frac{\mathbb{Z}_2[x]}{(x^2)},\ \mathbb{Z}_6,\ \mathbb{Z}_8,\ \mathbb{Z}_2\times\mathbb{Z}_4,\ \mathbb{Z}_2\times\mathbb{Z}_2\times\mathbb{Z}_2,\ \frac{\mathbb{Z}_2[x]}{(x^3)},\ \frac{\mathbb{Z}_4[x]}{(2x,x^2)},$$

$$\mathbb{Z}_2\times\mathbb{F}_4,\ \mathbb{Z}_2\times\frac{\mathbb{Z}_2[x]}{(x^2)},\ \frac{\mathbb{Z}_4[x]}{(x^2-2,2x)},\ \frac{\mathbb{Z}_2[x,y]}{(x,y)^2},\ \mathbb{Z}_9,\ \mathbb{Z}_3\times\mathbb{Z}_3,\ \frac{\mathbb{Z}_3[x]}{(x^2)}.$$

The study of the Jacobson graph of a commutative ring was continued by Azimi et al. in [125]. They concentrated on isomorphisms between Jacobson graphs. It is easy to see that the subgraph of $\mathfrak{J}_R$ induced by all vertices $x$ such that $1 - x^2$ is a unit is the co-normal product of some graphs, each of which is a union of isomorphic complete bipartite graphs. Specifically, they determined the structure of the automorphism group of Jacobson graphs and gave some conditions under which two rings are isomorphic whenever the automorphism group of their Jacobson graphs are isomorphic.

In 2016, Amraei et al. [57], classified, up to isomorphism, the commutative rings whose Jacobson graph is toroidal. More precisely, they proved the following theorem.

**Theorem 10.1.21.** ([57, Theorem 1.1]) *Let $R$ be a finite commutative ring. Then the Jacobson graph $\mathfrak{J}_R$ is a toroidal graph if and only if $R$ is isomorphic to*

$$\frac{\mathbb{F}_4[x]}{(x^2)},\ \frac{\mathbb{Z}_4[x]}{(x^2+x+1)},\ \mathbb{Z}_{10},\ \mathbb{Z}_{12},\ \mathbb{Z}_3\times\frac{\mathbb{Z}_2[x]}{(x^2)},\ \mathbb{Z}_3\times\mathbb{F}_4, \textit{or } \mathbb{Z}_{14}.$$

Asir et al. [94] obtained some bounds for the genus of Jacobson graphs; a few are listed below: The first result of this type provides a lower bound for the genus of $\mathfrak{J}_R$ through the cardinality of a maximal ideal of $R$.

**Theorem 10.1.22.** ([94, Theorem 2.5]) *Let $R$ be a commutative ring with maximal ideal $\mathfrak{m}$ such that $|\mathfrak{m}| = \lambda \geq 3$ and $|R/\mathfrak{m}| = \mu$ ($\lambda$ and $\mu$ may be infinite cardinals, in which case $g(\mathfrak{J}_R) = \infty$). Then*

$$g(\mathfrak{J}_R) \geq \begin{cases} 2\left\lceil \frac{(\lambda-3)(\lambda-4)}{12} \right\rceil + \left(\frac{\mu-3}{2}\right)\left\lceil \frac{(\lambda-2)^2}{4} \right\rceil & \textit{if } \mu \textit{ is odd,} \\ \left\lceil \frac{(\lambda-3)(\lambda-4)}{12} \right\rceil + \left(\frac{\mu-2}{2}\right)\left\lceil \frac{(\lambda-2)^2}{4} \right\rceil & \textit{if } \mu \textit{ is even.} \end{cases}$$

*Moreover, if $R$ is local, then equality holds.*

The next result exhibits a lower bound for the genus of $\mathfrak{J}_R$, where $R$ is isomorphic to the direct product of local rings, in terms of the cardinalities of the local rings. Recall that Artinian rings have such a decomposition into local rings.

In the sequel, for a graph $G$, $V(G)$ denotes its vertex set.

**Theorem 10.1.23.** ([94, Theorem 2.8]) *Let $R_1, \ldots, R_n$ be commutative rings and $R \cong R_1 \times \cdots \times R_n$ ($n \geq 2$).*

(1) *Suppose that $R_i$ is finite for $1 \leq i \leq n$. Let $min\{|R_1|, \ldots, |R_n|\} = |R_j|$, and set $r = |R_1| \cdots \widehat{|R_j|} \cdots |R_n|$, where the notation $\widehat{|R_j|}$ means that $|R_j|$ is deleted. Then $g(\mathfrak{J}_R) \geq \left\lceil \frac{(r-3)(r-4)}{12} \right\rceil$ if $r \geq 3$, and $g(\mathfrak{J}_R) \geq 0$ if $r = 2$;*

(2) *Suppose that at least one of the is $R_i$ is infinite. Then $g(\mathfrak{J}_R) = \infty$.*

Next, we state another lower bound for the genus of the Jacobson graph of a finite commutative ring.

**Theorem 10.1.24.** ([94, Theorem 2.11]) *Let $R$ be a finite commutative ring, and write $R \cong R_1 \times \cdots \times R_n$, where every $R_i$ is local. Set $k_i = g(\mathfrak{J}_{R_i})$, $\ell_i = |R_i \backslash J(R_i)|$, $m_1 = max\{\ell_1 k_2 + k_1, \ell_2 k_1 + k_2\}$, and $m_j = max\{\ell_1 \cdots \ell_j k_{j+1} + m_{j-1}, \ell_{j+1} m_{j-1} + k_{j+1}\}$ for $j = 2, \ldots, n-1$. Then $g(\mathfrak{J}_R) \geq m_{n-1}$.*

In the following theorem, we give an upper bound for $g(\mathfrak{J}_R)$.

**Theorem 10.1.25.** ([94, Theorem 2.13]) *Let $R$ be a finite commutative ring with $R \setminus U(R) = \bigcup_{i=1}^{m} \mathfrak{m}_i$, where the $\mathfrak{m}_i$'s are maximal ideals of $R$. Let $|\mathfrak{m}_i| = \lambda_i \geq 3$*

*and* $|R/\mathfrak{m}_i| = \mu_i$ *for* $i = 1, \ldots, m$. *Rearrange the* $\mathfrak{m}_i$*'s so that* $\mu_i$ *is odd for* $1 \leq i \leq j$ *and* $\mu_i$ *is even for* $j+1 \leq i \leq m$. *Then*

$$\begin{aligned} g(\mathfrak{J}_R) \leq {} & (m-1)(|R \setminus J(R)| - 1) \\ & + \sum_{i=1, |\lambda_i|>2}^{j} \left\{ 2\left\lceil \frac{(\lambda_i - 3)(\lambda_i - 4)}{12} \right\rceil + \left(\frac{\mu_i - 3}{2}\right) \left\lceil \frac{(\lambda_i - 2)^2}{4} \right\rceil \right\} \\ & + \sum_{i=j+1, |\lambda_i|>2}^{m} \left\{ \left\lceil \frac{(\lambda_i - 3)(\lambda_i - 4)}{12} \right\rceil + \left(\frac{\mu_i - 1}{2}\right) \left\lceil \frac{(\lambda_i - 2)^2}{4} \right\rceil \right\}. \end{aligned}$$

Next, we state an isomorphism relationship between the Jacobson graphs $\mathfrak{J}_R$ and $\mathfrak{J}_S$ of two finite rings $R$ and $S$. Note that $V(\mathfrak{J}_{R_1 \times R_2}) \supseteq V(\mathfrak{J}_{R_1}) \times V(\mathfrak{J}_{R_2})$ as $J(R_1 \times R_2) = J(R_1) \times J(R_2)$. However, equality does not hold in general. In order to see this, it is enough to note that $(0,1) \in (\mathbb{Z}_4 \times \mathbb{Z}_4) \setminus J(\mathbb{Z}_4 \times \mathbb{Z}_4)$, but $(0,1) \notin (\mathbb{Z}_4 \setminus J(\mathbb{Z}_4)) \times (\mathbb{Z}_4 \setminus J(\mathbb{Z}_4))$. In what follows, for a given local ring $R$, $\kappa(R)$ denotes the residue field of $R$.

**Theorem 10.1.26.** ([94, Theorem 3.3]) *Let* $R \cong R_1 \times \cdots \times R_n$ *and* $S \cong S_1 \times \cdots \times S_n$ *be finite commutative rings, where the* $R_i$*'s and* $S_i$*'s are local.*

(1) *If* $|R_i| = |S_i|$ *and* $\mathfrak{J}_{R_i} \cong \mathfrak{J}_{S_i}$ *for* $1 \leq i \leq n$, *then* $\mathfrak{J}_R \cong \mathfrak{J}_S$;

(2) *If* $|R_i| = |S_i|$ *and* $|\kappa(R_i)| = |\kappa(S_i)|$ *for* $1 \leq i \leq n$, *then* $\mathfrak{J}_R \cong \mathfrak{J}_S$.

As an application of bounds for the genus of Jacobson graphs, Asir et al. [94] characterized all commutative Artinian rings whose Jacobson graphs are toroidal using a similar result for the finite case.

Recently, Amraei et al. [58] studied the nonorientable genus of Jacobson graphs. In [58], the nonorientable genus of some Jacobson graphs is either computed or estimated by a lower bound. As an application, they classified, up to isomorphism, the finite commutative rings whose Jacobson graphs are projective (i.e., nonorientable genus one).

Next, we look at the intersection graph of ideals of a ring, which was first defined and studied by Chakrabarty et al. in [157].

**Definition 10.1.27.** ([157]) Let $R$ be a commutative ring and $I(R)^*$ be the set of nontrivial ideals of $R$. The *intersection graph of ideals* of $R$, denoted by $IG(R)$, is the graph with vertex set $I(R)^*$, and two distinct vertices $I$ and $J$ are adjacent if and only if $I \cap J \neq 0$.

For example, consider the ring $R = \mathbb{Z}_{p^2 q}$. Then $I(R)^* = \{(p), (q), (pq), (p^2)\}$, where $p$ and $q$ are distinct primes. The corresponding intersection graph is given in Figure 10.6.

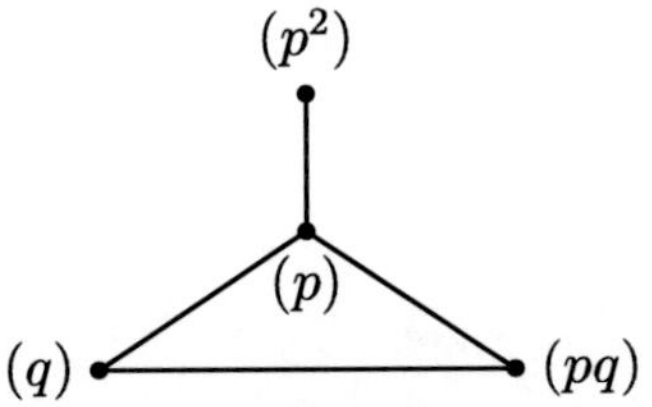


Figure 10.6: The intersection graph of $\mathbb{Z}_{p^2q}$

Note that if $R$ is an integral domain, but not a field, then $IG(R)$ is complete, and the graph $IG(\mathbb{Z}_n)$ is complete if and only if $n = p^m$, where $p$ is a prime number and $2 \leq m \in \mathbb{N}$. In [157], they determined the values of $n$ for which $IG(\mathbb{Z}_n)$ is connected, complete, bipartite, or Hamiltonian. Further, Devhare et al. [184] obtained a characterization for $IG(\mathbb{Z}_n)$ to be Eulerian after pinpointing an error in a previous characterization for the same. The results obtained in this regard by Devhare et al. [184] are given below. Recall that a commutative ring $R$ is a *special principal ideal ring* (or, SPIR for short) if $R$ is a local Artinian principal ideal ring. The following theorem characterizes Eulerian graphs $IG(R)$ for special principal ideal rings $R$.

**Theorem 10.1.28.** ([184, Corollary 3.8]) *Suppose that $R$ is a commutative Artinian principal ideal ring, and let $R_1, \ldots, R_k (k \in \mathbb{N})$ be special principal ideal rings such that $R \cong R_1 \times \cdots \times R_k$. If $M_i$ is the maximal ideal of $R_i (i \in \{1, \ldots, k\})$, then $IG(R)$ is Eulerian if and only if one of the following statements hold:*

(a) *$k \geq 3$ and $R_i$ is a field for every $i \in \{1, \ldots, k\}$;*

(b) *The index of nilpotency of $M_i$ is even for every $i \in \{1, \ldots, k\}$.*

*In particular, if $R$ is a reduced commutative Artinian ring with at least three prime ideals, then $IG(R)$ is Eulerian.*

The next theorem provides the correction to a characterization for $IG(\mathbb{Z}_n)$ to be Eulerian in [158] and generalizes the result by relaxing the distinct primes condition.

**Theorem 10.1.29.** ([184, Corollary 3.9]) *If $p_1, \ldots, p_k \in \mathbb{N}$ are (not necessarily distinct) prime numbers, then the graph $IG(\mathbb{Z}_{{p_1}^{n_1}} \times \cdots \times \mathbb{Z}_{{p_k}^{n_k}})$ is Eulerian if and only if either $k \geq 3$ and $n_1 = \cdots = n_k = 1$, or $n_i \in \mathbb{N}$ is even for every $i \in \{1, \ldots, k\}$.*

Since the structure of a ring is closely tied to the behavior of its ideals, one may expect that the intersection graph of ideals reflects many properties of the ring.

The intersection graph of submodules of a module was first introduced in [439].

**Definition 10.1.30.** ([439]) Let $M$ be a left $R$-module, where $R$ is a (not necessarily commutative) ring with unit. The *intersection graph* $IG(M)$ of proper $R$-submodules of $M$ is the undirected graph without loops and multiple edges with vertex set the set of all proper $R$-submodules of $M$, and distinct vertices $U$ and $W$ are adjacent if and only if $U \cap W \neq 0$.

Yaraneri [439] studied this graph to relate the combinatorial properties of $IG(M)$ to the algebraic properties of the $R$-module $M$. For instance, he studied the connectedness, domination, finiteness, coloring, and planarity for $IG(M)$. The intersection graph of rings and modules have also been studied in [46, 42, 313].

In continuation of above, Devhare et al. [183] generalized the connectedness characterization of intersection graphs to $R$-modules ($R$ not necessarily commutative). Actually, the following result on connectedness is a generalization of a characterization proved in [158] to every $R$-module ($R$ not necessarily commutative).

**Theorem 10.1.31.** ([183, Corollary 3.1]) *Let $R$ be a (not necessarily commutative) ring and $M$ an $R$-module. Then the intersection graph $IG(M)$ is disconnected if and only if $M$ contains at least two minimal submodules and every proper nonzero submodule of $M$ is minimal (as well as maximal).*

Further, Devhare et al. [182] showed that the intersection graphs of modules (over not necessarily commutative rings) are weakly perfect. The relevant result is given below.

**Theorem 10.1.32.** ([182, Corollary 4.3]) *If $R$ is a (not necessarily commutative) ring and $M$ is an $R$-module such that $\omega(IG(M)) < \infty$, then $IG(M)$ and its complement $CIG(M)$ are weakly perfect.*

## 10.2 Annihilator Graphs

The concept of annihilator graph of a commutative ring was introduced by A. Badawi [128] in 2014.

**Definition 10.2.1.** ([128]) The *annihilator graph* of a commutative ring $R$ is the simple graph $AG(R)$ with vertices $Z(R)^*$, and two distinct vertices $x$ and $y$ are adjacent if and only if $ann_R(xy) \neq ann_R(x) \cup ann_R(y)$ (i.e., if and only if $ann_R(x) \cup ann_R(y) \subsetneq ann_R(xy)$).

By definition, it follows that the zero-divisor graph $\Gamma(R)$ of a commutative ring $R$ is a subgraph of $AG(R)$. Figure 10.7 shows the annihilator graphs of the rings $\mathbb{Z}_3 \times \mathbb{Z}_3$, $\mathbb{Z}_8$, and $\mathbb{Z}_4 \times \mathbb{Z}_3$. It is clear that if $R \cong F_1 \times F_2$, where $F_1$ and $F_2$ are fields, then $AG(R) \cong K_{|F_1^*|,|F_2^*|}$ is a complete bipartite graph.

Note that if $R$ is a commutative ring, then $AG(R)$ is connected and $diam(AG(R)) \leq 2$. The following theorem answers the question of when the annihilator graph $AG(R)$ is identical to the zero-divisor graph $\Gamma(R)$.

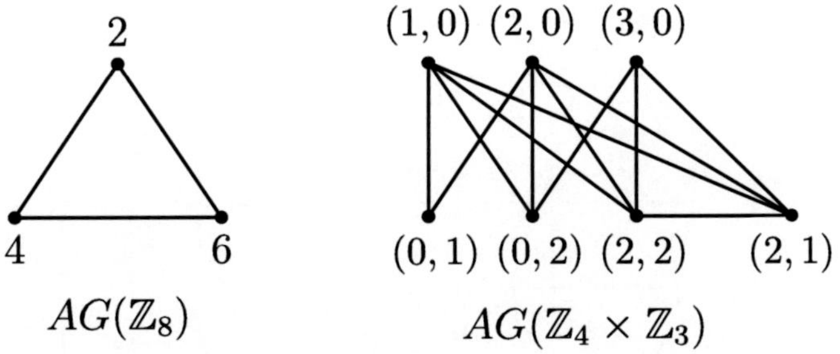


Figure 10.7: The annihilator graph of $\mathbb{Z}_8$ and $\mathbb{Z}_4 \times \mathbb{Z}_3$

**Theorem 10.2.2.** ([128, Theorem 3.6], for a proof, see Section 10.2.1 Theorem 10.2.17) *Let $R$ be a reduced commutative ring that is not an integral domain. Then $AG(R) = \Gamma(R)$ if and only if $R$ has exactly two distinct minimal prime ideals.*

Let $R = F_1 \times F_2$, where $F_1$ and $F_2$ are finite fields. Then $R$ is reduced with exactly two distinct minimal prime ideals. Hence, by Theorem 10.2.2, $AG(R) = \Gamma(R) = K_{|F_1^*|,|F_2^*|}$. Let $R \neq \mathbb{Z}_2 \times \mathbb{Z}_2$ be a local ring that is not a field and $|Z(R)^*| \geq 3$. Then $AG(R)$ is complete, and hence $gr(AG(R)) = 3$. On the other hand, let $R$ be a finite commutative ring, but not a field. Since $R$ is finite, $R \cong R_1 \times \cdots \times R_n$, where every $R_i$ is a local ring. If $n \geq 3$, then $(1,0,\ldots,0) - (0,1,0,\ldots,0) - (0,0,1,0,\ldots,0) - (1,0,\ldots,0)$ is a cycle in $AG(R)$, and hence $gr(AG(R)) = 3$. Note that $\Gamma(R)$ is a subgraph of $AG(R)$.

Anderson et al. [75] gave all zero-divisor graphs of order $\leq 4$. Using this, Table 10.1 provides all commutative local rings $R$ for which $1 \leq |Z(R)^*| \leq 4$ and $AG(R)$ is complete. Note that only two rings $\mathbb{Z}_2 \times \mathbb{F}_4$ and $\mathbb{Z}_3 \times \mathbb{Z}_3$ with $|Z(R)^*| \leq 4$ are not mentioned in Table 10.1 because $\Gamma(\mathbb{Z}_2 \times \mathbb{F}_4) = K_{1,3}$ and $\Gamma(\mathbb{Z}_3 \times \mathbb{Z}_3) = K_{2,2}$.

Redmond [325, 326] has given all local commutative rings $R$ with $|Z(R)^*| \leq 14$ (see Section 2.1). Using the list given in [325, 326], Table 10.2 also provides all commutative local rings for which $6 \leq |Z(R)^*| \leq 7$ and $AG(R)$ is complete.

Now, we see properties of annihilator graphs

**Lemma 10.2.3.** ([128, Lemma 2.3]) *Let $R$ be a commutative ring and $x - y$ an edge of $AG(R)$ that is not an edge of $\Gamma(R)$ for distinct $x, y \in Z(R)^*$. If there is a $w \in ann_R(xy) \setminus \{x, y\}$ such that $wx \neq 0$ and $wy \neq 0$, then $x - w - y$ is a path in $AG(R)$ that is not a path in $\Gamma(R)$. Thus $C : x - w - y - x$ is a cycle in $AG(R)$ of length three, and each edge of $C$ is not an edge of $\Gamma(R)$.*

*Proof.* Suppose that $x - y$ is an edge of $AG(R)$ that is not an edge of $\Gamma(R)$. Then $xy \neq 0$. Assume there is a $w \in ann_R(xy) \setminus \{x, y\}$ such that $wx \neq 0$ and $wy \neq 0$.

Table 10.1: Complete annihilator graphs with $1 \leq |Z(R)| \leq 4$

| $\lvert Z(R)^*\rvert$ | Local ring $R$ | $AG(R)$ |
|---|---|---|
| 1 | $\mathbb{Z}_4, \dfrac{\mathbb{Z}_2[x]}{(x^2)}$ | $K_1$ |
| 2 | $\mathbb{Z}_9, \dfrac{\mathbb{Z}_3[x]}{(x^2)}$ | $K_2$ |
| 3 | $\mathbb{Z}_8, \dfrac{\mathbb{Z}_2[x]}{(x^3)}, \dfrac{\mathbb{Z}_4[x]}{(2x, x^2-2)}$ | $K_3$ |
| 3 | $\dfrac{\mathbb{F}_4[x]}{(x^2)}, \dfrac{Z_4[x]}{(x^2+x+1)}$ | $K_3$ |
| 3 | $\dfrac{\mathbb{Z}_4[x]}{(x,2)^2}, \dfrac{\mathbb{Z}_2[x,y]}{(x,y)^2}$ | $K_3$ |
| 4 | $\mathbb{Z}_{25}, \dfrac{\mathbb{Z}_5[x]}{(x^2)}$ | $K_4$ |

Table 10.2: Complete annihilator graphs with $6 \leq |Z(R)| \leq 7$

| | | |
|---|---|---|
| 6 | $\mathbb{Z}_{49}, \frac{\mathbb{Z}_7[x]}{(x^2)}$ | $K_6$ |
| 7 | $\mathbb{Z}_{16}, \frac{\mathbb{Z}_2[x]}{(x^4)}, \frac{\mathbb{Z}_4[x]}{(x^2+2)}, \frac{\mathbb{Z}_4[x]}{(x^2+2x+2)}$ | $K_7$ |
| 7 | $\frac{\mathbb{Z}_4[x]}{(x^3-2,2x^2,2x)}, \frac{\mathbb{Z}_2[x,y]}{(x^3,xy,y^2)}, \frac{\mathbb{Z}_8[x]}{(2x,x^2)}$ | $K_7$ |
| 7 | $\frac{\mathbb{Z}_4[x]}{(x^3,2x^2,2x)}, \frac{\mathbb{Z}_4[x,y]}{(x^2-2,xy,y^2,2x,2y)}, \frac{\mathbb{Z}_4[x]}{(x^2+2x)}$ | $K_7$ |
| 7 | $\frac{\mathbb{Z}_8[x]}{(2x,x^2+4)}, \frac{\mathbb{Z}_2[x,y]}{(x^2,y^2-xy)}, \frac{\mathbb{Z}_4[x,y]}{(x^2,y^2-xy,xy-2,2x,2y)}$ | $K_7$ |
| 7 | $\frac{\mathbb{Z}_4[x,y]}{(x^2,y^2,xy-2,2x,2y)}, \frac{\mathbb{Z}_2[x,y]}{(x^2,y^2)}, \frac{\mathbb{Z}_4[x]}{(x^2)}$ | $K_7$ |
| 7 | $\frac{\mathbb{Z}_2[x,y,z]}{(x,y,z)^2}, \frac{\mathbb{Z}_4[x,y]}{(x^2,y^2,xy,2x,2y)}, \frac{\mathbb{F}_8[x]}{(x^2)}, \frac{\mathbb{Z}_4[x]}{(x^3+x+1)}$ | $K_7$ |

Since $y \in ann_R(xw) \setminus (ann_R(x) \cup ann_R(w))$, we conclude that $x - w$ is an edge of $AG(R)$. Since $x \in ann_R(yw) \setminus (ann_R(y) \cup ann_R(w))$, we conclude that $y - w$ is an edge of $AG(R)$. Hence $x - w - y$ is a path in $AG(R)$. Since $xw \neq 0$ and $yw \neq 0$, we have $x - w - y$ is not a path in $\Gamma(R)$. It is clear that $x - w - y - x$ is a cycle in $AG(R)$ of length three, and each edge of $C$ is not an edge of $\Gamma(R)$. □

**Theorem 10.2.4.** ([128, Theorem 2.4]) *Let $R$ be a commutative ring and $x - y$ an edge of $AG(R)$ that is not an edge of $\Gamma(R)$ for distinct $x, y \in Z(R)^*$. If $xy^2 \neq 0$ and $x^2y \neq 0$, then there is a $w \in Z(R)^*$ such that $x - w - y$ is a path in $AG(R)$ that is not a path in $\Gamma(R)$. Thus $C : x - w - y - x$ is a cycle in $AG(R)$ of length three, and each edge of $C$ is not an edge of $\Gamma(R)$.*

*Proof.* Suppose that $x - y$ is an edge of $AG(R)$ that is not an edge of $\Gamma(R)$. Then $xy \neq 0$, and there is a $w \in ann_R(xy) \setminus (ann_R(x) \cup ann_R(y))$. We show $w \notin \{x, y\}$. Assume $w \in \{x, y\}$. Then either $x^2y = 0$ or $y^2x = 0$, which is a contradiction. Thus $w \notin \{x, y\}$. Hence $x - w - y$ is the desired path in $AG(R)$ by Lemma 10.2.3. □

**Corollary 10.2.5.** ([128, Corollary 2.5]) *Let $R$ be a reduced commutative ring and $x - y$ an edge of $AG(R)$ that is not an edge of $\Gamma(R)$ for distinct $x, y \in Z(R)^*$. Then there is a $w \in ann_R(xy) \setminus \{x, y\}$ such that $x - w - y$ is a path in $AG(R)$ that is not a path in $\Gamma(R)$. Thus $C : x - w - y - x$ is a cycle in $AG(R)$ of length three, and each edge of $C$ is not an edge of $\Gamma(R)$.*

*Proof.* Suppose that $x - y$ is an edge of $AG(R)$ that is not an edge of $\Gamma(R)$ for some distinct $x, y \in Z(R)^*$. Since $R$ is reduced, we have $(xy)^2 \neq 0$. Hence $xy^2 \neq 0$ and $x^2y \neq 0$, and thus the claim is now clear by Theorem 10.2.4. □

In light of Corollary 10.2.5, we have the following result.

**Theorem 10.2.6.** ([128, Theorem 2.6]) *Let $R$ be a reduced commutative ring with $AG(R) \neq \Gamma(R)$. Then $gr(AG(R)) = 3$. Furthermore, there is a cycle $C$ of length three in $AG(R)$ such that each edge of $C$ is not an edge of $\Gamma(R)$.*

In view of Theorem 10.2.4, the following is an example of a nonreduced commutative ring $R$ where $x - y$ is an edge of $AG(R)$ that is not an edge of $\Gamma(R)$ for some distinct $x, y \in Z(R)^*$, but every path in $AG(R)$ of length two from $x$ to $y$ is also a path in $\Gamma(R)$.

**Example 10.2.7.** ([128, Example 2.7]) Let $R = \mathbb{Z}_8$. Then $2 - 6$ is an edge of $AG(R)$ that is not an edge of $\Gamma(R)$. Now $2 - 4 - 6$ is the only path in $AG(R)$ of length two from 2 to 6, and it is also a path in $\Gamma(R)$. Note that $AG(R) = K_3$, $\Gamma(R) = K_{1,2}$, $gr(\Gamma(R)) = \infty$, $gr(AG(R)) = 3$, $diam(\Gamma(R)) = 2$, and $diam(AG(R)) = 1$.

The following is an example of a nonreduced commutative ring $R$ with $AG(R) \neq \Gamma(R)$, and if $x - y$ is an edge of $AG(R)$ that is not an edge of $\Gamma(R)$ for some distinct $x, y \in Z(R)^*$, then there is no path in $AG(R)$ of length two from $x$ to $y$.

**Example 10.2.8.** ([128, Example 2.8])

(1) Let $R = \mathbb{Z}_2 \times \mathbb{Z}_4$, $a = (0,1), b = (1,2)$, and $c = (0,3)$. Then $a - b$ and $c - b$ are the only two edges of $AG(R)$ that are not edges of $\Gamma(R)$, but there is no path in $AG(R)$ of length two from $a$ to $b$ and there is no path in $AG(R)$ of length two from $c$ to $b$. Note that $AG(R) = K_{2,3}$, $\Gamma(R) = \overline{K}_{1,3}$, $gr(AG(R)) = 4$, $gr(\Gamma(R)) = \infty$, $diam(AG(R)) = 2$, and $diam(\Gamma(R)) = 3$.

(2) Let $R = \mathbb{Z}_2 \times \mathbb{Z}_2[x]/(x^2)$. Let $X = x + (x^2) \in \mathbb{Z}_2[x]/(x^2)$, $a = (0,1), b = (1, X)$, and $c = (0, 1 + X)$. Then $a - b$ and $c - b$ are the only two edges of $AG(R)$ that are not edges of $\Gamma(R)$, but there is no path in $AG(R)$ of length two from $a$ to $b$ and there is no path in $AG(R)$ of length two from $c$ to $b$. Again, note that $AG(R) = K_{2,3}$, $\Gamma(R) = \overline{K}_{1,3}$, $gr(AG(R)) = 4$, $gr(\Gamma(R)) = \infty$, $diam(AG(R)) = 2$, and $diam(\Gamma(R)) = 3$.

**Theorem 10.2.9.** ([128, Theorem 2.9]) *Let $R$ be a commutative ring with $AG(R) \neq \Gamma(R)$. Then the following statements are equivalent:*

(1) $gr(AG(R)) = 4$;

(2) $gr(AG(R)) \neq 3$;

(3) *If $x - y$ is an edge of $AG(R)$ that is not an edge of $\Gamma(R)$ for distinct $x, y \in Z(R)^*$, then there is no path in $AG(R)$ of length two from $x$ to $y$;*

(4) *There are some distinct $x, y \in Z(R)^*$ such that $x - y$ is an edge of $AG(R)$ that is not an edge of $\Gamma(R)$, and there is no path in $AG(R)$ of length two from $x$ to $y$;*

(5) *$R$ is ring-isomorphic to $\mathbb{Z}_2 \times \mathbb{Z}_4$ or $\mathbb{Z}_2 \times \mathbb{Z}_2[x]/(x^2)$.*

*Proof.* (1) $\Rightarrow$ (2) No comments.

(2) $\Rightarrow$ (3) Suppose that $x - y$ is an edge of $AG(R)$ that is not an edge of $\Gamma(R)$ for distinct $x, y \in Z(R)^*$. Since $gr(AG(R)) \neq 3$, there is no path in $AG(R)$ of length two from $x$ to $y$.

(3) $\Rightarrow$ (4) Since $AG(R) \neq \Gamma(R)$ by hypothesis, there are distinct $x, y \in Z(R)^*$ such that $x - y$ is an edge of $AG(R)$ that is not an edge of $\Gamma(R)$, and hence there is no path in $AG(R)$ of length two from $x$ to $y$ by (3).

(4) $\Rightarrow$ (5) Suppose there are distinct $x, y \in Z(R)^*$ such that $x - y$ is an edge of $AG(R)$ that is not an edge of $\Gamma(R)$ and there is no path in $AG(R)$ of length two from $x$ to $y$. Then $ann_R(x) \cap ann_R(y) = \{0\}$. Since $xy \neq 0$ and $ann_R(x) \cap ann_R(y) = \{0\}$, by Lemma 10.2.3 we conclude that $ann_R(xy) = ann_R(x) \cup ann_R(y) \cup \{y\}$ such that $y^2 \neq 0$ or $ann_R(xy) = ann_R(x) \cup ann_R(y) \cup \{x\}$ such that $x^2 \neq 0$ (note that if $\{x, y\} \subseteq ann_R(xy)$, then $x - xy - y$ is a path in $AG(R)$ of length two). Without lost of generality, we may assume that $ann_R(xy) =$

$ann_R(x) \cup ann_R(y) \cup \{y\}$ and $y^2 \neq 0$. Let $0 \neq a \in ann_R(x)$ and $0 \neq b \in ann_R(y)$. Since $ann_R(x) \cap ann_R(y) = \{0\}$, we have $a+b \in ann_R(xy) \setminus (ann_R(x) \cup ann_R(y))$, and hence $a + b = y$. Thus $|ann_R(x)| = |ann_R(y)| = 2$. Since $xy^2 = 0$, we have $ann_R(x) = \{0, y^2\}$ and $ann_R(y) = \{0, xy\}$. Since $y^2 + xy = y$, we have $(y^2+xy)^2 = y^2$. Since $xy^3 = 0$ and $xy^2 = x^2y^2 = 0$, we have $(y^2+xy)^2 = y^2$ implies that $y^4 = y^2$. Since $y^2 \neq 0$ and $y^4 = y^2$, we have $y^2$ is a nonzero idempotent of $R$. Hence $ann_R(xy) = ann_R(x) \cup ann_R(y) \cup \{y\} = \{0, y^2, xy, y\}$. Thus $ann_R(xy) \subseteq yR$, and since $yR \subseteq ann_R(xy)$, we conclude that $ann_R(xy) = yR = \{0, y^2, xy, y\}$. Since $y^2 + xy = y$ and $y^4 = y^2$, we have $(y^2 + xy)^3 = y^3$, and hence $y^3 = y^2$. Thus $y^2R = y(yR) = \{0, y^2\}$. Since $y^2$ is a nonzero idempotent of $R$ and $y^2R$ is a ring with two elements, we conclude that $y^2R$ is ring-isomorphic to $\mathbb{Z}_2$. Let $f \in ann_R(y^2)$. Then $y^2f = y(yf) = 0$, and thus $yf \in ann_R(y)$. Hence $yf = 0$ or $yf = yx$. Suppose $yf = 0$. Since $ann_R(y) = \{0, xy\}$, either $f = 0$ or $f = xy$. Suppose that $yf = yx$. Then $y(f - x) = 0$, and thus $f - x = 0$ or $f - x = xy$. Hence $f = x$ or $f = x + xy$. It is clear that $0, x, xy, x + xy$ are distinct elements of $R$, and thus $ann_R(y^2) = \{0, x, xy, x + xy\}$. Since $ann_R(y^2) = (1 - y^2)R$, we have $(1 - y^2)R = \{0, x, xy, x + xy\}$. Since $(1 - y^2)R$ is a ring with four elements, we conclude that $(1 - y^2)R$ is ring-isomorphic to $\mathbb{Z}_4$, $\mathbb{Z}_2 \times \mathbb{Z}_2$, $\mathbb{F}_4$, or $\mathbb{Z}_2[x]/(x^2)$. Since $x - y$ is an edge of $AG(R)$ that is not an edge of $\Gamma(R)$ and there is no path in $AG(R)$ of length two from $x$ to $y$ by hypothesis, $R$ is not reduced by Corollary 10.2.5. Since $R$ is ring-isomorphic to $y^2R \times (1 - y^2)R$ and not reduced, $R$ is ring-isomorphic to $\mathbb{Z}_2 \times \mathbb{Z}_4$ or $\mathbb{Z}_2 \times \mathbb{Z}_2[x]/(x^2)$.

(5) $\Rightarrow$ (1) See Example 10.2.8. □

**Corollary 10.2.10.** ([128, Corollary 2.10]) *Let $R$ be a commutative ring with $AG(R) \neq \Gamma(R)$, and assume that $R$ is not ring-isomorphic to $\mathbb{Z}_2 \times B$, where $B = \mathbb{Z}_4$ or $B = \mathbb{Z}_2[x]/(x^2)$. If $E$ is an edge of $AG(R)$ that is not an edge of $\Gamma(R)$, then $E$ is an edge of a cycle of length three in $AG(R)$.*

**Corollary 10.2.11.** ([128, Corollary 2.11]) *Let $R$ be a commutative ring with $AG(R) \neq \Gamma(R)$. Then $gr(AG(R)) \in \{3, 4\}$.*

### 10.2.1 Equal Annihilator and Zero-divisor Graphs

We recall the following results.

**Lemma 10.2.12.** (1) (Theorem 2.2.5) *Let $R$ be a reduced commutative ring that is not an integral domain. Then $\Gamma(R)$ is complete if and only if $R$ is ring-isomorphic to $\mathbb{Z}_2 \times \mathbb{Z}_2$;*

(2) ([269, Theorem 2.6(3)]) *Let $R$ be a commutative ring. Then $diam(\Gamma(R)) = 2$ if and only if either* (i) *$R$ is reduced with exactly two minimal primes and at least three nonzero zero-divisors, or* (ii) *$Z(R)$ is an ideal whose square is not $\{0\}$ and each pair of distinct zero-divisors has a nonzero annihilator.*

We first study the case when $R$ is reduced.

**Lemma 10.2.13.** ([128, Lemma 3.2]) *Let $R$ be a reduced commutative ring that is not an integral domain and let $z \in Z(R)^*$.*

(1) $ann_R(z) = ann_R(z^n)$ *for every positive integer* $n \geq 2$;

(2) *If* $c+z \in Z(R)$ *for* $c \in ann_R(z)\setminus\{0\}$, *then* $ann_R(z+c)$ *is properly contained in* $ann_R(z)$ *(i.e.,* $ann_R(c+z) \subsetneq ann_R(z)$*). In particular, if* $Z(R)$ *is an ideal of* $R$ *and* $c \in ann_R(z) \setminus \{0\}$, *then* $ann_R(z+c)$ *is properly contained in* $ann_R(z)$.

*Proof.* (1) Let $n \geq 2$. It is clear that $ann_R(z) \subseteq ann_R(z^n)$. Let $f \in ann_R(z^n)$. Since $fz^n = 0$ and $R$ is reduced, we have $fz = 0$. Thus $ann_R(z^n) = ann_R(z)$.

(2) Let $c \in ann_R(z) \setminus \{0\}$, and suppose that $c + z \in Z(R)$. Since $z^2 \neq 0$, we have $c + z \neq 0$, and hence $c + z \in Z(R)^*$. Since $c \in ann_R(z)$ and $R$ is reduced, we have $c \notin ann_R(c + z)$. Hence $ann_R(c + z) \neq ann_R(z)$. Since $ann_R(c + z) \subsetneq ann_R(z(c + z)) = ann_R(z^2)$ and $ann_R(z^2) = ann_R(z)$ by (1), it follows that $ann_R(c+z) \subsetneq ann_R(z)$. □

**Theorem 10.2.14.** ([128, Theorem 3.3]) *Let $R$ be a reduced commutative ring that is not an integral domain. Then the following statements are equivalent:*

(1) $AG(R)$ *is complete;*

(2) $\Gamma(R)$ *is complete (and hence* $AG(R) = \Gamma(R)$*);*

(3) $R$ *is ring-isomorphic to* $\mathbb{Z}_2 \times \mathbb{Z}_2$.

*Proof.* (1) $\Rightarrow$ (2) Let $a \in Z(R)^*$. Suppose that $a^2 \neq a$. Since $ann_R(a^3) = ann_R(a)$ by Lemma 10.2.13(1) and $a^3 \neq 0$, we have that $a - a^2$ is not an edge of $AG(R)$, a contradiction. Thus $a^2 = a$ for each $a \in Z(R)$. Let $x, y$ be distinct elements in $Z(R)^*$. We show that $x - y$ is an edge of $\Gamma(R)$. Suppose that $xy \neq 0$. Since $x - y$ is an edge of $AG(R)$, we have $ann_R(xy) \neq ann_R(x)$, and thus $xy \neq x$. Since $x^2 = x$, we have $ann_R(x(xy)) = ann_R(x^2y) = ann_R(xy)$, and thus $x - xy$ is not an edge of $AG(R)$, a contradiction. Hence $xy = 0$ and $x - y$ is an edge of $\Gamma(R)$.

(2) $\Rightarrow$ (3) It follows from Lemma 10.2.12(1).

(3) $\Rightarrow$ (1) It is easily verified. □

**Theorem 10.2.15.** ([128, Theorem 3.4]) *Let $R$ be a reduced commutative ring that is not an integral domain, and assume that $Z(R)$ is an ideal of $R$. Then $AG(R) \neq \Gamma(R)$ and $gr(AG(R)) = 3$.*

*Proof.* Let $z \in Z(R)^*$, $c \in ann_R(z) \setminus \{0\}$, and $m \in ann_R(c + z) \setminus \{0\}$. Then $m \in ann_R(c + z) \subsetneq ann_R(z)$ by Lemma 10.2.13(2), and thus $mc = 0$. Since $c^2 \neq 0$, we have $m \neq c$, and hence $c + z \neq m + z$. Since $\{c, m\} \subseteq ann_R(z)$ and $z^2 \neq 0$, we have $c + z$ and $m + z$ are distinct nonzero elements of $Z(R)$. Since $(m + z)(c + z) = z^2 \neq 0$, we have $(c + z) - (m + z)$ is not an edge of $\Gamma(R)$. Since $c^2 \neq 0$ and $m^2 \neq 0$, it follows that $(c+m) \in ann_R(z^2)\setminus(ann_R(c+z) \cup ann_R(m+z))$, and thus $(c+z)-(m+z)$ is an edge of $AG(R)$. Since $(c+z)-(m+z)$ is an edge of $AG(R)$ that is not an edge of $\Gamma(R)$, we have $AG(R) \neq \Gamma(R)$. Since $R$ is reduced and $AG(R) \neq \Gamma(R)$, we have $gr(AG(R)) = 3$ by Theorem 10.2.6. □

**Theorem 10.2.16.** ([128, Theorem 3.5]) *Let $R$ be a reduced commutative ring with $|Min(R)| \geq 3$ (possibly $Min(R)$ is infinite). Then $AG(R) \neq \Gamma(R)$ and $gr(AG(R)) = 3$.*

*Proof.* If $Z(R)$ is an ideal of $R$, then $AG(R) \neq \Gamma(R)$ by Theorem 10.2.15. Hence assume that $Z(R)$ is not an ideal of $R$. Since $|Min(R)| \geq 3$, we have $diam(\Gamma(R)) = 3$ by Lemma 10.2.12(2), and thus $AG(R) \neq \Gamma(R)$ by Theorem 8.1.2. Since $R$ is reduced and $AG(R) \neq \Gamma(R)$, we have $gr(AG(R)) = 3$ by Theorem 10.2.6. □

**Theorem 10.2.17.** ([128, Theorem 3.6]) *Let $R$ be a reduced commutative ring that is not an integral domain. Then $AG(R) = \Gamma(R)$ if and only if $|Min(R)| = 2$.*

*Proof.* Suppose that $AG(R) = \Gamma(R)$. Since $R$ is a reduced commutative ring that is not an integral domain, $|Min(R)| = 2$ by Theorem 10.2.16. Conversely, suppose that $|Min(R)| = 2$. Let $P_1$, $P_2$ be the minimal prime ideals of $R$. Since $R$ is reduced, we have $Z(R) = P_1 \cup P_2$ and $P_1 \cap P_2 = \{0\}$. Let $a, b \in Z(R)^*$. Assume that $a, b \in P_1$. Since $P_1 \cap P_2 = \{0\}$, neither $a \in P_2$ nor $b \in P_2$, and thus $ab \neq 0$. Since $P_1P_2 \subseteq P_1 \cap P_2 = \{0\}$, it follows that $ann_R(ab) = ann_R(a) = ann_R(b) = P_2$. Thus $a - b$ is not an edge of $AG(R)$. Similarly, if $a, b \in P_2$, then $a - b$ is not an edge of $AG(R)$. If $a \in P_1$ and $b \in P_2$, then $ab = 0$, and thus $a - b$ is an edge of $AG(R)$. Hence each edge of $AG(R)$ is an edge of $\Gamma(R)$, and therefore $AG(R) = \Gamma(R)$. □

**Theorem 10.2.18.** ([128, Theorem 3.7]) *Let $R$ be a reduced commutative ring. Then the following statements are equivalent:*

(1) $gr(AG(R)) = 4$;

(2) $AG(R) = \Gamma(R)$ *and* $gr(\Gamma(R)) = 4$;

(3) $gr(\Gamma(R)) = 4$;

(4) $T(R)$ *is ring-isomorphic to* $K_1 \times K_2$, *where each* $K_i$ *is a field with* $|K_i| \geq 3$;

(5) $|Min(R)| = 2$ *and every minimal prime ideal of $R$ has at least three distinct elements;*

(6) $\Gamma(R) = K_{m,n}$ *with* $m, n \geq 2$;

(7) $AG(R) = K_{m,n}$ *with* $m, n \geq 2$.

*Proof.* (1) ⇒ (2) Since $gr(AG(R)) = 4$, $AG(R) = \Gamma(R)$ by Theorem 10.2.6, and thus $gr(\Gamma(R)) = 4$.

(2) ⇒ (3) No comments.

(3) ⇔ (4) ⇔ (5) ⇔ (6) are clear by Theorem 2.3.8.

(6) $\Rightarrow$ (7) Since (6) implies $|Min(R)| = 2$ by Theorem 2.3.8, we conclude that $AG(R) = \Gamma(R)$ by Theorem 10.2.17, and thus $gr(AG(R)) = gr(\Gamma(R)) = 4$.

(7) $\Rightarrow$ (1) This is clear since $AG(R)$ is a complete bipartite graph and $m, n \geq 2$. □

**Theorem 10.2.19.** ([128, Theorem 3.8]) *Let $R$ be a reduced commutative ring that is not an integral domain. Then the following statements are equivalent:*

(1) $gr(AG(R)) = \infty$;

(2) $AG(R) = \Gamma(R)$ *and* $gr(AG(R)) = \infty$;

(3) $gr(\Gamma(R)) = \infty$;

(4) $T(R)$ *is ring-isomorphic to* $\mathbb{Z}_2 \times K$, *where* $K$ *is a field;*

(5) $|Min(R)| = 2$ *and at least one minimal prime ideal of* $R$ *has exactly two distinct elements;*

(6) $\Gamma(R) = K_{1,n}$ *for some* $n \geq 1$;

(7) $AG(R) = K_{1,n}$ *for some* $n \geq 1$.

*Proof.* (1) $\Rightarrow$ (2) Since $gr(AG(R)) = \infty$, $AG(R) = \Gamma(R)$ by Theorem 10.2.6, and thus $gr(\Gamma(R)) = \infty$.

(2) $\Rightarrow$ (3) No comments.

(3) $\Leftrightarrow$ (4) $\Leftrightarrow$ (5) $\Leftrightarrow$ (6) are clear by Theorem 2.3.10.

(6) $\Rightarrow$ (7) Since (6) implies $|Min(R)| = 2$ by Theorem 2.3.10, we conclude that $AG(R) = \Gamma(R)$ by Theorem 10.2.17, and thus $gr(AG(R)) = gr(\Gamma(R)) = \infty$. (7) $\Rightarrow$ (1) It is clear. □

In view of Theorem 10.2.18 and Theorem 10.2.19, we have the following result.

**Corollary 10.2.20.** ([128, Corollary 3.9]) *Let $R$ be a reduced commutative ring. Then* $AG(R) = \Gamma(R)$ *if and only if* $gr(AG(R)) = gr(\Gamma(R)) \in \{4, \infty\}$.

For the remainder of this section, we study the case when $R$ is not reduced.

**Theorem 10.2.21.** ([128, Theorem 3.10]) *Let $R$ be a nonreduced commutative ring with* $|Nil(R)^*| \geq 2$, *and let* $AG_N(R)$ *be the (induced) subgraph of* $AG(R)$ *with vertices* $Nil(R)^*$. *Then* $AG_N(R)$ *is complete.*

*Proof.* Suppose there are distinct $0 \neq a, b \in Nil(R)$ such that $ab \neq 0$. Assume that $ann_R(ab) = ann_R(a) \cup ann_R(b)$. Hence $ann_R(ab) = ann_R(a)$ or $ann_R(ab) = ann_R(b)$. Without lost of generality, we may assume that $ann_R(ab) = ann_R(a)$. Let $n$ be the least positive integer such that $b^n = 0$. Suppose that $ab^k \neq 0$ for every $k$, $1 \leq k < n$. Then $b^{n-1} \in ann_R(ab) \setminus ann_R(a)$, a contradiction. Thus assume that $k$, $1 \leq k < n$ is the least positive integer such that $ab^k = 0$. Since $ab \neq 0$, $1 < k < n$. Hence $b^{k-1} \in ann_R(ab) \setminus ann_R(a)$, a contradiction. Thus $a - b$ is an edge of $AG_N(R)$. □

In view of Theorem 10.2.21, we have the following result.

**Corollary 10.2.22.** ([128, Corollary 3.11]) *Let $R$ be a nonreduced local commutative ring with maximal ideal $Nil(R)$ such that $|Nil(R)^*| \geq 2$. Then $AG(R)$ is complete. In particular, $AG(\mathbb{Z}_{2^n})$ is complete for every $n \geq 3$, and if $q > 2$ is a prime number, then $AG(\mathbb{Z}_{q^n})$ is complete for every $n \geq 2$.*

The following is an example of a local commutative ring $R$ with maximal ideal $Nil(R)$ such that $w^2 = 0$ for every $w \in Nil(R)$, $diam(\Gamma(R)) = 2$, $diam(AG(R)) = 1$, and $gr(AG(R)) = gr(\Gamma(R)) = 3$.

**Example 10.2.23.** ([128, Example 3.12]) Let $R = \mathbb{Z}_2[x, y]/(x^2, y^2)$, $a = x + (x^2, y^2) \in R$, and $b = y + (x^2, y^2) \in R$. Then $R$ is a local commutative ring with maximal ideal $Nil(R) = (a, b)R$. It is clear that $w^2 = 0$ for every $w \in Nil(R)$ and $diam(AG(R)) = 1$ by Corollary 10.2.22. Since $Nil(R)^2 \neq \{0\}$ and $abNil(R) = \{0\}$, we have $diam(\Gamma(R)) = 2$ by Lemma 10.2.12(2). Since $a - ab - (ab + a) - a$ is a cycle of length three in $\Gamma(R)$, we have $gr(AG(R)) = gr(\Gamma(R)) = 3$.

**Theorem 10.2.24.** ([128, Corollary 3.13]) *Let $R$ be a nonreduced commutative ring with $|Nil(R)^*| \geq 2$, and let $\Gamma_N(R)$ be the induced subgraph of $\Gamma(R)$ with vertices $Nil(R)^*$. Then $\Gamma_N(R)$ is complete if and only if $Nil(R)^2 = \{0\}$.*

*Proof.* If $Nil(R)^2 = \{0\}$, then it is clear that $\Gamma_N(R)$ is complete. Hence assume that $\Gamma_N(R)$ is complete. We need only show that $w^2 = 0$ for every $w \in Nil(R)^*$. Let $w \in Nil(R)^*$ and assume that $w^2 \neq 0$. Let $n$ be the least positive integer such that $w^n = 0$. Then $n \geq 3$. Thus $w, w^{n-1} + w$ are distinct elements of $Nil(R)^*$. Since $w(w^{n-1} + w) = 0$ and $w^n = 0$, we have $w^2 = 0$, a contradiction. Thus $w^2 = 0$ for each $w \in Nil(R)$. □

**Theorem 10.2.25.** ([128, Theorem 3.14]) *Let $R$ be a nonreduced commutative ring and suppose that $Nil(R)^2 \neq \{0\}$. Then $AG(R) \neq \Gamma(R)$ and $gr(AG(R)) = 3$.*

*Proof.* Since $Nil(R)^2 \neq \{0\}$, $AG(R) \neq \Gamma(R)$ by Theorem 10.2.21 and Theorem 10.2.24. Thus $gr(AG(R)) \in \{3, 4\}$ by Corollary 10.2.11. Let $F = \mathbb{Z}_2 \times B$, where $B$ is $\mathbb{Z}_4$ or $\mathbb{Z}_2[x]/(x^2)$. Since $Nil(F)^2 = \{0\}$ and $Nil(F) \neq \{0\}$, we have $gr(AG(R)) \neq 4$ by Theorem 10.2.9. Thus $gr(AG(R)) = 3$. □

**Theorem 10.2.26.** ([128, Theorem 3.15]) *Let $R$ be a nonreduced commutative ring such that $Z(R)$ is not an ideal of $R$. Then $AG(R) \neq \Gamma(R)$.*

*Proof.* Since $R$ is not reduced and $Z(R)$ is not an ideal of $R$, $diam(\Gamma(R)) = 3$ by Corollary 2.4.27. Hence $AG(R) \neq \Gamma(R)$ by Theorem 8.1.2. □

**Theorem 10.2.27.** ([128, Theorem 3.16]) *Let $R$ be a nonreduced commutative ring. Then the following statements are equivalent:*

(1) $gr(AG(R)) = 4$;

(2) $AG(R) \neq \Gamma(R)$ *and* $gr(AG(R)) = 4$;

(3) $R$ *is ring-isomorphic to either* $\mathbb{Z}_2 \times \mathbb{Z}_4$ *or* $\mathbb{Z}_2 \times \mathbb{Z}_2[x]/(x^2)$;

(4) $\Gamma(R) = \overline{K}_{1,3}$;

(5) $AG(R) = K_{2,3}$.

*Proof.* (1) $\Rightarrow$ (2) Suppose $AG(R) = \Gamma(R)$. Then $gr(\Gamma(R)) = 4$, and $R$ is ring-isomorphic to $D \times B$, where $D$ is an integral domain with $|D| \geq 3$ and $B = \mathbb{Z}_4$ or $\mathbb{Z}_2[x]/(x^2)$ by Theorem 2.3.9. Assume that $R$ is ring-isomorphic to $D \times \mathbb{Z}_4$. Since $|D| \geq 3$, there is an $a \in D \setminus \{0, 1\}$. Let $x = (0, 1), y = (1, 2), w = (a, 2) \in R$. Then $x, y, w$ are distinct elements in $Z(R)^*$, $w(xy) = (0, 0)$, $wx \neq (0, 0)$, and $wy \neq (0, 0)$. Thus $x - w - y - x$ is a cycle of length three in $AG(R)$ by Lemma 10.2.3, a contradiction. Similarly, assume that $R$ is ring-isomorphic to $D \times \mathbb{Z}_2[x]/(x^2)$. Again, since $|D| \geq 3$, there is an $a \in D \setminus \{0, 1\}$. Let $b = x + (x^2) \in \mathbb{Z}_2[X]/(X^2)$. Then it is easily verified that $(0, 1) - (a, b) - (1, b) - (0, 1)$ is a cycle of length three in $AG(R)$, a contradiction. Thus $AG(R) \neq \Gamma(R)$.

(2) $\Rightarrow$ (3) It is clear by Theorem 10.2.9.

(3) $\Leftrightarrow$ (4) It is clear by Theorem 2.3.11.

(4) $\Rightarrow$ (5) Since (4) implies (3) by Theorem 2.3.11, it is easily verified that the annihilator graph of the two rings in (3) is $K_{2,3}$. (4) $\Rightarrow$ (5) Since $AG(R)$ is a $K_{2,3}$, it is clear that $gr(AG(R)) = 4$. □

We observe that $gr(\Gamma(\mathbb{Z}_8)) = \infty$, but $gr(AG(\mathbb{Z}_8)) = 3$. We have the following result.

**Theorem 10.2.28.** [128, Theorem 3.17]) *Let $R$ be a commutative ring with $AG(R) \neq \Gamma(R)$. Then the following statements are equivalent:*

(1) $\Gamma(R)$ *is a star graph;*

(2) $\Gamma(R) = K_{1,2}$;

(3) $AG(R) = K_3$.

*Proof.* (1) $\Rightarrow$ (2) Since $gr(\Gamma(R)) = \infty$ and $\Gamma(R) \neq AG(R)$, we have that $R$ is not reduced by Corollary 10.2.20 and $|Z(R)^*| \geq 3$. Since $\Gamma(R))$ is a star graph, there are two sets $A, B$ such that $Z(R)^* = A \cup B$ with $|A| = 1$, $A \cap B = \emptyset$, $AB = \{0\}$, and $b_1b_2 \neq 0$ for every $b_1, b_2 \in B$. Since $|A| = 1$, let $A = \{w\}$ for $w \in Z(R)^*$. Since every edge of $\Gamma(R)$ is an edge of $AG(R)$ and $AG(R) \neq \Gamma(R)$, there are $x, y \in B$ such that $x - y$ is an edge of $AG(R)$ that is not an edge of $\Gamma(R)$. Since $ann_R(c) = w$ for every $c \in B$ and $ann_R(xy) \neq ann_R(x) \cup ann_R(y)$, we have $ann_R(xy) \neq w$. Thus $ann_R(xy) = B$ and $xy = w$. Since $A = \{xy\}$ and $AB = \{0\}$, we have $x(xy) = x^2y = 0$ and $y(xy) = y^2x = 0$. We show that $B = \{x, y\}$, and hence $|B| = 2$. Thus assume there is a $c \in B$ such that $c \neq x$ and $c \neq y$. Then $wc = xyc = 0$. We show that $(xc + xy) \neq x$ and $(xc + xy) \neq xy$ (note that $xy = w$). Suppose that $(xc + xy) = x$. Then $y(xc + xy) = yx$.

But $y(xc + xy) = yxc + xy^2 = 0 + 0 = 0$ and $xy \neq 0$, a contradiction. Hence $x \neq (xc + xy)$. Since $x, c \in B$, we have $xc \neq 0$, and thus $(xc + xy) \neq xy$. Hence $x, (xc + xy), xy$ are distinct elements of $Z(R)^*$. Since $x^2y = 0$ and $y \in B$, either $x^2 = 0$, $x^2 = xy$, or $x^2 = y$. Suppose that $x^2 = y$. Since $xy = w \neq 0$, we have $xy = x(x^2) = x^3 = w \neq 0$. Since $x^2y = 0$, we have $x^4 = 0$. Since $x^4 = 0$ and $x^3 \neq 0$, we have $x^2, x^3, x^2 + x^3$ are distinct elements of $Z(R)^*$, and thus $x^2 - x^3 - (x^2 + x^3) - x^2$ is a cycle of length three in $\Gamma(R)$, a contradiction. Hence, we assume that either $x^2 = 0$ or $x^2 = xy = w$. In both cases, we have $x^2c = 0$. Since $x, (xc + xy), xy$ are distinct elements of $Z(R)^*$ and $xy^2 = yx^2 = x^2c = 0$, we have $x - (xc + xy) - xy - x$ is a cycle of length three in $\Gamma(R)$, a contradiction. Thus $B = \{x, y\}$ and $|B| = 2$. Hence $\Gamma(R) = K_{1,2}$.

(2) $\Rightarrow$ (3) Since every edge of $\Gamma(R)$ is an edge of $AG(R)$ and $\Gamma(R) \neq AG(R)$ and $\Gamma(R) = K_{1,2}$, it is clear that $AG(R)$ must be $K_3$.

(3) $\Rightarrow$ (1) Since $|Z(R)^*| = 3$, $\Gamma(R)$ is connected, and $AG(R) \neq \Gamma(R)$, exactly one edge of $AG(R)$ is not an edge of $\Gamma(R)$. Thus $\Gamma(R)$ is a star graph. □

**Theorem 10.2.29.** ([128, Theorem 3.18]) *Let $R$ be a nonreduced commutative ring with $|Z(R)^*| \geq 2$. Then the following statements are equivalent:*

(1) *$AG(R)$ is a star graph;*

(2) *$gr(AG(R)) = \infty$;*

(3) *$AG(R) = \Gamma(R)$ and $gr(\Gamma(R)) = \infty$;*

(4) *$Nil(R)$ is a prime ideal of $R$ and either $Z(R) = Nil(R) = \{0, w, -w\}$ ($w \neq -w$) for some $0 \neq w \in R$, or $Z(R) \neq Nil(R)$ and $Nil(R) = \{0, w\}$ for some $0 \neq w \in R$ (and hence $wZ(R) = \{0\}$);*

(5) *Either $AG(R) = K_{1,1}$ or $AG(R) = K_{1,\infty}$;*

(6) *Either $\Gamma(R) = K_{1,1}$ or $\Gamma(R) = K_{1,\infty}$.*

*Proof.* (1) $\Rightarrow$ (2) It is clear by the definition of a star graph.

(2) $\Rightarrow$ (3) Since $gr(AG(R)) = \infty$, $AG(R) = \Gamma(R)$ by Corollary 10.2.11, and thus $gr(\Gamma(R)) = \infty$.

(3) $\Rightarrow$ (4) Suppose that $|Nil(R)^*| \geq 3$. Since $AG_N(R)$ is complete by Theorem 10.2.21 and $|Nil(R)^*| \geq 3$, we have $gr(AG(R)) = gr(\Gamma(R)) = 3$, a contradiction. Thus $|Nil(R)^*| \in \{1, 2\}$. Suppose $|Nil(R)^*| = 2$. Then $Nil(R) = \{0, w, -w\}$ ($w \neq -w$) for some $0 \neq w \in R$. We show that $Z(R) = Nil(R)$. Assume there is a $k \in Z(R) \setminus Nil(R)$. Suppose that $wk = 0$. Since $Nil(R)^2 = \{0\}$, $w - k - (-w) - w$ is a cycle of length three in $\Gamma(R)$, a contradiction. Thus assume that $wk \neq 0$. Hence there is an $f \in Z(R)^* \setminus \{w, -w, k\}$ such that $w - f - z$ is a path of length two in $\Gamma(R)$ by Theorem 8.1.2 (note that we are assuming that $AG(R) = \Gamma(R)$). Thus $w - f - (-w) - w$ is a cycle of length three in $\Gamma(R)$, a contradiction. Hence if $|Nil(R)^*| = 2$, then $Z(R) = Nil(R)$. Thus assume that $Nil(R) = \{0, w\}$ for some $0 \neq w \in R$. We show $Nil(R)$ is a prime ideal of $R$. Since $gr(AG(R)) = gr(\Gamma(R)) = \infty$, we have $AG(R) = \Gamma(R)$

is a star graph by Theorem 2.3.11 and Theorem 10.2.27. Since $|Z(R)^*| \geq 2$ by hypothesis and $|Nil(R)^*| = 1$, we have $Z(R) \neq Nil(R)$. Let $c \in Z(R)^* \setminus Nil(R)^*$. We show $wc = 0$. Suppose that $wc \neq 0$. Since $|Nil(R)^*| = 1$ and $wc \neq 0$, we have $wc = w$. Thus $w(c-1) = 0$. Since $w + 1 \in U(R)$ and $c \notin U(R)$, we have $c - 1 \neq w$. Since $\Gamma(R)$ is a star graph, $w(c-1) = 0$, and $wc \neq 0$, we have $(c-1)j = 0$ for every $j \in Z(R)^* \setminus \{c-1\}$. In particular, $(c-1)[(c-1)+w] = 0$, and therefore $w - (c-1) - (c-1+w) - w$ is a cycle of length three in $\Gamma(R)$, a contradiction. Hence $wc = 0$. Since $wZ(R) = \{0\}$ and $\Gamma(R)$ is a star graph, we have $Nil(R) = \{0, w\}$ is a prime ideal of $R$.

(4) $\Rightarrow$ (5) Suppose that $Nil(R)$ is a prime ideal of $R$. If $Z(R) = Nil(R)$ and $|Nil(R)^*| = 2$, then $AG(R) = K_{1,1}$. Hence, assume that $Nil(R) = \{0, w\}$ for some $0 \neq w \in R$. We show that $Z(R)$ is infinite. Let $c \in Z(R) \setminus Nil(R)$ and let $n > m \geq 1$. We show that $c^m \neq c^n$. Suppose that $c^m = c^n$. Then $c^m(1 - c^{n-m}) = 0$. Since $Nil(R) = \{0, w\}$ is a prime ideal of $R$, we have $(1 - c^{n-m}) = w$. Since $1 - w \in U(R)$, we have $1 - w = c^{n-m} \in U(R)$, a contradiction. Thus $c^m \neq c^n$, and hence $Z(R)$ is infinite. Since $Nil(R) = \{0, w\}$ is a prime ideal of $R$ and $wZ(R) = \{0\}$, we have $AG(R) = K_{1,\infty}$.

(5) $\Rightarrow$ (6) It is clear.

(6) $\Rightarrow$ (1) Since $\Gamma(R)$ is a star graph and $\Gamma(R) \neq K_{1,2}$, we have $AG(R) = \Gamma(R)$ by Theorem 10.2.28, and thus $gr(AG(R)) = \infty$. □

**Corollary 10.2.30.** ([128, Corollary 3.19]) *Let $R$ be a nonreduced commutative ring with $|Z(R)^*| \geq 2$. Then $\Gamma(R)$ is a star graph if and only if $\Gamma(R) = K_{1,1}$, $\Gamma(R) = K_{1,2}$, or $\Gamma(R) = K_{1,\infty}$*

*Proof.* The proof is a direct consequence of Theorem 10.2.28 and Theorem 10.2.29. □

In the following example, we construct two nonreduced commutative rings $R_1$ and $R_2$ with $AG(R_1) = K_{1,1}$ and $AG(R_2) = K_{1,\infty}$.

**Example 10.2.31.** ([128, Example 3.20])

(1) Let $R_1 = \mathbb{Z}_3[x]/(x^2)$ and $a = x + (x^2) \in R_1$. Then $Z(R_1) = Nil(R_1) = \{0, -a, a\}$ and $AG(R_1) = \Gamma(R_1) = K_{1,1}$. Also note that $AG(\mathbb{Z}_9) = \Gamma(\mathbb{Z}_9) = K_{1,1}$.

(2) Let $R_2 = \mathbb{Z}_2[x, y]/(xy, x^2)$, $a = x + (xy + x^2)$, and $b = y + (xy + x^2) \in R_2$. Then $Z(R_2) = (a, b)R_2$, $Nil(R_2) = \{0, a\}$, and $Z(R_2) \neq Nil(R_2)$. It is clear that $AG(R_2) = \Gamma(R_2) = K_{1,\infty}$.

**Remark 10.2.32.** ([128, Example 3.21]) Let $R$ be a nonreduced commutative ring. In view of Theorem 10.2.26, Theorem 10.2.27, and Theorem 10.2.29, if $AG(R) = \Gamma(R)$, then $Z(R)$ is an ideal of $R$ and $gr(AG(R)) = gr(\Gamma(R)) \in \{3, \infty\}$. The converse is true if $gr(AG(R) = gr(\Gamma(R)) = \infty$ (see Theorem 10.2.26 and 10.2.29). However, if $Z(R)$ is an ideal of $R$ and $gr(AG(R)) = gr(\Gamma(R)) = 3$, then it is possible to have all the following cases:

(1) It is possible to have a commutative ring $R$ with $Z(R)$ an ideal of $R$, $Z(R) \neq Nil(R)$, $AG(R) = \Gamma(R)$, and $gr(AG(R)) = 3$. See Example 10.2.33.

(2) It is possible to have a commutative ring $R$ with $Z(R)$ an ideal of $R$, $Z(R) \neq Nil(R)$, $Nil(R)^2 = \{0\}$, $AG(R) \neq \Gamma(R)$, $diam(AG(R)) = diam(\Gamma(R)) = 2$, and $gr(AG(R)) = gr(\Gamma(R)) = 3$. See Example 10.2.34.

(3) It is possible to have a commutative ring $R$ with $Z(R)$ an ideal of $R$, $Z(R) \neq Nil(R)$, $Nil(R)^2 = \{0\}$, $AG(R)$ is a complete graph (i.e., $diam(AG(R)) = 1$), $AG(R) \neq \Gamma(R)$, $diam(\Gamma(R)) = 2$, and $gr(AG(R)) = gr(\Gamma(R)) = 3$. See Theorem 10.2.35.

**Example 10.2.33.** ([128, Example 3.22]) Let $D = \mathbb{Z}_2[x, y, w]$, $I = (x^2, y^2, xy, xy)D$, $R = D/I$, and $a = x + I, b = y + I, c = w + I \in R$. Then $Nil(R) = (a, b)R$, and $Z(R) = (a, b, c)R$ is an ideal of $R$. By construction, we have $Nil(R)^2 = \{0\}$, $AG(R) = \Gamma(R)$, $diam(AG(R)) = diam(\Gamma(R)) = 2$, and $gr(AG(R)) = gr(\Gamma(R)) = 3$ (for example, $a - (a + b) - b - a$ is a cycle of length three).

**Example 10.2.34.** ([128, Example 3.23]) Let $D = \mathbb{Z}_2[x, y, w]$, $I = (x^2, y^2, xy, xw, yw^3)D$, $R = D/I$, and $a = x + I, b = y + I, c = w + I \in R$. Then $Nil(R) = (a, b)R$, and $Z(R) = (a, b, c)R$ is an ideal of $R$. By construction, $Nil(R)^2 = \{0\}$, $diam(AG(R)) = diam(\Gamma(R)) = 2$, $gr(AG(R)) = gr(\Gamma(R)) = 3$. However, since $c^3 \neq 0$ and $b \in ann_R(c^3) \setminus (ann_R(c) \cup ann_R(c^2))$, we have $c - c^2$ is an edge of $AG(R)$ that is not an edge of $\Gamma(R)$, and hence $AG(R) \neq \Gamma(R)$.

Recall that, for a given commutative ring $R$ and an $R$-module $M$, the *idealization* of $M$ is the ring $R(+)M = R \times M$ with addition defined by $(r, m) + (s, n) = (r + s, m + n)$ and multiplication defined by $(r, m)(s, n) = (rs, rn + sm)$ for all $r, s \in R$ and $m, n \in M$. Note that $\{0\}(+)M \subseteq Nil(R(+)M)$ since $(\{0\}(+)M)^2 = \{(0, 0)\}$. We have the following result.

**Theorem 10.2.35.** ([128, Theorem 3.24]) *Let $D$ be a principal ideal domain that is not a field with quotient field $K$ (for example, $D = \mathbb{Z}$ or $D = F[X]$ for a field $F$), and $Q = (p)$ be a nonzero prime ideal of $D$ for some prime (irreducible) element $p \in D$. Set $M = K/D_Q$ and $R = D(+)M$. Then $Z(R) \neq Nil(R)$, $AG(R)$ is a complete graph, $AG(R) \neq \Gamma(R)$, and $gr(AG(R)) = gr(\Gamma(R)) = 3$.*

*Proof.* By construction of $R$, $Z(R) = Q(+)M$, $Nil(R) = \{0\}(+)M$, and $Nil(R)^2 = \{(0, 0)\}$. Let $x, y$ be distinct elements of $Z(R)^*$ and suppose that $xy \neq 0$. Since $Nil(R)^2 = \{(0, 0)\}$, to show that $AG(R)$ is complete, we consider two cases.
**Case I:** Assume that $x \in Nil(R)^*$ and $y \in Z(R) \setminus Nil(R)$. Then $x = (0, \frac{a}{cp^m} + D_Q)$ for some $0 \neq a \in D$, $c \in D \setminus Q$, and some positive integer $m \geq 1$ such that $\gcd(a, cp^m) = 1$, and $y = (hp^n, f)$ for some positive integer $n \geq 1$, a $0 \neq h \in D$, and $f \in M$. Since $xy \neq 0$, we have $n < m$. Thus $xy = (0, \frac{ha}{cp^{m-n}} + D_Q) \in Nil(R)^*$.

Since $(p^{m-n}, 0) \in ann_R(xy) \setminus (ann_R(x) \cup ann_R(y))$, we have $x - y$ is an edge of $AG(R)$.
**Case II:** Assume that $x, y \in Z(R)^* \setminus Nil(R)^*$. Then $x = (dp^u, g)$ and $y = (vp^r, w)$ for some positive integers $u, r \geq 1$, nonzero $d, v \in D \setminus Q$, and $g, w \in M$. Hence $xy = (dvp^{u+r}, dp^u w + vp^r g)$. Since $(0, \frac{1}{p^{u+r}} + D_Q) \in ann_R(xy) \setminus (ann_R(x) \cup ann_R(y))$, we have $x - y$ is an edge of $AG(R)$. Since $(0, \frac{1}{p} + D_Q) - (0, \frac{1}{p^2} + D_Q) - (0, \frac{1}{p^3} + D_Q) - (0, \frac{1}{p} + D_Q)$ is a cycle of length three in $\Gamma(R)$, we have $gr(AG(R)) = gr(\Gamma(R)) = 3$. □

The following example shows that the hypothesis "$Q$ is principal" is crucial in the above theorem.

**Example 10.2.36.** ([128, Example 3.25]) Let $D = \mathbb{Z}[x]$ with quotient field $K$ and $Q = (2, x)D$. Then $Q$ is a nonprincipal prime ideal of $D$. Set $M = K/D_Q$ and $R = D(+)M$. Then $Z(R) = Q(+)M$, $Nil(R) = \{0\}(+)M$, and $Nil(R)^2 = \{(0,0)\}$. Let $a = (2, 0)$ and $b = (0, \frac{1}{x} + D_Q)$. Then $ab = (0, \frac{2}{X} + D_Q) \in Nil(R)^*$. Since $ann_R(ab) = ann_R(b)$, we have $a - b$ is not an edge of $AG(R)$. Thus $AG(R)$ is not a complete graph.

The concept of 2-absorbing ideals, a generalization of prime ideals, was introduced in [127] (this concept was generalized in [69]). A proper ideal $I$ of $R$ is called a *2-absorbing ideal* of $R$ if whenever $a, b, c \in R$ and $abc \in I$, then $ab \in I$, $ac \in I$, or $bc \in I$.

The authors in [308] established the following connection between $\Gamma(R)$ and $AG(R)$ in terms of 2-absorbing ideals (the more detailed proof given below is from [84, Theorem 2.8]).

**Theorem 10.2.37.** ([308, Theorem 3.1]) *Let $R$ be a commutative ring. Then $\Gamma(R) = AG(R)$ if and only if $0$ is a 2-absorbing ideal of $R$.*

*Proof.* Suppose that 0 is a 2-absorbing ideal of $R$. Let $x, y \in Z(R)^*$ and $x - y$ be an edge of $AG(R)$. Then $\operatorname{ann}(x) \cup \operatorname{ann}(y) \subsetneq \operatorname{ann}(xy)$. Therefore, there is a $z \in \operatorname{ann}(xy) \setminus \operatorname{ann}(x) \cup \operatorname{ann}(y)$, and so $zxy = 0$, $zx \neq 0$, and $zy \neq 0$. Thus $xy = 0$ since 0 is a 2-absorbing ideal of $R$. Hence, $x - y$ is an edge of $\Gamma(R)$.

Conversely, assume that $\Gamma(R) = AG(R)$. We show that 0 is a 2-absorbing ideal of $R$. Let $x, y, z \in R$, $xyz = 0$ with $xy \neq 0$ and $xz \neq 0$; we show that $yz = 0$. First, suppose that $x = y = z$, Then $x$ and $w = x(1 + x)$ are distinct adjacent vertices in $AG(R)$, and thus $xw = 0$. Hence $yz = x^2 = 0$. Next, suppose that $x = y$ and $z$ are distinct. Then $x^2 \neq 0$ and $xz \neq 0$; so $x$ and $z$ are adjacent in $AG(R)$. Hence $yz = xz = 0$. Finally, assume that $x, y, z$ are all distinct. Then $x \in \operatorname{ann} yz \setminus \operatorname{ann} y \cup \operatorname{ann} z$, and thus $y - z$ is an edge of $AG(R)$. Therefore, $y - z$ is an edge of $\Gamma(R)$, and so $yz = 0$. □

Let $I$ be an ideal of a commutative ring $R$. In [17], Afkhami, Hoseni, and Khashyarmanesh defined the *ideal-based annihilator graph of $R$* (with respect to $I$)

as the simple graph $AG_I(R)$ with vertices $\{x \in R\backslash I \mid xy \in I \text{ for some } y \in R\backslash I\}$, and distinct vertices $x$ and $y$ are adjacent if and only if $(I : xy) \neq (I : x) \cup (I : y)$ (i.e., $((I : x) \cup (I : y) \subsetneq (I : xy))$. Note that $AG_{\{0\}}(R) = AG(R)$.

Let $S$ be a semigroup with 0 element. In [22], Afkhami, Khashyarmanesh, and Sakhdari defined the *annihilator graph of* $S$ to be the simple graph $AG(S)$ with vertices $Z(S)^*$ and distinct vertices $x$ and $y$ are adjacent if and only if $ann_S(xy) \neq ann_S(x) \cup ann_S(y)$ (i.e., $(ann_S(x) \cup ann_S(y) \subsetneq ann_S(xy))$.

For other graphs related to $AG(R)$, we recommend the survey article in [131]. More properties of $AG(R)$ can be found in [418] and [84].

## 10.3 Compressed Annihilator Graphs

Throughout this section, let $Ass(R)$ be the set of associated prime ideals and $Min(R)$ the set of minimal prime ideals of $R$. If $n \geq 3$ is a positive integer, then $C_n$ denotes a cycle with $n$ vertices. For $x, y \in R$, we say that $x \sim y$ if and only if $ann_R(x) = ann(y)$. As noted in [295], $\sim$ is an (congruence) equivalence relation. Let $[x]$ denote the equivalence class of $x$. Note that $[0] = \{0\}$, $[1] = R \setminus Z(R)$ and the other equivalence classes form a partition of $Z(R)^*$. Payrovi and Babaei in [308] introduced the concept of the compressed annihilator graph of a commutative ring. The compressed annihilator graph is a special case of a congruence-based graph (see Section 10.5 and Example 10.5.1(d)).

**Definition 10.3.1.** ([308, Definition 2.1]) The *compressed annihilator graph* of $R$, denoted $\mathrm{AG_E}(R)$, is the simple graph associated to $Z(R)$ whose vertices are the classes of elements in $Z(R)^*$, and two distinct classes $[x]$ and $[y]$ are adjacent if and only if $ann_R(x) \cup \mathrm{ann}_R(y) \subsetneq ann_R(xy)$.

**Remark 10.3.2.** ([308, Remark 2.2]) Adjacency is well-defined: let $[x_1] = [x_2]$ and $[y_1] = [y_2]$; that is $ann_R(x_1) = ann_R(x_2)$ and $ann_R(y_1) = ann_R(y_2)$. If $z \in ann_R(x_1y_1)\backslash(ann_R(x_1) \cup ann_R(y_1))$, then $zx_1y_1 = 0$, and so $zy_1 \in ann_R(x_1) = ann_R(x_2)$. Thus $zy_1x_2 = 0$, and so $zx_2 \in ann_R(y_1) = ann_R(y_2)$. Therefore, $z \in ann_R(x_2y_2) \setminus (ann_R(x_2) \cup ann_R(y_2))$. Hence, $ann_R(x_1) \cup ann_R(y_1) \subsetneq ann_R(x_1y_1)$ if and only if $ann_R(x_2) \cup ann_R(y_2) \subsetneq ann_R(x_2y_2)$.

The graph $\mathrm{AG_E}(R)$ has the same vertices as $\Gamma_E(R)$ (recall from Section 5.9 that $\Gamma_E(R)$ is the compressed zero-divisor graph whose vertices are the classes of elements in $Z(R)^*$, and two distinct classes $[x]$ and $[y]$ are adjacent if and only if $[x][y] = [0]$, if and only if $xy = 0$. It is obvious that every edge of $\Gamma_E(R)$ is an edge of $\mathrm{AG_E}(R)$.

**Lemma 10.3.3.** ([308, Lemma 2.3]) *Let $R$ be a commutative ring and $y_1, y_2 \in R$ such that $ann_R(y_1)$ and $ann_R(y_2)$ are distinct associated primes of $R$. Then $[y_1]$ and $[y_2]$ are adjacent in $\mathrm{AG_E}(R)$. If $R$ is Noetherian and $[x]$ is a vertex of $\mathrm{AG_E}(R)$, then either $ann_R(x) \in Ass(R)$ or there is a vertex $[x']$ adjacent to $[x]$ such that $ann_R(x') \in Ass(R)$.*

**Theorem 10.3.4.** ([308, Theorem 2.4]) *Let $R$ be a commutative ring. Then* $\mathrm{AG_E}(R)$ *is connected with* $diam(\mathrm{AG_E}(R)) \leq 2$.

Since $AG_E(R)$ is a congruence-based graph, the following theorem actually holds without the Noetherian hypothesis by Theorem 10.5.5(d).

**Theorem 10.3.5.** ([308, Theorem 2.5]) *If $R$ be a commutative Noetherian ring. Then* $gr(\mathrm{AG_E}(R)) \in \{3, 4, \infty\}$.

**Example 10.3.6.** ([308, Example 2.6]) Let $R = \mathbb{Z}_{p^2q}$, where $p, q$ are two distinct prime numbers. Then $[p]$, $[p^2]$, $[q]$, and $[pq]$ are vertices of $\mathrm{AG_E}(R)$ and $\mathrm{AG_E}(R) = C_4$. Therefore, $gr(\mathrm{AG_E}(R)) = 4$.

**Example 10.3.7.** ([308, Example 2.7]) Let $R = \mathbb{Z}_4[x, y]/(x^2, yx, 2x)$, and let $a$ and $b$ denote the images of $x$ and $y$ in $R$, respectively. Then we have $\Gamma_E(R)$ is the path $[2] - [a] - [b]$. It is clear that $ann_R(2y) = ann_R(2)$, and so $[2] - [b]$ is not an edge of $\mathrm{AG_E}(R)$. Therefore, $\mathrm{AG_E}(R) = \Gamma_E(R)$ and $gr(\mathrm{AG_E}(R)) = \infty$.

It is shown that there are no complete graphs $\Gamma_E(R)$ with three or more vertices, see Proposition 5.9.4. But the following example shows that for every positive integer $n$, there is a commutative ring for which the compressed annihilator graph is complete with $n$ vertices.

**Example 10.3.8.** ([308, Example 2.8]) Let $p$ be a prime number and $n \geq 2$. Then $[p], [p^2], \ldots, [p^n]$ are all distinct vertices of $\mathrm{AG_E}(\mathbb{Z}_{p^{n+1}})$. Assume that $1 \leq i, j \leq n$. It is obvious that $[p^i][p^j] = 0$ whenever $i + j \geq n$, so that $[p^i], [p^j]$ are adjacent. If $i + j < n$, then $p^{n-(i+j)} \in ann_R(p^{i+j}) \setminus (ann_R(p^i) \cup ann_R(p^j))$ so that $[p^i], [p^j]$ are adjacent. Hence, $\mathrm{AG_E}(\mathbb{Z}_{p^{n+1}})$ is a complete graph with $n$ vertices.

**Theorem 10.3.9.** ([308, Theorem 2.9]) *Let $R$ be a commutative ring such that* $\mathrm{AG_E}(R)$ *is a complete graph with at least three vertices. Then $Z(R)$ is an ideal of $R$.*

**Lemma 10.3.10.** ([308, Lemma 2.10]) *Let $R$ be a commutative ring, $x \in R$, $P = ann_R(x)$ a prime ideal of $R$, and $y \in Z(R)^*$ with $[x] \neq [y]$. If $[x] - [y]$ is not an edge of $\Gamma_E(R)$, then $[x] - [y]$ is not an edge of* $\mathrm{AG_E}(R)$.

**Corollary 10.3.11.** ([308, Corollary 2.11]) *Let $R$ be a commutative ring with* $\mathrm{AG_E}(R)$ *a complete graph on $n$ vertices. If $ann_R(x)$is a prime ideal of $R$ for $x \in Z(R)^*$, then $deg([x]) = n - 1$ in $\Gamma_E(R)$.*

**Theorem 10.3.12.** ([308, Theorem 2.12]) *Let $R$ be a commutative Noetherian ring with* $\mathrm{AG_E}(R)$ *a complete graph on $n \geq 3$ vertices. Then $Z(R)$ is a prime ideal of $R$ and $|Ass(R)| = 1$.*

**Theorem 10.3.13.** ([308, Theorem 2.13]) *Let $R$ be a commutative Noetherian ring such that $\Gamma_E(R)$ is star graph with at least four vertices. Then* $\mathrm{AG_E}(R)$ *is a complete graph.*

**Example 10.3.14.** ([308, Example 2.14] and [357, Example 2.6]) Let $R = \mathbb{Z}_2[x,y,z]/(x^2,y^2)$. Then $\Gamma_E(R)$ is an infinite star graph. Therefore, $\mathrm{AG_E}(R)$ is an infinite complete graph.

**Corollary 10.3.15.** ([308, Corollary 2.15]) *Let $R$ be a commutative Noetherian ring. Then there is no star graph $\mathrm{AG_E}(R)$ with more than three vertices.*

**Proposition 10.3.16.** ([308, Proposition 2.16]) *Let $R$ be Noetherian. Then $gr(\mathrm{AG_E}(R)) = 4$ if and only if $\mathrm{AG_E}(R) = C_4$.*

**Theorem 10.3.17.** ([308, Theorem 2.17]) *Let $R$ be a reduced commutative ring. Then $|\mathrm{AG_E}(R)| = 2$ if and only if $\Gamma(R)$ is a complete bipartite graph.*

**Proposition 10.3.18.** (proof of [357, Corollary 1.6]) *Let $R$ be a commutative Noetherian ring with $|\mathrm{AG_E}(R)| = 3$. Then $\Gamma_E(R)$ is the graph $[x] - [y] - [z]$, where $\{[x],[y],[z]\}$ is the set of all vertices of $\Gamma_E(R)$.*

**Theorem 10.3.19.** ([308, Theorem 2.18]) *Let $R$ be a commutative Noetherian ring with $|\mathrm{AG_E}(R)| = 3$. Then $R$ is not reduced.*

*Proof.* Let $[x]$, $[y]$, and $[z]$ be the vertices of $\mathrm{AG_E}(R)$. It follows that $\Gamma_E(R)$ is the graph $[x] - [y] - [z]$ by Proposition 10.3.18. So we have $\mathrm{AG_E}(R) = \Gamma_E(R)$ or $\mathrm{AG_E}(R) = K_3$. If $\mathrm{AG_E}(R) = \Gamma_E(R)$, then $ann_R(x) \cup ann_R(z) = ann_R(xz)$. Without loss of generality, assume that $ann_R(x) = ann_R(xz)$. Thus $ann_R(z) \subsetneq ann_R(x)$. Let $s \in ann_R(x) \setminus ann_R(z)$. Then $ann_R(s) = ann_R(x)$, and so $x^2 = 0$; so $R$ is not reduced. Now, assume that $\mathrm{AG_E}(R) = K_3$. Thus $[x] - [z]$ is an edge of $\mathrm{AG_E}(R)$. Hence $ann_R(x) \cup ann_R(z) \subsetneq ann_R(xz)$, and therefore $ann_R(xz) = ann_R(y)$. We have $xy = 0$ and $x^2z = 0$. Thus $(xz)^2 = 0$; so $R$ is not reduced. □

**Corollary 10.3.20.** ([308, Corollary 2.19]) *Let $R$ be a commutative Noetherian ring with $gr(\mathrm{AG_E}(R)) = \infty$. Then $\mathrm{AG_E}(R)$ is a path of length $\leq 2$.*

**Proposition 10.3.21.** ([172, Proposition 4.1]) *Let $R$ and $S$ be commutative rings and $\varphi : R \longrightarrow S$ a flat ring-monomorphism. Then $\Gamma_E(R)$ is isomorphic to an induced subgraph of $\Gamma_E(S)$.*

**Proposition 10.3.22.** ([308, Proposition 2.20]) *Let $R$ and $S$ be commutative rings and $\varphi : R \longrightarrow S$ be a flat ring-monomorphism. Then $\mathrm{AG_E}(R)$ is isomorphic to an induced subgraph of $\mathrm{AG_E}(S)$.*

**Corollary 10.3.23.** ([308, Corollary 2.21]) *Let $R$ be a commutative ring and $T = R \setminus Z(R)$. Then $\mathrm{AG_E}(R) \cong \mathrm{AG_E}(R_T)$.*

Recall that a proper ideal $I$ of $R$ ([127]) is called *2-absorbing* if whenever $abc \in I$ for $a,b,c \in R$, then $ab \in I$, $ac \in I$, or $bc \in I$.

**Proposition 10.3.24.** ([308, Proposition 3.5]) *Let $R$ be a reduced commutative ring. Then $\Gamma_E(R) = \mathrm{AG_E}(R)$ if and only if $\Gamma(R) = AG(R)$.*

*Proof.* In view of Theorem 10.2.37, it is enough to show that $\Gamma_E(R) = \mathrm{AG_E}(R)$ if and only if $\{0\}$ is a 2-absorbing ideal of $R$. Let $\Gamma_E(R) = \mathrm{AG_E}(R)$ and $x, y, z \in R$ such that $xyz = 0$, $xy \neq 0$, and $xz \neq 0$. Then $y, z \in Z(R)^*$. Since $R$ is reduced, we have $[y] \neq [z]$. If $yz \neq 0$, then $[y] - [z]$ is not an edge of $\Gamma_E(R)$. Therefore, $[y] - [z]$ is not an edge of $\mathrm{AG_E}(R)$. Then $ann_R(y) \cup ann_R(z) = ann_R(yz)$, and so $ann_R(y) = ann_R(yz)$ or $ann_R(z) = ann_R(yz)$. It follows that $xy = 0$ or $xz = 0$, a contradiction. Hence $yz = 0$ and $\{0\}$ is a 2-absorbing ideal of $R$. The converse is clear. □

If $R_1, R_2$ are commutative rings such that either $Z(R_1)^* \neq \emptyset$ or $Z(R_2)^* \neq \emptyset$, then 0 is not a 2-absorbing ideal of $R_1 \times R_2$.

**Theorem 10.3.25.** ([308, Theorem 3.6]) *Let $R_1$ and $R_2$ be commutative rings with $Z(R_1)^* \neq \emptyset$ or $Z(R_2)^* \neq \emptyset$, and let $R = R_1 \times R_2$. Then*

(i) $\mathrm{AG_E}(R)$ *is not complete;*

(ii) $\mathrm{AG_E}(R) \neq \Gamma_E(R)$.

**Theorem 10.3.26.** ([308, Theorem 3.7]) *Let $R_1$ and $R_2$ be commutative rings such that $\mathrm{AG_E}(R_1)$ or $\mathrm{AG_E}(R_2)$ has at least one edge, and let $R = R_1 \times R_2$. Then $gr(\mathrm{AG_E}(R)) = 3$ and $diam(\mathrm{AG_E}(R)) = 2$.*

*Proof.* Let $a, b \in Z(R)^*$ and $[a] - [b]$ be an edge of $\mathrm{AG_E}(R_1)$. Then there is $c \in ann_R(ab) \setminus (ann_R(a) \cup ann_R(b))$. Thus $(c, 0) \in ann_R((a, 1)(b, 1))$, but $(c, 0) \notin ann_R(a, 1) \cup ann_R(b, 1)$. It follows that $[(a, 1)] - [(b, 1)]$ is an edge of $\mathrm{AG_E}(R)$. Now, the proof of Theorem 10.3.25(ii) shows that $[(1, 0)] - [(a, 1)] - [(b, 1)] - [(1, 0)]$ is a cycle in $\mathrm{AG_E}(R)$. Therefore, $gr(\mathrm{AG_E}(R)) = 3$. Furthermore, $\mathrm{AG_E}(R)$ is not complete by Theorem 10.3.25(i). Hence, $diam(\mathrm{AG_E}(R)) = 2$ by Theorem 10.3.4. □

## 10.4 $\Gamma_I(R), \overline{\Gamma}_I(R)$, and $AG_I(R)$

For a commutative ring $R$ with nonzero identity, we have defined three "zero-divisor" graphs with vertex set $Z(R)^*$, but different adjacency relations: the zero-divisor graph $\Gamma(R)$ ([83]) in Section 2.1, the extended zero-divisor graph $\overline{\Gamma}(R)$ ([143]) in Section 5.10, and the annihilator graph $AG(R)$ ([128]) in Section 10.2. More generally, for an ideal $I$ of $R$, we have defined three analogous "ideal-based zero-divisor" graphs with vertex set $Z_I(R) = \{x \in R \setminus I \mid xy \in I$ for some $y \in R \setminus I\}$, but different adjacency relations: the ideal-based zero-divisor graph $\Gamma_I(R)$ ([323]) in Section 5.1, the ideal-based extended zero-divisor graph $\overline{\Gamma}_I(R)$ ([84]) in Section 5.10, and the ideal-based annihilator graph $AG_I(R)$ ([17]) in Section 10.2. Recall that $\Gamma_{\{0\}}(R) = \Gamma(R), \overline{\Gamma}_{\{0\}}(R) = \overline{\Gamma}(R)$, and $AG_{\{0\}}(R) = AG(R)$ since $Z_{\{0\}}(R) = Z(R)^*$; $Z(R)^* = \emptyset$ if and only if $R$ is an integral domain; and $Z_I(R) = \emptyset$ if and only if $I = R$ or $I$ is a prime ideal of $R$.

In this section, we present a unified treatment of these six zero-divisor graphs as given by Anderson and McClurkin in [84]. We give results for ideal-based graphs, and then recover the analogous results for zero-divisor graphs by setting $I = \{0\}$. Clearly $\Gamma_I(R)$ is a subgraph of $\overline{\Gamma}_I(R)$; and $\overline{\Gamma}_I(R)$ is also a subgraph of $AG_I(R)$.

**Theorem 10.4.1.** ([84, Theorem 2.1]) *Let $R$ be a commutative ring with $1 \neq 0$ and $I$ an ideal of $R$. Then $\Gamma_I(R) \subseteq \overline{\Gamma}_I(R) \subseteq AG_I(R)$.*

*Proof.* Clearly $\Gamma_I(R) \subseteq \overline{\Gamma}_I(R)$. We show that $\overline{\Gamma}_I(R) \subseteq AG_I(R)$. Let $x$ and $y$ be adjacent vertices in $\overline{\Gamma}_I(R)$. Then $x^m y^n \in I$ for positive integers $m$ and $n$ with $x^m \notin I$ and $y^n \notin I$. We may assume that $x^{m-1}y^n \notin I$ and $x^m y^{n-1} \notin I$. Let $z = x^{m-1}y^{n-1}$. Then $z(xy) = x^m y^n \in I$, but $zx = x^m y^{n-1} \notin I$ and $zy = x^{m-1}y^n \notin I$; so $z \in (I : xy) \setminus ((I : x) \cup (I : y))$. Thus $x$ and $y$ are also adjacent in $AG_I(R)$; so $\overline{\Gamma}_I(R) \subseteq AG_I(R)$. □

**Corollary 10.4.2.** ([84, Corollary 2.2], [135, Theorem 2.2]) *Let $R$ be a commutative ring with $1 \neq 0$. Then $\Gamma(R) \subseteq \overline{\Gamma}(R) \subseteq AG(R)$.*

The following example shows that all possible inclusions may occur.

**Example 10.4.3.** ([84, Example 2.3]) Let $R = \mathbb{Z}_2 \times \mathbb{Z}_8$. Then $(0, 2)$ and $(1, 2)$ are adjacent in $\overline{\Gamma}(R)$, but not in $\Gamma(R)$; and $(1, 2)$ and $(1, 4)$ are adjacent in $AG(R)$, but not in $\overline{\Gamma}(R)$. Thus $\Gamma(R) \subsetneq \overline{\Gamma}(R) \subsetneq AG(R)$.

It is easy to check that (1) $\Gamma(R) = \overline{\Gamma}(R) = AG(R) = K_1$ for $R = \mathbb{Z}_4$, (2) $\Gamma(R) \subsetneq \overline{\Gamma}(R) = AG(R)$ for $R = \mathbb{Z}_2 \times \mathbb{Z}_4$, and (3) $\Gamma(R) = \overline{\Gamma}(R) \subsetneq AG(R)$ for $R = \mathbb{Z}_2 \times \mathbb{Z}_2 \times \mathbb{Z}_2$.

The next theorem gives conditions for equality in Theorem 10.4.1. As in [127], an ideal $I$ of $R$ is a 2-absorbing ideal if whenever $xyz \in I$ for $x, y, z \in R$, then $xy \in I$, $xz \in I$, or $yz \in I$.

**Theorem 10.4.4.** ([84, Theorem 2.8]) *Let $R$ be a commutative ring with $1 \neq 0$ and $I$ an ideal of $R$.*

(a) $\overline{\Gamma}_I(R) = \Gamma_I(R)$ *if and only if*

(1) $x \in \sqrt{I}$ *implies* $x^2 \in I$, *and*

(2) $x \in R \setminus \sqrt{I}$ *implies* $(I : x^2) = (I : x)$.

*In particular,* $\overline{\Gamma}_I(R) = \Gamma_I(R)$ *when $I$ is a radical ideal of $R$;*

(b) $AG_I(R) = \Gamma_I(R)$ *if and only if $I$ is a 2-absorbing ideal of $R$.*

**Corollary 10.4.5.** ([84, Corollary 2.9]) *Let $R$ be a commutative ring with $1 \neq 0$.*

(a) ([143, Theorem 2.1]) $\overline{\Gamma}(R) = \Gamma(R)$ *if and only if*

(1) $x \in Nil(R)$ *implies* $x^2 = 0$, *and*

(2) $x \in R \setminus Nil(R)$ *implies* $ann_R(x^2) = ann_R(x)$.

*In particular,* $\overline{\Gamma}(R) = \Gamma(R)$ *when* $R$ *is reduced;*

(b) ([308, Theorem 3.1]) $AG(R) = \Gamma(R)$ *if and only if* $\{0\}$ *is a 2-absorbing ideal of* $R$.

Distinct, nonzero nilpotent elements $x$ and $y$ need not be adjacent in $\Gamma(R)$, but $d(x, y) \leq 2$ in $\Gamma(R)$, and they are adjacent in $AG(R)$. This fact generalizes to $\Gamma_I(R)$ and $AG_I(R)$.

**Theorem 10.4.6.** ([84, Theorem 2.5]) *Let* $R$ *be a commutative ring with* $1 \neq 0$, $I$ *an ideal of* $R$, *and* $x, y \in \sqrt{I} \setminus I$ *with* $x \neq y$.

(a) $d(x, y) \leq 2$ *in* $\Gamma_I(R)$;

(b) $x$ *and* $y$ *are adjacent in* $AG_I(R)$.

**Corollary 10.4.7.** ([84, Corollary 2.6]) *Let* $R$ *be a commutative ring with* $1 \neq 0$ *and* $0 \neq x, y \in Nil(R)$ *with* $x \neq y$.

(a) (cf. [64, Lemma 3.1]) $d(x, y) \leq 2$ *in* $\Gamma(R)$;

(b) ([128, Theorem 3.10]) $x$ *and* $y$ *are adjacent in* $AG(R)$.

In [81, Theorem 2.2] (resp., [21, Theorem 4.10]), it was proved that $\Gamma(T(R)) \cong \Gamma(R)$ (resp., $AG(T(R)) \cong AG(R)$); the analogous result also holds for $\overline{\Gamma}(T(R))$.

**Theorem 10.4.8.** ([84, Theorem 2.10]) *Let* $R$ *be a commutative ring with* $1 \neq 0$ *and total quotient ring* $T(R)$. *Then* $\Gamma(T(R)) \cong \Gamma(R), \overline{\Gamma}(T(R)) \cong \overline{\Gamma}(R)$, *and* $AG(T(R)) \cong AG(R)$.

Several of the following results concerning the diameter and girth have been given in earlier sections.

**Theorem 10.4.9.** ([84, Theorem 2.11]) *Let* $R$ *be a commutative ring with* $1 \neq 0$ *and* $I$ *an ideal of* $R$.

(a) $\Gamma_I(R), \overline{\Gamma}_I(R)$, *and* $AG_I(R)$ *are connected;*

(b) $0 \leq diam(AG_I(R)) \leq diam(\overline{\Gamma}_I(R)) \leq diam(\Gamma_I(R)) \leq 3$;

(c) $diam(AG_I(R)) \in \{0, 1, 2\}$;

(d) $\Gamma_I(R), \overline{\Gamma}_I(R)$, *and* $AG_I(R)$ *have girth* $3, 4$, *or* $\infty$. *Moreover,* $gr(AG_I(R)) \leq gr(\overline{\Gamma}_I(R)) \leq gr(\Gamma_I(R))$.

*Proof.* (a) and (b) are clear since $\Gamma_I(R)$ is connected with $diam(\Gamma_I(R)) \in \{0, 1, 2, 3\}$ ([323, Theorem 2.4]), all three graphs have the same vertex set $Z_I(R)$, and $\Gamma_I(R) \subseteq \overline{\Gamma}_I(R) \subseteq AG_I(R)$ by Theorem 10.4.1.

(c) By part (b), we need only show that if $d(x, y) = 3$ in $\Gamma_I(R)$, then $x$ and $y$ are adjacent in $AG_I(R)$. Since $d(x, y) = 3$ in $\Gamma_I(R)$, there are $a, b \in R \setminus I$ such that $xa, ab, by \in I$ and $xb, ay \notin I$. Thus $(I : x) \subsetneq (I : y)$ and $(I : y) \subsetneq (I : x)$; so $(I : xy) \neq (I : x) \cup (I : y)$. Hence $x$ and $y$ are adjacent in $AG_I(R)$.

(d) We have $gr(\Gamma_I(R)) \in \{3, 4, \infty\}$ by [323, Lemma 5.1] and $gr(AG_I(R)) \in \{3, 4, \infty\}$ by [17, Theorem 1.7]. If $I = \{0\}$, then $\overline{\Gamma}_I(R) = \overline{\Gamma}(R)$ has girth $3, 4$, or $\infty$ by [143, Theorem 4.1]. So we may assume that $I$ is a nonzero proper ideal of $R$ that is not prime. If $gr(\Gamma_I(R)) \in \{3, 4\}$, then $gr(\overline{\Gamma}_I(R)) \in \{3, 4\}$ since $\Gamma_I(R) \subseteq \overline{\Gamma}_I(R)$. If $gr(\Gamma_I(R)) = \infty$, then $\Gamma_I(R) = K_2$ by (the proof of) [110, Theorem 3.1]. Thus $\overline{\Gamma}_I(R) = K_2$; so $gr(\overline{\Gamma}_I(R)) = \infty$. Hence $gr(\overline{\Gamma}_I(R)) \in \{3, 4, \infty\}$.

The "moreover" statement is clear. □

**Corollary 10.4.10.** ([84, Corollary 2.12 and Theorem 2.13]) *Let $R$ be a commutative ring with $1 \neq 0$.*

(a) *$\Gamma(R), \overline{\Gamma}(R)$, and $AG(R)$ are connected;*

(b) *$0 \leq diam(AG(R)) \leq diam(\overline{\Gamma}(R)) \leq diam(\Gamma(R)) \leq 3$ and $diam(\Gamma(R)) \leq diam(AG(R)) + 1$;*

(c) *$diam(AG(R)) \in \{0, 1, 2\}$;*

(d) *$\Gamma(R), \overline{\Gamma}(R)$, and $AG(R)$ have girth $3, 4$, or $\infty$. Moreover, $gr(AG(R)) \leq gr(\overline{\Gamma}(R)) \leq gr(\Gamma(R))$.*

The next theorem gives a more careful comparison of the possible values for the diameter and girth of the three zero-divisor graphs.

**Theorem 10.4.11.** ([84, Theorems 2.14 and 2.15])

(a) *Let $a, b, c \in \mathbb{N} \cup \{0\}$. Then there is a commutative ring $R$ with $1 \neq 0$ such that $(a, b, c) = (diam(\Gamma(R)), diam(\overline{\Gamma}(R)), diam(AG(R)))$ if and only if $(a, b, c)$ is $(0,0,0), (1,1,1), (2,1,1), (2,2,1), (2,2,2), (3,2,2)$, or $(3,3,2)$;*

(b) *Let $a, b, c \in \mathbb{N} \cup \{\infty\}$. Then there is a commutative ring $R$ with $1 \neq 0$ such that $(a, b, c) = (gr(\Gamma(R)), gr(\overline{\Gamma}(R)), gr(AG(R)))$ if and only if $(a, b, c)$ is $(3,3,3), (4,4,3), (4,4,4), (\infty,3,3), (\infty,4,4)$, or $(\infty,\infty,\infty)$.*

## 10.5 Congruence-based Graphs

Let $R$ be a commutative ring with nonzero identity. In this section, we give a unified approach using congruence relations on $R$ to the many various "zero-divisor" graphs associated to a commutative ring based on a special case of the "zero-divisor" semigroup graphs: $\Gamma(S)$, $\overline{\Gamma}(S)$, and $AG(S)$ from Section 6.2. In [82], Anderson and Lewis introduced and studied $\Gamma_\sim(R)$, and in [84], Anderson and McClurkin extended these ideas to $\overline{\Gamma}_\sim(R)$ and $AG_\sim(R)$. This approach clarifies the many isolated results concerning the various "zero-divisor" (e.g., ideal-based and compressed) graphs spread throughout the literature.

Let $\sim$ be a multiplicative congruence relation on $R$ (i.e., $\sim$ is an equivalence relation on $R$ and $x \sim y$ implies $xz \sim yz$ for $x, y, z \in R$). Then $R/\!\sim\, = \{[x]_\sim \mid$

$x \in R\}$, the set of $\sim$-congruence classes of $R$, is a commutative monoid under the induced multiplication $[x]_\sim[y]_\sim = [xy]_\sim$ with identity $[1]_\sim$ and zero element $[0]_\sim$. As in [82], the *congruence-based zero-divisor graph* of $R$ (or *$\sim$-zero-divisor graph* of $R$ for short) is the simple graph $\Gamma_\sim(R) = \Gamma(R/\sim)$ (i.e., the semigroup zero-divisor graph) with vertices $Z(R/\sim)^* = Z(R/\sim) \setminus \{[0]_\sim\}$, and distinct vertices $[x]_\sim$ and $[y]_\sim$ are adjacent if and only if $[x]_\sim[y]_\sim = [xy]_\sim = [0]_\sim$, if and only if $xy \sim 0$. As in [84], the *extended $\sim$-zero-divisor graph* of $R$ is the simple graph $\overline{\Gamma}_\sim(R) = \overline{\Gamma}(R/\sim)$ with vertices $Z(R/\sim)^*$, and distinct vertices $[x]_\sim$ and $[y]_\sim$ are adjacent if and only if $[x]_\sim^m[y]_\sim^n = [x^my^n]_\sim = [0]_\sim$ for positive integers $m$ and $n$ with $[x]_\sim^m \neq [0]_\sim$ and $[y]_\sim^n \neq [0]_\sim$; and the *$\sim$-annihilator graph* of $R$ is the simple graph $AG_\sim(R) = AG(R/\sim)$ with vertices $Z(R/\sim)^*$, and distinct vertices $[x]_\sim$ and $[y]_\sim$ are adjacent if and only if $ann_{R/\sim}([x]_\sim[y]_\sim) \neq ann_{R/\sim}([x]_\sim) \cup ann_{R/\sim}([y]_\sim)$ (i.e., $(ann_{R/\sim}([x]_\sim) \cup ann_{R/\sim}([y]_\sim)) \subsetneq ann_{R/\sim}([x]_\sim[y]_\sim)$). These three graphs have the same vertex set $Z(R/\sim)^*$, and $\Gamma_\sim(R) \subseteq \overline{\Gamma}_\sim(R) \subseteq AG_\sim(R)$ (by Theorem 10.5.3). Let $I = [0]_\sim$. Then $[x]_\sim[y]_\sim = [0]_\sim$ if and only if $xy \in I$; so $[x]_\sim$ and $[y]_\sim$ are adjacent in $\Gamma_\sim(R)$ (resp., $\overline{\Gamma}_\sim(R)$, $AG_\sim(R)$) if and only if $xy \in I$ (resp., $x^my^n \in I$ for positive integers $m$ and $n$ with $x^m \notin I$ and $y^n \notin I$, $(I : xy) \neq (I : x) \cup (I : y)$).

Let $\mathcal{C}(R)$ be the set of all multiplicative congruence relations on $R$. Then $\mathcal{C}(R)$ is partially ordered by inclusion, i.e., for $\sim_1, \sim_2 \in \mathcal{C}(R)$, $\sim_1 \leq \sim_2$ if and only if $\sim_1 \subseteq \sim_2$ as subsets of $R \times R$, if and only if $x \sim_1 y$ implies $x \sim_2 y$ for $x, y \in R$, if and only if $[x]_{\sim_1} \subseteq [x]_{\sim_2}$ for every $x \in R$. Moreover, $\mathcal{C}(R)$ has a least element $=_R = \{(x,x) \mid x \in R\}$ and a greatest element $R \times R$. For $\sim \in \mathcal{C}(R)$, $I = [0]_\sim$ is a semigroup ideal of $R$, where $R$ is considered to be a monoid under multiplication. Note that an (ring) ideal of $R$ is always a semigroup ideal, but not conversely.

For $I$ a semigroup ideal of $R$, let $\mathcal{C}_I(R) = \{\sim \in \mathcal{C}(R) \mid [0]_\sim = I\}$. Then $\mathcal{C}(R) = \sqcup\{\mathcal{C}_I(R) \mid I \text{ is a semigroup ideal of } R\}$, and each $\mathcal{C}_I(R)$ is partially ordered by inclusion. Note that $[0]_\sim = R$ if and only if $\sim = R \times R$; so $\mathcal{C}_R(R) = \{R \times R\}$. Moreover, for $\sim, \sim_1, \sim_2 \in \mathcal{C}(R)$, $[0]_{\sim 1} \subseteq [0]_{\sim 2}$ if $\sim_1 \leq \sim_2$; so $\sim_1 \leq \sim \leq \sim_2$ with $\sim_1, \sim_2 \in \mathcal{C}_I(R)$ implies $\sim \in \mathcal{C}_I(R)$.

Let $\sim \in \mathcal{C}(R)$. Many properties of the congruence-based graphs are determined by the semigroup ideal $I = [0]_\sim$ and happen in "levels" given by $\mathcal{C}_I(R)$. For example, given $\sim_1, \sim_2 \in \mathcal{C}_I(R)$, we have $[x]_{\sim_1} \in Z(R/\sim_1)^*$ if and only if $[x]_{\sim_2} \in Z(R/\sim_2)^*$ ([82, Lemma 2.4]). Of special interest are the cases when either $I = \{0\}$ or $I$ is an ideal of $R$.

Every $\mathcal{C}_I(R)$ has a least element $\sim_1$ given by $x \sim_1 y \Leftrightarrow x = y$ or $x, y \in I$ and a greatest element $\sim_3$ given by $x \sim_3 y \Leftrightarrow (I : x) = (I : y)$ by [82, Theorem 2.8]. Thus $\mathcal{C}_I(R) = \{\sim \in \mathcal{C}(R) \mid \sim_1 \leq \sim \leq \sim_3\}$. For $I = \{0\}$, $x \sim_1 y \Leftrightarrow x = y$ and $x \sim_3 y \Leftrightarrow ann_R(x) = ann_R(y)$; so $\mathcal{C}_{\{0\}}(R) = \{\sim \in \mathcal{C}(R) \mid x \sim y \Rightarrow ann_R(x) = ann_R(y)\}$.

We next give several examples (cf. Example 6.2.1).

**Example 10.5.1.** (cf. [82, Example 2.1]) Let $R$ be a commutative ring with $1 \neq 0$.

(a) Let $\sim$ be $=_R$. Then $R/{\sim} = R$, $\Gamma_\sim(R) = \Gamma(R)$, $\overline{\Gamma}_\sim(R) = \overline{\Gamma}(R)$, and $AG_\sim(R) = AG(R)$, the usual zero-divisor graph, extended zero-divisor graph, and annihilator graph of $R$, respectively (we identify $[x]_\sim = \{x\}$ with $x$), and $[0]_\sim = \{0\}$. Thus $\sim \in \mathcal{C}_{\{0\}}(R)$.

(b) Let $I$ be a proper ideal of $R$, and let $\sim_I \in (R)$ be defined by $x \sim_I y \Leftrightarrow x - y \in I$ for $x, y \in R$. Then $R/{\sim_I} = R/I$, $\Gamma_\sim(R) = \Gamma(R/I)$, $\overline{\Gamma}_\sim(R) = \overline{\Gamma}(R/I)$, $AG_\sim(R) = AG(R/I)$, and $[0]_\sim = I$. Thus $\sim_I \in \mathcal{C}_I(R)$, and $\sim_I \leq \sim_J \Leftrightarrow I \subseteq J$. (Note that for $I$ a semigroup ideal of $R$, $\sim_I$ defines a multiplicative congruence relation on $R$ if and only if $I$ is an ideal of $R$.)

(c) Let $I$ be a semigroup ideal of $R$, and let $\sim_I$ be defined by $x \sim_I y \Leftrightarrow x = y$ or $x, y \in I$ for $x, y \in R$. Then $R/{\sim_I} = R_I$ is the Rees semigroup of $R$ with respect to $I$, and $[0]_{\sim_I} = I$. (The Rees semigroup of $R$ with respect to $I$ is usually denoted by $R/I$, where the semigroup ideal $I$ collapses to 0.) Thus $\sim_I \in \mathcal{C}_I(R)$, and $\sim_I \leq \sim_J$ if and only if $I \subseteq J$. If $I$ is an ideal of $R$, then $\Gamma_\sim(R) = \Gamma_I(R)$, $\overline{\Gamma}_\sim(R) = \overline{\Gamma}_I(R)$, and $AG_\sim(R) = AG_I(R)$, the usual ideal-based zero-divisor graph, ideal-based extended zero-divisor graph, and ideal-based annihilator graph of $R$ with respect to $I$, respectively (again, we identify $[x]_\sim = \{x\}$ with $x$ for $x \in R \setminus I$). Hence for any semigroup ideal $I$ of $R$, we will denote $\Gamma_\sim(R)$ (resp., $\overline{\Gamma}_\sim(R)$, $AG_\sim(R)$) by $\Gamma_I(R)$ (resp., $\overline{\Gamma}_I(R)$, $AG_I(R)$).

(d) Let $I$ be a semigroup ideal of $R$, and let $\sim$ $\mathcal{C}(R)$ be defined by $x \sim y \Leftrightarrow (I : x) = (I : y)$ for $x, y \in R$. Then $[0]_\sim = I$; so $\sim \in \mathcal{C}_I(R)$. If $I$ is a proper ideal of $R$, then $\Gamma_\sim(R) = \Gamma_E(R/I)$, $\overline{\Gamma}_\sim(R) = \overline{\Gamma}_E(R/I)$, and $AG_\sim(R) = AG_E(R/I)$, the compressed zero-divisor graph, compressed extended zero-divisor graph, and compressed annihilator graph of $R/I$, respectively (we identify $[x]_\sim$ in $R/\sim$ with $[x + I]$ in $(R/I)_E$, see Corollary 10.5.10). If $I = \{0\}$, then $(I : x) = ann_R(x)$; so $\Gamma_\sim(R) = \Gamma_E(R)$, $\overline{\Gamma}_\sim(R) = \overline{\Gamma}_E(R)$, and $AG_\sim(R) = AG_E(R)$, the usual compressed zero-divisor graph, compressed extended zero-divisor graph, and compressed annihilator graph of $R$, respectively.

The following theorem also holds for $\overline{\Gamma}_\sim(R)$ and $AG_\sim(R)$ since all three graphs have the same vertex set $Z(R/{\sim})^*$.

**Theorem 10.5.2.** ([82, Theorem 2.3]) *Let $R$ be a commutative ring with $1 \neq 0$, and let $\sim \in \mathcal{C}(R)$.*

(a) *$\Gamma_\sim(R)$ is the empty graph if and only if $[0]_\sim = R$ or $[0]_\sim$ is a prime ideal of $R$;*

(b) *$\Gamma_\sim(R)$ is the empty graph for every $\sim \in \mathcal{C}(R)$ if and only if $R$ is a field.*

The following theorem generalizes Theorems 10.4.1 and 10.4.4.

**Theorem 10.5.3.** ([84, Theorem 3.1]) *Let $R$ be a commutative ring with $1 \neq 0$, $\sim$ a multiplicative congruence relation on $R$, and $I = [0]_\sim$.*

(a) *$\Gamma_\sim(R) \subseteq \overline{\Gamma}_\sim(R) \subseteq AG_\sim(R)$;*

(b) $\overline{\Gamma}_{\sim}(R) = \Gamma_{\sim}(R)$ *if*

(1) $x \in \sqrt{I}$ *implies* $x^2 \in I$, *and*

(2) $x \in R \setminus \sqrt{I}$ *implies* $(I : x^2) = (I : x)$.

*In particular,* $\overline{\Gamma}_{\sim}(R) = \Gamma_{\sim}(R)$ *when* $I$ *is a radical semigroup ideal of* $R$;

(c) $AG_{\sim}(R) = \Gamma_{\sim}(R)$ *if* $I$ *is a 2-absorbing semigroup ideal of* $R$.

The next example shows that parts (b) and (c) of Theorem 10.5.3 are not "if and only if" as in Theorem 10.4.4.

**Example 10.5.4.** ([84, Example 3.2]) Let $R = \mathbb{Z}_8$ and $x \sim y \Leftrightarrow ann_R(x) = ann_R(y)$ (equivalently, $x \sim y \Leftrightarrow y = ux$ for some $u \in U(R)$). Then $\sim \in C_I(R)$, $I = [0]_{\sim} = \{0\}$, and $Z(R/\sim)^* = \{2,4\}$; so $\Gamma_{\sim}(R) = \overline{\Gamma}_{\sim}(R) = AG_{\sim}(R) = K_2$. However, $2 \in \sqrt{I} = Nil(R) = \{0,2,4,6\}$ and $2^2 = 4 \notin I$; so the converse of Theorem 10.5.3(b) fails. And $I$ is not a 2-absorbing ideal of $R$; so the converse of Theorem 10.5.3(c) fails.

The next theorem concerning the diameter and girth of the congruence-based graphs generalizes Theorem 10.4.9.

**Theorem 10.5.5.** ([84, Theorem 3.3]) *Let* $R$ *be a commutative ring with* $1 \neq 0$ *and* $\sim$ *a multiplicative congruence relation on* $R$.

(a) $\Gamma_{\sim}(R), \overline{\Gamma}_{\sim}(R)$, *and* $AG_{\sim}(R)$ *are connected;*

(b) $0 \leq diam(AG_{\sim}(R)) \leq diam(\overline{\Gamma}_{\sim}(R)) \leq diam(\Gamma_{\sim}(R)) \leq 3$;

(c) $diam(AG_{\sim}(R)) \in \{0,1,2\}$;

(d) $\Gamma_{\sim}(R), \overline{\Gamma}_{\sim}(R)$, *and* $AG_{\sim}(R)$ *have girth* $3, 4$, *or* $\infty$. *Moreover,* $gr(AG_{\sim}(R)) \leq gr(\overline{\Gamma}_{\sim}(R)) \leq gr(\Gamma_{\sim}(R))$.

The next theorem generalizes Theorem 5.2.29 from $\Gamma_I(R)$ to $\Gamma_{\sim}(R)$.

**Theorem 10.5.6.** ([84, Theorem 3.5]) *Let* $R$ *be a commutative ring with* $1 \neq 0$, $\sim$ *a multiplicative congruence relation on* $R$, *and* $I = [0]_{\sim}$.

(a) *If* $I$ *is a radical semigroup ideal of* $R$, *then* $\Gamma_{\sim}(R)$ *is uniquely complemented if and only if* $\Gamma_{\sim}(R)$ *is complemented;*

(b) *If* $I$ *is a radical ideal of* $R$, *then the following statements are equivalent:*

(1) $\Gamma_{\sim}(R)$ *is uniquely complemented;*

(2) $\Gamma_{\sim}(R)$ *is complemented;*

(3) $\Gamma(R/I)$ *is uniquely complemented;*

(4) $\Gamma(R/I)$ *is complemented;*

(5) $T(R/I)$ *is von Neumann regular.*

We next consider maps between congruence-based graphs. This work was initiated in [82] for $\Gamma_{\sim}(R)$, and then extended to $\overline{\Gamma}_{\sim}(R)$ and $AG_{\sim}(R)$ in [84]. For $\sim_1, \sim_2 \in \mathcal{C}(R)$ with $\sim_1 \leq \sim_2$, the map $1_R : R \longrightarrow R$ induces a surjective monoid homomorphism $f : R/\sim_1 \longrightarrow R/\sim_2$ given by $f([x]_{\sim_1}) = [x]_{\sim_2}$ with $f([0]_{\sim_1}) = [0]_{\sim_2}$ and $f([1]_{\sim_1}) = [1]_{\sim_2}$. If $[0]_{\sim_1} = [0]_{\sim_2}$, then $f$ induces a surjective function $F : \Gamma_{\sim_1}(R) \longrightarrow \Gamma_{\sim_2}(R)$ given by $F([x]_{\sim_1}) = [x]_{\sim_2}$. Since $F$ is surjective, there is an (not necessarily unique) injective function $G : \Gamma_{\sim_2}(R) \longrightarrow \Gamma_{\sim_1}(R)$ such that $FG = 1_{\Gamma_{\sim_2}(R)}$. Moreover, $G$ is an injective graph homomorphism and embeds $\Gamma_{\sim_2}(R)$ as an induced subgraph of $\Gamma_{\sim_1}(R)$. Similarly, we have maps $F' : \overline{\Gamma}_{\sim_1}(R) \longrightarrow \overline{\Gamma}_{\sim_2}(R)$, $F'' : AG_{\sim_1}(R) \longrightarrow AG_{\sim_2}(R)$, and inverses $G', G''$ . Note that the functions $F, F', F''$ (resp., $G, G', G''$) take the same values on $Z(R/\sim_1)^*$ (resp., $Z(R/\sim_2)^*$). The following theorem collects the above observations.

**Theorem 10.5.7.** ([82, Theorem 3.1] and [84, Theorem 5.2]) *Let $R$ be a commutative ring with $1 \neq 0$, and let $\sim_1, \sim_2 \in \mathcal{C}(R)$ with $\sim_1 \leq \sim_2$. If $[0]_{\sim_1} = [0]_{\sim_2}$, then the surjective monoid homomorphism $f : R/\sim_1 \longrightarrow R/\sim_2$ given by $f([x]_{\sim_1}) = [x]_{\sim_2}$ induces a surjective function $F : \Gamma_{\sim_1}(R) \longrightarrow \Gamma_{\sim_2}(R)$ (resp., $F' : \overline{\Gamma}_{\sim_1}(R) \longrightarrow \overline{\Gamma}_{\sim_2}(R)$, $F'' : AG_{\sim_1}(R) \longrightarrow AG_{\sim_2}(R)$) given by $F([x]_{\sim_1}) = [x]_{\sim_2}$ (resp., $F'([x]_{\sim_1}) = [x]_{\sim_2}$, $F''([x]_{\sim_1}) = [x]_{\sim_2}$) and an injective graph homomorphism $G : \Gamma_{\sim_2}(R) \longrightarrow \Gamma_{\sim_1}(R)$ (resp., $G' : \overline{\Gamma}_{\sim_2}(R) \longrightarrow \overline{\Gamma}_{\sim_1}(R)$, $G'' : AG_{\sim_2}(R) \longrightarrow AG_{\sim_2}(R)$) such that $FG = 1_{\Gamma_{\sim_1}(R)}$ (resp., $F'G' = 1_{\overline{\Gamma}_{\sim_2}(R)}$, $F''G'' = 1_{AG_{\sim_2}(R)}$). Moreover, for adjacent vertices $[x]_{\sim_1}$ and $[y]_{\sim_1}$ in $\Gamma_{\sim_1}(R)$ (resp., $\overline{\Gamma}_{\sim_1}(R)$, $AG_{\sim_1}(R)$), either $F([x]_{\sim_1}) = F([y]_{\sim_1})$ (resp., $F'([x]_{\sim_1}) = F'([y]_{\sim_1})$, $F''([x]_{\sim_1}) = F''([y]_{\sim_1})$) or $F([x]_{\sim_1})$ and $F([y]_{\sim_1})$ (resp., $F'([x]_{\sim_1})$ and $F'([y]_{\sim_1})$, $F''([x]_{\sim_1})$ and $F''([y]_{\sim_1})$) are adjacent in $\Gamma_{\sim_2}(R)$ (resp., $\overline{\Gamma}_{\sim_2}(R)$, $AG_{\sim_2}(R)$); and $\Gamma_{\sim_2}(R)$ (resp., $\overline{\Gamma}_{\sim_2}(R)$, $AG_{\sim_2}(R)$) is isomorphic to an induced subgraph of $\Gamma_{\sim_1}(R)$ (resp., $\overline{\Gamma}_{\sim_1}(R)$, $AG_{\sim_1}(R)$).*

By Theorem 10.5.7, for a fixed semigroup ideal $I$ of $R$, there is a largest and a smallest $\sim$-zero-divisor (resp., extended $\sim$-zero-divisor, $\sim$-annihilator) graph for $\sim \in \mathcal{C}_I(R)$. The largest is $\Gamma_I(R)$ (resp., $\overline{\Gamma}_I(R)$, $AG_I(R)$), and the smallest is $\Gamma_{\sim}(R)$ (resp., $\overline{\Gamma}_{\sim}(R)$, $AG_{\sim}(R)$), where $x \sim y \Leftrightarrow (I : x) = (I : y)$. Moreover, $\Gamma_{\sim}(R) = \Gamma_E(R/I)$, $\overline{\Gamma}_{\sim}(R) = \overline{\Gamma}_E(R/I)$, and $AG_{\sim}(R) = AG_E(R/I)$ when $I$ is a proper ideal of $R$ (see Corollary10.5.10). In particular, for $I = \{0\}$, $\Gamma(R)$ (resp., $\overline{\Gamma}(R)$, $AG(R)$) is the largest and $\Gamma_E(R)$ (resp., $\overline{\Gamma}_E(R)$, $AG_E(R)$) is the smallest zero-divisor (resp., extended zero-divisor, annihilator) graph of $R$.

The functions $F$ (resp., $F'$, $F''$) need not be graph homomorphisms since distinct vertices in $\Gamma_{\sim_1}(R)$ (resp., $\overline{\Gamma}_{\sim_1}(R)$, $AG_{\sim_1}(R)$) may collapse to the same vertex in $\Gamma_{\sim_2}(R)$ (resp., $\overline{\Gamma}_{\sim_2}(R)$, $AG_{\sim_2}(R)$). The next theorem gives a sufficient condition for $F, F'$, and $F''$ to be graph homomorphisms. For example, let $R = \mathbb{Z}_9$, $\sim_1 = =_R$, and $x \sim_2 y \Leftrightarrow ann_R(x) = ann_R(y)$. Then $\sim_1, \sim_2 \in \mathcal{C}_{\{0\}}(R)$ with $\sim_1 \leq \sim_2$, $\Gamma_{\sim_1}(R) = \Gamma(R) = \overline{\Gamma}(R) = AG(R) = K_2$ and $\Gamma_{\sim_2}(R) = \Gamma_E(R) =$

$\overline{\Gamma}_E(R) = AG_E(R) = K_1$; so $F = F'' = F''$ are not graph homomorphisms. Note that $\overline{\Gamma}_\sim(R) = \Gamma_\sim(R)$ when $I = [0]_\sim$ is a radical semigroup ideal of $R$.

**Theorem 10.5.8.** ([84, Theorem 5.3]) *Let $R$ be a commutative ring with $1 \neq 0$, $I$ a radical semigroup ideal of $R$, and $\sim_1, \sim_2 \in \mathcal{C}_I(R)$ with $\sim_1 \leq \sim_2$. Then $F : \Gamma_{\sim_1}(R) \longrightarrow \Gamma_{\sim_2}(R)$ (resp., $F'' : AG_{\sim_1}(R) \longrightarrow AG_{\sim_2}(R)$) given by $F([x]_\sim) = [x]_{\sim_2}$ (resp., $F''([x]_{\sim_1}) = [x]_{\sim_2}$) is a surjective graph homomorphism.*

In some cases, a $\sim \in \mathcal{C}_I(R)$ induces a $\sim' \in \mathcal{C}_{\{0\}}(R)$ such that $[x]_\sim \mapsto [x+I]_{\sim'}$ gives graph isomorphisms $F : \Gamma_\sim(R) \longrightarrow \Gamma_{\sim'}(R/I)$, $F' : \overline{\Gamma}_\sim(R) \longrightarrow \overline{\Gamma}_{\sim'}(R/I)$, and $F'' : AG_\sim(R) \longrightarrow AG_{\sim'}(R/I)$. Thus a congruence-based graph may come from different base rings.

**Theorem 10.5.9.** ([84, Theorem 5.4]) *Let $R$ be a commutative ring with $1 \neq 0$, and let $I$ be a proper ideal of $R$. Define $\sim_1 \in \mathcal{C}_I(R)$ by $x \sim_1 y \Leftrightarrow x - y \in I$. Given $\sim \in \mathcal{C}_I(R)$ with $\sim_1 \leq \sim$, define $\sim' \in \mathcal{C}_{\{0\}}(R)$ by $x + I \sim' y + I \Leftrightarrow x \sim y$. Then $F : \Gamma_\sim(R) \longrightarrow \Gamma_{\sim'}(R/I)$ (resp., $F' : \overline{\Gamma}_\sim(R) \longrightarrow \overline{\Gamma}_{\sim'}(R/I)$, $F'' : AG_\sim(R) \longrightarrow AG_{\sim'}(R/I)$) given by $F([x]_\sim) = [x+I]_{\sim'}$ (resp., $F'([x]_\sim) = [x+I]_{\sim'}$, $F''([x]_\sim) = [x+I]_{\sim'}$) is a graph isomorphism.*

**Corollary 10.5.10.** ([84, Corollary 5.6]) *Let $R$ be a commutative ring with $1 \neq 0$, and let $I$ be a proper ideal of $R$. Define $\sim_1, \sim_2 \in \mathcal{C}_I(R)$ by $x \sim_1 y \Leftrightarrow x - y \in I$ and $x \sim_2 y \Leftrightarrow (I : x) = (I : y)$. Then $\Gamma_{\sim_1}(R) = \Gamma(R/I)$, $\overline{\Gamma}_{\sim_1}(R) = \overline{\Gamma}(R/I)$, $AG_{\sim_1}(R) = AG(R/I)$, and $\Gamma_{\sim_2}(R) = \Gamma_E(R/I)$, $\overline{\Gamma}_{\sim_2}(R) = \overline{\Gamma}_E(R/I)$, $AG_{\sim_2}(R) = AG_E(R/I)$.*

There are many other results concerning maps between congruence-based graphs in [82] and [84].

## 10.6 Element Graphs

In this section, we give several "zero-divisor" graphs defined by certain subsets of $Z(R)$. In [71], Anderson and Badawi defined five "zero-divisor" graphs based on certain types of elements of $R$. For details of the results mentioned below, see [71, Section 7]. Let $Idem(R) = \{a \in R \mid a^2 = a\}$ be the set of idempotent elements of $R$, $vnr(R) = \{a \in R \mid a^2x = a \text{ for some } x \in R\}$ the set of von Neumann regular elements of $R$, $\pi - r(R) = \{a \in R \mid a^{2n}x = a^n \text{ for some } n \in \mathbb{N}$ and $x \in R\}$ the set of $\pi$-regular elements of $R$, $vnl(R) = \{a \in R \mid a \in vnr(R)$ or $1 - a \in vnr(R)\}$ the set of von Neumann local elements of $R$, and $cln(R) = \{a \in R \mid a = u + e \text{ for some } u \in U(R), e \in Idem(R)\}$ the set of clean elements of $R$. These five sets yield five induced subgraphs $\Gamma(Idem(R))$, $\Gamma(vnr(R))$, $\Gamma(\pi - r(R))$, $\Gamma(vnl(R))$, and $\Gamma(cln(R))$ of $\Gamma(R)$ with vertices $Idem(R) \cap Z(R)^*$, $vnr(R) \cap Z(R)^*$, $\pi - r(R) \cap Z(R)^*$, $vnl(R) \cap Z(R)^*$, and $cln(R) \cap Z(R)^*$, respectively.

For $Z(R)^* \neq \emptyset$, we have $\Gamma(Idem(R)) = \Gamma(R)$, (resp., $\Gamma(vnr(R)) = \Gamma(R)$) if and only if $R$ is a Boolean (resp., von Neumann regular) ring, and for $Nil(R) \subsetneq Z(R)$, we have $\Gamma(\pi - r(R)) = \Gamma(R)$ if and only if $R$ is a $\pi$-regular ring. We also have $\Gamma(Idem(R)) \subseteq \Gamma(vnr(R)) \subseteq \Gamma(\pi - r(R)) \subseteq \Gamma(cln(R)) \subseteq \Gamma(R)$ and $\Gamma(vnr(R)) \subseteq \Gamma(vnl(R)) \subseteq \Gamma(cln(R))$ for any commutative ring $R$. Since a zero-dimensional commutative (e.g., finite) ring is $\pi$-regular, $\Gamma(\pi - r(R)) = \Gamma(vnl(R)) = \Gamma(cln(R)) = \Gamma(R)$ when $R$ is zero-dimensional. However, for $T = \mathbb{Z} \times \mathbb{Z}_4 \times \mathbb{Z}_4$, we have $\Gamma(Idem(T)) \subsetneq \Gamma(vnr(T)) \subsetneq \Gamma(\pi - r(T)) \subsetneq \Gamma(cln(T)) \subsetneq \Gamma(T)$, $\Gamma(vnr(T)) \subsetneq \Gamma(vnl(T)) \subsetneq \Gamma(cln(T))$, and $\Gamma(\pi - r(T))$ and $\Gamma(vnl(T))$ are not comparable. Moreover, $\Gamma(T)$ is infinite, while the other five subgraphs are all finite.

Recall that $diam(\Gamma(R)) \in \{0, 1, 2, 3\}$ and $gr(\Gamma(R)) \in \{3, 4, \infty\}$. These two results also hold for the subgraphs $\Gamma(Idem(R))$, $\Gamma(vnr(R))$, and $\Gamma(\pi - r(R))$ of $\Gamma(R)$. However, the subgraphs $\Gamma(vnl(R))$ and $\Gamma(cln(R))$ need not be connected.

Let $R$ be a commutative ring with nonzero identity graded by an abelian group $G$, and let $Z_G(R)^*$ be the set of nonzero homogeneous zero-divisors of $R$. In [166], Cooper and Johnson defined the *$G$-graded zero-divisor graph* of $R$ to be the simple graph $\Gamma_G(R)$ with vertices $Z_G(R)^*$, and distinct vertices $x$ and $y$ are adjacent if and only if $xy = 0$. They proved that $\Gamma_G(R)$, like $\Gamma(R)$, is connected with $diam(\Gamma_G(R)) \in \{0, 1, 2, 3\}$ and $gr(\Gamma_G(R)) \in \{3, 4, \infty\}$. However, unlike $\Gamma(R)$, every star graph, complete graph, and complete bipartite graph can be realized as a graded zero-divisor graph.

## 10.7 Dot Product Graphs

Let $A$ be a commutative ring, $n$ a positive integer, and $R = A \times A \times \cdots \times A$ ($n$ times). Let $x = (x_1, \ldots, x_n), y = (y_1, \ldots, y_n) \in R$. Then the dot product $x \cdot y = x_1y_1 + x_2y_2 + \cdots + x_ny_n \in A$. Badawi[130] introduced the *total dot product graph* of $R$ to be the (simple) graph $TD(R)$ with vertices $R^* = R \setminus \{(0, 0, \ldots, 0)\}$, and distinct vertices $x$ and $y$ are adjacent if and only if $x \cdot y = 0 \in A$. Then the *zero-divisor dot product graph* of $R$ is the induced subgraph $ZD(R)$ of $TD(R)$ with vertices $Z(R)^* = Z(R) \setminus \{(0, 0, \ldots, 0)\}$. It follows that every edge (resp., path) of the classical zero-divisor graph $\Gamma(R)$ is an edge (resp., path) of $ZD(R)$. We observe that if $n = 1$, then $TD(R)$ is a disconnected graph, where $ZD(R)$ is identical to $\Gamma(R)$, and hence is connected.

**Theorem 10.7.1.** ([130, Theorem 2.1]) *Let $A$ be an integral domain and $R = A \times A$. Then $TD(R)$ is disconnected and $ZD(R) = \Gamma(R)$ is connected. In particular, if $A$ is ring-isomorphic to $\mathbb{Z}_2$, then $ZD(R)$ is complete (i.e., $diam(ZD(R)) = 1$) and $gr(ZD(R)) = \infty$. If $A$ is not ring-isomorphic to $\mathbb{Z}_2$, then $diam(ZD(R)) = 2$ and $gr(ZD(R)) = 4$.*

*Proof.* Let $B = \{(a, a), (-a, a), (a, -a) \mid a \in A^*\}$ and $x \in B$. Suppose that

$y \in R^*$ and $x \cdot y = 0$. Since $A$ is an integral domain, one can easily see that $y \in B$. Let $M = \{(a, 0), (0, a) \mid a \in A^*\}$ and $w \in M$. Suppose that $w \cdot s = 0$ for some $s \in R^*$. Again, since $A$ is an integral domain, we have $s \in M$. Thus the vertices $(1, 1)$ and $(0, 1)$ are not connected by a path in $TD(R)$. Hence $TD(R)$ is disconnected. Since $A$ is an integral domain, $Z(R)^* = M$. Let $x, y \in M$. Then $x \cdot y = 0$ if and only if $xy = (0, 0)$. Thus $ZD(R) = \Gamma(R)$. Suppose that $A$ is ring-isomorphic to $\mathbb{Z}_2$. Then it is clear that $diam(ZD(R)) = 1$ and $gr(ZD(R)) = \infty$. Suppose $A$ is not ring-isomorphic to $\mathbb{Z}_2$. Since $ZD(R) = \Gamma(R)$ and $A$ is an integral domain, $diam(ZD(R)) = 2$ by Theorem 2.4.28 and $gr(ZD(R)) = 4$ by Theorem 2.3.8. □

In view of Theorem 10.7.1, we have the following result.

**Theorem 10.7.2.** ([130, Theorem 2.2]) *Let $2 \leq n < \infty$, $A$ a commutative ring, and $R = A \times A \times \cdots \times A$ ($n$ times). Then $ZD(R) = \Gamma(R)$ if and only if either $n = 2$ and $A$ is an integral domain or $R$ is ring-isomorphic to $\mathbb{Z}_2 \times \mathbb{Z}_2 \times \mathbb{Z}_2$.*

*Proof.* If $n = 2$ and $A$ is an integral domain, then $ZD(R) = \Gamma(R)$ by Theorem 10.7.1. Suppose that $R$ is ring-isomorphic to $\mathbb{Z}_2 \times \mathbb{Z}_2 \times \mathbb{Z}_2$. Then by simple hand-calculations, for every $x, y \in Z(R)^*$, we have $x \cdot y = 0$ if and only if $xy = (0, 0, 0)$, and hence $ZD(R) = \Gamma(R)$.

Conversely, suppose that $ZD(R) = \Gamma(R)$. Assume that $A$ is not an integral domain. Then there is an $a \in Z(A)^*$. Hence $x = (1, a, 0, 0, \ldots, 0), y = (a, -1, 0, 0, \ldots, 0) \in Z(R)^*$, and $x \cdot y = 0$, but $xy \neq (0, 0, ..., 0)$. Thus $x - y$ is an edge of $ZD(R)$ that is not an edge of $\Gamma(R)$, a contradiction. Thus $A$ must be an integral domain. Now assume that $n = 3$ and $A$ is not ring-isomorphic to $\mathbb{Z}_2$. Then there is an $a \in A \setminus \{0, 1\}$. Let $x = (1, a, 0)$ and $y = (-a, 1, 0)$. Then $x \neq y$, and it is clear that $x - y$ is an edge of $ZD(R)$ that is not an edge of $\Gamma(R)$, a contradiction again. Hence assume that $n \geq 4$. Let $x = (1, 1, 0, 1, 0, 0, \ldots, 0)$ and $y = (-1, 1, 1, 0, 0, \ldots, 0)$. Then $x \neq y$, $x \cdot y = 0$, but $xy \neq (0, 0, \ldots, 0)$, a contradiction. Thus either $n = 2$ and $A$ is an integral domain or $R$ is ring-isomorphic to $\mathbb{Z}_2 \times \mathbb{Z}_2 \times \mathbb{Z}_2$. □

**Theorem 10.7.3.** ([130, Theorem 2.3]) *Let $A$ be a commutative ring that is not an integral domain and $R = A \times A$. Then the following statements hold:*

(1) *$TD(R)$ is connected and $diam(TD(R)) = 3$;*

(2) *$ZD(R)$ is connected, $ZD(R) \neq \Gamma(R)$, and $diam(ZD(R)) = 3$;*

(3) *$gr(ZD(R)) = gr(TD(R)) = 3$.*

*Proof.* (1) Let $x = (a, b), y = (c, d) \in R^*$, where $x \neq y$, and assume that $x \cdot y \neq 0$. Since $A$ is not an integral domain, there are $f, g \in A^*$ (not necessarily distinct) such that $fg = 0$. Let $w = (-bf, af)$ and $v = (-dg, cg)$. Note that $w, v \in Z(R)$. Clearly $x \cdot w = w \cdot v = v \cdot y = 0$. Since $x \cdot y \neq 0$, $w \neq y$ and $v \neq x$. First, assume that $v, w \in Z(R)^*$. If $x \cdot y = 0$ or $y \cdot w = 0$, then $x - v - y$ or $x - w - y$ is a path of length 2 in $TD(R)$ from $x$ to $y$. Assume that neither $x \cdot y = 0$ nor $y \cdot w = 0$. Then $x, w, v, y$ are distinct, and hence $x - w - v - y$ is a path of length 3 in $TD(R)$ from $x$ to $y$. Now assume that $w = (0,0)$ or $v = (0,0)$. If $w = (0,0)$, then replace $w$ by $(f, -f) \in Z(R)^*$, and hence $x \cdot w = (a,b) \cdot (f,-f) = 0$. Similarly, if $v = (0,0)$, then replace $v$ by $(g, -g) \in Z(R)^*$. Hence if $w = (0,0)$ or $v = (0,0)$, then we are able to redefine $w$ and $v$ so that $w, v \in Z(R)^*$ and $x \cdot w = w \cdot v = v \cdot y = 0$. Thus as in the earlier argument, we conclude that there is a path of length at most 3 in $TD(R)$ from $x$ to $y$. Thus $TD(R)$ is connected and $d_T(x,y) \leq 3$ for every $x, y \in R^*$. Now, let $x = (1,1)$ and $y = (1,0)$. We show $d_T(x,y) = 3$, and hence $diam(TD(R)) = 3$. Let $w \in R^*$ such that $x \cdot w = 0$. Then $w = (a, -a)$ for some $a \in A^*$. Since $w \cdot y = a \neq 0$, $d_T(x,y) > 2$. Hence $d_T(x,y) = 3$. In particular, let $k, t \in A^*$ with $kt = 0$, $w = (k, -k)$, and $v = (0,t)$. Then $x - w - v - y$ is a path of length 3 in $TD(R)$ from $x$ to $y$.

(2) Since $A$ is not an integral domain, $ZD(R) \neq \Gamma(R)$ by Theorem 10.7.2. Let $x, y \in Z(R)^*$, and assume that $x \cdot y \neq 0$. In view of the proof of (1), there are $w, v \in Z(R)^*$ such that either $x - w - y$ is a path in $ZD(R)$, $x - v - y$ is a path in $ZD(R)$, or $x - w - v - y$ is a path in $ZD(R)$. Hence $diam(ZD(R)) \leq 3$. Let $a \in Z(A)^*$. Then $x = (1,a), y = (0,1) \in Z(R)^*$. We show $d_Z(x,y) = 3$, and thus $diam(ZD(R)) = 3$. Since $x \cdot y \neq 0$, $d_Z(x,y) > 1$. Suppose there is a $v = (g,h) \in Z(R)^*$ such that $x - v - y$ is a path of length 2 in $ZD(R)$ from $x$ to $y$. Since $v \cdot y = 0$, we have $h = 0$, and hence $v = (g, 0)$. Since $x \cdot y = 0$, we have $g = 0$ , and thus $v = (0,0)$, a contradiction. Thus $d_Z(x,y) = 3$, and hence $diam(ZD(R)) = 3$.

(3) Since $A$ is not an integral domain, there are $a, b \in A^*$ (not necessarily distinct) with $ab = 0$. Then $x = (a,0), y = (0,b), w = (b,a) \in Z(R)^*$. Hence $x - y - w - x$ is a cycle of length 3 in $ZD(R)$. Thus $gr(TD(R)) = gr(ZD(R)) = 3$. □

**Theorem 10.7.4.** ([130, Theorem 2.4]) *Let $A$ be a commutative ring with $1 \neq 0$, $3 \leq n < \infty$, and $R = A \times A \times \cdots \times A$ ($n$ times). Then $TD(R)$ is connected with $diam(TD(R)) = 2$.*

*Proof.* Let $x = (x_1, \ldots, x_n), y = (y_1, \ldots, y_n) \in R^*$, and suppose that $x \cdot y \neq 0$. Then let $M = \{i \mid x_i = y_i = 0, 1 \leq i \leq n\}$. Suppose that $M \neq \emptyset$. Choose $k \in M$, and let $w = (w_1, \ldots, w_n) \in R^*$, where $w_k = 1$ and $w_i = 0$ if $i \neq k$. Then $x - w - y$ is a path of length 2 in $TD(R)$ from $x$ to $y$. Thus suppose that $M = \emptyset$. Let $f(x) = min\{i \mid x_i \neq 0, 1 \leq i \leq n\}$ and $f(y) = min\{i \mid y_i \neq 0, 1 \leq i \leq n\}$. Since $M = \emptyset$, we have $f(x) = 1$ or $f(y) = 1$. We may assume that $f(x) = 1$. Let $v = (x_2y_3 - x_3y_2,\ x_3y_1 - x_1y_3,\ x_1y_2 - x_2y_1, 0, 0, 0, \ldots, 0) \in R$. Suppose that $v \neq (0, \ldots, 0)$. Then it is easy to check that $x \cdot y = v \cdot y = 0$. Since $x \cdot y \neq 0$, $v \neq x$ and $v \neq y$. Thus $x - v - y$ is a path of length 2 in $TD(R)$ from $x$ to $y$. Suppose that $v = (0, \ldots, 0)$. Then $x_1y_2 - x_2y_1 = 0$. Let $w = (-x_2,\ x_1, 0, 0, \ldots, 0) \in R$. Since $x_1 \neq 0$, $w \in R^*$. Hence $x \cdot w = -x_1x_2 + x_1x_2 = 0$ and $w \cdot y = x_1y_2 - x_2y_1 = 0$. Since $x \cdot w = w \cdot y = 0$ and $x \cdot y \neq 0$, $x \neq w$ and $y \neq w$. Thus $x - w - y$

is a path of length 2 in $TD(R)$ from $x$ to $y$. Hence $TD(R)$ is connected with $diam(TD(R)) = 2$. □

**Theorem 10.7.5.** ([130, Theorem 2.5]) *Let $A$ be a commutative ring. Then the following statements hold:*

(1) *If $A$ is an integral domain and $R = A \times A \times A$, then $ZD(R)$ is connected ($ZD(R) \neq \Gamma(R)$ by Theorem 10.7.2) with $diam(ZD(R)) = 3$;*

(2) *If $A$ is not an integral domain and $R = A \times A \times A$, then $ZD(R)$ is connected ($ZD(R) \neq \Gamma(R)$ by Theorem 10.7.2) with $diam(ZD(R)) = 2$;*

(3) *If $4 \leq n < \infty$ and $R = A \times A \times \cdots \times A$ ($n$ times), then $ZD(R)$ is connected ($ZD(R) \neq \Gamma(R)$ by Theorem 10.7.2) with $diam(ZD(R)) = 2$.*

*Proof.* (1) Since $\Gamma(R)$ is connected and every path in $\Gamma(R)$ is a path in $ZD(R)$, we conclude that $ZD(R)$ is connected. Since $diam(ZD(R)) \leq diam(\Gamma(R))$ and $diam(\Gamma(R)) = 3$ by Theorem 2.4.28, we have $diam(ZD(R)) \leq 3$. Let $x = (1, 0, -1), y = (0, 1, -1) \in Z(R)^*$. Then $x \cdot y = 1 \neq 0$. We show that $d_Z(x, y) = 3$. Let $w = (w_1, w_2, w_3) \in R$ such that $x \cdot w = w \cdot y = 0$. Then a trivial calculation leads to $w_1 = w_2 = w_3$. Since $A$ is an integral domain, $w \in Z(R)$ if and only if $w = (0, 0, 0)$. Thus $d_z(x, y) = 3$, and hence $diam(ZD(R)) = 3$.

(2) (Similar to the proof of Theorem 10.7.4). Let $x = (x_1, x_2, x_3), y = (y_1, y_2, y_3) \in Z(R)^*$, and suppose that $x \cdot y \neq 0$. Then let $M = \{i \mid x_i = y_i = 0, 1 \leq i \leq 3\}$. Suppose that $M \neq \emptyset$. Choose a $k \in M$, and let $w = (w_1, w_2, w_3) \in Z(R)^*$, where $w_k = 1$ and $w_i = 0$. If $i \neq k$, then $x - w - y$ is a path of length 2 in $ZD(R)$ from $x$ to $y$. Now suppose that $M = \emptyset$. Let $f(x) = min\{i \mid x_i \neq 0, 1 \leq i \leq n\}$ and $f(y) = min\{i \mid y_i \neq 0, 1 \leq i \leq 3\}$. Since $M = \emptyset$, we have $f(x) = 1$ or $f(y) = 1$. We may assume that $f(x) = 1$. Let $v = (x_2y_3 - x_3y_2,\ x_3y_1 - x_1y_3,\ x_1y_2 - x_2y_1) \in R$. Suppose that $v \in Z(R)^*$. Then it is easy to check that $x \cdot y = v \cdot y = 0$. Since $x \cdot y \neq 0$, $v \neq x$ and $v \neq y$. Thus $x - v - y$ is a path of length 2 in $ZD(R)$ from $x$ to $y$. Suppose that $v \notin Z(R)$. Then choose an $a \in Z(A)^*$. Then $av \in Z(R)^*$ and $x - av - y$ is a path of length 2 in $ZD(R)$ from $x$ to $y$. Suppose that $v = (0, 0, 0)$. Then $x_1y_2 - x_2y_1 = 0$. Let $w = (-x_2,\ x_1, 0) \in Z(R)$. Since $x_1 \neq 0$, $w \in Z(R)^*$. Hence $x \cdot w = -x_1x_2 + x_1x_2 = 0$ and $w \cdot y = x_1y_2 - x_2y_1 = 0$. Since $x \cdot w = w \cdot y = 0$ and $x \cdot y \neq 0$, $x \neq w$ and $y \neq w$. Thus $x - w - y$ is a path of length 2 in $ZD(R)$ from $x$ to $y$. Hence $ZD(R)$ is connected with $diam(ZD(R)) = 2$.

(3) The proof is similar to the proof of Theorem 10.7.4. Just observe that if $n \geq 4$, then $v$, as in the proof of Theorem 10.7.4, is in $Z(R)$. □

**Theorem 10.7.6.** ([130, Theorem 2.6]) *Let $A$ be a commutative ring, $3 \leq n < \infty$, and $R = A \times A \times \cdots \times A$ ($n$ times). Then $gr(ZD(R)) = gr(TD(R)) = 3$.*

*Proof.* Let $a = (1, 0, \ldots, 0), b = (0, 1, 0, \ldots, 0)$, and $c = (0, 0, 1, 0 \ldots, 0)$. Then $a - b - c - a$ is a cycle of length 3. □

**Corollary 10.7.7.** ([130, Corollary 2.7]) *Let $A$ be a commutative ring, $2 \leq n < \infty$, and $R = A \times A \times \cdots \times A$ ($n$ times). Then the following statements are equivalent:*

(1) $gr(ZD(R)) = 3$;

(2) $gr(TD(R)) = 3$;

(3) *Either $A$ is not an integral domain and $n = 2$ or $n \geq 3$.*

*Proof.* This is clear by Theorem 7.2.7 and Theorem 10.7.6. □

**Corollary 10.7.8.** ([130, Corollary 2.8]) *Let $A$ be a commutative ring, $2 \leq n < \infty$, and $R = A \times A \times \cdots \times A$ ($n$ times). Then the following statements are equivalent:*

(1) $gr(ZD(R)) = \infty$;

(2) *$A$ is ring-isomorphic to $\mathbb{Z}_2$ and $n = 2$;*

(3) $diam(ZD(R)) = 1$.

*Proof.* (1) $\Rightarrow$ (2) Suppose that $gr(ZD(R)) = \infty$. Then $n = 2$ by Theorem 10.7.6. Hence $A$ is an integral domain by Corollary 10.7.7. Thus $ZD(R) = \Gamma(R)$ by Theorem 10.7.1. Thus $A$ is ring-isomorphic to $\mathbb{Z}_2$ by Theorem 2.3.10. (2) $\Rightarrow$ (3) It is clear. (3) $\Rightarrow$ (1) Since $diam(ZD(R)) = 1$, we have $n = 2$ and $A$ is an integral domain by Theorems 10.7.3 and 10.7.5. Thus $A$ is ring-isomorphic to $\mathbb{Z}_2$ by Theorem 10.7.1. Hence $gr(ZD(R)) = \infty$. □

**Corollary 10.7.9.** ([130, Corollary 2.9]) *Let $A$ be a commutative ring not ring-isomorphic to $\mathbb{Z}_2$, $0 \leq n < \infty$, and $R = A \times A \times \cdots \times A$ ($n$ times). Then the following statements are equivalent:*

(1) $gr(ZD(R)) = 4$;

(2) $ZD(R) = \Gamma(R)$;

(3) *$TD(R)$ is disconnected;*

(4) *$A$ is an integral domain and $n = 2$.*

*Proof.* This is clear by Theorem 10.7.1, Theorem 10.7.2, Corollary 10.7.7, and Corollary 10.7.8. □

**Corollary 10.7.10.** ([130, Corollary 2.10]) *Let $A$ be a commutative ring, $2 \leq n < \infty$, and $R = A \times A \times \cdots \times A$ ($n$ times). Then the following statements are equivalent:*

(1) $diam(ZD(R)) = 3$;

(2) *Either $A$ is not an integral domain and $n = 2$, or $A$ is an integral domain and $n = 3$.*

**Corollary 10.7.11.** ([130, Corollary 2.11]) *Let $A$ be a commutative ring, $2 \leq n < \infty$, and $R = A \times A \times \cdots \times A$ ($n$ times). Then the following statements are equivalent:*

(1) *$diam(ZD(R)) = 2$;*

(2) *Either $A$ is an integral domain not ring-isomorphic to $\mathbb{Z}_2$ and $n = 2$, $A$ is is not an integral domain and $n = 3$, or $n \geq 4$.*

*Proof.* This is clear by Theorem 10.7.1, Theorem 10.7.3, and Theorem 10.7.5. □

**Corollary 10.7.12.** ([130, Corollary 2.12]) *Let $A$ be a commutative ring with $1 \neq 0$, $2 \leq n < \infty$, and $R = A \times A \times \cdots \times A$ ($n$ times). Then $diam(TD(R)) = 3$ if and only if $A$ is not an integral domain and $n = 2$.*

*Proof.* This is clear by Theorem 10.7.1, Theorem 10.7.5, and Corollary 10.7.10. □

**Corollary 10.7.13.** ([130, Corollary 2.13]) *Let $A$ be a commutative ring with $1 \neq 0$, $2 \leq n < \infty$, and $R = A \times A \times \cdots \times A$ ($n$ times). Then the following statements are equivalent:*

(1) *$diam(TD(R)) = 2$;*

(2) *$TD(R)$ is connected and $n \geq 3$;*

(3) *$n \geq 3$.*

*Proof.* This is clear by Theorem 10.7.1, Theorem 10.7.3, and Theorem 10.7.4. □

**Corollary 10.7.14.** ([130, Corollary 2.14]) *Let $A$ be a commutative ring, $2 \leq n < \infty$, and $R = A \times A \times \cdots \times A$ ($n$ times). Then $diam(TD(R)) = diam(ZD(R)) = 3$ if and only if $A$ is not an integral domain and $n = 2$.*

*Proof.* This is clear by Corollary 10.7.10 and Corollary 10.7.12. □

### 10.7.1 Subgraphs of Dot Product Graphs

Let $R$, $TD(R)$, and $ZD(R)$ be as in Section 10.7. The group of units of $R$ is denoted by $U(R)$. The *unit dot product graph* of $R$ is the induced subgraph $UD(R)$ of $TD(R)$ with vertices $U(R)$. Let $p$ be a positive prime number, $n \geq 2$, $A = GF(p^n)$ the finite field with $p^n$ elements, and $R = A \times A$. Then $TD(R)$ is not connected by Theorem 10.7.1. The first two results give a complete description of the structure of $UD(R)$ and $TD(R)$.

**Theorem 10.7.15.** ([11, Theorem 2.1]) *Let $n \geq 1$, $m = 2^n - 1$, and $R = GF(2^n) \times GF(2^n)$. Then*

(1) *$ZD(R) = \Gamma(R) = K_{m,m}$;*

(2) *$UD(R)$ is the disjoint union of one $K_m$ and $2^{(n-1)} - 1$ $K_{m,m}$'s;*

(3) *$TD(R)$ is the disjoint union of one $K_m$ and $2^{(n-1)}$ $K_{m,m}$'s.*

*Proof.* (1) The result is clear by Theorem 10.7.1 and Theorem 2.3.8.

(2) Let $A = GF(2^n)$. Then $R = A \times A$. Let $v_1, v_2 \in U(R)$. Since $R$ is a vector space over $A$, $v_1 = u(1,a) \in R$ and $v_2 = v(1,b) \in R$ for some $u, v, a, b \in A^*$. Hence $v_1$ is adjacent to $v_2$ if and only if $v_1 \cdot v_2 = uv + uvab = 0$ in $A$, if and only if $b = -a^{-1} = a^{-1}$ in $A$ . Thus for every $a \in U(A) = A^*$, let $X_a = \{u(1,a) \mid u \in A^*\}$ and $Y_a = \{u(1,a^{-1}) \mid u \in A^*\}$. It is clear that $|X_a| = |Y_a| = 2^n - 1$. Let $a = 1$. Since $char(A) = char(R) = 2$, $X_a = Y_a$ and the dot product of every two distinct vertices in $X_a$ is zero. Hence every two distinct vertices in $X_a$ are adjacent. Thus the vertices in $X_a$ form the graph $K_m$ that is a complete subgraph of $TD(R)$. Let $a \in U(A)$ such that $a \neq 1$. Since $a^2 \neq 1$ for every $a \in U(A) \setminus \{1\}$, we have $X_a \cap Y_a = \emptyset$, every two distinct vertices in $X_a$ are not adjacent, and every two distinct vertices in $Y_a$ are not adjacent. Since $char(A) = char(R) = 2$, it is clear that every vertex in $X_a$ is adjacent to every vertex in $Y_a$. Thus the vertices in $X_a \cup Y_a$ form the graph $K_{m,m}$ that is a complete bipartite subgraph of $TD(R)$. By construction, there are exactly $(2^n - 2)/2 = 2^{n-1} - 1$ disjoint complete bipartite $K_{m,m}$ subgraphs of $TD(R)$. Hence $UD(R)$ is the disjoint union of one complete subgraph $K_m$ and $2^{n-1} - 1$ complete bipartite $K_{m,m}$ subgraphs.

(3) The claim follows from (1) and (2). □

**Theorem 10.7.16.** ([11, Theorem 2.2]) *Let $p \geq 3$ be a prime integer, $n \geq 1$, $m = p^n - 1$, and $R = GF(p^n) \times GF(p^n)$. Then*

(1) *$ZD(R) = \Gamma(R) = K_{m,m}$;*

(2) *If $4 \nmid m$, then $UD(R)$ is the disjoint union of $m/2$ $K_{m,m}$'s;*

(3) *If $4 \mid m$, then $UD(R)$ is the disjoint union of two $K_m$'s and $(m-2)/2$ $K_{m,m}$'s;*

(4) *If $4 \nmid m$, then $TD(R)$ is the disjoint union of $(m+2)/2$ $K_{m,m}$'s;*

(5) *If $4 \mid m$, then $TD(R)$ is the disjoint union of two $K_m$'s and $m/2$ $K_{m,m}$'s.*

*Proof.* (1) The result is clear by Theorem 10.7.1 and Theorem 2.3.8.

(2) Let $A = GF(p^n)$. Then $R = A \times A$. Let $v_1, v_2 \in U(R)$. Since $R$ is a vector space over $A$, $v_1 = u(1,a) \in R$ and $v_2 = v(1,b) \in R$ for some $u, v, a, b \in A^*$. Hence $v_1$ is adjacent to $v_2$ if and only if $v_1 \cdot v_2 = uv + uvab = 0$ in $A$, if and only if $b = -a^{-1}$ in $A$. Since $R$ is a vector space over $A$, for every $a \in U(A) = A^*$, let $X_a = \{u(1,a) \mid u \in A^*\}$ and $Y_a = \{u(1,-a^{-1}) \mid u \in A^*\}$. It is clear that $|X_a| = |Y_a| = m = p^n - 1$. Since $4 \nmid m$, $U(A) = A^*$ has no elements of order 4. Thus $a^2 \neq -1$ for every $a \in U(A)$. Hence $X_a \cap Y_a = \emptyset$; so every two distinct vertices in $X_a$ are not adjacent, and every two distinct vertices in $Y_a$ are not adjacent. By construction of $X_a$ and $Y_a$, it is clear that every vertex in $X_a$ is adjacent to every vertex in $Y_a$. Thus the vertices in $X_a \cup Y_a$ form the graph $K_{m,m}$ that is a complete bipartite subgraph of $TD(R)$. By construction, there

are exactly $m/2$ disjoint complete bipartite $K_{m,m}$ subgraphs of $TD(R)$. Hence $UD(R)$ is the disjoint union of $m/2$ $K_{m,m}$'s.

(3) Note that $|U(A)| = m$. Since $U(A) = A^*$ is cyclic and $4 \mid m$, $U(A)$ has exactly one subgroup of order 4. Thus $U(A)$ has exactly two elements of order 4, say $b, c$. Since $a \in U(A)$ is of order 4 if and only if $a^2 = -1$, it is clear that $x^2 = -1$ for some $x \in U(A)$ if and only if $x = b, c$. Let $X_b = \{u(1,b) \mid u \in U(A)\}$ and $X_c = \{u(1,c) \mid u \in U(A)\}$. It is clear that $|X_b| = |X_c| = m$. Let $H = \{b, c\}$. Then the dot product of every two distinct vertices in $X_h$ is zero for every $h \in H$. Thus every two distinct vertices in $X_h$ are adjacent for every $h \in H$. Thus for each $h \in H$, the vertices in $X_h$ form the graph $K_m$ that is a complete subgraph of $TD(R)$. Let $a \in U(A) \setminus H$, $X_a = \{u(1,a) \mid u \in A^*\}$, and $Y_a = \{u(1, -a^{-1}) \mid u \in A^*\}$. It is clear that $|X_a| = |Y_a| = m$. Since $a \notin H$, we have $X_a \cap Y_a = \emptyset$, every two distinct vertices in $X_a$ are not adjacent, and every two distinct vertices in $Y_a$ are not adjacent. By construction, it is clear that every vertex in $X_a$ is adjacent to every vertex in $Y_a$. Thus the vertices in $X_a \cup Y_a$ form the graph $K_{m,m}$ that is a complete bipartite subgraph of $TD(R)$. By construction, there are $(m-2)/2$ disjoint $K_{m,m}$ subgraphs. Hence $UD(R)$ is the disjoint union of two $K_m$'s and $(m-2)/2$ $K_{m,m}$'s.

(4) The claim follows from (1) and (2).

(5) The claim follows from (1) and (3). □

In view of Theorem 10.7.16, we have the following corollary.

**Corollary 10.7.17.** ([11, Corollary 2.3]) *Let $p \geq 3$ be a prime integer and $R = \mathbb{Z}_p \times \mathbb{Z}_p$. Then*

(1) $ZD(R) = \Gamma(R) = K_{p-1,p-1}$;

(2) *If $4 \nmid p-1$, then $UD(R)$ is the disjoint union of $(p-1)/2$ $K_{p-1,p-1}$'s;*

(3) *If $4 \mid p-1$, then $UD(R)$ is the disjoint union of two $K_{p-1}$'s and $(p-3)/2$ $K_{p-1,p-1}$'s;*

(4) *If $4 \nmid p-1$, then $TD(R)$ is the disjoint union of $(p+1)/2$ $K_{p-1,p-1}$'s;*

(5) *If $4 \mid p-1$, then $TD(R)$ is the disjoint union of two $K_{p-1}$'s and $(p-1)/2$ $K_{p-1,p-1}$'s.*

**Example 10.7.18.** ([11, Example 2.4]) Let $A = \mathbb{Z}_2[X]/(X^2+X+1)$. Then $A$ is a finite field with 4 elements. Let $v = X + (X^2+X+1) \in A$. Since $(A^*, \cdot)$ is a cyclic group and $A^* = \langle v \rangle$, we have $A = \{0, v, v^2, v^3 = 1 + (X^2+X+1)\}$. Let $R = A \times A$. Then $UD(R)$ is the disjoint union of one $K_3$ and one $K_{3,3}$ by Theorem 10.7.15(2). The following is the graph of $UD(R)$.

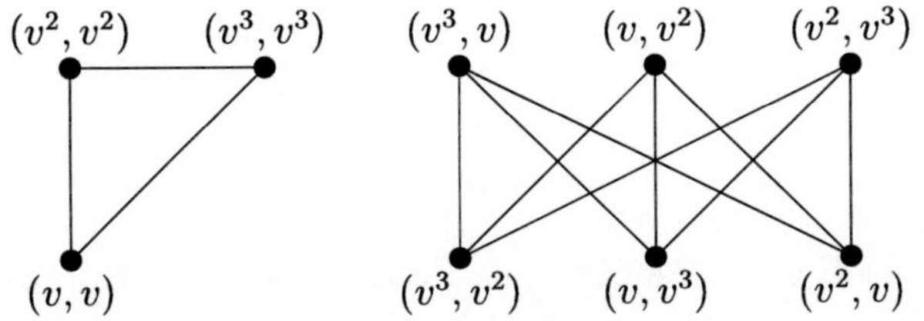


**Example 10.7.19.** ([11, Example 2.5]) Let $A = \mathbb{Z}_5$ and $R = A \times A$. Then $UD(R)$ is the disjoint union of two $K_4$'s and one $K_{4,4}$ by Corollary 10.7.17(3). The following is the graph of $UD(R)$.

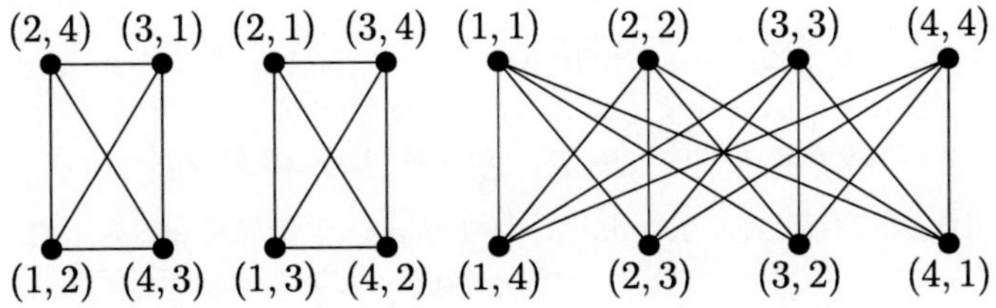


Let $n > 1$ and write $n = p_1^{k_1} \cdots p_m^{k_m}$, where the $p_i$'s are distinct positive prime integers. Then $U(\mathbb{Z}_n) = \{1 \leq a < n \mid a \text{ is an integer and } \gcd(a, n) = 1\}$. It is known that $U(\mathbb{Z}_n)$ is a group under multiplication modulo $n$ and $|U(\mathbb{Z}_n)| = \varphi(n) = (p_1 - 1)p_1^{k_1-1}(p_2 - 1)p_2^{k_2-1} \cdots (p_m - 1)p_m^{k_m-1}$.

The following lemma is needed.

**Lemma 10.7.20.** ([11, Lemma 2.1]) *Let $n$ be a positive integer and write $n = p_1^{k_1} p_2^{k_2} \cdots p_r^{k_r}$, where the $p_i$'s are distinct positive prime integers. Then*

(1) *If $4 \mid n$, then $a^2 \not\equiv n - 1 (mod\ n)$ for every $a \in U(\mathbb{Z}_n)$;*

(2) *If $4 \nmid n$, then $x^2 \equiv n - 1 (mod\ n)$ has a solution in $U(\mathbb{Z}_n)$ if and only if $4 \mid (p_i - 1)$ for every odd prime factor $p_i$ of $n$. Furthermore, if $x^2 \equiv n - 1 (mod\ n)$ has a solution in $U(\mathbb{Z}_n)$, then it has exactly $2^{r-1}$ distinct solutions in $U(\mathbb{Z}_n)$ if $n$ is even and it has exactly $2^r$ distinct solutions in $U(\mathbb{Z}_n)$ if $n$ is odd.*

*Proof.* (1) Suppose that $4 \mid n$. Then $n \geq 4$. Since $4 \nmid (n - 2)$, $n - 1 \not\equiv 1 (mod\ 4)$, and thus $a^2 \not\equiv n - 1 (mod\ n)$ for every $a \in U(\mathbb{Z}_n)$ by [259, Theorem 5.1].

(2) Suppose that $4 \nmid n$. Then $a^2 \equiv n - 1 (mod\ n)$ for some $a \in U(Z_n)$ if and only if $a^2 \equiv n - 1 (mod\ p_i)$ for every odd prime factor $p_i$ of $n$ by [259, Theorem 5.1]. Thus $a^2 \equiv n - 1 (mod\ n)$ for some $a \in U(\mathbb{Z}_n)$ if and only if $(a\ mod\ p_i)^2 \equiv p_i - 1 (mod\ p_i)$ for every odd prime factor $p_i$ of $n$. Since $U(\mathbb{Z}_{p_i}) = \mathbb{Z}_{p_i}^* = \{1, \ldots, p_i - 1\}$ for every prime factor $p_i$ of $n$, we have $|U(\mathbb{Z}_{p_i})| = p_i - 1$. For every $x \in U(\mathbb{Z}_{p_i})$, $1 \leq i \leq r$, let $|x|$ denote the order of $x$ in $U(\mathbb{Z}_{p_i})$. Let

$p_i$, $1 \leq i \leq r$, be an odd prime factor of $n$. Since $|p_i - 1| = 2$ in $U(\mathbb{Z}_{p_i})$, $b^2 = p_i - 1$ in $U(\mathbb{Z}_{p_i})$ for some $b \in U(\mathbb{Z}_{p_i})$ if and only if $|b| = 4$ in $U(\mathbb{Z}_{p_i})$. Since $|U(\mathbb{Z}_{p_i})| = p_i - 1$, we conclude that $b^2 = p_i - 1$ in $U\mathbb{Z}_{P_i})$ for some $b \in U(\mathbb{Z}_{p_i})$ if and only if $4 \mid (p_i - 1)$. Thus $x^2 \equiv n - 1(mod\ n)$ has a solution in $U(\mathbb{Z}_n)$ if and only if $4 \mid (p_i - 1)$ for every odd prime $p_i$ factor of $n$. Suppose that $x^2 \equiv n - 1(mod\ n)$ has a solution in $U(\mathbb{Z}_n)$. We consider two cases: Case 1. Suppose that $n$ is an even integer. Then there are exactly $r - 1$ distinct odd prime factors of $n$. Since $4 \nmid n$, $x^2 \equiv n - 1(mod\ n)$ has exactly $2^{r-1}$ distinct solutions in $U(\mathbb{Z}_n)$ by [259, Theorem 5.2]. Case 2. Suppose that $n$ is an odd integer. Then there are exactly $r$ distinct odd prime factors of $n$. Thus $x^2 \equiv n - 1(mod\ n)$ has exactly $2^r$ distinct solutions in $U(\mathbb{Z}_n)$ by [259, Theorem 5.2]. □

Let $A = \mathbb{Z}_n$, where $n$ is not prime. Then $TD(A \times A)$ is connected by [130, Theorem 2.3]. In the following result, we show that $UD(A \times A)$ is disconnected, and we give a complete description of the structure of $UD(A \times A)$.

**Theorem 10.7.21.** ([11, Theorem 3.2]) *Let $n \geq 3$ be an integer, $R = \mathbb{Z}_n \times \mathbb{Z}_n$, and $\varphi(n) = m$. Write $n = p_1^{k_1} p_2^{k_2} \cdots p_r^{k_r}$, where the $p_i$'s are distinct positive prime integers, $1 \leq i \leq r$. Then*

(1) *If $4 \mid n$, then $UD(R)$ is the disjoint union of $m/2$ $K_{m,m}$'s;*

(2) *If $4 \nmid n$ and $4 \nmid (p_i - 1)$ for at least one of the $p_i$'s in the prime factorization of $n$, then $UD(R)$ is the disjoint union of $m/2$ $K_{m,m}$'s;*

(3) *If $4 \nmid n$ and $4 \mid (p_i - 1)$ for all the odd $p_i$'s in the prime factorization of $n$, then we consider two cases:*

*Case I. If $n$ is even, then $UD(R)$ is the disjoint union of $m/2 - 2^{r-2}$ $K_{m,m}$'s and $2^{r-1}$ $K_m$'s;*

*Case II. If $n$ is odd, then $UD(R)$ is the disjoint union of $m/2 - 2^{r-1}$ $K_{m,m}$'s and $2^r$ $K_m$'s.*

*Proof.* Let $A = \mathbb{Z}_n$. Then $R = A \times A$. Note that $UD(R)$ has exactly $m^2$ vertices. Let $v_1, v_2 \in U(R)$. Since $R$ is a vector space over $A$, $v_1 = u(1, a) \in R$ and $v_2 = v(1, b) \in R$ for some $u, v, a, b \in U(A)$. Hence $v_1$ is adjacent to $v_2$ if and only if $v_1 \cdot v_2 = uv + uvab = 0$ in $A$, if and only if $b = -a^{-1}$ in $A$. Thus for every $a \in U(A)$, let $X_a = \{u(1, a) \mid u \in U(A)\}$ and $Y_a = \{u(1, -a^{-1}) \mid u \in U(A)\}$. It is clear that $|X_a| = |Y_a| = m$.

(1) Since $4 \mid n$, $a^2 \not\equiv n - 1(mod\ n)$ for each $a \in U(\mathbb{Z}_n)$ by Lemma 10.7.20(1). Hence $X_a \cap Y_a = \emptyset$; so every two distinct vertices in $X_a$ are not adjacent, and every two distinct vertices in $Y_a$ are not adjacent. By construction of $X_a$ and $Y_a$, it is clear that every vertex in $X_a$ is adjacent to every vertex in $Y_a$. Thus the vertices in $X_a \cup Y_a$ form the graph $K_{m,m}$ that is a complete bipartite subgraph of $TD(R)$. By construction, there are exactly $m/2$ disjoint complete bipartite $K_{m,m}$ subgraphs of $TD(R)$. Hence $UD(R)$ is the disjoint union of $m/2$ $K_{m,m}$'s.

(2) Write $n = p_1^{k_1} p_2^{k_2} \cdots p_r^{k_r}$, where the $p_i$'s are distinct positive prime integers. Since $4 \nmid n$ and $4 \nmid (p_i - 1)$ for at least one of the $p_i$'s, $a^2 \not\equiv n-1 (mod\ n)$ for every $a \in U(\mathbb{Z}_n)$ by Lemma 10.7.20. Thus by the same argument as in (1), $UD(R)$ is the disjoint union of $m/2$ $K_{m,m}$'s.

(3) Write $n = p_1^{k_1} p_2^{k_2} \cdots p_r^{k_r}$, where the $p_i$'s are distinct positive prime integers. Suppose that $4 \nmid n$ and $4 \mid p_i - 1$ for all the odd $p_i$'s in the prime factorization of $n$. Let $B = \{b \in U(\mathbb{Z}_n) \mid b^2 = n-1 \text{ in } U(\mathbb{Z}_n)\}$ and $C = \{c \in U(Z_n) \mid c^2 \neq n-1 \text{ in } U(\mathbb{Z}_n)\}$. We consider two cases: Case I. Suppose that $n$ is even. Then $|B| = 2^{r-1}$ by Lemma 10.7.20(2), and hence $|C| = m - 2^{r-1}$. For every $a \in B$, we have $X_a = Y_a$, and hence the dot product of every two distinct vertices in $X_a$ is zero. Thus the vertices in $X_a$ form the graph $K_m$ that is a complete subgraph of $TD(R)$. Hence $UD(\mathbb{Z}_n)$ has exactly $2^{r-1}$ disjoint $K_m$'s. For each $a \in C$, we have $X_a \cap Y_a = \emptyset$; so every two distinct vertices in $X_a$ are not adjacent, and every two distinct vertices in $Y_a$ are not adjacent. By construction, it is clear that every vertex in $X_a$ is adjacent to every vertex in $Y_a$. Thus the vertices in $X_a \cup Y_a$ form the graph $K_{m,m}$ that is a complete bipartite subgraph of $TD(R)$. Thus $UD(\mathbb{Z}_n)$ has exactly $(m - 2^{r-1})2 = m/2 - 2^{r-2}$ disjoint $K_{m,m}$'s. Case II. Suppose that $n$ is odd. Then $|B| = 2^r$ by Lemma 10.7.20(2), and hence $|C| = m - 2^r$. For every $a \in B$, we have $X_a = Y_a$, and hence the dot product of every two distinct vertices in $X_a$ is zero. Thus the vertices in $X_a$ form the graph $K_m$ that is a complete subgraph of $TD(R)$. Hence $UD(Z_n)$ has exactly $2^r$ disjoint $K_m$'s. For every $a \in C$, we have $X_a \cap Y_a = \emptyset$, every two distinct vertices in $X_a$ are not adjacent, and every two distinct vertices in $Y_a$ are not adjacent. By construction, it is clear that every vertex in $X_a$ is adjacent to every vertex in $Y_a$. Thus the vertices in $X_a \cup Y_a$ form the graph $K_{m,m}$ that is a complete bipartite subgraph of $TD(R)$. Thus $UD(\mathbb{Z}_n)$ has exactly $(m - 2^r)/2 = m/2 - 2^{r-1}$ disjoint $K_{m,m}$'s. □

Recall that a graph $G$ is called *completely disconnected* if every two vertices of $G$ are not connected by an edge in $G$.

**Theorem 10.7.22.** ([11, Theorem 3.3]) *Let $n \geq 4$ be an even integer and $R = \mathbb{Z}_n \times \mathbb{Z}_n \times \cdots \times \mathbb{Z}_n$ ($k$ times), where $k$ is an odd positive integer. Then $UD(R)$ is completely disconnected.*

*Proof.* Let $x = (x_1, \ldots, x_k), y = (y_1, \ldots, y_k) \in U(R)$. Then $x_i, y_i \in U(\mathbb{Z}_n)$ for every i, $1 \leq i \leq k$. Since $n$ is an even integer, $x_i$ and $y_i$ are odd integers for every $i$, $1 \leq i \leq k$. Hence, since $k$ is an odd integer, $x \cdot y = x_1y_1 + \cdots + x_ky_k$ is an odd integer, and thus $x \cdot y = x_1y_1 + \cdots + x_ky_k \neq 0$ in $\mathbb{Z}_n$ since $n$ is even. Hence $UD(R)$ is completely disconnected. □

**Theorem 10.7.23.** ([11, Theorem 3.4]) *Let $n \geq 4$ be an even integer and $R = \mathbb{Z}_n \times \mathbb{Z}_n$. Then the vertex $(n/2, n/2)$ in $ZD(R)$ is adjacent to every vertex in $UD(R)$.*

*Proof.* It is clear that $(n/2, n/2)$ is a vertex of $ZD(R)$. Let $u \in U(\mathbb{Z}_n)$. Since $n$ is even, $u$ is an odd integer. Thus $u - 1 = 2m$ for some integer $m$. Hence $(n/2)(u-1) = (n/2)(2m) = mn = 0 \in \mathbb{Z}_n$. Thus $(n/2)u = n/2$ in $\mathbb{Z}_n$. Now let $(a, b) \in U(R)$. Then $a, b \in U(\mathbb{Z}_n)$ are odd integers. Hence $(a, b)(n/2, n/2) = n/2 + n/2 = n = 0 \in \mathbb{Z}_n$. Thus the vertex $(n/2, n/2)$ in $ZD(R)$ is adjacent to every vertex in $UD(R)$. □

**Example 10.7.24.** ([11, Example 3.5]) Let $A = \mathbb{Z}_8$ and $R = A \times A$. Then $UD(R)$ is the disjoint union of two $K_{4,4}$'s by Theorem 10.7.21(1). The following is the graph of $UD(R)$.

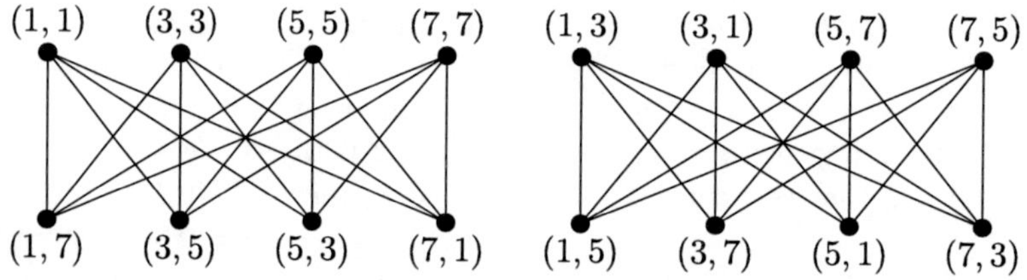


**Example 10.7.25.** ([11, Example 3.6]) Let $A = \mathbb{Z}_{10}$ and $R = A \times A$. Then $UD(R)$ is the disjoint union of two $K_4$'s and one $K_{4,4}$ by Theorem 10.7.21(3, case I). The following is the graph of $UD(R)$.

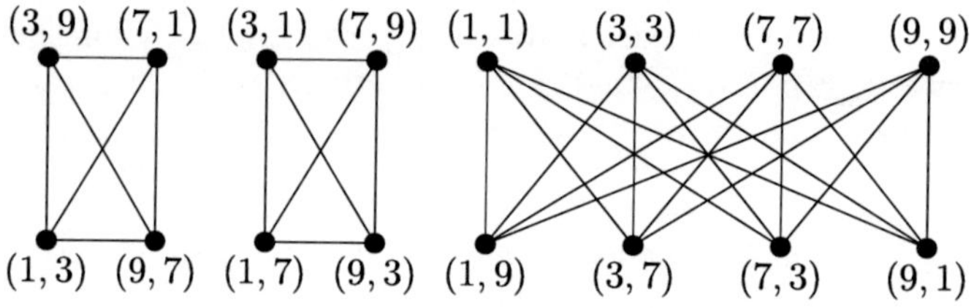


For an integer $n \geq 2$, let $R_1 = \{(u_1, z_1) \mid u_1 \in U(\mathbb{Z}_n) \text{ and } z_1 \in Z(\mathbb{Z}_n)\}$ and $R_2 = \{(z_2, u_2) \mid u_2 \in U(\mathbb{Z}_n) \text{ and } z_2 \in Z(\mathbb{Z}_n)\}$. It is clear that $R_1 \subsetneq Z(\mathbb{Z}_n \times \mathbb{Z}_n)$ and $R_2 \subsetneq Z(\mathbb{Z}_n \times \mathbb{Z}_n)$. We study the induced subgraph $ZD(R_1 \cup R_2)$ of $ZD(\mathbb{Z}_n \times \mathbb{Z}_n)$ with vertices $R_1 \cup R_2$.

**Theorem 10.7.26.** *Let $n \geq 2$, $R = \mathbb{Z}_n \times \mathbb{Z}_n$, and $\varphi(n) = m$. Then*

(1) *If $n$ is prime, then $ZD(R_1 \cup R_2) = ZD(\mathbb{Z}_n \times \mathbb{Z}_n) = \Gamma(R) = K_{n-1,n-1}$;*

(2) *If $n$ is not prime, then $ZD(R_1 \cup R_2)$ is the disjoint union of of $n - m$ $K_{m,m}$'s.*

*Proof.* (1) Suppose that $n$ is prime. Then it is clear that $R_1 \cup R_2 = Z(\mathbb{Z}_n \times \mathbb{Z}_n)$. If $n = 2$, then it is trivial to see that $ZD(R_1 \cup R_2) = ZD(\mathbb{Z}_n \times \mathbb{Z}_n) = \Gamma(R) = K_{1,1}$. If $n \geq 3$, then the claim is clear by Corollary 10.7.17(1).

(2) Let $A = \mathbb{Z}_n$. Suppose that $n$ is not prime. It is clear that every two vertices in $R_i$ are not adjacent for every $i \in \{1, 2\}$. Let $v_1 \in R_1$ and $v_2 \in R_2$. Then $v_1 = u(1, a) \in R_1$ and $v_2 = v(b, 1) \in R_2$ for some $u, v \in U(A)$ and $a, b \in Z(A)$. Then $v_1$ is adjacent to $v_2$ if and only if $v_1 \cdot v_2 = uvb + uva = 0$ in $A$, if and only if $b = -a$ in $A$. Hence for every $a \in Z(A)$, let $X_a = \{u(1, a) \mid u \in U(A)\}$ and $Y_a = \{u(-a, 1) \mid u \in U(A)\}$. It is clear that $|X_a| = |Y_a| = m$. For every $a \in Z(A)$, $X_a \cap Y_a = \emptyset$, every two distinct vertices in $X_a$ are not adjacent, and every two distinct vertices in $Y_a$ are not adjacent. By construction, it is clear that every vertex in $X_a$ is adjacent to every vertex in $Y_a$. Thus the vertices in $X_a \cup Y_a$ form the graph $K_{m,m}$ that is a complete bipartite subgraph of $ZD(R)$. Since $|R_1| = |R_2| = m(n - m)$ and $R_1 \cap R_2 = \emptyset$, we have $|R_1 \cup R_2| = 2m(n - m)$. Thus $ZD(R_1 \cup R_2)$ is the disjoint union of of $n - m$ $K_{m,m}$'s. □

### 10.7.2 Equivalence Dot Product Graphs

Let $A = \mathbb{Z}_n$ and $R = A \times A$. Define a relation $\sim$ on $U(R)$ such that $x \sim y$, where $x, y \in U(R)$, if $x = (c, c)y$ for some $(c, c) \in U(R)$. It is clear that $\sim$ is an equivalence relation on $U(R)$. If $S$ is an equivalence class of $U(R)$, then there is an $a \in U(A)$ such that $S = \overline{(1, a)} = \{u(1, a) \mid u \in U(\mathbb{Z}_n)\}$. Let $E(U(R))$ be the set of all distinct equivalence classes of $U(R)$. We define the *equivalence unit dot product graph* of $U(R)$ to be the (undirected) graph $EUD(R)$ with vertices $E(U(R))$, and two distinct vertices $X$ and $Y$ are adjacent if and only if $a \cdot b = 0 \in A$ for every $a \in X$ and every $b \in Y$ (where $a \cdot b$ denote the normal dot product of $a$ and $b$). We have the following results.

**Theorem 10.7.27.** ([11, Theorem 5.1]) *Let $n \geq 1$, $m = 2^n - 1$, and $R = GF(2^n) \times GF(2^n)$. Then $EUD(R)$ is the disjoint union of one $K_1$ and $2^{(n-1)} - 1$ $K_{1,1}$'s.*

*Proof.* Let $A = GF(2^n)$. For every $a \in U(A)$, let $X_a$ and $Y_a$ be as in the proof of Theorem 10.7.15. Then $X_a, Y_a \in E(U(R))$. Since $|X| = m$ for every $X \in E(U(R))$, we conclude that every $K_m$ of $UD(R)$ is a $K_1$ of $EUD(R)$ and every $K_{m,m}$ of $UD(R)$ is a $K_{1,1}$ of $EUD(R)$. Hence the claim follows by the proof of Theorem 10.7.15. □

**Theorem 10.7.28.** ([11, Theorem 5.2]) *Let $p \geq 3$ be a prime integer, $n \geq 1$, $m = p^n - 1$, and $R = GF(p^n) \times GF(p^n)$. Then*

(1) *If $4 \nmid m$, then $EUD(R)$ is the disjoint union of $m/2$ $K_{1,1}$'s;*

(2) *If $4 \mid m$, then $EUD(R)$ is the disjoint union of two $K_1$'s and $(m - 2)/2$ $K_{1,1}$'s.*

*Proof.* Let $A = GF(p^n)$. For every $a \in U(A)$, let $X_a$ and $Y_a$ be as in the proof of Theorem 10.7.16. Then $X_a, Y_a \in E(U(R))$. Since $|X| = m$ for every $X \in E(U(R))$, we conclude that every $K_m$ of $UD(R)$ is a $K_1$ of $EUD(R)$ and every $K_{m,m}$ of $UD(R)$ is a $K_{1,1}$ of $EUD(R)$. Hence the claim follows by the proof of Theorem 10.7.16. □

**Theorem 10.7.29.** ([11, Theorem 5.3]) *Let $n \geq 3$ be an integer, $R = \mathbb{Z}_n \times \mathbb{Z}_n$, and $\varphi(n) = m$. Write $n = p_1^{k_1} p_2^{k_2} \cdots p_r^{k_r}$, where the $p_i$'s are distinct positive prime integers, $1 \leq i \leq r$. Then*

(1) *If $4 \mid n$, then $EUD(R)$ is the disjoint union of $m/2$ $K_{1,1}$'s;*

(2) *If $4 \nmid n$ and $4 \nmid (p_i - 1)$ for at least one of the $p_i$'s in the prime factorization of $n$, then $EUD(R)$ is the disjoint union of $m/2$ $K_{1,1}$'s;*

(3) *If $4 \nmid n$ and $4 \mid (p_i - 1)$ for all the odd $p_i$'s in the prime factorization of $n$, then we consider two cases:*

*Case I. If $n$ is even, then $EUD(R)$ is the disjoint union of $m/2 - 2^{r-2}$ $K_{1,1}$'s and $2^{r-1}$ $K_1$'s;*

*Case II. If $n$ is odd, then $EUD(R)$ is the disjoint union of $m/2 - 2^{r-1}$ $K_{1,1}$'s and $2^r$ $K_1$'s.*

*Proof.* Let $A = \mathbb{Z}_n$. For every $a \in U(A)$, let $X_a$ and $Y_a$ be as in the proof of Theorem 10.7.21. Then $X_a, Y_a \in E(U(R))$. Since $|X| = m$ for every $X \in E(U(R))$, we conclude that every $K_m$ of $UD(R)$ is a $K_1$ of $EUD(R)$ and every $K_{m,m}$ of $UD(R)$ is a $K_{1,1}$ of $EUD(R)$. Hence the claim follows by the proof of Theorem 10.7.21. □

Let $R_1 = \{(u_1, z_1) \mid u_1 \in U(\mathbb{Z}_n) \text{ and } z_1 \in Z(\mathbb{Z}_n)\}$ and $R_2 = \{(z_2, u_2) \mid u_2 \in U(\mathbb{Z}_n) \text{ and } z_2 \in Z(\mathbb{Z}_n)\}$. We define a relation $\sim$ on $R_1 \cup R_2$ such that $x \sim y$, where $x, y \in R_1 \cup R_2$, if $x = (c, c)y$ for some $(c, c) \in U(\mathbb{Z}_n \times \mathbb{Z}_n)$. It is clear that $\sim$ is an equivalence relation on $R_1 \cup R_2$. By construction of $R_1$ and $R_2$, it is clear that if $x \sim y$ for some $x, y \in R_1 \cup R_2$, then $x, y \in R_1$ or $x, y \in R_2$. Hence if $S$ is an equivalence class of $R_1 \cup R_2$, then there is an $a \in Z(\mathbb{Z}_n)$ such that either $S = \overline{(1, a)} = \{u(1, a) \mid u \in U(\mathbb{Z}_n)\}$ or $S = \overline{(a, 1)} = \{u(a, 1) \mid u \in U(\mathbb{Z}_n)\}$. Let $E(R_1 \cup R_2)$ be the set of all distinct equivalence classes of $R_1 \cup R_2$. We define the *equivalence zero-divisor dot product graph* $R_1 \cup R_2$ to be the (undirected) graph $EZD(R_1 \cup R_2)$ with vertices $E(R_1 \cup R_2)$, and two distinct vertices $X$ and $Y$ are adjacent if and only if $a \cdot b = 0 \in A$ for every $a \in X$ and every $b \in Y$, where $a \cdot b$ denotes the normal dot product of $a$ and $b$. We have the following result.

**Theorem 10.7.30.** ([11, Theorem 5.4]) *Let $n \geq 2, R = \mathbb{Z}_n \times \mathbb{Z}_n$, and $\varphi(n) = m$. Then*

(1) *If $n$ is prime, then $EZD(R_1 \cup R_2) = K_{1,1}$;*

(2) *If $n$ is not prime, then $EZD(R_1 \cup R_2)$ is the disjoint union of of $n - m$ $K_{1,1}$'s.*

*Proof.* (1) If $n$ is prime, then $E = \{\overline{(1,0)}, \overline{(0,1)}\}$. Thus $EZD(R_1 \cup R_2) = K_{1,1}$.

(2) Suppose that $n$ is not prime, and let $A = \mathbb{Z}_n$. For every $a \in Z(A)$, let $X_a$ and $Y_a$ be as in the proof of Theorem 10.7.26. Then $X_a, Y_a \in E(R_1 \cup R_2)$. Since $|X| = m$ for every $X \in E(R_1 \cup R_2)$, we conclude that each $K_{m,m}$ of $ZD(R_1 \cup R_2)$ is a $K_{1,1}$ of $EZD(R_1 \cup R_2)$. Hence the claim follows by the proof of Theorem 10.7.26. □

**Remark 10.7.31.** ([11, Remark 5.5])

(1) Let $A = \mathbb{Z}_n$ and $R = \mathbb{Z}_n \times \mathbb{Z}_n$. Since for every $X \in E(U(R))$, there is an $a \in U(A)$ such that $X = \overline{(1,a)} = \{u(1,a) \mid u \in U(A)\}$, note that we can recover the graph $UD(R)$ from the graph $EUD(R)$. However, drawing $EUD(R)$ is much simpler than drawing $UD(R)$.

(2) Since for every $X \in E(R_1 \cup R_2)$ there exists an $a \in Z(Z_n)$ such that either $X = \overline{(1,a)} = \{u(1,a) \mid u \in U(Z_n)\}$ or $X = \overline{(a,1)} = \{u(a,1) \mid u \in U(Z_n)\}$, note that we can recover the graph $ZD(R_1 \cup R_2)$ from the graph $EZD(R_1 \cup R_2)$. However, drawing $EZD(R_1 \cup R_2)$ is much simpler than drawing $ZD(R_1 \cup R_2)$.

**Example 10.7.32.** ([11, Example 5.6]) Let $A = \mathbb{Z}_{20}$ and $R = A \times A$. Then $EUD(R)$ is the disjoint union of 4 $K_{1,1}$'s by Theorem 10.7.29(1), and thus $UD(R)$ is the disjoint union of 4 $K_{8,8}$'s. The following is the graph of $EUD(R)$.

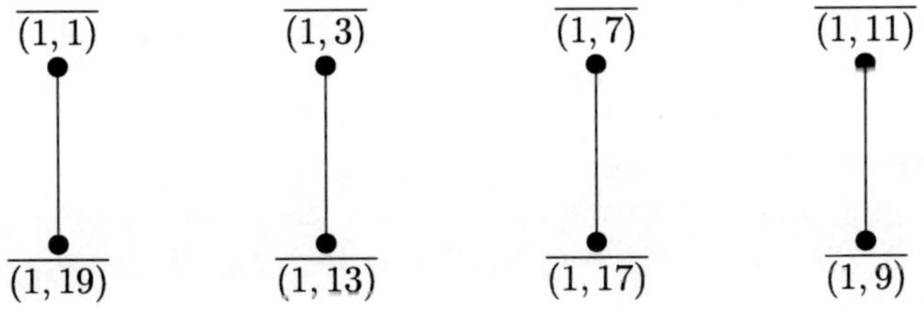


**Example 10.7.33.** ([11, Example 5.7]) Let $A = \mathbb{Z}_{34}$ and $R = A \times A$. Then $EUD(R)$ is the disjoint union of 7 $K_{1,1}$'s and 2 $K_1$'s by Theorem 10.7.29(3, Case I), and thus $UD(R)$ is the disjoint union of 7 $K_{16,16}$'s and 2 $K_8$'s. The following is the graph of $EUD(R)$.

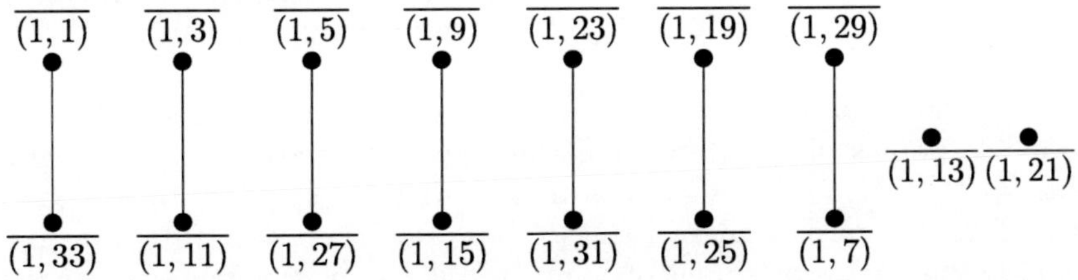

## 10.8 Trace Graph of Matrices

### 10.8.1 Basic Properties of Trace Graphs

In a manner analogous to the commutative case, research on graphs from noncommutative rings are also taken under consideration. The connection between matrices and graph theory is growing enormously and is studied by many researchers. The study of graphs from the ring of matrices is especially important. The Artin-Wedderburn theorem is a classification theorem for semisimple rings and semisimple algebras. The theorem states that an Artinian semisimple ring $R$ is isomorphic to the product of finitely many $n_i \times n_i$ matrix rings over division rings $D_i$, for some integers $n_i$, both of which are uniquely determined up to permutation of the index $i$. In particular, any simple left or right Artinian ring is isomorphic to the $n \times n$ matrix ring over a division ring $D$, where both $n$ and $D$ are uniquely determined. Hence several authors [26, 92, 151, 263, 349, 408] have extensively studied graphs from matrices.

For an integer $n \geq 2$, let $M_n(R)$ be the ring of $n \times n$ matrices over the ring $R$ and $M_n(R)^*$ be the set of nonzero matrices of $M_n(R)$. A matrix $A$ over ring $R$ is usually specified as $A = (a_{ij})$. $E_{ij}$ represents the matrix with 1 as the $ij^{th}$ entry and 0 elsewhere. The trace of a square matrix $A \in M_n(R)$, denoted by $\text{Tr}(A)$, is defined to be the sum of elements on the principal diagonal of $A$. A square matrix $A = (a_{ij}) \in M_n(R)$ is said to be *anti-diagonal* if $a_{ij} = 0$ for all $i, j \in \{1, \ldots, n\}$ such that $i + j \neq n + 1$.

For $A = (a_{ij})_{n \times m}$ over $R$, if every entry $a_{ij}$ has nonzero annihilator, then the generalized rank of $A$, denoted by $rank_M A$, is defined to be zero. If $\text{rank}_M A \neq 0$, its generalized rank is the greatest positive integer $r \leq n$ such that the set of all $r \times r$ subdeterminants of $A$ does not have a common nonzero annihilator.

If $R$ is a ring and $I$ is an ideal of $R$, then the natural map $\varphi : M_n(R) \to M_n(R/I)$ is a ring homomorphism and $ker(\varphi) = M_n(I)$. Hence, we have the following theorem.

**Theorem 10.8.1.** *Let $n \geq 2$ be an integer, $R$ a ring, and $I$ an ideal of $R$. Then $M_n(R)/M_n(I) \cong M_n(R/I)$.*

Let $I$ be a nontrivial ideal of $R$. If $A = (a_{ij}) \in M_n(R) \setminus M_n(I)$, then the corresponding matrix in $M_n(R/I)$ is $(a_{ij} + I)$. If $A = (a_{ij}) \in M_n(I)$, then the corresponding matrix in $M_n(R/I)$ is the zero matrix in $M_n(R/I)$. For convenience, we denote the matrix $(a_{ij} + I) \in M_n(R/I)$ corresponding to the matrix $A = (a_{ij})$ as $\overline{A}$.

Akbari and Mohammadian [26] studied the problem of determining when the zero-divisor graphs of rings are isomorphic given that the zero-divisor graphs of their matrix rings are isomorphic. In [26], they investigated some properties of zero-divisor graphs of matrix rings. Bozic and Petrovic [151] investigated the relationship between graphs of commutative and noncommutative rings and proved several theorems about zero-divisor graphs of matrix rings. In particular, they investigated the possible diameters of zero-divisor graphs of matrix rings over a commutative ring $R$ in terms of the diameter of $\Gamma(R)$. Li and Tucci [263] studied the zero-divisor graphs of upper triangular matrix rings over commutative rings with identity. In [92], Ashrafi and Tadayyonfar completely determined the structure of the zero-divisor graph of $2 \times 2$ matrices over a field.

Mostly, research work on graphs from matrices has focused on construction of graphs with edge set based on the determinant of the matrices. As an exception, the trace graph of matrices is a study of graphs with edge set representing the trace of the matrices. Almahdi et al. [55] introduced the concept of trace graph of matrices over a finite commutative ring as given below.

**Definition 10.8.2.** ([55]) Let $R$ be a commutative ring. The *trace graph* of the matrix ring $M_n(R)$, denoted by $\Gamma_t(M_n(R))$, is the simple undirected graph with vertex set $\{A \in M_n(R)^* \mid \text{there exists } B \in M_n(R)^* \text{ such that } \mathrm{Tr}(AB) = 0\}$ and two distinct vertices $A$ and $B$ are adjacent if and only if $\mathrm{Tr}(AB) = 0$.

The following example illustrates the definition.

**Example 10.8.3.** The trace graph $\Gamma_t(M_2(\mathbb{Z}_2))$ is given below in Fig. 10.8. Note that it is biregular.
In Fig. 10.8,

$$A_1 = \begin{pmatrix} 1 & 0 \\ 0 & 1 \end{pmatrix} \quad A_2 = \begin{pmatrix} 0 & 1 \\ 0 & 0 \end{pmatrix} \quad A_3 = \begin{pmatrix} 1 & 0 \\ 1 & 0 \end{pmatrix} \quad A_4 = \begin{pmatrix} 1 & 1 \\ 0 & 1 \end{pmatrix} \quad A_5 = \begin{pmatrix} 1 & 1 \\ 0 & 0 \end{pmatrix}$$

$$A_6 = \begin{pmatrix} 0 & 0 \\ 1 & 0 \end{pmatrix} \quad A_7 = \begin{pmatrix} 0 & 0 \\ 1 & 1 \end{pmatrix} \quad A_8 = \begin{pmatrix} 1 & 0 \\ 1 & 1 \end{pmatrix} \quad A_9 = \begin{pmatrix} 0 & 1 \\ 0 & 1 \end{pmatrix} \quad A_{10} = \begin{pmatrix} 1 & 1 \\ 1 & 1 \end{pmatrix}$$

$$A_{11} = \begin{pmatrix} 1 & 0 \\ 0 & 0 \end{pmatrix} \quad A_{12} = \begin{pmatrix} 0 & 1 \\ 1 & 0 \end{pmatrix} \quad A_{13} = \begin{pmatrix} 0 & 1 \\ 1 & 1 \end{pmatrix} \quad A_{14} = \begin{pmatrix} 1 & 1 \\ 1 & 0 \end{pmatrix} \quad A_{15} = \begin{pmatrix} 0 & 0 \\ 0 & 1 \end{pmatrix}.$$

**Example 10.8.4.** The subgraph of $\Gamma_t(M_2(\mathbb{Z}_2))$ induced by the matrices with nonzero trace value in $M_2(\mathbb{Z}_2)$ is given in Fig. 10.9.

The matrices with nonzero trace value in $M_2(\mathbb{Z}_2)$ are as follows.

$$A_1 = \begin{pmatrix} 1 & 0 \\ 1 & 0 \end{pmatrix} \quad A_2 = \begin{pmatrix} 1 & 1 \\ 0 & 0 \end{pmatrix} \quad A_3 = \begin{pmatrix} 1 & 1 \\ 1 & 0 \end{pmatrix} \quad A_4 = \begin{pmatrix} 0 & 0 \\ 0 & 1 \end{pmatrix}$$

$$A_5 = \begin{pmatrix} 1 & 0 \\ 0 & 0 \end{pmatrix} \quad A_6 = \begin{pmatrix} 0 & 1 \\ 1 & 1 \end{pmatrix} \quad A_7 = \begin{pmatrix} 0 & 1 \\ 0 & 1 \end{pmatrix} \quad A_8 = \begin{pmatrix} 0 & 0 \\ 1 & 1 \end{pmatrix}$$

Except in the $n = 1$ case, for every matrix $A \in M_n(R)^*$, there is a matrix $B \in M_n(R)^*$, $B \neq A$, such that $\mathrm{Tr}(AB) = 0$. Hence, whenever the case $n \geq 2$

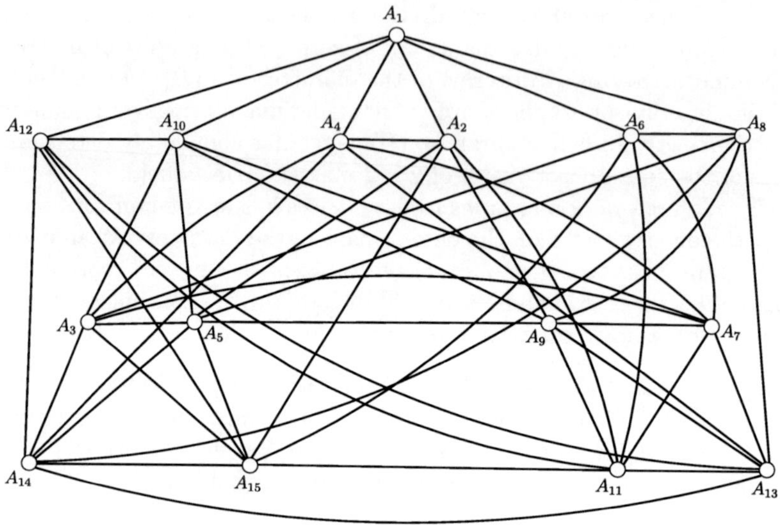


Figure 10.8: The trace graph $\Gamma_t(M_2(\mathbb{Z}_2))$

is assumed, the vertex set is $M_n(R)^*$. When $n = 1$, the trace graph $\Gamma_t(M_n(R))$ is the same as the zero-divisor graph $\Gamma(M_n(R)) = \Gamma(R)$. Hence in the discussion on trace graph of matrices, we assume that $n \geq 2$. Having introduced the trace graph of matrices, Almahdi et al. [55] studied the trace graph of matrices over a commutative ring. More specifically, they studied the degree of vertices and irregularity index of the trace graph.

**Theorem 10.8.5.** ([55, Theorem 2.2]) *Let $R$ be a ring with identity $1 \neq 0$ and $n \geq 2$ an integer. Then $\Gamma_t(M_n(R))$ is connected with $diam(\Gamma_t(M_n(R))) = 2$ and $gr(\Gamma_t(M_n(R))) = 3$.*

For $A = (a_{ij}) \in M_n(R)$, $I(A) = \sum_{1 \leq i,j \leq n} Ra_{ij}$ is the ideal of $R$ generated by all entries of $A$.

**Proposition 10.8.6.** ([55, Proposition 2.3]) *Let $R$ be a finite commutative ring with identity $1 \neq 0$ and $n \geq 2$ an integer.*

(1) *For any vertex $A$ of $\Gamma_t(M_n(R))$, we have:*

(a) $deg(A) = \dfrac{|R|^{n^2}}{|I(A)|} - 1$ *if* $Tr(A^2) \neq 0$;

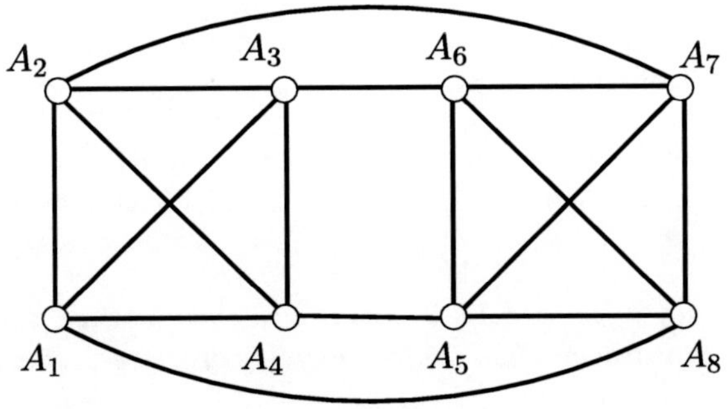


Figure 10.9: Induced subgraph of $\Gamma_t(M_2(\mathbb{Z}_2))$

(b) $deg(A) = \dfrac{|R|^{n^2}}{|I(A)|} - 2$ *if* $Tr(A^2) = 0$;

(2) $\delta(\Gamma_t(M_n(R))) = |R|^{n^2-1} - 2$;

(3) *If* $R$ *is a field, then* $\Delta(\Gamma_t(M_n(R))) = |R|^{n^2-1} - 1$.

A graph $G$ is said to be *sub-Eulerian* if it is a spanning subgraph of some Eulerian graph and a non-Eulerian graph $G$ is said to be super-Eulerian if it has a spanning Eulerian subgraph. In the following theorem, it was proved that the trace graph is not sub Eulerian.

**Theorem 10.8.7.** ([148]) *The following are equivalent for a connected graph* $G$*:*

(1) $G$ *is not sub-Eulerian;*

(2) $G$ *is spanned by* $K_{2m+1,2n+1}$ *for some* $m$ *and* $n$*;*

(3) $G$ *has an even number of points and* $CG$ *has at least two components with an odd number of points.*

The following are some observations on $\Gamma_t(M_n(R))$.

**Observation 10.8.8.** (i) The underlying undirected graph $\Gamma(M_n(R))$ of the directed zero-divisor graph of $M_n(R)$ is a subgraph of the trace graph of $M_n(R)$, i.e., $\Gamma(M_n(R)) \subseteq \Gamma_t(M_n(R))$, since if $A, B \in M_n(R)$ such that $AB = O$ or $BA = O$ (here $O$ represents the zero matrix of $M_n(R)$), then clearly $\text{Tr}(AB) = 0$.

(ii) The graph $\Gamma_t(M_n(R))$ is the null graph if and only if $n = 1$ and $R$ is an integral domain.

(iii) For $n \geq 2$, by Theorem 10.8.5, $gr(\Gamma_t(M_n(R))) = 3$, and hence $\Gamma_t(M_n(R))$ can never be a bipartite graph.

Using the above observation and by definitions, Tamizh Chelvam et al. [349] proved the following theorem regarding the eccentricity of vertices in the trace graph.

**Theorem 10.8.9.** ([349, Theorem 3.1]) *Let $R$ be a commutative ring with identity $1 \neq 0$ and $n \geq 2$ an integer. Then the eccentricity of every vertex of $\Gamma_t(M_n(R))$ is 2.*

**Remark 10.8.10.** From Theorem 10.8.13, for a commutative ring $R$ with identity $1 \neq 0$ and an integer $n \geq 2$, we have the following.

(i) $rad(\Gamma_t(M_n(R))) = 2$.

(ii) $\Gamma_t(M_n(R))$ is self-centered.

Note that for graph $G$ with diameter 2, $\kappa(G) = \delta(G)$. By combining this with Theorem 10.8.5 and Proposition 10.8.6(2), we obtain the following corollary.

**Corollary 10.8.11.** ([349, Corollary 3.3]) *Let $R$ be a finite commutative ring with identity $1 \neq 0$ and $n \geq 2$ an integer. Then $\kappa(\Gamma_t(M_n(R))) = |R|^{n^2-1} - 2$.*

Note that 2-connectivity of a graph is also called biconnectivity, and it is a part of the necessary conditions for several graph-theoretical properties of graphs. In the following theorem, it is proved that $\Gamma_t(M_n(R))$ is 2-connected.

**Theorem 10.8.12.** ([349, Theorem 3.4]) *Let $R$ be a commutative ring with identity $1 \neq 0$ and $n \geq 2$ an integer. Then $\Gamma_t(M_n(R))$ is 2-connected. In particular, $\Gamma_t(M_n(R))$ has no cut-vertices.*

In the following theorem,the authors proved a relationship between $\Gamma_t(M_{n_1}(R))$ and $\Gamma_t(M_{n_2}(R))$ when $n_1 \leq n_2$.

**Theorem 10.8.13.** ([349, Theorem 3.5]) *Let $R$ be a commutative ring and $n_1, n_2$ positive integers such that $n_1 \leq n_2$. Then $\Gamma_t(M_{n_1}(R))$ is isomorphic to a subgraph of $\Gamma_t(M_{n_2}(R))$.*

The following theorem concerns the existence of odd degree vertices in the trace graph.

**Theorem 10.8.14.** ([349, Theorem 4.1]) *Let $R$ be a finite commutative ring with identity and $n \geq 2$ be a positive integer. Then $\Gamma_t(M_n(R))$ contains at least one vertex of odd degree.*

From the above theorem, it is obvious that for $n \geq 2$, and a finite commutative ring $R$, $\Gamma_t(M_n(R))$ is never Eulerian. In the following lemma, it is proved that $C\Gamma_t(M_n(R))$ is connected, which is used to provide a characterization for $\Gamma_t(M_n(R))$ to be sub-Eulerian.

**Lemma 10.8.15.** ([349, Lemma 4.2]) *Let $R$ be a commutative ring with identity and $n \geq 2$ an integer. Then the complement, $C\Gamma_t(M_n(R))$, of the trace graph is connected.*

The following theorem provides a characterization for $\Gamma_t(M_n(R))$ to be sub-Eulerian.

**Theorem 10.8.16.** ([349, Theorem 4.3]) *Let $R$ be a commutative ring with identity and $n \geq 2$ an integer. Then $\Gamma_t(M_n(R))$ is sub-Eulerian.*

**Theorem 10.8.17.** ([349, Theorem 4.4]) *Let $R$ be a commutative ring with identity and $n \geq 3$ an integer. Then every edge of $\Gamma_t(M_n(R))$ lies on a triangle.*

**Theorem 10.8.18.** ([349, Theorem 4.5]) *Let $R$ be a commutative ring with identity. Then every edge of $\Gamma_t(M_2(R))$ lies on a triangle.*

Using Theorems 10.8.17 and 10.8.18, it is proved that $\Gamma_t(M_n(R))$ is super-Eulerian.

**Theorem 10.8.19.** ([349, Theorem 4.6]) *Let $n \geq 2$ be an integer and $R$ a finite commutative ring with identity. Then $\Gamma_t(M_n(R))$ is super-Eulerian.*

**Theorem 10.8.20.** ([349, Theorem 4.8]) *Let $R$ be a commutative ring with identity and $n \geq 2$ an integer. Then every vertex of $\Gamma_t(M_n(R))$ lies on a cycle of length 4.*

### 10.8.2 Planarity and Thickness of Trace Graphs

In this section, we obtain certain bounds for the genus of $\Gamma_t(M_n(R))$ and characterize the trace graph of matrices over a finite commutative ring with identity having thickness 2. We start with an observation of the existence of a maximal clique.

**Theorem 10.8.21.** ([349, Theorem 6.1]) *Let $R$ be a commutative ring $R$ with identity and $n \geq 2$ an integer. Let $X = \{A = (a_{ij}) \in M_n(R)^* \mid A$ is upper triangular with $a_{ii} = 0$ for every $i = 1, \ldots, n\}$ and for $1 \leq i \leq n$, $E_{ii}$ be the $n \times n$ matrix with 1 as the $(i,i)^{th}$ entry and 0 as the other entries. Then the set $X \bigcup \{E_{ii}\}_{i=1}^{n}$ is a maximal clique for $\Gamma_t(M_n(R))$ of order $|R|^{\frac{n(n-1)}{2}} + n - 1$.*

The statement in Theorem 10.8.21 also holds for lower triangular matrices instead of upper triangular matrices. In the following theorem, it is observed that the trace graph $\Gamma_t(M_n(R))$ is not planar when $n \geq 2$.

**Theorem 10.8.22.** ([349, Theorem 6.2]) *Let $R$ be a commutative ring with identity and $n \geq 2$ an integer. Then $\Gamma_t(M_n(R))$ is not planar.*

In the following theorem, the authors obtained certain lower bounds on the genus of the trace graph.

**Theorem 10.8.23.** ([349, Theorem 6.3]) *Let $R$ be a commutative ring with identity. Then the following are true:*

(i) *If $n \geq 2$ and $|R| \geq 3$, then $g(\Gamma_t(M_n(R))) \geq 9$;*

(ii) *If $n \geq 3$ and $|R| = 2$, then $g(\Gamma_t(M_n(R))) \geq 4$.*

The *graph-theoretical thickness* (thickness, for short) of a graph $G$, denoted by $\Theta(G)$, is the minimum number of planar subgraphs into which the graph can be decomposed. The thickness of a planar graph is one. In the following, a characterization is obtained for the thickness of the trace graph to be 2.

**Theorem 10.8.24.** ([349, Theorem 6.4]) *Let $R$ be a finite commutative ring with identity and $n \geq 2$ an integer. Then $\Theta(\Gamma_t(M_n(R))) = 2$ if and only if $n = 2$ and $|R| = 2$.*

### 10.8.3 Domination in Trace Graphs of Matrices

In this section, we present results involving the domination number of $\Gamma_t(M_n(R))$ for a commutative semisimple ring. Further, we present an upper bound for the domination number of $\Gamma_t(M_n(R))$ for $n \geq 2$. Let $R, S$ be commutative rings and $n \geq 2$ an integer. Then $M_n(R \times S) \cong M_n(R) \times M_n(S)$; so $\text{Tr}(A, B) = (\text{Tr}(A),\text{Tr}(B))$ for $(A, B) \in M_n(R) \times M_n(S)$. From this, we have

$$\text{Tr}\left(((a_{ij})_R, (b_{ij})_S) \cdot ((c_{ij})_R, (d_{ij})_S)\right) = (0,0) \text{ if and only if}$$
$$\text{Tr}\left(((a_{ij}, b_{ij}))_{R\times S} \cdot ((c_{ij}, d_{ij}))_{R\times S}\right) = (0,0).$$

Thus $\Gamma_t(M_n(R)\times M_n(S)) \cong \Gamma_t(M_n(R\times S))$. This gives the domination number of the trace graph of matrices over a product as the minimum of the domination numbers of the matrices over the component rings.

**Theorem 10.8.25.** ([349, Theorem 5.1]) *Let $R, S$ be commutative rings and $n \geq 2$ an integer. Then*

$$\gamma(\Gamma_t(M_n(R \times S))) = \min\left\{\gamma(\Gamma_t(M_n(R))), \gamma(\Gamma_t(M_n(S)))\right\}.$$

From the above, the domination number of the trace graph of a commutative semisimple ring is obtained as stated below.

**Theorem 10.8.26.** ([349, Theorem 5.2]) *Let $R$ be a commutative semisimple ring and $R \cong F_1 \times F_2 \times \cdots \times F_m$, where the $F_k$'s are fields. Then*

$$\gamma(\Gamma_t(M_n(R))) = \min\left\{\gamma(\Gamma_t(M_n(F_1))), \ldots, \gamma(\Gamma_t(M_n(F_m)))\right\}.$$

**Theorem 10.8.27.** ([349, Theorem 5.3]) *Let $R$ be a commutative ring and $n \geq 2$ an integer. Then $\gamma(\Gamma_t(M_n(R))) \leq (|R|-1)^2+2$.*

**Remark 10.8.28.** ([349, Remark 5.4]) The bound obtained in Theorem 10.8.27 is sharp. For, one can check that

$$X = \left\{ \begin{pmatrix} 0 & 0 \\ 0 & 1 \end{pmatrix}, \begin{pmatrix} 1 & 0 \\ 0 & 0 \end{pmatrix}, \begin{pmatrix} 1 & 0 \\ 0 & 1 \end{pmatrix} \right\}$$

is a $\gamma$-set of the trace graph of $\mathbb{Z}_2$. Thus $\gamma(\Gamma_t(M_2(\mathbb{Z}_2))) = 3 = (|\mathbb{Z}_2|-1)^2+2$.

### 10.8.4 Ideal-based Trace Graphs of Matrices

n 2001, S. P. Redmond [322] introduced the concept of ideal-based zero-divisor graph of a commutative ring $R$ with respect to an ideal $I$ of $R$. This turns out to be a very natural generalization of the zero-divisor graph. Redmond noted a strong connection between $\Gamma_I(R)$ and $\Gamma(R/I)$ and described a three-step construction method for $\Gamma_I(R)$ based on $\Gamma(R/I)$ (see the three paragraphs after Lemma 5.1.3).

For an ideal $I$ of $R$, the trace graph of the matrix ring $M_n(R)$ with respect to an ideal $I$ of $R$, denoted by $\Gamma_{I^t}(M_n(R))$, is the simple undirected graph with vertex set $M_n(R) \setminus M_n(I)$ and two distinct vertices $A$ and $B$ are adjacent if and only if $\mathrm{Tr}(AB) \in I$.

Recall that if $A = (a_{ij}) \in M_n(R) \setminus M_n(I)$ the corresponding matrix in $M_n(R/I)$ is $(a_{ij}+I)$. If $A = (a_{ij}) \in M_n(I)$, then the corresponding matrix in $M_n(R/I)$ is the zero matrix in $M_n(R/I)$. For convenience, we denote the matrix $(a_{ij}+I) \in M_n(R/I)$ corresponding to the matrix $A = (a_{ij})$ as $\overline{A}$.

First, we list some properties of the trace graph of a matrix ring with respect to an ideal $I$ of $R$ that can be proved by similar arguments as in the case of the trace graph of matrix rings over commutative rings. For $A = (a_{ij}) \in M_n(R)$, we set $J_I(A) = \sum_{1\leq i,j\leq n} (R/I)(a_{ij}+I) \in (R/I)$; the sum of the ideals of $R/I$ generated by all entries of $A = (a_{ij}+I)$ over $R/I$. Note that $J_I(A)$ is an ideal of $R/I$.

**Proposition 10.8.29.** ([408, Proposition 2.1]) *For a nonzero ideal $I$ of $R$ and an integer $n \geq 2$, $\Gamma_{I^t}(M_n(R)))$ contains no isolated vertices.*

In the following, it is proved that no vertex in $\Gamma_{I^t}(M_n(R))$ is adjacent to all other vertices.

**Proposition 10.8.30.** ([408, Proposition 2.2]) *For a nonzero ideal $I$ of $R$ and an integer $n \geq 2$, no vertex of $\Gamma_{I^t}(M_n(R))$ is adjacent to every other vertex of $\Gamma_{I^t}(M_n(R))$.*

Now we obtain the degree of vertices in $\Gamma_{I^t}(M_n(R))$.

**Proposition 10.8.31.** ([408, Proposition 2.3]) *Let $R$ be a finite commutative ring and $n \geq 2$ an integer.*

1. *For any vertex $A$ of $\Gamma_{I^t}(M_n(R))$, we have:*

   a. $deg(A) = \frac{|R|^{n^2}}{|J_I(A)|} - 1$ *if* $Tr(A^2) \notin I$, *and*

   b. $deg(A) = \frac{|R|^{n^2}}{|J_I(A)|} - 2$ *if* $Tr(A^2) \in I$;

2. $\delta(\Gamma_t(M_n(R))) = |R|^{n^2-1}|I| - 2$.

From Proposition 10.8.31, for a finite commutative ring $R$, $\Gamma_{I^t}(M_n(R))$ can never be Eulerian. For, consider the matrices $E_{11}$ and $E_{1n}$ where $n \neq 1$. $\mathrm{Tr}(E_{11}^2) = 1 \notin I$ and $\mathrm{Tr}(E_{1n}^2) = 0 \in I$. Hence, by the Proposition 10.8.31, either $E_{11}$ and $E_{1n}$ must have odd degree.

**Proposition 10.8.32.** ([408, Proposition 2.4]) *Let $R$ be a commutative ring, $n \geq 2$ an integer, and $I$ a nontrivial ideal of $R$. Then $\Gamma_{I^t}(M_n(R))$ is connected with $diam(\Gamma_{I^t}(M_n(R))) = 2$ and $gr(\Gamma_{I^t}(M_n(R))) = 3$.*

**Remark 10.8.33.** ([408, Remark 2.5]) (1) By Theorem 10.8.30 and Proposition 10.8.32, the eccentricity of every vertex in $\Gamma_{I^t}(M_n(R))$ is 2, and hence the radius of $\Gamma_{I^t}(M_n(R))$ is 2, i.e., the graph $\Gamma_{I^t}(M_n(R))$ is self-centered.

(2) By Proposition 10.8.32 $\Gamma_{I^t}(M_n(R))$ contains an odd cycle, and so $\Gamma_{I^t}(M_n(R))$ can never be a bipartite graph.

Now we present certain connections between the graphs $\Gamma_{I^t}(M_n(R))$ and $\Gamma_t(M_n(R/I))$. This explores the relationship between the structure of $\Gamma_t(M_n(R/I))$ and $\Gamma_{I^t}(M_n(R))$.

**Remark 10.8.34.** Given an ideal $I$ of $R$ and a matrix $A \in M_n(R)$, we can view the trace of the coset $A + M_n(I)$ in $M_n(R)/M_n(I)$ as the trace of $\overline{A}$ in $M_n(R/I)$. Thus, the trace graph of $M_n(R)/M_n(I)$ is the trace graph of $M_n(R/I)$.

**Theorem 10.8.35.** ([408, Theorem 3.3 ]) *Let $I$ be an ideal of a commutative ring $R$, $n \geq 2$ an integer, and $A = (a_{ij}), B = (b_{ij}) \in M_n(R) \setminus M_n(I)$ Then the following are true:*

(i) *If $\overline{A}$ is adjacent to $\overline{B}$ in $\Gamma_t(M_n(R/I))$, then $A$ and $B$ are adjacent in $\Gamma_{I^t}(M_n(R))$;*

(ii) *If $A$ is adjacent to $B$ in $\Gamma_{I^t}(M_n(R))$ and $\overline{A} \neq \overline{B}$, then $\overline{A}$ is adjacent to $\overline{B}$ in $\Gamma_t(M_n(R/I))$;*

(iii) *If $A$ is adjacent to $B$ in $\Gamma_{I^t}(M_n(R))$ and $\overline{A} = \overline{B}$, then $Tr(A^2)$, $Tr(B^2) \in I$;*

(iv) *If $Tr(A^2) \in I$ and $\overline{A} = \overline{B}$, then $A$ is adjacent to $B$ in $\Gamma_{I^t}(M_n(R))$ and $Tr(B^2) \in I$;*

(v) *If $A$ and $B$ are (distinct) adjacent vertices in $\Gamma_{I^t}(M_n(R))$, then all (distinct) elements of $\overline{A}$ are adjacent to all elements of $\overline{B}$ in $\Gamma_{I^t}(M_n(R))$. In particular, if $Tr(A^2) \in I$, then all the distinct elements of $\overline{A}$ are adjacent in $\Gamma_{I^t}(M_n(R))$.*

**Corollary 10.8.36.** ([408, Corollary 3.4]) *Let $I$ be an ideal of a commutative ring $R$ and $n \geq 2$ an integer. Then $\Gamma_{I^t}(M_n(R))$ contains $|M_n(I)|$ disjoint subgraphs each isomorphic to $\Gamma_t(M_n(R/I))$.*

**Remark 10.8.37.** ([408, Remark 3.5]) The following are true:

(i) $\Gamma_t(M_n(R)/M_n(I))$ is a graph with $|M_n(R/I)| - 1$ vertices;

(ii) $\Gamma_{I^t}(M_n(R))$ is a graph with $|M_n(R)| - |M_n(I)|$ vertices;

(iii) Let $R$ be a finite commutative ring. Note that Corollary 10.8.36 exhibits a partition of $\Gamma_{I^t}(M_n(R))$ into vertex disjoint subgraphs. Thus

$$|M_n(I)|\,|V(\Gamma_t(M_n(R/I)))| = |V(\Gamma_{I^t}(M_n(R)))|.$$

The following theorem gives a partition of $\Gamma_{I^t}(M_n(R))$ into edge-disjoint subgraphs. In view of Theorem 10.8.35(iv), if $\mathrm{Tr}(A^2) \in I$ and $\overline{A} = \overline{B}$, then $A$ is adjacent to $B$ in $\Gamma_{I^t}(M_n(R))$ and $\mathrm{Tr}(B^2) \in I$. This means that if $\mathrm{Tr}(A^2) \in I$ for a matrix $A$, then the same is true for all matrices in the coset of $A$.

**Theorem 10.8.38.** ([408, Theorem 3.6]) *Let $R$ be a commutative ring with identity, $I$ a nontrivial ideal of $R$, $n \geq 2$ an integer, and $\lambda = |\{\overline{A} \in V(\Gamma_t(M_n(R/I))) \mid Tr(A^2) \in I$ and $A$ is a coset representative of $\overline{A}\}|$. Then $\Gamma_{I^t}(M_n(R)) \cong |M_n(I)|^2\Gamma_t(M_n(R/I)) \oplus \lambda K_{|M_n(I)|}$.*

Now we present results which exhibit some bounds for the clique, chromatic, and independence numbers of $\Gamma_{I^t}(M_n(R))$ and obtain a condition for the chromatic and clique numbers of $\Gamma_{I^t}(M_n(R))$ to be equal.

**Theorem 10.8.39.** ([408, Theorem 4.1]) *Let $n \geq 2$ be an integer, $R$ a commutative ring, and $I$ a nontrivial ideal of $R$. Then the following hold:*

(i) *$\omega(\Gamma_t(M_n(R/I))) \leq \omega(\Gamma_{I^t}(M_n(R))) \leq |M_n(I)|\omega(\Gamma_t(M_n(R/I)))$. Moreover, the equality $\omega(\Gamma_{I^t}(M_n(R))) = |M_n(I)|\omega(\Gamma_t(M_n(R/I)))$ holds if there exists a clique of maximum order in $\Gamma_t(M_n(R/I))$ such that $Tr(A^2) \in I$ for every vertex $\overline{A}$ in the clique;*

(ii) *$\chi(\Gamma_t(M_n(R/I))) \leq \chi(\Gamma_{I^t}(M_n(R))) \leq |M_n(I)|\chi(\Gamma_t(M_n(R/I)))$.*

The following theorem is a generalization of the "moreover" case of Theorem 10.8.39(i).

**Theorem 10.8.40.** ([408, Theorem 4.2]) *Let $n \geq 2$ be an integer, $R$ a commutative ring, and $I$ a nontrivial ideal of $R$. Let $S$ be a clique of maximum order in $\Gamma_t(M_n(R/I))$ and $S$ have the largest number of elements $\overline{A}$ with $Tr(A^2) \in I$. Let $X = \{\overline{A} \in S \mid Tr(A^2) \notin I\}$. Then*

$$\omega(\Gamma_{I^t}(M_n(R))) = |X| + |M_n(I)|(\omega(\Gamma_t(M_n(R/I))) - |X|).$$

**Theorem 10.8.41.** ([408, Theorem 4.3]) *Let $n \geq 2$ be an integer, $R$ a commutative ring, and $I$ a nontrivial ideal of $R$. Let $\Gamma_t(M_n(R/I))$ contain a clique of maximum order such that $Tr(A^2) \in I$ for every $\overline{A}$ in the clique. If*

$$\chi(\Gamma_t(M_n(R/I))) = \omega(\Gamma_t(M_n(R/I))),$$

*then*

$$\chi(\Gamma_{I^t}(M_n(R))) = \omega(\Gamma_{I^t}(M_n(R))).$$

**Theorem 10.8.42.** ([408, Theorem 4.4]) *Let $n \geq 2$ be an integer, $R$ a commutative ring, and $I$ a nontrivial ideal of $R$. Then*

(i) $\alpha(\Gamma_t(M_n(R/I))) \leq \alpha(\Gamma_{I^t}(M_n(R))) \leq |M_n(I)|\alpha(\Gamma_t(M_n(R/I)))$;

(ii) *In particular, if there is an independent set of maximum order in $\Gamma_t(M_n(R/I))$ such that $Tr(A^2) \notin I$ for every vertex $A$ in the independent set, then $\alpha(\Gamma_{I^t}(M_n(R))) = |M_n(I)|\alpha(\Gamma_t(M_n(R/I)))$.*

## 10.9 Graphs of Noncommutative Rings

In this section, we give some definitions and basic concepts for graphs constructed from noncommutative rings.

### 10.9.1 Commuting and Non-commuting Graphs

In 2004, Akbari et al. [27] introduced a new graph called the commuting graph of $R$.

**Definition 10.9.1.** ([27]) Let $R$ be a noncommutative ring and $\mathcal{Z}(R)$ the center of $R$. The *commuting graph* of $R$, denoted by $\Gamma_C(R)$, is the graph with vertex set $R \setminus \mathcal{Z}(R)$, and distinct vertices $x$ and $y$ are adjacent if and only if $xy = yx$.

Akbari et al. [27] obtained the minimum degree, maximum degree, and clique number of $\Gamma_C(M_n(F))$, where $F$ is a field and $M_n(F)$ is the ring of $n \times n$ matrices over $F$. The commuting graph of a noncommutative ring has been studied in [6, 41, 43, 291, 303]. In [26], it was shown that $\Gamma_C(M_n(F))$ is connected if and only if every field extension of $F$ of degree $n$ contains a proper intermediate field.

In [27], it was shown that for any two finite semisimple rings $R$ and $S$, if $\Gamma_C(R) \cong \Gamma_C(S)$, then there are commutative semisimple rings $R_1$ and $S_1$ and a semisimple ring $T$ such that $R \cong T \times R_1$, $S \cong T \times S_1$, and $|R_1| = |S_1|$. In general, the isomorphism between the commuting graphs of two rings does not imply an isomorphism between these rings. In [6], two rings with distinct cardinality and the same commuting graphs were given. In [303], it was shown that the commuting graph of every ring $R$ of order $p^i$ for $i = 2, 3$ and prime $p$ is a disjoint union of complete graphs, and so if $\Gamma_C(R) \cong \Gamma_C(S)$, then $|R| = |S|$.

The following conjecture was proposed by Akbari et al. [27].

**Conjecture:** Let $R$ be a ring with identity, $F$ a finite field, and $n \geq 2$. If $\Gamma_C(R) \cong \Gamma_C(M_n(F))$, then $R \cong M_n(F)$.

Mohammadian [291] proved this conjecture when $n = 2$. Moreover, in [6] it was shown that for every integer $n \geq 2$, the rings $R$ in the above conjecture must have the same order $|M_n(F)|$.

Note that in the complement of the commuting graph, denoted by $\Gamma_C(R)^c$, two distinct vertices $x$ and $y$ are adjacent whenever $xy \neq yx$. This graph is called the *non-commuting graph*. In [27], it has been shown that for a noncommutative ring $R$, the graph $\Gamma_C(R)^c$ is Hamiltonian and $diam(\Gamma_C(R)^c) \leq 2$. In [6], Abdollahi showed that, up to isomorphism, there are only two finite noncommutative rings $R$ with $diam(\Gamma_C(R)^c) = 1$. Also, he identified the rings where the complements of their commuting graphs are planar.

### 10.9.2 Nilpotent and Idempotent Graphs

In [162], Chen defined a graph structure on a ring $R$ whose vertices are all the elements of $R$, and two distinct vertices $x$ and $y$ are adjacent if and only if $xy \in Nil(R)$, where $Nil(R)$ denotes the set of all nilpotent elements of $R$. The nilpotent graph of $R$ was studied by Li et al. in [262, 261].

**Definition 10.9.2.** ([262]) Let $R$ be a ring. The *nilpotent graph* of $R$, denoted by $\Gamma_N(R)$, is the simple graph with vertex set $Z_N(R)^*$, where $Z_N(R) = \{x \in R \mid xy \in Nil(R) \text{ for some } y \in R^*\}$, and two distinct vertices $x$ and $y$ are adjacent if and only if $xy \in Nil(R)$, or equivalently, $yx \in Nil(R)$.

For a commutative ring $R$, it is easy to see that the usual zero-divisor graph $\Gamma(R)$ is a subgraph of $\Gamma_N(R)$. Also, if $R$ is a reduced commutative ring, then $Z_N(R) = Z(R)$, and so $\Gamma_N(R) = \Gamma(R)$.

For example, let $R = \mathbb{Z}_2 \times \mathbb{Z}_2 \times \mathbb{Z}_2$. Then $Z_N(R) = Z(R) = \{(0,0,0), (1,0,0), (0,1,0), (0,0,1), (1,1,0), (0,1,1), (1,0,1)\}$. Figure 10.10 shows the nilpotent graphs of $\mathbb{Z}_2 \times \mathbb{Z}_2 \times \mathbb{Z}_2$ and $\mathbb{Z}_8$.

Note that for a finite commutative ring $R$, the graph $\Gamma_N(R)$ is complete if and only if $R \cong \mathbb{Z}_2 \times \mathbb{Z}_2$. Let $R = F_1 \times F_2$ be a finite commutative ring, where

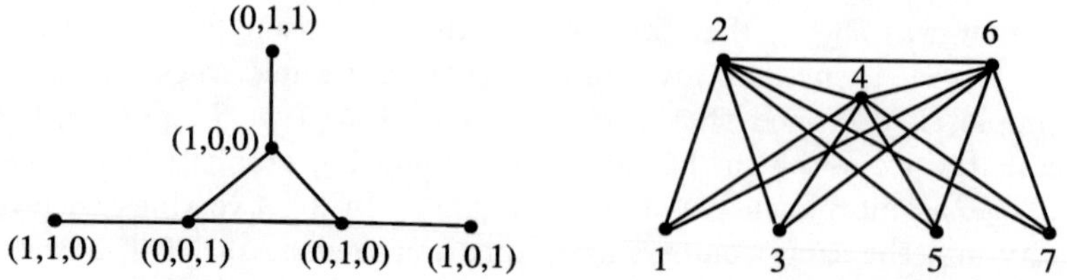


$\Gamma_N(\mathbb{Z}_2 \times \mathbb{Z}_2 \times \mathbb{Z}_2)$ $\Gamma_N(\mathbb{Z}_8)$

Figure 10.10: Nilpotent graph of $\mathbb{Z}_2 \times \mathbb{Z}_2 \times \mathbb{Z}_2$ and $\mathbb{Z}_8$

$F_1$ and $F_2$ are fields. Then $\Gamma_N(R)$ is a complete bipartite graph, and hence $\Gamma_N(R) \cong K_{|F_1^*|,\, |F_2^*|}$. The following two results deal with the split graph nature of $\Gamma_N(R)$. Recall that a graph $G$ is said to be *split* if $V(G)$ can be partitioned into two subsets $X$ and $Y$ such that $\langle X \rangle$ is complete and $Y$ is an independent set.

**Theorem 10.9.3.** *Let $R \cong R_1 \times \cdots \times R_n \times F_1 \times \cdots \times F_m$ be a finite commutative ring, but not a field, where every $R_i$ is a commutative local ring and every $F_j$ a field. Then $\Gamma_N(R)$ is a split graph if and only if $R$ is local or isomorphic to $\mathbb{Z}_2 \times F$ or $\mathbb{Z}_2 \times \mathbb{Z}_2 \times \mathbb{Z}_2$.*

The interplay between the ring-theoretic properties of $R$ and the graph-theoretic properties of $\Gamma_N(R)$ was investigated in [262, 261]. It was shown that if $R$ is Artinian or commutative, then $\Gamma_N(R)$ is connected, the diameter of $\Gamma_N(R)$ is at most 3; and if $\Gamma_N(R)$ contains a cycle, then the girth of $\Gamma_N(R)$ is at most 4; moreover, if $R$ is not reduced, then the girth of $\Gamma_N(R)$ is 3. For a finite commutative ring $R$, it is proved that the edge chromatic number of $\Gamma_N(R)$ is equal to the maximum degree of $\Gamma_N(R)$ unless $R$ is a nilpotent ring of even order. It is also shown that, with two exceptions, if $R$ is a reduced finite commutative ring, $S$ is a commutative ring which is not an integral domain, and $\Gamma_N(R) \cong \Gamma_N(S)$, then $R \cong S$. If $R$ and $S$ are nonreduced finite commutative rings and $\Gamma_N(R) \cong \Gamma_N(S)$, then $|R| = |S|$ and $|Nil(R)| = |Nil(S)|$.

In 2013, Akbari et al. [28] introduced the concept of the idempotent graph of a ring.

**Definition 10.9.4.** ([28]) The *idempotent graph* of a ring $R$, denoted by $I(R)$, is a graph with vertices all nontrivial idempotents of $R$, and distinct vertices $x$ and $y$ are adjacent if and only if $xy = yx = 0$.

The authors of [28] showed that if $D$ is a division ring, then the clique number of $I(M_n(D))$ $(n \geq 2)$ is $n$, and for any commutative Artinian ring $R$, the clique number and the chromatic number of $I(R)$ are equal to the number of maximal ideals of $R$. They proved that for every left Noetherian ring $R$, the clique number of $I(R)$ is finite. For every finite field $F$, they also determined an independent set of $I(M_n(F))$ with maximum size. If $F$ is an infinite field, then they proved that the domination number of $I(M_n(F))$ is infinite. Furthermore, they showed that the idempotent graph of a reduced ring is connected, and if $n \geq 3$ and $D$ is a division ring, then $I(M_n(D))$ is connected with $diam(I(M_n(D))) \leq 5$.

The study of the idempotent graph was continued by Dorbidi et al. in [186]. They showed that $diam(I(M_n(D))) = 4$, for every integer $n \geq 4$ and $diam(I(M_3(D))) = 5$ when $D$ is a division ring. Also, they provided some classes of rings whose idempotent graphs are connected. Moreover, the regularity, clique number, and chromatic number of idempotent graphs are studied in [186].

In 2008, Cvetko-Vah [174] defined a subgraph of the zero-divisor graph of a ring associated to the ring idempotents. Readers may refer to [174] for more details on this graph.

## 10.10 Graphs of Ideals

In 2012, Amini et al. [56] introduced a new graph over the ideals of a ring. They associate a graph $\Gamma_+(R)$ to a ring $R$ whose vertices are the nonzero proper right ideals of $R$, and distinct vertices $I$ and $J$ are adjacent if and only if $I + J = R$. They translated properties of this graph into algebraic properties of $R$ and vice versa. For example, they characterized the rings $R$ for which $\Gamma_+(R)$ is connected, complete, planar, complemented, or a forest. Also, they found the dominating number of $\Gamma_+(R)$.

The inclusion ideal graph was introduced by Akbari et al. [29].

**Definition 10.10.1.** ([29]) The *inclusion ideal graph* of a ring $R$, denoted by $In(R)$, is the graph with vertices the nontrivial left ideals of $R$, and distinct left ideals $I$ and $J$ are adjacent if and only if $I \subseteq J$ or $J \subseteq I$.

We give two examples of inclusion ideal graphs in Figure 10.11.

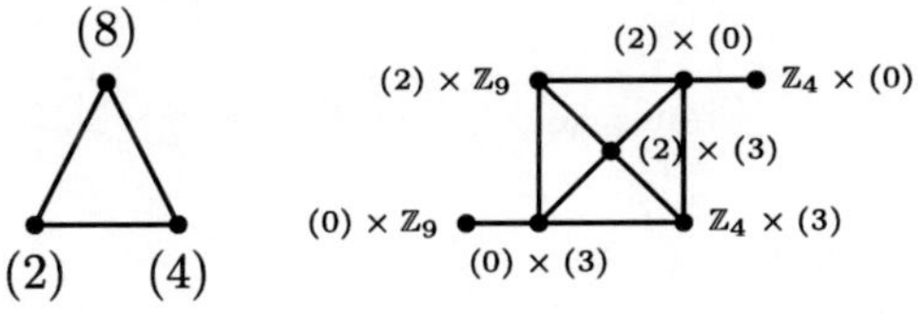


$In(\mathbb{Z}_{16})$ $In(\mathbb{Z}_4 \times \mathbb{Z}_9)$

Figure 10.11: The Inclusion ideal graphs

In [29], they showed that $In(R)$ is not connected if and only if $R \cong M_2(D)$ or $D_1 \times D_2$, for division rings $D, D_1$, and $D_2$. Moreover, if $R$ is connected, then

$diam(In(R)) \leq 3$. Also, they proved that $In(R)$ is a complete graph if and only if $R$ is a uniserial ring and that $In(M_n(D))$ is not regular when $D$ is a division ring and $n > 3$.

In 2015, Akbari et al. [30] further studied the properties of the inclusion ideal graph. They proved that if $In(R)$ is connected, then $diam(In(R)) = 3$. Also it is shown that if $In(R)$ is a tree, then $In(R)$ is a caterpillar with $diam(In(R)) = 3$. Also, they proved that $gr(In(R)) \in \{3, 6, \infty\}$ and determined the clique number and chromatic number of the inclusion ideal graph for some classes of rings.

The concept of the Cayley graph of ideals in a commutative ring was introduced by Afkhami et al. [14].

**Definition 10.10.2.** ([14]) The *Cayley ideal graph* $\overrightarrow{I\Gamma}(R)$ of a commutative ring $R$ is the digraph with vertices the set of nontrivial ideals $\mathbb{I}^*(R)$ of $R$, and for distinct vertices $I$ and $J$, there is an arc from $I$ to $J$ if and only if there is a nontrivial ideal $L$ with $J = IL$.

In fact, the ideal digraph $\overrightarrow{I\Gamma}(R)$ is isomorphic to the Cayley graph $Cay(\mathbb{I}^*, \mathbb{I}^*)$. The *undirected ideal graph* of $R$, denoted by $I\Gamma(R)$, is the simple graph with vertices the set $\mathbb{I}^*$, and distinct vertices $I$ and $J$ are adjacent if and only if either $J = IL$ or $I = JL$ for some nontrivial ideal $L$ of $R$. In Figure 10.12, two Cayley graph of ideals of rings are given.

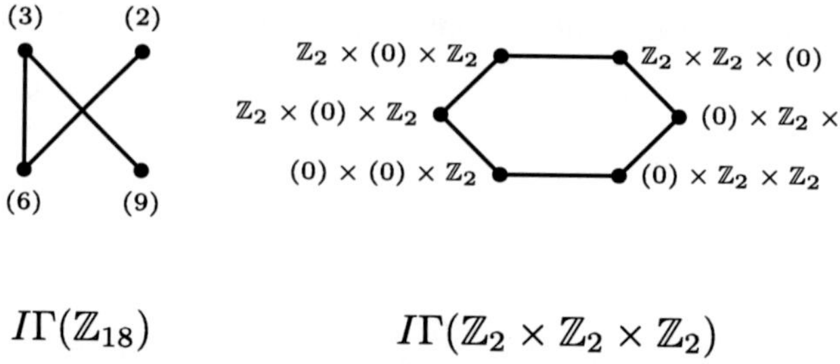


Figure 10.12: The Cayley graph of ideals

The ideal digraph $\overrightarrow{I\Gamma}(R)$ contains no directed Hamiltonian cycle or directed Euler path. The Cayley graph $I\Gamma(R)$ consists of a collection of maximal ideals of $R$ if and only if either $R = F_1 \times F_2$, where $F_1$ and $F_2$ are fields, or $R$ has only one nonzero proper ideal. If $R$ is a field, then $I\Gamma(R)$ is the empty graph.

In a nonempty graph $G$, the maximum length of the paths in $G$ is called the *graph height* of $G$. We denote the graph height of the ideal graph $\overrightarrow{I\Gamma}(R)$ by $h_{\overrightarrow{\Gamma}}(R)$, and the graph height of the ideal graph $I\Gamma(R)$ by $h_{\Gamma}(R)$. If $I \to J$ is in $\overrightarrow{I\Gamma}(R)$, then by the definition, there is an $L \in \mathbb{I}^*(R)$ with $J = IL$. Also, it is proved that $gr(\overrightarrow{I\Gamma}(R)) = \infty$, and so $\overrightarrow{I\Gamma}(R)$ contains no directed Hamilton cycles or directed Euler paths. Moreover, for every maximal ideal $\mathfrak{m}$ and every minimal ideal $I$ of $R$, we have that $deg^-(\mathfrak{m}) = deg^+(I) = 0$. The next result deals with the completeness of $I\Gamma(R)$.

**Theorem 10.10.3.** [14, Proposition 3.1] *Let $R$ be an integral domain that is not a field. Then $I\Gamma(R)$ is connected with* $\text{diam}(I\Gamma(R)) \leq 2$. *In particular, $I\Gamma(R)$ is complete if and only if $R$ is a multiplicative ring and the set of nontrivial ideals of $R$ is totally ordered with respect to inclusion.*

We observe that a multiplicative ring is usually called a multiplication ring, and a multiplication domain is just a Dedekind domain. Thus, the "in particular" statement in the above theorem can be restated as: $I\Gamma(R)$ is complete if and only if $R$ is a DVR.

Thus, if the ideal graph $I\Gamma(R)$ is complete, then $R$ is a local ring. Note that if $R$ is not an integral domain, then $I\Gamma(R)$ may be connected. For example, $\mathbb{Z}_4$ is not an integral domain and $I\Gamma(\mathbb{Z}_4)$ is connected. Moreover, it is proved ([14, Proposition 3.7]) that if every ideal of $R$ is multiplicative and prime, then $I\Gamma(R)$ is complete. (However, it is well known that every proper ideal of a commutative ring $R$ is prime if and only if $R$ is a field.)

## 10.11 Divisor Graphs

Let $R$ be a commutative ring. In this section, we discuss several "divisor" graphs associated to an element $x \in R$. These graphs are due to Coykendall and Maney [171] (also see [283] and [150]) and LaGrange [253] (also see [236] and [252]). LaGrange's article [253] gives many more applications, examples, and references to these types of graphs.

For an integral domain $R$, let $Irr(R)$ be the set of irreducible elements of $R$, and let $\overline{Irr}(R)$ be a (chosen) set of coset representatives for $\{xU(R) \mid x \in Irr(R)\}$. For a nonzero nonunit $x \in R$ that can be factored into irreducibles, Coykendall and Maney [171] defined the *irreducible divisor graph of* $x$ to be the graph $G(x)$ with vertices $\{y \in \overline{Irr}(x) \mid y|x\}$, distinct vertices $y$ and $z$ are adjacent if and only if $yz \mid x$, and $n-1$ loops are attached to the vertex $y$ if $y^n \mid x$. The *reduced irreducible divisor graph of* $x$ is the subgraph $\overline{G}(x)$ of $G(x)$ containing no loops. They then used these graphs to characterize several factorization properties; we give two of them below. (Recall that an integral domain $R$ is atomic if every nonzero nonunit is a product of irreducibles, and $R$ is an FFD (finite factorization domain) if $R$ is atomic and every nonzero nonunit of $R$ has only finitely many nonassociate irreducible divisors, see [59].)

**Theorem 10.11.1.** *Let $R$ be an atomic integral domain.*

(a) ([171, Proposition 3.1]) *$R$ is an FFD if and only if $G(x)$ is finite for every nonzero nonunit $x \in R$;*

(b) ([171, Theorem 5.1]) *R is a UFD if and only if $G(x)$ is connected for every nonzero nonunit $x \in R$, if and only if $\overline{G}(x)$ is complete for every nonzero nonunit $x \in R$.*

These concepts have also been investigated in the context of commutative rings with zero-divisors by Axtell and Stickles in [121]. See [117] for a survey article on irreducible divisor graphs.

For a commutative ring $R$, not necessarily with identity, and $x \in R$, let $d(x) = \{r \in R \mid rs = x \text{ for some } s \in R\}$ be the set of divisors of $x$. In [253], LaGrange defined the *$x$-divisor graph of $R$* to be the simple graph $\Gamma_x(R)$ with vertices $d(x)$ and distinct vertices $r$ and $s$ are adjacent if and only if $rs = x$. If $x = 0$, then $\Gamma_x(R)$ is just Beck's zero-divisor graph $\Gamma_0(R)$. Also, if $x$ is not a zero-divisor, then $\Gamma_x(R)$ is either empty or a disjoint union of complete graphs $K_1$ and $K_2$ [253, Proposition 1].

Several other variants of $\Gamma_x(R)$ are given in [253]. Let $\Gamma_x^d(R)$ be the simple graph with vertices $d(x)$, and distinct vertices $r$ and $s$ are adjacent if and only if $rs \in d(x)$. If $R$ is an integral domain and $x \in R$ can be factored into irreducibles, then $\overline{G}(x)$ is an induced subgraph of $\Gamma_x^d(R)$. For $R$ an integral domain and $0 \neq x \in R$, the *compressed divisor graph* associated with $x$ is the simple graph $(\Gamma_E)_x^{d^\times}(R)$ with vertices the associate-equivalence classes $\overline{r} = rU(R)$ of elements $r \in d(x)^\times = d(x) \setminus (xU(R) \cup U(R))$, and distinct vertices $\overline{r}$ and $\overline{s}$ are adjacent if and only if $rs \in d(x)$. More generally, for a partially ordered commutative monoid $M$ with least element equal to its identity 1, let $\Gamma_{\leq x}(M)$ be the simple graph with vertices $\{m \in M \mid 1 < m < x\}$, and two distinct vertices $m$ and $n$ are adjacent if and only if $mn \leq x$. Then for an integral domain $R$ and $0 \neq x \in R$, $(\Gamma_E)_x^{d^\times}(R) = \Gamma_{xU(R)}^{d^\times}(G(R)^+) = \Gamma_{\leq xU(R)}(G(R)^+)$, where $G(R)^+ = (R \setminus \{0\})/U(R)$ and $rU(R) \leq sU(R)$ in $G(R)^+$ if and only if $r$ divides $s$ in $R$ [253, Proposition 3].

In [253], these graphs were used to study integral domains and von Neumann regular rings and to give several characterizations of UFDs and FFDs as above, but without assuming that the domain $R$ is atomic. For example, if $R$ is von Neumann regular, then the components of $\Gamma_x(R)$ are completely characterized [253, Corollary 1]. We give three factorization characterizations below.

**Theorem 10.11.2.** *Let $R$ be an integral domain.*

(a) ([253, Theorem 13]) *R is an FFD if and only if $(\Gamma_E)_x^{d^\times}(R)$ is finite for every $0 \neq x \in R$;*

(b) ([253, Proposition 7]) *R is a UFD if and only if, for every $0 \neq x \in R \setminus (Irr(R) \cup U(R))$, $(\Gamma_E)_x^{d^\times}(R)$ is finite and every set of nonassociate irreducible divisors of $x$ forms a clique;*

(c) ([253, Theorem 7]) *R is a UFD if and only if $(\Gamma_E)_x^{d^\times}(R)$ is either a null graph or a finite graph that contains a dominant clique for every $0 \neq x \in R$.*

## 10.12 Subgroup Graphs of a Group

In [74], Anderson, Fasteen, and LaGrange introduced and studied the *subgroup graph of a group*. Let $H$ be a subgroup of a group $G$. Then $\vec{\Gamma}_H(G)$ is the directed graph with vertex set $G$ such that $x$ is an initial vertex and $y$ is a terminal vertex if and only if $x \neq y$ and $xy \in H$. If $xy, yx \in H$, then we view $x$ and $y$ to be connected by a single undirected edge. Thus $\vec{\Gamma}_H(G)$ is undirected if $H$ is a normal subgroup of $G$. If $G = R$ is a commutative ring with nonzero identity (viewed as a group under $+$), and $H = Z(R)$ is closed under addition (i.e., $Z(R)$ is a prime ideal of $R$), then $\vec{\Gamma}_{Z(R)}(R) = T(\Gamma(R))$, the total graph of $R$ as in Chapter 7. We may have $\vec{\Gamma}_H(G) \cong \vec{\Gamma}_{H'}(G')$ with $G$ and $G'$ abelian, $|G| = |G'|$, and $|H| = |H'| > 1$, but $G \not\cong G'$ and $H \not\cong H'$ (see [74, Example 6.1]).

They investigated the structure of the connected components of $\vec{\Gamma}_H(G)$ and classified all possible components when $|H| \leq 3$ ([74, Theorems 4.2 and 5.2]). For example, we have the following results concerning $\vec{\Gamma}_H(G)$ ($N(H)$ is the normalizer of $H$ in $G$).

**Theorem 10.12.1.** ([74, Theorem 2.3]) *Let $H$ be a subgroup of a group $G$ and $g \in G$. Then the connected component of $\vec{\Gamma}_H(G)$ containing $g$ is $\vec{\Gamma}_H(HgH \cup Hg^{-1}H)$. Furthermore, the (directed) distance from any element of $HgH \cup Hg^{-1}H$ to $g$ is at most three, and the (directed) distance from $g$ to any element of $HgH \cup Hg^{-1}H$ is also at most three.*

**Theorem 10.12.2.** ([74, Theorem 2.5]) *Let $H$ be a subgroup of a group $G$ and $g \in G$. Then $g \in N(H)$ if and only if the component of $\vec{\Gamma}_H(G)$ containing $g$ is undirected. In particular, $H$ is normal in $G$ if and only if $\vec{\Gamma}_H(G)$ is undirected.*

**Theorem 10.12.3.** ([74, Theorem 2.7]) *Let $H$ be a subgroup of a group $G$, and set $K = \{x \in N(H) \mid x^2 \in H\}$. Then the component of $\vec{\Gamma}_H(N(H))$ containing $x$ is either a complete graph with vertex set $xH$ or is a complete bipartite graph with vertex set $xH \cup x^{-1}H$. In particular, every component of $\vec{\Gamma}_H(N(H))$ is isomorphic to either $K_{|H|}$ or $K_{|H|,|H|}$. Furthermore, if $N(H)$ is finite, then the number of complete graph components of $\vec{\Gamma}_H(N(H))$ is given by $\alpha(N(H), H) = |\{xH \mid x \in K\}|$ and the number of complete bipartite components of $\vec{\Gamma}_H(N(H))$ is given by $\beta(N(H), H) = |\{xH \mid x \in N(H) \setminus K\}|/2$.*

**Theorem 10.12.4.** ([74, Theorem 2.8]) *Let $H$ be a subgroup of a group $G$. Then any component of $\vec{\Gamma}_H(G)$ is either isomorphic to $K_1$ or $K_2$, or it contains a cycle of length 3 or 4. In particular, if $g \notin N(H)$, then the component of $\vec{\Gamma}_H(G)$ containing $g$ has a cycle.*

**Theorem 10.12.5.** ([74, Corollary 2.12]) *Let $H$ be a subgroup of a finite group $G$. Then $|H|$ divides the order of any component of $\vec{\Gamma}_H(G)$.*

Let $H$ be a normal subgroup of $G$ such that $G/H$ is abelian. Then $K = \{x \in G \mid x^2 \in H\}$ is a subgroup of $G$ containing $H$, and $\overrightarrow{\Gamma}_H(G)$ has two types of components: (i) a complete graph of the form $xH$ for $x \in K$ (so a $K_{|H|}$), and (ii) a complete bipartite graph of the form $xH \cup x^{-1}H$ for $x \in G \setminus K$ (so a $K_{|H|,|H|}$) by Theorem 10.12.3. The graph $\overrightarrow{\Gamma}_H(G)$ is then completely classified in [74, Theorems 6.2 and 6.3].

# Bibliography


[1] G. Aalipour, S. Akbari, *Application of some combinatorial arrays in coloring of total graph of a commutative ring*, arXiv:1305.4315v1, May (2013).

[2] G. Aalipour, S. Akbari, *On the Cayley graph of a commutative ring with respect to its zero-divisors*, Comm. Algebra, **44(4)** (2016), 1443–1459.

[3] G. Aalipour, S. Akbari, R. Nikandish, M. J. Nikmehr, F. Shaveisi, *On the coloring of the annihilating-ideal graph of a commutative ring*, Discrete Math., **312** (2012), 2620–2626.

[4] G. Aalipour, S. Akbari, R. Nikandish, M. J. Nikmehr, F. Shahsavari, *Minimal prime ideals and cycles in annihilating-ideal graphs*, Rocky Mountain J. Math., **43(5)** (2013), 1415–1425.

[5] G. Aalipour, S. Akbari, M. Behboodi, R. Nikandish, M. J. Nikmehr, F. Shaveisi, *The classification of the annihilating-ideal graphs of commutative rings*, Algebra Colloq., **21** (2014), 249–256.

[6] A. Abdollahi, *Commuting graphs of full matrix rings over finite fields*, Linear Algebra Appl., **422** (2008), 654–658.

[7] A. Abdollahi, S. Akbari, H. R. Maimani, *Non-commuting graph of a group*, J. Algebra, **298** (2006), 468–492.

[8] A. Abbasi, S. Habibi, *The total graph of a commutative ring with respect to proper ideals*, J. Korean Math. Soc. **49** (2012), 85–98.

[9] A. Abbasi, S. Habibi, *The total graph of a module over a commutative ring with respect to proper submodules*, J. Algebra Appl., **11(3)** (2012), # 1250048 [13 pages].

[10] C. Abdioğlu, *Zero-divisor graph of matrix rings and Hurwitz rings*, Turk. J. Math., **40(1)** (2016), 201–209.

[11] M. Abdulla, A. Badawi, *On the dot product graph of a commutative ring II*, Int. Electron. J. Algebra, **28** (2020), 61–175.

[12] O. A. Abu-Ghneim, E. E. Abd-Al-Jawad, H. Al-Ezeh, *The clique number of* $\Gamma(\mathbb{Z}_{p^n}(\alpha))$, Rocky Mountain J. Math., **42(1)** (2012), 1–13.

[13] O. A. Abu-Ghneim, E. E. Abd-Al-Jawad, H. Al-Ezeh, *The independence number of* $\Gamma(\mathbb{Z}_{p^n}(x))$, Bull. Malaysian Math. Soc., **37(2)** (2014), 345–357.




D. F. Anderson et al., *Graphs from Rings*, https://doi.org/10.1007/978-3-030-88410-9

[14] M. Afkhami, M. R. Ahmadi, R. Jahani-Nezand, K. Khashyarmanesh, *Cayley graphs of ideals in a commutative ring*, Bull. Malays. Math. Sci. Soc., **37(3)** (2014), 833–843.

[15] M. Afkhami, Z. Barati, K. Khashyarmanesh, *Planar zero divisor graphs of partially ordered sets*, Acta Math. Hungar., **137(1–2)** (2012), DOI: 10.1007/s10474-012-0231-6.

[16] M. Afkhami, Z. Barati, K. Khashyarmanesh, *When the unit, unitary and total graphs are ring graphs and outer planar*, Rocky Mountain J. Math., **44(3)** (2014), 705–716.

[17] M. Afkhami, N. Hoseini, K. Khashyarmanesh, *The annihilator ideal graph of a commutative ring*, Note. Mat., **36(1)** (2016), 1–10.

[18] M. Afkhami, K. Khashyarmanesh, *The cozero-divisor graph of a commutative ring*, South-East Asian Bull. Math., **35** (2011), 753–762.

[19] M. Afkhami, K. Khashyarmanesh, *On the cozero-divisor graphs of commutative rings and their complements*, Bull. Malays. Math. Sci. Soc., **35(4)** (2012), 935–944.

[20] M. Afkhami, K. Khashyarmanesh, K. Nafar, *Generalized Cayley graphs associated to commutative rings*, Linear Algebra Appl., **437** (2012), 1040–1049.

[21] M. Afkhami, K. Khashyarmanesh, Z. Rajabi, *Some results on the annihilator graph of a commutative ring*, Czechoslovak Math. J., **67(1)** (2017), 151–169.

[22] M. Afkhami, K. Khashyarmanesh, S.M. Sakhdari, *The annihilator graph of a commutative semigroup*, J. Algebra Appl., **14(2)** (2015), # 1550015 [14 pages].

[23] M. Afkhami, K. Khashyarmanesh, S. Salehifar, *On the annihilator graph of group rings*, Bull. Korean Math. Soc., **54(1)** (2017), 331–342.

[24] S. Akbari, F. Alizadeh, S. Khojasteh, *Some results on cozero-divisor graph of a commutative ring*, J. Algebra Appl., 13(3) (2014), # 1350113 [14 pages].

[25] S. Akbari, M. Aryapoor, M.Jamaali, *Chromatic number and clique number of subgraphs of regular graph of matrix algebras*, Linear Algebra Appl., **436** (2012), 2419–2424.

[26] S. Akbari, H. Bidkhori, A. Mohammadian, *Commuting graphs of matrix algebras*, Comm. Algebra, **36** (2008), 4020–4031.

[27] S. Akbari, M. Ghandehari, M. Hadian, A. Mohammadian, *On commuting graphs of semisimple rings*, Linear Algebra Appl., **390** (2004), 345–355.

[28] S. Akbari, M. Habibi, A. Majidinya, R. Manaviyat, *On the idempotent graph of a ring*, J. Algebra Appl., **12(6)** (2013), # 1350003 [14 pages].

[29] S. Akbari, M. Habibi, A. Majidinya, R. Manaviyat, *The inclusion ideal graph of rings*, Electron. Notes Discrete Math., **45(6)** (2014), 73–78.

[30] S. Akbari, M. Habibi, A. Majidinya, R. Manaviyat, *The inclusion ideal graph of rings*, Comm. Algebra, **43(6)** (2015), 2457–2465.

[31] S. Akbari, F. Heydari, *The regular graph of a commutative ring*, Period. Math. Hungar., **67(2)** (2013), 211–220.

[32] S. Akbari, F. Heydari, *The regular graph of a noncommutative ring*, Bull. Austral. Math. Soc., **89(1)** (2014), 132–140.

[33] S. Akbari, F. Heydari, *The regular graph of a noncommutative ring*, Electron. Notes Discrete Math., **45** (2014), 79–85.

[34] S. Akbari, M. Jamaali, S. A. Seyed Fakhari, *The clique numbers of regular graphs of matrix algebras are finite*, Linear Algebra Appl., **431** (2009), 1715–1718.

[35] S. Akbari, M. Jamaali, S.A. Seyed Fakhari, in: P.J. Cameron (Ed.), *Research Problems from the BCC22*, Discrete Math., **311** (2011), 1074–1083.

[36] S. Akbari, D. Kiani, F. Mohammadi, S. Moradi, *The total graph and regular graph of a commutative ring*, J. Pure Appl. Algebra, **213** (2009), 2224–2228.

[37] S. Akbari, H. R. Maimani, S. Yassemi, *When a zero–divisor graph is planar or a complete r–partite graph*, J. Algebra, **270** (2003), 169–180.

[38] S. Akbari, A. Mohammadian, *On the zero–divisor graph of a commutative ring*, J. Algebra, **274** (2004), 847–855.

[39] S. Akbari,, A. Mohammadian, *Zero-divisor graphs of non-commutative rings*, J. Algebra, **296** (2006), 462–479.

[40] S. Akbari, A. Mohammadian, *On the zero–divisor graphs of finite rings*, J. Algebra, **314** (2007), 168–184.

[41] S. Akbari, A. Mohammadian, H. Radjavi, P. Raja, *On the diameters of commuting graphs*, Linear Algebra Appl., **418** (2006), 161–176.

[42] S. Akbari, R. Nikandish, *Some results on the intersection graph of ideals of matrix algebras*, Linear and Multilinear Algebra, **62(2)** (2013), 195–206.

[43] S. Akbari, P. Raja, *Commuting graphs of some subsets in simple rings*, Linear Algebra Appl., **416** (2006), 1038–1047.

[44] R. Akhtar, M. Boggess, T. Jackson-Henderson, I. Jiménez, R. Karpman, A. Kinzel, D. Pritikin, *On the unitary Cayley graph of a finite ring*, Electron. J. Combin., **16** (2009), # R117.

[45] R. Akhtar, L. Lee, *Connectivity of the zero-divisor graph for finite rings*, Involve, **9(3)** (2016), 415–422.

[46] S. Akbari, H. A. Tavallaee, S. K. Ghezelahmad, *Intersection graph of submodules of a module*, J. Algebra Appl., **11(1)** (2012), # 1250019 [8 pages].

[47] F. Aliniaeifard, M. Behboodi, *Rings whose annihilating-ideal graphs have positive genus*, J. Algebra Appl., **11(3)** (2012), # 1250049 [13 pages].

[48] F. Aliniaeifard, M. Behboodi, *Commutative rings whose zero–divisor graphs have positive genus*, Comm. Algebra, **41** (2013), 3629–3634.

[49] F. Aliniaeifard, M. Behboodi, E. Mehdi-Nezhad, A. M. Rahimi, *The annihilating-ideal graph of a commutative ring with respect to an ideal*, Comm. Algebra, **42** (2014), 2269–2284.

[50] F. Aliniaeifard, M. Behboodi, E. Mehdi-Nezhad, Amir M. Rahimi, *On the diameter and girth of an annihilating-ideal graph*, arXiv:1411.4163v1, November (2014).

[51] F. Aliniaeifard, M. Behboodi, Y. Li, *The annihilating-ideal graph of a ring*, arXiv:1411.4159, November (2014).

[52] F. Aliniaeifard, M. Behboodi, Y. Li, *Noetherian rings whose annihilating-ideal graphs have finite genus*, arXiv:1501.04329, January (2015).

[53] M. Alizadeh, A.K. Das, H. R. Maimani, M. R. Pournaki, S. Yassemi, *On the diameter and girth of zero-divisor graphs of posets*, Discrete Appl. Math., **160** (2012), 1319–1324.

[54] M. Alizadeh, H. R. Maimani, M. R. Pournaki, S. Yassemi, *An ideal theoretic approach to complete partite zero-divisor graphs of posets*, J. Algebra Appl., **12(2)** (2013), # 1250148 [11 pages].

[55] F. A. A. Almahdi, K. Louartiti, M. Tamekkante, *The trace graph of the matrix ring over a finite commutative ring*, Acta Math. Hungar., **156(1)**(2018), 132–144.

[56] A. Amini, B. Amini, E. Momtahan, M. H. Shirdareh Haghighi, *On a graph of ideals*, Acta Math. Hungar., **134(3)** (2012), 369–384.

[57] H. Amraei, H. R. Maimani, M. R. Pournaki, A. Zaeembashi, *Classification of the Toroidal Jacobson Graphs*, Bull. Malays. Math. Sci. Soc., **41** (2018), 321-334.

[58] H. Amraei, H. R. Maimani, M. R. Pournaki, A. Zaeembashi, *The nonorientable genus of some Jacobson graphs and classification of the projective ones*, Publ. Math. Debrecen, **88(3–4)** (2016), 425–437.

[59] D. D. Anderson, D. F. Anderson, M. Zafrullah, *Factorization in integral domains*, J. Pure Appl. Algebra, **69** (1990), 1–19.

[60] D. D. Anderson, M. Naseer, *Beck's coloring of a commutative ring*, J. Algebra, **159** (1993), 500–514.

[61] D. D. Anderson, E. Smith, *Weakly prime ideals*, Houston J. Math., **29(4)** (2003), 831–840.

[62] D. D. Anderson, J. Stickles, *Commutative rings with finitely generated multiplicative semigroup*, Semigroup Forum, **60(3)** (2000), 436–443.

[63] D. D. Anderson, M. Winders, *Idealization of a module*, J. Commutative Algebra, **1** (2009), 3–56.

[64] D. F. Anderson, *On the diameter and girth of a zero-divisor graph II*, Houston J. Math., **34(2)** (2008), 361–372.

[65] D. F. Anderson, S. E. Atani, M. S. Kohan, Z. E. Sarvandi, *The ideal-based zero-divisor graph of commutative chained rings*, Sarajevo J. Math., **10 (22)** (2014), 3–12.

[66] D. F. Anderson, M. Axtell, J. Stickles, *Zero–divisor graphs in commutative rings.* In : M. Fontana, S. E. Kabbaj, B. Olberding and I. Swanson(eds.), Commutative Algebra Noetherian and Non-Noetherian Perspectives, Springer-Verlag, New York (2010), 23–45.

[67] D. F. Anderson, A. Badawi, *The total graph of a commutative ring*, J. Algebra, **320** (2008), 2706–2719.

[68] D. F. Anderson, A. Badawi, *On the zero-divisor graph of a ring*, Comm. Algebra, **36** (2008), 3073–3092.

[69] D. F. Anderson, A. Badawi, *On n-absorbing ideals of commutative rings*, Comm. Algebra, **39** (2011), 1646–1672.

[70] D. F. Anderson, A. Badawi, *On the total graph of a commutative ring without the zero element*, J. Algebra Appl., **11(4)** (2012), # 1250074 [18 pages].

[71] D. F. Anderson, A. Badawi, *Von Neumann regular and related elements in commutative rings,* Algebra Colloq., **19 (Spec 1)** (2012), 1017–1040.

[72] D. F. Anderson, A. Badawi, *The generalized total graph of a commutative ring*, J. Algebra Appl., **12(5)** (2013), # 1250212 [18 pages].

[73] D. F. Anderson, A. Badawi, *The zero-divisor graph of a commutative semigroup: A survey,* In groups, modules, and model theory-surveys and recent developments, edited by Manfred Droste, et al. (2017), 23–39, Germany/New York: Springer.

[74] D. F. Anderson, J. Fasteen, J. D. LaGrange, *The subgroup graph of a group*, Arab. J. Math., **1** (2012), 17–27.

[75] D. F. Anderson, A. Frazier, A. Lauve, P. S. Livingston, *The zero–divisor graph of a commutative ring-II*, Lecture Notes in Pure and Appl. Math., **202** (2001), 61–72.

[76] D. F. Anderson, S. Ghalandarzadeh, S. Shirinkam, P. M. Rad, *On the diameter of the graph* $\Gamma_{Ann(M)}(R)$, Filomat, **26(3)** (2012), 623–629.

[77] D.F. Anderson, J.D. LaGrange, *Commutative Boolean monoids, reduced rings and the compressed zero-divisor graph*, J. Pure Appl. Algebra, **216** (2012), 1626–1636.

[78] D.F. Anderson, J.D. LaGrange, *Abian's poset and the ordered monoid of annihilator classes in a reduced commutative ring*, J. Algebra Appl., **13(8)** (2014), # 1450070 [18 pages].

[79] D.F. Anderson, J.D. LaGrange, *The semilattice of annihilator classes in a reduced commutative ring*, Comm. Algebra, **43** (2015), 29–42.

[80] D. F. Anderson, J. D.LaGrange, *Some remarks on the compressed zero-divisor graph*, J. Algebra, **447** (2016), 297–321.

[81] D. F. Anderson, R. Levy, J. Shapiro, *Zero-divisor graphs, von Neumann regular rings and Boolean algebras*, J. Pure Appl. Algebra, **180** (2003) 221–241.

[82] D. F. Anderson, E. F. Lewis, *A general theory of zero-divisor graphs over a commutative ring*, Int. Electron. J. Algebra, **20** (2016), 111–135.

[83] D. F. Anderson, P.S. Livingston, *The zero–divisor graph of a commutative ring*, J. Algebra, **217** (1999), 434–447.

[84] D. F. Anderson, G. McClurkin, *Generalizations of the zero-divisor graph*, Int. Electron. J. Algebra, **27** (2020), 237–262.

[85] D. F. Anderson, S.B. Mulay, *On the diameter and girth of a zero-divisor graph*, J. Pure Appl. Algebra, **210(2)** (2007), 543–550.

[86] D. F. Anderson, S. Shirinkam, *Some remarks on the Graph* $\Gamma_I(R)$, Comm. Algebra, **42** (2014), 545–562.

[87] D. F. Anderson, D. Weber, *The zero-divisor graph of a commutative ring without identity*, Int. Electron. J. Algebra, **23** (2018), 176–202.

[88] H. Ansari-Toroghy, S. Habibi, *The annihilating-submodule graph over a commutative ring II*, Arab. J. Math., **5(4)** (2016), 187–194.

[89] D. Archdeacon, *Topological graph theory: A survey*, Congr. Numer., **115** (1996), 115–135.

[90] D. Archdeacon, *A Kuratowski Theorem for the projective plane*, J. Graph Theory, **5** (1981), 243–246.

[91] N. Ashrafi, H. R. Maimani, M. R. Pournaki, S. Yassemi, *Unit graphs associated with rings*, Comm. Algebra, **38** (2010), 2851–2871.

[92] A.R. Ashrafi, A. Tadayyonfar, *The zero divisor graph of* $2 \times 2$ *matrices over a field*, Quaest. Math., **39(7)** (2016), 977–990.

[93] T. Asir, *The genus two class of graphs arising from rings*, J. Algebra Appl., **17(10)** (2018), #1850193 [9 pages].

[94] T. Asir, H. R. Maimani, M. R. Pournaki, T. Tamizh Chelvam, *Some bounds for the genus of a class of graphs arising from rings*, Houston J. Math., **45(2)** (2019), 371–384.

[95] T. Asir, K. Mano, *Classification of rings with crosscap two class of graphs*, Discrete Appl. Math., **256** (2019), 13–21.

[96] T. Asir, K. Mano, *Bounds for the genus of generalized total graph of a commutative ring*, Algebra Colloq. **26(3)** (2019), 519–528.

[97] T. Asir, K. Mano, *The classification of rings with its genus of class of graphs*, Turkish J. Math., **42(3)** (2018), 1424–1435.

[98] T. Asir, K. Mano, *Classification of non-local rings with genus two zero-divisor graphs*, Soft Comput., **24** (2020), 237–245.

[99] T. Asir, K. Mano, T. Tamizh Chelvam, *Correction to: Classification of non-local rings with genus two zero-divisor graphs*, Soft Computing, **25**(2021), 3355–3356, https://doi.org/ 10.1007/s00500-020-05533-z.

[100] T. Asir, T. Tamizh Chelvam, *On the total graph and its complement of a commutative ring*, Comm. Algebra, **41(10)** (2013), 3820–3835.

[101] T. Asir, T. Tamizh Chelvam, *On the intersection graph of gamma sets in the total graph II*, J. Algebra Appl., **12(4)** (2013), # 1250199 [14 pages].

[102] T. Asir, T. Tamizh Chelvam, *On the genus of generalized unit and unitary Cayley graphs of a commutative ring*, Acta Math. Hungar., **142(2)** (2014), 444–458, DOI: 10.1007/s10474-013-0365-1.

[103] T. Asir, T. Tamizh Chelvam, *Genus of total graphs of commutative rings: A survey*, Electron. Notes Discrete Math., **63** (2017), 59–68.

[104] T. Asir, T. Tamizh Chelvam, *On the genus two characterizations of unit, unitary Cayley and co-maximal graphs*, ARS Combin., **138** (2018), 77–91.

[105] M. F. Atiyah, I. G. Macdonald, *Introduction to Commutative Algebra*, Addison-Wesley Publishing Company (1969).

[106] S. E. Atani, *The zero-divisor graph with respect to ideals of a commutative semiring*, Glas. Mat. **43(63)** (2008), 309–320.

[107] S. E. Atani, *An ideal-based zero-divisor graph of a commutative semiring*, Glas. Mat., **44(64)** (2009), 141–153.

[108] S. E. Atani, A. Y. Darani, *On weakly primal ideals (I)*, Demonstratio Math., **40(1)** (2007), 23–32.

[109] S. E. Atani, A. Y. Darani, *Zero–divisor graphs with respect to primal and weakly primal ideals*, J. Korean Math. Soc., **46(2)** (2009), 313–325.

[110] S. E. Atani, A. Y. Darani, E. R. Puczylowski, *On the diameter and girth of ideal-based zero-divisor graphs*, Publ. Math., **78(3)** (2011), 607–612.

[111] S. E. Atani, Sh. Habibi, *The total torsion element graph of a module over a commutative ring*, An. St. Univ. Ovidius Constanta, **19(1)** (2011), 23–34.

[112] S. E. Atani, F. E. Khalil Sarael, *The total graph of a commutative semiring*, An. St. Univ. Ovidius Constant, **21(2)** (2013), 21–33.

[113] S. E. Atani, M. S. Kohan, *The diameter of a zero-divisor graph for finite direct product of commutative rings*, Sarajevo J. Math. **3(16)** (2007), 149–156.

[114] S. E. Atani, S. D. Pish Hesari, M. Khoramdel, *The identity-summand graph of commutative semirings*, J. Korean Math. Soc., **51(3)** (2014), 189–202.

[115] S. E. Atani, S. D. Pish Hesari, M. Khoramdel, *Total graph of a commutative semiring with respect to identity-summand elements*, J. Korean Math. Soc., **51(1)** (2014), 593–607.

[116] J. Aubert, B. Schneider, *Decomposition of $K_m + K_n$ in Hamilton cycles*, Discrete Math., **37** (1981), 19–27.

[117] M. Axtell, M. Baeth, J. Stickles, *Survey article - graphical representations of factorizations in commutative rings*, Rocky Mt. J. Math., **43** (2013), 1–36.

[118] M. Axtell, N. Baeth, J. Stickles, *Cut vertices in zero-divisor graphs of finite commutative rings*, Comm. Algebra, **39** (2011), 2179–2188.

[119] M. Axtell, J. Coykendall, J. Stickles, *Zero-divisor graphs of polynomials and power series over commutative rings*, Comm. Algebra, **33(6)** (2005), 2043–2050.

[120] M. Axtell, J. Stickles, *Zero-divisor graphs of idealizations*, J. Pure Appl. Algebra, **204** (2006), 235–243.

[121] M. Axtell, J. Stickles, *Irreducible divisor graphs in commutative rings with zero-divisors*, Comm. Algebra, **36** (2008), 1883–1893.

[122] M. Axtell, J. Stickles, W. Trampbachls, *Zero-divisor ideals and realizable zero-divisor graphs*, Involve, **2(1)** (2009), 17–27.

[123] M. Axtell, J. Stickles, J. Warfel, *Zero-divisor graphs of direct products of commutative rings*, Houston J. Math., **32(4)** (2006), 985–994.

[124] A. Azimi, A. Erfanian, M. Farrokhi, *The Jacobson graph of commutative rings*, J. Algebra Appl., **12(3)** (2013), # 1250179 [18 pages].

[125] A. Azimi, A. Erfanian, M. Farrokhi, *Isomorphisms between Jacobson graphs*, Rend. Circ. Mat. Palermo, **63(2)** (2014), 277–286.

[126] A. Badawi, *On domains which have prime deals that are linearly ordered*, Comm. Algebra, **23(12)** (1995), 4365–4373.

[127] A. Badawi, *On 2-absorbing ideals of commutative rings*, Bull. Austral. Math. Soc., **75** (2007), 417–429.

[128] A. Badawi, *On the annihilator graph of a commutative ring*, Comm. Algebra, **42(1)** (2014), 108–121.

[129] A. Badawi, *On the total graph of a ring and its related graphs: A survey*, in: M. Fontana et al. (Eds.) Commutative Algebra: Recent advances in commutative rings, integer-valued polynomials, and polynomial functions, Springer Science and Business Media, New York (2014), 39–54.

[130] A. Badawi, *On dot-product graph of a commutative ring*, Comm. Algebra, **43(1)** (2015), 43–50.

[131] A. Badawi, *Recent results on annihilator graph of a commutative ring: A survey.* In nearrings, nearfields, and related topics, edited by K. Prasad et al, (2017), New Jersey: World Scientific, 170–185.

[132] R. Balakrishnan, K. Ranganathan, *A text book of Graph Theory*, Springer, (2000).

[133] A. Barnard, *Multiplication Modules*, J. Algebra, **71** (1981), 174–178.

[134] Z. Barati, K. Khashyarmanesh, F. Mohammadi, K. Nafar, *On the associated graphs to a commutative ring, J. Algebra Appl.*, **11(2)** (2012), # 1250037 [17 pages].

[135] M. Bataineh, D. Bennis, J. Mikram,, F. Taraza, *On two extensions of the classical zero-divisor graph*, São Paulo J. Math. Sci., **14(1)** (2020), 341–350.

[136] M. Baziar, E. Momtahan, S. Safaeeyan, *A zero-divisor graph for modules with respect to their (first) dual*, J. Algebra Appl., **12(2)** (2013), # 1250151, [11 pages].

[137] I. Beck, *Coloring of commutative ring*, J. Algebra, **116** (1988), 208–226.

[138] M. Behboodi, *Zero divisor graphs for modules over commutative rings*, J. Commut. Algebra, **4(2)** (2012), 175–197.

[139] M. Behboodi, Z. Rakeei, *The annihilating-ideal graph of commutative ring I*, J. Algebra Appl., **10(4)** (2011), 727–739.

[140] M. Behboodi, Z. Rakeei, *The annihilating-ideal graph of commutative ring II*, J. Algebra Appl., **10(4)** (2011), 741–753.

[141] L. W. Beineke, F. Harary, *The thickness of the complete graph*, Canad. J. Math., **17** (1965), 850–859.

[142] R. Belshoff, J. Chapman, *Planar zero-divisor graphs*, J. Algebra **316(1)** (2007), 471–480.

[143] D. Bennis, J. Mikram, F. Taraza, *On the extended zero divisor graph of commutative rings*, Turkish J. Math., **40(2)** (2016), 376–388.

[144] D. Bennis, J. Mikram, F. Taraza, *Extended zero-divisor graphs of idealizations*, Commun. Korean Math. Soc., **32(1)** (2017), 7–17.

[145] S. H. Bhatwadekar, M. N. Dumaldar, P. K. Sharma, *Some non–chromatic rings*, Comm. Algebra, **26(2)** (1998), 477–505.

[146] N. Biggs, *Algebraic Graph Theory*, Cambridge University Press, Cambridge, (1974).

[147] N. Bloomfield, C. Wickham, *Local rings with genus two zero-divisor graph*, Comm. Algebra, **38** (2010), 2965–2980.

[148] F. Boesch, C. Suffel, R. Tindell, *The spanning subgraphs of Eulerian graphs*, J. Graph Theory, **1(1)** (1977), 69–75.

[149] B. Bollaboás, *Graph Theory*, An Introductory Course, Springer-Verlag, New York (1979).

[150] J. G. Boynton, J. Coykendall, *On the graph of divisibility of an integral domain*, Canad. Math. Bull., **58(3)** (2015), 449–458.

[151] I. Bozic, Z. Petrovic, *Zero-divisor graphs of matrices over commutative rings*, Comm. Algebra, **37** (2009), 1186–1192.

[152] T. Brand, M. Jameson, M. Mcgowan, J.D. Mckeel, *Zero-divisor and ideal-divisor graphs*, Involve, **2(1)** (2009), 17–27.

[153] M. Brickel, *Complements of zero-divisor graphs*, Plan B Project, Miami University, (2004).

[154] P. J. Cameron, S. Ghosh, *The power graph of a finite group*, Discrete Math., **311(13)** (2011), 1220–1222.

[155] G. A. Cannon, K. M. Neuerburg, S. P. Redmond, *Zero-divisor graphs of near–rings and semigroups*, Near rings and Near fields, Dordrecht: Springer (2005), 189–200.

[156] A. Cayley, *Desiderata and Suggestions: No. 2. The Theory of Groups: Graphical Representation*, Amer. J. Math., **1(2)** (1878), 174–176.

[157] I. Chakrabarty, S. Ghosh, T. K. Mukherjee, M. K. Sen, *Intersection graphs of ideals of rings*, Electron. Notes Discrete Math., **23** (2005), 23–32.

[158] I. Chakrabarty, S. Ghosh, T. K. Mukherjee, M. K. Sen, *Intersection graphs of ideals of rings*, Discrete Math., **309** (2009), 5381–5392.

[159] G. Chartrand, Lesniak, *Graphs and Digraphs*, Chapman & Hall/CRC, (2000).

[160] G. Chartrand, P. Zhang, *Introduction to Graph Theory*, Tata McGraw-Hill Edition, (2006).

[161] G. Chartrand, P. Zhang, Chromatic Graph Theory, *CRC press*, (2009).

[162] P. W. Chen, *A kind of graph structure of rings*, Algebra Colloq., **10(2)** (2003), 229–238.

[163] H.-J. Chiang-Hsieh, P.-F. Lee, H.-J. Wang, *The embedding of line graphs associated to the zero-divisor graphs of commutative rings*, Israel J. Math., **180** (2010), 193–222.

[164] H.-J. Chiang-Hsieh, N. O. Smith, H-J. Wang, *Commutative rings with toroidal zero-divisor graphs*, Houston J. Math., **36(1)** (2010), 1–31.

[165] H.-J. Chiang-Hsieh, *Classification of rings with projective zero-divisor graphs*, J. Algebra, **319** (2008), 2789–2802.

[166] K. Cooper, B. Johnson, *Zero divisor graphs of commutative graded rings*, Involve, **11(2)** (2018), 283–295.

[167] B. Corbas, G.D Williams, *Rings of order $p^5$ II. Local rings*, J. Algebra, **231(2)** (2000), 691–704.

[168] N. I. Córdova, C. Gholston, H. A. Hauser, *The structure of zero-divisor graphs*, SUMSRI J., Miami University, (2005) # 3.

[169] B. Coté, C. Ewing, M. Huhn, C. M. Plaut, D. Weber, *Cut-sets and cut vertices in the zero divisor graph of* $\prod_{i=1}^{n} \mathbb{Z}_{n_i}$, Rose-Hulman Undergraduate Math. J., **11(1)** (2010), 1–8.

[170] B. Coté, C. Ewing, M. Huhn, C. M. Plaut, D. Weber, *Cut-sets in zero-divisor graphs of finite commutative rings*, Comm. Algebra, **39** (2011), 2849–2861.

[171] J. Coykendall, J. Maney, *Irreducible divisor graphs*, Comm. Algebra, **35** (2007) 885–895.

[172] J. Coykendall, S. Sather-Wagstaff, L. Sheppardson, S. Spiroff, *On zero divisor graphs*, Progress in Commutative Algebra **2**: Closures, finiteness and factorization, edited by (C. Francisco et al. Eds.), Walter Gruyter, Berlin, (2012), 241–299.

[173] D. L. Craft, *On the genus of joins and composition of graphs*, Discrete Math., **178** (1998) 25–50.

[174] K. Cvetko-Vah, D. Dolžan, *Indecomposability graphs of rings*, Bull. Austral. Math. Soc., **77** (2008), 151–159.

[175] A. Y. Darani, *Notes on the ideal-based zero-divisor graph*, J. Math. Appl., **32** (2010), 103–107.

[176] F. DeMeyer, L. DeMeyer, *Zero divisor graphs of semigroups*, J. Algebra, **283** (2005), 190–198.

[177] F. R. DeMeyer, T. McKenzie, K. Schneider, *The zero-divisor graph of a commutative semigroup*, Semigroup Forum, **65(2)** (2002), 206–214.

[178] F. DeMeyer, K. Schneider, *Automorphisms and zero-divisor graphs of commutative rings*, Internat. J. Commutative Rings, **1(3)** (2002), 93–105.

[179] L. DeMeyer, L. Greve, A. Sabbaghi,, J. Wang, *The zero-divisor graph associated to a semigroup*, Comm. Algebra, **38** (2010), 3370–3391.

[180] L. DeMeyer, Y. Jiang, C. Loszewski,, E. Purdy, *Classification of commutative zero-divisor semigroup graphs*, Rocky Mountain J. Math., **40** (2010), 1481–1503.

[181] L. DeMeyer, A. Schneider, *The annihilating-ideal graph of commutative semigroups*, J. Algebra, **469** (2017), 402–420.

[182] S. Devhare, V. Joshi, J. D. LaGrange, *On the complement of the zero- divisor graph of a partially ordered set*, Bull. Aust. Math. Soc., **97** (2018), 185–193.

[183] S. Devhare, V. Joshi, J. D. LaGrange, *On the connectedness of the complement of the zero-divisor graph of a poset*, Quaest. Math., **42(7)** (2019), 939–951.

[184] S.Devhare, V. Joshi, J. D. LaGrange, *Eulerian and Hamiltonian complements of zero-divisor graphs of pseudocomplemented posets*, Palest. J. Math., **8(2)** (2019), 30–39.

[185] R. Diestel, *Graph Theory*, Springer-Verlag, New York (1997).

[186] H. R. Dorbidi, R. Manaviyat, S. Mirvakili, *Some Properties of the idempotent graph of a ring*, Mediterranean J. Math., (2015), 1–9.

[187] A. Duane, *Proper colorings and p-partite structures of the zero-divisor graph*, Undergraduate Math. J., **7(2)** (2006), 1–7.

[188] M. N. Dumaldar, P. K. Sharma, *Comments over some non-chromatic rings*, Comm. Algebra, **26(11)** (1998), 3871–3883.

[189] D. S. Dummit, R. M. Foote, *Abstract Algebra*, Second Edition, Wiley Student Edition, (2008).

[190] Z. A. El-Bast, P. F. Smith, *Multiplication modules*, Comm. Algebra, **16(4)** (1988), 755–779.

[191] S. Endo, *Note on p.p. rings*, Nagoya Math. J., **17** (1960), 167–170.

[192] D. Endean, K. Henry, E. Manlove, *Zero-divisor graphs of $\mathbb{Z}_n$ and polynomial quotient rings over $\mathbb{Z}_n$*, Undergraduate Math. J., **8(2)** (2007), # 10.

[193] N. Epstein, P. Nasehpour, *Zero-divisor graphs of nilpotent-free semigroups*, J. Algebraic Combin., **37** (2013), 523–543.

[194] C. H. Eslahchi, A. M. Rahimi, *The k-zero-divisor hypergraph of a commutative ring*, Int. J. Math. Math. Sci. Art., **15** (2007), # 50875 [15 pages].

[195] E. Estaji, K. Khashyarmanesh, *The zero-divisor graph of a lattice*, Results. Math., **61(1)** (2012), 1–11.

[196] D. E. Fields, *Zero-divisors and nilpotent elements in power series rings*, Proc. Amer. Math. Soc., **27(3)** (1971), 427–433.

[197] E. Fuchs, *Longest induced cycles in circulant graphs*, Electron. J. Combin., **14(1)** (2005), # N52.

[198] L. Fuchs, *On primal ideals*, Proc. Amer. Math. Soc., **1** (1950), 1–6.

[199] J. A. Gallian, *Contemporary Abstract Algebra*, D. C. Heath and Company, (1990).

[200] N. Ganesan, *Properties of rings with a finite number of zero-divisors*, Math. Ann., **157** (1964), 215–218.

[201] N. Ganesan, *Properties of rings with a finite number of zero-divisors II*, Math. Ann., **161** (1965), 241–246.

[202] Sh. Ghalandarzadeh, P. Malakooti Rad, *Torsion graph over multiplication modules*, Extracta Math., **24(3)** (2009), 281–299.

[203] Sh. Ghalandarzadeh, P. Malakooti Rad, *Torsion graph of modules*, Extracta Math., **26(1)** (2011), 153–163.

[204] Sh. Ghalandarzadeh, P. Malakooti Rad,, S. Shirinkam, *The connected subgraph of the torsion graph of a module*, J. Korean Math. Soc., **49(5)** (2012), 1031–1051.

[205] Sh. Ghalandarzadeh, S. Shirinkam, P. Malakooti Rad, *Annihilator ideal-based zero-divisor graphs over multiplication modules*, Comm. Algebra, **41** (2013), 1134–1148.

[206] R. Gilmer, *Zero divisors in commutative rings*, Amer. Math. Monthly, **93(5)** (1986), 382–387.

[207] R. Gilmer, A. Grams, T. Parker, *Zero-divisors in power series rings*, J. Reine Angew. Math., **278/279** (1975), 145–164.

[208] I. Gitler, E. Reyes, R.H. Villarreal, *Ring graphs and complete intersection toric ideals*, Discrete Math., **310** (2010), 430–441.

[209] H. Glover, J.P. Huneke, C. S. Wang, *103 graphs that are irreducible for the projective plane*, J. Combin. Theory Ser.B, 27(1979), 332–370.

[210] J. S. Golan, *The theory of semirings (with applications in mathematics and theoretical computer science)*, Pitman Monographs and Surveys in Pure and Appl. Math. **54** (1992), Longman Scientific & Tech., Essex.

[211] R. P. Grimaldi, *Graphs from rings*, Proceedings of the twentieth Southeastern Conference on Combinatorics, Graph Theory, and Computing (Boca Raton, FL, 1989), Congr. Numer., **71** (1990), 95–103.

[212] M. Hadian, *Unit action and the geometric zero-divisor ideal graph*, Comm. Algebra, **40** (2012), 2920–2931.

[213] R. Halaš, M. Jukl, *On Beck's coloring of posets*, Discrete Math., **309(13)** (2009), 4584–4589.

[214] T. W. Haynes, S. T. Hedetniemi, P. J. Slatar, *Fundamental of domination in graphs*, Marcel Dekker. Inc. (1998).

[215] T. W. Haynes, S. T. Hedetniemi, P. J. Slatar, *Domination in graphs–advanced topics*, Marcel Dekker. Inc. (2000).

[216] S. Heidar Jafari, N. Jafari Rad, *On domination of zero-divisor graphs of matrix rings*, Canad. Math. Bull., **58(2)** (2015), 271–275.

[217] Huadong Su, Pailing Li, *On the genus of the zero-divisor graph of* $\mathbb{Z}_n$, Internat. J. Combin., (2014), # 390732 [5 pages].

[218] Houyi Yu, Tongsuo Wu, Weiping Gu, *Artinian local rings whose annihilating-ideal graphs are star graphs*, Algebra Colloq., **22(1)** (2015), 73–82.

[219] J. A. Huckaba, *Commutative rings with zero-divisors*, Marcel Dekker, New York/Basel (1988).

[220] T. W. Hungerford, *Algebra*, Springer (2005).

[221] A. Ilić, *The energy of unitary Cayley graphs*, Linear Algebra Appl., **431(10)** (2009), 1881–1889.

[222] Ivana Božić, Zoran Petrović, *Zero-divisor graph of matrices over commutative rings*, Comm. Algebra, **37(4)** (2009), 1186–1192.

[223] N. Jafari Rad, S. H. Jafari, D. A. Mojdeh, *On domination in zero-divisor graphs*, Canad. Math. Bull., **56(2)** (2013), 407–411.

[224] M. I. Jinnah, S. C. Mathew, *When is the comaximal graph split?*, Comm. Algebra, **40(7)** (2012), 2400–2404.

[225] V. Joshi, *Zero divisor graph of a poset with respect to an ideal*, Order, **29(3)** (2012), 499–506.

[226] G. Kalaimurugan, P. Vignesh, T. Tamizh Chelvam, *On zero-divisor graphs of commutative rings without identity*, J. Algebra Appl., **19(12)**(2020), # 2050226 [20 pages].

[227] G. Kalaimurugan, P. Vignesh, T. Tamizh Chelvam, *Toroidal zero-divisor graphs of decomposable rings without identity*, Bol. Soc. Mat. Mex., **26(3)** (2020), 807-829.

[228] I. Kaplansky, *Commutative rings*, Washington, NJ: Polygonal Publishing House, (1974).

[229] A.V. Kelarev, *Graph Algebras and Automata*, Marcel Dekker, New York, (2003).

[230] A.V. Kelarev, C.E. Praeger, *On transitive Cayley graphs of groups and semi-groups*, European J. Combin., **24(1)** (2003), 59–72.

[231] A.V. Kelarev, J. Ryan, J. Yearwood, *Cayley graphs as classifiers for datamining: The influence of asymmetries*, Discrete Math., **309(17)** (2009), 5360–5369.

[232] K. Khashyarmanesh, M. R. Khorsandi, *A generalization of the unit and unitary Cayley graphs of a commutative ring*, Acta Math. Hungar., **137(4)** (2012), 242–253.

[233] K. Khashyarmanesh, M. R. Khorsandi, *Projective total graphs of commutative rings*, Rocky Mountain J. Math., **43(4)** (2013), 1207–1213.

[234] S. Kiani, H.R. Maimani, R. Nikandish, *Some results on the domination number of a zero-divisor graph*, Canad. Math. Bull., **57(3)** (2014), 573–578.

[235] D. Kiani, M. M. Haji Aghaei, *On the unitary Cayley graph of a ring*, Electron. J. of Combin., **19(2)** (2012), # P10 [10 pages].

[236] C. F. Kimball, J. D. LaGrange, *The idempotent-divisor graphs of a commutative ring*, Comm. Algebra, **46(9)** (2018), 3899–3912.

[237] W. Klotz, T. Sander, *Some properties of unitary Cayley graphs*, Electron. J. Combin., **14(1)** (2007), # P45 [12 pages].

[238] A. S. Kuz'mina, *Description of finite nonnilpotent rings with planar zero-divisor graphs*, Discrete. Mat. Appl., **19(6)** (2009), 601–617.

[239] A. S. Kuz'mina, *On structure of rings with planar zero-divisor graphs*, Izv. Altaisk. Gos. Univ., **1** (2009), 17–26.

[240] A. S. Kuzmina, *Finite rings with Eulerian zero-divisor graphs*, J. Algebra Appl., **11(3)** (2012), # 1250055 [9 pages].

[241] A. S. Kuz'mina, Yu. N. Maltsev, *Nilpotent finite rings with planar zero-divisor graphs*, Asian-European J. Math., **1(4)** (2008), 565–574.

[242] A. S. Kuz'mina, Yu. N. Maltsev, *Finite rings with complete bipartite zero-divisor graphs*, Russian Math., **56(3)** (2012), 20–25.

[243] J. D. LaGrange, *Zero-divisor graphs, commutative rings of quotients, and Boolean algebras*, Ph.D. Thesis, University of Tennessee, Knoxville.

[244] J. D. LaGrange, *Complemented zero-divisor graphs and Boolean rings*, J. Algebra, **315(2)** (2007), 600–611.

[245] J.D. LaGrange, *The cardinality of an annihilator class in a von Neumann regular ring*, Int. Electron. J. Algebra, **4** (2008), 63–82.

[246] J. D. LaGrange, *On realizing zero-divisor graphs*, Comm. Algebra, **36(12)** (2008), 4509–4520.

[247] J. D. LaGrange, *Weakly central-vertex complete graphs with applications to commutative rings*, J. Pure Appl. Algebra, **214(7)** (2010), 1121–1130.

[248] J. D. LaGrange, *Characterizations of three classes of zero-divisor graphs*, Canad. Math. Bull., **55(1)** (2012), 127–137.

[249] J. D. LaGrange, *Boolean rings and reciprocal eigenvalue properties*, Linear Algebra Appl., **436(7)** (2012), 1863–1871.

[250] J. D. LaGrange, *Invariants and isomorphism theorems for zero-divisor graphs of commutative rings of quotients*, J. Commut. Algebra, **6(3)** (2014), 407–438.

[251] J. D. LaGrange, *Annihilators in zero-divisor graphs of semilattices and reduced commutative semigroups*, J. Pure Appl. Algebra, **220(8)** (2016), 2955–2968.

[252] J. D. LaGrange, *The $x$-divisor pseudographs of a commutative groupoid*, Int. Electron. J. Algebra, **22** (2017), 62–77.

[253] J. D. LaGrange, *Divisor graphs of a commutative ring*, Advances in commutative algebra, Trends Math., Birkhauser/Springer, Singapore, 2019, 217–244.

[254] J. D. LaGrange, K. A. Roy, *Poset graphs and the lattice of graph annihilators*, Discrete Math., **313(10)** (2013), 1053–1062.

[255] J. Lambek, *Lectures on rings and modules*, Blaisdell Publishing Company, Waltham (1966).

[256] C. Lanski, A. Maroti, *Rings elements as sums of units*, Cent. Eur. J. Math., **7(3)** (2009), 395–399.

[257] J. Lauri, R. Scapellato, *Topics in graphs automorphisms and reconstruction*, London Mathematical Society Lecture Note Series (2003).

[258] S. C. Lee, Rezvan Varmazyar, *Zero-divisor graphs of multiplication modules*, Honam Math. J., **34(4)** (2012), 571–584.

[259] W. J. LeVeque, Fundamentals of Number Theory, Addison-Wesley Publishing Company, Reading, Massachusetts (1977).

[260] R. Levy, J. Shapiro, *The zero-divisor graph of von Neumann regular rings*, Comm. Algebra, **30(2)** (2002), 745–750.

[261] A. H. Li, Q. H. Li, *A kind of graph structure on non-reduced rings*, Algebra Colloq., **17(1)** (2010), 173–180.

[262] A. H. Li, Q. H. Li, *A kind of graph structure on von Neumann regular rings*, Int. J. Algebra, **4(6)** (2010), 291–302.

[263] A. Li, R. P. Tucci, *Zero divisor graphs of upper triangular matrix rings*, Comm. Algebra, **41**(2013), 4622–4636.

[264] P. S. Livingston, *Structure in zero-divisor graphs*, Master's Thesis, The University of Tennessee, Knoxville, TN, December (1997).

[265] C. Long, *Elementary Introduction to Number Theory*, Prentice-Hall, Englewood Cliffs, New Jersey (1987).

[266] L. Lovász, *Combinatorial structures and their applications*, Proc. Calgary Internat. Conf., Calgary, Alberta, Gordon and Breach, New York, Problem 11, (1970), 243–246.

[267] D. C. Lu, T. S. Wu, *The zero-divisor graphs which are uniquely determined by neighborhoods*, Comm. Algebra, **35(12)** (2007), 3855–3864.

[268] D. C. Lu, T. S. Wu, *The zero-divisor graphs of posets and an application to semigroups*, Graphs Combin., **26(6)** (2010), 793–804.

[269] T. G. Lucas, *The diameter of a zero-divisor graph*, J. Algebra, **301(1)** (2006), 174–193.

[270] T. G. Lucas, *Two annihilator conditions: Property (A) and (A.C.)*, Comm. Algebra, **14(3)** (1986), 557–580.

[271] H. R. Maimani, M. R. Pouranki, A. Tehranian, S. Yassemi, *Graphs attached to rings revisited*, AJSE-Mathematics, Springer, **36(6)** (2011), 997–1011.

[272] H. R. Maimani, M. R. Pournaki, S. Yassemi, *Zero-divisor graph with respect to an ideal*, Comm. Algebra, **34(3)** (2006), 923–929.

[273] H. R. Maimani, M. R. Pournaki, S. Yassemi, *Rings which are generated by their units: A graph theoretical approach*, Elem. Math., **65(1)** (2010), 17–25.

[274] H. R. Maimani, M. R. Pournaki, S. Yassemi, *Weakly perfect graphs arising from rings*, Glasg. Math. J., **52(3)** (2010), 417–425.

[275] H. R. Maimani, M. R. Pournaki, S. Yassemi, *Necessary and sufficient conditions for unit graphs to be hamiltonian*, Pacific J. Math., **249(2)** (2011), 419–429.

[276] H. R. Maimani, M. Salimi, A. Sattari, S. Yassemi, *Comaximal graph of commutative rings*, J. Algebra, **319(4)** (2008), 1801–1808.

[277] H. R. Maimani, C. Wickham, S. Yassemi, *Rings whose total graphs have genus at most one*, Rocky Mountain J. Math., **42(5)** (2012), 1551–1560.

[278] H. R. Maimani, S. Yassemi, *Zero-divisor graphs of amalgamated duplication of a ring along an ideal*, J. Pure Appl. Algebra, **212(1)** (2008), 168–174.

[279] H. R. Maimani, S. Yassemi, *On the zero-divisor graphs of commutative semigroups*, Houston J. Math., **372(3)** (2011), 733–740.

[280] A. Mallika, R. Kala, K. Selvakumar, *A note on ideal based zero-divisor graph of a commutative ring*, Discuss. Math. Gen. Algebra Appl. **37**(2017), 177–187.

[281] A. Mallika, R. Kala, K. Selvakumar, *The edge chromatic number of* $\Gamma_I(R)$, Algebra Discrete Math., **24(2)** (2017), 250–261.

[282] Yu. N. Maĺtsev, A. S. Kuz'mina, *Describing ring varieties in which all finite rings have Hamiltonian zero-divisor graphs*, Algebra Logic, **52(2)** (2013), 137–146.

[283] J. Maney, *Irreducible divisor graphs, II*, Comm. Algebra, **36**(2008), 3496–3513.

[284] W.S. Massey, *Algebraic Topology: An introduction*, Harcourt, Brace and World, New York (1967).

[285] N.H. McCoy, *Remarks on divisors of zero*, Amer. Math. Monthly, **49(5)**(1942), 286–295.

[286] W. W. McGovern, *Clean semiprime f-rings with bounded inversion*, Comm. Algebra, **31(7)** (2003), 3295–3304.

[287] T. A. McKee, F. R. McMorris, *Topics in intersection graph theory*, SIAM Monogr. Discrete Math. Appl., USA (1999).

[288] E. Mehdi-Nezhad, A. M. Rahimi, *On some graphs associated to commutative semirings*, Results Math., **68(3-4)** (2015), 293–312.

[289] C. Mihaila, *Comaximal ideal graphs of commutative rings*, Honors Thesis Collection (2012).

[290] S. M. Moconja, Z. Z. petrović, *On the structure of comaximal graphs of commutative rings with identity*, Bull. Aust. Math. Soc., **83(1)** (2011), 11–21.

[291] A. Mohammadian, *On commuting graphs of finite matrix rings*, Comm. Algebra, **38(3)** (2010), 988–994.

[292] B. Mohar, C. Thomassen, *Graphs on Surfaces*, Baltimore, MD: Johns Hopkins (2001).

[293] D. A. Mojdeh, A. M. Rahimi, *Dominating sets of some graphs associated to commutative rings*, Comm. Algebra, **40(9)** (2012), 3389–3396.

[294] A. Moussavi, K. Paykan, *Zero divisor graphs of skew generalized power series rings*, Commun. Korean Math. Soc., **30(4)** (20152), 363–377.

[295] S.B. Mulay, *Cycles and symmetries of zero-divisors*, Comm. Algebra, **30(7)** (2002), 3533–3558.

[296] S.B. Mulay, *Rings having zero-divisor graphs of small diameter or large girth*, Bull. Austral. Math. Soc., **72(3)** (2005), 481–490.

[297] K. Nazzal, *Total graphs associated to a commutative ring*, Palest. J. Math., **5(Special Issue 1)** (2016), 108–126.

[298] R. Nikandish, H. R. Maimani, *Dominating sets of the annihilating-ideal graphs*, Electron. Notes Discrete Math., **45** (2014), 17–22.

[299] M. J. Nikmehr, S. Khojasteh, *A generalized ideal-based zero-divisor graph*, J. Algebra Appl. **14(6)** (2015), # 1550079 [11 pages].

[300] M. J. Nikmehr, R. Nikandish, M. Bakhtyiari, *More on the annihilator graph of a commutative ring*, Hokkaido Math. J., **46(1)** (2017), 107–118.

[301] S. K. Nimbhorkar, M. P. Wasadikar, L. DeMeyer, *Coloring of meet-semilattices*, Ars Combin., **84** (2007), 97–104.

[302] S. K. Nimbhokar, M. P. Wasadikar, M. M. Pawar, *Coloring of lattices*, Math. Slovaca, **60(4)** (2010), 419–434.

[303] G. R. Omidi, E. Vatandoost, *On the commuting graph of rings*, J. Algebra Appl., **10(3)** (2011), 521–527.

[304] E. A. Osba, *The complement graph for Gaussian integers modulo n*, Comm. Algebra, **40(5)** (2012), 1886–1892.

[305] E. A. Osba, S. Al-Addasi, B. Al-Khamaiseh, *Some properties of the zero-divisor graph for the ring of Gaussian integers modulo n*, Glasg. Math. J., **53(2)** (2011), 391–399.

[306] E. A. Osba, S. Al-Addasi, N. A. Jaradeh, *Zero-divisor graph for the ring of Gaussin integers modulo n*, Comm. Algebra, **36(10)** (2008), 3865–3877.

[307] K. Paykan, H. Daneshmand, *Zero divisor graphs of skew Hurwitz series rings*, Matematiche(Catania), **73(1)** (2018), 25–40.

[308] Sh. Payrovi, S. Babaei, *The compressed annihilator graph of a commutative ring*, Indian J. Pure Appl. Math., **49(1)** (2018), 177–186.

[309] A. Phillips, J. Rogers, K. Tolliver, F. Worek, *Uncharted territory of zero-divisor graphs and their complements*, Miami University, SUMSRI (2004).

[310] Pranjali, D. F. Anderson, B. D. Acharya,, P. Gupta, *Zero-divisor labellings of graphs*, Comm. Algebra, **44(11)** (2016), 4697–4708.

[311] Z. Pucanović, *The total graph of a module*, Math. Vesnik, **63(4)** (2011), 305–312.

[312] Z. Pucanović, Z. Petrović, *On the radius and the relation between the total graph of a commutative ring and its extensions*, Publ. Inst. Math.(Beograd)(N.S., **89(103)** (2011), 1–9.

[313] Z. S. Pucanović, Z. Z. Petrović, *Toroidality of intersection graphs of ideals of commutative rings*, Graphs Combin., **30(3)** (2013), 707–716.

[314] Z. Pucanović, Z. Petrović, *The line graph associated to the total graph of a commutative ring*, Ars Combin., **127** (2016), 185–195.

[315] P. M. Rad, *Planar torsion graph of modules*, Filomat, **30(2)** (2016), 367–372.

[316] P. M. Rad, Sh. Ghalandarzadeh, S. Shirinkam, *On the torsion graph and von Neumann regular rings*, Filomat, **26(2)** (2012), 259–253.

[317] P. M. Rad, S. Yassemi, Sh. Ghalandarzadeh, P. Safari, *Diameter and girth of torsion graph*, An. Stiint. Univ. Ovidius Constanta Ser. Mat., **22(3)** (2014), 127–136.

[318] H. N. Ramaswamy, C. R. Veena, *On the energy of unitary Cayley graphs*, Electron. J. Combin., **16(1)** (2009), # N24 [8 pages].

[319] A. Ramin, *The total graph of a finite commutative ring*, Turkish J. Math., **37(3)** (2013), 391–397.

[320] R. Raphael, *Rings which are generated by their units*, J. Algebra **28** (1974), 199–205.

[321] S.P. Redmond, *Generalizations of the zero-divisor graph of a ring, Dissertation*, The University of Tennessee, Knoxville, TN, May (2001).

[322] S. P. Redmond, *The zero-divisor graph of a noncommutative ring*, Int. J. Commutative Rings, **1(4)** (2002), 203–211.

[323] S. P. Redmond, *An ideal-based zero-divisor graph of a commutative ring*, Comm. Algebra, **31(9)** (2003), 4425–4443.

[324] S. P. Redmond, *Central sets and radii of the zero-divisor graphs of commutative rings*, Comm. Algebra, **34(7)** (2006), 2389–2401.

[325] S. P. Redmond, *On zero-divisor graphs of small finite commutative rings*, Discrete Math., **307(9-10)** (2007), 1155–1166.

[326] S. P. Redmond, *Corrigendum to "On zero-divisor graphs of small finite commutative rings"* , Discrete Math., **307(21)** (2007), 2449–2452.

[327] S. P. Redmond, *Counting zero-divisors*, Commutative Rings: New Research, Nova Science Publishers, Hauppauge, NY (2009), 7–12.

[328] S. P. Redmond, *Cut vertices and degree one vertices of zero-divisor graphs*, Comm. Algebra, **40(8)** (2012), 2749–2756.

[329] S. P. Redmond, *Recovering rings from zero-divisor graphs*, J. Algebra Appl., **12(8)** (2013), # 1350047 [9 pages].

[330] G. Ringel, *On the genus of the graph $K_n \times K_2$ or the n-prism*, Discrete Math., **20(3)** (1977), 287–294.

[331] S. Safaeeyan M. Baziar, E. Momtahan, *A generalization of the zero-divisor graph for modules*, J. Korean Math. Soc., **51(1)** (2014), 87–98.

[332] K. Selvakumar, *A note on the annihilating ideal graph of commutative rings*, Ars Combin., **CXXXVII**(2018), 113–122.

[333] K. Selvakumar, V.C. Amritha, *On the genus of the k-maximal hypergraph of commutative rings*, Discrete Math. Algorithms Appl., **11(1)** (2019), # 1950010 [9 pages].

[334] K. Selvakumar, S. Karthik, *On the genus of the co-annihilating graph of commutative rings*, Discuss. Math. Gen. Algebra Appl., **32(2)**(2019), 1–16.

[335] K. Selvakumar, V. Ramanathan, *Classification of nonlocal rings with genus one 3-zero-divisor hypergraphs*, Comm. Algebra, **45(1)**(2016), 275–284.

[336] K. Selvakumar, V. Ramanathan, *Classification of nonlocal rings with projective 3-zero-divisor hypergraph*, Ricerche di Mat., **66(2)** (2017), 457–468.

[337] K.Selvakumar, V. Ramanathan, *On the genus of the k-annihilating ideal hypergraph of commutative ring*, Indian J. Pure Appl. Math., **50(2)** (2019), 461–475.

[338] K. Selvakumar, V. Ramanathan, *The k-annihilating hypergraph of commutative ring*, AKCE Inter. J. of Graphs and Combim., **16** (2019), 241–252.

[339] K. Selvakumar, M. Subajini, *Crosscap of the Non-cyclic graph of groups*, AKCE Inter. J. of Graphs and Combin. **13**(2016), 235–240.

[340] K. Selvakumar, M. Subajini, *Classification of rings with toroidal Jacobson graph*, Czechoslovak Math. J., **66(2)**(2016), 307–316.

[341] K. Selvakumar, M. Subajini, *Classification of groups with toroidal coprime graph*, Australas. J. Combin., **69(2)** (2017), 174–183.

[342] K. Selvakumar, M. Subajini, *Commutative rings with genus two annihilator graphs*, Comm. Algebra, **46(1)** (2018), 28–37.

[343] K. Selvakumar, M. Subajini, M.J. Nikmehr, *Finite commutative ring with genus two essential graph*, J. Algebra App., **16(2)** (2018), # 1850121[11 pages].

[344] K. Selvakumar, M. Subajini, *On the genus of the essential graph of commutative rings*, Australas. J. Combin., **74(1)** (2019), 74–85.

[345] K. Selvakumar, P. Subbulakshmi, J. Amjadi, *On the genus of the graph associated to a commutative ring*, Discrete Math. Algorithms Appl., **9(5)** (2017), # 1750058 [11 pages].

[346] A. Shariatinia, A. Tehranian, *Domination number of total graph of module*, Algebr. Struct. Appl., **2(1)** (2015), 1–9.

[347] P. K. Sharma, S. M. Bhatwadekar, *A note on graphical representation of rings*, J. Algebra, **176(1)** (1995), 124–127.

[348] M. H. Shekarriz, M. H. Shiradareh Haghighi, H. Sharif, *On the total graph of a finite commutative ring*, Comm. Algebra, **40(8)** (2012), 2798–2807.

[349] M. Sivagami, T. Tamizh Chelvam, *On the trace graph of matrices*, Acta Math. Hungar., **158(1)** (2019), 235–250.

[350] J. Skowronek-Kaziów, *Some digraphs arising from number theory and remarks on the zero-divisor graph of ring $\mathbb{Z}_n$*, Inform. Process. Lett., **108(3)** (2008), 165–169.

[351] J. G. Smith Jr, *Properties of ideal-based zero-divisor graphs of commutative rings*, ProQuest LLC, Ann Arbor, MI, Thesis (Ph.D.), The University of Tennessee (2014).

[352] J. G. Smith Jr, *When ideal-based zero-divisor graphs are complemented or uniquely complemented*, Int. Electron. J. Algebra, **21** (2017), 198–204.

[353] J. G. Smith Jr, *Isomorphisms and planarity of zero-divisor graphs*, Advances in commutative algebra, Trends Math., Birkhauser/Springer, Singapore, 2019, 245–263.

[354] N. O. Smith, *Graph-theoretic properties of the zero-divisor graph of a ring*, ProQuest LLC, Ann Arbor, MI, Thesis (Ph.D.), The University of Tennessee (2004).

[355] N. O. Smith, *Planar zero-divisor graphs*, Int. J. Commut. Rings, **2(4)** (2003), 177–188.

[356] N. O. Smith, *Infinite planar zero-divisor graphs*, Comm. Algebra, **35(1)** (2007), 171–180.

[357] S. Spiroff, C. Wickham, *A zero divisor graph determine by equivalence classes of zero divisors*, Comm. Algebra, **39(7)** (2011), 2338–2348.

[358] T. Tamizh Chelvam, S. Anukumar Kathirvel, *Generalized unit and unitary Cayley graphs of finite rings*, J. Algebra Appl., **18(1)** (2019), # 1950006 [21 pages].

[359] T. Tamizh Chelvam, S. Anukumar Kathirvel, *Note on generalized Cayley graph of finite rings and its complement*, J. Anal., **27(2)** (2019), 555–566.

[360] T. Tamizh Chelvam, S. Anukumar Kathirvel, M. Balamurugan, *Intersection graph of gamma sets in generalized unit and unitary Cayley graphs*, Houston J. Math., **46(3)** (2020), 561–582.

[361] T. Tamizh Chelvam, S. Anukumar Kathirvel, M. Balamurugan, *Domination in generalized unit and unitary Cayley graphs of finite rings*, Indian J. Pure Appl. Math., **51(2)** (2020), 533–556.

[362] T. Tamizh Chelvam, T. Asir, *A note on total graph of $\mathbb{Z}_n$*, J. Discrete Math. Sci. Cryptogr., **14(1)** (2011), 1–7.

[363] T. Tamizh Chelvam, T. Asir, *Domination in the total graph on $\mathbb{Z}_n$*, Discrete Math. Algorithms Appl., **3(4)** (2011), 413–421.

[364] T. Tamizh Chelvam, T. Asir, *Intersection graph of gamma sets in the total graph*, Discuss. Math. Graph Theory, **32(2)** (2012), 341–356.

[365] T. Tamizh Chelvam, T. Asir, *On the genus of the total graph of a commutative ring*, Comm. Algebra, **41(1)** (2013), 142–153.

[366] T. Tamizh Chelvam, T. Asir, *The intersection graph of gamma sets in the total graph of a commutative ring I*, J. Algebra Appl., **12(4)** (2013), # 1250198 [14 pages].

[367] T. Tamizh Chelvam, T. Asir, *Domination in the total graph of a commutative ring*, J. Combin. Math. Combin. Comput., **87** (2013), 147–158.

[368] T. Tamizh Chelvam, T. Asir, *Distances in zero-divisor and total graphs from commutative rings - A survey*, AKCE Int. J. Graphs Comb., **13(3)** (2016), 290–298.

[369] T. Tamizh Chelvam, T. Asir, *Generalized total graphs of commutative rings - A survey*, Algebra and its Applications, Ed. by Ashraf, Mohammad / De Filippis, Vincenzo / Rizvi, Syed Tariq, 37-55, ISBN: 978-3-11-054240-0, De Gruyter, Germany (2018).

[370] T. Tamizh Chelvam, T. Asir, *Genus of total graphs from rings: A survey*, AKCE Int. J. Graphs Comb., **15(1)** (2018), 97–104.

[371] T. Tamizh Chelvam, T. Asir, K. Selvakumar, *On the domination in graphs from commutative rings*, Algebra and its Applications: Springer Proceedings in Mathematics and Statistics 174, S. Tariq Rizvi et al.(eds), Singapore (2016), 363–379.

[372] T. Tamizh Chelvam, M. Balamurugan, *On the generalized total graph of fields and its complement*, Palest. J. Math., **7(2)**(2018), 450-457.

[373] T. Tamizh Chelvam, M. Balamurugan, *Complement of the generalized total graph*, Filomat, **33(18)** (2019), 6103–6113.

[374] T. Tamizh Chelvam, M. Balamurugan, *Complement of the generalized total graph of commutative rings*, J. Anal., **27** (2019), 539–553.

[375] T. Tamizh Chelvam, G. Kalaimurugan, Well Y. Chou, *Signed star domination number of Cayley graphs* Discrete Math. Algorithms Appl., **4(2)** (2012), # 1250017 [10 pages].

[376] T. Tamizh Chelvam, N. Mohammed Rilwan, G. Kalaimurugan, *Antimagic and magic labeling in Cayley digraphs*, Australas. J. Combin., **55** (2013), 65–71.

[377] T. Tamizh Chelvam, M.S. Mutharasu, *Efficient domination in Cartesian product of cycles*, Journal of Advanced Research in Pure Mathematics, **3(3)** (2011), 42–49.

[378] T. Tamizh Chelvam, M.S. Mutharasu, *Subgroups as efficient dominating set in Cayley Graphs on $\mathbb{Z}_n$*, Discrete Appl. Math., **161(8)** (2013), 1187–1190.

[379] T. Tamizh Chelvam, M.S. Mutharasu, I. Rani, *Domination in generalized circulant graphs*, Indian J. Math., **52(3)** (2010), 531–541.

[380] T. Tamizh Chelvam, S. Nithya, *Zero-divisor graph of an ideal in a near-ring*, Discrete Math. Algorithms Appl., **5(1)** (2013), # 1350007 [11 pages].

[381] T. Tamizh Chelvam, S. Nithya, *Domination in the zero-divisor graph of an ideal of a near-ring*, Taiwanese J. Math., **17(5)**(2013), 1613–1625.

[382] T. Tamizh Chelvam, S. Nithya, *A note on the zero-divisor graph of a lattice*, Trans. Comb., **3(3)** (2014), 51–59.

[383] T. Tamizh Chelvam, S. Nithya, *Crosscap of the ideal based zero-divisor graph*, Arab J. Math. Sci., **22(1)** (2016), 29–37.

[384] T. Tamizh Chelvam, K. Prabha Ananthi, *The genus of graphs associated with vector spaces*, J. Algebra Appl., **19(5)** (2020), # 2050086 [11 pages].

[385] T. Tamizh Chelvam, S. Raja, I. Gutman, *Strongly regular integral Circulant graphs and their energies*, Bulletin of International Mathematical Virtual Institute, **2** (2012), 9–16.

[386] T. Tamizh Chelvam, S. Raja, *Integral circulant graphs with four distinct eigenvalues*, Discrete Math. Algorithms Appl., **10(5)** (2018) # 1850062 [10 pages].

[387] T. Tamizh Chelvam, I. Rani, *Dominating Sets in Cayley graphs on $\mathbb{Z}_n$*, Tamkang J. Math., *37(4)* (2007), 341–345.

[388] T. Tamizh Chelvam, I. Rani, *Independent Domination number of a Cayley Graph on $\mathbb{Z}_n$*, J. Combin. Math. Combin. Comput., **69** (2009), 251–255.

[389] T. Tamizh Chelvam, I. Rani, *Total and Connected Domination number for Cayley graphs on $\mathbb{Z}_n$*, Advanced Studies in Contemporary Mathematics, **20(1)** (2010), 57–61.

[390] T. Tamizh Chelvam, I. Rani, *Locating dominating sets in circulant graphs II*, J. Combin. Math. Combin. Comput., **107** (2018), 115–125.

[391] T. Tamizh Chelvam, M. Sattanathan, *Subgroup Intersection graph of a group*, Journal of Advanced Research in Pure Mathematics, **3(4)** (2011), 44–49.

[392] T. Tamizh Chelvam, M. Sattanathan, *On the subgroup intersection graph of finite abelian groups*, Transactions on Combinatorics, **1(3)** (2012), 5–10.

[393] T. Tamizh Chelvam, M. Sattanathan, *Power graph of finite abelian groups*, Algebra and Discrete Mathe., **16(1)** (2013), 33–41.

[394] T. Tamizh Chelvam, Sivagnanam Mutharasu, *Bounds for Domination Parameters in Circulant Graphs*, Advanced Studies in Contemporary Mathematics, **22(4)** (2012), 525–529.

[395] T. Tamizh Chelvam, Sivagnanam Mutharasu *Efficient open domination in Cayley graphs*, Appl. Math. Lett., **25** (2012), 1560–1564.

[396] T. Tamizh Chelvam, K. Selvakumar, *Domination in zero-divisor graph of matrices*, J. Adv. Res. Pure. Math., **3(2)** (2011), 31–39.

[397] T. Tamizh Chelvam, K. Selvakumar, *On the intersection graph of gamma sets in the zero-divisor graph*, J. Adv. Res. Pure. Math. **4(4)** (2012), 57–66.

[398] T. Tamizh Chelvam, K. Selvakumar, *Domination in the directed zero-divisor graph of ring of matrices*, J. Combin. Math. Combin. Comput., **91** (2014), 155–163.

[399] T. Tamizh Chelvam, K. Selvakumar, *Central sets in the annihilating-ideal graph of a commutative ring*, J. Combin. Math. Combin. Comput., **88** (2014), 277–288.

[400] T. Tamizh Chelvam, K. Selvakumar, *On the connectivity of the annihilating-ideal graphs*, Discuss. Math. Gen. Algebra Appl., **35(2)** (2015), 195–204.

[401] T. Tamizh Chelvam, K. Selvakumar, *On the intersection graph of gamma sets in the zero-divisor graph*, Discrete Math. Algorithms Appl., **7(1)** (2015), # 1450067 [14 pages].

[402] T. Tamizh Chelvam, K. Selvakumar, V. Ramanathan, *On the planarity of the k-zero-divisor hypergraphs*, AKCE Int. J. Graphs Comb., **12(2)** (2015), 169–176.

[403] T. Tamizh Chelvam, K. Selvakumar, *On the genus of the annihilator graph of a commutative ring*, Algebra Discrete Math., **24(2)** (2017), 191–208.

[404] T. Tamizh Chelvam, K. Selvakumar, V. Ramanathan, *Cayley sum graph of ideals of commutative rings*, J. Algebra Appl. **16(2)** (2018), # 1850125 [14 pages].

[405] T. Tamizh Chelvam, K. Selvakumar, P. Subbulakshmi, *On the genus of the nil-graph of ideals of a commutative ring*, Arab J. Math. Sci., **23(2)** (2017), 186–195.

[406] T. Tamizh Chelvam, K. Selvakumar, P. Subbulakshmi, *On the nil-graph of ideals of commutative Artinian rings*, Discrete Math. Algorithms Appl., **10(5)** (2018), # 1850057 [10 pages].

[407] T. Tamizh Chelvam, M. Sivagami, *Structure and substructure connectivity of Cayley graphs*, Arab J. Math., **27(1)**(2021), 94–103.

[408] T. Tamizh Chelvam, M. Sivagami, *On the ideal based trace graph of matrices*, Hacet. J. Math. Stat., **49(2)** (2020), 608–616.

[409] G. H. Tang, H. D. Su,, B. S. Ren, *Commutative zero-divisor semigroups of graphs with at most four vertices*, Algebra Colloq., **16(2)** (2009), 341–350.

[410] G. H. Tang, H. D. Su,, B. S. Ren, *Zero-divisor semigroups of star graphs and two-star graphs*, Ars Comb., **119** (2015), 3–11.

[411] M. Taylor, *Zero-divisor graphs with looped vertices*, Unpublished manuscript.

[412] A. Tehranian, H. R, Maimani, *A study of the total graph*, Iran. J. Math. Sci. Inform., **6(2)** (2011), 75–80.

[413] Tengxia Ju, Meiyun Wu, *On iteration digraph and zero-divisor graph of the ring* $\mathbb{Z}_n$, Czechoslovak Math. J., **64 (139)** (2014), 611–628.

[414] I. Tomon, *On the chromatic number of regular graphs of matrix algebras*, Linear Algebra Appl., **475** (2015), 154–162.

[415] U. Vishne, *The graph of zero-divisor ideals*, Unpublished Manuscript, dated July 11, 2002.

[416] S. Visweswaran, Hiren D. Patel, *When is the annihilating ideal graph of a zero-dimensional quasi semilocal commutative ring complemented?*, Arab J. Math. Sci., **22(1)** (2016), 1–21.

[417] S. Visweswaran, Hiren D. Patel, *On the clique number of the complement of the annihilating ideal graph of a commutative ring*, Beitr. Algebra Geom., **57(2)** (2016), 307–320.

[418] S. Visweswaran, P. Sarman, *Some properties of the complement of the annihilator graph of a commutative reduced ring*, Palest. J. Math., **6(2)** (2017), 434–447.

[419] K. Wagner, *Über eine Erweiterung eines Satzes von Kuratowski*, Deutsche Math., **2** (1937), 280–285.

[420] H.-J. Wang, *Zero-divisor graphs of genus one*, J. Algebra, **304(2)** (2006), 666–678.

[421] H.-J. Wang, *Coloring of graphs associated to zero-divisors*, arXiv:math/0703051, March (2007).

[422] H.-J. Wang, *Genera of zero-divisor graphs with respect to ideals*, arXiv:math/0703049, March (2007).

[423] H.-J. Wang, *Graphs associated to co-maximal ideals of commutative rings*, J. Algebra, **320(7)** (2008), 2917–2933.

[424] H.-J. Wang, *Co-maximal graph of non-commutative rings*, Linear Algebra Appl., **430(2-3)** (2009), 633–641.

[425] D. Weber, *Zero-divisor graphs and lattices of finite commutative rings*, Rose-Hulman Undergrad. Math. J., **12(1)** (2011), 57–70.

[426] G. Wendt, *On zero divisors in near-rings*, Int. J. Algebra, **3(1)** (2009), 21–32.

[427] D.B. West, *Introduction to Graph Theory*, Prentice Hall, Inc., Upper Saddle River, NJ (1996).

[428] A. T. White, *The genus of the cartesian product of two graphs*, J. Combinatorial. Theory Ser., **11** (1971), 89–94.

[429] A. T. White, *Graphs, Groups and Surfaces*, North-Holland: Amsterdam (1973).

[430] C. Wickham, *Classification of rings with genus one zero-divisor graphs*, Comm. Algebra, **36(2)** (2008), 325–345.

[431] C. Wickham, *Rings whose zero-divisor graphs have positive genus*, J. Algebra, **321(2)** (2009), 377–383.

[432] R. Wilson, *Introduction to Graph Theory*, 4th edn, Pearson Prentice Hall, Malaysia (1996).

[433] S. E. Wright, *Lengths of paths and cycles in zero-divisor graphs and digraphs of semigroups*, Comm. Algebra, **35(6)** (2007), 1987–1991.

[434] T. Wu, *On directed zero-divisor graphs of finite rings*, Discrete Math. **296(1)** (2005), 73–86.

[435] T. Wu, F. Cheng, *The structure of zero-divisor semigroups with graph $K_n o K_2$*, Semigroup Forum, **76(2)** (2008), 330–340.

[436] T. Wu, Q. Liu, L. Chen, *Zero-divisor semigroups and refinements of a star graph*, Discrete Math., **309(8)**(2009), 2510–2518.

[437] Z. Xue, S. Liu, *Zero-divisor graphs of partially ordered sets*, Appl. Math. Lett., **23(4)** (2010), 449–452.

[438] Tian Yanzhao, Wei Qijiao, *Remarks on the zero-divisor graph of a commutative ring*, Adv. Appl. Math., **00** (2015), 1–3.

[439] E. Yaraneri, *Intersection graph of a module*, J. Algebra Appl., **12(5)** (2013), # 1250218 [30 pages].

[440] M. Ye, T. Wu, *Co-maximal ideal graphs of commutative rings*, J. Algebra Appl., **11(6)** (2012), # 1250114 [14 pages].

[441] Yongwei Yao, *Infinite rings with planar zero-divisor graphs*, Comm. Algebra, **36(11)** (2008), 4068–4077.

[442] J. W. T. Youngs, *Solution of the map-coloring problem - cases 3, 5, 6 and 9*, J. Combinatorial. Theory, **8** (1970), 175–219.

[443] M. Zuo, T. S. Wu, *A new graph structure of commutative semigroups*, Semigroup, Forum **70** (2005), 71–80.

# Index




D. F. Anderson et al., *Graphs from Rings*, https://doi.org/10.1007/978-3-030-88410-9